P9-CJX-536

The Seismic Design Handbook

The Seismic Design Handbook

Edited by

Farzad Naeim, Ph.D., P.E.

Director of Research and Development
John A. Martin and Associates, Los Angeles, California

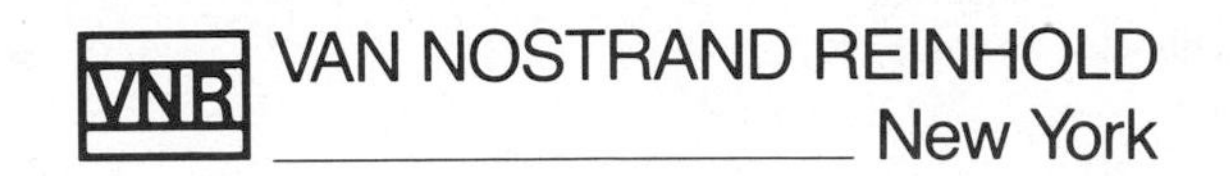

VAN NOSTRAND REINHOLD
New York

Library of Congress Catalog Card Number 89-30616
ISBN 0-442-26922-6

Printed in the United States of America

Van Nostrand Reinhold
115 Fifth Avenue
New York, New York 10003

Van Nostrand Reinhold International Company Limited
11 New Fetter Lane
London EC4P 4EE, England

Van Nostrand Reinhold
480 La Trobe Street
Melbourne, Victoria 3000, Australia

Nelson Canada
1120 Birchmount Road
Scarborough, Ontario, M1K 5G4, Canada

16 15 14 13 12 11 10 9 8 7 6 5 4 3 2

Library of Congress Cataloging-in-Publication Data

The seismic design handbook/edited by Farzad Naeim.
p. cm.
Includes bibliographies and index.
ISBN 0-442-26922-6
1. Earthquake resistant design—Handbooks, manuals, etc.
2. Buildings—Earthquake effects—Handbooks, manuals, etc.
I. Naeim, Farzad.
TA658.44.H36 1989
624.1′762—dc19 89-30616
CIP

—— **VNR STRUCTURAL ENGINEERING SERIES** ——

Structural Engineering Theory

STRUCTURES: FUNDAMENTAL THEORY AND BEHAVIOR by Richard Gutkowski
STRUCTURAL DYNAMICS, 2nd Ed., by Mario Paz
MICROCOMPUTER-AIDED ENGINEERING: STRUCTURAL DYNAMICS by Mario Paz
EARTHQUAKE RESISTANT BUILDING DESIGN AND CONSTRUCTION, 2nd Ed., by Norman Green
THE SEISMIC DESIGN HANDBOOK edited by Farzad Naeim

Steel Design

STEEL DESIGN FOR ENGINEERS AND ARCHITECTS by Rene Amon, Bruce Knobloch and Atanu Mazumder

Concrete Design

HANDBOOK OF CONCRETE ENGINEERING by Mark Fintel
STRUCTURAL DESIGN GUIDE TO THE ACI BUILDING CODE, 3rd Ed., by Paul F. Rice, Edward S. Hoffman, David P. Gustafson and Albert J. Gouwens
TORSION OF REINFORCED CONCRETE by T. Hsu
MODERN PRESTRESSED CONCRETE, 3rd Ed., by James R. Libby

Masonry

SIMPLIFIED MASONRY SKILLS by R. Kreh

Wood

MECHANICS OF WOOD AND WOOD COMPOSITES by Jozsef Bodig and Benjamin Jayne
STRUCTURAL USE OF WOOD IN ADVERSE ENVIRONMENTS edited by Robert W. Meyer and Robert M. Kellogg
STRUCTURAL DESIGN IN WOOD by Judith J. Stalnaker and Ernest C. Harris

Tall Buildings

DEVELOPMENTS IN TALL BUILDINGS by the Council on Tall Buildings
ADVANCES IN TALL BUILDINGS by the Council on Tall Buildings
SECOND CENTURY OF THE SKYSCRAPER by the Council on Tall Buildings

Other Structures-Related Books

ICE INTERACTION WITH OFFSHORE STRUCTURES by A.B. Cammaert and D.B. Muggeridge
FOUNDATION ENGINEERING HANDBOOK by Hans Winterkorn and H.Y. Fang
FOUNDATION ENGINEERING FOR DIFFICULT SUBSOIL CONDITIONS by Leonardo Zeevaert
ADVANCED DAM ENGINEERING edited by Robert Jansen
TUNNEL ENGINEERING HANDBOOK by John Bickel and T.R. Kuesel
SURVEYING HANDBOOK edited by Russell Brinker and Roy Minnick

CONTRIBUTORS

James C. Anderson, Ph.D.
Professor of Civil Engineering, University of Southern California, Los Angeles, California.
(*Seismic Response of Buildings*)

Christopher Arnold, A.I.A.
President, Building Systems Development, Inc., San Mateo, California.
(*Architectural Considerations*)

Bruce A. Bolt, Ph.D.
Professor of Seismology and Director of Seismographic Stations, University of California, Berkeley, California.
(*The Nature of Earthquake Ground Motion*)

Rao Boppana, Ph.D., S.E.
President, Boppana and Associates, Los Angeles, California.
(*Seismic Design of Floor Diaphragms*)

Arnaldo T. Derecho, Ph.D.
Senior Consultant, Wiss, Janney, Elstner & Associates, Northbrook, Illinois.
(*Seismic Design of Reinforced Concrete Structures*)

Roger M. Di Julio, Jr., Ph.D., P.E.
Professor of Engineering, California State University, Northridge, California.
(*Static Lateral Force Procedures*)

Fawzi E. Elghadamsi, Ph.D.
Assistant Professor of Civil Engineering, Southern Methodist University, Dallas, Texas.
(*Earthquake Ground Motion and Response Spectra*)

Gary C. Hart, Ph.D.
Professor of Civil Engineering, University of California, Los Angeles and President, Englekirk & Hart, Inc., Los Angeles, California.
(*Seismic Design of Masonry Structures*)

Marshall Lew, Ph.D., P.E.
Director of Earthquake Engineering, Leroy Crandall and Associates, Glendale, California.
(*Geotechnical and Foundation Design Considerations*)

Ronald L. Mayes, Ph.D.
President, Dynamic Isolation Systems, Inc., Berkeley, California.
(*Design of Systems with Seismic Isolation*)

Bijan Mohraz, Ph.D.
Professor of Civil Engineering, Southern Methodist University, Dallas, Texas.
(*Earthquake Ground Motion and Response Spectra*)

Farzad Naeim, Ph.D., P.E.
Director of Research and Development, John A. Martin & Associates, Los Angeles, California.
(*Design for Drift and Lateral Stability; Seismic Design of Floor Diaphragms*)

Joseph P. Nicoletti, S.E.
Senior Vice President, URS/John A. Blume & Associates, San Francisco, California.
(*Seismic Design of Steel Structures*)

John M. Nissen, S.E.
Project Manager, John A. Martin & Associates, Los Angeles, California.
(*Geotechnical and Foundation Design Considerations*)

Thomas A. Sabol, Ph.D., S.E.
Project Manager, Englekirk & Hart, Inc., Los Angeles, California.
(*Design of Nonstructural Systems and Components*)

Preface

This handbook contains up-to-date information on planning, analysis, and design of earthquake-resistant building structures. Its intention is to provide engineers, architects, developers, and students of structural engineering and architecture with authoritative, yet practical, design information. It represents an attempt to bridge the persisting gap between advances in the theories and concepts of earthquake-resistant design and their implementation in seismic design practice.

The distinguished panel of contributors is composed of 15 experts from industry and universities, recognized for their knowledge and extensive practical experience in their fields. They have aimed to present clearly and concisely the basic principles and procedures pertinent to each subject and to illustrate with practical examples the application of these principles and procedures in seismic design practice. Where applicable, the provisions of various seismic design standards such as ANSI-82, BOCA-87, ATC 3-06, NEHRP-85, UBC-85, and UBC-88 are thoroughly explained and their differences are highlighted.

A number of chapters are of particular interest because they cover subjects that have never before been published in book form. Seismic base isolation, floor diaphragm design, and design for drift and lateral stability are among these subjects.

One should realize that seismic design is still as much an art as it is a science. Hence, no matter how helpful the material in this handbook might prove to be, it cannot replace or substitute sound engineering judgment. Furthermore, one must recognize that on some seismic design and detailing issues, a general consensus on the appropriate approaches does not yet exist. As an eminent engineer once said: "No two design offices completely agree on all aspects of seismic design or proper detailing." It is the editor's belief, however, that it is through the publication of books like this one, and continuation of research and development, that a general consensus on these issues will finally be reached.

The primary purpose of this handbook is to serve practicing engineers and architects. However, its scope and its treatment of both theory and practice should also make it valuable to both teachers and students of earthquake-resistant design.

The editor gratefully acknowledges the efforts of contributors in preparing excellent manuscripts. Thanks are also due to the management and staff at John A. Martin and Associates, especially Jack and Trailer Martin, who graciously understood and accommodated my shift of emphasis from everyday office practice to preparation of this handbook. If it had not been for their encouragement and support, this project would have not been completed. Last, but not least, the efforts of Hamid Fahmian, Marvin Mittelstaedt, Dawn Fajt, and Tibor Fajt in preparing the artwork is gratefully appreciated.

FARZAD NAEIM

Contents

Chapter 1

The Nature of Earthquake Ground Motion

*Bruce A. Bolt, Ph.D.**

1.1 INTRODUCTION

On the average, 10,000 people die each year from earthquakes (see Figure 1-1). A UNESCO study gives damage losses amounting to $10,000,000,000 from 1926 to 1950 from earthquakes. In Central Asia in this interval two towns and 200 villages were destroyed. Since then several towns, including Ashkhabad (1948), Agadir (1960), Skopje (1963), Managua (1972), Tangshan (1976), Mexico City (1985), Spitaka (1988), and hundreds of villages have been essentially razed to the ground by ground shaking. Historical writings testify to man's long concern about earthquake hazards.

The first modern stimulus for seismology came from the extensive field work of the Irish engineer Robert Mallett, after the great Neopolitan earthquake of 1857 in southern Italy. He set out to explain the "masses of dislocated stone and mortar" in terms of mechanical principles and in doing so established basic vocabulary such as "seismology," "hypocenter," and "isoseismal." The close links between engineering and seismology have continued ever since.

It is part of strong-motion seismology to explain and predict the large-amplitude, long-duration shaking observed in damaging earthquakes. In the first 60 years of the century, however, the great seismological advances occurred in studying waves from distant earthquakes using very sensitive seismographs. Because the wave amplitudes in even a nearby magnitude-5 earthquake would exceed the dynamic range of the usual seismographs, not much fundamental work was done by seismologists on the rarer large earthquakes of engineering importance.

Nowadays, the situation has changed. After the 1971 San Fernando earthquake, hundreds of strong-motion records were available for this magnitude-6.5 earthquake. The $1.2g$ acceleration at Pacoima Dam led to questions on topographic amplification and the construction of realistic models of fault rupture and travel path that could explain the strong-motion patterns. Digital recorders and fast computers mean that both seismologists and engineers can tackle more fundamental and realistic problems of earthquake generation and ground shaking.

Knowledge of strong ground shaking is now advancing rapidly, largely because of the growth of appropriately sited strong-motion accelerographs in seismic areas of the world. For example, in the Strong Motion Instrumentation Program in California, by 1986 there were 325 instruments in the free field, and 85 buildings and 30 other structures were instrumented. Over 200 records had been digitized and were available for use in research or practice.

In earthquake-prone regions, structural design of large or critical engineered structures such as high-rise build-

*Professor of Seismology and Director of Seismographic Stations, University of California, Berkeley, California.

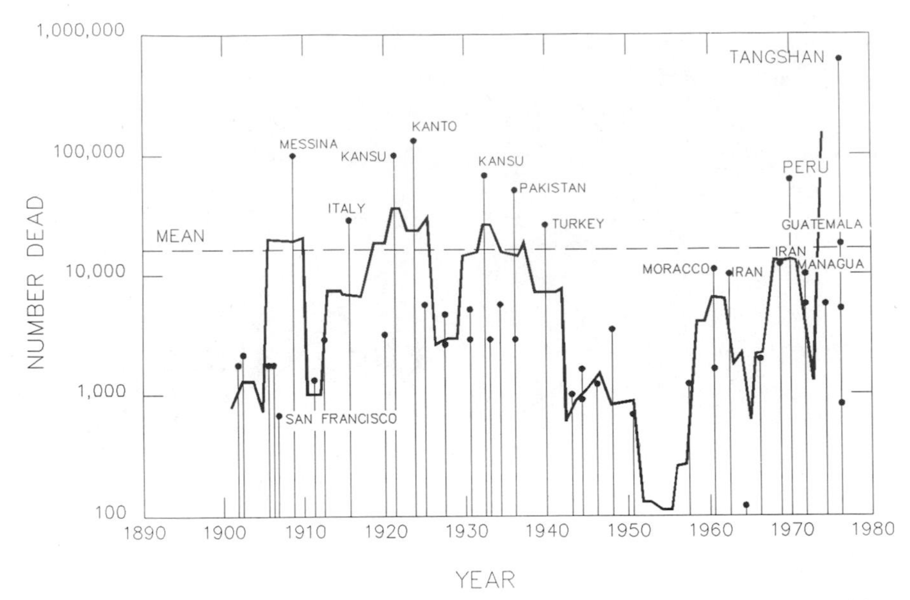

Figure 1-1 Loss of life caused by major earthquakes. (After Hiroo Kanamori (1-5).)

ings, dams, and bridges now often involves quantitative dynamic analysis; engineers ask penetrating questions on the likely seismic intensity for construction sites and require input motions or spectra or defining parameters. Predicted time histories for dynamic modeling in structural design or vulnerability assessments are often needed.

The aim of this chapter is to provide a basic understanding about earthquakes, their worldwide distribution, what causes them, their likely damage mechanisms, earthquake measuring scales, and current efforts on the prediction of strong seismic ground motions. Additional helpful background on the subject may be found in References 1-1 through 1-26.

1.2 SEISMICITY OF THE WORLD

From the earthquake-wave readings at different seismographic observatories, the position of the center of an earthquake can be calculated.[1-1] In this way, a uniform picture of the earthquake distribution around the world has been obtained (see Figure 1-2). Definite belts of seismic activity separate large oceanic and continental regions, themselves almost devoid of earthquake centers. Other concentrations of earthquake sources can be seen in the oceanic areas, for example, along the center of the Atlantic and Indian Oceans. These are the sites of gigantic submarine mountain ranges called mid-oceanic ridges. The geological strains that prevail throughout this global ridge system are evidenced by great mountain peaks and deep rift valleys. Volcanic eruptions are frequent, and earthquakes originating along these ridges often occur in "swarms," so that many hundreds of shocks are concentrated in a small area in a short time.

Dense concentrations of earthquake centers, some as much as 680 km beneath the surface, also coincide with island arcs, such as those of the Pacific and the eastern Caribbean.

On the eastern side of the Pacific Ocean, the whole west coast of Central and South America is agitated by many earthquakes, great and small. High death tolls have ensued from the major ones. In marked contrast, the eastern part of South America is almost entirely free from earthquakes, and can be cited as an example of low seismic risk country. Other seismically quiet continental areas can be seen in Figure 1-2.

In Europe, earthquake activity is quite widespread. To the south, Turkey, Greece, Yugoslavia, Italy, Spain, and Portugal suffer from it, and large numbers of people have died in disasters throughout the years. An earthquake off southwest Iberia on November 1, 1755 produced a great tsunami, which caused many of the 50,000 to 70,000 deaths occurring in Lisbon, Portugal, and surrounding areas; the shaking was felt in Germany and the Netherlands. In Alicante, Spain, on March 21, 1829, a shock killed about 840 persons and injured many hundred more. Total or partial destruction of more than 5000 houses was reported in and near Torrevieja and Murcia. On December 28, 1908, a devastating earthquake hit Messina, Italy, causing 120,000 deaths and widespread damage. A much studied earthquake to affect that country struck on May 6, 1976 in the Friuli region near Gemona; about 965 persons were killed and 2280 injured.

On December 27, 1939, in Erzincan, Turkey, 23,000 lives were lost from a major earthquake. Similar killer earthquakes have occurred in Turkey and Iran in recent years.

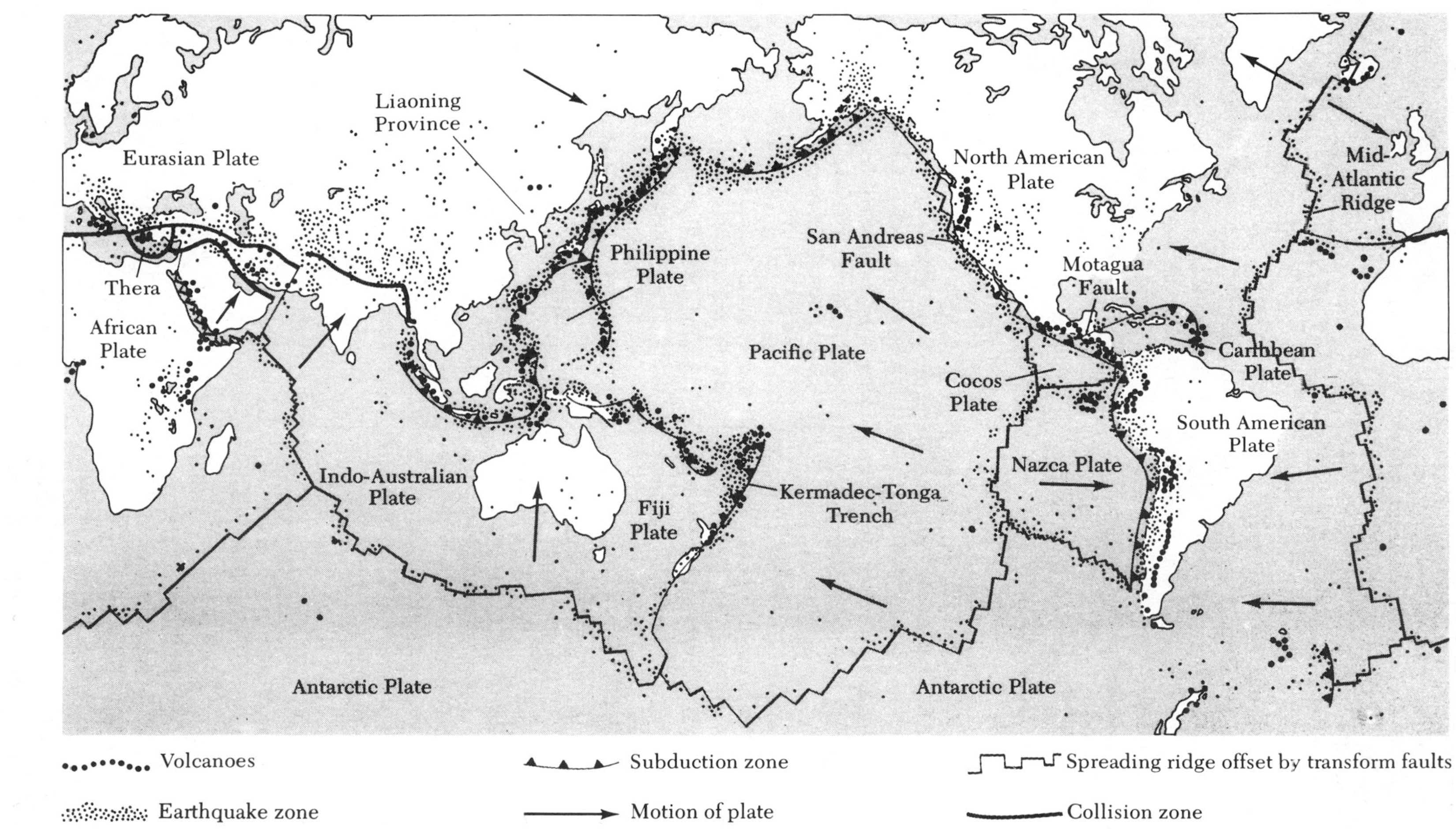

Figure 1-2 Tectonic plates and worldwide distribution of earthquakes. (From *Earthquakes*, by Bruce A. Bolt. Copyright 1978, 1988 W. H. Freeman and Company. Used with permission.)

North of the Mediterranean margin, Europe is much more stable. However, destructive earthquakes do occur from time to time in Romania, Germany, Austria, and Switzerland, and even in the North Sea region and Scandinavia. For example, on October 8, 1927, an earthquake occurred near Schwadorf in Austria and caused damage in an area southeast of Vienna. This earthquake was felt in Hungary, Germany, and Czechoslovakia at distances of 250 km from the center of the disturbance. The seismicity in the North Sea is sufficiently significant to require attention to earthquake-resistant design of oil platforms there.

Even in Great Britain, damaging earthquakes have occurred in historical times. On December 17, 1896, an earthquake series occurred, the largest of which caused some damage in Hereford, a city of 4565 inhabited houses. About 200 chimneys had to be repaired or rebuilt, and the cathedral was slightly damaged. The area affected was just over 1000 km^2. The most recent moderate earthquake there was in north Wales on July 19, 1984.

An example of infrequent and dispersed seismicity is the occurrence of earthquakes in Australia. Nevertheless, this country does have some areas of significant present-day seismicity. Of particular interest is a damaging earthquake of moderate size that was centered near Meckering, Western Australia, on October 14, 1968 and was associated with fresh surface faulting about 30 km long.

During an earthquake, seismic waves radiate from the earthquake source somewhere below the ground surface. Although in natural earthquakes this source is spread out through a volume of rock, it is often convenient to imagine an earthquake source as a point from which the waves first emanate. This point is called the *earthquake focus*. The point on the ground surface directly above the focus is called the *earthquake epicenter*.

Although many foci are situated at shallow depths, in some regions they are hundreds of kilometers deep. Such regions include the South American Andes, the Tonga Islands, Samoa, the New Hebrides chain, the Japan Sea, Indonesia, and the Caribbean Antilles. On the average, the frequency of occurrence of earthquakes in these regions declines rapidly below a depth of 200 km, but some foci are as deep as 680 km. Rather arbitrarily, earthquakes with foci from 70–300 km deep are called *intermediate-focus*, and those below this depth are called *deep-focus*. Some intermediate- and deep-focus earthquakes are located away from the Pacific region: in the Hindu Kush, in Romania, in the Aegean Sea, and under Spain.

The shallow-focus earthquakes (focus depth less than 70 km) wreak the most devastation, and they contribute about three-quarters of the total energy released in earthquakes throughout the world. In California, for example, all of the known earthquakes to date have been shallow-focus. In fact, it has been shown that the great majority of earthquakes occurring in central California originate from foci in the upper 10 km of the Earth, and only a few are even as deep as 15 km.

Most moderate to large shallow earthquakes are followed, in the ensuing hours and even over the next several months, by numerous smaller earthquakes in the same vicinity. These earthquakes are called *aftershocks*, and large earthquakes are sometimes followed by remarkable numbers of them. The great Rat Island earthquake in the Aleutian Islands on February 4, 1965 was followed, within the next 24 days, by more than 750 aftershocks large enough to be recorded by distant seismographs. A few earthquakes are preceded by smaller *foreshocks* from the source area, and it has been suggested that these might be used to predict the main shock.

1.3 CAUSES OF EARTHQUAKES

1.3.1 Tectonic Earthquakes

In the time of the ancient Greeks it was natural to link the Aegean volcanoes with the earthquakes of the Mediterranean. As time went on, however, it became clear that most damaging earthquakes were in fact not caused by volcanic activity.

One recent and coherent explanation of the majority of earthquakes is in terms of what is called plate tectonics. The basic idea is that the Earth's outermost part (called the lithosphere) consists of several large and fairly stable slabs called plates. The ten largest plates are mapped in Figure 1-2. Each plate extends to a depth of about 80 km.

Moving plates of the Earth's surface (see Figures 1-2 and 1-3) provide an explanation for a great deal of the seismic activity of the world. Collisions between adjacent lithospheric plates, destruction of a slablike plate as it descends, or *subducts*, into a dipping zone beneath island arcs (see Figure 1-4), and spreading along mid-oceanic ridges are all mechanisms that produce significant straining and fracturing of crustal rocks. Thus, the earthquakes in these tectonically active boundary regions are called plate-edge earthquakes. The very hazardous shallow earthquakes of Chile, Peru, the eastern Caribbean, Central America, Southern Mexico, California, Southern Alaska, the Aleutians, the Kuriles, Japan, Taiwan, the Philippines, Indonesia, New Zealand, and the Alpine–Caucasian–Himalayan belt are of plate-edge type.

As the mechanics of the lithospheric plates becomes better understood, long-term predictions may be possi-

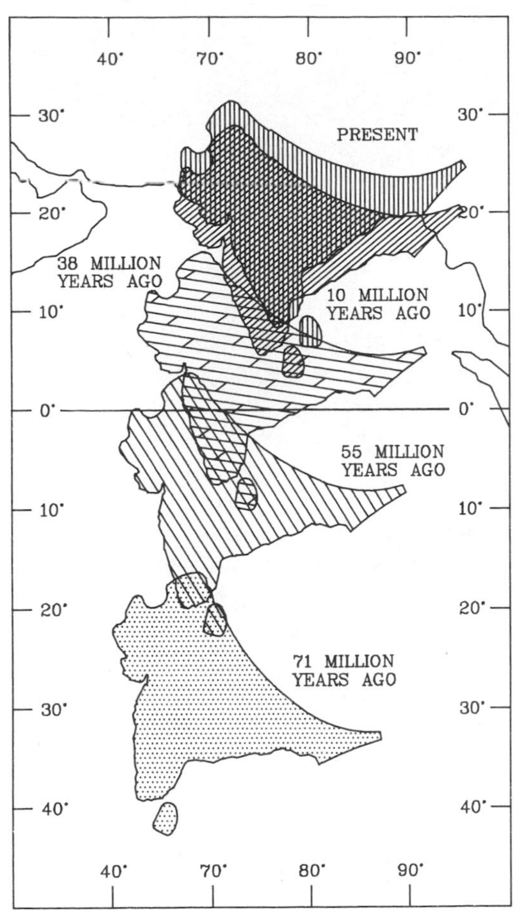

Figure 1-3 Continued drift of the Indian plate towards Asian plate causes major Himalayan earthquakes. (From *The Collision Between India and Eurasia*, by Molnar and Tapponnier. Copyright 1977 by Scientific American, Inc. All rights reserved.)

ble for plate-edge earthquakes. For example, many plates spread toward the subduction zones at rates of from 2–5 cm (about 1–2 in.) per year. Therefore in active arcs like the Aleutian and Japanese Islands and subduction zones like Chile and western Mexico, knowledge of the history of large earthquake occurrence might flag areas that currently lag in earthquake activity.

While the simple plate-tectonic theory is an important one for a general understanding of earthquakes and volcanoes, it does not explain all seismicity in detail, for within continental regions—away from boundries—large, devastating earthquakes sometimes occur. These intraplate earthquakes can be found on nearly every continent.

One example of such earthquakes is the Dashte-e-Bayaz earthquake of August 31, 1968 in northeastern Iran. In the United States, the most famous are the major earthquake series of 1811–1812, which occurred in the New Madrid area of Missouri along the Mississippi River, and the 1886 Charleston (South Carolina) earthquake. One important group, which seems to bear no simple mechanical relation to the present plate edges, has occurred in northern China.

Such major internal seismic activity indicates that lithospheric plates are not rigid or free of internal rupture. The occurrence of intraplate earthquakes makes the prediction of earthquake occurrence difficult in many regions where there is a significant seismic risk.

1.3.2 Dilatancy in the Crustal Rocks

At a depth in the crust of 5 km or so, the lithostatic pressure (due to the weight of the overlying rocks) is already about equal to the strength of typical uncracked rock samples at the temperature (500°C) and pressure appropriate for that depth. If no other factors entered, the shearing forces required to bring about sudden brittle failure and frictional slip along a crack would never be attained; rather, the rock would deform plastically. A way around this problem was the discovery that the presence of water provides a mechanism for sudden rupture by reduction of the effective friction along crack boundaries.

Soviet seismologists in the early 1960s, working with earthquakes in the Garm region, reported that P-wave velocities in the crust appeared to decrease and then increase in a region before a sizable earthquake occurred in it. Similar observations subsequently were obtained in a number of places in the United States.

Studies of the time of travel of P and S waves before the 1971 San Fernando earthquake indicated that four years before, the ratio of the velocity of the P waves to that of the S waves decreased rather suddenly by 10 percent from its average value of 1.75. Thereafter there was a steady increase in this ratio back to a more normal value. One explanation is the dilatancy model. This states that as the crustal rocks become strained, cracking occurs locally and the volume of rock increases (the rock *dilates*). Cracking may occur too quickly for ground water to flow into the dilated volume to fill the spaces, so the cracks become vapor-filled. The consequent fall in pore pressure leads to a reduction mainly in P-wave velocities. Subsequent diffusion of ground water into the dry cracks increases the pore pressure and provides water for lubrication along the walls of the cracks, so that the P-wave velocity increases again (see Figure 1-31 in Section 1.10 below).

The full implications and relevance of the dilatancy theory of earthquake genesis are not yet clear, but the hypothesis is attractive in that it is consistent with precursory changes in ground levels, electrical conductivity, and other physical properties which have been noted in the past before earthquakes. The theory has a potential for forecasting earthquakes under certain circumstances. For example, measurement of the P veloc-

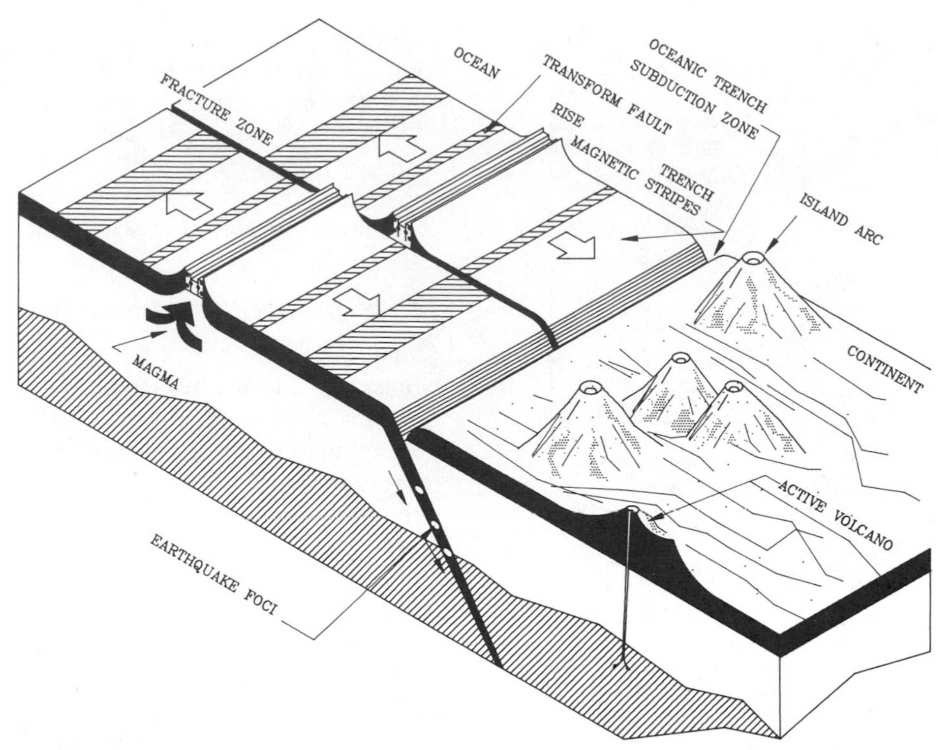

Figure 1-4 A sketch of the Earth's crust showing mid-oceanic ridges and active continental margin along a deep trench.

ity in the vicinity of large reservoirs before and after impounding of water might provide a more direct method of indicating an approaching seismic event near dams than is now available.

1.3.3 Explosions

Earthquakes may be produced by the underground detonation of chemical or nuclear devices. When a nuclear device is detonated in a borehole underground, enormous nuclear energy is released. Underground nuclear explosions fired during the past several decades at a number of test sites around the world have produced substantial earthquakes (up to magnitude 7.0). Resultant seismic waves have traveled throughout the Earth's interior to be recorded at distant seismographic stations.

1.3.4 Volcanic Earthquakes

As Figure 1-4 shows, volcanoes and earthquakes often occur together along the margins of plates around the world. As with earthquakes, there are also intraplate volcanic regions, such as the Hawaiian volcanoes.

Despite these tectonic connections between volcanoes and earthquakes, there is no evidence that moderate to major shallow earthquakes are not essentially all of tectonic, elastic-rebound type. Those earthquakes that can be reasonably associated with volcanoes are relatively rare and fall into three categories: (i) volcanic explosions, (ii) shallow earthquakes arising from magma movements, and (iii) sympathetic tectonic earthquakes.

Among the three, category (iii), tectonically associated with volcanoes, is the most difficult to tie down, as cases which may fit this category are rare. There is no report of significantly increased volcanic activity in the great 1964 Alaska earthquake, but Puyehue Volcano in the Andes erupted 48 h after the great 1960 Chilean earthquake. One might suppose that in a large earthquake the ground shaking would set up waves in reservoirs of magma; the general compression and dilatation of the gas–liquid melt may trigger volcanic activity.

1.3.5 Collapse Earthquakes

Collapse earthquakes are small earthquakes occurring in regions of underground caverns and mines. The immediate cause of ground shaking is the sudden collapse of the roof of the mine or cavern. An often observed variation is the *mine burst*. This rock rupture happens when the induced stress around the mine workings causes large masses of rock to fly off the mine face explosively, producing seismic waves. Mine bursts have been observed, for example, in Canada, and are especially common in deep mines of South Africa.

An intriguing variety of collapse earthquakes is sometimes produced by massive landsliding. For example, a

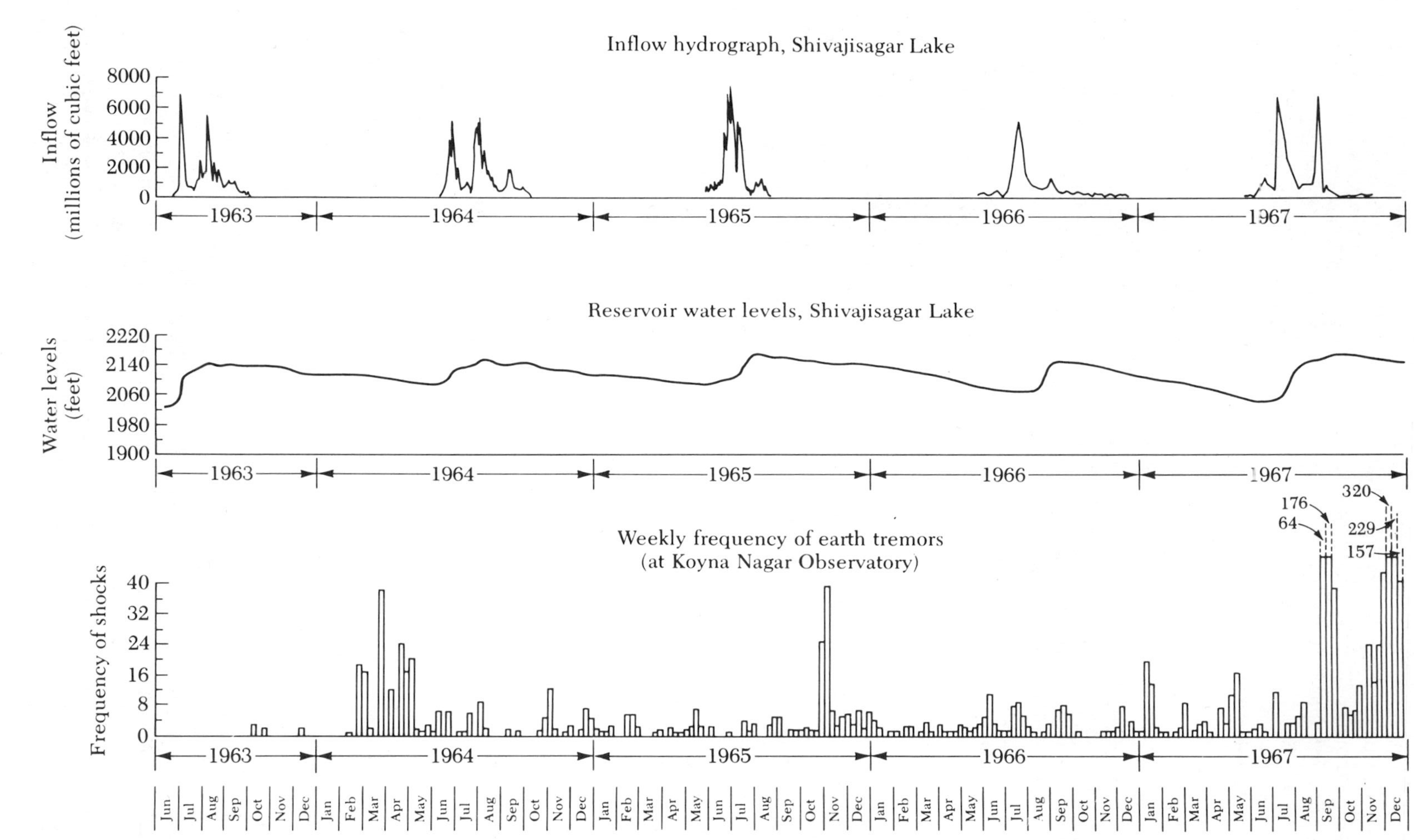

Figure 1-5 The relationship between reservoir level and local seismic activity at Koyna Dam. (From *Earthquakes*, by Bruce A. Bolt. Copyright 1978, 1988 W. H. Freeman and Company. Used with permission.)

spectacular landslide on April 25, 1974 along the Mantaro River, Peru, produced seismic waves equivalent to a magnitude-4.5 earthquake. The slide had a volume of 1.6×10^9 m^3 and killed about 450 people.

1.3.6 Large Reservoir-Induced Earthquakes

The idea that earthquakes might be triggered by impounding surface water is not new. In the 1870s the U.S. Corps of Engineers rejected proposals for major water storage in the Salton Sea in southern California on the grounds that such action might cause earthquakes. The first detailed evidence of such an effect came with the filling of Lake Mead behind Hoover Dam (height 221 m), Nevada–Arizona, beginning in 1935. Although there may have been some local seismicity before 1935, after 1936 earthquakes were much more common. Nearby seismographs subsequently showed that after 1940 the seismicity declined. The foci of hundreds of detected earthquakes cluster on steeply dipping faults on the east side of the lake and have focal depths of less than 8 km.

In Koyna, India, an earthquake (magnitude 6.5) centered close to the dam (height 103 m) caused significant damage on December 11, 1967. After impounding began in 1962, reports of local shaking became prevalent in a previously almost aseismic area. Seismographs showed that foci were concentrated at shallow depths under the lake. In 1967 a number of sizable earthquakes occurred, leading up to the principal earthquake of magnitude 6.5 on December 11. This ground motion caused significant damage to buildings nearby, killed 177 persons, and injured more than 1500. A strong-motion seismograph in the dam gallery registered a maximum acceleration of $0.63g$. The series of earthquakes recorded at Koyna has a pattern that seems to follow the rhythm of the rainfall (see Figure 1-5). A comparison of the frequency of earthquakes with the water level suggests that seismicity increases a few months after each rainy season when the reservoir level is highest. Such correlations are not so clear in other examples now known.

In the ensuing years, suggestive case histories have been accumulated for several dozen large dams, but only a few are well documented. Most of these dams are more than 100 m high, and although the geological framework at the sites varies, the most convincing examples of reservoir-induced earthquakes occur in tectonic regions with at least some history of earthquakes. Indeed, most of the thousands of large dams around the world give no sign of earthquake induction. A poll in 1976 showed that for only 4 percent of large dams had an earthquake been reported with magnitude greater than 3.0 within 16 km of the dam.

Figure 1-6 Normal fault at the Corinth Canal, Greece. (Photo courtesy of L. Weiss.)

1.4 EARTHQUAKE FAULT SOURCES

Field observations show that abrupt changes in the structure of rocks are common. In some places one type of rock can be seen butting up against rock of quite another type along a plane of contact. Such offsets of geological structure are called *faults*. Clear vertical offset of layers of rock along an exposed fault in the wall of the Corinth canal, Greece, can be seen in Figure 1-6.

Faults may range in length from a few meters to many kilometers and are drawn on a geological map as continuous or broken lines (see Figure 1-7). The presence of such faults indicates that, at some time in the past, movement took place along them. Such movement could have been either slow slip, which produces no ground shaking, or sudden rupture (an earthquake). Figure 1-8 shows one of the most famous examples of sudden fault rupture slips: that of the San Andreas fault in April 1906. In contrast, the observed surface faulting of most shallow-focus earthquakes is much shorter and shows much less offset. Indeed, in the majority of earthquakes, fault rupture does not reach the surface and consequently is not directly visible. Geological mappings and geophysical work show that faults seen at the surface sometimes extend to depths of tens of kilometers in the Earth's crust.

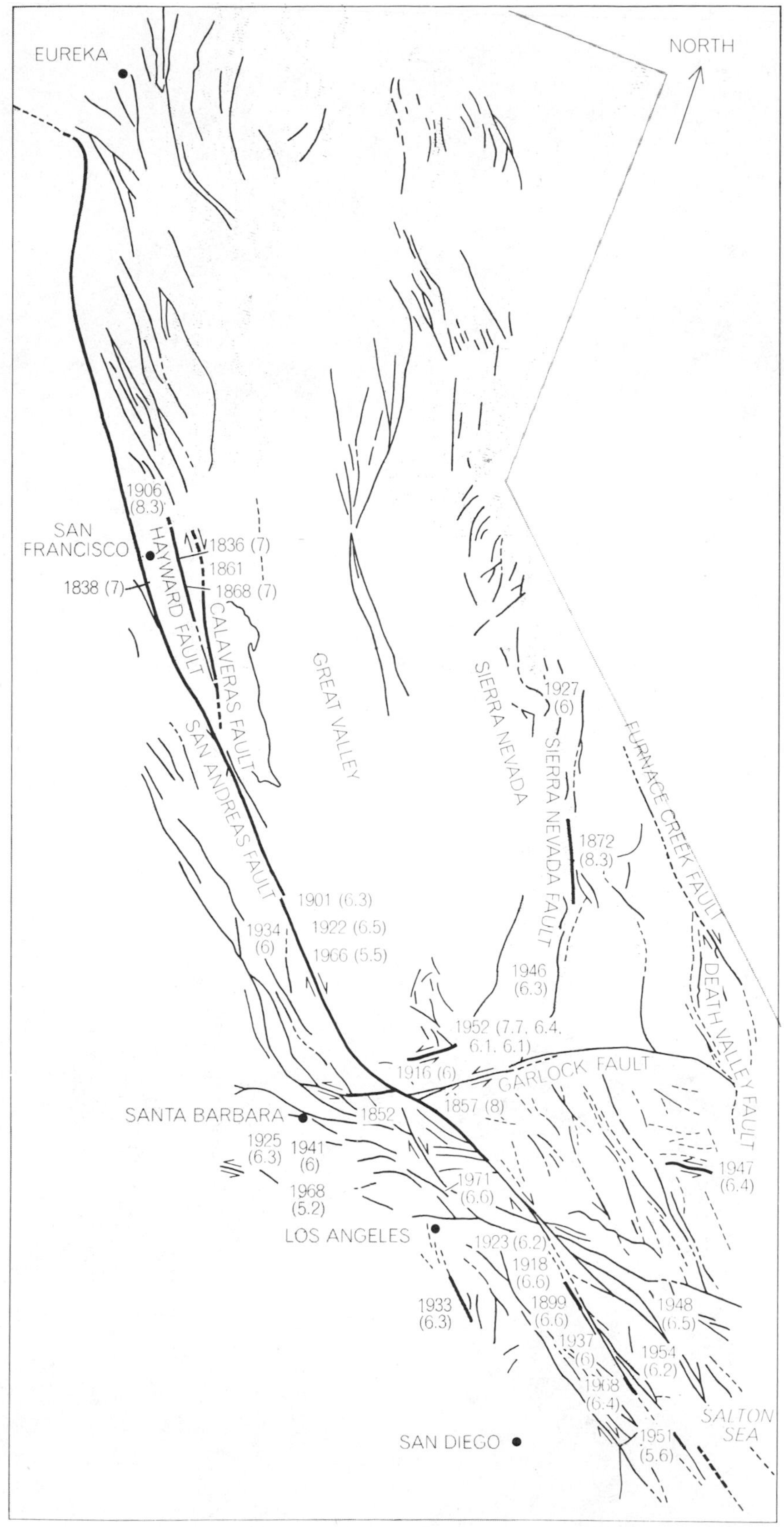

Figure 1-7 A simplified fault map of California. (From *The San Andreas Fault*, by Don L. Anderson. Copyright 1971 by Scientific American, Inc. All rights reserved.)

Figure 1-8 Right-lateral horizontal movement of the San Andreas Fault in the 1906 earthquake across the Old Sir Francis Highway. (Photo by G. K. Gilbert; courtesy of USGS.)

It must be emphasized that most faults plotted on geological maps are inactive. However, sometimes previously unrecognized active faults are discovered from fresh ground breakage during an earthquake. Thus, a fault was delineated by a line of cracks in open fields south of Oroville after the Oroville earthquake of August 1, 1975. The last displacement to occur along a typical fault may have taken place tens of thousands or even millions of years ago. The local disruptive forces in the earth nearby may have subsided long ago, and chemical processes involving water movement may have healed the ruptures, particularly at depth. Such an inactive fault is not now the site of earthquakes and may never be again.

Figure 1-9 Normal fault scarp associated with the Dixie Valley – Fairview Park earthquake (Nevada) of December 1954. (Photo courtesy of P. Byerly.)

In seismology and earthquake engineering, the primary interest is of course in active faults, along which rock displacements can be expected to occur. Many of these faults are in well-defined plate-edge regions of the Earth, such as the mid-oceanic ridges and young mountain ranges. However, sudden fault displacements can also occur away from regions of clear present tectonic activity.

Fault displacement in an earthquake may be almost entirely horizontal, as it was in the 1906 San Francisco earthquake along the San Andreas fault, but often large vertical motions occur, such as were evident in the 1954 Dixie Valley (Nevada) earthquake (Figure 1-9). In California in the 1971 San Fernando earthquake, an elevation change of 3 m occurred across the ruptured fault in some places.

The classification of faults depends only on the geometry and direction of relative slip. Various types are sketched in Figure 1-10. The *dip* of a fault is the angle that the fault surface makes with a horizontal plane, and the *strike* is the direction of the fault line exposed at the ground surface relative to the north.

A *strike-slip* fault, sometimes called a transcurrent fault, involves displacements of rock laterally, parallel to the strike. If when we stand on one side of a fault we see the motion on the other side is from left to right, the fault is *right-lateral* strike-slip. Similarly, we can identify *left-lateral* strike-slip.

A *dip-slip* fault is one in which the motion is largely parallel to the dip of the fault and thus has vertical components of displacement. A *normal* fault is one in which the rock above the inclined fault surface moves downward relative to the underlying crust. Faults with almost vertical slip are also included in this category.

A *reverse* fault is one in which the crust above the inclined fault surface moves upward relative to the

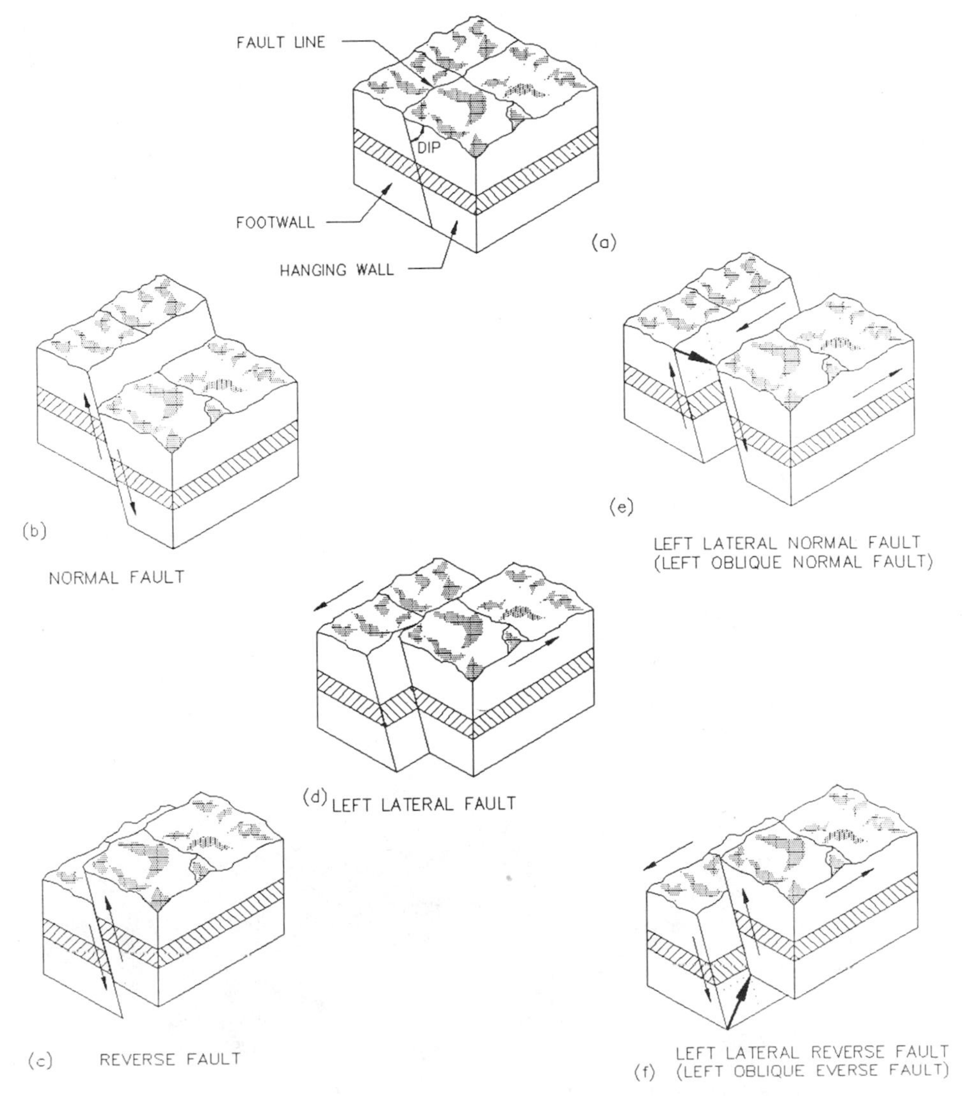

Figure 1-10 Diagrammatic sketches of fault types.

block below the fault. Thrust faults are included in this category but are generally restricted to cases when the dip angle is small.

In most cases, fault slip is a mixture of strike-slip and dip-slip and is called *oblique* faulting.

For many decades it has been known that displacement in fault zones occurs not only by sudden rupture in an earthquake but also by slow differential slippage of the sides of the fault. The fault is said to be undergoing *tectonic creep*. Slippage rates range from a few millimeters to several centimeters per year.

The best examples of fault creep come from the San Andreas zone near Hollister, California, where a winery built straddling the fault trace is being slowly deformed; in the town, sidewalks, curbs, fences and homes are being offset. On the Hayward fault, on the east side of San Francisco Bay, many structures are being deformed and even seriously damaged by slow slip, including a large water-supply tunnel, a drainage culvert, and railroad tracks that intersect the zone.

Horizontal fault slippage has now also been detected on other faults around the world, including the north Anatolian fault at Ismetpasa in Turkey and along the Jordan Valley rift in Israel. Usually, such episodes of fault slip are aseismic—that is, they do not produce local earthquakes.

It is sometimes argued that a large, damaging earthquake will not be generated along a fault that is undergoing slow fault slip, because the slippage allows the strain in the crustal rocks to be relieved periodically without sudden rupture. However, an alternative view is also plausible. It may be that, as the elastic crystalline rocks of the deeper crust strain elastically and accumulate the energy to be released in a nearby earthquake, the weak gouge material at the top of the fault zone is carried along by the adjacent stronger rock to the side and underneath. This would mean that the slow slip in the gouge seen at the surface is an indication that strain is being stored in the basement rocks. The implication of this view is that, on portions of the fault where slippage occurs, an earthquake at depth could result from sudden rupture, but surface offset would be reduced. On the portion where slippage is small or nonexistent, offsets would be maximum. A prediction of this kind can be checked after earthquakes occur near places where slippage is known to be taking place.

Sometimes aseismic slip is observed at the ground surface along a ruptured fault that has produced a substantial earthquake. For example, along the San Andreas fault break in the 1966 earthquake on June 27 near Parkfield, California, the offset of road pavement increased by a few centimeters in the days following the main earthquake. Such continued adjustment of the crustal rock after the initial major offset is probably caused partly by aftershocks and partly by the yielding of the weaker surface rocks and gouge in the fault zone as they accommodate to the new tectonic pressures in the region.

It is clear that slow slippage, when it occurs in built-up areas, may have unfortunate economic consequences. This is another reason why certain types of structures should not be built across faults if at all possible. When utility lines, roads, and railroads must be laid across active faults, they should have jointed or flexible sections in the fault zone.

1.5 SEISMIC WAVES

Three basic types of elastic waves make up the shaking that is felt and causes damage in an earthquake. These waves are similar in many important ways to the famil-

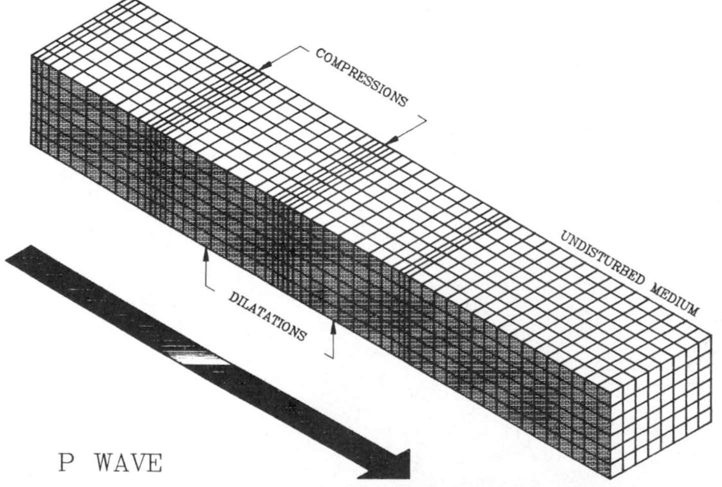

Figure 1-11 Ground motion near the ground surface due to P waves. (From *Nuclear Explosions and Earthquakes*, by Bruce A. Bolt. Copyright 1976 W. H. Freeman and Company. Used with permission.)

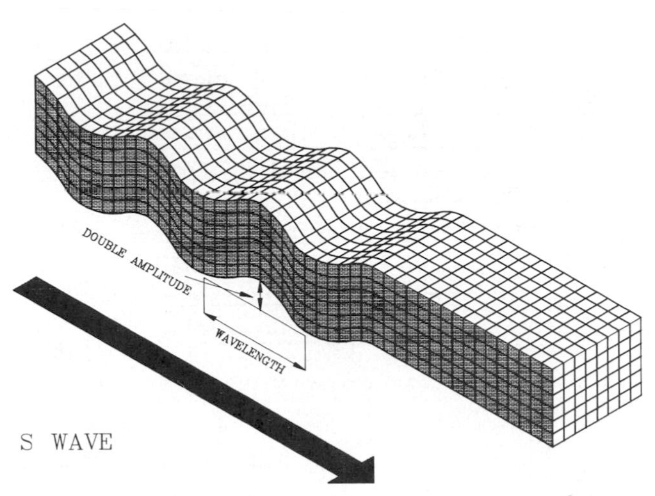

Figure 1-12 Ground motion near the ground surface due to S waves. (From *Nuclear Explosions and Earthquakes*, by Bruce A. Bolt. Copyright 1976 W. H. Freeman and Company. Used with permission.)

iar waves in air, water, and gelatin. Of the three, only two propagate within a body of solid rock. The faster of the seismic waves is appropriately called the primary or P wave. Its motion is the same as that of a sound wave in a fluid, in that, as it spreads out, it alternately pushes (compresses) and pulls (dilates) the rock (see Figure 1-11). These P waves, just like sound waves, are able to travel through both solid rock, such as granite mountains, and liquid material, such as volcanic magma or the water of the oceans.

The slower wave through the body of rock is called the secondary or S wave. As an S wave propagates, it shears the rocks sideways, at right angles to the direction of travel (see Figure 1-12). Thus, at the ground surface S waves can produce both vertical and horizontal motions. The S waves cannot propagate in the liquid parts of the Earth, such as the oceans, and their amplitude is significantly reduced in liquefied soil.

The actual speed of P and S seismic waves depends on the density and elastic properties of the rocks and soil through which they pass. In most earthquakes, the P waves are felt first. The effect is similar to a sonic boom that bumps and rattles windows. Some seconds later, the S waves arrive with their significant component of side-to-side motion, so that the ground shaking is both vertical and horizontal. This S-wave motion is most effective in damaging structures.

The third general type of earthquake wave is called a surface wave, because its motion is restricted to near the ground surface. Such waves correspond to ripples of water that travel across a lake. Most of the wave motion is located at the outside surface itself, and as the depth below this surface increases, the wave displacements become less and less.

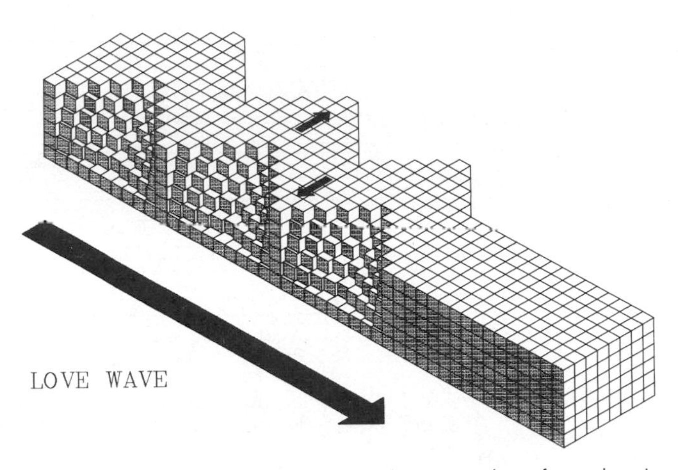

Figure 1-13 Ground motion near the ground surface due to Love waves. (From *Nuclear Explosions and Earthquakes*, by Bruce A. Bolt. Copyright 1976 W. H. Freeman and Company. Used with permission.)

Surface waves in earthquakes can be divided into two types. The first is called a Love wave. Its motion is essentially the same as that of S waves that have no vertical displacement; it moves the ground side to side in a horizontal plane parallel to the Earth's surface, but at right angles to the direction of propagation, as can be seen from the illustration in Figure 1-13. The second type of surface wave is known as a Rayleigh wave. Like rolling ocean waves, the pieces of rock disturbed by a Rayleigh wave move both vertically and horizontally in a vertical plane pointed in the direction in which the waves are traveling, as shown by the arrows in Figure 1-14. Each piece of rock moves in an ellipse as the wave passes.

Surface waves travel more slowly than body waves, and of the two kinds of surface waves, Love waves generally travel faster than Rayleigh waves. Thus, as the waves radiate outwards from the earthquake source into

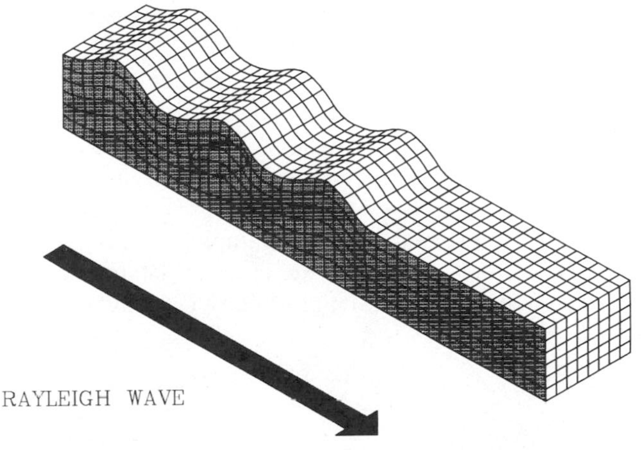

Figure 1-14 Ground motion near the ground surface due to Rayleigh waves. (From *Nuclear Explosions and Earthquakes*, by Bruce A. Bolt. Copyright 1976 W. H. Freeman and Company. Used with permission.)

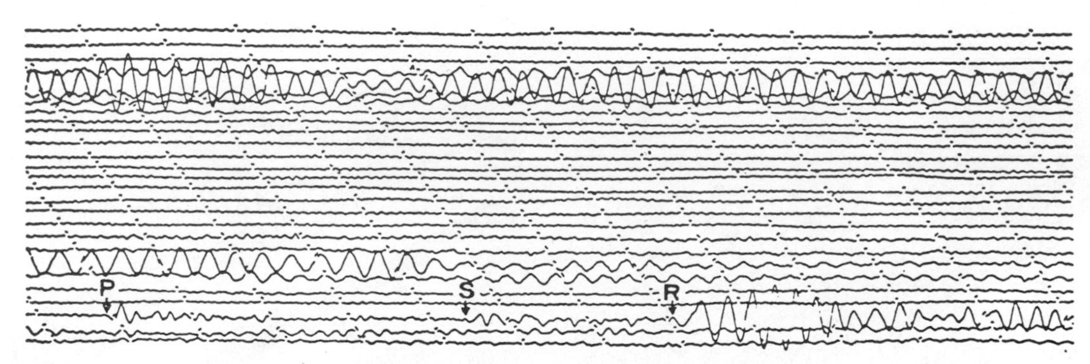

Figure 1-15 A seismograph record of the vertical component of a distant earthquake on which the arrival of P, S and Rayleigh waves are marked. (Time increases from left to right.)

the rocks of the Earth's crust, the different types of waves separate out from one another in a predictable pattern. An illustration of the pattern seen at a distant place is shown in Figure 1-15. In this example the seismograph has recorded only the vertical motion of the ground, and so the seismogram contains only P, S, and Rayleigh waves, because Love waves are not recorded by vertical instruments.

When the body waves (the P and S waves) move through the layers of rock in the crust, they are reflected or refracted at the interfaces between rock types, as illustrated in Figure 1-16a. Also, whenever either one is reflected or refracted, some of the energy of one type is converted to waves of the other type (see Figure 1-16b).

When P and S waves reach the surface of the ground, most of their energy is reflected back into the crust, so that the surface is affected almost simultaneously by upward- and downward-moving waves. For this reason considerable amplification of shaking typically occurs near the surface—sometimes doubling the amplitude of the upcoming waves. This surface amplification enhances the shaking damage produced at the surface of the earth. Indeed, in many earthquakes mine workers below ground report less shaking than people on the surface.

Seismic waves of all types are progressively damped as they travel because of the inelastic properties of the rocks and soils. The attenuation of S waves is greater than that of P waves, but for both types attenuation increases as wave frequency increases. One useful seismological quantity to measure damping is the parameter Q such that the amplitude A at a distance d of a wave frequency f (hertz) and velocity C is given by

$$A = A_0 e^{-(\pi f d/QC)}$$

For P and S waves in sediments, Q is about 500 and 200, respectively.

The above physical description is approximate, and while it has been verified closely for waves recorded by seismographs at a considerable distance from the wave source (the *far field*), it is not adequate to explain important details of the heavy shaking near the center of a large earthquake (the *near field*). Near a fault that is suddenly rupturing, the strong ground shaking in the associated earthquake consists of a mixture of various kinds of seismic waves that have not separated very

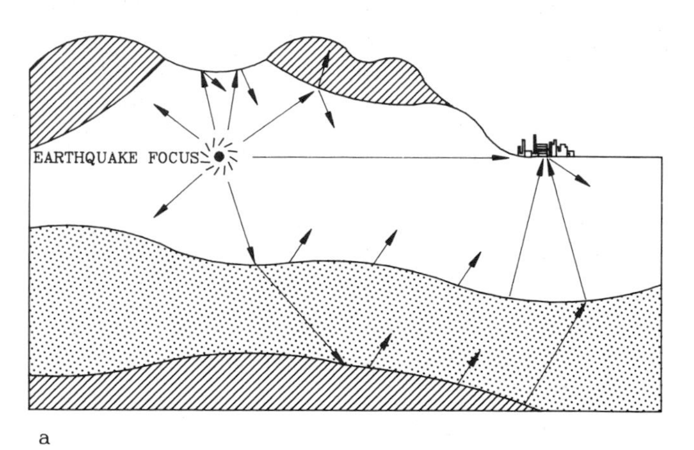

a

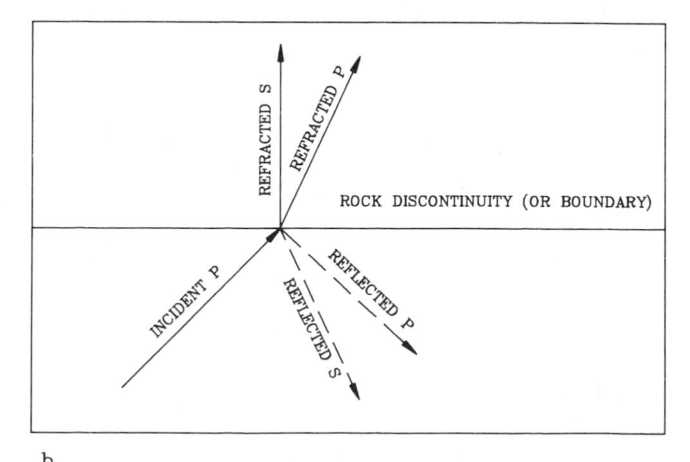

b

Figure 1-16 Reflection, refraction, and transformation of body waves. (From *Nuclear Explosions and Earthquakes*, by Bruce A. Bolt. Copyright 1976 W. H. Freeman and Company. Used with permission.)

distinctly. To complicate the matter, because the source of radiating seismic energy is itself spread out across an area, the types of ground motion may be further mixed together. This complication makes identification of P, S, and surface waves on strong-motion records obtained near the rupturing fault particularly difficult. However, much progress in this skill, based on intense study and theoretical modeling, has been made in recent years. This advance has made possible the computation of realistic ground motions at specified sites for engineering design purposes.

A final point about seismic waves is worth noting here. There is considerable evidence, observational and theoretical, that earthquake waves are affected by both soil conditions and topography. For example, in weathered surface rocks, in alluvium, and in water-filled soil, the size of seismic waves may be either increased or decreased, depending on their frequency, as they pass to the surface from the more rigid basement rock.

1.6 EARTHQUAKE DAMAGE MECHANISMS

Earthquakes can damage structures in various ways, such as:

1. by inertial forces generated by severe ground shaking;
2. by earthquake-induced fires;
3. by changes in the physical properties of the foundation soils (e.g., consolidation, settling, and liquefaction);
4. by direct fault displacement at the site of a structure;
5. by landslides, or other surficial movements;
6. by seismically induced water waves such as seismic sea waves (tsunamis) or fluid motions in reservoirs and lakes (seiches);
7. by large-scale tectonic changes in ground elevation.

Of the above categories, by far the most serious and widespread earthquake damage and accompanying loss of life are caused by severe ground shaking. The bulk of this handbook is devoted to design techniques and measures for reducing this type of hazard.

Fire hazards in earthquakes must also be emphasized. Vivid memories remain of the great conflagrations that followed the 1906 San Francisco earthquake and the 1923 Tokyo earthquake. In the former earthquake perhaps 20 percent of the total loss was due directly to ground motions. However, the fire, which in three days burned 12 km^2 and 521 blocks of downtown San Francisco, was the major property hazard.

The Managua (Nicaragua) earthquake of December 23, 1972 provides a more recent example of fire hazards associated with earthquakes. The second floor of the central fire station, built in 1964 to withstand earthquake damage, collapsed, crushing fire apparatus, killing two firemen, and injuring others. Fires soon began to break out in the city. All fire-fighting equipment and personnel continued operations on a 24-h basis for

Figure 1-17 Tilting of buildings due to soil liquefaction during the Niigata (Japan) earthquake of 1964.

seven days. On December 29, fires were still burning in the downtown section.

Soil-related problems have caused major economic loss in past earthquakes. One classic example of this type of damage happened in the 1964 earthquake at Niigata, Japan. The maximum ground acceleration was approximately 0.16*g*, which—considering the amount of damage—is not high. Expansion of the modern city of Niigata had involved reclamation of land along the Shinano river. In the newly deposited and reclaimed land areas many buildings tilted or subsided as a result of soil liquefaction (see Figure 1-17). Altogether 3018 houses were destroyed and 9750 were moderately or severely damaged in Niigata prefecture alone; most of the damage was caused by cracking and unequal settlement of the ground. About 15,000 houses in Niigata city were inundated by the collapse of a protective embankment along the Shinano river. The number of deaths was only 26. Precautionary and design measures against earthquake-induced soil problems are discussed in Chapter 11.

Perhaps surface fault displacements are the most frightening aspect of earthquakes to the general public. However, although severe local damage has occurred in this way, compared with damage caused by strong ground shaking, this type of damage is rather rare. Even in very large earthquakes, the area exposed to direct surface fault displacement is much smaller than the area affected by strong ground shaking. One of the clearest examples of damage caused by direct fault displacement occurred in the Managua, Nicaragua earthquake of 1972, where four distinct faults ruptured under the city (see Figure 1-18). The total length of fault rupture within the city was about 20 km, and the maximum fault displacement on two of the faults reached about 30 cm. Even in this case, the total area damaged by direct faulting was less than one percent of the area damaged by strong ground shaking.

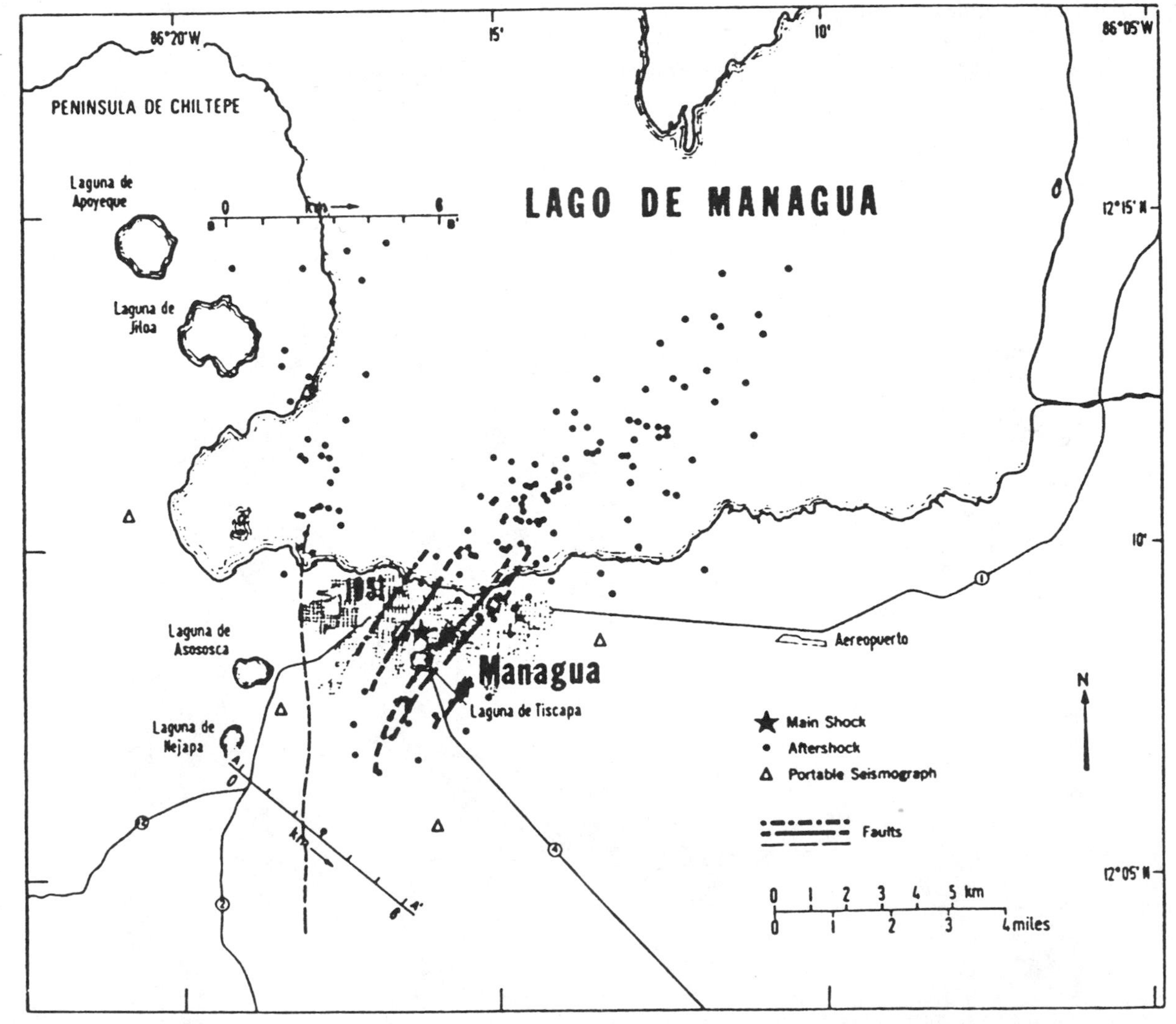

Figure 1-18 Faults rupturing under Managua (Nicaragua) during the earthquake of 1972.

Figure 1-19 Aerial view of Mt. Huascaran and the debris avalanche that destroyed Yungay and Ranrahirca in May 1970 Peru earthquake. (Photo courtesy of Servicio Aerofotografico National de Peru and L. Cluff.)

Figure 1-20 Surface effects of underlying thrust faulting in the Meckering earthquake of October 14, 1968, Western Australia. (Photo courtesy of West Australian Newspapers Limited.)

Earthquake-induced landslides and avalanches, although responsible for major devastation, are fortunately localized. The most pronounced example of this kind of damage occurred in the Peru earthquake of May 31, 1970. This magnitude-7.75 earthquake led to the greatest seismological disaster yet experienced in the Western Hemisphere. An enormous debris avalanche from the north peak of Huascaran Mountain (see Figure 1-19), amounting to 50,000,000 or more cubic meters of rock, snow, ice, and soil, traveled 15 km from the mountain to the town of Yungay with an estimated speed of 320 km/h. At least 18,000 people were buried under this avalanche, which covered the town of Ranrahirca and most of Yungay.

Earthquake-induced changes in ground elevations (see Figure 1-20) are not known to cause major injuries or loss of life. Their most important threat is the damage they can cause to agricultural irrigation works.

Seismic sea waves, or *tsunamis*, are long water waves generated by sudden displacements under water. The most common cause of significant tsunamis is the impulsive displacement along a submerged fault associated with a large earthquake. Because of the great earthquakes that occur around the Pacific, this ocean is particularly prone to seismic sea waves. For earthquakes to generate tsunamis, dip-slip faulting (see Figure 1-10) seems to be necessary; strike-slip faulting is almost never accompanied by tsunamis. History contains many accounts of great offshore earthquakes being accompanied by destructive tsunamis. On June 15, 1896, in the Honshu region of Japan, a tsunami with a visual runup height exceeding 20 m (65 ft) drowned about 26,000 people. More recently, the Chilean earthquake of 1960 caused a tsunami with a runup height of 10 m at Hilo, Hawaii (see Figure 1-21). A tsunami at Crescent City, California, caused by the great Alaskan earthquake of 1964, resulted in deaths and widespread local damage.

The most important scheme to prevent loss of life in the Pacific from tsunamis is the Seismic Sea Wave Warning System. The warning system is made up of several seismological observatories (including Berkeley, California; Tokyo, Japan; and Victoria, Canada) and about 30 tide stations around the Pacific Ocean. The time of travel of a tsunami wave from Chile to the Hawaiian islands, for example, is about 10 h, and from Chile to Japan about 20 h. Under this system, therefore,

Before tsunami

After tsunami

Figure 1-21 Damage at Hilo, Hawaii, due to tsunami of May 23, 1960. (Photos courtesy of R. L. Wiegel.)

there is ample time for alerts to be followed up by local police action along coastlines so that people can be evacuated.

Apart from the tsunami warning system, the hazard can be mitigated by adequate design of wharf, breakwater, and other facilities based on techniques of coastal engineering. Often, however, zoning around coastlines is desirable to prevent building in the lowest-lying areas where tsunamis are known to overwash the surface level. Sufficient information is nowadays usually available to allow local planners to make prudent decisions.

1.7 QUANTIFICATION OF EARTHQUAKES

1.7.1 Earthquake Intensity

The oldest useful measure of the "strength" of an earthquake is its *intensity*. Intensity is the measure of damage to works of man, to the ground surface, and of human reaction to the shaking. Because earthquake intensity assessments do not depend on instruments, but on the actual observation of effects in the meizoseismal zone, intensities can be assigned even to historical earthquakes. In this way, the historical record becomes of utmost importance in modern estimates of seismological risk.

The first intensity scale was developed by de Rossi of Italy and Forel of Switzerland in the 1880s. This scale, with values from I to X, was used for reports of the intensity of the 1906 San Francisco earthquake, for example. A more refined scale was devised in 1902 by the Italian volcanologist and seismologist Mercalli with a twelve-degree range from I to XII. A version is given in Table 1-1, as modified by H. O. Wood and Frank Neumann to fit conditions in California.[1-9] The descriptions in Table 1-1 allow the damage to places affected by an earthquake to be rated numerically. These spot intensity ratings can often be separated by lines which form an isoseismal map (see Figure 1-22). Such intensity maps provide crude, but valuable information on the distribution of strong ground shaking, the effect of surficial soil and underlying geological strata, the extent of the source, and other matters pertinent to insurance and engineering problems.

Because intensity scales are subjective and depend upon social and construction conditions of a country,

Table 1-1 Modified Mercalli Intensity Scale (MMI) of 1931

I	Not felt except by a very few under especially favorable circumstances.
II	Felt only by a few persons at rest, especially on upper floors of buildings. Delicately suspended objects may swing.
III	Felt quite noticeably indoors, especially on upper floors of buildings, but many people do not recognize it as an earthquake. Standing motor cars may rock slightly. Vibration like passing of truck. Duration estimated.
IV	During the day felt indoors by many, outdoors by few. At night some awakened. Dishes, windows, doors disturbed; walls make cracking sound. Sensation like heavy truck striking building. Standing motor cars rocked noticeably.
V	Felt by nearly everyone, many awakened. Some dishes, windows, etc., broken; a few instances of cracked plaster; unstable objects overturned. Disturbances of trees, poles, and other tall objects sometimes noticed. Pendulum clocks may stop.
VI	Felt by all, many frightened and run outdoors. Some heavy furniture moved; a few instances of fallen plaster or damaged chimneys. Damage slight.
VII	Everybody runs outdoors. Damage negligible in buildings of good design and construction; slight to moderate in well-built ordinary structures; considerable in poorly built or badly designed structures; some chimneys broken. Noticed by persons driving motor cars.
VIII	Damage slight in specially designed structures; considerable in ordinary substantial buildings, with partial collapse; great in poorly built structures. Panel walls thrown out of frame structures. Fall of chimneys, factory stacks, columns, monuments, walls. Heavy furniture overturned. Sand and mud ejected in small amounts. Changes in well water. Persons driving motor cars disturbed.
IX	Damage considerable in specially designed structures; well-designed frame structures thrown out of plumb; great in substantial buildings, with partial collapse. Buildings shifted off foundations. Ground cracked conspicuously. Underground pipes broken.
X	Some well-built wooden structures destroyed; most masonry and frame structures destroyed with foundations; ground badly cracked. Rails bent. Landslides considerable from river banks and steep slopes. Shifted sand and mud. Water splashed (slopped) over banks.
XI	Few, if any, (masonry) structures remain standing. Bridges destroyed. Broad fissures in ground. Underground pipelines completely out of service. Earth slumps and land slips in soft ground. Rails bent greatly.
XII	Damage total. Practically all works of construction are damaged greatly or destroyed. Waves seen on ground surface. Lines of sight and level are distorted. Objects are thrown into the air.

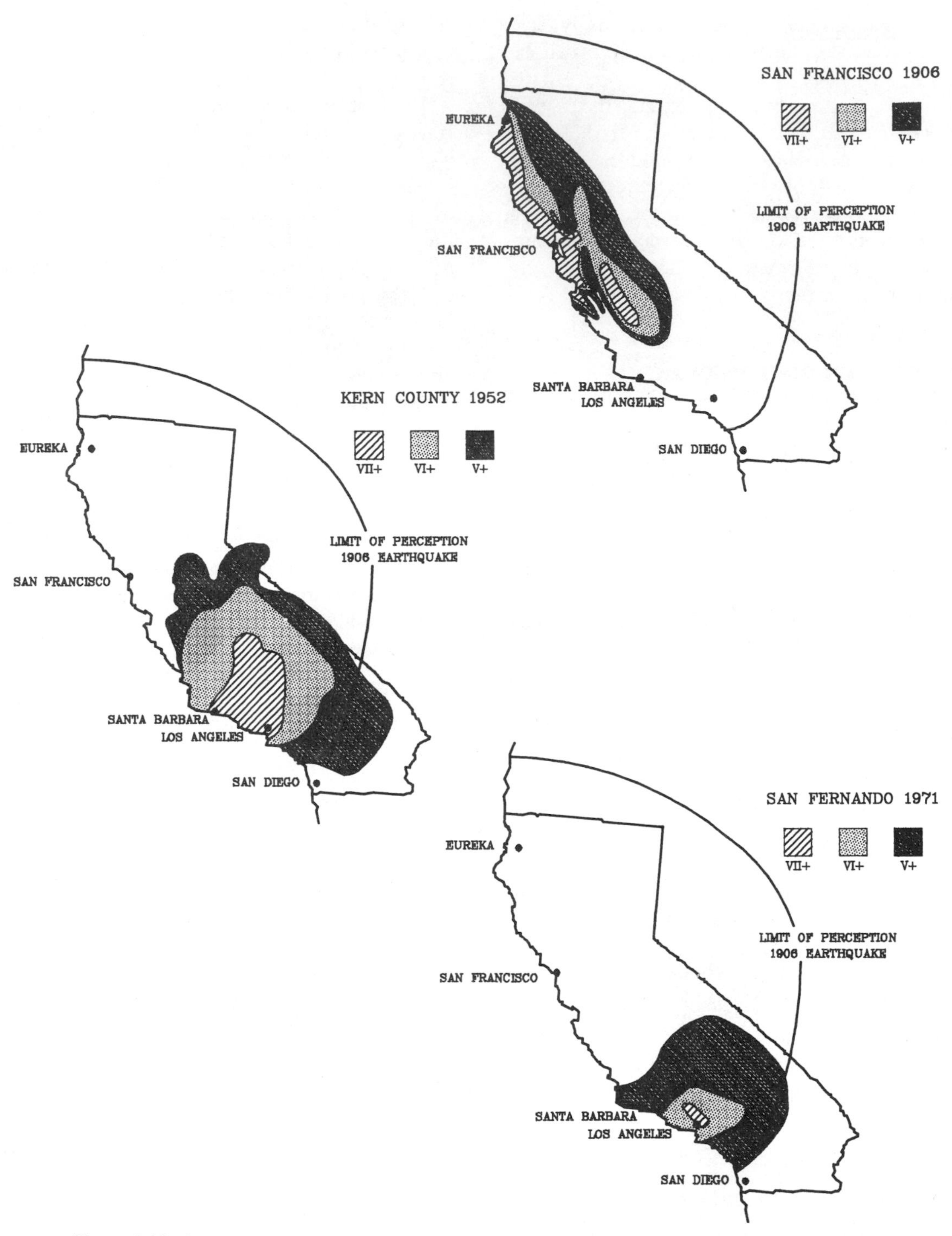

Figure 1-22 Isoseismal maps (modified Mercalli intensity scale) for three California earthquakes.

they need revising from time to time. Regional effects must be taken into account. In this respect, it is interesting to compare the Japanese scale (0 to VII) summarized in Table 1-2 with the modified Mercalli descriptions.

1.7.2 Earthquake Magnitude

If sizes of earthquakes are to be compared worldwide, a measure is needed that does not depend, as does intensity, on the density of population and type of construction. A strictly quantitative scale that can be applied to earthquakes in both inhabited and uninhabited regions was originated in 1931 by Wadati in Japan and developed by Charles Richter in 1935 in California.

Richter defined the magnitude of a local earthquake as the logarithm to base ten of the maximum seismic wave amplitude in microns (10^{-4} cm) recorded on a Wood–Anderson seismograph located at a distance of

Table 1-2 Japanese Seismic Intensity Scale

0	Not felt; too weak to be felt by humans; registered only by seismographs.
I	Slight: felt only feebly by persons at rest or by those who are sensitive to an earthquake.
II	Weak: felt by most persons, causing light shaking of windows and Japanese latticed sliding doors (shoji).
III	Rather strong: shaking of houses and buildings, heavy rattling of windows and Japanese latticed sliding doors, swinging of hanging objects, sometimes stopping pendulum clocks, and moving of liquids in vessels. Some persons are so frightened as to run out of doors.
IV	Strong: resulting in strong shaking of houses and buildings, overturning of unstable objects, spilling of liquid out of vessels.
V	Very strong: causing cracks in brick and plaster walls, overturning of stone lanterns and grave stones, etc. and damaging of chimneys and mud and plaster warehouses. Landslides in steep mountains are observed.
VI	Disastrous: causing demolition of more than 1% of Japanese wooden houses; landslides, fissures on flat ground accompanied sometimes by spouting of mud and water in low fields.
VII	Ruinous: causing demolition of almost all houses; large fissures and faults are observed.

100 km from the earthquake epicenter (see Figure 1-23). This means that every time the magnitude goes up by one unit, the amplitude of the earthquake waves increases 10 times. Since the fundamental period of the Wood–Anderson seismograph is 0.8 sec, it selectively amplifies those seismic waves with a period ranging approximately from 0.1 to 2 sec. Because the natural period of many building structures is within this range, the local Richter magnitude is of special value to engineers.

It follows from the definition of the magnitude that it has no theoretical upper or lower limits. However, the size of an earthquake is limited at the upper end by the strength of the rocks of the Earth's crust. Since 1935, only a few earthquakes have been recorded on seismographs that have had a magnitude over 8.0. At the other extreme, highly sensitive seismographs can record earthquakes with a magnitude of less than −2. See Table 1-3 for the average frequency of worldwide earthquakes of various magnitudes.

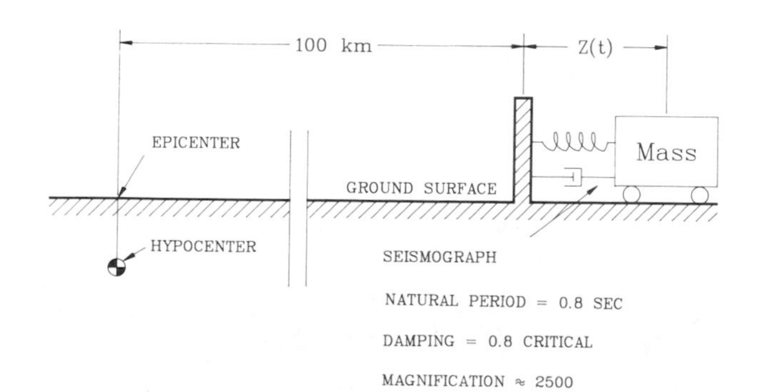

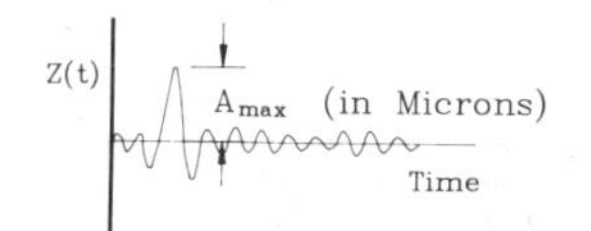

Figure 1-23 Definition of local Richter magnitude.

Generally speaking, shallow earthquakes have to attain Richter magnitudes of more than 5.5 before significant damage occurs near the source of the waves.

At its inception, the idea behind the Richter local magnitude scale (M_L) was a modest one. It was defined for Southern California, shallow earthquakes, and epicentral distances less than about 600 km. Today, the method has been extended to apply to a number of types of seismographs throughout the world (see Figure 1-24). Consequently a variety of magnitude scales, based on different formulas for epicentral distance and ways of choosing an appropriate wave amplitude, have emerged:

Surface-Wave Magnitude (M_s) Surface waves with a period around 20 sec are often dominant on the seismograph records of distant earthquakes (epicentral distances of more than 2000 km). To quantify these earthquakes, Gutenberg defined a magnitude scale (M_s) which is based on measuring the amplitude of surface waves with a period of 20 sec.

Body-Wave Magnitude (m_b) Deep-focus earthquakes have only small or insignificant trains of surface waves. Hence, it has become routine in seismology to measure the amplitude of the P wave, which is not affected by the focal depth of the source, and thereby determine a P-wave magnitude (m_b). This magnitude type has also been found useful in continental regions

Table 1-3 Worldwide Earthquakes per Year

Magnitude M_s	Average No. > M_s
8	1
7	15
6	100
5	3,000
4	15,000
3	> 100,000

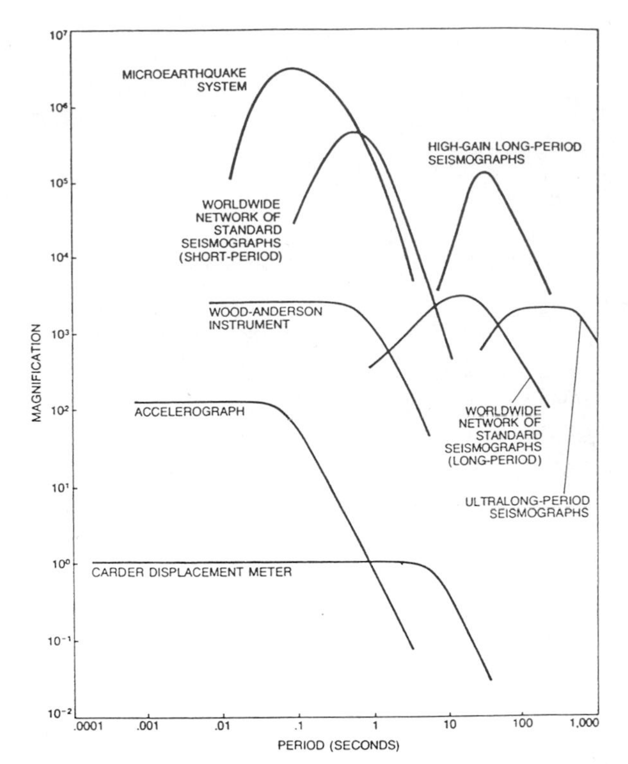

Figure 1-24 Amplification of seismic waves by various seismographs. (From *The Motion of Ground in Earthquakes*, by David M. Boore. Copyright 1977 by Scientific American, Inc. All rights reserved.)

like the eastern United States where no Wood–Anderson instruments have operated historically.

Duration Magnitude (M_D) This magnitude scale has come into wide use for rapidly measuring small earthquakes ($M_L \leqslant 3$). It assigns a magnitude to an earthquake based on total duration of the earthquake in seconds.(1-12) The duration magnitude for small events usually correlates with the measured Richter magnitude (M_L), but calibration tests are not always available. Because M_D is mainly intended for measuring small earthquakes, it is more important to seismologists than it is to engineers, except perhaps for seismic-risk studies.

Moment Magnitude (M_w) Because of significant shortcomings of M_L, m_b, and to a lesser degree M_s in distinguishing between great earthquakes, the moment-magnitude scale was devised.(1-13) This scale assigns a magnitude to the earthquake in accordance with its seismic moment M_0, which is directly related to the energy released by the earthquake:

$$M_w = (\log M_0/1.5) - 10.7$$

where M_0 is the seismic moment in dyn-cm.

Magnitude Saturation As described earlier, the Richter magnitude scale (M_L) measures the seismic waves in a period range of particular importance to structural engineers (about 0.1–2 sec). This range corresponds approximately to wavelengths of 300 m to 6 km. Hence, although theoretically there is no upper bound to Richter magnitude, it underestimates more seriously the strength of earthquakes produced by the longer fault rupture lengths. The saturation point for the Richter magnitude scale is about $M_L = 7$. The body-wave magnitude (m_b) saturates at about the same value.

In contrast, the surface-wave magnitude (M_s), which uses the amplitude of 20-sec surface waves (wavelength

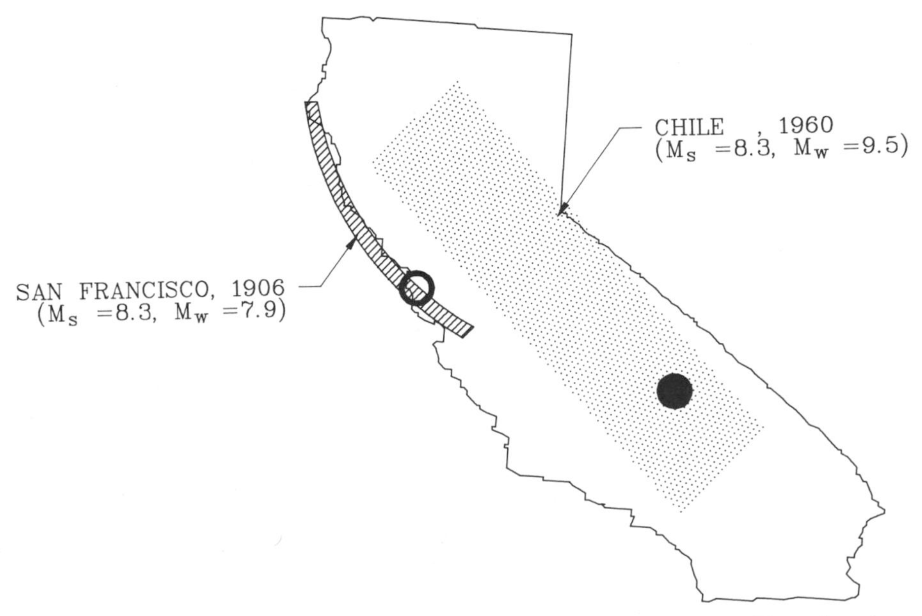

Figure 1-25 Fault rupture area for the San Francisco 1906 and Chile 1960 earthquakes. (Modified from *The Motion of Ground in Earthquakes*, By David M. Boore. Copyright 1977 by Scientific American, Inc. All rights reserved.)

of about 60 km), saturates at about $M_s = 8$. Its inadequacy in measuring size of greater earthquakes can be illustrated by comparing the San Francisco earthquake of 1906 and the Chilean earthquake of 1960 (see Figure 1-25). Both earthquakes had a magnitude M_s of 8.3. However, the area that ruptured in the San Francisco earthquake (the dashed area) was approximately 15 km deep and 400 km long, whereas the area that ruptured in the Chilean earthquake (the dotted area) had a width equal to half of the width of the state of California. The hollow circle shows the location of the epicenter of the San Francisco earthquake, and the solid circle the epicenter of the Chilean earthquake, superimposed on the map of California. Clearly the Chilean earthquake was a much larger event.

The moment-magnitude scale (M_w) is the only magnitude scale which does not suffer from the abovementioned saturation problem for great earthquakes. The reason is that it is directly based on the forces that work at the fault rupture to produce the earthquake and not on the recorded amplitude of specific types of seismic waves. Hence, as might be expected, when moment magnitudes were assigned to the San Francisco earthquake of 1906 and the Chilean earthquake of 1960, the magnitude of the San Francisco earthquake was about 7.9, whereas the magnitude of the Chilean earthquake was raised to 9.5. M_s and M_w for some great earthquakes are compared in Table 1-4.

In light of the above discussion, application of different scales have been suggested for measuring shallow earthquakes of various magnitudes:

M_D or M_L	for magnitudes less than 3
M_L or m_b	for magnitudes between 3 and 7
M_s	for magnitudes between 5 and 7.5
M_w	for magnitudes larger than 7.5

Table 1-4 Magnitudes of Some of the Great Earthquakes

Date	Region	M_s	M_w
July 9, 1905	Mongolia	$8\frac{1}{4}$	8.4
Jan. 31, 1906	Ecuador	8.6	8.8
April 18, 1906	San Francisco	$8\frac{1}{4}$	7.9
Jan. 3, 1911	Turkestan	8.4	7.7
Dec. 16, 1920	Kansu, China	8.5	7.8
Sept. 1, 1923	Kanto, Japan	8.2	7.9
March 2, 1933	Sanrika	8.5	8.4
May 24, 1940	Peru	8.0	8.2
April 6, 1943	Chile	7.9	8.2
Aug. 15, 1950	Assam	8.6	8.6
Nov. 4, 1952	Kamchatka	8	9.0
March 9, 1957	Aleutian Islands	8	9.1
Nov. 6, 1958	Kurile Islands	8.7	8.3
May 22, 1960	Chile	8.3	9.5
March 28, 1964	Alaska	8.4	9.2
Oct. 17, 1966	Peru	7.5	8.1
Aug. 11, 1969	Kurile Islands	7.8	8.2
Oct. 3, 1974	Peru	7.6	8.1
July 27, 1976	China	8.0	7.5
Aug. 16, 1976	Mindanao	8.2	8.1
March 3, 1985	Chile	7.8	7.5
Sep. 19, 1985	Mexico	8.1	8.0

1.8 EARTHQUAKE SOURCE MODELS

Field evidence in the 1906 California earthquake showed clearly that the strained rocks immediately west of the San Andreas fault had moved northwest relative to the rocks to the east. Displacements of adjacent points along the fault reached a maximum of 6 m near Olema in the Point Reyes region.

H. F. Reid[1-14] studied the triangulation surveys made by the U.S. Coast and Geodetic Survey across the region traversed by the 1906 fault break. These surveys, made in 1851–1865, in 1874–1892, and just after the earthquake, showed (i) small inconsistent changes in elevation along the San Andreas fault, (ii) significant horizontal displacements parallel to the fault trace, and (iii) movement of distant points on opposite sides of the fault of 3.2 m over the 50-year period, the west side moving north.

Based on geological evidence, geodetic surveys, and his own laboratory experiments, Reid put forth the elastic-rebound theory for a source mechanism that would generate seismic waves. This theory supposes that the crust of the Earth in many places is being slowly displaced by underlying forces. Differential displacements set up elastic strains that reach levels greater than can be endured by the rock. Ruptures (faults) then occur, and the strained rock rebounds along the fault under the elastic stresses until the strain is partly or wholly relieved (see Figures 1-26 and 1-27). This theory of the earthquake mechanism has been verified under many circumstances and has required only minor modification.

The strain slowly accumulating in the crust builds a reservoir of elastic energy, in the same way as a coiled spring does, so that at some place, the *focus*, within the strained zone, rupture suddenly commences, and spreads in all directions along the fault surface in a series of erratic movements due to the uneven strength of the rocks along the tear. This uneven propagation of the dislocation leads to bursts of high-frequency waves which travel into the earth to produce the seismic shaking that causes the damage to buildings. The fault rupture moves with a typical velocity of 2–3 km/sec, and the irregular steps of rupture occur in fractions of a second. Ground shaking away from the fault consists of all types of wave vibrations with different frequencies and amplitudes.

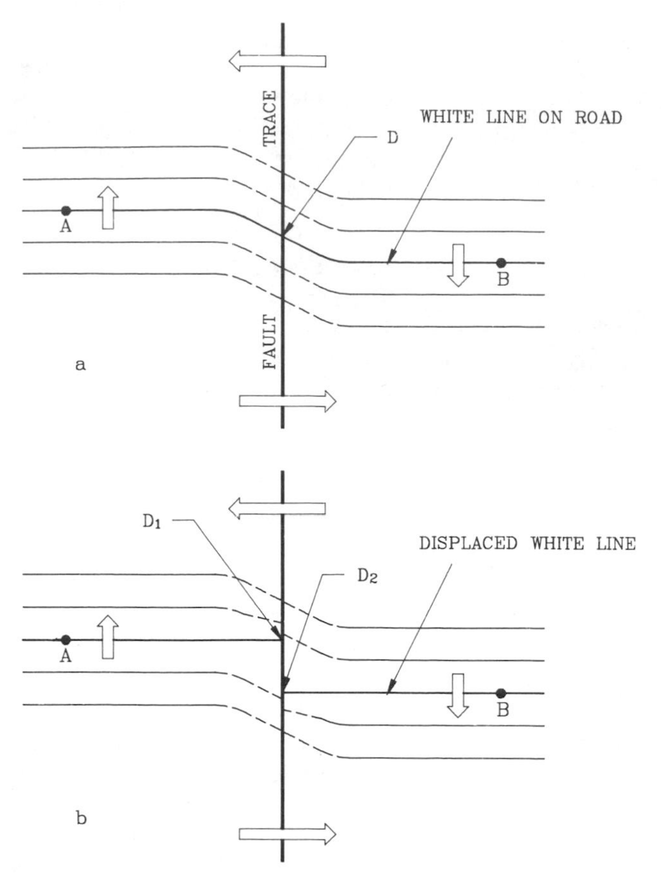

Figure 1-26 A bird's eye view of marker lines drawn along a road AB, which crosses a fault trace at the ground surface. (a) Elastic strain accumulation before fault rupture. (b) Final position after the fault rupture. (From *Earthquakes*, by Bruce A. Bolt. Copyright 1978, 1988 W. H. Freeman and Company. Used with permission.)

In 1964 and 1966, N. A. Haskell[1-15, 1-16] developed a model "in which the fault displacement is represented by a coherent wave only over segments of the fault and the radiations from adjacent sections are assumed to be statistically independent or incoherent." The physical situation in this model is that the rupture begins suddenly and then spreads with periods of acceleration and retardation along the weakly welded fault zone. In this model, the idea of statistical randomness of fault slip or "chattering" in irregular steps along the fault plane is introduced.

More recently, Das and Aki[1-17, 1-18] have considered a fault plane having various "barriers" distributed over it. They conceive that rupture would start near one of the barriers and then propagate over the fault plane until it is brought to rest or slowed at the next barrier. Sometimes the barriers are broken by the dislocation; sometimes the barriers remain unbroken but the dislocation reinitiates on the far side and continues; sometimes the barrier is not broken initially but, due to local repartitioning of the stresses and possible nonlinear effects, eventually breaks, perhaps with the occurrence of aftershocks.

The elastic-rebound model involving a moving dislocation along a fault plane segmented by barriers, over which roughnesses (or *asperities*) of various types are distributed stochastically, is thus the starting point for the modern interpretation of near-field records.[1-19] Based on this model, there have been recently a number of attempts to compute synthetic seismograms for points near the source, and comparisons have been made with observations (see Section 1.10).

As mentioned earlier, there are different kinds of fault ruptures. Some involve purely horizontal slip

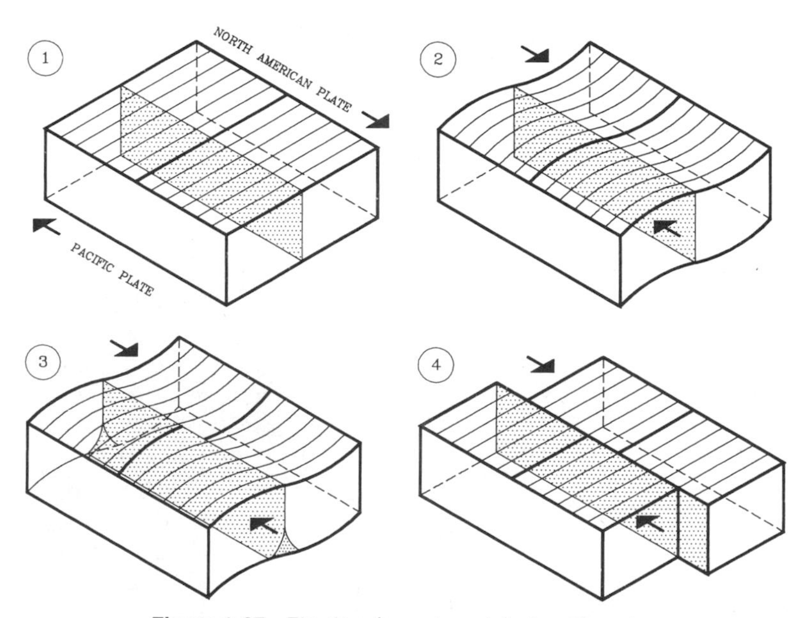

Figure 1-27 Elastic rebound model of earthquakes.

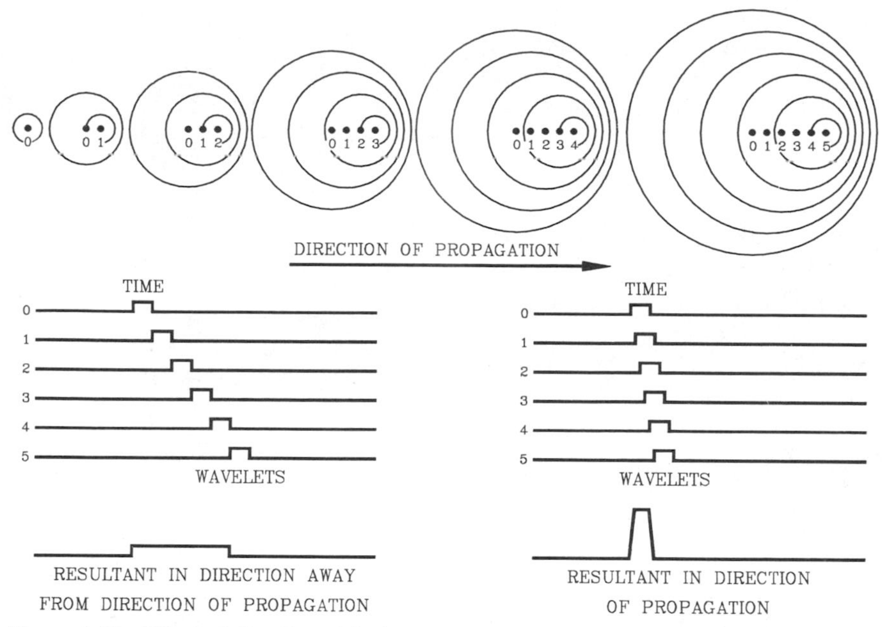

Figure 1-28 Effect of direction of fault rupture on ground motion experienced at a site. [After Benioff (1-20) and Singh (1-21).]

(strike-slip); some involve vertical slip (dip-slip). It might be expected that the wave patterns generated by fault geometries and mechanisms of different kinds would be different to a larger or lesser extent, because of the different radiation patterns produced. These different geometries can be modeled mathematically by appropriate radiation functions.

The theory must also incorporate effects of the motion of the source. The Doppler-like consequences will depend on the speed of fault rupture and the directions of faulting. The physical problem is analogous to the problem of sound emission from moving sources (but more difficult). The solution can be approached both kinematically and dynamically. The acoustic problem shows that in the far field the pressure is the same as when the source is at rest. However, in the near field the time dependence of both frequency and wave amplitude is a function of the azimuth of the site relative to the moving source (Figure 1-28).

We now summarize the main lines of approach to modeling the earthquake source mathematically. The first model is the kinematic approach in which the time history of the slip on the generating fault is known *a priori*. Several defining parameters may be specified, such as the shape, duration, and amplitude of the source (or source time-function and slip), the velocity of the slip over the fault surface, and the final area of the region over which the slip occurred. Numerous theoretical papers using this approach have been published (see the various discussions in Reference 1-22). The process is a kind of complicated curve fitting whereby the parameters of the source are varied in order to estimate by inspection the closeness of fit between recorded and computed near-field or far-field seismic waves. Once the seismic source is defined by this comparison process, then the estimated source parameters can be used to extrapolate from the known ground motions near a historical source to the future conditions required for engineering purposes.

A second approach is to use the differential equations involving the forces which produce the rupture. This dynamic procedure has received considerable emphasis lately. The basic model is a shear crack which is initiated in the preexisting stress field and which causes stress concentrations around the tip of the crack. These concentrations, in turn, cause the crack to grow. Many of the articles on this subject have been built on the work of Kostrov. For example, Burridge and Willis[1-23] obtained analytic expressions for particle accelerations in given directions from a uniformly growing elliptical crack, although they did not include the effect of crack stoppage (this unrealistic boundary condition is included in most work of this kind).

The key to the crack problem seems to be in modeling the physical processes of the typical crack, where there is interaction between the stress accumulation, the rate of crack growth, and the criterion of fracture. Most studies on dynamic shear cracks are concerned primarily with the actual rupture process, and so the crack is assumed to be embedded in an infinite homogeneous medium. Studies more concerned with the seismic waves that are recorded in the near field need a numerical approach, such as finite elements or finite differences, to handle realistic structural conditions.

The studies mentioned as kinematic and dynamic models are built around the elastic-rebound theory of slip on a fault. There are, however, more general studies that take a less specific view of the earthquake source. Work by Backus,[1-24] for example, has taken up the important idea of the uniqueness of the various source descriptions; the representation of an arbitrary source of seismic waves is given in terms of moment tensors. The idea is that any seismic source can, in principle, be expanded in terms of spatial moments; at wavelengths that are long compared to the fault dimensions, only the low-degree terms in the expansion need to be included. Thus, for small earthquakes or far-field problems, it is sufficient to represent a seismic source in terms of a single first-degree moment of the equivalent force, which is a symmetric second-rank tensor. Then, the waves calculated can be interpreted in terms of any specific model. In practice it turns out, however, that to represent the near field in this way, higher terms give very complicated tensor components and analytic evaluation may not be worthwhile.

It should be mentioned here that the scalar seismic moment (component of force couples along the fault ignored) is given by

$$M_0 = \mu A D$$

where μ is the rigidity of the material surrounding the fault, A is the slipped area, and D is the amount of slip (see Figure 1-29). The seismic moment is now the preferred parameter for specifying quantitatively the overall size of an earthquake source.

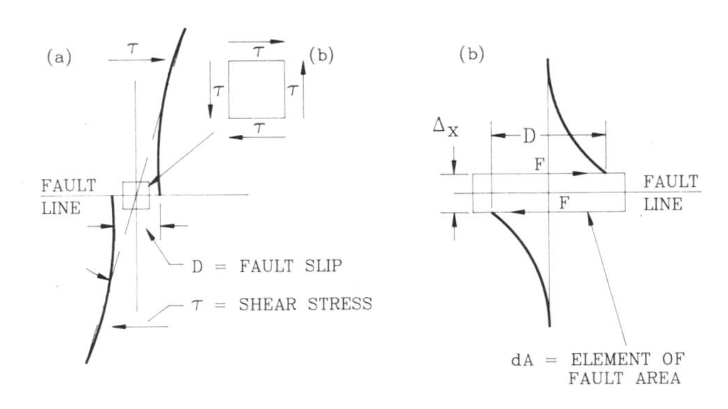

EQUIVALENT COUPLE MOMENT $\Delta M_0 = F \Delta x$

EQUIVALENT SHEAR STRAIN $= D/\Delta x$

EQUIVALENT SHEAR STRESS $= F/dA = \mu D/\Delta x$

μ = SHEAR MODULUS

$\Delta M_0 = \mu D dA;\ M_0 = \int \mu D dA;\ \underline{M_0 = \mu \bar{D} A}$

Figure 1-29 Definition of seismic moment.

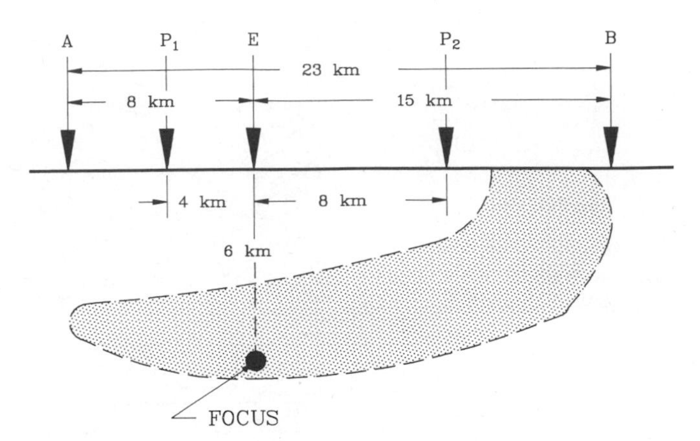

Figure 1-30 Simplified model of the vertical rupture surface for the Coyote Lake earthquake of 1979 in California.

Let us now summarize the physical model for the earthquake source generally accepted at present (see Figure 1-30). The source extends over a fault plane in the strained crustal rocks by a series of dislocations which initiate at the focus and spread out with various rupture velocities. The dislocation front changes speed as it passes through patches of roughness (asperities on the fault e.g. below P_1 and P_2).

At the dislocation itself, there is a finite time for a slip to take place, and the form of the slip is an elastic rebound of each side of the fault, leading to a decrease of overall strain. The slip can have vertical components, as well as horizontal components, and can vary along the fault. The waves are produced near the dislocation front as a result of the release of the strain energy in the slippage.

This model resembles in many ways the radiation of radio waves from a finite antenna. In the far field, the theory of radio propagation gives complete solutions for the reception of radio signals through stratified media. However, when the receiver is very near to the extended antenna, the signal becomes mixed because of the finiteness of the source and interference through end effects. The main parameters in the model are:

Rupture length	L
Rupture area	A
Fault slippage (offset)	D
Rupture velocity	V
Slip (rise) time	T
Roughness distribution density	$\phi(x)$

The main work in theoretical seismology on source properties today is to determine which of these parameters are essential, whether a set of them is an optimal one, and how best to estimate each parameter from both field observations and analysis of the seismograms made in the near and the far field.

A number of papers have now been published that demonstrate that, in certain important cases, synthetic seismograms for seismic waves near the source can be computed realistically.[1-22] The synthetic motions can be compared with the three observed orthogonal components of either acceleration, velocity, or displacement at a site. There remain difficulties, however, in modeling certain observed complexities, and there is a lack of uniqueness in the physical formulations which lead to acceptable fits with observations (see also Section 1.10).

1.9 SEISMIC RISK EVALUATION

Regional seismicity or risk maps recommended by seismic design codes (see Chapter 4) usually do not attempt to reflect geological conditions or to take into account variations due to soil properties. It is necessary, therefore, for critical construction in populated regions to make special geological–engineering studies for each site, the detail and the level of concern depending on the density of occupancy as well as the proposed structural type. In inhabited areas, more casualities are likely to result from a failed dam or a damaged nuclear reactor, for example, than from a damaged oil pipeline.

The factors which must be considered in assessment of seismic risk of a site have been well defined in recent times. Here a brief summary of these factors is listed.

Geological Input Any of the following investigations may be required:

1. Provision of a structural geologic map of the region, together with an account of recent tectonic movements.
2. Compilation of active faults in the region and the type of displacement (e.g., left-lateral, strike-slip, etc.). Field work is sometimes necessary here. Of particular importance are geological criteria for fault movements in Holocene time (the past 10,000 years), such as displacements in recent soils, dating by radiocarbon methods of organic material in trenches across the fault, and other methods.
3. Mapping of the structural geology around the site, with attention to scarps in bedrock, effects of differential erosion, and offsets in overlying sedimentary deposits. Such maps must show rock types, surface structures, and local faults, and include assessments of the probable length, continuity, and type of movement on such faults.
4. In the case of throughgoing faults near the site, geophysical exploration to define the location of recent fault ruptures and other lineaments. Geophysical work sometimes found useful includes measurement of electrical resistivity and gravity along a profile normal to the fault. Other key geological information is evidence for segmentation of the total fault length, such as stepover of fault strands, and changes in strike.
5. Reports of landslides, major settlements, ground warping, or inundation from floods or tsunamis at the site.
6. Checks of ground-water levels in the vicinity to determine if ground-water barriers are present which may be associated with faults or affect the soil response to the earthquake shaking.

Seismological Input Procedures for the estimation of ground-shaking parameters for optimum engineering design are still in the early stages, and many are untested. It is important, therefore, to state the uncertainties and assumptions employed in the following methods:

1. Detailed documentation of the earthquake history of the region around the site. Seismicity catalogs of historical events are needed in preparing lists of felt earthquakes. The lists should show the locations, magnitudes, and maximum modified Mercalli intensities for each earthquake. This information should be illustrated by means of regional maps.
2. Construction, where the record permits, of recurrence curves of the frequency of regional earthquakes, down to even small magnitudes. Estimates of the frequency of occurrence of damaging earthquakes can then be based on these statistics.
3. A review of available historic records of ground shaking, damage, and other intensity information near the site.
4. Estimation of the maximum modified Mercalli intensities on firm ground near the site from felt reports from each earthquake of significance.
5. Definition of the design earthquakes. The geological and seismological evidence assembled in the above sections should then be used to predict the earthquakes which would give the most severe ground shaking at the site. (Several such design earthquakes might be necessary or prudent.) Where possible, specific faults on which rupture might occur should be stated, together with the likely mechanism (strike-slip, thrust, and so on). The likely focal depth and length of rupture and estimated amount of fault displacement should be determined, with their uncertainties. These values are useful in estimating the possible magnitude of damaging earthquakes from standard curves that relate fault rupture to magnitude (see Table 1-5).

Soil Engineering Input When there is geological indication of the presence of structurally poor foundation material (such as in flood plains and filled tidelands), a field report on the surficial strata underlying the site is advisable. In addition, areas of subsidence and settlement (either natural or from groundwater withdrawal) and the stability of nearby slopes must be studied. We mention here only three factors that may require special scrutiny.

1. Study of engineering properties of foundation soils to the extent warranted for the type of building. Bor-

Table 1-5 Earthquake Magnitude versus Fault Rupture Length

Magnitude (Richter)	Rupture (km)
5.5	5-10
6.0	10-15
6.5	15-30
7.0	30-60
7.5	60-100
8.0	100-200
8.5	200-400

ings, trenchings, and excavations are important for such analyses, as well as a search for the presence of sand layers which may lead to liquefaction.

2. Measurements of the physical properties (density, water content, shear strength, behavior under cyclic loading, attenuation values) of the soil in situ or by laboratory tests of borehole core samples.

3. Determination of P- and S-wave speeds and Q attenuation values and in the overburden layers by geophysical prospecting methods (Section 1-5).

1.10 EARTHQUAKE AND GROUND-MOTION PREDICTION

Aspects of earthquake prediction that tend to receive the most publicity are the place, the size, and the time of the earthquake. For most people, prediction of earthquakes means prediction of the time of occurrence. A more important aspect for mitigation of hazard is the prediction of the strong ground motion likely at a particular site (see also Chapter 2).

First, consider the status of forecasting of the time and size of an earthquake. Prediction of the region where earthquakes are likely to occur has now been largely achieved by seismicity studies using earthquake observatories. Because empirical relations between the magnitude of an earthquake and the length of observed fault rupture have also been constructed (see Table 1-5), rough limits can be placed on the size of earthquakes for a region.

Many attempts have been made to find clues for forewarning. Some physical clues for earthquake prediction are shown in Figure 1-31. In 1975, Chinese officials, using, in part, increased seismicity (foreshocks) and animal restlessness, evacuated a wide area before the damaging Haicheng earthquake. However, in the 1976 Tangshan catastrophe no forewarnings were issued. Elsewhere, emphasis has been placed on geodetic data, such as geodimeter measurements of deformation of the Californian crust along the San Andreas fault. An ex post facto premonitory change in ground level was found after the Niigata earthquake, which, if it had been discovered beforehand, might have served as one indication of the coming earthquake.

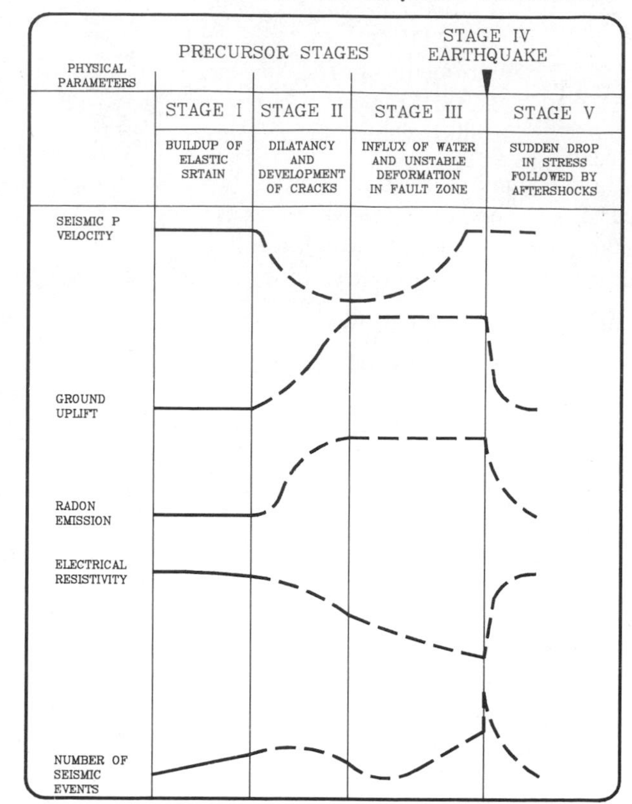

Figure 1-31 Physical clues for earthquake prediction. [After National Academy of Sciences (1-25).]

Another scheme is based on detecting spatial and temporal gaps in the seismicity of a tectonic region. In 1973, a prediction was made by seismologists of the U.S. Geological Survey that an earthquake with a magnitude of 4.5 would occur along the San Andreas fault south of Hollister within the next six months. The prediction was based on four principal shocks which had occurred within three years on both ends of a 25-km-long stretch of the San Andreas fault, bracketing a 6-km-long section free from earthquakes in that time interval. The assumption was that the midsection was still stressed but locked, ready to release the elastic energy in an earthquake. However, no earthquake occurred in the six months predicted. One difficulty with such methods is the assessment of a zero epoch with which to compare the average background occurrence rate.

The most promising prediction experiment in California to date depends on the detection of a 22-year periodicity in moderate-magnitude ($M_L = 5.5$) earthquakes centered on the San Andreas fault near Parkfield. Similar earthquakes were recorded in 1901, 1922, 1934, and 1966. Available seismograms have in addition allowed quantitative comparison of the source mechanisms for the 1922, 1934, and 1966 earthquakes. Many monitoring instruments are now in place to try to detect precursors for a possible 1988 repetition. These include changes in ground water tables, radon concentration, seismicity, and fault slippage.

It has often been pointed out that even if the ability to predict the time and size of an earthquake were achieved by seismologists, many problems would remain on the hazard side. Suppose that an announcement were made that there was a chance of one in two of a destructive earthquake occurring within a month. What would be the public response? Would the major industrial and commercial work in the area cease for a time, thus dislocating large segments of the local economy? Even with shorter-term prediction, there are difficulties if work is postponed until the earthquake warning period is over. Suppose that the predictive time came to an end and no earthquake occurred; who would take the responsibility of reopening schools and resuming other activities?

An allied approach to reduction of seismic danger may some day turn out to be rewarding. Is there a way in which earthquakes can be controlled, like floods and tsunami runup? A chance occurrence near Denver, Colorado suggests one possibility. There, when waste liquid was being pumped down a deep borehole for disposal, earthquakes which correlated quite closely in number with the amount of water injected began to occur in the area. Perhaps the water reduced friction along preexisting faults around the disposal well to a point where residual strain in the crust could be reduced by sudden slip, or perhaps an aquifer was inflated with consequent adjustments. As a followup, small earthquakes were produced by water injection by seismologists of the U.S. Geological Survey at the Rangely oil field.

The experiment has led to the proposal that the magnitude of earthquakes along an active fault might be restricted to, say, magnitude 5 or less by deliberately inducing rupture by pumping water down holes bored into fault zones. The length of fault rupture might be restricted by bounding the region with "seismovalves" at which points water would be pumped from the fault zone in order to "lock" it. The scheme supposes a better knowledge of the rate and direction of groundwater movement than is now available for most regions.

Secondly, let us consider the calculation of artificial (synthetic) seismic strong ground motions. The engineering demand is for the estimation of certain parameters which will be used for design and structural checking. There are two representations usually used. The first is the time history of the ground motion at the site, represented instrumentally by the seismogram or accelerogram. The second is the Fourier or response spectra for the whole motion at the site. These two representations are equivalent and are connected by appropriate transformations between the time and frequency domains.

In the time-history representation, the major interest is in the peak amplitudes of acceleration, velocity, and displacement as a function of frequency of the ground motion. Another parameter of great importance is the duration of the strong ground motion, usually given in terms of the interval of time above a certain acceleration threshold (say $0.05g$) in a particular frequency range. Typically the duration of a magnitude-7 earthquake at a distance of 10 km is about 25 sec. The pattern of wave motion is also important in earthquake engineering because the nonlinear response of structures is dependent on the sequence of arrival of the various types of waves. In other words, damage would be different if the ground motion were run backwards rather than in the actual sequence of arrival. In this respect, phasing of the ground motion becomes very important, and the phase spectra should be considered along with the amplitude spectrum.

The phasing of the various wave types on synthetic seismograms can be determined by estimation of times of arrival of the P, S, and surface waves. In this way, a realistic envelope of amplitudes in the time histories can be achieved.

The sequence of steps which go into the construction of a synthetic ground motion can be summarized as follows. First, the appropriate source type and magnitude for the controlling earthquake in the region is selected from the available geological and seismological evidence. Secondly, sample recorded ground motions for the selected source type as recorded on accelerometers at the relevant distance are selected. These are usually, of course, much smaller-magnitude sources than required. Such smaller recorded ground motions contain essential properties of the particular earthquake mechanism involved, however, as well as the effects of the particular geological structure between the source and the station. In terms of the theory of the response of mechanical structures, they are called "empirical Green's functions." They can be considered as the response of the local geological system to an approximate impulsive force of short duration applied at the rupturing fault. If such empirical Green's functions appropriate to the study of the site in question are not available, they must be constructed making certain mathematical assumptions and introducing appropriate tensor analysis (see Section 1.8).

The size of the earthquake is then selected in terms of its seismic moment, which, as was seen in Section 1.8, is given quantitatively in terms of the area of the fault slip and the amount of slip. An appropriate fault area is then mapped in terms of an elementary mesh of finite elements. The empirical Green's function mentioned above is then applied to each element of the mesh in the sequence required to achieve the appropriate rupture velocity across the whole fault surface, as well as maintain the specified overall moment for the larger earthquake. This superposition can be done in terms of amplitude and phase spectra by available computer programs, and the synthetic seismogram calculated at a point on the surface in the vicinity of the original recorded empirical Green's function. Various modifications of the process described above have been explored, and a number of test cases have been published.

As can be seen, the prediction of ground motions in this way involves a number of assumptions and extrapolations. A particular value of the method is that it permits the exploration of the effect of changing some of the basic parameters on the expected ground motions. A major difficulty is often the lack of knowledge of the appropriate wave attenuation for the region in question (see Section 1.5). Because of the importance of the application of attenuation factors in calculation of predicted ground motion at arbitrary distances, a great deal of work has been recently done on empirical attenuation forms. The usual form for the peak value at distance x is given by

$$y = \frac{ae^{bM}}{ce^{dM} + x^2}$$

where a, b, c, and d are constants and M is the magnitude.

It is usual that attenuation changes significantly from one geological province to another, and local regional studies need to be made to obtain the parameters involved. A discussion is given in the book by Bullen and Bolt.[1-26]

1.11 CONCLUSIONS

The state of the art in strong-motion seismology is now such that prediction of key parameters, such as the peak ground acceleration and the duration of the significant portion of shaking at a given site, is relatively reliable. A number of checks have been carried out in recent earthquakes which have provided many recordings of the strong ground motion and various site conditions of rock and soil. In addition, the great earthquakes of 1985 in Chile and Mexico ($M_s \approx 8$) have provided accelerograms within a wide region for large-subduction-zone earthquakes. There are still, however, no clear record-

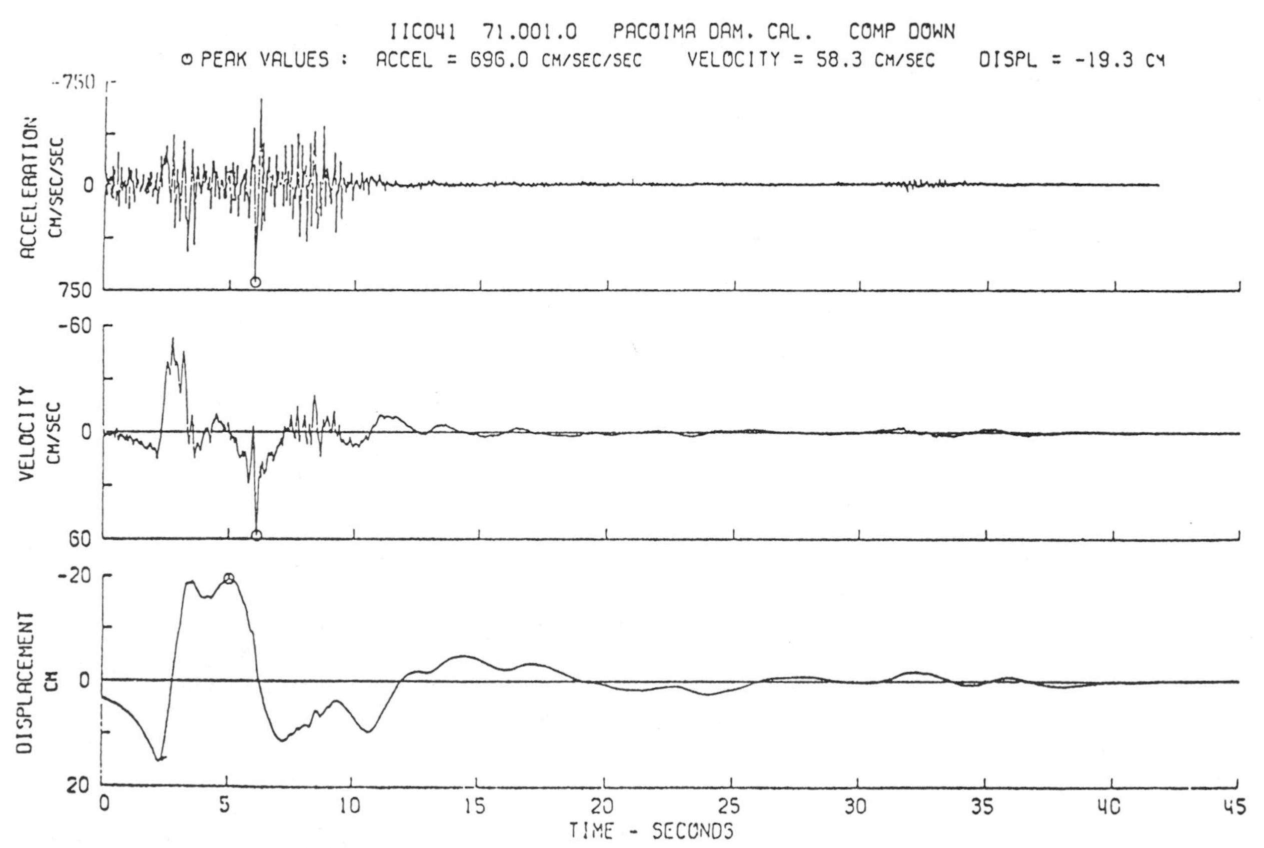

Figure 1-32 San Fernando earthquake of February 9, 1971 as recorded at the Pacoima Dam site in California.

ings of ground motion in the near field from earthquakes with $M_s > 7.5$, so that the validation of synthetic ground motions in extreme cases of wide engineering interest are not available.

To meet this need and others of engineering importance, more strong-motion instruments are being placed in highly seismic areas of the world. Of special interest is the recent operation of clusters of such instruments, called strong-motion arrays.[1-26] Such arrays allow the mixture of seismic waves involved in strong-motion shaking in the near field to be analyzed and the complexities properly understood (see Figure 1-32).

Of special importance is the rigorous seismic analysis of large structures (such as long bridges). Such analysis requires realistic predictions of free-field surface motions at all interface points on the supporting foundations under design earthquake conditions. In the past, engineers have normally carried out seismic analyses under the incorrect assumption that the motions of all support points are fully correlated, i.e., "rigid foundation" inputs are assumed.

Already some strong-motion array data have come to hand that have provided results on the cross-correlation characteristics of free-field surface motions as a function of separation distance and wave direction. These observations allow the study of the effects of magnitude, epicentral distance, focal depth, etc. on such characteristics. At the present time, considerable effort is directed towards measuring the coherency of seismic waves as a function of spatial separation, frequency, wave type, and foundation conditions, in studies of the 1971 San Fernando earthquake and of earthquakes recorded by the SMART 1 array in Taiwan. Incoherent-phase arrivals in velocity and acceleration time histories have been measured across distances of hundreds of kilometers, suggesting progressive loss of coherency as a function of increased wave frequency.

REFERENCES

1-1 Bolt, B. A., *Earthquakes*, W. H. Freeman and Company, New York, 1987.

1-2 Bolt, B. A., Horn, W. L., Macdonald, G. A., and Scott, R. F., *Geological Hazards*, 2nd edition, Springer-Verlag, New York, 1977.

1-3 Bolt, B. A., *Nuclear Explosions and Earthquakes*, W. H. Freeman and Company, San Francisco, 1975.

1-4 Boore, D. M., "The Motion of Ground in Earthquakes," *Scientific American*, Dec. 1977.

1-5 Kanamori, H., "Quantification of Earthquakes," *Nature*, Vol. 271, Feb. 2, 1978.

1-6 Anderson, D. L., "The San Andreas Fault," *Scientific American*, Nov. 1971.

1-7 Press, F., "Earthquake Prediction," *Scientific American*, May 1975.

1-8 Molnar, P., and Tapponnier, P., "The Collision between India and Eurasia," *Scientific American*, Apr. 1977.

1-9 Richter, C. F., *Elementary Seismology*, W. H. Freeman and Company, San Francisco, 1958.

1-10 Gutenberg, B., and Richter, C. F., *Seismicity of the Earth and Associated Phenomena*, Princeton University Press, Princeton, 1954.

1-11 Wiegel, R. L. (ed.), *Earthquake Engineering*, Prentice-Hall, Englewood Cliffs, NJ, 1970.

1-12 Real, C. R., and Teng, T., "Local Richter Magnitude and Total Signal Duration in Southern California," *Bull. Seism. Soc. Am.*, 63, No. 5, Oct. 1973.

1-13 Kanamori, H., "The Energy Release in Great Earthquakes," *J. Geol. Res.*, 82, 20, 1977.

1-14 Reid, H. F., "The Elastic Respond Theory of Earthquakes," *Bull. Dept. Geol. Univ. Calif.*, 6, 413–444, 1911.

1-15 Haskell, N. A., "Total Energy and Energy Spectral Density of Elastic Wave Radiation from Propagating Faults," *Bull. Seism. Soc. Am.*, 54, 1811–1841, 1964.

1-16 Haskell, N. A., "Total Energy and Energy Spectral Density of Elastic Wave Radiation from Propagating Faults—Part II, A Statistical Source Model," *Bull. Seism. Soc. Am.*, 56, 125–140, 1966.

1-17 Das, S. and Aki, K., "A Numerical Study of Two-Dimensional Spontaneous Rupture Propagation," *Geophys. J. Roy. Astr. Soc.*, 50, 643–668, 1977.

1-18 Das, S., and Aki, K., "Fault Plane with Barriers: A Versatile Earthquake Model," *J. Geophys. Res.*, 82, 5658–5670, 1977.

1-19 Boore, D. M., Joyner, W. B., Oliver, A. A. III, and Page, R. A., "Estimation of Ground Motion Parameters," United States Geological Survey, Circular 795, Arlington, VA, 1987.

1-20 Benioff, H., "Mechanism and Strain Characteristics of the White Wolf Fault as Indicated by the Aftershock Sequence," California Division of Mines and Geology, Bulletin 171, 199–202, 1955.

1-21 Singh, J. P., "Earthquake Ground Motions: Implications for Designing Structures and Reconciling Structural Damage," *Earthquake Spectra*, *EERI*, Vol. 1, No. 2, Feb. 1985.

1-22 Bolt, B. A. (ed.), *Seismic Strong Motion Synthetics*, Academic Press, 1987.

1-23 Burridge, R. and Willis, J., "The Self-Similar Problem of the Expanding Elliptical Crack in an Anisotropic Solid," *Proceedings of the Cambridge Philosophical Society*, 66, 443–468, 1969.

1-24 Backus, G. E., "Interpreting the Seismic Glut Moment of Total Degree Two or Less," *Geophys. J. Roy. Astr. Soc.*, 51, 1–25, 1977.

1-25 Panel on Earthquake Prediction of the Committee of Seismology, *Predicting Earthquakes*, National Academy of Sciences, Washington, 1976.

1-26 Bullen, K. E. and Bolt, B. A., *An Introduction to the Theory of Seismology*, Cambridge University Press, 1985.

Chapter 2

Earthquake Ground Motion and Response Spectra

Bijan Mohraz, Ph.D. and Fawzi E. Elghadamsi, Ph.D.**

2.1 INTRODUCTION

Ground vibrations during an earthquake can severely damage buildings and equipment housed in them. The ground acceleration, velocity, and displacements (referred to as ground motion), when transmitted through a structure, are in most cases amplified. The amplified motion can produce forces and displacements which may exceed those the structure can sustain. Many factors influence ground motion and its amplification. An understanding of how these factors influence the response of structures and equipment is essential for a safe and economical design.

Earthquake ground motion is usually measured by a strong-motion accelerograph which records the acceleration of the ground at a particular location. The record accelerograms, after they are corrected for instrumented errors and adjusted for baseline (see next section), are integrated to obtain the velocity and displacement time histories. The maximum values of the ground motion (peak ground acceleration, peak ground velocity, and peak ground displacement) are of interest in seismic analysis and design. These parameters, however, do not by themselves describe the intensity of shaking that structures or equipment experience. Other factors, such as the earthquake magnitude, distance from fault or epicenter, duration of strong shaking, soil condition of the site, and frequency content of the motion also influence the response of a structure. Some of these effects, such as the amplitude of motion, duration of strong shaking, frequency content, and local soil conditions are best represented through the response spectrum,[2-1 to 2-4] which describes the maximum response of a damped single-degree-of-freedom oscillator to various frequencies or periods. The response spectra from a number of records are often averaged and smoothed to obtain design spectra, which also represent the amplification of ground motion at various frequencies or periods of structure.

This chapter discusses earthquake ground motion, response spectra, and response amplifications. The influence of earthquake parameters such as earthquake magnitude, duration of strong motion, soil condition, and epicentral distance on ground motion and response spectra is presented and discussed, as well as the evaluation of seismic risk at a given site. Earthquake design spectra proposed by several investigators and those recommended by various agencies are presented. The latter part of the chapter includes the inelastic earthquake spectra and deamplification factors and a brief discussion of artificially generated ground motion.

*Civil and Mechanical Engineering Department, Southern Methodist University, Dallas, Texas.

2.2 RECORDED GROUND MOTION

Ground motion during an earthquake is measured by a strong motion accelerograph which records the acceleration of the ground at a particular location. Three orthogonal components of the motion, two in the horizontal direction and one in the vertical, are recorded by the instrument. The instruments may be located on free field or mounted in structures. Typical strong-motion accelerograms, recorded on free field and ground floor of the Imperial County Services Building during the Imperial Valley earthquake of October 15, 1979,(2-5) are shown in Figure 2-1.

Accelerograms are generally recorded on photographic paper or film. The records are digitized for engineering applications, and during the digitization process, errors associated with instrument and digitization (transverse play of the recording paper or film, warping of the paper, enlargement of the trace, etc.) are removed. In addition, the unknown zero-acceleration line must be established before the velocity and displacement are computed. Small errors in establishing the zero-acceleration baseline can result in appreciable velocity and displacement errors. To minimize the errors, a correction is carried out by assuming linear zero-acceleration and -velocity baselines and then using a least-squares fit to determine the parameters of the lines. For a detailed description of the procedure used in digitizing and correcting accelerograms, one should refer to Trifunac,(2-6) Hudson et al.,(2-7) and Hudson.(2-8) Another procedure, which assumes a second-degree polynomial for the zero-acceleration baseline, has also been used in the past.(2-9) However, the Trifunac–Hudson procedure is more automated and has been used extensively to correct accelerograms. The corrected accelerograms are integrated to obtain the velocity and displacement time histories. The digitization, correction, and processing of accelerograms have been carried out

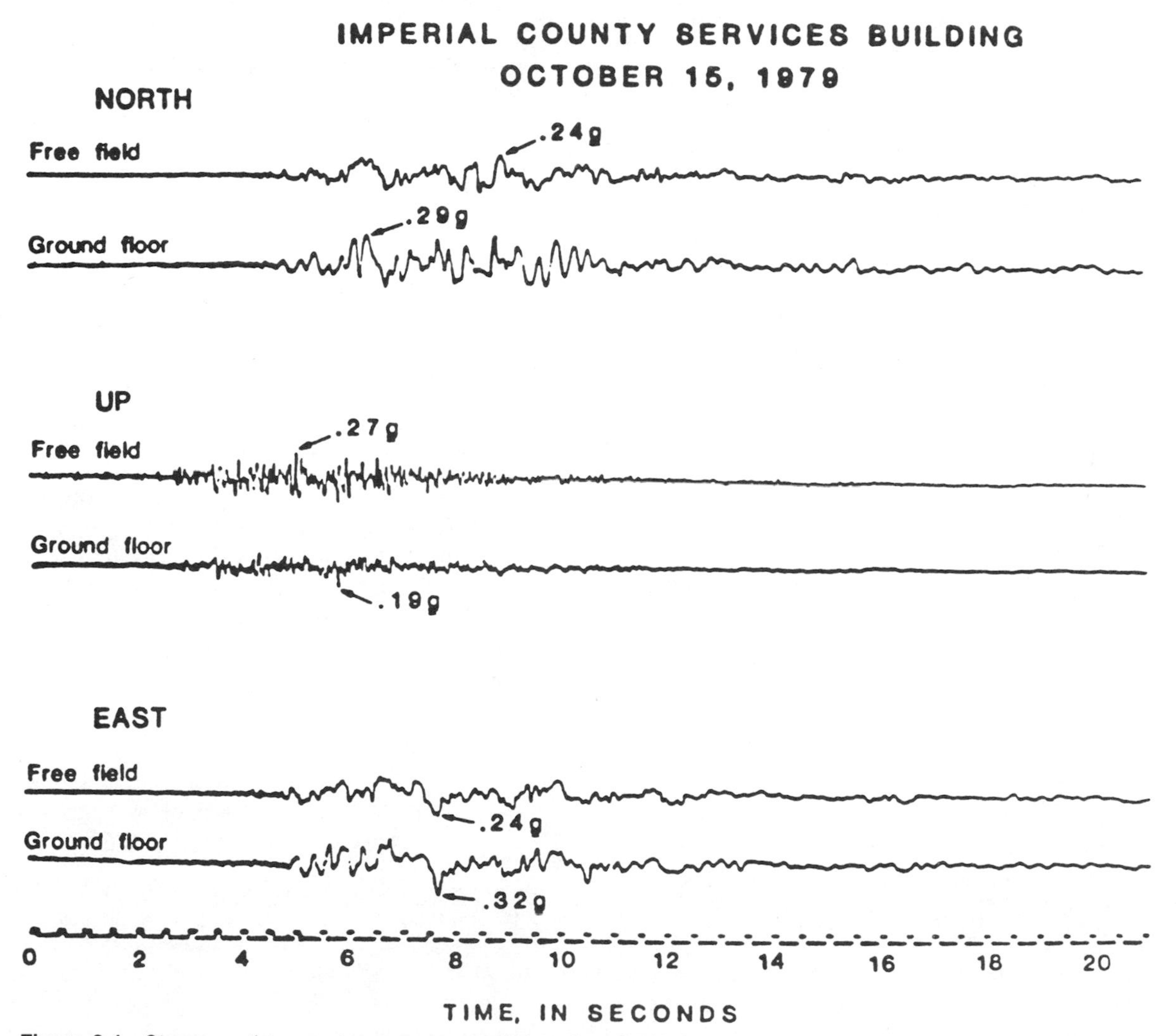

Figure 2-1 Strong motion accelerograms recorded on free field and ground floor of the Imperial County Service Building during the Imperial Valley earthquake of October 15, 1979. [After Rojahn and Mork (2-5).]

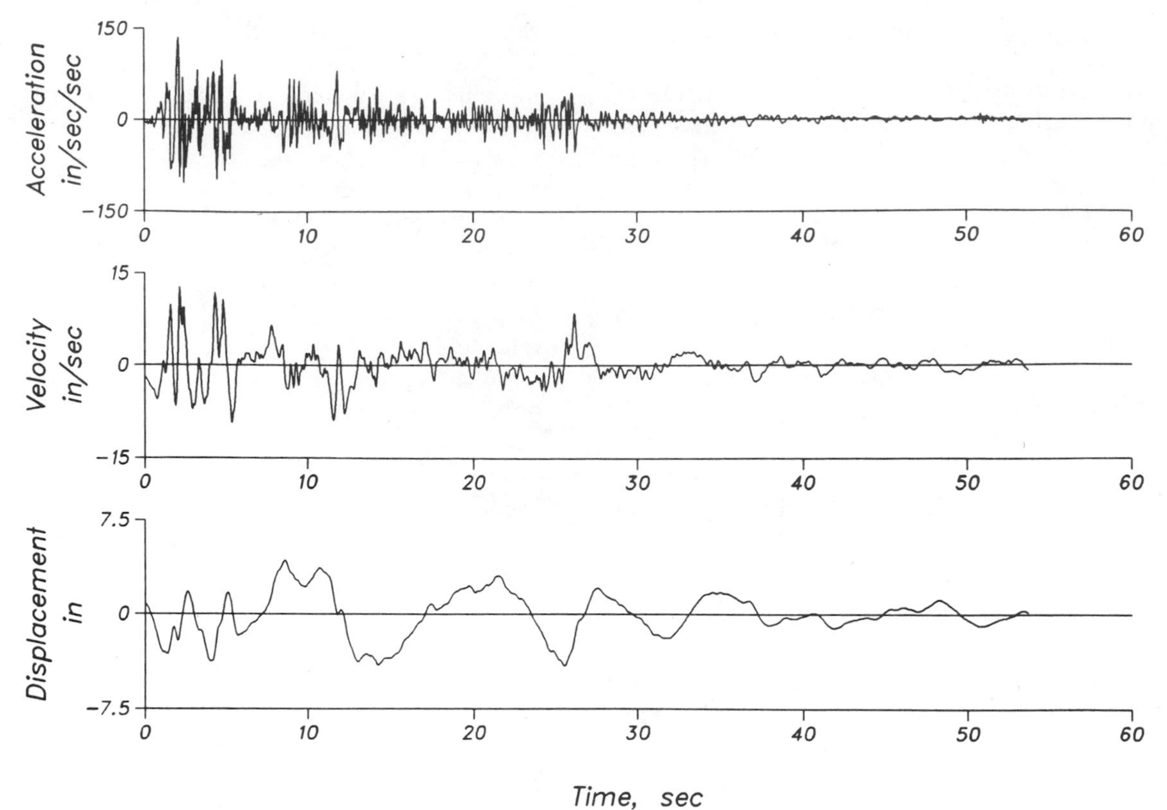

Figure 2-2 Corrected accelerogram and integrated velocity and displacement time-histories for the S00E component of El Centro, the Imperial Valley Earthquake of May 18, 1940.

by the Earthquake Engineering Research Laboratory of the California Institute of Technology in the past and now by the United States Geological Survey. A typical corrected accelerogram and the integrated velocity and displacement for the S00E component of El Centro, the Imperial Valley earthquake of May 18, 1940, are shown in Figure 2-2.

2.3 CHARACTERISTICS OF EARTHQUAKE GROUND MOTION

The characteristics of earthquake ground motion which are important in earthquake engineering applications are:

1. peak ground motion (peak ground acceleration, peak ground velocity, and peak ground displacement),
2. duration of strong motion,
3. frequency content.

Each of these parameters influences the response of a structure. Peak ground motion primarily influences the vibration amplitudes. Duration of strong motion has a pronounced effect on the severity of shaking. A ground motion with a moderate peak acceleration and long duration may cause more damage than a ground motion with a larger acceleration and a shorter duration. Frequency content and spectral shapes relate to frequencies or periods of vibration of a structure. In a structure, ground motion is amplified the most when the frequency content of the motion and the vibration frequencies of the structure are close to each other. Each of these characteristics is briefly discussed below.

2.3.1 Peak ground motion

Table 2-1 gives the peak ground acceleration, velocity, displacement, earthquake magnitude, epicentral distance, and site description for typical records from a number of seismic events. Some of these records are frequently used in earthquake engineering applications. Peak ground acceleration has been widely used to scale earthquake design spectra and acceleration time histories. As will be discussed later, recent studies recommend that in addition to peak ground acceleration, peak ground velocity and displacement should be used for scaling purposes. Relationships between ground motion parameters are discussed in Section 2-6.

2.3.2 Duration of Strong Motion

Several investigators have proposed procedures for computing the strong-motion duration of an accelerogram. Page et al.[2-10] and Bolt[2-11] proposed the *bracketed duration*, which is the time interval between the first and the last acceleration peaks greater than a specified value

Table 2-1 Peak Ground Motion, Earthquake Magnitude, Epicentral Distance and Site Description for Typical Recorded Accelerograms

Earthquake and location	Comp.	Epicentral distance (km)	Mag.	Peak Acc. (g)	Peak Vel. (in./sec)	Peak Dis. (in.)	Site description
Northwest California, 10/7/51	S44W			0.104	1.89	0.94	Deep cohesionless soil,
Ferndale City Hall	N46W	56.2	5.8	0.112	2.91	1.08	500 ft
	Vert			0.027	0.87	0.64	Seed (2-68)
Eureka, 12/21/54	N44E			0.159	14.04	5.58	Deep cohesionless soil,
Ferndale City Hall	N46W	40.0	6.5	0.201	10.25	3.79	500 ft
	Vert			0.043	2.99	1.54	Seed (2-68)
San Fernando, 2/9/71	N21E			0.315	6.76	1.66	Sandstone
Castaic Old Ridge	N69W	29.5	6.4	0.271	10.95	3.74	Hudson (2-87)
Route	Vert			0.156	2.54	1.38	
San Francisco, 3/22/57	N10E			0.083	1.94	0.89	Rock
San Francisco Golden	S80E	11.5	5.3	0.105	1.82	0.33	Seed (2-68)
Gate Park	Vert			0.038	0.48	0.27	
Helena, 10/31/35	S00W			0.146	2.89	0.56	Rock
Helena, Montana	S90W	6.3	6.0	0.145	5.25	1.47	Seed (2-68)
Carroll College	Vert			0.089	3.82	1.11	
San Fernando, 2/9/71	S15W			1.170	44.58	14.83	Highly jointed
Pacoima Dam	N75W	7.2	6.4	1.075	22.73	4.26	Diorite gneiss
	Vert			0.709	22.96	7.61	Hudson (2-87)
San Fernando, 2/9/71	S00W			0.180	8.08	2.87	Granitic
Griffith Park Observatory	S90W	32.5	6.4	0.171	5.73	2.15	Hudson (2-87)
	Vert			0.123	2.92	1.33	
Imperial Valley, 5/18/40	S00E			0.348	13.17	4.28	Alluvium, several 1000 ft
El Centro site	S90W	11.5	6.7	0.214	14.54	7.79	Hudson (2-87)
	Vert			0.210	4.27	2.19	
Kern County, 7/21/52	N21E			0.156	6.19	2.64	40 ft of alluvium
Taft Lincoln School	S69E	41.4	7.7	0.179	6.97	3.60	over poorly cemented
Tunnel	Vert			0.105	2.63	1.98	sandstone. Hudson (2-87)
Eureka, 12/21/54	N11W			0.168	12.44	4.89	Deep cohesionless soil
Eureka Federal Bldg.	N79E	24.0	6.5	0.258	11.57	5.53	250 ft deep
	Vert			0.083	3.23	1.83	Seed (2-68)
Hollister, 4/8/61	S01W			0.065	3.06	1.12	Unconsolidated alluvium over
Hollister City Hall	N89W	22.2	5.7	0.179	6.75	1.51	partly consolidated gravels
	Vert			0.050	1.85	0.85	Trifunac (2-88)
Borrego Mountain, 4/8/68	S00W			0.130	10.16	4.82	Alluvium, several 1000 ft
El Centro site	S90W	67.3	6.4	0.057	5.78	4.33	Hudson (2-87)
	Vert			0.030	1.36	1.52	
Western Washington, 4/13/49	N04W			0.165	8.43	3.38	Deep cohesionless soil,
Olympia, Washington	N86E	16.9	7.1	0.280	6.73	4.09	420 ft
Highway Test Lab	Vert			0.092	2.77	1.59	Seed (2-68)
Parkfield, 6/27/66	N05W			0.355	9.12	2.09	Alluvium
Cholame, Shandom,	N85E	56.1	5.6	0.434	10.01	2.80	Hudson (2-87)
California Array No. 5	Vert			0.119	2.87	1.35	
San Fernando, 2/9/71	N00W			0.255	11.81	5.87	Alluvium
8244 Orion Blvd.	S90W	21.1	6.4	0.134	9.42	5.45	Hudson (2-87)
1st floor	Vert			0.171	12.58	5.76	

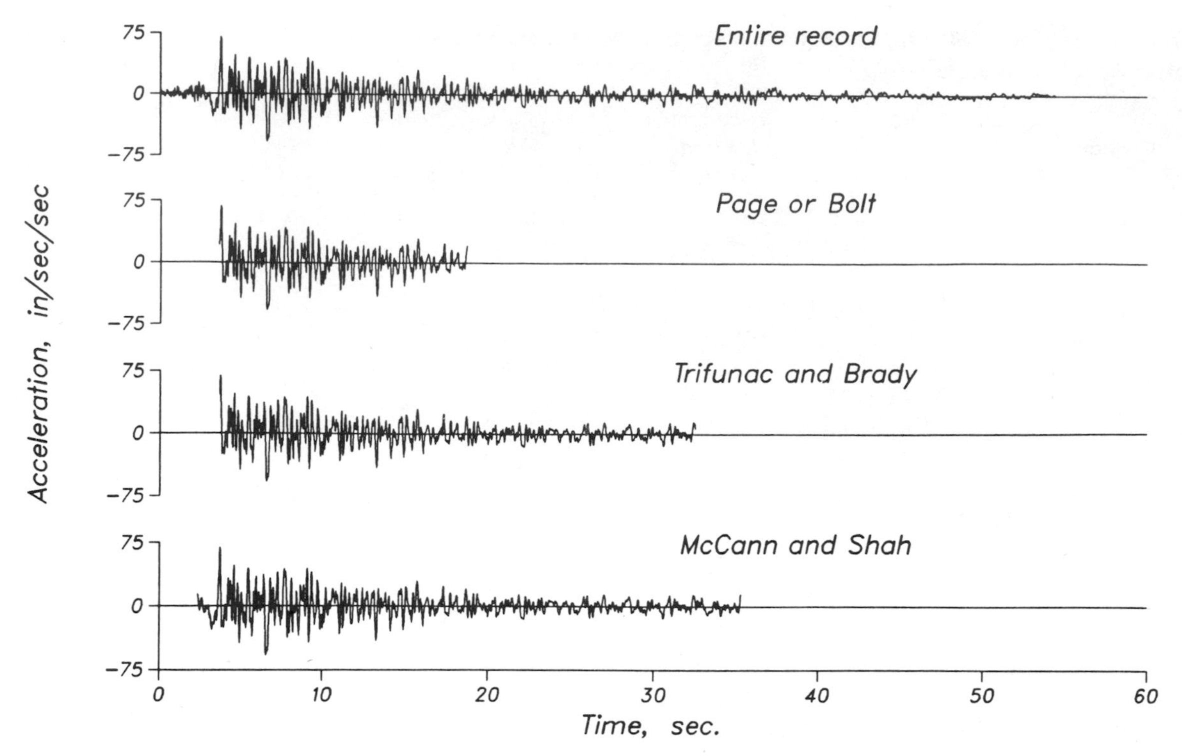

Figure 2-3 Comparison of strong motion duration for the S69E component of the Taft, California earthquake of July 21, 1982 using different procedures.

(usually 0.05g). Trifunac and Brady[(2-12)][a] defined the duration of the strong motion as the time interval in which a significant contribution to the integral of the square of the acceleration ($\int a^2\,dt$)—referred to as the *accelerogram intensity*—takes place. They selected the time interval between the 5% and the 95% contributions as the duration of strong motion. A third procedure, suggested by McCann and Shah,[(2-14)] is based on the average energy arrival rate. The duration is obtained by examining the cumulative root-mean-square (rms)[b] acceleration of the accelerogram. A search is performed on the rate of change of the cumulative rms to determine the two cutoff times. The final cutoff time T_2 is obtained when the rate of change of the cumulative rms acceleration becomes negative and remains so for the remainder of the record. The initial time T_1 is obtained in the same manner except that the search is performed starting from the "tail end" of the record.

Figure 2-3 shows comparison between the strong-motion duration extracted from an original record using different procedures. Table 2-2 gives the initial time T_1, the final time T_2, the duration of strong motion ΔT, the rms acceleration, and the percentage contribution to ($\int a^2\,dt$) for several records. The comparisons show that these procedures result in different values of the duration of strong motion. This is to be expected, since the procedures are based on different criteria. Since there is no standard definition of strong-motion duration, the selection of a procedure for computing it for a study depends on the purpose of the study. For example, it seems reasonable to use McCann and Shah's definition, which is based on rms acceleration, when studying the stationary characteristics of earthquake records and in computing power spectral density. The bracketed duration proposed by Page et al.[(2-10)] and Bolt[(2-11)] may be more appropriate when computing elastic and inelastic response.

2.3.3 Frequency Content

The frequency content of ground motion can be examined by transforming the motion from time domain to frequency domain through a Fourier transform. The Fourier amplitude spectrum and power spectral density, which are based on this transformation, may be used to characterize the frequency content. They are briefly discussed below:

Fourier Amplitude Spectrum The finite Fourier transform $F(\omega)$ of an accelerogram $a(t)$ is obtained as

$$F(\omega) = \int_0^T a(t)\, e^{-i\omega t}\, dt, \qquad i = \sqrt{-1} \tag{2-1}$$

[a]An earlier study by Husid et al.[(2-13)] used a similar definition for the duration of strong motion.
[b]See next section for definition of rms.

Table 2.2 Comparison of Strong-Motion Duration for Eight Earthquake Records

Record	Comp.	Method*	*T1* (sec)	*T2* (sec)	ΔT (sec)	rms Accel. (cm/sec²)	Percentage $\int a^2\,dt$
El Centro, 1940	S00E	a	0.00	53.74	53.74	46.01	100
		b	0.88	26.74	25.86	65.16	97
		c	1.68	26.10	24.42	64.75	90
		d	0.88	26.32	25.44	65.60	96
	S90W	a	0.00	53.46	53.46	38.85	100
		b	1.24	26.64	25.40	54.88	95
		c	1.66	26.20	24.54	54.39	90
		d	0.80	26.62	25.82	54.73	96
Taft, 1952	N21E	a	0.00	54.34	54.34	25.03	100
		b	3.44	22.94	19.50	38.50	85
		c	3.70	34.24	30.54	31.70	90
		d	2.14	36.46	34.32	30.85	96
	S69E	a	0.00	54.38	54.38	26.10	100
		b	3.60	18.72	15.12	44.61	82
		c	3.66	32.52	28.86	33.96	90
		d	2.34	35.30	32.96	32.71	95
El Centro, 1934	S00W	a	0.00	90.28	90.28	19.48	100
		b	1.92	14.78	12.86	46.89	83
		c	2.82	23.92	21.10	38.27	90
		d	1.92	23.88	21.96	38.38	94
	S90W	a	0.00	90.22	90.22	20.76	100
		b	1.98	20.10	18.12	44.58	93
		c	2.86	23.14	20.28	41.57	90
		d	1.62	20.10	18.48	44.26	93
Olympia, 1949	N04W	a	0.00	89.06	89.06	22.98	100
		b	0.74	23.04	22.30	44.25	93
		c	1.78	27.58	25.80	40.51	90
		d	0.08	23.02	22.94	43.73	93
	N86E	a	0.00	89.02	89.02	28.10	100
		b	1.00	22.04	21.04	56.00	94
		c	4.34	22.42	18.08	59.22	90
		d	0.28	21.80	21.52	55.48	94

* a, entire record; b, Page et al.(2-10) or Bolt(2-11); c, Trifunac and Brady(2-12); d, McCann and Shah.(2-14)

where T is the duration of the accelerogram. The Fourier amplitude spectrum $FS(\omega)$ is defined as the square root of the sum of the squares of the real and imaginary parts of $F(\omega)$. Thus,

$$FS(\omega) = \left\{\left[\int_0^T a(t)\sin\omega t\,dt\right]^2 + \left[\int_0^T a(t)\cos\omega t\,dt\right]^2\right\}^{1/2} \qquad (2\text{-}2)$$

Since $a(t)$ has units of acceleration, $FS(\omega)$ has units of velocity. The Fourier amplitude spectrum is of interest to seismologists in characterizing ground motion. Figure 2-4 shows a typical Fourier amplitude spectrum for the S00E component of El Centro, the Imperial Valley earthquake of May 18, 1940. The figure indicates that most of the energy in the accelerogram is in the frequency range of 0.1 to 10 Hz, and that the largest amplitude is at a frequency of approximately 1.5 Hz.

When subjecting an undamped single-degree-of-freedom system to a base acceleration $a(t)$, the response of the system and the Fourier amplitude spectrum of the acceleration are closely related. The equation of motion of the system can be written as

$$\ddot{x} + \omega_n^2 x = -a(t) \qquad (2\text{-}3)$$

in which x and $\ddot{x}$ are the relative displacement and acceleration, and ω_n is the natural frequency of the system. Using Dahamel's integral, the steady-state re-

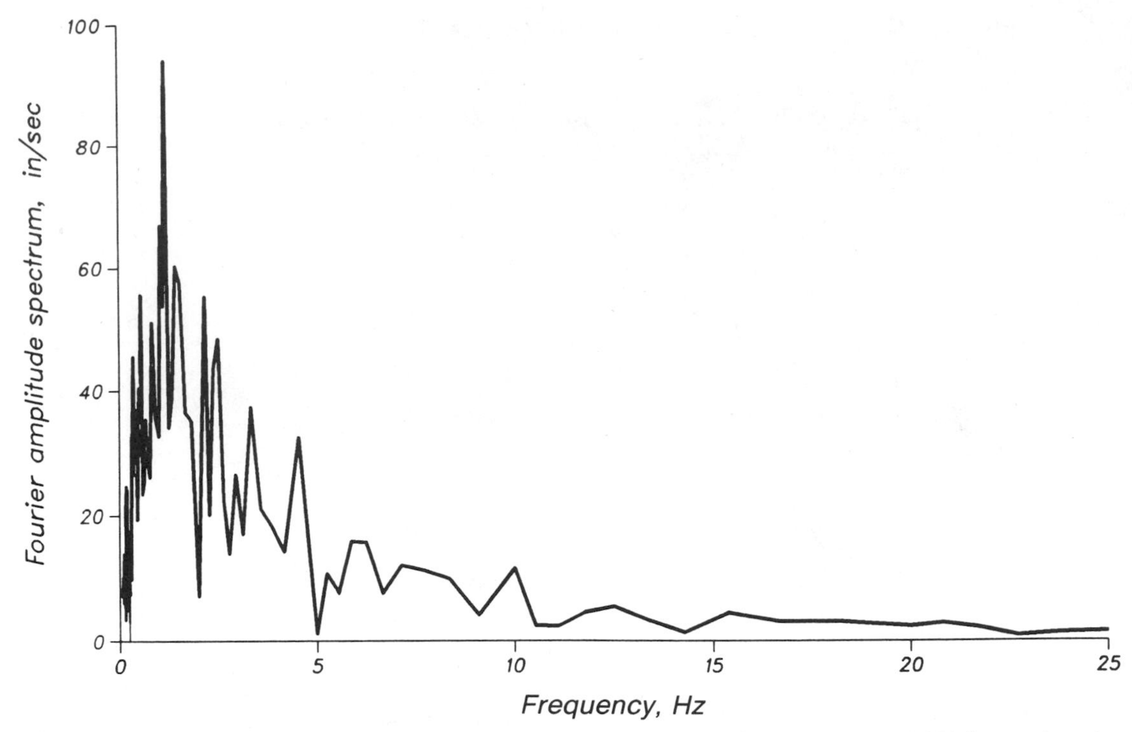

Figure 2-4 Fourier amplitude spectrum for the S00E component of El Centro, the Imperial Valley earthquake of May 18, 1940.

sponse can be obtained as

$$x(t) = \frac{1}{\omega_n}\int_0^t - a(\tau)\sin\omega_n(t-\tau)\,d\tau \qquad (2\text{-}4)$$

and the relative velocity $\dot{x}(t)$ follows directly from Equation 2-4 as

$$\dot{x}(t) = \int_0^t a(\tau)\cos\omega_n(t-\tau)\,d\tau \qquad (2\text{-}5)$$

Denoting the sum of the kinetic and strain energies per unit mass at the end of the duration of motion T by $E(\omega_n)$, then

$$E(\omega_n) = \int_0^T \ddot{x}\,dt + \omega_n^2\frac{x^2(T)}{2} = \tfrac{1}{2}\left[\dot{x}^2(T) + \omega_n^2 x^2(T)\right] \qquad (2\text{-}6)$$

Substituting from Equations 2-4 and 2-5 into Equation 2-6 leads to

$$E(\omega_n) = \frac{1}{2}\left\{\left[\int_0^T a(\tau)\cos\omega_n(T-\tau)\,d\tau\right]^2 + \left[\int_0^T a(\tau)\sin\omega_n(T-\tau)\,d\tau\right]^2\right\} \qquad (2\text{-}7)$$

Expanding the terms in Equation 2-7 and replacing the dummy variable τ by t, one gets

$$E(\omega_n) = \frac{1}{2}\left\{\left[\int_0^T a(t)\sin\omega_n t\,dt\right]^2 + \left[\int_0^T a(t)\cos\omega_n t\,dt\right]^2\right\} \qquad (2\text{-}8)$$

Comparing Equations 2-2 and 2-8, one concludes that

$$E(\omega) = \tfrac{1}{2}\left[FS(\omega)\right]^2 \qquad (2\text{-}9)$$

Equation 2-9 implies that at the end of the duration of motion the total energy per unit mass in an undamped single-degree-of-freedom system having a frequency ω is equal to one-half the square of the Fourier amplitude spectrum evaluated at that frequency.

Another relationship between the response of the single-degree-of-freedom system and the Fourier amplitude spectrum of the base acceleration can be obtained if Equation 2-5 is expanded as

$$\dot{x}(t) = \left[\int_0^t a(\tau)\cos\omega_n\tau\,d\tau\right]\cos\omega_n t + \left[\int_0^t a(\tau)\sin\omega_n\tau\,d\tau\right]\sin\omega_n t \qquad (2\text{-}10)$$

Denoting the maximum relative velocity (spectral velocity) of a system with frequency ω by $SV(\omega)$ and assuming that it occurs at time t_v, one can write

$$SV(\omega) = \left\{\left[\int_0^{t_v} a(\tau)\sin\omega\tau\,d\tau\right]^2 + \left[\int_0^{t_v} a(\tau)\cos\omega\tau\,d\tau\right]^2\right\}^{1/2} \quad (2\text{-}11)$$

The pseudo-velocity $PSV(\omega)$, defined as the product of the natural frequency ω and the maximum relative displacement or the spectral displacement $SD(\omega)$, is close to the maximum relative velocity (see Section 2-7). If $SD(\omega)$ occurs at t_d, then

$$PSV(\omega) = \omega\,SD(\omega) = \left\{\left[\int_0^{t_d} a(\tau)\sin\omega\tau\,d\tau\right]^2 + \left[\int_0^{t_d} a(\tau)\cos\omega\tau\,d\tau\right]^2\right\}^{1/2} \quad (2\text{-}12)$$

Comparison between Equations 2-2 and 2-11 shows that for zero damping the maximum relative velocity and the Fourier amplitude spectrum are equal when $t_v = T$. A similar comparison between Equations 2-2 and 2-12 shows that the pseudovelocity and the Fourier amplitude spectrum are equal if $t_d = T$. Figure 2-5 shows a comparison between $FS(\omega)$ and $SV(\omega)$ for zero damping for the S00E component of El Centro, the Imperial Valley Earthquake of May 18, 1940. The figure shows the close relationship between the two functions. It should be noted that, in general, the Fourier amplitude spectrum is less than the pseudovelocity spectrum.

Power Spectral Density The inverse Fourier transform of $F(\omega)$ is

$$a(t) = \frac{1}{\pi}\int_0^{\omega_0} F(\omega)\,e^{i\omega t}\,d\omega \quad (2\text{-}13)$$

where ω_0 is the maximum frequency detected in the data (referred to as the Nyquist frequency). Equations 2-1 and 2-13 are called Fourier transform pairs. As mentioned previously, the intensity or total energy of an accelerogram is

$$I = \int_0^T a^2(t)\,dt \quad (2\text{-}14)$$

Based on Parseval's theorem, the intensity I can also be expressed in the frequency domain as

$$I = \frac{1}{\pi}\int_0^{\omega_0} |F(\omega)|^2\,d\omega \quad (2\text{-}15)$$

The intensity per unit of time, or the temporal mean square acceleration ψ^2, can be obtained by diving Equa-

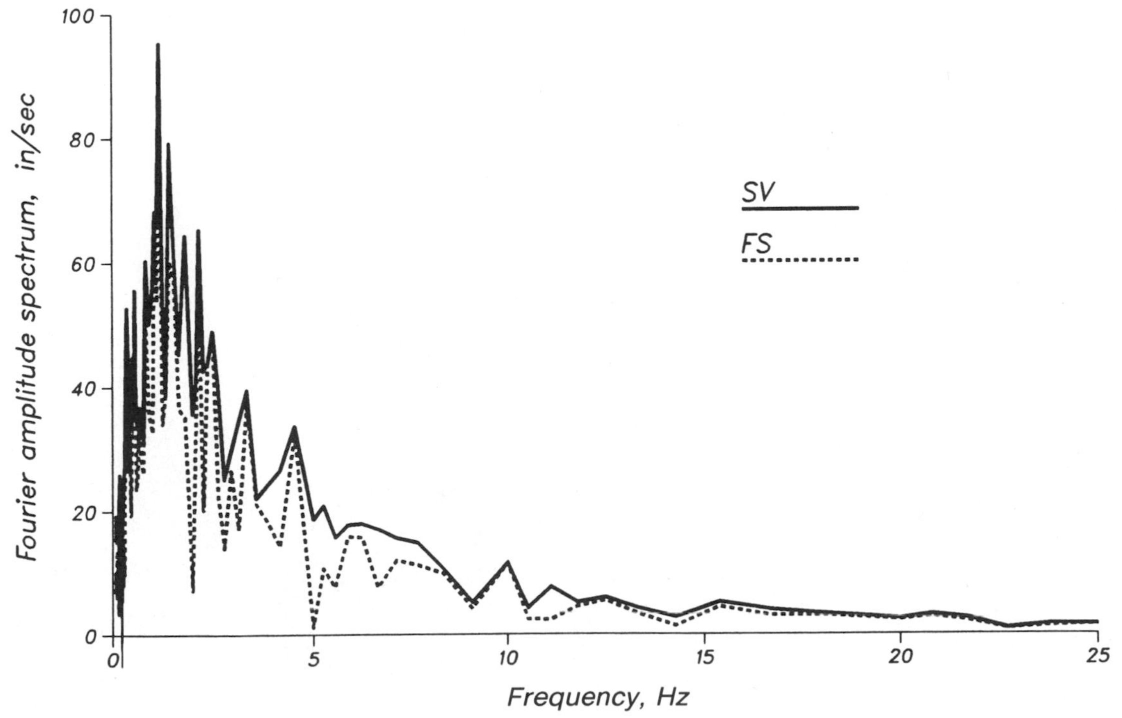

Figure 2-5 Comparison of Fourier amplitude spectrum and velocity spectrum for an undamped single-degree-of-freedom system for the S00E component of El Centro, the Imperial Valley earthquake of May 18, 1940.

tion 2-14 or 2-15 by the duration T. Therefore,

$$\psi^2 = \frac{1}{T}\int_0^T a^2(t)\,dt = \frac{1}{\pi T}\int_0^{\omega_0} |F(\omega)|^2\,d\omega \quad (2\text{-}16)$$

The temporal power spectral density is defined as

$$G(\omega) = \frac{1}{\pi T}|F(\omega)|^2 \quad (2\text{-}17)$$

Combining Equations 2-16 and 2-17, the mean square value can be obtained as

$$\psi^2 = \int_0^{\omega_0} G(\omega)\,d\omega \quad (2\text{-}18)$$

In practice, the power spectral density of ground motion is computed by averaging across the temporal power spectral densities of the representative ensemble of N accelerograms (see for example Reference 2-15). Therefore

$$G(\omega) = \frac{1}{N}\sum_{i=1}^{N} G_i(\omega) \quad (2\text{-}19)$$

where $G_i(\omega)$ is the power spectral density of the ith record.

The power spectral density is frequently presented as the product of a normalized power spectral density $G^{\langle n\rangle}(\omega)$ (area = 1.0) and a mean square acceleration as

$$G(\omega) = \psi^2 G^{\langle n\rangle}(\omega) \quad (2\text{-}20)$$

Figure 2-6 shows a typical example of a normalized power spectral density computed for an ensemble of 161 accelerograms recorded on alluvium. Studies[2-15, 2-16] have shown that the strong-motion segment of accelerograms constitutes a locally stationary random process and that the power spectral density can be presented as a time-dependent function $G(t, \omega)$ in the form

$$G(t,\omega) = \psi^2 S(t) G^{\langle n\rangle}(\omega) \quad (2\text{-}21)$$

where $S(t)$ is a slowly varying time-scale factor which accounts for the local variation of the mean square acceleration with time.

The power spectral density is useful not only as a measure of the frequency content of ground motion but also in estimating its statistical properties. Among such properties are the rms acceleration ψ, the central frequency f_c, and the shape factor δ, defined as

$$\psi = \sqrt{\lambda_0} \quad (2\text{-}22)$$

$$f_c = \sqrt{\lambda_2/\lambda_0} \quad (2\text{-}23)$$

and

$$\delta = \sqrt{1 - \frac{\lambda_1^2}{\lambda_0\lambda_2}} \quad (2\text{-}24)$$

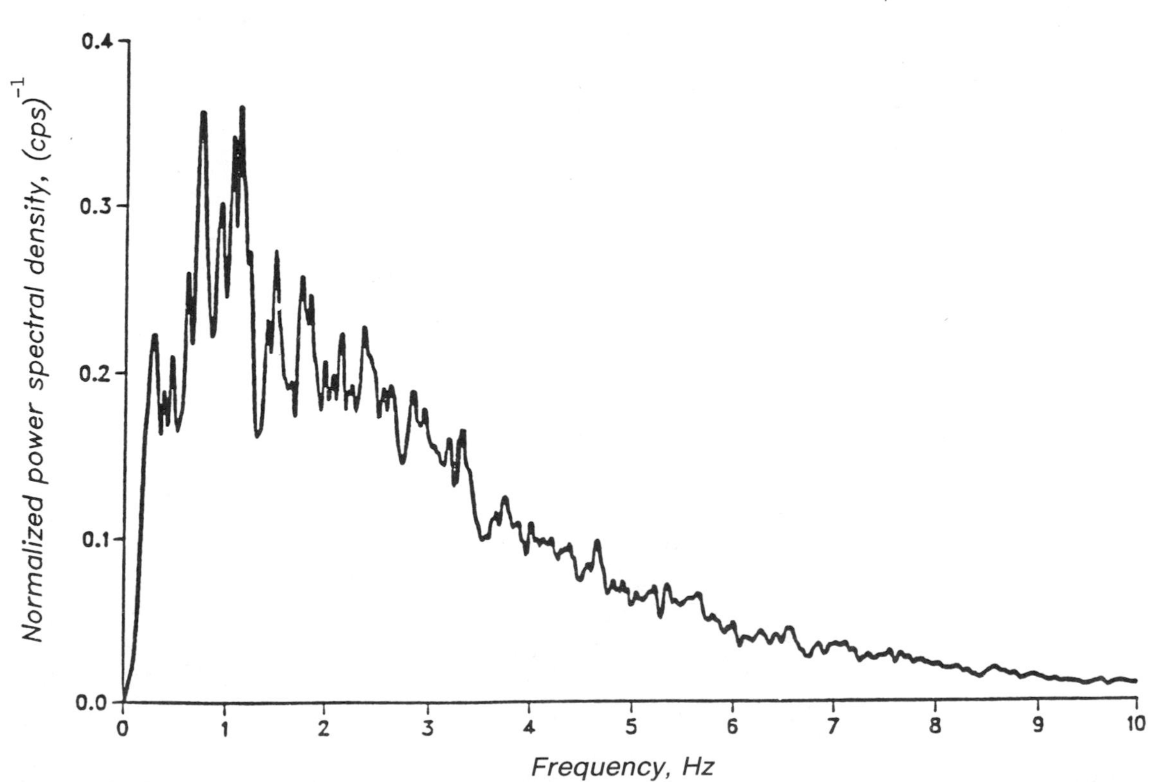

Figure 2-6 Normalized power spectral density of an ensemble of 161 horizontal components of accelerograms recorded on alluvium.

where λ_r is the rth spectral moment defined as

$$\lambda_r = \int_0^{\omega_0} \omega^r G(\omega)\, d\omega \tag{2-25}$$

Smooth power spectral densities of the ground acceleration have been commonly presented in the form proposed by Kanai[(2-17)] and Tajimi[(2-18)] in the form

$$G(\omega) = \frac{1 + 4\xi_g^2(\omega/\omega_g)^2}{\left[1 - (\omega/\omega_g)^2\right]^2 + (2\xi_g\omega/\omega_g)^2} G_0 \tag{2-26}$$

The Kanai–Tajimi parameters ξ_g, ω_g, and G_0 represent ground damping, ground frequency, and ground intensity. These parameters are computed by equating the rms acceleration, the central frequency, and the shape factor, Equations 2-22, 2-23 and 2-24, of the smooth and the raw (unsmooth) power spectral densities.[(2-16, 2-19)] Table 2-3 gives the values of ω_g, ξ_g, and G_0 for the normalized power spectral densities on different soil conditions. Also shown are the central frequency ω_c and the shape factor δ. Using the Kanai–Tajimi parameters in Table 2-3, normalized power spectral densities for horizontal and vertical motion on various soil conditions were computed and are presented in Figures 2-7 and 2-8. The figures indicate that as the site gets stiffer, the predominant frequency increases and the power spectral densities spread over a wider frequency range. The figures also show that the power spectral densities for horizontal motion have a sharper peak and span over a narrower frequency region than the corresponding ones for vertical motion.

Lai[(2-19)] presents empirical relationships for estimating ground frequency ω_g and central frequency ω_c for a given epicentral distance R or a given local magnitude

Table 2-3 Central Frequency, Shape Factor, Ground Frequency, Ground Damping, and Ground Intensity for Different Soil Conditions*

Site category	No. of Records	Central freq. ω_c	Shape factor δ	Ground freq. ω_g	Ground damping ξ_g	Ground intensity G_0
Horizontal						
Alluvium	161	25.8	0.65	18.4	0.34	0.102
Alluvium on rock	60	28.9	0.59	22.9	0.30	0.078
Rock	26	34.0	0.59	27.0	0.34	0.070
Vertical						
Alluvium	78	39.4	0.63	26.2	0.46	0.080
Alluvium on rock	29	42.0	0.62	29.1	0.46	0.072
Rock	13	47.3	0.55	38.8	0.46	0.053

*After Elghadamasi et al.[(2-16)]

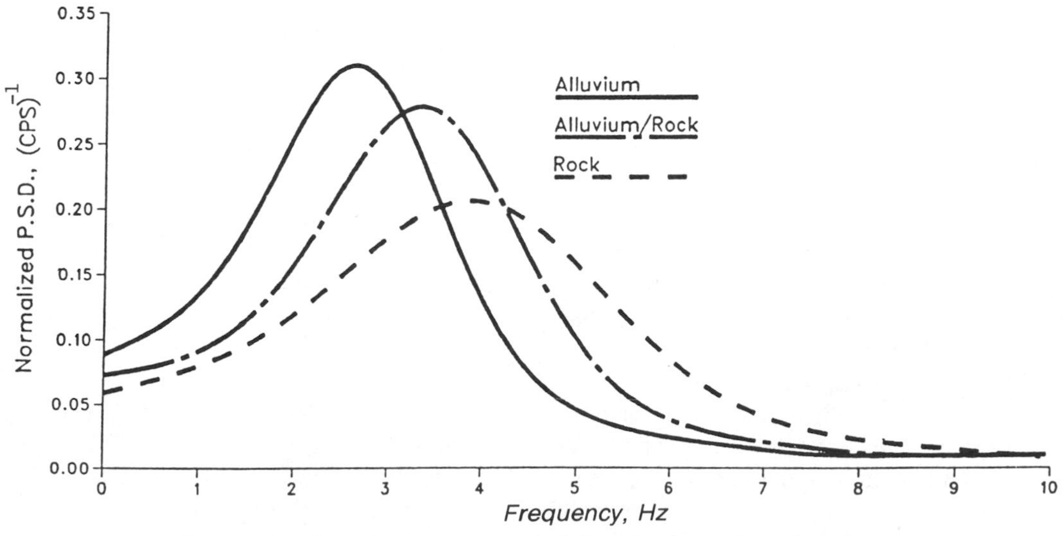

Figure 2-7 Normalized power spectral densities for horizontal motion.

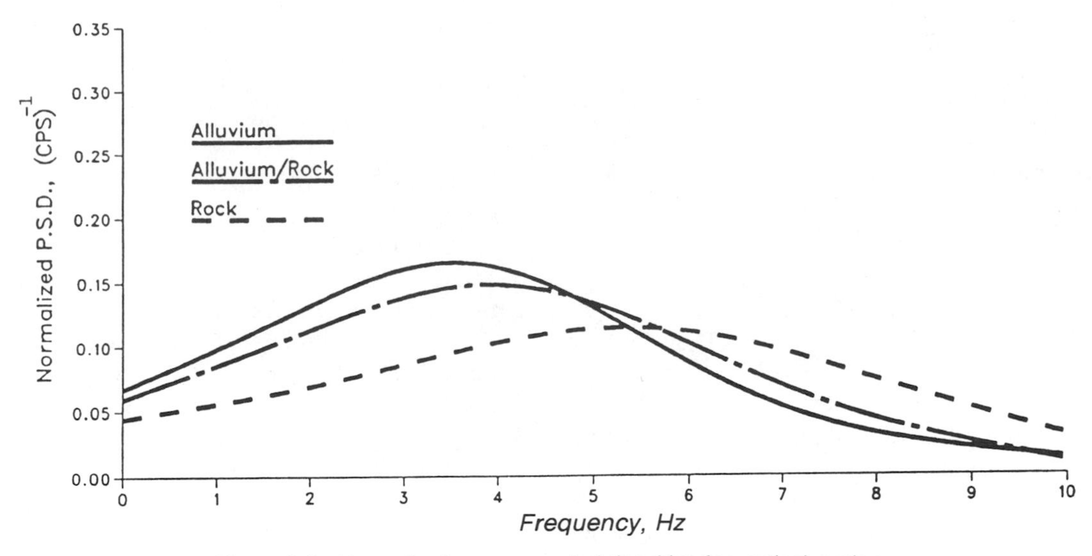

Figure 2-8 Normalized power spectral densities for vertical motion.

m. These relationships are

$$\omega_g = 27 - 0.09R, \qquad 10 \text{ km} \leqslant R \leqslant 160 \text{ km} \qquad (2\text{-}27)$$

$$\omega_g = 65 - 7.5m, \qquad 5 \leqslant m \leqslant 7 \qquad (2\text{-}28)$$

$$\omega_c = \omega_g + 5.15 \qquad (2\text{-}29)$$

Using these relationships and the acceleration attenuation equations (see Section 2.4), Lai proposes a procedure for estimating a smooth power spectral density for a given strong-motion duration and ground damping.

Once the power spectral density of ground motion at a site is established, random-vibration methods may be used to formulate probabilistic procedures for computing the response of structures. In addition the power spectral density of ground motion may also be used for other applications such as generating artificial accelerograms.

2.4 FACTORS INFLUENCING GROUND MOTION

Earthquake ground motion and its duration at a particular location are influenced by a number of factors, the most important being: (i) earthquake magnitude, (ii) distance of the source of energy release (epicentral distance or distance from causative fault), (iii) local soil conditions, (iv) variation in geology and propagation velocity along the travel path, and (v) earthquake-source conditions and mechanism (fault type, stress conditions, stress drop). Past earthquake records have been used to study some of these influences. While the effect of some of these parameters, such as local soil conditions and distance from source of energy release, are fairly well understood and documented, the influence of the source mechanism and the variation of geology along the travel path are more complex and difficult to quantify.[2-20] Several of these influences are interrelated, and consequently it is difficult to discuss them individually without incorporating the others. Some of these influences are discussed below.

2.4.1 Distance

The variation of ground motion with distance to the source of energy release has been studied by many investigators. In most studies, peak ground motion (usually peak ground acceleration) is plotted as a function of distance. A smooth curve based on a regression analysis is fitted to the data, and the curve or its equation is used to predict the expected ground motion as a function of distance. These relationships, referred to as motion attenuation, are sometimes plotted independently of the earthquake magnitude. This was the case in the earlier studies because of the lack of sufficient number of earthquake records. However, with the availability of a large number of records, particularly during the 1971 San Fernando earthquake and subsequent seismic events, the data base for attenuation studies was increased and a number of investigators reexamined their earlier studies, modified their proposed relationships for estimating peak accelerations, and included earthquake magnitude as a parameter. Donovan[2-21] compiled a data base of more than 500 recorded accelerations from seismic events in the United States, Japan, and elsewhere, and later increased it to more than 650.[2-22] The plot of peak ground acceleration vs. fault distance for different earthquake magnitudes from his data base is shown in Figure 2-9. Even though there is considerable scatter in the data, the figure shows that the peak acceleration decreases as the distance from the source of energy release increases. Also shown in the figure are the least-squares fit be-

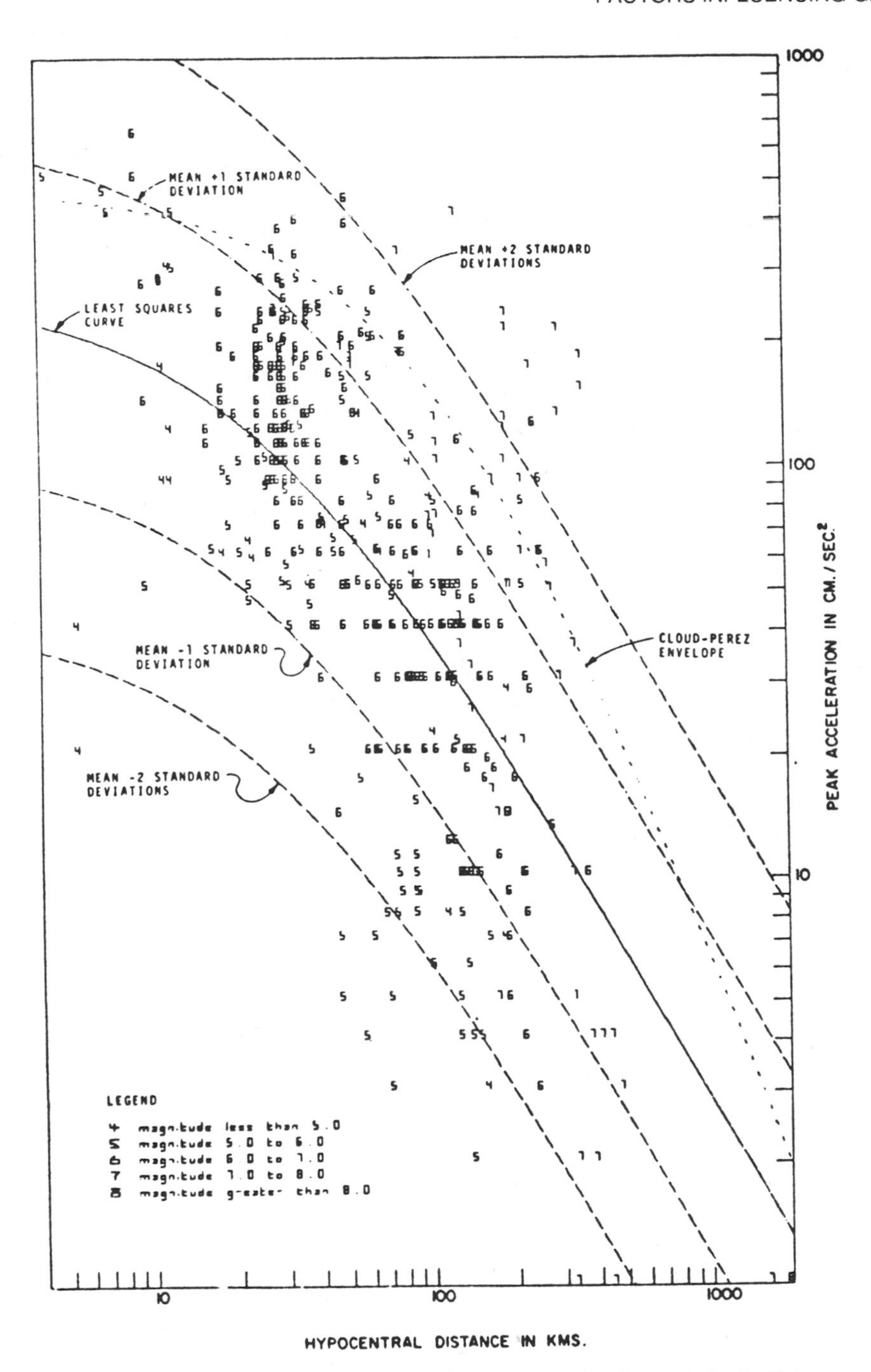

Figure 2-9 Peak ground acceleration plotted as a function of fault distance obtained from worldwide set of 515 strong motion records without normalization of magnitude. [After Donovan (2-21).]

tween acceleration and distance and the curves corresponding to mean plus—and mean minus—one and two standard deviations. The dashed curve in the figure represents the envelope curve proposed by Cloud and Perez.[(2-23)]

Other investigators have also proposed attenuation relationships for peak ground acceleration which are similar to Figure 2-9. A summary of some of the relationships, compiled by Donovan[(2-21)] and updated by the authors, is shown in Table 2-4. A comparison of various relationships[(2-22)] for an earthquake magnitude of 6.5 with the data from the 1971 San Fernando earthquake is shown in Figure 2-10. This figure is significant because it shows the comparison of various attenuation relationships with actual data from a single seismic event and a single magnitude. While the figure shows the differences in various attenuation relationships, it indicates that they all follow the same trend.

Table 2-4 Typical Attenuation Relationships

Data source	Relationship*	Reference
1. San Fernando earthquake February 9, 1971	$a = 190/R^{1.83}$	Donovan[2-21]
2. California earthquake	$a = \dfrac{y_0}{1 + (R'/h)^2}$ where $\log y_0 = -(\bar{b} + 3) + 0.81m - 0.027m^2$; $\bar{b}$ is a site factor	Blume[2-89]
3. California and Japanese earthquakes	$a = \dfrac{0.0051}{\sqrt{T_G}} 10^{0.61m - P\log R + Q}$ where $P = 1.66 + 3.60/R$; $Q = 0.167 - 1.83/R$; T_G = fundamental period of site	Kanai[2-29]
4. Cloud (1963)	$a = \dfrac{0.0069e^{1.64m}}{1.1e^{1.1m} + R^2}$	Milne and Davenport[2-91]
5. Cloud (1963)	$a = 1.254e^{0.8m}/(R + 25)^2$	Esteva[2-91]
6. U.S.C. & G.S.	$\log a = \dfrac{6.5 - 2\log(R' + 80)}{981}$	Cloud & Perez[2-23]
7. 303 instrumental values	$a = 1.325e^{0.67m}/(R + 25)^{1.6}$	Donovan[2-21]
8. Western U.S. records	$a = 0.0193e^{0.8m}/(R^2 + 400)$	Donovan[2-21]
9. U.S., Japan, etc.	$a = 1.35e^{0.58m}/(R + 25)^{1.52}$	Donovan[2-21]
10. Western U.S. records; USSR; Iran	$\ln a = 3.99 + 1.28m - 1.75\ln[R + 0.147\exp(0.732m)]$ where $m = \begin{cases} \text{surface-wave magnitude}, & m \geqslant 6 \\ \text{local magnitude}, & m < 6 \end{cases}$	Campbell[2-26]
11. Western U.S. records; world wide	$\log a = -1.02 + 0.249m - \log(R^2 + 7.3^2)^{1/2} - 0.00255(R^2 + 7.3^2)^{1/2}$ where m = moment magnitude	Joyner and Boore[2-27]
12. Western U.S. records	$\ln a = \ln\alpha(m) - \beta(m)\ln(R + 20)$ where $m = \begin{cases} \text{surface-wave magnitude}, & m \geqslant 6 \\ \text{local magnitude}, & m < 6 \end{cases}$; $R = \begin{cases} \text{smallest distance to source}, & m \geqslant 6 \\ \text{hypocentral distance}, & m < 6 \end{cases}$; $\alpha(m), \beta(m)$ are magnitude dependent coefficients	Idriss[2-39]
13. Italian records	$\log a = -1.562 + 0.306m - \log(R^2 + 5.8^2)^{1/2} + 0.169S$ where $S = 1.0$ for soft sites and 0 for rock	Sabetta and Pugliese[2-92]

*Acceleration a in g; distance to causative fault, R in kilometers; epicentral distance R' in miles; local depth h in miles; magnitude m.

It has been reported by Housner,[2-24] Donovan,[2-21] and Seed and Idriss[2-20] that at distances away from the fault or the source of energy release (far field), the earthquake magnitude influences the attenuation, whereas at distances close to the fault (near field), the attenuation is affected by smaller earthquake magnitudes and not by the larger ones. This can be observed from the earthquake data in Figure 2-9.

The majority of attenuation relationships for predicting peak ground motion are presented in terms of earthquake magnitude. Prior to the Imperial Valley Earthquake of 1979, the vast majority of available accelerograms were recorded at distances of greater than approximately 10–15 km from the source of energy release. An array of accelerographs which were placed at both sides of the Imperial Fault[2-25] prior to this earthquake (see Figure 2-11) provided excellent acceleration data at small distances from the fault. This data made it possible to study attenuation relationships very close to the fault. The attenuation relationship from this array, presented by Campbell,[2-26] is shown in Figure 2-12. The figure shows clearly the flatness of the acceleration attenuation curve at distances close to the source, a phenomenon which is not observed in the attenuation curves for far-field data. Similar observations can also be made from the attenuation curves proposed by Joyner

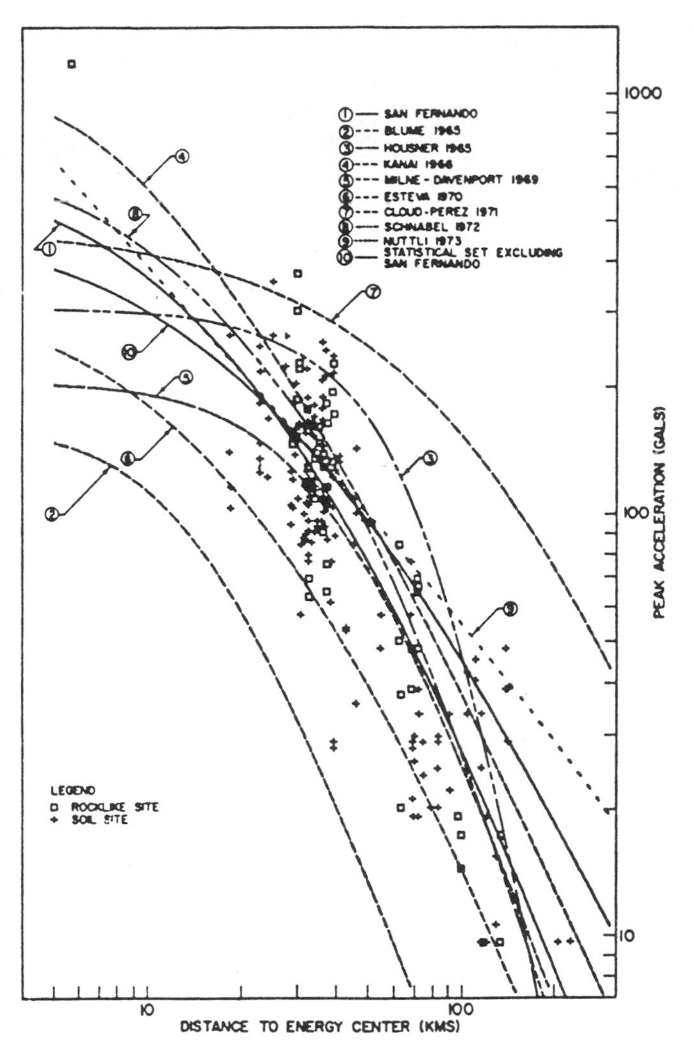

Figure 2-10 Comparison of attenuation relations with data from the 1971 San Fernando earthquake. [After Donovan (2-22).]

and Boore,[2-27] presented in Figure 2-13. The majority of attenuation studies and the relationships presented in Table 2-4 are mainly from data in the western United States. It is believed by several seismologists that ground acceleration attenuates more slowly in the eastern United States and eastern Canada. A comparison of the attenuation curves for the western United States and a curve for the eastern United States recommended by Nuttli and Herrmann[2-28] is shown in Figure 2-14. Another comparison for eastern North America, prepared by Milne and Davenport,[2-29] is presented in Figure 2-15. Both of these figures show the slower attenuation of earthquakes in the eastern United States and eastern Canada. According to Donovan[2-21] a similar phenomenon also cxists for Japanese earthquakes.

The variation of peak ground velocity with distance from the source of energy relase (velocity attenuation) has also been studied by several investigators (see for example Page et al.,[2-10] Boore et al.,[2-30, 2-31] Joyner and Boore,[2-27] Seed and Idriss[2-20]). Velocity attenuation curves have similar shapes and follow similar trends to the acceleration attenuation. Typical velocity attenuation curves proposed by Joyner and Boore are shown in Figure 2-16. Comparisons between Figures 2-13 and 2-16 indicate that velocity attenuates somewhat faster than acceleration, and unlike acceleration attenuation, velocity attenuation depends on soil condition.

The variation of peak ground displacement with the fault distance or the distance from the source of energy release (displacement attenuation) can also be plotted. Boore et al.[2-30, 2-31] have presented displacement attenuations for different ranges of earthquake magnitude. Only a few studies have addressed displacement attenuations, probably because of their limited use and the uncertainties in computing displacements accurately.

Distance also influences the duration of strong motion. Correlation of the duration of strong motion with epicentral distance has been studied by Page et al.,[2-10] Trifunac and Brady,[2-12] Chang and Krinitzsky,[2-32] and others. Page et al., using the bracketed duration, conclude that for a given magnitude, the duration decreases with an increase in the distance from the source. Chang and Krinitzsky, also using the bracketed duration, present the curves shown in Figures 2-17 and 2-18 for estimating durations for soil and rock as a function of distance. These figures show that for a given magnitude the duration of strong motion in soil is greater (approximately two times) than that in rock.

Using the 90% contribution of the acceleration intensity ($\int a^2\,dt$) as a measure of duration, Trifunac and Brady[2-12] conclude that the average duration in soil is approximately 10–12 sec longer than that in rock. They also observe that the duration increases by approximately 1.0–1.5 sec for every 10-km increase in distance. Although there seems to be a contradiction between their finding and those of Page et al. and of Chang and Krinitzsky, it stems from using two different definitions. The bracketed duration is based on the absolute acceleration level ($0.05g$). At longer epicentral distances, the acceleration peaks are smaller and a shorter duration is to be expected. The acceleration-intensity definition of duration is based on the relative measure of the percentile contribution to the acceleration intensity. Conceivably, a more intense shaking within a shorter time may result in a shorter duration than a much less intense shaking over a longer time. According to Housner,[2-24] at greater distances from the fault the duration of strong shaking may be longer, but it will be less intense than closer to the fault.

2.4.2 Site Geology

Soil conditions influence ground motion and its attenuation. Several investigators, such as Boore et al.[2-30, 2-31] and Seed and Idriss,[2-20] have presented attenuation curves for soil and rock. According to Boore et al., the peak horizontal acceleration is not appreciably affected

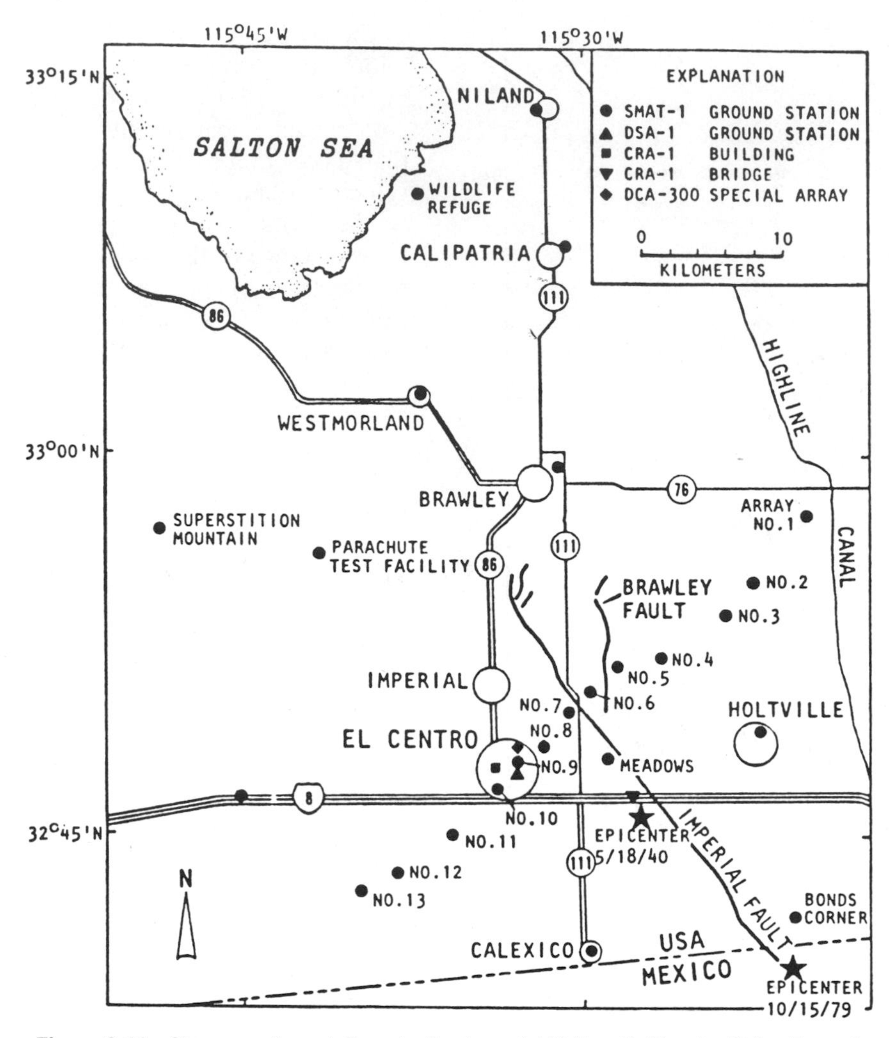

Figure 2-11 Strong motion stations in the Imperial Valley, California. [After Porcella and Matthiesen (2-25); reproduced from 2-20.]

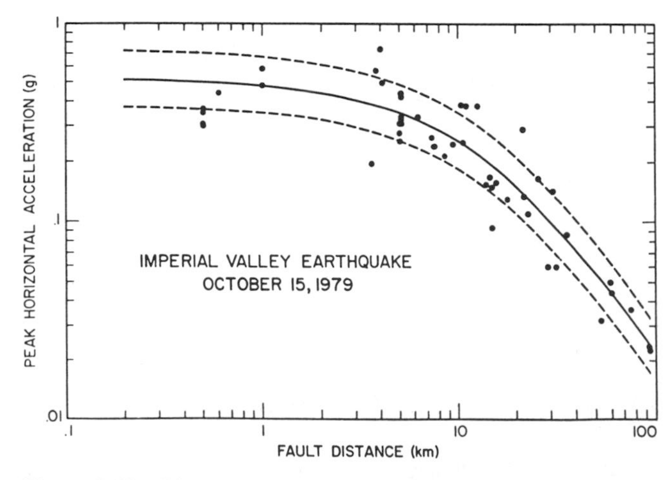

Figure 2-12 Observed and predicted mean horizontal peak accelerations for the October 15, 1979, Imperial Valley earthquake plotted as a function of distance from the fault. The solid curve represents the median predictions based on the observed values and the dashed curves represent the standard error bounds for the regression. [After Campbell (2-26).]

by soil condition (it is nearly the same for rock and for soil). Seed and Idriss compare the acceleration attenuation for rock from earthquakes with magnitudes of approximately 6.6 with that for alluvium from the 1979 Imperial Valley earthquake (magnitude 6.8). Their comparison, shown in Figure 2-19 indicates that at a given distance from the source of energy release, peak accelerations on rock are somewhat greater than those on alluvium. Studies from other earthquakes indicate that this is generally the case for acceleration levels greater than approximately $0.10g$. At levels smaller than this value, accelerations on deep alluvium are slightly greater than those on rock. The effect of soil condition on peak acceleration is illustrated by Seed and Idriss in Figure 2-20. According to this figure, the difference in acceleration on rock and on stiff soil is practically negligible. Even though in specific cases, particularly soft soil, soil condition can affect peak accelerations, Seed and Idriss conclude that the influence of soil condition can gener-

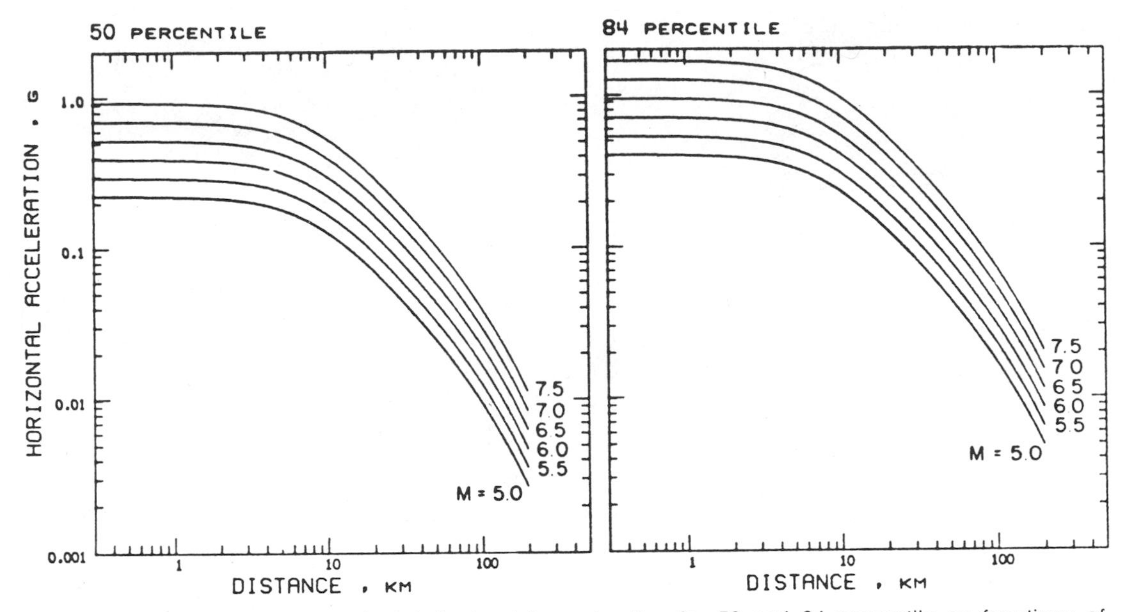

Figure 2-13 Predicted values of peak horizontal acceleration for 50 and 84 percentile as functions of distance and moment magnitude. [After Joyner and Boore (2-27).]

ally be neglected when using acceleration-attenuation curves.

There seems to be general agreement among various investigators that soil condition has a pronounced influence on velocities and displacements. According to Boore et al.,(2-30, 2-31) Joyner and Boore,(2-27) and Seed and Idriss,(2-20) larger peak horizontal velocities are more likely to be expected on soil than on rock. It should be noted that in a statistical study of earthquake ground motion and response spectra by Mohraz,(2-33) it was shown that the average velocity-acceleration ratio

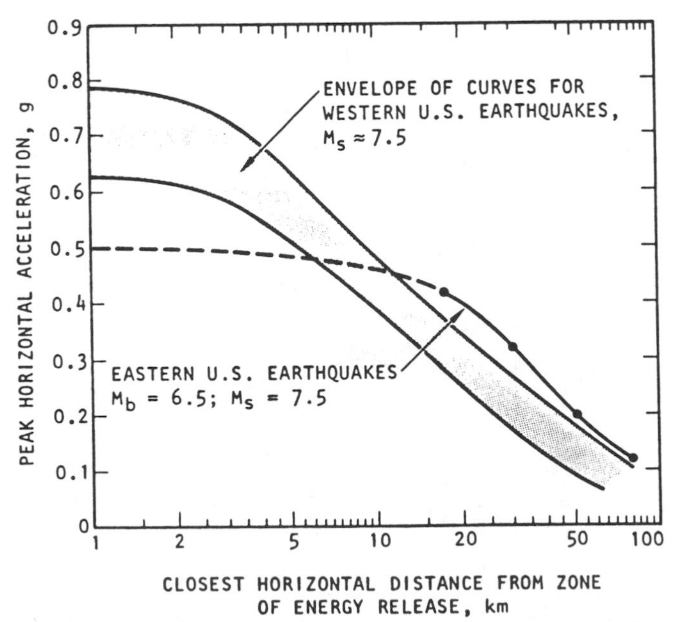

Figure 2-14 Comparison of attenuation curves for the western U.S. earthquakes. (Reproduced from 2-20.)

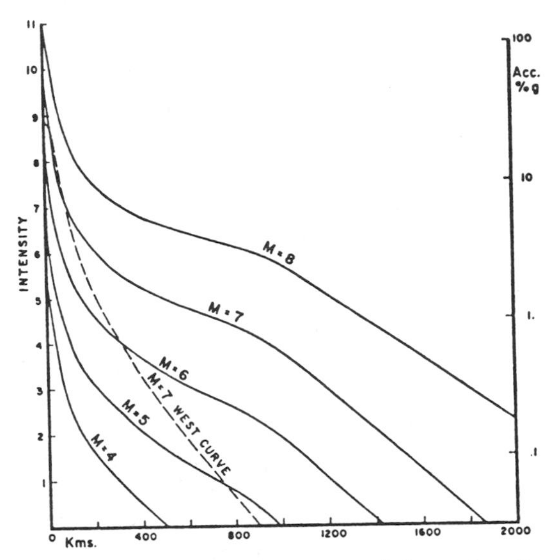

Figure 2-15 Intensity versus distance graph for western Canada. [After Milne and Davenport (2-29).]

for records on alluvium is greater than the corresponding ratio for rock, indicating that soil condition influences the ground velocity.

The influence of soil condition on duration of strong motion has already been discussed in the preceding section.

2.4.3 Magnitude

As expected, at a given distance from the source of energy release, larger earthquake magnitudes result in larger peak ground accelerations, velocities, and displacements. Because of the lack of adequate data for earthquake magnitudes greater than approximately 7.5,

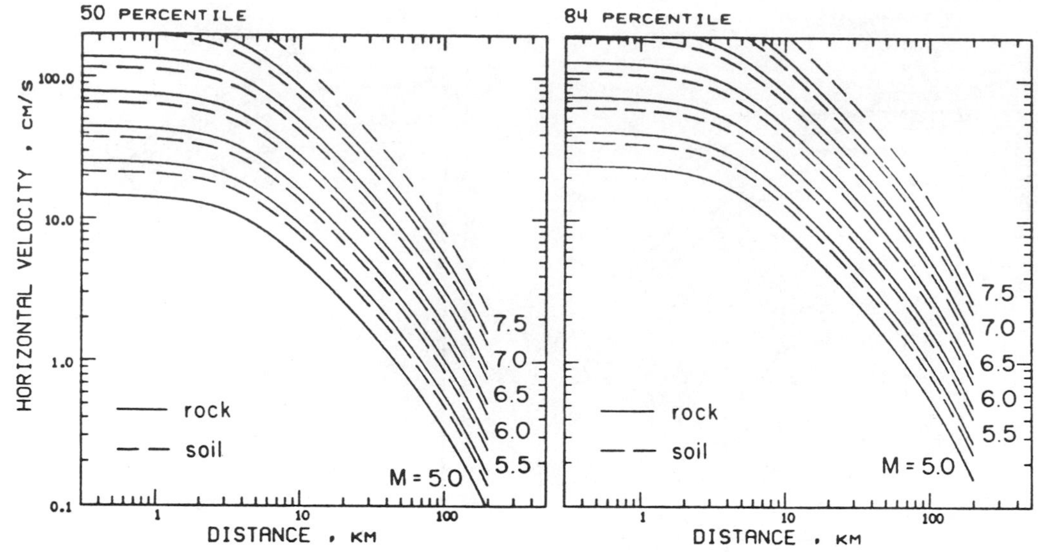

Figure 2-16 Predicted values of peak horizontal velocity for 50 and 84 percentile as functions of distance, moment magnitude, and geologic site condition. [After Joyner and Boore (2-27).]

the effect of magnitude on peak ground motion and duration is generally determined through extrapolation of data from earthquake magnitudes smaller than 7.5. Attenuation relationships can also be plotted as a function of magnitude for a given distance to source.

The influence of earthquake magnitude on the duration of strong motion has been studied by several investigators. Housner[2-24, 2-34] presents values for maximum accelerations and durations of strong phase of shaking in the vicinity of the fault for different earthquake magnitudes (Table 2-5). Donovan[2-21] presents the linear relationships in Figure 2-21 for estimating duration in terms of magnitude. His estimates compare closely with those presented by Housner in Table 2-5. Using the bracketed duration (0.05g), Page et al.[2-10] gives estimates of duration for various earthquake magni-

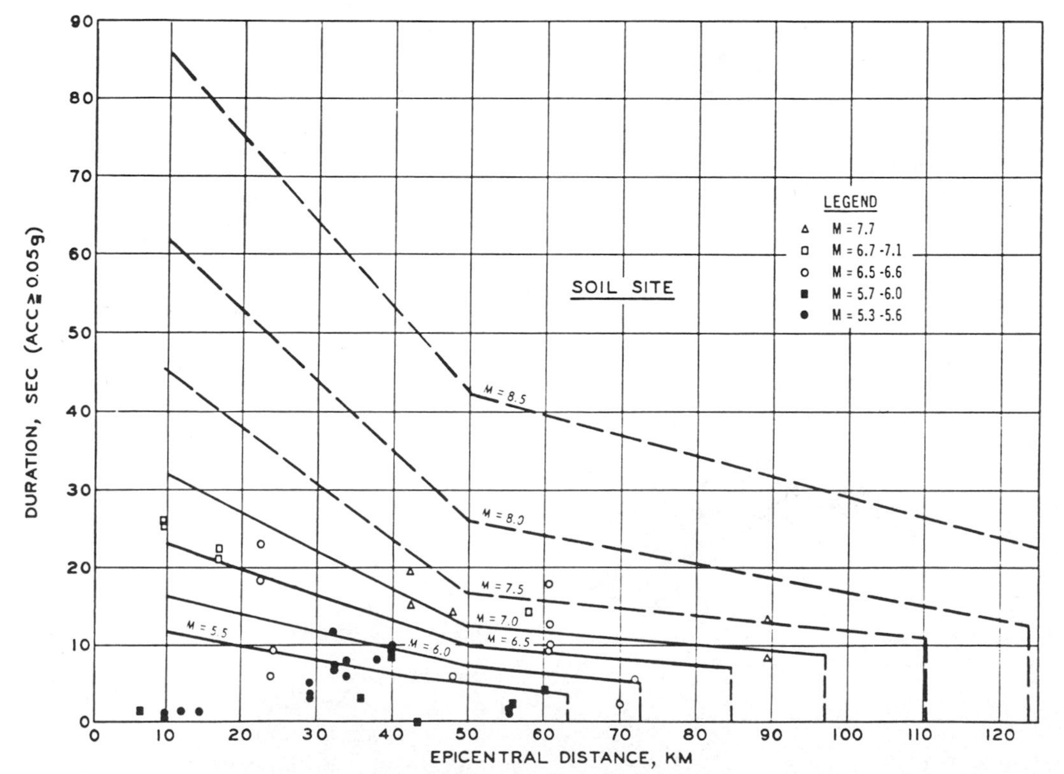

Figure 2-17 Duration versus epicentral distance and magnitude for soil. [After Chang and Krinitzsky (2-32).]

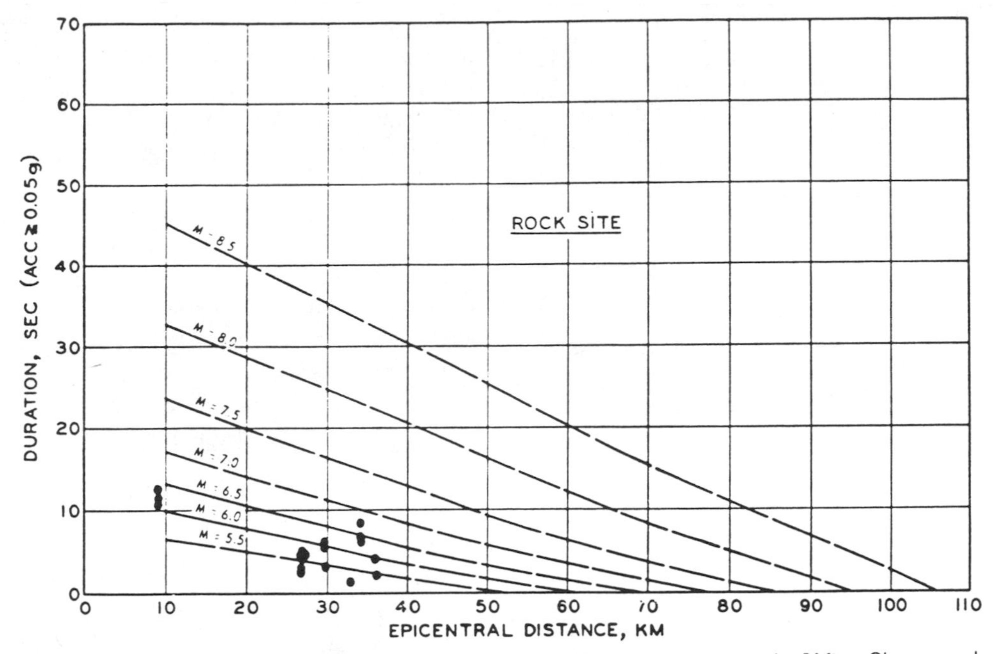

Figure 2-18 Duration versus epicentral distance and magnitude for rock. [After Chang and Krinitzsky (2-32).]

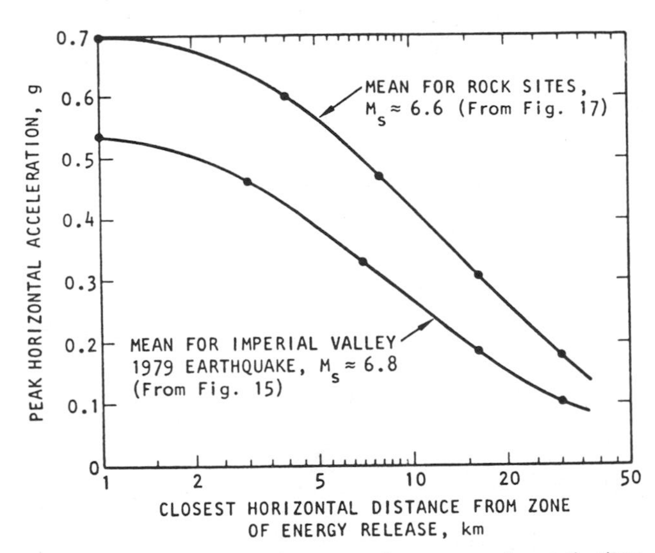

Figure 2-19 Comparison of attenuation curves for rock sites and the Imperial Valley earthquake of 1979. [After Seed and Idriss (2-20).]

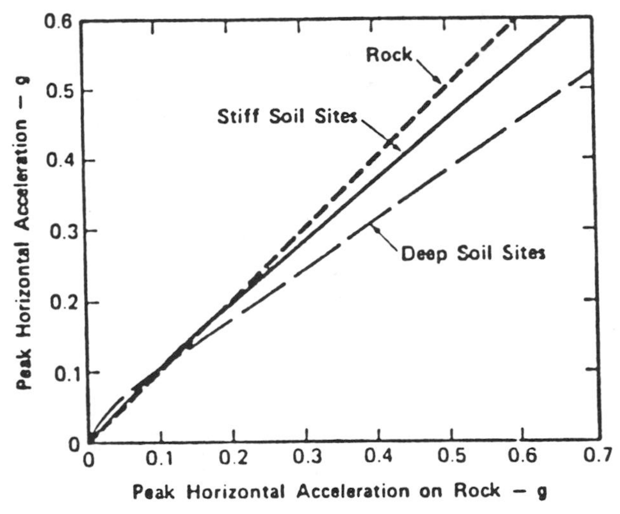

Figure 2-20 Relationship between peak accelerations on rock and soil. [After Seed and Idriss (2-20).]

tudes near the fault (Table 2-6). Chang and Krinitzsky[(2-32)] give approximate upper bounds for duration for soil and rock (Table 2-7). Their values for soil are close to those presented by Page, and the ones for rock are consistent with those given by Housner and Donovan.

2.5 EVALUATING SEISMIC RISK AT A SITE

Evaluating seismic risk is based on information from three sources: (i) the recorded ground motion, (ii) the history of seismic events in the vicinity of the site, and

Table 2-5 Maximum Ground Accelerations and Durations of Strong Phase of Shaking*

Magnitude	Maximum acceleration (%*g*)	Duration (sec)
5.0	9	2
5.5	15	6
6.0	22	12
6.5	29	18
7.0	37	24
7.5	45	30
8.0	50	34
8.5	50	37

*After Housner.[(2-34)]

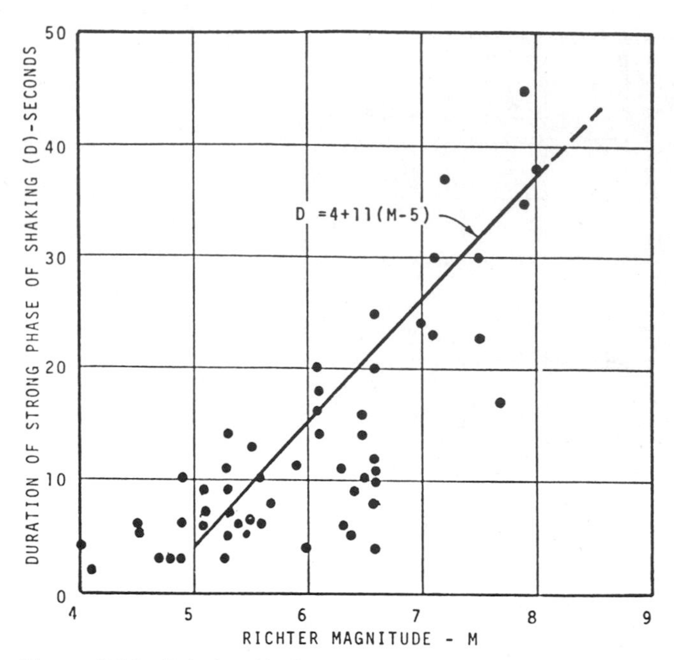

Figure 2-21 Relationship between magnitude and duration of strong phase of shaking. [After Donovan (2-21).]

(iii) the geological data and fault activities of the region. For most regions of the world this information, particularly from the first source, is very limited and may not be sufficient to predict the size and recurrence intervals of future earthquakes. Nevertheless, earthquake engineers have relied on this limited information to establish some acceptable levels of risk.

Table 2-6 Duration of Strong Motion Near Fault*

Magnitude	Duration (sec)
5.5	10
6.5	17
7.0	25
7.5	40
8.0	60
8.5	90

*After Page et al.[2-10]

Table 2-7 Duration for Different Earthquake Magnitudes*

	Duration† (sec)	
Magnitude	Rock	Soil
5.0	4	8
5.5	6	12
6.0	8	16
6.5	11	23
7.0	16	32
7.5	22	45
8.0	31	62
8.5	43	86

*After Chang and Krinitzsky.[2-32]
†Using bracketed duration with 0.05-g amplitude.

The seismic-risk analysis usually begins by developing mathematical models which are used to estimate the recurrence intervals of future earthquakes with certain magnitude and/or intensity. These models, together with the appropriate attenuation relationships, are commonly utilized to estimate ground-motion parameters such as the peak acceleration and velocity corresponding to a specified probability and return period. Among the earthquake recurrence models most used in practice is the Gutenberg–Richter relationship,[2-35, 2-36] also known as the Richter law of magnitude, which states that there exists an approximate linear relationship between the logarithm of the average number of annual earthquakes and earthquake magnitude in the form

$$\log N(m) = a - bm \tag{2-30}$$

where $N(m)$ is the average number of earthquakes per annum with magnitude greater than or equal to m, and a and b are constants determined from a regression analysis of data from the seismological and geological studies of the region over a period of time. The Gutenberg–Richter relationship is highly sensitve to magnitude intervals and the fitting procedure used in the regression analysis.[2-37–2-39] Figure 2-22 shows a typical plot of the Gutenberg–Richter relationship presented by Schwartz and Coppersmith[2-38] for the south–central

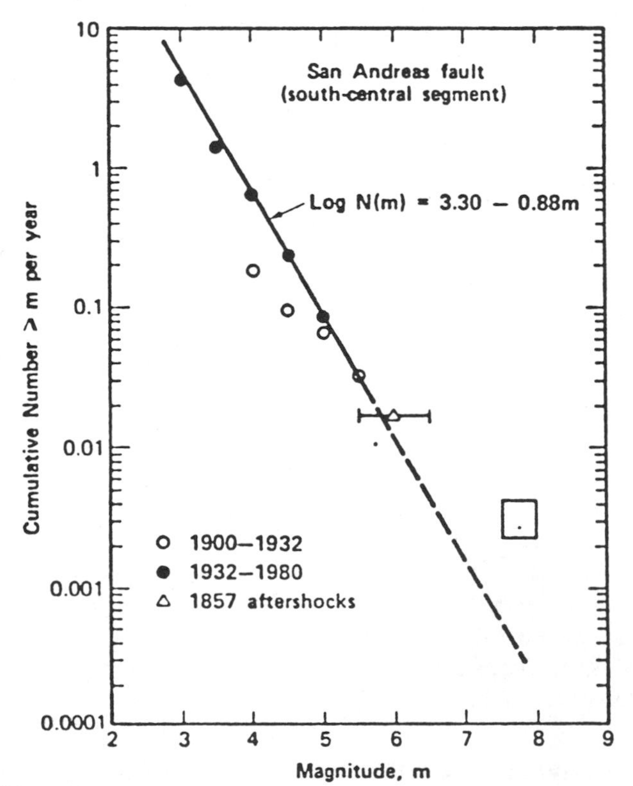

Figure 2-22 Cumulative frequency–magnitude plot of instrumental seismicity. The box in the figure represents range of recurrence based on geological data for earthquake magnitudes of 7.5–8. [After Schwartz and Coppersmith (2-38); reproduced from Idriss (2-39).]

segment of the San Andreas Fault. The relationship was obtained from historical and instrumental data in the period 1900–1980 for a 40 km-wide strip centered on the fault. The box shown in the figure represents recurrence intervals based on geological data for earthquakes of magnitudes 7.5 8.0.[2-40] It is apparent from the figure that the extrapolated portion of the Gutenberg–Richter equation (dashed line) underestimates the frequency of occurrence of earthquakes with large magnitudes, and therefore the model requires modification of the b value in Equation 2-30 for magnitudes greater than approximately 6.0.[2-39]

Cornell[2-41] introduced a simplified method for evaluating seismic risk. The method incorporates the influence of all potential sources of earthquakes. His procedure, as described by Vanmarcke,[2-42] is as follows:

1. The potential sources of seismic activity are identified and divided into smaller subsources (point sources).
2. The average number of earthquakes per annum $N_i(m)$ of magnitudes greater than or equal to m from the ith subsource is determined from the Gutenberg–Richter relationship (Equation 2-30) as

$$\log N_i(m) = a_i - b_i m \tag{2-31}$$

where a_i and b_i are known constants for the ith subsource.

3. Assuming that the design ground motion is specified in terms of the peak ground acceleration a and the epicentral distance from the ith subsource to the site is R_i, the magnitude $m_{a,i}$ of an earthquake initiated at this subsource may be estimated from

$$m_{a,i} = f(R_i, a) \tag{2-32}$$

where $f(R_i, a)$ is a function which can be obtained from the attenuation relationships. Substituting Equation 2-32 into Equation 2-31, one obtains

$$\log N_i(m_{a,i}) = a_i - b_i[f(R_i, a)] \tag{2-33}$$

Assuming the seismic events are independent (no overlapping), the total number of earthquakes per annum, N_a, which may result in a peak ground acceleration greater than or equal to a is obtained from the contribution of each subsource as

$$N_a = \sum_{\text{all}} N_i(m_{a,i}) \tag{2-34}$$

4. Finally, the mean return period T_a in years is obtained as

$$T_a = \frac{1}{N_a} \tag{2-35}$$

In the above expression, N_a can also be interpreted as the average annual probability λ_a that the peak ground acceleration exceeds a certain acceleration a. In a typical design situation, the engineer is interested in the probability that such a peak exceeds a during the life of the structure t_L. This probability can be estimated using the Poisson distribution

$$P = 1 - e^{-\lambda_a t_L} \tag{2-36}$$

Another distribution, based on a Bayesian procedure,[2-43] was proposed by Donovan.[2-21] It is more conservative than the Poisson distribution, and therefore more appropriate when additional uncertainties such as those associated with the long return periods of large magnitude earthquakes are encountered. It should be noted that other ground-motion parameters may be used, in lieu of acceleration, for evaluating seismic risk. Other similar procedures for seismic-risk analysis, based on more sophisticated models, have also been proposed (see for example Der Kiureghian and Ang).[2-44]

The evaluation of seismic risk at a site is demonstrated by Donovan,[2-21] who uses as an example the downtown area of San Francisco. The epicentral data and earthquake magnitudes he considers in the evaluation were obtained over a period of 163 years and are depicted in Figure 2-23. Most of the data are associated with three major faults, San Andreas, Hayward, and Calaveras. Using attenuation relationships for competent soil and rock, Donovan computed the return periods for different peak accelerations (see Table 2-8). He then computed the probability of exceeding various peak ground accelerations during a 50-year life of the structure, which is shown in Figure 2-24. Plots such as those in Figure 2-24 may be used to estimate the peak acceleration for various probabilities. For example, if the structure is to be designed to resist a moderate earthquake with a probability of 0.6 and a severe earthquake with a probability between 0.1 and 0.2 of occurring at least once during the life of the structure, the peak accelerations using Figure 2-24b for the rock, are $0.15g$ and $0.4g$, respectively.

Using the seismic-risk principles of Cornell,[2-41] Algermissen and Perkins[2-45, 2-46] developed isoseismal maps for peak ground accelerations and velocities. Figure 2-25 is a map which shows contours of peak accelerations on rock having 90% probability of not being exceeded in 50 years. The Applied Technology Council[2-47] has used this map to develop similar maps for effective peak acceleration (Figure 2-26) and effective peak velocity-related acceleration (Figure 2-27). The effective peak acceleration and the effective peak velocity-related acceleration are defined by the Applied Technology Council[2-47] based on a study by McGuire.[2-48] They are obtained by dividing the spectral accelerations at periods of 0.1–0.5 sec and the

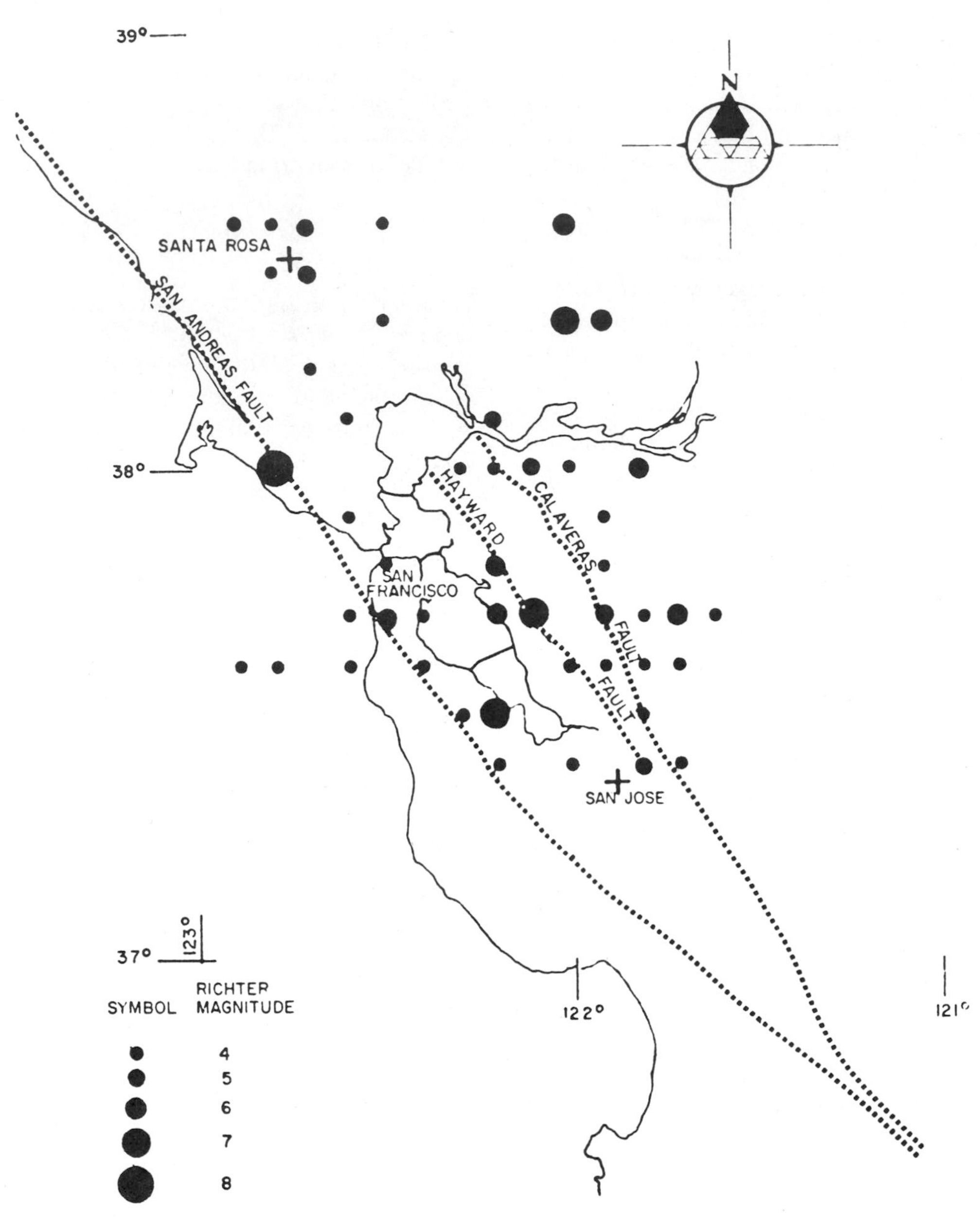

Figure 2-23 Historical instrumental or estimated epicentral locations within 100 km of San Francisco. [After Donovan (2-21).]

Table 2-8 Return Periods for Peak Ground Acceleration*

	Return period (years)	
Peak acceleration (*g*)	**Soil**	**Rock**
0.05	4	8
0.10	20	30
0.15	50	60
0.20	100	100
0.25	250	200
0.30	450	300
0.40	2000	700

*After Donovan.(2-21)

spectral velocity at a period of approximately 1.0 sec by a constant factor (the amplification 2.5 for a 5% damping spectrum).

2.6 ESTIMATING GROUND MOTION

Seismic-risk procedures and attenuation relationships are mostly developed for estimating the expected peak horizontal acceleration at the site.[c] Although the struc-

[c]The use of peak ground acceleration for design has been questioned by several investigators who believe effective peak acceleration rather than peak acceleration should be used.

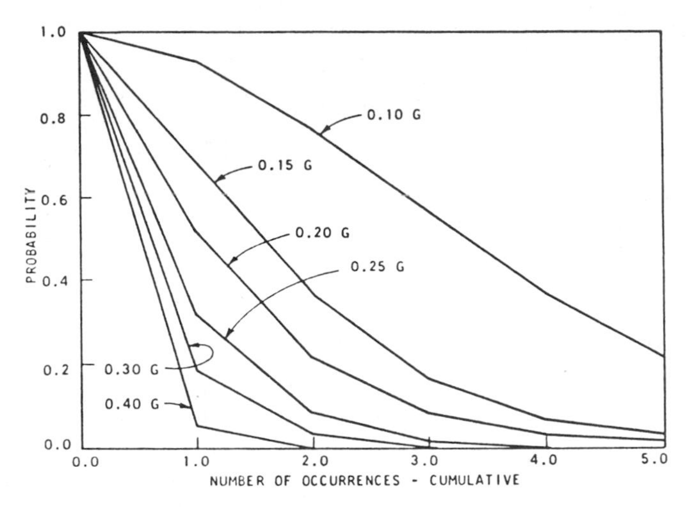

(a) Based on return periods using relationship 9 in Table 2-4

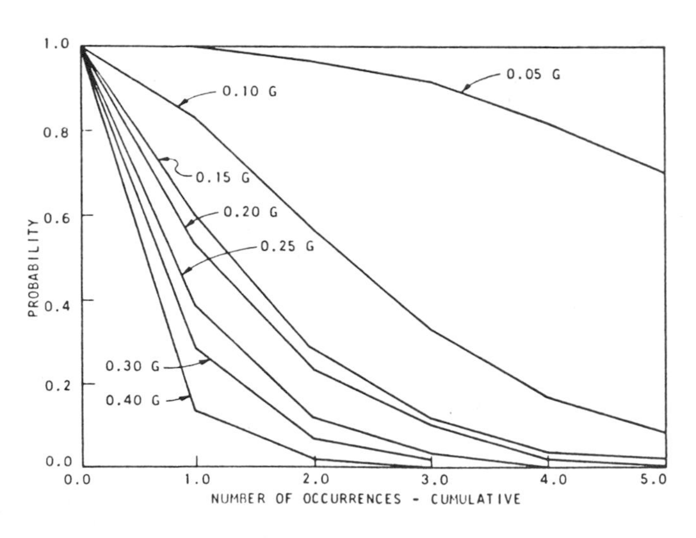

(b) Based on return periods using attenuation equation for rock

Figure 2-24 Estimated probabilities for a 50-yr project life. [After Donovan (2-21).]

tural response and damage potential to buildings are closely related to peak ground acceleration, certain response characteristics may relate more appropriately to ground velocity and displacement. Studies by Mohraz et al.,(2-9) Mohraz,(2-33) Newmark and Hall,(2-49) and Newmark et al.(2-50) have recommended using ground velocity and displacement, in addition to ground acceleration, in defining spectral shapes and ordinates.

Earlier studies by Newmark and Hall(2-51, 2-52) specify a maximum horizontal ground velocity of 48 in./sec and a maximum horizontal ground displacement of 36 in., corresponding to a maximum horizontal ground acceleration of $1.0g$. In addition, Newmark recommends that the maximum vertical ground motion can be taken as $\frac{2}{3}$ of the horizontal values for constructing smooth spectrum for vertical motion.

With the availability of a large number of recorded earthquake ground motions, particularly during the 1971 San Fernando earthquake, a number of statistical studies(2-9, 2-53, 2-33) were carried out to determine average values of ground velocity and displacement corresponding to a given acceleration. These studies recommend two ratios—the ratio of the peak velocity to peak acceleration, v/a, and the ratio of the peak-acceleration—displacement product to the square of the peak velocity, ad/v^2—be used to estimate peak velocity and displacement. The reason for selecting ad/v^2 instead of d/a or d/v is that certain ground-motion and response-spectrum characteristics can be related to ad/v^2. For example, the ratio ad/v^2 is a measure of sharpness or flatness of the response spectrum. According to Newmark and Rosenbleuth,(2-54) for most earthquakes of practical interest ad/v^2 ranges from approximately 5–15. They state that for harmonic oscillations ad/v^2 is 1, and for steady-state square acceleration waves it is $\frac{1}{2}$.

A statistical study of v/a and ad/v^2 ratios was carried out by Mohraz,(2-33) who used a total of 162 components of 54 earthquake records from 46 stations in 16 seismic events. A summary of v/a and ad/v^2 for records on alluvium, on rock, on less than 30 ft of alluvium underlain by rock, and on 30–200 ft of alluvium underlain by rock from Mohraz's study is given in Table 2-9.[d] The results for each soil category are presented in three groups: the first includes the components with the larger of the two peak horizontal accelerations, the second includes the components with the smaller of the two peak horizontal accelerations, and the third includes the vertical components. The reason for separating the two horizontal components is to exclude the influence of the components with the smaller peak acceleration on the ground-motion ratios.

The results in Table 2-9 indicate that v/a ratios for rock are substantially lower than those for alluvium, and those for the two intermediate categories fall between alluvium and rock. It is noted that for each soil category, the v/a for the horizontal components with the larger of the two peak ground accelerations is lower than for the components with the smaller of the two peak accelerations. Since the maximum horizontal ground acceleration is the parameter usually specified for design, a realistic estimate of the ground velocity should be based on the v/a for the group with the larger peak acceleration. Table 2-9 also shows that the v/a ratios for the vertical components are close to those for the horizontal components with the larger of the two peak accelerations. It is noted that the 50-per-

[d]Both a normal and a lognormal distribution were used, and it was found that slightly better correlations were obtained for the lognormal distribution.

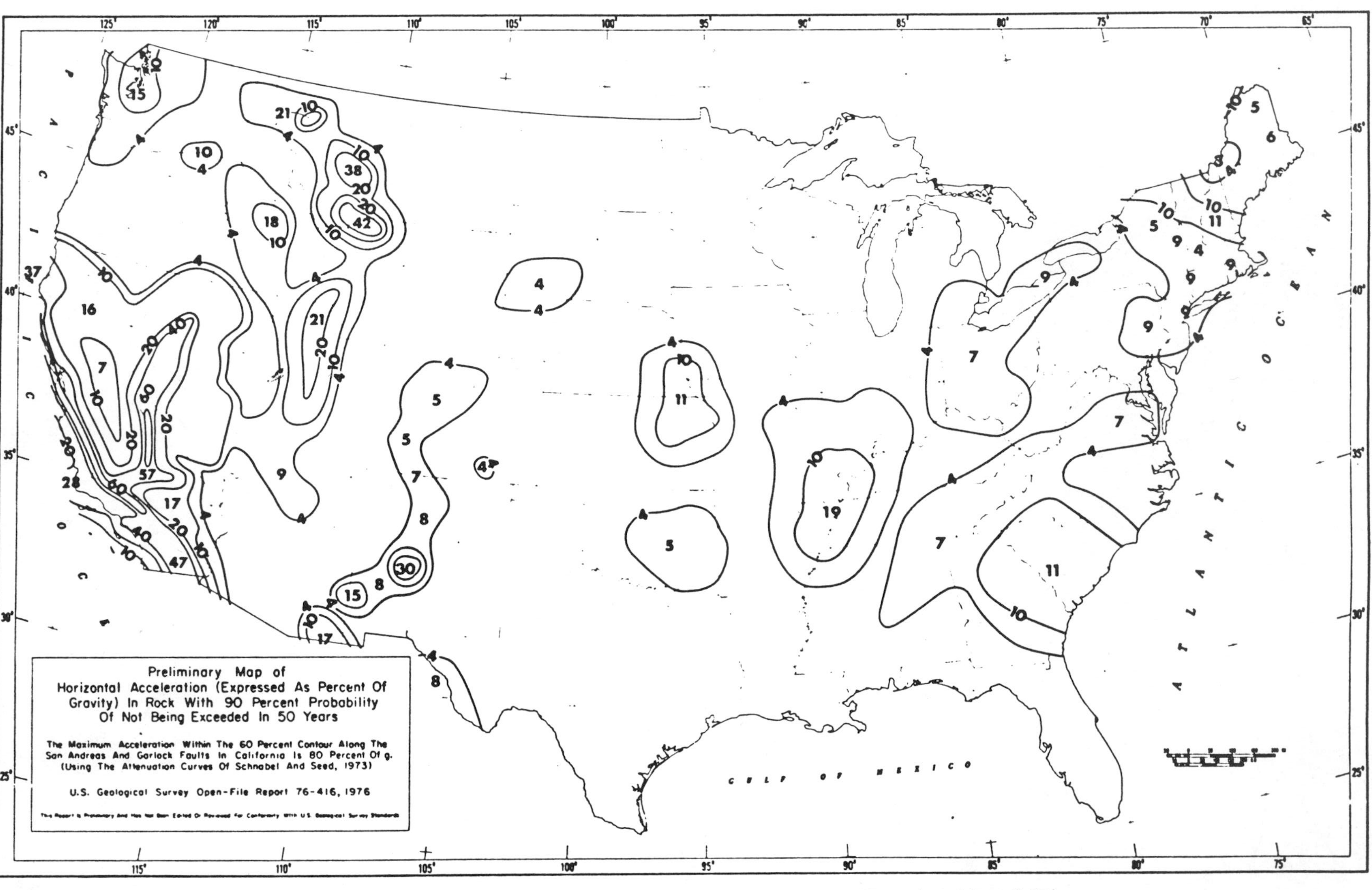

Figure 2-25 Seismic risk map developed by Algermissen and Perkins. (Reproduced from 2-47.)

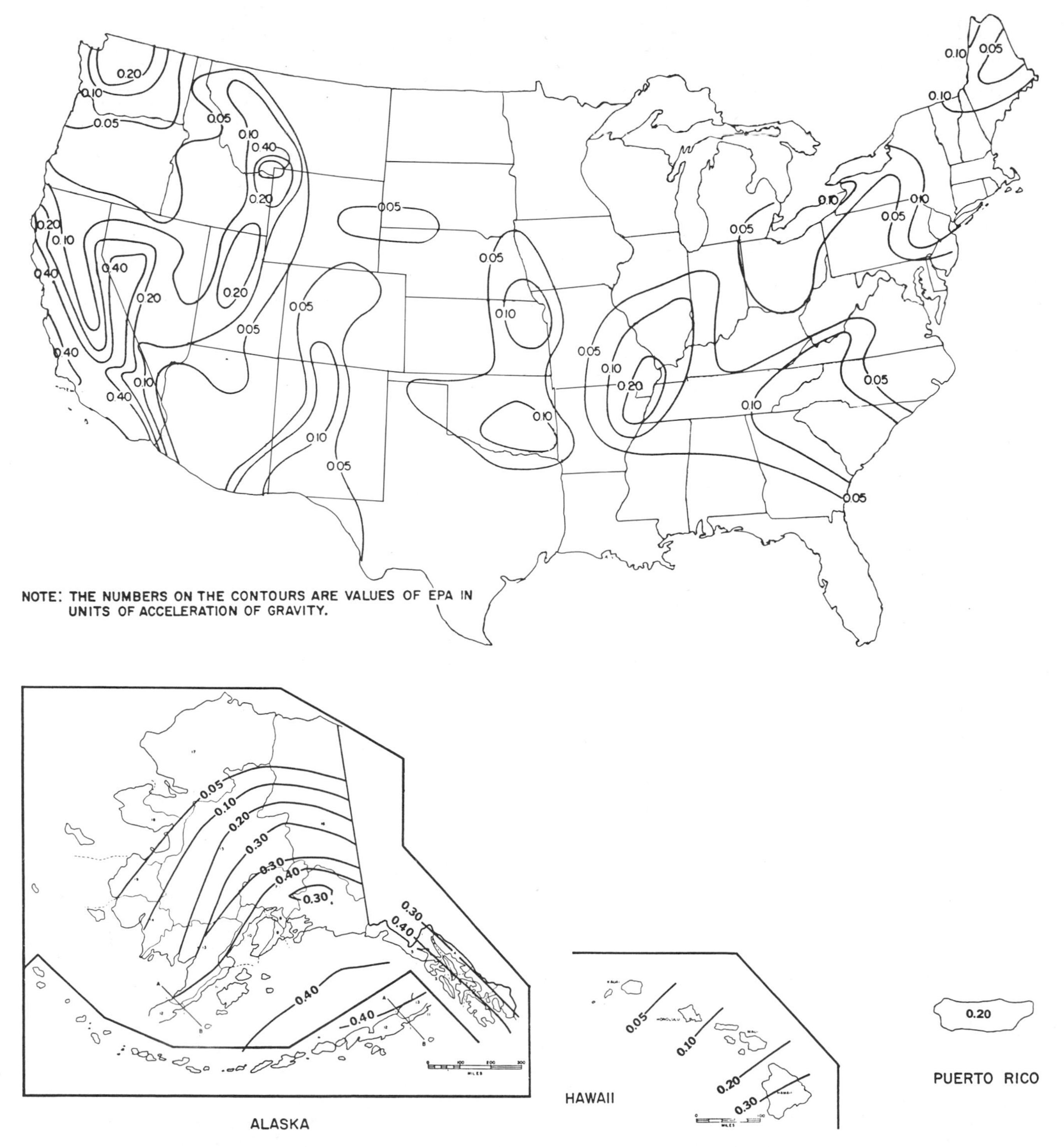

Figure 2-26 Contour map for effective peak acceleration (ATC, 2-47).

centile v/a ratios for the larger of the two peak accelerations from Table 2-9 [24 (in./sec)/g for rock and 48 (in./sec)/g for alluvium] and those given by Seed and Idriss [22 (in./sec)/g for rock and 43 (in./sec)/g for alluvium] are in close agreement. The ad/v^2 ratios in Table 2-9 indicate that, in general, the ratios for alluvium are smaller than those for rock and those for alluvium layers underlain by rock. The d/a ratios computed using the 50-percentile v/a and ad/v^2 ratios are also shown in Table 2-9. The values indicate that for a given acceleration, the displacements for alluvium are 2 to 3 times those for rock.

A statistical summary of the peak ground acceleration ratios for the four soil categories is presented in

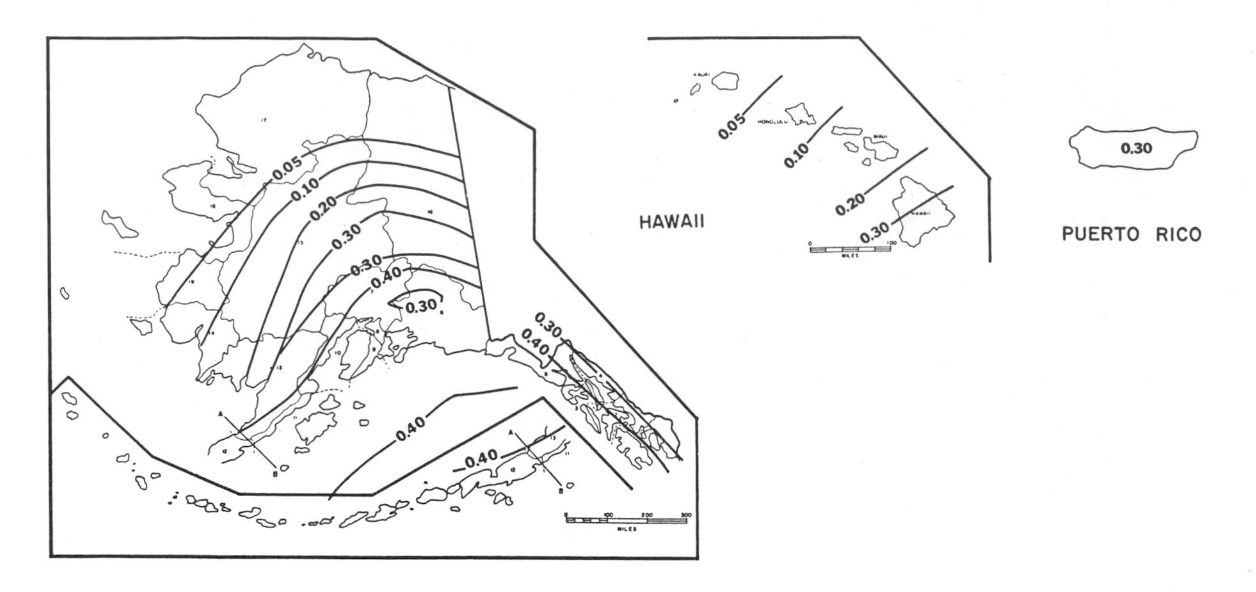

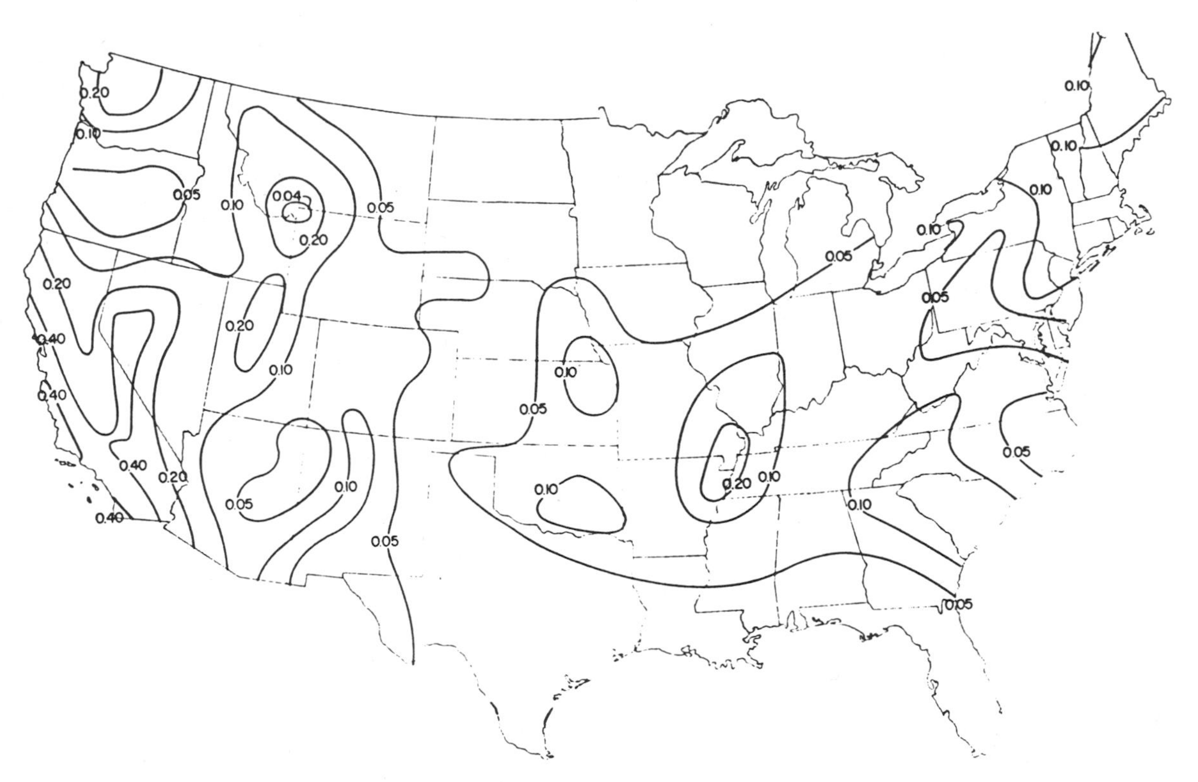

Figure 2-27 Contour map for effective peak velocity-related acceleration coefficient (ATC, 2-47).

Table 2-10. The table includes the ratio of the smaller to the larger of the two peak horizontal accelerations and the ratio of the vertical to the larger of the two peak horizontal accelerations. In each column, the ratios are generally close to each other, indicating that the soil condition does not influence the acceleration ratios. The ratio $\frac{2}{3}$ of the vertical to horizontal acceleration, which has been recommended by Newmark and has been employed in seismic design, is closer to the 84.1-percentile than to the 50-percentile ratio. Although the ratio $\frac{2}{3}$ is conservative, its use has been justified as taking account of variations greater than median and uncertainties in the ground motion in the vertical direction.[2-53]

Table 2-9 Summary of v/a and ad/v^2 Ratios* (Lognormal Distribution)

		v/a, (in./sec)/g		ad/v^2		d/a, in./g
		Percentile		Percentile		Percentile
Soil category	Group†	50	84.1	50	84.1	50
Rock	L	24	38	5.3	11.0	8
	S	27	44	5.2	11.2	10
	V	28	45	6.1	11.8	12
< 30 ft of alluvium underlain by rock	L	30	57	4.5	7.7	11
	S	39	62	4.2	8.2	17
	V	33	53	6.8	13.3	19
30 – 200 ft of alluvium underlain by rock	L	30	46	5.1	7.8	12
	S	36	58	3.8	6.4	13
	V	30	46	7.6	13.7	18
Alluvium	L	48	69	3.9	6.0	23
	S	57	85	3.5	4.9	29
	V	48	70	4.6	7.0	27

*After Mohraz.(2-33)
†L: Horizontal components with the larger of the two peak accelerations; S: Horizontal components with the smaller of the two peak accelerations; V: Vertical components.

Table 2-10 Summary of Peak Ground Acceleration Ratios* (Lognormal Distribution)

	a_s/a_l			a_v/a_l		
	Percentile			Percentile		
Soil category	50	84.1	Mean	50	84.1	Mean
Rock	0.81	0.99	0.82	0.48	0.69	0.52
< 30 ft of alluvium underlain by rock	0.89	0.98	0.89	0.47	0.62	0.46
30 – 200 ft of alluvium underlain by rock	0.82	0.96	0.83	0.40	0.66	0.46
Alluvium	0.75	0.96	0.79	0.42	0.61	0.45

†After Mohraz.(2-33)
*a_l = the larger peak acceleration of the two horizontal components; a_s = the smaller peak acceleration of the two horizontal components; a_v = the peak acceleration of the vertical component.

2.7 EARTHQUAKE RESPONSE SPECTRA

The importance of the response spectrum in seismic analysis and design of structures and equipment is well known to earthquake design engineers. The response spectrum introduced by Biot[2-1, 2-2] and Housner[2-3] describes the maximum response of a damped single-degree-of-freedom oscillator at various frequencies or periods. The detailed procedure for computing and constructing the response spectrum is discussed in Chapter 3 and in a number of publications (see for example References 2-34, 2-55, 2-56, 2-57). As discussed in Chapter 3, it is convenient to plot response spectra on tripartite paper (four-way logarithmic paper) so that at a given frequency or period, the maximum relative displacement SD, the pseudovelocity PSV, and the pseudoacceleration PSA can all be read from the plot simultaneously. The parameters PSV and PSA, which are expressed in terms of SD and the circular natural frequency ω as $PSV = \omega SD$ and $PSA = \omega^2 SD$, have certain characteristics that are of practical interest.[2-49] The pseudovelocity PSV is close to the maximum relative velocity SV for high frequencies, and practically equal for intermediate frequencies, but different for low frequencies (see Figure 2-28). For zero damping, the pseudoacceleration PSA is equal to the maximum absolute acceleration SA, but for dampings other than zero, the two are slightly different. However, for damping levels encountered in most engineering applications, the two can be considered practically equal (see Figure 2-29). Arithmetic and semilogarithmic plots have also been used to represent response spectra. It is sometimes convenient to plot the ratio of the response to the appropriate ground motion (amplification). In many

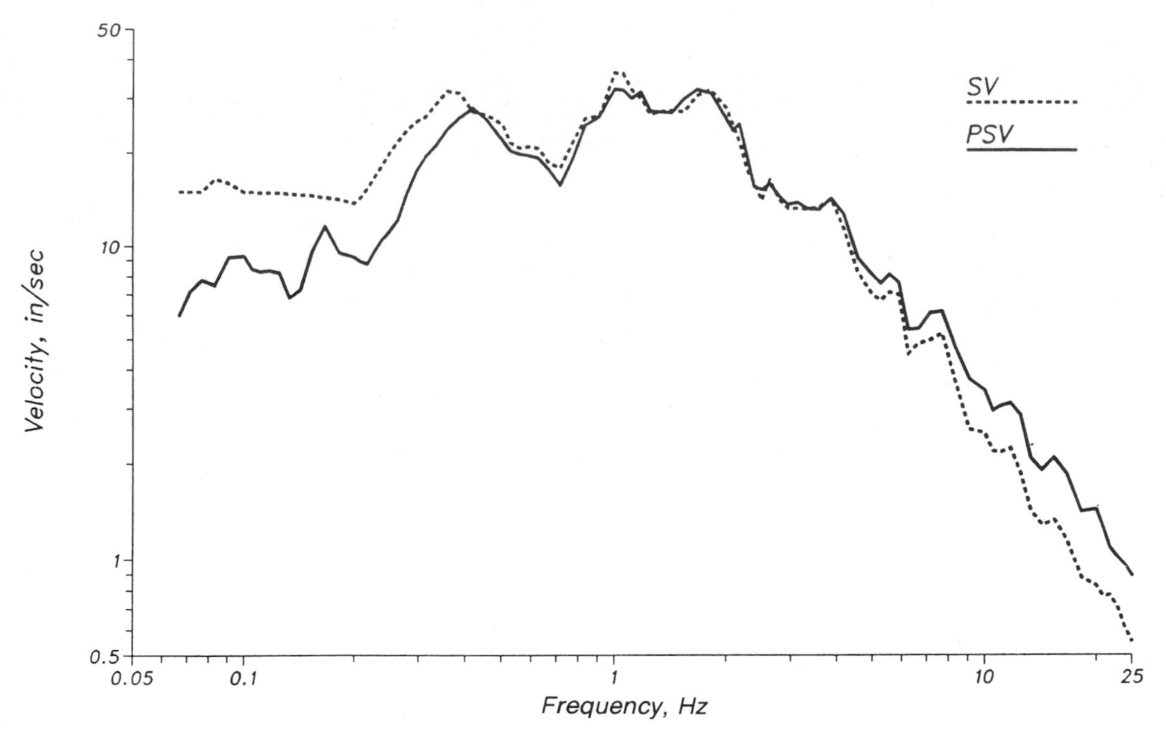

Figure 2-28 Comparison of pseudovelocity and maximum relative velocity for 5% damping for the S00E component of El Centro, the Imperial Valley earthquake of May 18, 1940.

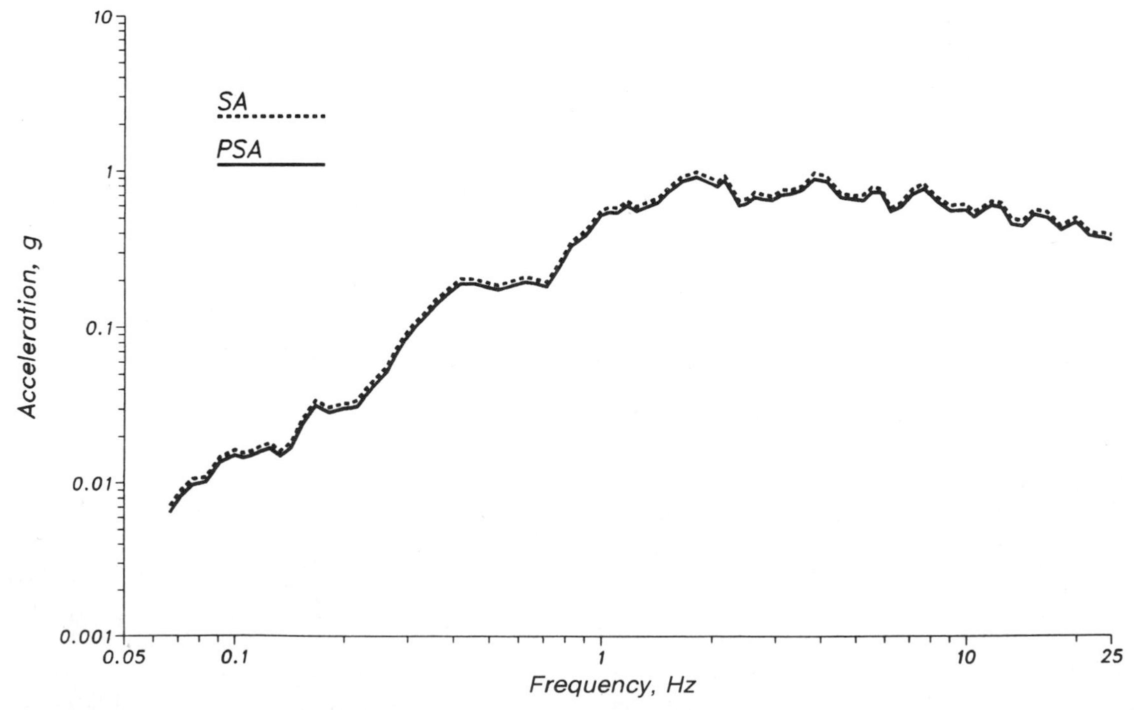

Figure 2-29 Comparison of pseudoacceleration and maximum absolute acceleration for 5% damping for the S00E component of El Centro, the Imperial Valley Earthquake of May 18, 1940.

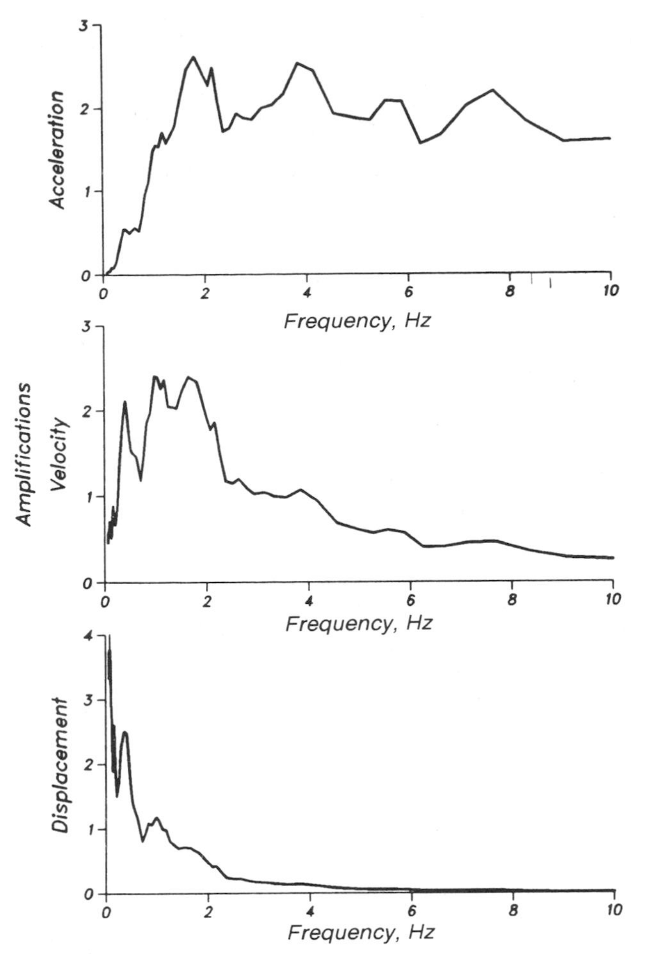

Figure 2-30 Acceleration, velocity, and displacement amplifications plotted as a function of frequency for 5% damping for the S00E component of El Centro, the Imperial Valley earthquake of May 18, 1940.

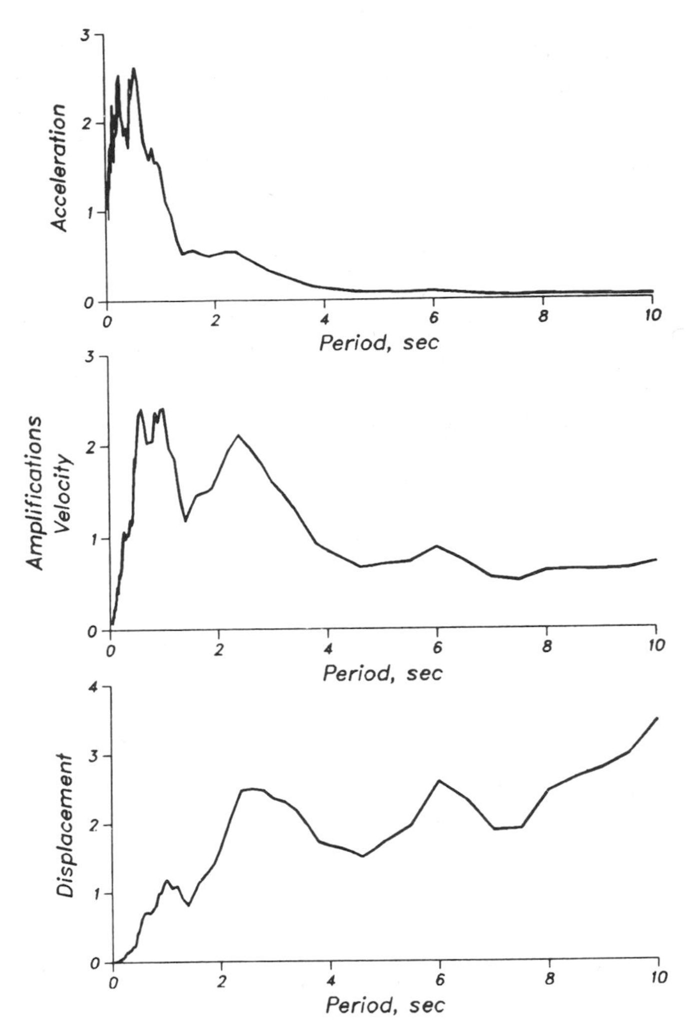

Figure 2-31 Acceleration, velocity, and displacement amplifications plotted as a function of period for 5% damping for the S00E component of El Centro, the Imperial Valley earthquake of May 18, 1940.

cases, the ratio of spectral acceleration to peak ground acceleration (acceleration amplification) is plotted as a function of either frequency or period. Typical acceleration, velocity, and displacement amplifications for the S00E component of El Centro, the Imperial Valley earthquake of May 18, 1940, are shown in Figures 2-30 and 2-31—one plotted as a function of frequency and the other as a function of period.

To show how ground motion is amplified in various regions of the spectrum, the peak ground displacement, peak ground velocity, and peak ground acceleration for the S00E component of El Centro are plotted together with the response spectra in Figure 2-32. Several observations can be made from this figure. At small frequencies, the maximum relative displacement is large, whereas the pseudoacceleration is very small. At large frequencies, the relative displacement is extremely small, whereas the pseudoacceleration is relatively large. At intermediate frequencies, the pseudovelocity is substantially larger than at either end of the spectrum. For this reason, three regions are usually identified in a response spectrum—the low-frequency or displacement region, the intermediate-frequency or velocity region, and the high-frequency or acceleration region. In each region the corresponding ground motion is amplified the most. Figure 2-32 also shows that at small frequencies, (0.05 Hz or less), the spectral displacement approaches the peak ground displacement, indicating that for very flexible systems (systems with small frequencies or long periods) the maximum distortion in the spring or column is equal to the ground displacement. At large frequencies, such as 25–30 Hz, the pseudoacceleration approaches the peak ground acceleration, indicating that for rigid systems (systems with large frequencies or short periods) the absolute acceleration of the mass is the same as that of the ground. Figure 2-32 shows that response spectrum for a given earthquake record is quite irregular and has a number of peaks and valleys. The irregularities are sharp for small dampings, and they get somewhat smoother as damping increases.

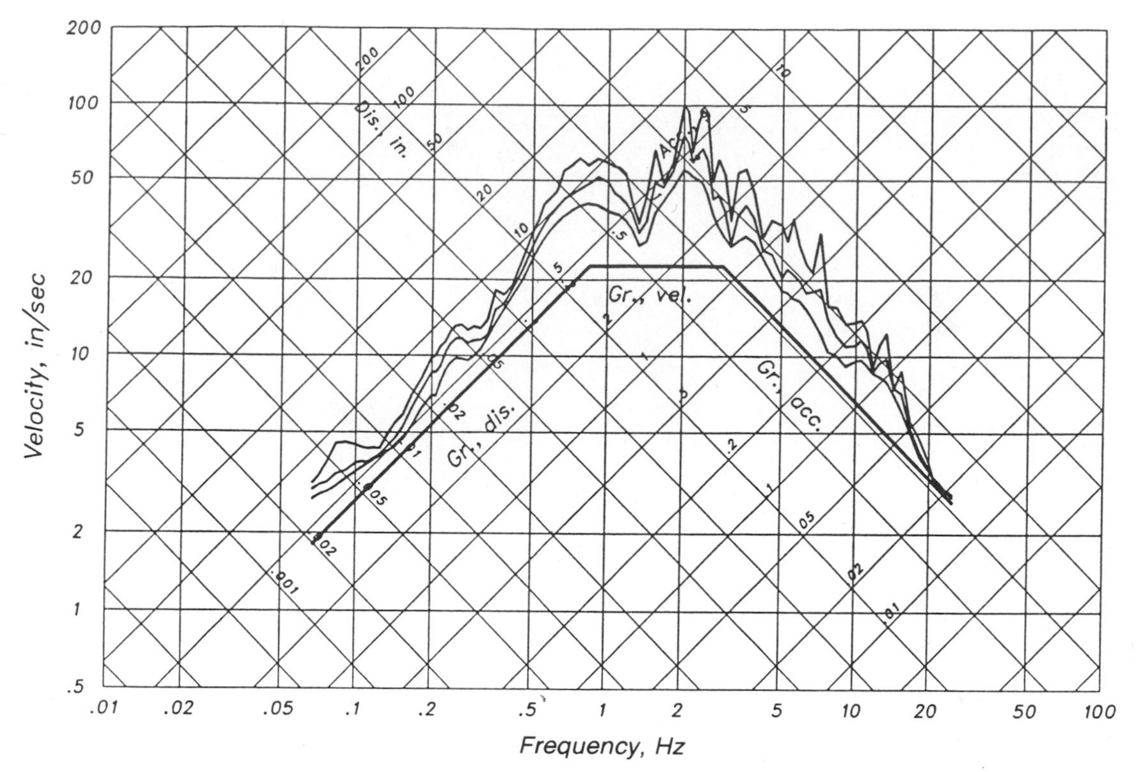

Figure 2-32 Response spectra for 2, 5, and 10% damping for the S00E component of El Centro, the Imperial Valley earthquake of May 18, 1940, together with the peak ground motion.

While response spectra for a specified earthquake record may be used to obtain the response of a structure to an earthquake ground motion with similar characteristics, they cannot be used for design, because the response of the same structure to another earthquake will undoubtedly be different. Nevertheless, the recorded ground motion and computed response spectra of past earthquakes exhibit certain similarities. For example, studies have shown that the response spectra from accelerograms recorded on similar soil conditions reflect similarities in shape and amplifications. For this reason, response spectra from records with common characteristics are averaged and then smoothed before they are used in design.

2.8 EARTHQUAKE DESIGN SPECTRA

Because the detailed characteristics of future earthquakes are not known, the majority of earthquake design spectra are obtained by averaging a set of response spectra from records with common characteristics. For practical applications, design spectra are presented as smooth curves or straight lines. Smoothing is justified because of the difficulties in determining the exact frequencies and mode shapes of structures during severe earthquakes when the behavior is most likely nonlinear. It should be noted that in some cases, determining the shape of design spectra for a particular site is complicated, and caution should be used in arriving at a representative set of records. For example, long-period components of strong motion influence the response of structures in certain frequency ranges. Recent strong-motion data indicate that long-period components are influenced by factors such as source type, rupture propagation, travel path, and local soil conditions.[2-59, 2-60] In addition, the direction and spread of rupture propagation can affect motion in the near field. For these reasons, the selection of an appropriate set of records in arriving at representative design spectra is important and may involve the selection of different sets of records for different regions of the spectrum.

The difference between response spectra and design spectra should be kept in mind whenever a design spectrum is used. A response spectrum is a plot of the maximum response of a single-degree-of-freedom oscillator with different frequencies and damping ratios to a specific ground motion, whereas a smooth, or design, spectrum is a specification of the seismic-design force or displacement of a structure having a certain frequency or period of vibration and damping.[2-58]

Since the peak ground acceleration, velocity, and displacement for various earthquake records differ, the computed response cannot be averaged on an absolute basis. Various procedures are used to normalize response spectra before averaging is carried out. Among these procedures, two which have been most commonly

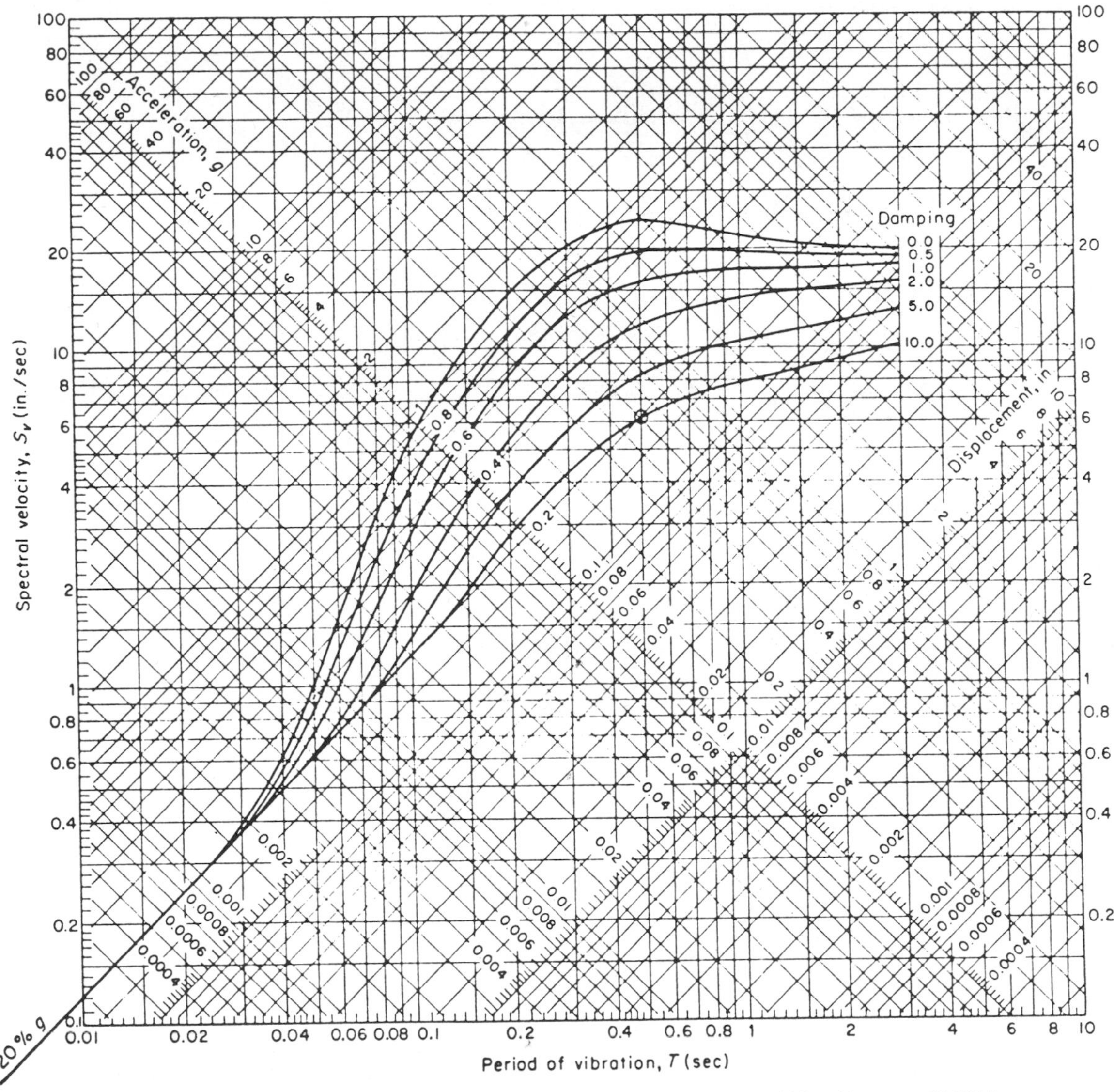

Figure 2-33 Design spectral scaled to 20% ground acceleration. [After Housner (2-63).]

used are: (i) normalization according to spectrum intensity,[2-61] where the areas under the spectra between two given frequencies or periods are set equal to each other, and (ii) normalization to peak ground motion, where the spectral ordinates are divided by the peak ground acceleration, velocity or displacement. Other normalizations, based on peak effective acceleration and rms acceleration, have also been suggested.

The first earthquake design spectrum was developed by Housner.[2-62, 2-63] His design spectra, shown in Figure 2-33, are based on the characteristics of the two horizontal components of four earthquake ground motions recorded at El Centro, California in 1934 and 1940, Olympia, Washington in 1949, and Taft, California in 1952. The plots are normalized to 20% acceleration ($0.2g$) at zero period (ground acceleration). For any other acceleration, the plots or the information read from them are simply scaled up or down by multiplying them by the ratio of the desired acceleration to $0.2g$.

In the late sixties, Newmark and Hall[2-51, 2-52] recommended that straight lines be used to represent earthquake design spectra. They suggested that three amplifications (acceleration, velocity, and displacement) which are constant in the high, intermediate, and low frequency regions of the spectrum (Table 2-11), together with peak ground acceleration, velocity, and displacement of $1.0g$, 48 in./sec, and 36 in., be used to construct normalized design spectra. Their recommended ground motion and the amplifications were based on the characteristics of several earthquake records. The spectral ordinates (amplified motion), which are obtained by multiplying the three ground motions by the corresponding amplifications, are plotted on tripartite (four-way logarithmic) paper as shown in Figure 2-34.

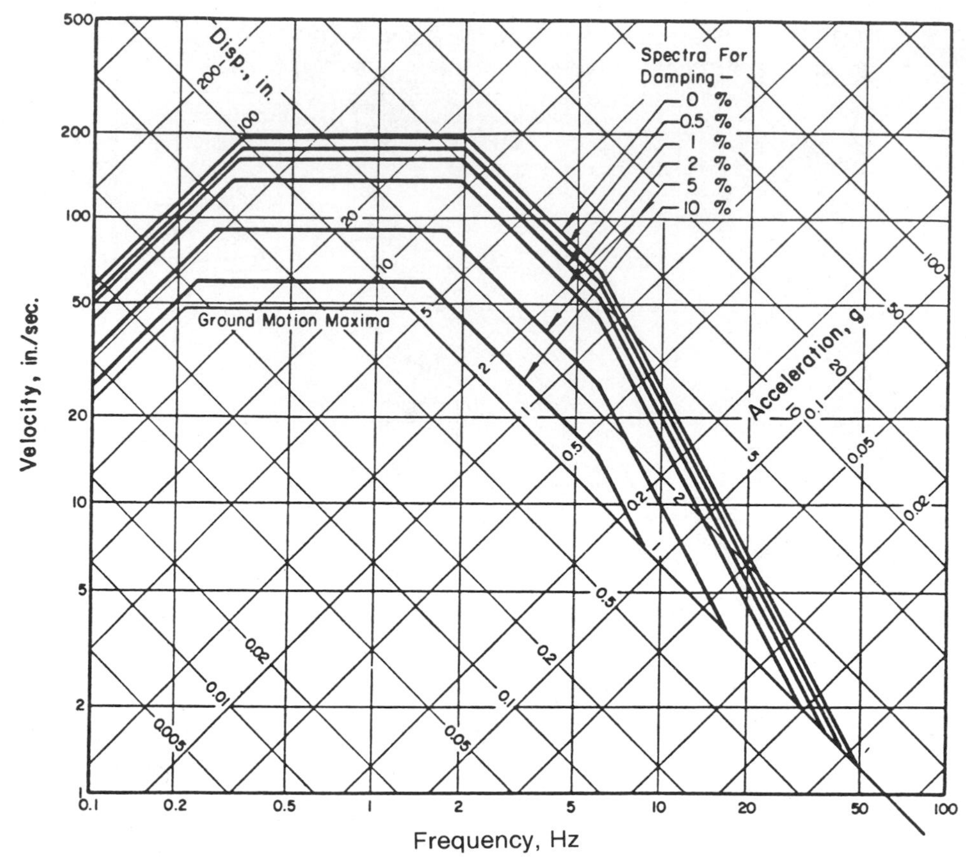

Figure 2-34 Design spectra normalized to 1.0*g*. [After Newmark and Hall (2-52).]

Table 2-11 Relative Values of Spectrum Amplification Factors*

Percent of critical damping	Amplification factor for Displacement	Velocity	Acceleration
0	2.5	4.0	6.4
0.5	2.2	3.6	5.8
1	2.0	3.2	5.2
2	1.8	2.8	4.3
5	1.4	1.9	2.6
7	1.2	1.5	1.9
10	1.1	1.3	1.5
20	1.0	1.1	1.2

*After Newmark and Hall.(2-52)

The spectral displacement, spectral velocity, and spectral acceleration are plotted parallel to maximum ground displacement, ground velocity, and ground acceleration, respectively. The frequencies at the intersections of the spectral displacement and velocity and of the spectral velocity and acceleration define the three amplified regions of the spectrum. At a frequency of approximately 6 Hz, the spectral accelerations are tapered down to the maximum ground acceleration. It is assumed that the spectral acceleration for 2% damping intersects the maximum ground acceleration at a frequency of 30 Hz. The tapered spectral-acceleration lines for other dampings are parallel to the one for 2%. The normalized response spectral in Figure 2-34 can be used for design by scaling the ordinates to the design acceleration at the site. Newmark and Hall also recommend the damping coefficients in Table 2-12 for various structural types, conditions, and stress levels.

In the early seventies, with increased activity in the design of nuclear-reactor facilities, two studies—one by John A. Blume and Associates,(2-64) and the other by N. M. Newmark Consulting Engineering Services(2-9) —were carried out with the objective of developing recommendations for horizontal and vertical design response spectra for nuclear-reactor facilities. These studies, which used a statistical analysis of a number of recorded earthquake ground-motion and computed response spectra, were the basis for the Nuclear Regulatory Commission (NRC) Regulatory Guide 1.60.(2-50, 2-65) The studies recommended that the mean-plus-one-standard-deviation (84.1-percentile) response be used for the design of nuclear-reactor facilities. The NRC horizontal and vertical design spectra are constructed using a set of amplifications corresponding to four control frequencies (Figures 2-35 and 2-36). The spectra are normalized to 1.0-*g* horizontal ground accel-

Table 2-12 Recommended Damping Coefficients for Different Structural Types, Conditions, and Stress Levels*

Stress level	Type and condition of structure	Percentage of critical damping
Working stress, no more than about $\frac{1}{2}$ yield point	(a) Vital piping	0.5 – 1.0
	(b) Welded steel, prestressed concrete, well reinforced concrete (only slight cracking)	2
	(c) Reinforced concrete with considerable cracking	3 – 5
	(d) Bolted and/or riveted steel, wood structures with nailed or bolted joints	5 – 7
At or just below yield point	(a) Vital piping	2
	(b) Welded steel, prestressed concrete (without complete loss in prestress)	5
	(c) Prestressed concrete with no prestress left	7
	(d) Reinforced concrete	7 – 10
	(e) Bolted and/or riveted steel, wood structures, with bolted joints	10 – 15
	(f) Wood structures with nailed joints	15 – 20

*After Newmark and Hall.(2-52)

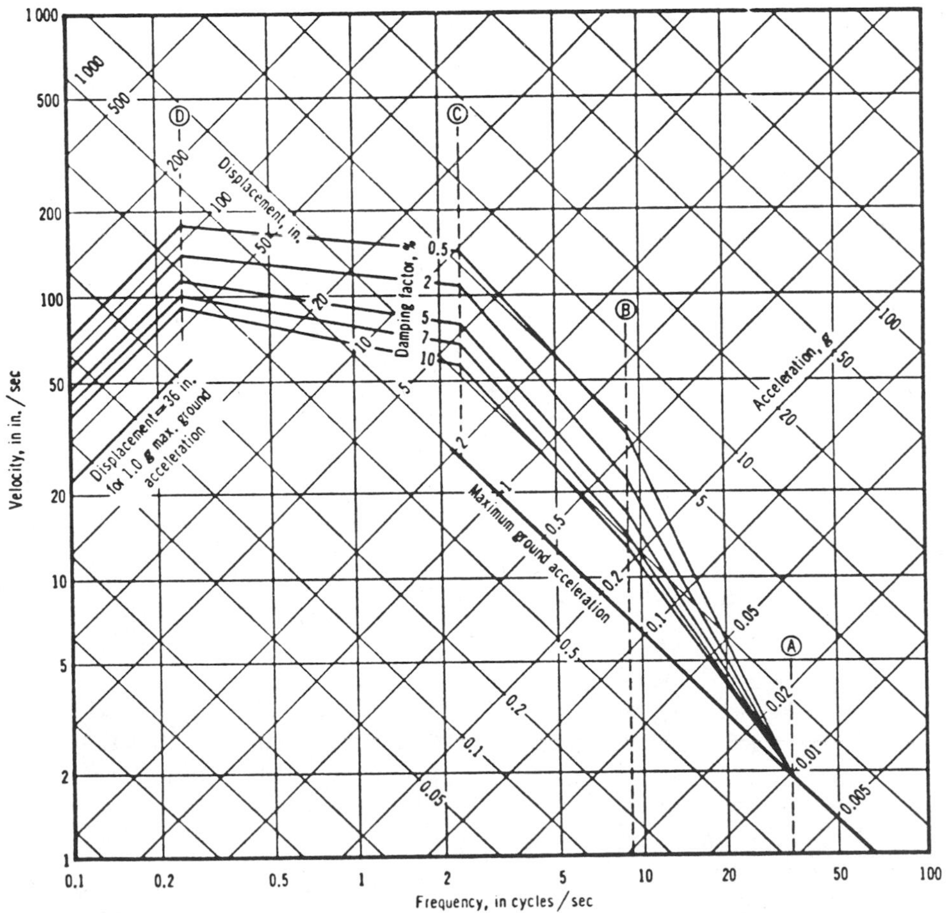

Figure 2-35 NRC horizontal design response spectra scaled to 1.0*g* horizontal ground acceleration. *A*, *B*, *C*, and *D* are control frequencies corresponding to 33, 9, 2.5, and 0.25 Hz, respectively.

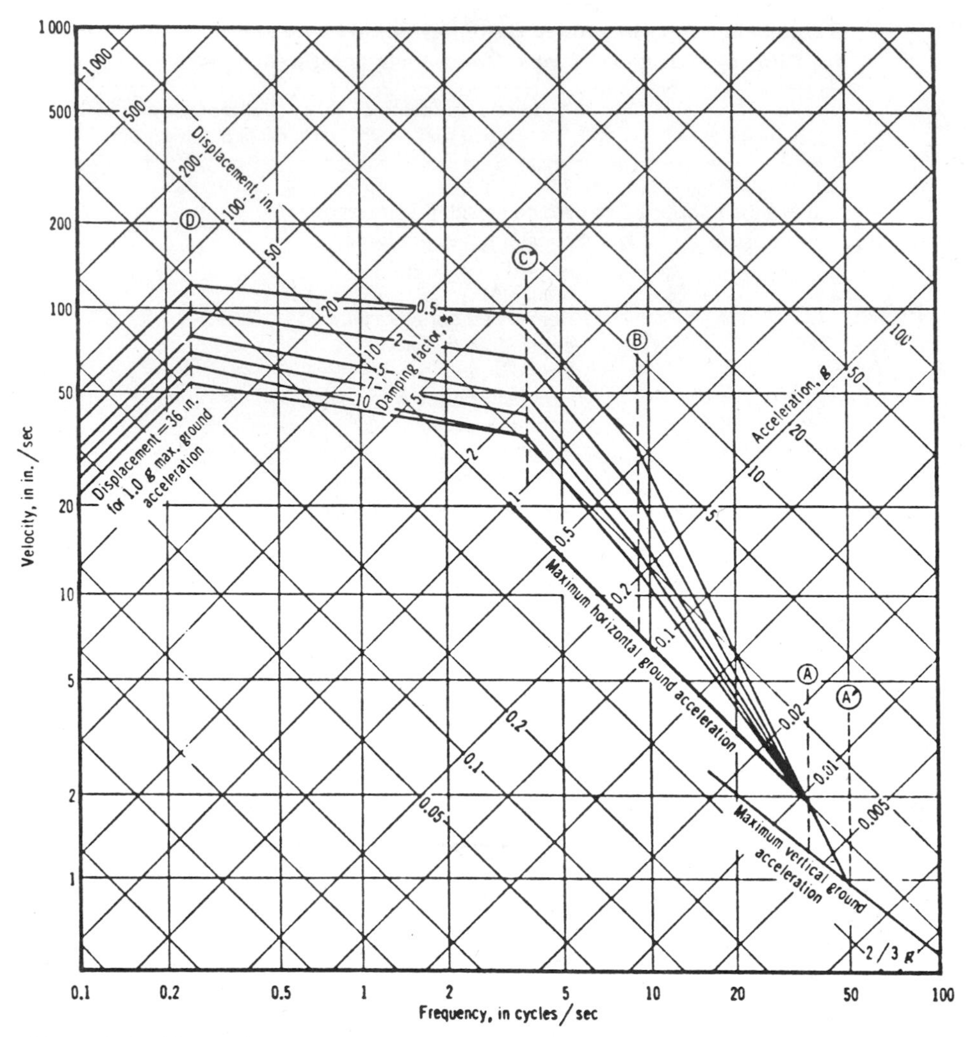

Figure 2-36 NRC vertical design spectra scaled to 1.0g horizontal ground acceleration. A, B, C, and D are control frequencies corresponding to 33, 9, 3.5, and 0.25 Hz, respectively.

eration. The figures show that the design spectra for vertical motion span a wider frequency range in the velocity region than the spectra for the horizontal motion. While the NRC spectra were developed for the design of nuclear power plants, which are generally considered as stiff structures, they have been used to develop and compare design spectra for other applications.

2.9 INFLUENCE OF SOIL CONDITIONS ON RESPONSE AND DESIGN SPECTRA

Prior to the San Fernando earthquake of 1971, accelerograms from previous earthquakes were limited in number, and the majority were recorded on alluvium. Consequently, design spectra such as those in Figures 2-33 to 2-36 were based on records on alluvium and did not consider soil condition as a parameter. Studies by Hayashi et al.[(2-66)] and Kuribayashi et al.[(2-67)] on the effect of soil conditions on Japanese earthquakes had shown that soil conditions substantially affect the spectral shapes. In addition, studies by Mohraz et al.[(2-9)] and Hall et al.[(2-53)] refer to the influence of soil conditions on spectral shapes.

The San Fernando earthquake provided a large data base and made it possible to study the influence of soil conditions among other parameters on earthquake ground motion and response spectra. In 1976, two independent studies, one by Seed, Ugas, and Lysmer[(2-68)] and the other by Mohraz,[(2-33)] considered the influence of soil conditions on response spectra. The study by Seed used a total of 104 horizontal components of earthquake records with peak ground accelerations greater than 0.05g from 23 earthquakes. The records were divided into four categories: rock, stiff soils less than about 150 ft deep, deep cohesionless soil with depths greater than 250 ft, and soft to medium clay and sand. The response spectra for 5% damping[e] were nor-

[e]They limited their study to 5% damping, but the conclusions can be extended to other damping coefficients.

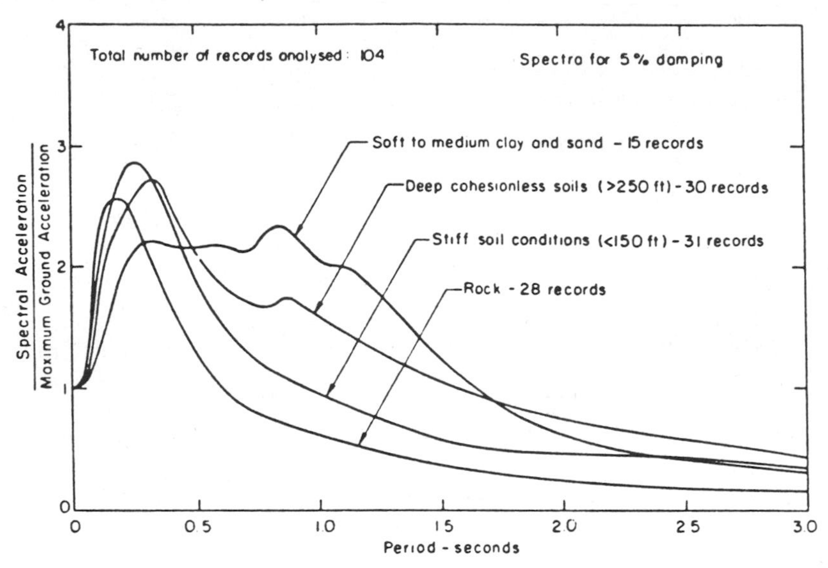

Figure 2-37 Average acceleration spectra for different site conditions. [After Seed et al. (2-68).]

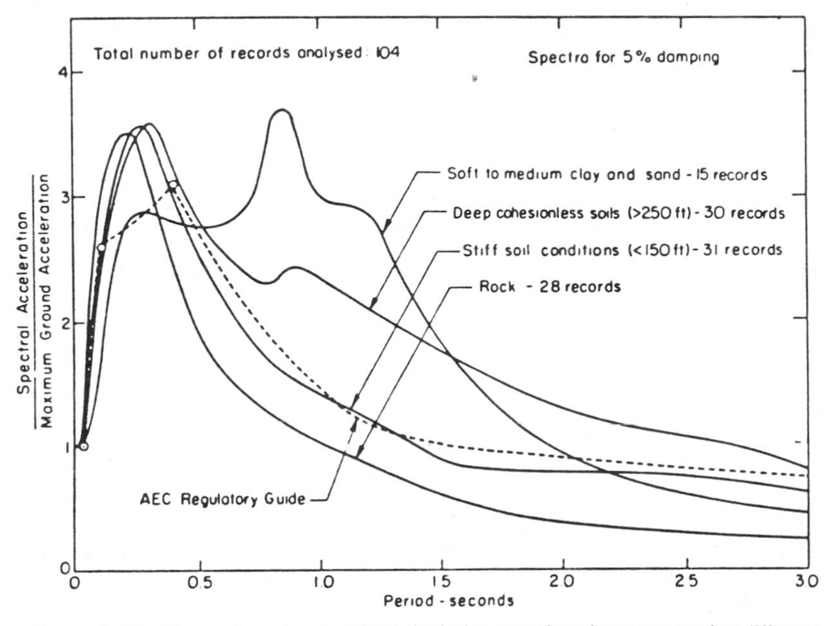

Figure 2-38 Mean plus one standard deviation acceleration spectra for different site conditions. [After Seed et al. (2-68).]

malized to the peak ground acceleration of the records and averaged at various periods. The average and the mean-plus-one-standard-deviation (84.1-percentile) spectra for the four categories from their study are presented in Figures 2-37 and 2-38. The ordinates in these plots represent the acceleration amplifications. Also shown in Figure 2-38 is the AEC or NRC spectra (Figure 2-35) proposed by Newmark et al.[2-50, 2-65] It is seen that the soil condition affects the average spectra to the significant degree. The figures show that for periods greater than approximately 0.4–0.5 sec, the normalized spectral ordinates (amplifications) for rock are substantially below those for soft to medium clay and for deep cohesionless soil. This indicates that using the spectra from the latter two groups may overestimate the design amplifications for rock.

The study by Mohraz[2-33] considered a total of 162 components of earthquake records divided into four soil categories: alluvium, rock, less than 30 ft of alluvium underlain by rock, and 30–200 ft of alluvium underlain by rock. To exclude the influence of the horizontal components of records with the smaller peak ground acceleration on amplification, Mohraz divided the response spectra for each soil category into three sets:

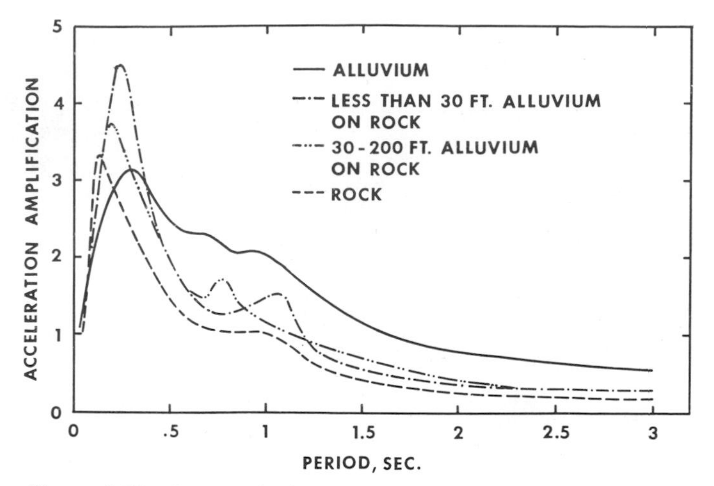

Figure 2-39 Average horizontal acceleration amplifications for 2% damping for four soil categories. [After Mohraz (2-33).]

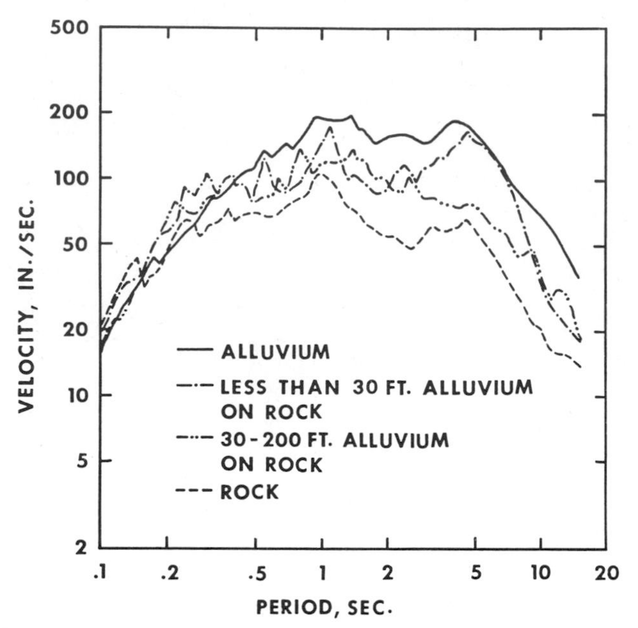

Figure 2-41 Mean plus one standard deviation response spectra for 2% damping for four soil categories, normalized to 1.0g horizontal ground acceleration. [After Mohraz (2-33).]

horizontal components with the larger of the two peak ground accelerations, horizontal components with the smaller of the two peak ground accelerations, and vertical components. Figure 2-39 shows the average acceleration amplifications (ratio of spectral ordinates to peak ground acceleration) for 2% damping for the horizontal components with the larger of the two peak ground accelerations. Consistent with Seed's study, the figure shows that the soil condition influences the spectral shapes to a significant degree. The acceleration amplification for alluvium extends over a larger frequency region than the amplifications for the other three soil categories. It is noted that the maximum acceleration amplification for less than 30 ft of alluvium on rock is approximately 40% greater than that for alluvium and about 33% greater than that for rock. A comparison of acceleration amplifications for 5% damping from Seed's and Mohraz's studies is shown in Figure 2-40. The figure indicates remarkably close agreement, even though the records used in the two studies are somewhat different. Normalized response spectra corresponding to the mean plus one standard deviation (84.1 percentile) for the four soil categories are given in Figure 2-41. The plot indicates that for short periods (high frequencies) the spectral ordinates for alluvium are lower than the others, whereas for intermediate and long periods they are higher.

Design amplifications were obtained by computing the mean-plus-one-standard-deviation amplifications at different frequencies and averaging them in each frequency region. The design amplifications in the three frequency regions, together with the ground motion ratios in Table 2-9, were used to compute spectral ordinates and to construct design spectra for unit ground acceleration. The site-dependent spectral ordinates for various dampings are given in Table 2-13 and presented graphically in Figures 2-42 to 2-44. For design, these ordinates should be scaled to the expected acceleration at the site, which may be estimated using seismic-risk procedures. Typical design spectra for horizontal motion and 5% damping obtained from Table 2-13 or Figures 2-42 to 2-44 are given in Figure 2-45. It should be noted that in constructing the design spectra, the

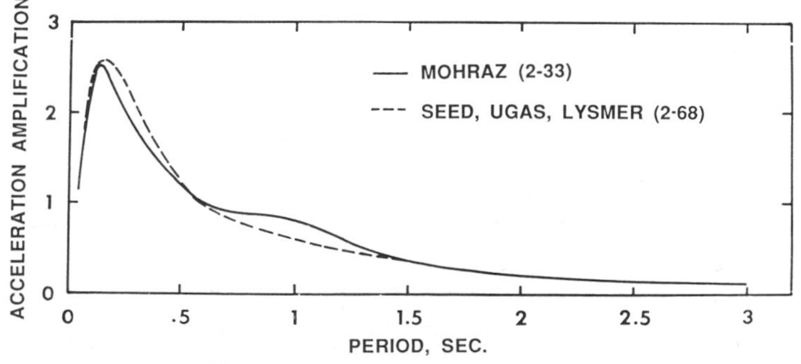

Figure 2-40 Comparison of the average horizontal acceleration amplifications for 5% damping for rock. [After Mohraz (2-33).]

Table 2-13 Spectral Ordinates for Unit Ground Acceleration*

		Damping = 0%‡			2%‡			5%‡			10%‡			20%‡		
Site category	Group†	Dis., in.	Vel., in./sec	Acc., *g*	Dis., in.	Vel., in./sec	Acc., *g*	Dis., in.	Vel., in./sec	Acc., *g*	Dis., in.	Vel., in./sec	Acc., *g*	Dis., in.	Vel., in./sec	Acc., *g*
Rock	L	54	97	7.02	43	66	3.80	35	51	2.82	28	40	2.11	21	30	1.54
	S	71	118	8.14	53	75	4.16	44	58	3.04	36	46	2.29	27	34	1.65
	V	87	115	8.15	67	74	3.81	54	57	2.70	44	45	2.03	30	35	1.65
< 30 ft of alluvium underlain by rock	L	93	174	10.42	75	106	4.76	59	77	3.38	46	58	2.54	32	42	1.81
	S	127	158	10.17	106	101	4.73	87	77	3.37	68	59	2.49	49	44	1.73
	V	151	140	10.55	123	88	4.28	101	68	2.93	81	52	2.09	61	37	1.46
30 – 200 ft of alluvium underlain by rock	L	68	167	8.48	53	96	4.13	44	72	2.94	35	54	2.19	26	39	1.60
	S	80	201	8.85	62	117	4.11	50	89	2.88	41	68	2.17	31	49	1.62
	V	110	198	10.04	84	110	4.22	68	80	2.95	53	59	2.18	38	41	1.60
Alluvium	L	132	242	7.52	99	141	3.55	81	106	2.58	63	81	1.99	47	59	1.53
	S	160	288	9.18	122	169	4.00	99	128	2.86	78	98	2.19	58	70	1.65
	V	146	246	10.91	111	143	4.43	92	109	3.08	73	81	2.32	55	57	1.69

*After Mohraz.(2-33)

†L: horizontal components with the larger of the two peak ground accelerations; S: horizontal components with the smaller of the two peak ground accelerations; V: vertical components.

‡Of critical damping.

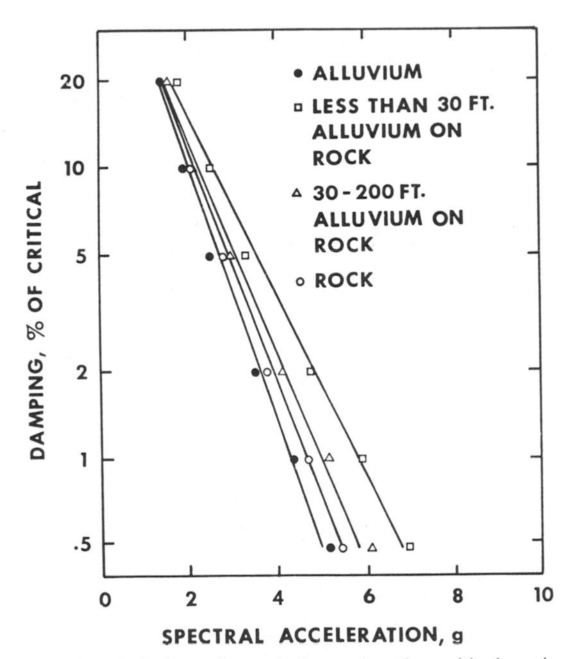

Figure 2-42 Variation of spectral acceleration with damping for unit ground acceleration for horizontal components with the larger of the two peak ground accelerations. [After Mohraz (2-33).]

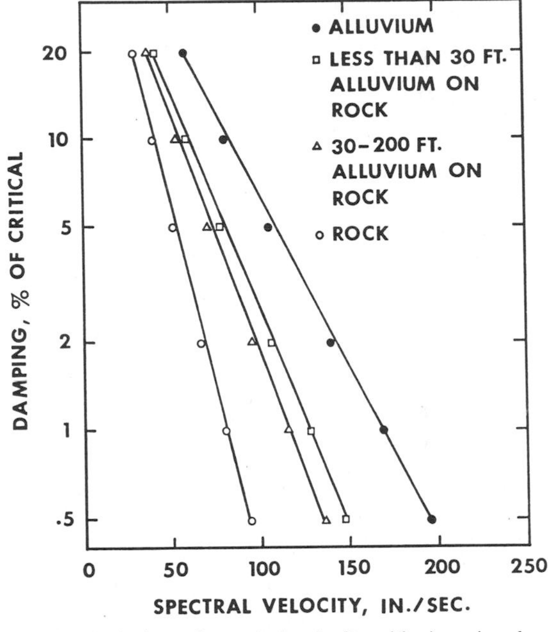

Figure 2-43 Variation of spectral velocity with damping for unit ground acceleration for horizontal components with the larger of the two peak ground accelerations. [After Mohraz (2-33).]

acceleration spectral ordinates which remain constant up to a frequency of 8 Hz for horizontal motion and 10 Hz for vertical, are tapered to peak ground acceleration at a frequency of 33 Hz.

In 1978, the Applied Technology Council(2-47) recommended that a smooth version of the normalized spectral shapes proposed by Seed et al.(2-68) be used for developing seismic regulations for buildings. The spectral shapes in Figures 2-37 and 2-38 were smoothed using four control periods.(2-20) In addition, the four soil categories were reduced to three: rock and stiff soils (soil type 1), deep cohesionless or stiff clay soils (soil type 2), and soft to medium clays and sands (soil type 3). The ATC spectra, which have also been adopted by the Seismology Committee of the Structural Engineers Association of California,(2-69) are presented in Figure 2-46. A comparison of the spectral shapes from Mohraz's study and those proposed by the ATC is shown in Figure 2-47.

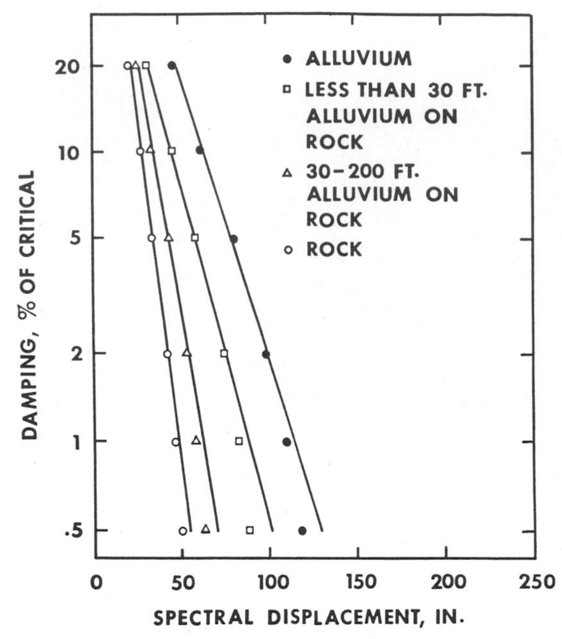

Figure 2-44 Variation of spectral displacement with damping for a unit ground acceleration for horizontal components with the larger of the two peak ground accelerations. [After Mohraz (2-33).]

2.10 INFLUENCE OF MAGNITUDE AND DURATION ON RESPONSE AND DESIGN SPECTRA

The influence of earthquake magnitude and epicentral distance on response and design spectra is generally taken into consideration when specifying the peak ground acceleration at a site. Consequently, the spectral shapes in Figures 2-37 and 2-38 and the spectral ordinates in Table 2-13 and Figures 2-42 to 2-44, which were obtained independently of earthquake magnitude, can be used to estimate design spectra for any site. Earthquake magnitude does, however, influence spectral amplifications to the certain degree. A study by Mohraz(2-70) on the influence of earthquake magnitude on response amplifications for alluvium shows larger acceleration amplifications for records with magnitudes between 6 and 7 than for those with magnitudes between 5 and 6 (see Figure 2-48). While the study used a limited number of records and no specific recommendation was made, the figure shows that earthquake magnitude can influence spectral shapes, and it may need to be considered when developing design spectra for a specific site, particularly for critical structures.

The influence of duration of strong motion on spectral shapes has recently been studied by Peng et al.,(2-71) who used a random-vibration approach to estimate site-dependent probabilistic response spectra. A comparison of the mean-plus-one-standard-deviation acceleration amplification for two different durations from that study, with the response spectral shapes recommended by the Applied Technology Council(2-47) and by the Structural Engineers Association of California,(2-69) is shown in Figures 2-49 and 2-50. It should be noted that the results in Figure 2-49 from the study by Peng et al. are for deep alluvium (deeper than 200 ft), whereas

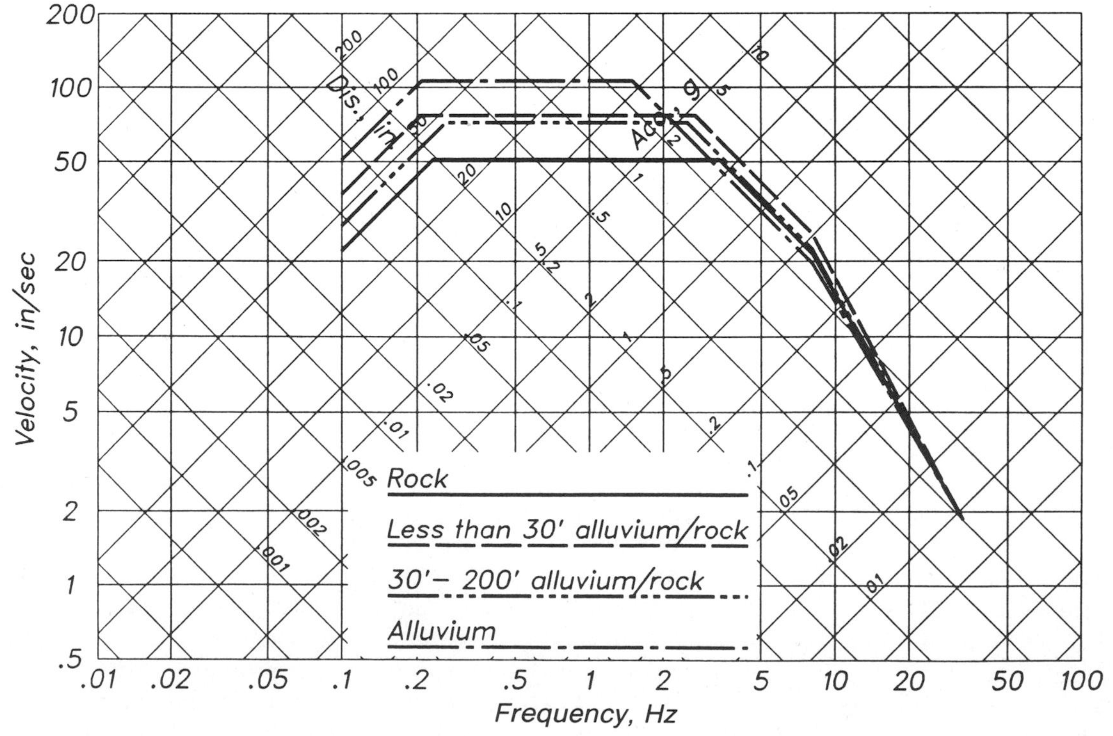

Figure 2-45 Design spectra for horizontal motion for 5% damping. [After Mohraz (2-33).]

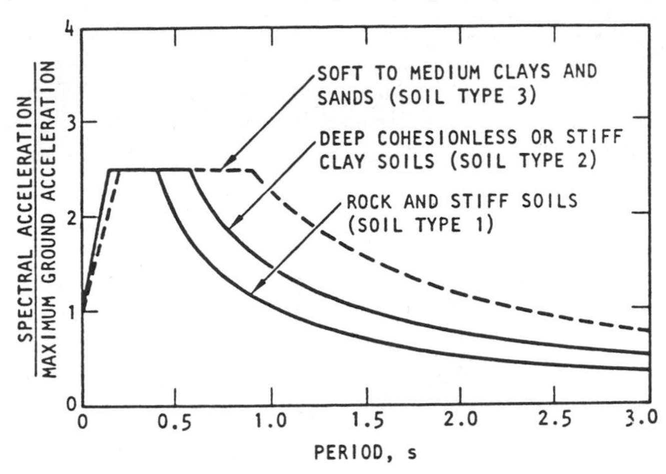

Figure 2-46 Normalized spectral curves recommended for use in building code. (Reproduced from 2-20.)

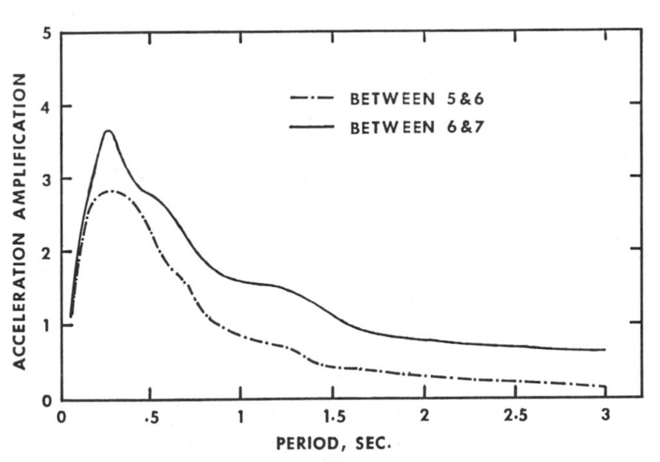

Figure 2-48 Effect of earthquake magnitude on spectral shapes. [After Mohraz (2-70).]

soil types 2 and 3 are for deep cohesionless or stiff clay and soft to medium clays and sand, respectively. The results in Figures 2-49 and 2-50 show that a longer duration of strong motion increases the response in the low and intermediate frequency regions. This is consistent with the fact that accelerograms with long duration of strong motion have a greater probability of containing long-period waves which can result in a higher response in the long-period (low-frequency) region of the spectrum.

2.11 INELASTIC RESPONSE SPECTRA

Structures subjected to severe earthquake ground motion experience deformations beyond the elastic range. Inelastic deformations to a large extent depend on the load–deformation characteristics of the structure and often result in stiffness deterioration. Because of the cyclic nature of ground motion, structures experience successive loadings and unloadings, and the force–displacement or resistance–deformation relationship fol-

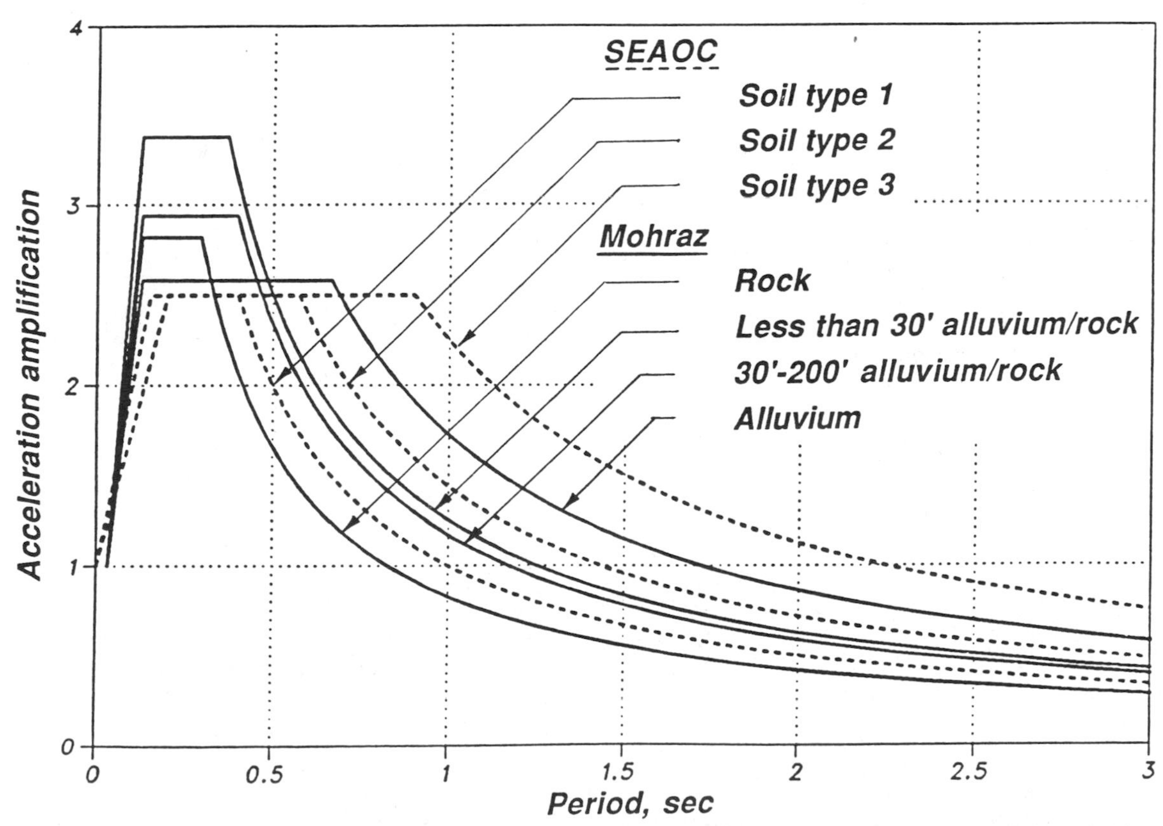

Figure 2-47 Comparison of spectral shapes for 5% damping proposed by Mohraz with those recommended by SEAOC.

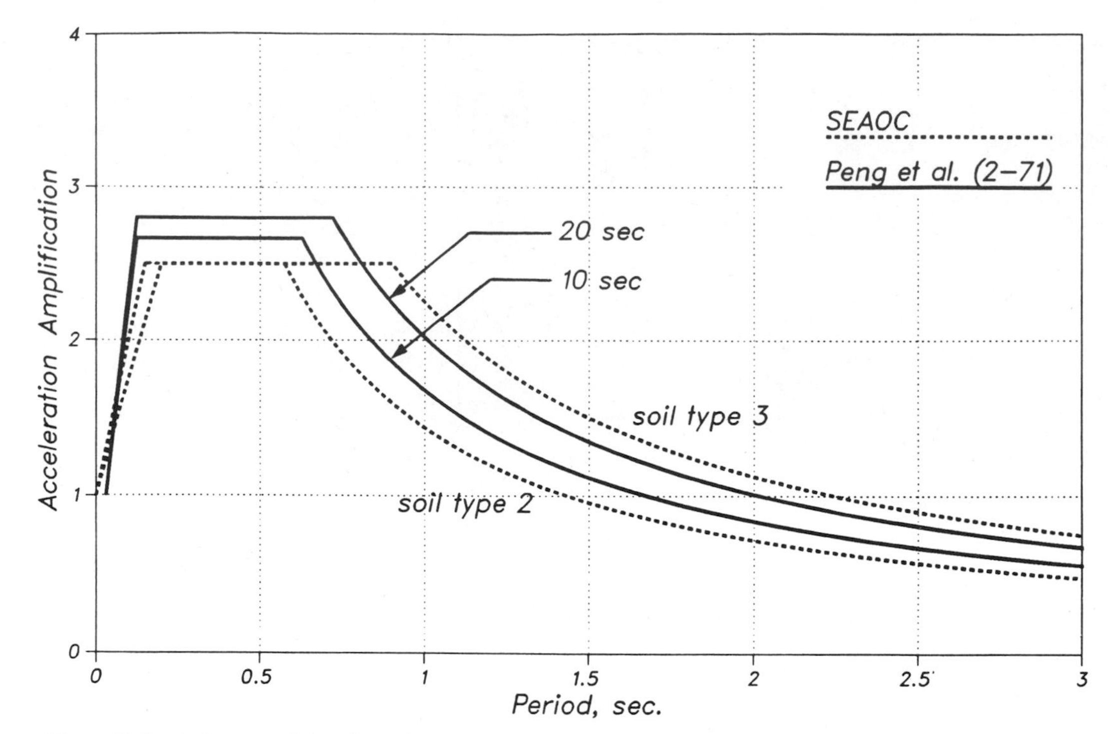

Figure 2-49 Influence of duration of strong motion on spectral shapes. The plots are for alluvium and 5% damping. The spectral shapes recommended by SEAOC are shown for comparison.

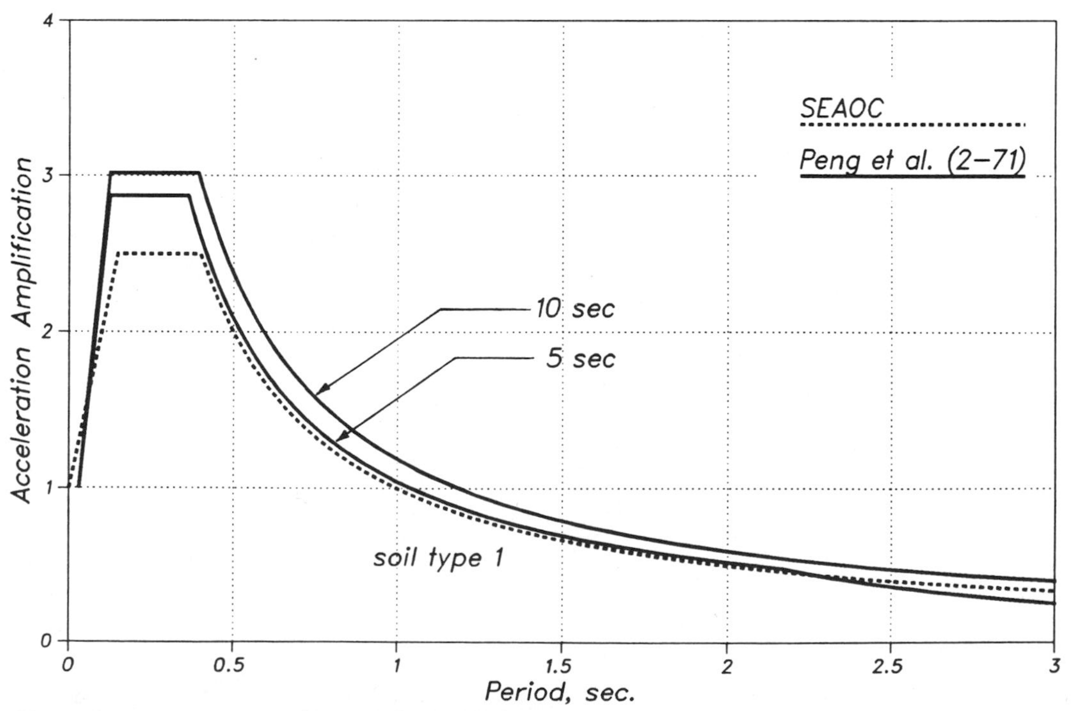

Figure 2-50 Influence of duration of strong motion on spectral shapes. The plots are for rock and 5% damping. The spectral shapes recommended by SEAOC are shown for comparison.

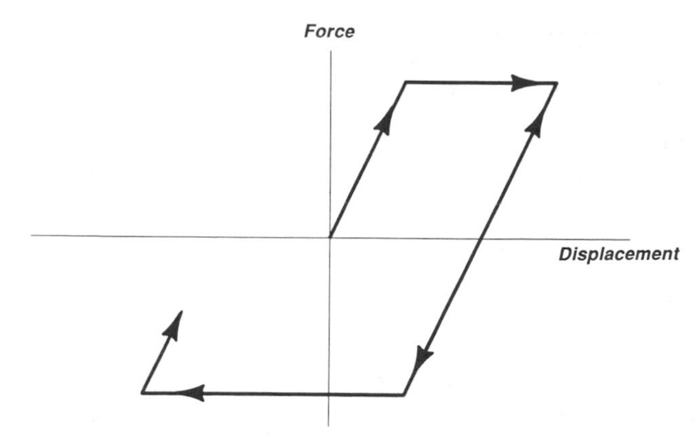

Figure 2-51 Elastic-plastic force-displacement relationship.

lows a sequence of loops known as hysteresis loops. The loops are a measure of the structure's capacity to dissipate energy. The shape and orientation of the hysteresis loops depend primarily on the structure's stiffness and yield displacement. Factors such as structural material, structural system, and connection configuration influence the hysteretic behavior of a structure. Consequently, arriving at an appropriate mathematical model to describe the inelastic behavior of structures during earthquakes is a difficult task. A simple model which has extensively been used to approximate the inelastic behavior of structural systems and components is the elastic-plastic (elastoplastic) model shown in Figure 2-51. In this model unloadings and subsequent loadings are assumed to be parallel to the original loading curve. Though elastic-plastic models are used to predict the response of structures, the computations are time-consuming and costly, especially when they have to be repeated with a number of accelerograms to arrive at representative response values for design.

Response spectra modified to account for inelastic behavior, commonly referred to as inelastic spectra, have been proposed for analysis and design when inelastic deformations are permitted. The use of inelastic spectra has, however, been limited to systems which can be modeled as single-degree-of-freedom. Procedures for utilizing inelastic spectra in the analysis and design of multi-degree-of-freedom systems have not yet been developed to the extent that they can be implemented in design. Similar to elastic spectra, inelastic spectra are usually plotted on tripartite paper for a given damping and various ductilities[f] or various yield deformations. When the spectra are plotted for various ductilities, computations are repeated for several yield deformations to achieve the preassigned ductility. Depending on the parameter plotted, different names have been used to identify the spectrum (Riddell and Newmark[(2-72)]). In the inelastic yield spectrum, the yield displacement is plotted on the displacement axis; in the inelastic acceleration spectrum, the maximum force per unit mass is plotted on the acceleration axis; and in the inelastic total deformation spectrum, the absolute maximum total displacement is plotted on the displacement axis. For elastic-plastic behavior, the inelastic yield spectrum and the inelastic acceleration spectrum are identical. Examples of inelastic spectra for a 5% damped elastic-plastic system for the S90W component of El Centro, the Imperial Valley earthquake of May 18, 1940, are shown in Figures 2-52 and 2-53. The figures show that for inelastic yield and acceleration spectra, the curves for various ductilities fall below the elastic curve (ductility of one), whereas for the inelastic total deformation spectra, they primarily fall above the elastic, particularly in the acceleration region.

A different presentation of inelastic spectra has been proposed by the authors.[(2-73)] The spectrum, referred to as the yield-displacement spectrum (YDS), is similar to the inelastic total-deformation spectrum except that it is plotted for a given yield displacement instead of a given ductility. In such a presentation, the ductility is obtained as the ratio of the maximum displacement to the yield displacement for which the spectrum is plotted. This procedure offers an efficient computational technique, particularly when statistical studies are used to obtain inelastic design spectra. It requires approximately 5–10% of the time needed for computing inelastic spectra in terms of ductility. An inelastic spectral zone is computed from the accelerograms normalized to 1.0-*g* ground acceleration. The lower bound of the zone is the elastic spectrum, and the upper bound constitutes the yield-displacement spectrum for a yield level of 0.05 in. (Figure 2-54). Interpolation relationships are then used to estimate the inelastic smooth spectrum for any given yield displacement.

Before Riddell and Newmark's statistical study of inelastic response, the most common procedure for estimating inelastic earthquake design spectra was that proposed by Newmark et al.[(2-74, 2-75)] and Newmark and Hall.[(2-51)] Based on results similar to those in Figures 2-52 and 2-53, and studies by Housner[(2-76)] and Blume,[(2-77 – 2-79)] Newmark observed that: (i) at low frequencies, an elastic and an inelastic system have the same total displacement, (ii) at intermediate frequencies, an elastic and an inelastic system absorb the same total energy, and (iii) at high frequencies, an elastic and an inelastic system have the same force. These observations resulted in the recommendation by Newmark for constructing inelastic spectra from the elastic ones by dividing the ordinates of the elastic spectrum by two coefficients depending on the ductility μ. Figure 2-55 shows the construction of the inelastic spectrum from the elastic. The solid lines $DVAA_0$ represent the elastic

[f]Ratios of maximum deformation to yield deformation.

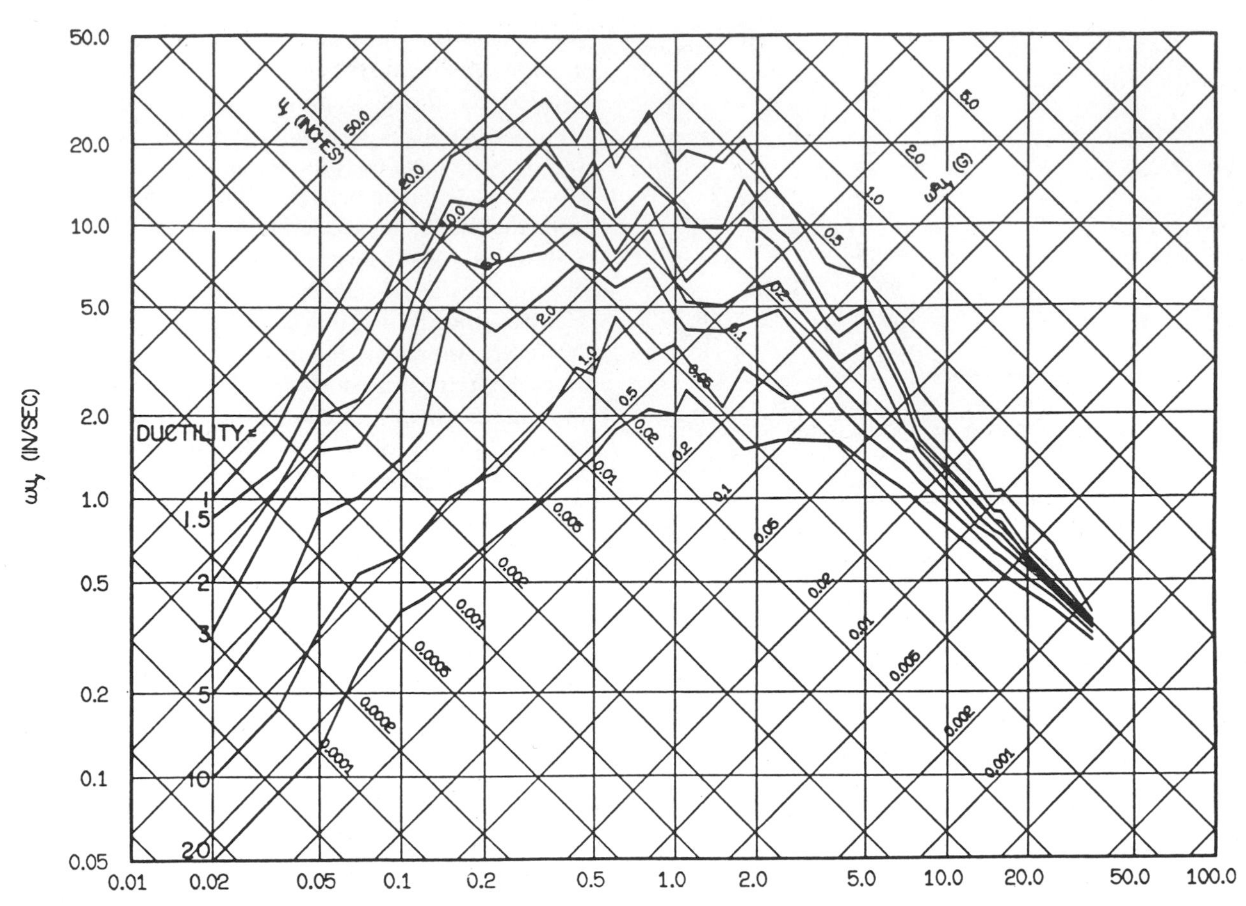

Figure 2-52 Inelastic yield spectra for the S90W component of El Centro, the Imperial Valley earthquake of May 18, 1940. Elastic-plastic systems with 5% damping. [After Riddell and Newmark (2-72).]

response spectrum. The solid circles at the intersections of the lines correspond to frequencies which remain constant in obtaining the inelastic spectrum. The lines $D'V'A'A_0$ represent the inelastic acceleration spectrum, whereas the lines $DVA''A_0''$ show the total displacement spectrum. D' and V' are obtained by dividing D and V by μ. A' is obtained by dividing A by $\sqrt{2\mu - 1}$ (to insure that the same energy is absorbed by the elastic and the inelastic systems). A'' and A_0'' are obtained by multiplying A' and A_0 by μ.

The Riddell-Newmark study[(2-72)] also considered bilinear and stiffness-degrading models, in addition to elastic-plastic, and concluded that using the elastic-plastic spectrum for inelastic analysis is generally on the conservative side.

2-12 DEAMPLIFICATION FACTORS

When inelastic deformations are permitted in design, the elastic forces can be reduced if adequate ductility is provided. Riddell and Newmark[(2-72)] presented a set of coefficients, referred to as *deamlification factors*, by which the ordinates of the elastic design spectrum are multiplied to obtain the inelastic yield spectrum. Lai and Biggs,[(2-80)] using artificial accelerograms with varied durations of strong motion, presented a set of coefficients, referred to as *inelastic acceleration response ratios*, by which the ordinates of the elastic spectrum are divided to obtain the inelastic yield spectrum. Since these two approaches are the inverse of one another, the reciprocals of the Lai-Biggs coefficients represent deamplification factors. Deamplification factors can also be obtained from the Newmark-Hall and from the Elghadamsi-Mohraz procedures for estimating inelastic spectra. Comparisons of the deamplification factors from the four procedures are shown in Figures 2-56 and 2-57 for two damping ratios and two ductilities. The figures indicate that the Riddell-Newmark deamplification factors are in general the smallest (largest reduction in the elastic force) of the four. Both Riddell-Newmark and Newmark-Hall deamplification factors remain constant over certain frequency segments, whereas the Lai-Biggs and Elghadamsi-Mohraz factors follow parallel patterns. While the deamplification factors are affected by ductility, they are practically not influenced by damping. Since the elastic spectral ordinates decrease significantly with increase in damp-

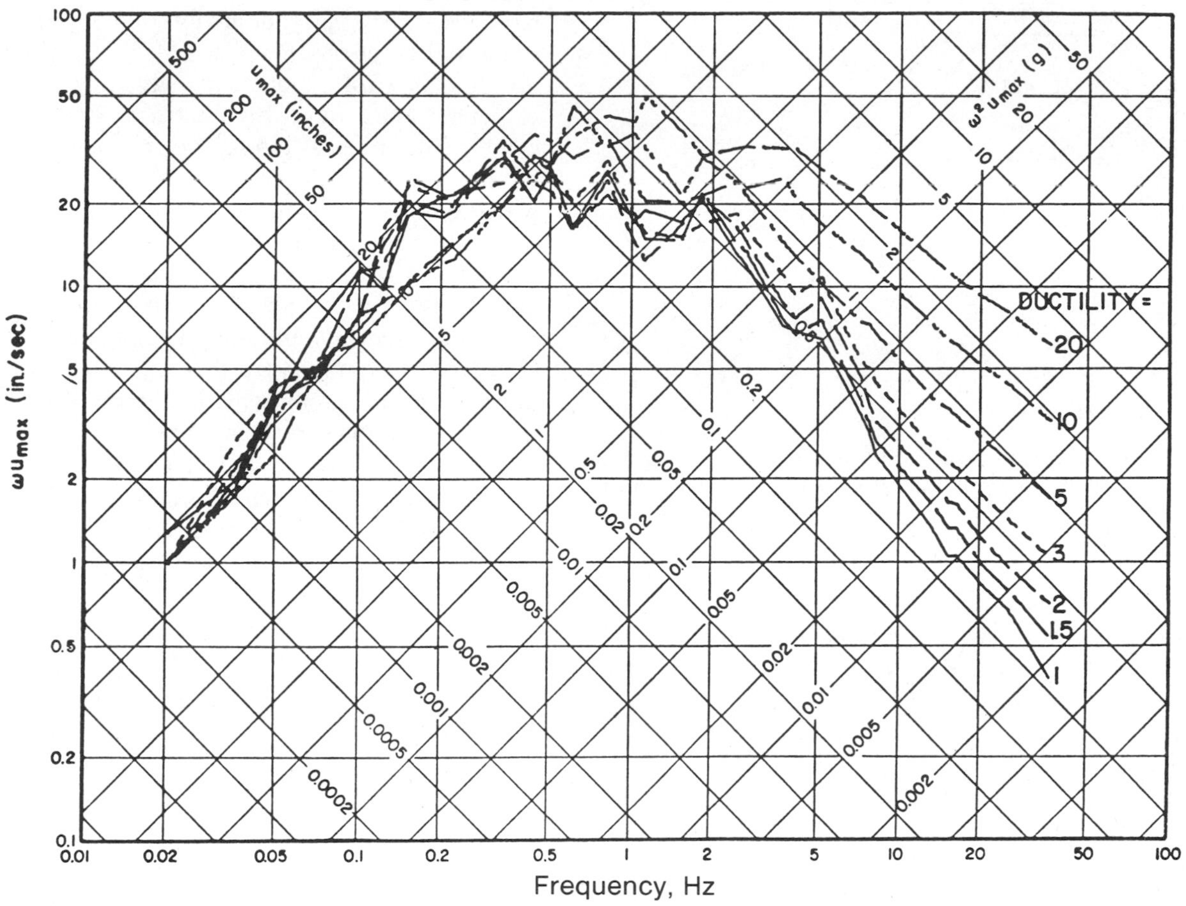

Figure 2-53 Total deformation spectra for the S90W component of El Centro, the Imperial Valley earthquake of May 18, 1940. Elastic–plastic systems with 5% damping. [After Riddell and Newmark (2-72).]

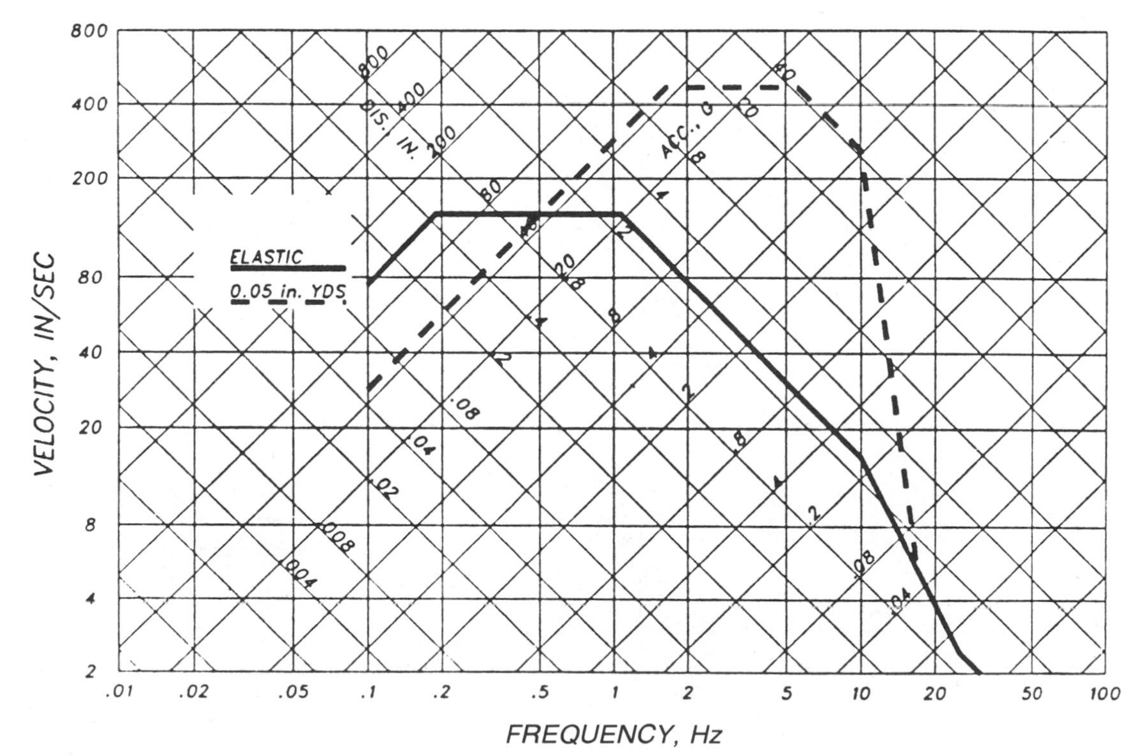

Figure 2-54 Elastic and inelastic spectral bounds for 1.0*g* horizontal acceleration on alluvium for 5% damping. [After Elghadamsi and Mohraz (2-73).]

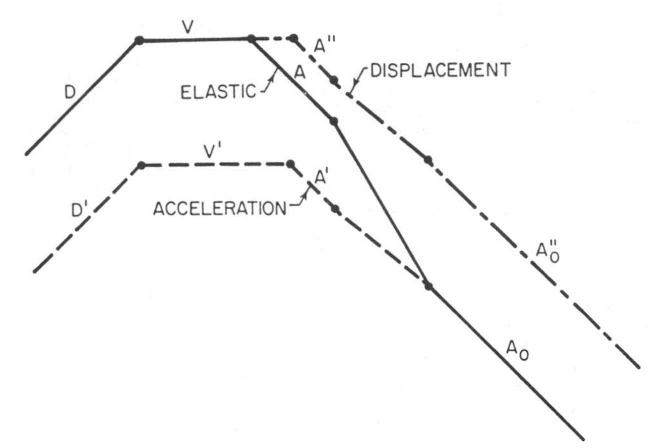

Figure 2-55 Construction of inelastic acceleration and inelastic total displacement spectra from the elastic spectrum. [After Newmark and Hall (2-51).]

ing, the decrease in inelastic spectral ordinates stems primarily from the elastic spectral ordinates.

The authors(2-73) have also presented deamplification factors for alluvium and rock. Typical results for 5% damping are presented in Figure 2-58 which shows that the deamplifications are not significantly affected by soil condition.

It has been recognized that the duration of strong motion influences the inelastic behavior of structures. In a nondeterministic study of nonlinear structures, Penzien and Liu(2-81) conclude that structures with elastic–plastic and stiffness-degrading behavior are more sensitive to the duration of strong motion than elastic structures. Using a random-vibration approach and extreme-value theory, Peng et al.(2-82) incorporated the duration of strong motion in estimating the maximum response of structures with elastic–plastic behavior. The effect of duration of strong motion on deamplification factors from Peng's study is shown in Figure 2-59, which indicates that for longer duration of strong motion one should use a larger deamplification (small reduction in elastic force). It should be noted that Lai and Biggs(2-80) conclude that inelastic response spectra are not significantly affected by strong-motion duration. However, they emphasize that this conclusion is valid only when the ground motions with various strong-motion durations are compatible with the same prescribed elastic response spectrum.

For elastic–plastic behavior (Figure 2-60) the deamplification factor can be expressed as the ratio of the yield force F_y to the elastic force F_e. Thus,

$$F_y = \phi(\mu) F_e \tag{2-37}$$

where $\phi(\mu)$ is the deamplification factor as a function of the ductility μ. In seismic design, particularly for critical and important structures, two earthquake levels are usually specified. According to Housner and Jennings,(2-58) it is uneconomical to design structures to resist the strongest possible earthquake without damage. The philosophy in building codes is to design structures to resist moderate earthquakes without damage (elastic) but to allow yielding and limited structural damage

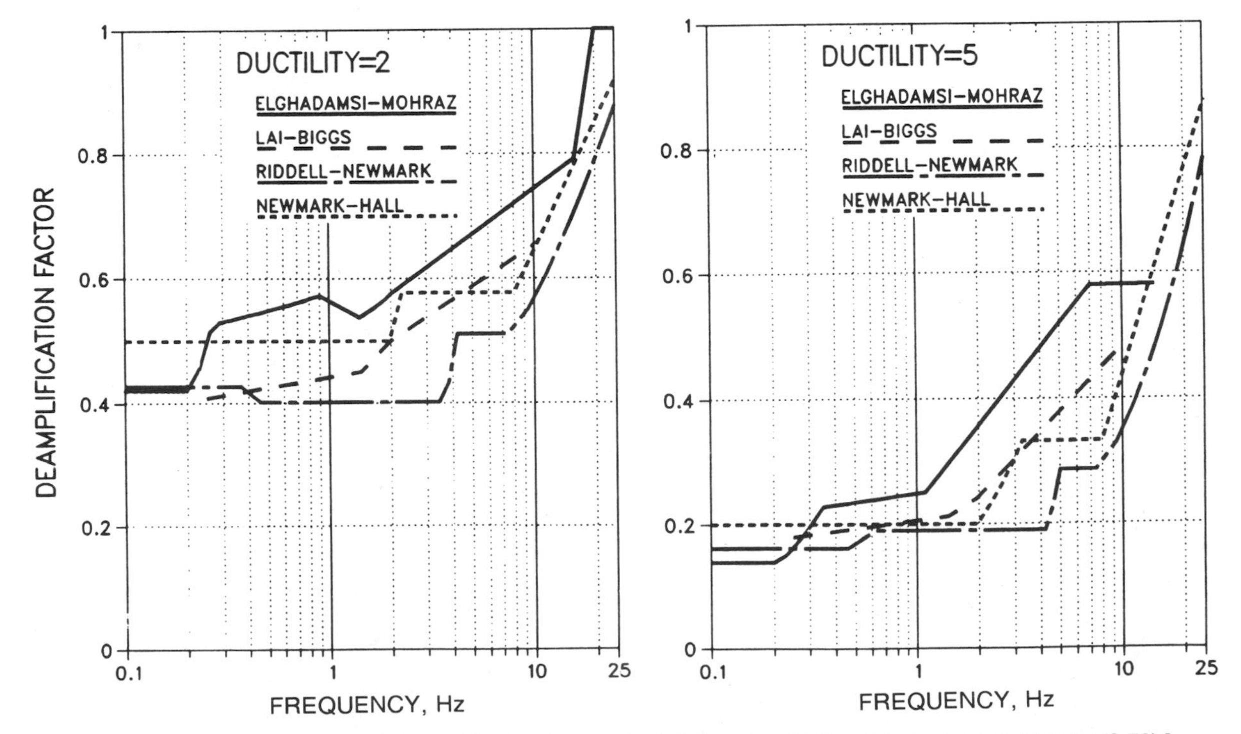

Figure 2-56 Comparison of deamplification factors for 2% damping. [After Elghadamsi and Mohraz (2-73).]

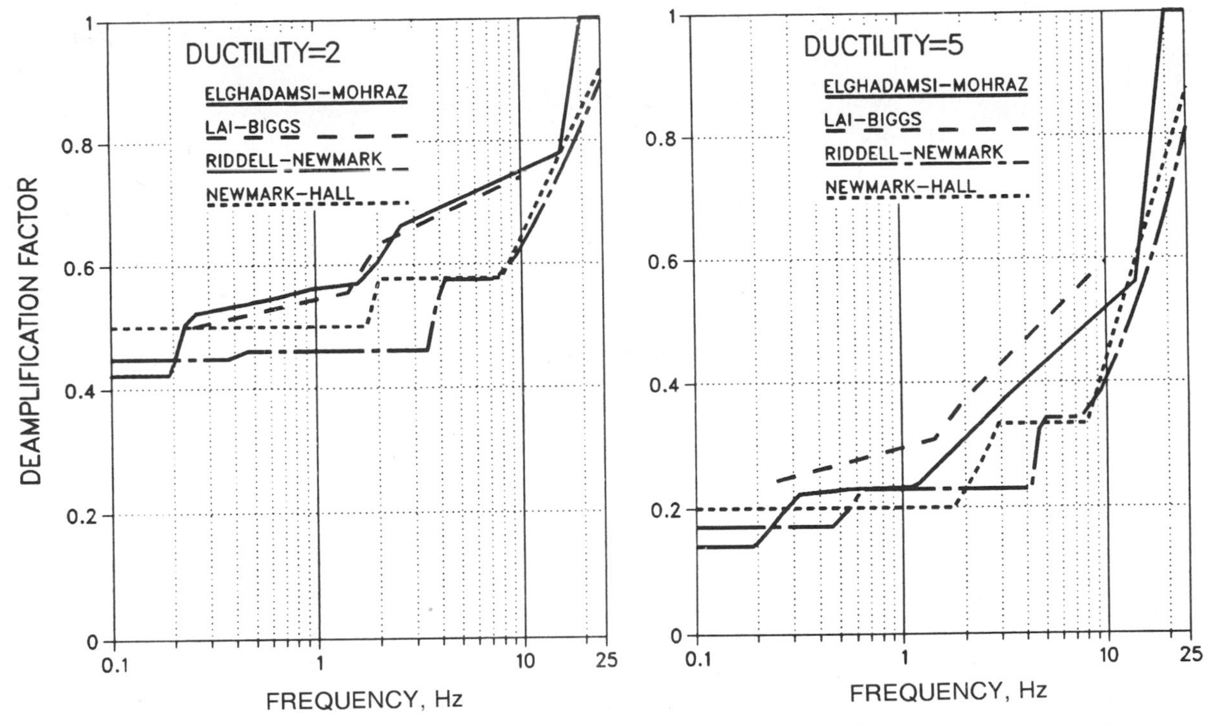

Figure 2-57 Comparison of deamplification factors for 5% damping. [After Elghadamsi and Mohraz (2-73).]

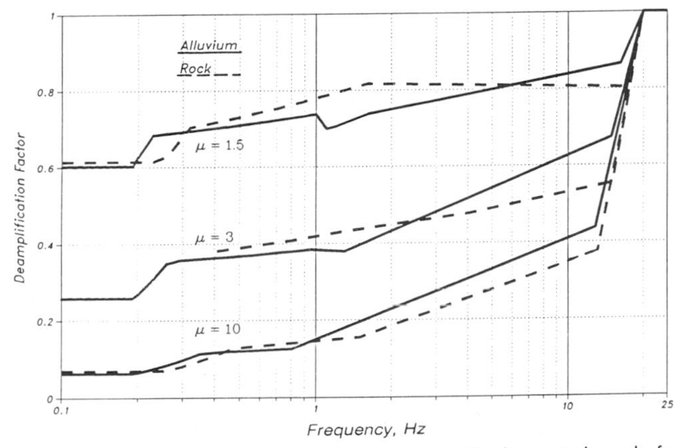

Figure 2-58 Deamplification factors for alluvium and rock for 5% damping. [After Elghadamsi and Mohraz (2-73).]

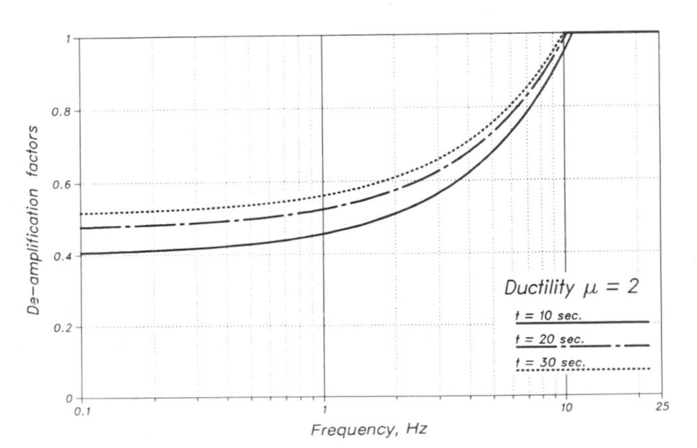

Figure 2-59 Effect of strong motion duration on deamplification factor for systems with 2% damping. [After Peng et al. (2-82).]

during severe earthquakes as long as the damage is not hazardous to life and limb. A question which often arises is what ductility should be provided so that the structure can resist severe earthquakes without collapse. Suppose the design accelerations corresponding to moderate and severe earthquakes are a_m and a_s, respectively. If the structure were to remain elastic during the severe earthquake, the amplified forces for the two levels of earthquake would be proportional to the two specified accelerations. If, for a given ductility, it is desirable to reduce the force corresponding to the severe earthquake to the level of that for the moderate earthquake, the deamplification factor can be expressed as the ratio of the two accelerations as

$$\phi(\mu) = a_m/a_s \tag{2-38}$$

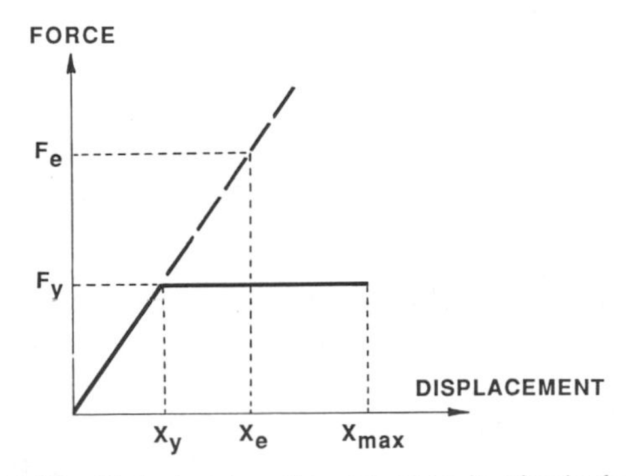

Figure 2-60 Plot showing the reduction in elastic force when elastic–plastic behavior is permitted.

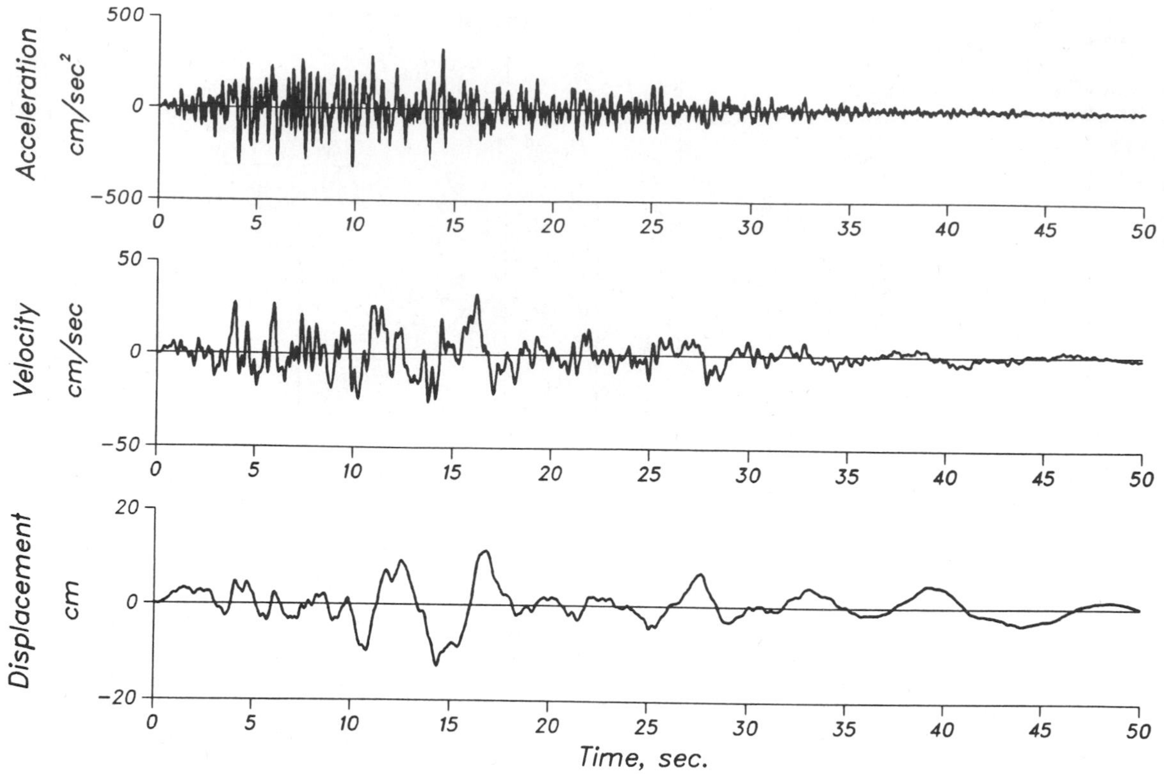

Figure 2-61 Acceleration – time history and integrated velocity and displacement generated from the Kanai – Tajimi power spectral density for alluvium using the peak ground acceleration and the duration of S00E component of El Centro, the Imperial Valley earthquake of May 18, 1940.

Using the above deamplification factor, an estimate of the required ductility μ for the structure can be obtained from plots similar to those in Figures 2-56 to 2-58. For example, suppose a structure which can be modeled as a single-degree-of-freedom has a frequency of 2 Hz (period of 0.5 sec) and a damping of 5 percent. If it is assumed that the structure is to be located on alluvium and that it behaves elastically up to a ground acceleration $a_m = 0.1g$. It is necessary to estimate the required ductility at a ground acceleration of $a_s = 0.2g$.

The deamplification factor in this case is

$$\phi(\mu) = \frac{0.1}{0.2} = 0.5$$

Referring to Figure 2-58, the deamplification of 0.5 and the frequency of 2 Hz corresponds to a ductility of approximately 2.5 (by interpolating between ductilities 1.5 and 3).

2.13 ARTIFICIALLY GENERATED GROUND MOTION

One major drawback to the response-spectrum method in the analysis and design of structures lies in its inability to provide temporal information on structural response and behavior. Such information is sometimes necessary in arriving at a satisfactory design. For example, the response-spectrum procedure can be used to estimate the maximum response in each mode of vibration, and procedures such as the square root of the sum of the squares can be used to combine the modal responses. However, when the natural frequencies are close to each other and the square root of the sum of the squares can result in inaccurate estimates of the response, the *CQC*[g] procedure or a time-history analysis may be used. If inelastic deformation is permitted in design, the inelastic spectra and the deamplification factors presented in the previous section can only be used if the structure is modeled as a single degree of freedom. For multi-degree-of-freedom systems, one still relies on the time-history analysis for inelastic response. In many cases, structures house sensitive equipment which will be subjected to the floor vibrations during an earthquake. It is sometimes necessary to develop floor response spectra from the time-history response of the floor. In addition, when designing critical facilities or major structures such as nuclear power plants, dams, or even high-rise buildings, the final design is usually based on a complete time-history analysis. The problem which

[g]An improved procedure for computing modal responses, referred to as complete quadrature combination (see Chapter 3).

often arises is what representative accelerogram should be used. Artificially generated accelerograms which represent earthquake characteristics such as a given magnitude, epicentral distance, and soil condition of the site have been used for this purpose as well as in research. For example, Penzien and Liu[2-81] used artificial accelerograms to investigate the statistical characteristics of inelastic systems, and Lai and Biggs[2-80] used them to obtain inelastic acceleration and displacement response ratios.

Different random models have been used to simulate earthquake ground motion and generate artificial accelerograms. Both stationary and nonstationary random processes have been suggested (see for example References 2-83 to 2-86). Recent studies have proposed site-dependent power spectral densities from recorded ground motion which can be utilized in generating artificial accelerograms. One of the first attempts to generate artificial accelerograms was one by Housner and Jennings,[2-83] which modeled ground motion as a stationary Gaussian random process with a power spectral density from undamped velocity spectra of recorded accelerograms. They developed a procedure for generating a random function that has the same properties of strong earthquake ground motion and obtained eight artificial accelerograms of 30-sec duration which exhibit the same appearance and statistical properties of real ground motion.

The detailed description of the procedures for generating artificial accelerograms is beyond the scope of this chapter. It may, however, be useful to mention the basic elements which are usually needed to generate artificial ground motion. In most cases, these elements consist of power spectral density or a zero-damped response spectrum, a random-phase-angle generator, and an envelope function. The simulated motion is then obtained as a finite sum of several harmonic excitations. Usually an iterative procedure is used to examine the consistency of the artificial motion by examining its frequency content through its response spectrum or its power spectral density. A typical artificial accelerogram and integrated velocity and displacement generated from the Kanai–Tajimi[2-17, 2-18] power spectral density for alluvium, using the peak acceleration and the duration of strong motion of the S00E component of El Centro the Imperial Valley earthquake of May 18, 1940, is shown in Figure 2-61.

2.14 CONCLUSION

The state of the art in strong motion seismology and earthquake engineering has advanced considerably in the last two decades. It is now possible to estimate the peak ground motion and the spectral shapes for a given site. Earthquake parameters such as magnitude, epicentral distance, site geology, and duration of strong motion influence ground motion and spectral shapes and should be taken into account when arriving at representative response values for design.

While the response spectrum is used extensively in seismic analysis and design of structures, its use has been limited to elastic cases. The application of inelastic spectra (elastic spectra modified to account for inelastic behavior) has been restricted to structures which can be modeled as a single degree of freedom. Research in utilizing inelastic spectra in the design of multi-degree-of-freedom systems is in preliminary stages. Future findings will undoubtedly provided simplified procedures for estimating the reduction in elastic forces and the ductilities required for such systems.

In many cases, a complete time-history analysis is warranted. Determination of a representative accelerogram for this purpose is very difficult, and often it is obtained through artificially generated ground motion. Statistical procedures and random vibration have been used to propose mathematical models for generating artificial accelerograms. The artificial accelerogram should possess the earthquake characteristics expected at the site.

It should be noted that in most cases, particularly for critical and important structures, the advice of geologists, seismologists, geotechnical engineers, and structural engineers should be considered before ground-motion and spectral-shape estimates are finalized for design.

REFERENCES

2-1 Biot, M. A., "A Mechanical Analyzer for Prediction of Earthquake Stresses," *Bull. Seism. Soc. Am.* 31, 151–171, 1941.

2-2 Biot, M. A., "Analytical and Experimental Methods in Engineering Seismology," *Proc. ASCE* 68, 49–69, 1942.

2-3 Housner, G. W., "An Investigation of the Effects of Earthquakes on Buildings," Ph.D. Thesis, California Institute of Technology, Pasadena, 1941.

2-4 Hudson, D. E., "Response Spectrum Techniques in Engineering Seismology," Proc. World Conf. Earthquake Eng., 4-1 to 4-12, Berkeley, CA, 1956.

2-5 Rojahn, C. and Mork, P., "An Analysis of Strong Motion Data from a Severely Damaged Structure," the Imperial Services Building, El Centro, California," USGS Open File Rep. 81-194, 1981.

2-6 Trifunac, M. D., "Low Frequency Digitization Errors and a New Method for Zero Baseline Correction of Strong-Motion Accelerograms," Earthquake Eng. Research Laboratory, EERL 70-07, California Institute of Technology, Pasadena, 1970.

2-7 Hudson, D. E., Brady, A. G., Trifunac, M. D., and Vijayaraghavan, A., "Analysis of Strong-Motion Earthquake Accelerograms —Digitized and Plotted Data, Vol. II, Corrected Accelerograms and Integrated Ground Velocity and Displacement Curves, Parts A through Y," Earthquake Research Laboratory, California Institute of Technology, Pasadena, 1971–1975.

2-8 Hudson, D. E., "Reading and Interpreting Strong Motion Accelerograms," Earthquake Eng. Research Institute, Berkeley, California, 1979.

2-9 Mohraz, B., Hall, W. J., and Newmark, N. M., "A Study of Vertical and Horizontal Earthquake Spectra," Nathan M. Newmark Consulting Engineering Services, Urbana, IL, AEC Report WASH-1255, 1972.

2-10 Page, R. A., Boore, D. M., Joyner, W. B., and Caulter, H. W., "Ground Motion Values for Use in the Seismic Design of the Trans-Alaska Pipeline System," USGS Circular 672, 1972.

2-11 Bolt, B. A., "Duration of Strong Motion," Proc. 4th World Conf. Earthquake Eng., 1304–1315, Santiago, Chile, 1969.

2-12 Trifunac, M. D. and Brady, A. G., "A Study of the Duration of Strong Earthquake Ground Motion," *Bull. Seism. Soc. Am.* 65, 581–626, 1975.

2-13 Husid R., Median, H., and Rios, J., "Analysis de Terremotos Norteamericanos y Japonesses," *Rivista del IDIEM* 8, No. 1, 1969.

2-14 McCann, W. M. and Shah, H. C., "Determining Strong-Motion Duration of Earthquakes," *Bull. Seism. Soc. Am.* 69, No. 4, 1253–1265, 1979.

2-15 Moayyad, P. and Mohraz, B., "A Study of Power Spectral Density of Earthquake Accelerograms," NSF Report PFR 8004824, Civil and Mechanical Engineering Dept., Southern Methodist University, Dallas, TX, 1982.

2-16 Elghadamsi, F. E., Mohraz, B., Lee, C. T., and Moayyad, P., "Time-Dependent Power Spectral Density of Earthquake Ground Motion," *Int. J. Soil Dynamics and Earthquake Eng.* 7, No. 1, 15–21, 1988.

2-17 Kanai, K., "Semi-Imperical Formula for the Seismic Characteristics of the Ground," Bull. Earthquake Research Inst. 35, University of Tokyo, 1957.

2-18 Tajimi, H., "A Statistical Method of Determining the Maximum Response of a Building Structure During an Earthquake," Proc. 2nd World Conf. Earthquake Eng., Vol. II, 781–797, Tokyo, 1960.

2-19 Lai, S. P., "Statistical Characterization of Strong Motions Using Power Spectral Density Function," *Bull. Seism. Soc. Am.* 72, 259–274, 1982.

2-20 Seed, H. B. and Idriss, I. M., "Ground Motions and Soil Liquefaction During Earthquakes," Earthquake Engineering Research Inst., Berkeley, CA, 1982.

2-21 Donovan, N. C., "Earthquake Hazards for Buildings," Building Practices for Disaster Mitigation, National Bureau of Standards, U.S. Department of Commerce, Building Research Services, 46, 82–111, 1973.

2-22 Donovan, N. C. and Bornstein, A. E., "Uncertainties in Seismic Risk Procedures," *J. Geotechnical Eng. Div., ASCE*, 104, No. GT 7, 869–887, 1978.

2-23 Cloud, W. K. and Perez, V., "Unusual Accelerograms Recorded at Lima, Peru," *Bull. Seism. Soc. Am.* 61, No. 3, 633–640, 1971.

2-24 Housner, G. W., "Intensity of Earthquake Ground Shaking Near the Causative Fault," Proc. 3rd. World Conf. Earthquake Eng., Vol. 1, III, 94–115, New Zealand, 1965.

2-25 Porcella, R. L. and Matthiesen, R. B., USGS Open-File Report 79-1654, 1979.

2-26 Campbell, K. W., "Near-Source Attenuation of Peak Horizontal Acceleration," *Bull. Seism. Soc. Am.* 71, No. 6, 2039–2070, 1981.

2-27 Joyner, W. B. and Boore, D. M., "Peak Horizontal Acceleration and Velocity from Strong Motion Records Including Records from the 1979 Imperial Valley, California, Earthquake," *Bull. Seism. Soc. Am.* 71, No. 6, 2011–2038, 1981.

2-28 Nuttli, O. W. and Herrmann, R. B., "Consequences of Earthquakes in Mississippi Valley," ASCE Preprint 81-519, ASCE National Convention, St. Louis, 1981.

2-29 Milne, W. G. and Davenport, A. G., "Distribution of Earthquake Risk in Canada," *Bull. Seism. Soc. Am.* 59, No. 2, 754–779, 1969.

2-30 Boore, D. M., Joyner, W. B., Oliver, A. A., and Page, R. A., "Estimation of Ground Motion Parameters," USGS, Circular 795, 1978.

2-31 Boore, D. M., Joyner, W. B., Oliver, A. A., and Page, R. A., "Peak Acceleration Velocity and Displacement from Strong Motion Records," *Bull. Seism. Soc. Am.* 70, No. 1, 305–321, 1980.

2-32 Chang, F. K. and Krinitzsky, E. L., "Duration, Spectral Content, and Predominant Period of Strong Motion Earthquake Records from Western United States," U.S. Army Engineer Waterways Experiment Station, Miscellaneous Paper S-73-1, Vicksburg, MS, 1977.

2-33 Mohraz, B., "A Study of Earthquake Response Spectra for Different Geological Conditions," *Bull. Seism. Soc. Am.* 66, No. 3, 915–935, 1976.

2-34 Housner, G. W., Strong Ground Motion, Chapter 4 in *Earthquake Engineering*, R. L. Wiegel (ed.), Prentice-Hall, Englewood Cliffs, N.J., 1970.

2-35 Gutenberg, B. and Richter, C. F. "Earthquake Magnitude, Intensity, Energy, and Acceleration," *Bull. Seism. Soc. Am.* 46, No. 2, 143–145, 1956.

2-36 Richter, C. F., *Elementary Seismology*, W. H. Freeman, San Francisco, 1958.

2-37 Bender, B., "Maximum Likelihood Estimation of *b* Values for Magnitude Grouped Data," *Bull. Seism. Soc. Am.* 73, No. 3, 831–851, 1983.

2-38 Schwartz, D. P. and Coppersmith, K. J., "Fault Behavior and Characteristic Earthquakes: Examples from the Wasatch and San Andreas Fault Zones," *J. Geophs. Res.* 89, No. B7, 5681–5698, 1984.

2-39 Idriss, I. M., "Evaluating Seismic Risk in Engineering Practice," Chapter 6, *Proc. 11th International Conf. Soil Mechanics and Foundation Eng.*, 255–320, San Francisco, 1985.

2-40 Sieh, K. E., "Prehistocial Large Earthquakes Produced by Slip on the San Andreas Fault at Pallett Creek, California," *J. Geophys. Res.*, 83, No. B8, 3907–3939, 1978.

2-41 Cornell, C. A., "Engineering Seismic Risk Analysis," *Bull. Seism. Soc. Am.* 58, No. 5, 1583–1606, 1968.

2-42 Vanmarcke, E. H., Seismic Safety Assessment, *Random Excitation of Structures by Earthquakes and Atmospheric Turbulence*, H. Parkus (ed.), International Center for Mechanical Sciences, Course and Lectures No. 225, Springer-Verlag, 1–76, 1977.

2-43 Benjamin, J. R., "Probabilistic Models for Seismic Force Design," *J. Structural Div., ASCE*, 94, ST5, 1175–1196, 1968.

2-44 Der-Kiureghian, A. and Ang, A. H.-S., "A Fault–Rupture Model for Seismic Risk Analysis," *Bull. Seism. Soc. Am.* 67, No. 4, 1173–1194, 1977.

2-45 Algermissen, S. T. and Perkins, D. M., "A Technique for Seismic Risk Zoning, General Considerations and Parameters," Proc. Microzonation Conf., 865–877, Seattle, WA, 1972.

2-46 Algermissen, S. T. and Perkins, D. M., "A Probabilistic Estimate of Maximum Acceleration in Rock in Contiguous United States," USGS Open File Report, 76-416, 1976.

2-47 Applied Technology Council, National Bureau of Standards, and National Science Foundation, "Tentative Provisions for the Development of Seismic Regulations for Buildings," ATC Publication 3-06, NBS Publication 510, NSF Publication 78-8, 1978.
2-48 McGuire, R. K., "Seismic Structural Response Risk Analysis, Incorporating Peak Response Progressions on Earthquake Magnitude and Distance," Report R74-51, Dept. of Civil Engineering, Mass. Inst. of Technology, Cambridge, MA, 1975.
2-49 Newmark, N. M. and Hall, W. J., "Earthquake Spectra and Design," Earthquake Engineering Research Inst., Berkeley, CA, 1982.
2-50 Newmark, N. M., Blume, J. A., and Kapur, K. K., "Seismic Design Criteria for Nuclear Power Plants," *J. Power Div., ASCE*, 99, No. PO2, 287–303, 1973.
2-51 Newmark, N. M. and Hall, W. J., "Seismic Design Criteria for Nuclear Reactor Facilities," Proc. 4th World Conf. Earthquake Eng., B-4, 37–50, Santiago, Chile, 1969.
2-52 Newmark, N. M. and Hall, W. J., "Procedures and Criteria for Earthquake Resistant Design," Building Practices for Disaster Mitigation, National Bureau of Standards, U.S. Department of Commerce, Building Research Series, 46, 209–236, 1973.
2-53 Hall, W. J., Mohraz, B., and Newmark, N. M., *Statistical Studies of Vertical and Horizontal Earthquake Spectra*, Nathan M. Newmark Consulting Engineering Services, Urbana, IL, 1975.
2-54 Newmark, N. M. and Rosenblueth, *Fundamentals of Earthquake Engineering*, Prentice-Hall, Englewood Cliffs, NJ, 1971.
2-55 Clough, R. W. and Penzien, J., *Dynamics of Structures*, McGraw-Hill, New York, 1975.
2-56 Trifunac, M. D., Brady, A. G., and Hudson, D. E., "Analysis of Strong-Motion Earthquake Accelerograms, Vol. III, Response Spectra, Parts A through Y," Earthquake Eng. Research Lab., California Inst. of Technology, Pasadena, 1972–1975.
2-57 Chopra, A. K., "Dynamics of Structures—A Primer," Earthquake Engineering Research Inst., Berkeley, CA, 1981.
2-58 Housner, G. W. and Jennings, P. C., "Earthquake Design Criteria," Earthquake Engineering Research Inst., Berkeley, CA, 1982.
2-59 Singh, J. P., "Earthquake Ground Motions: Implications for Designing Structures and Reconciling Structural Damage," *Earthquake Spectra* 1, No. 2, 239–270, 1985.
2-60 "Reducing Earthquake Hazards: Lessons Learned from Earthquakes," Earthquake Engineering Research Inst., Publication 86-02, Berkeley, CA, 1986.
2-61 Housner, G. W., "Spectrum Intensities of Strong-Motion Earthquakes," Proc. Symposium on Earthquakes and Blast Effects on Structures, Earthquake Engineering Research Inst., 1952.
2-62 Housner, G. W., "Behavior of Structures During Earthquakes," *J. Eng. Mech. Div.*, ASCE, 85, No. EM4, 109–129, 1959.
2-63 Housner, G. W., Design Spectrum, Chapter 5 in *Earthquake Engineering*, R. L. Wiegel (ed.), Prentice-Hall, Englewood Cliffs, NJ, 1970.
2-64 Blume, J. A., Sharpe, R. L., and Dalal, J. S., "Recommendations for Shape of Earthquake Response Spectra," John A. Blume & Associates, San Francisco, AEC Report WASH-1254, 1972.
2-65 Atomic Energy Commission, "Design Response Spectra for Seismic Design of Nuclear Power Plants," Regulatory Guide 1.60, Directorate of Regulatory Standards, Washington, 1973.
2-66 Hayashi, S., Tsuchida, H., and Kurata, E., "Average Response Spectra for Various Subsoil Conditions," Third Joint Meeting, U.S.–Japan Panel on Wind and Seismic Effects, UJNR, Tokyo, 1971.
2-67 Kuribayashi, E., Iwasaki, T., Iida, Y., and Tuji, K., "Effects of Seismic and Subsoil Conditions on Earthquake Response Spectra," Proc. International Conf. Microzonation, Seattle, WA, 499–512, 1972.
2-68 Seed, H. B., Ugas, C., and Lysmer, J., "Site-Dependent Spectra for Earthquake-Resistance Design," *Bull. Seism. Soc. Am.* 66, No. 1, 221–243, 1976.
2-69 Seismology Committee, Structural Engineers Association of California," Recommended Lateral Force Requirements," 1986.
2-70 Mohraz, B., "Influences of the Magnitude of the Earthquake and the Duration of Strong Motion on Earthquake Response Spectra," Proc. Central Am. Conf. on Earthquake Eng., San Salvadore, El Salvador, 1978.
2-71 Peng, M. H., Elghadamsi, F. E., and Mohraz, B., "A Simplified Procedure for Constructing Probabilistic Response Spectra," *Earthquake Spectra*, Vol. 5, No. 2, April 1989.
2-72 Riddell, R., and Newmark, N. M., "Statistical Analysis of the Response of Nonlinear Systems Subjected to Earthquakes," Civil Engineering Studies, Structural Research Series 468, Dept. of Civil Engineering, Univ. of Illinois, Urbana, 1979.
2-73 Elghadamsi, F. E. and Mohraz, B., "Inelastic Earthquake Spectra," *J. Earthquake Engineering and Structural Dynamics* 15, 91–104, 1987.
2-74 Blume, J. A., Newmark, N. M., and Corning, L. H., "Design of Multistory Reinforced Concrete Buildings for Earthquake Motions," Portland Cement Assoc., 1961.
2-75 Newmark, N. M., "Current Trends in the Seismic Analysis and Design of High-Rise Structures," Chapter 16 in *Earthquake Engineering*, R. L. Wiegel (ed.), Prentice-Hall, Englewood Cliffs, NJ, 1970.
2-76 Housner, G. W., "Limit Design of Structures to Resist Earthquakes," Proc. 1st World. Conf. Earthquake Engineering, 5-1 to 5-13, Berkeley, CA, 1956.
2-77 Blume, J. A., "A Reserve Energy Technique for the Earthquake Design and Rating of Structures in the Inelastic Range," Proc. 2nd World Conf. Earthquake Engineering, Vol. II, 1061–1084, Tokyo, 1960.
2-78 Blume, J. A., "Structural Dynamics in Earthquake-Resistant Design," *Trans. ASCE* 125, 1088–1139, 1960.
2-79 Blume, J. A., Discussion of "Electrical Analog for Earthquake Yield Spectra," *J. Eng. Mech. Div., ASCE*, 86, No. EM3, 177–184, 1960.
2-80 Lai, S. P. and Biggs, J. M., "Inelastic Response Spectra for Aseismic Building Design," *J. Structural Div., ASCE*, 106, No. ST6, 1295–1310, 1980.
2-81 Penzein, J. and Liu, S. C., "Nondeterministic Analysis of Nonlinear Structures Subjected to Earthquake Excitations," Proc. 4th World Conf. Earthquake Engineering, A-1, 114–129, Santiago, Chile, 1969.
2-82 Peng, M. H., Elghadamsi, F. E., and Mohraz, B., "A Stochastic Procedure for Seismic Analysis of SDOF Structures," Civil and Mechanical Engineering Dept., School of Engineering and Applied Science, Southern Methodist Univ., Dallas, TX, 1987.
2-83 Housner, G. W. and Jennings, P. C., "Generation of Artificial Earthquakes," *J. Eng. Mech. Div., ASCE*, 90, 113–150, 1964.
2-84 Shinozuka, M. and Sato, Y., "Simulation of Nonstationary Random Process," *J. Eng. Mech. Div., ASCE*, 93, 11–40, 1967.
2-85 Amin, M. and Ang, A. H.-S., "Nonstationary Stochastic Model of Earthquake Ground Motion," *J. Eng. Mech. Div., ASCE*, 74, No. EM2, 559–583, 1968.
2-86 Iyengar, R. N. and Iyengar, K. T. S., "A Nonstationary Random Process Model for Earthquake Accelerograms," *Bull. Seism. Soc. Am.* 59, 1163–1188, 1969.
2-87 Hudson, D. E. (ed.), "Strong Motion Instrumental Data on the San Fernando Earthquake of Feb. 9, 1971," Earthquake Engineering Research Lab., California Institute of Technology, Pasadena, 1971.
2-88 Trifunac, M. D. and Brady, A. G., "On the Correlation of Seismic Intensity Scales with the Peaks of Recorded Strong

Ground Motion," *Bull. Seism. Soc. Am.* 65, No. 1, 139–162, 1975.

2-89 Blume, J. A., "Earthquake Ground Motion and Engineering Procedures for Important Installations Near Active Faults," Proc. 3rd World Conf. Earthquake Eng., Vol. IV, 53–67, New Zealand, 1965.

2-90 Kanai, K., "Improved Empircal Formula for Characteristics of Stray Earthquake Motions," Proc. Japan Earthquake Symposium, 1–4, 1966. (In Japanese.)

2-91 Esteva, K., Seismic Risk and Seismic Design Decisions, *Seismic Design for Nuclear Power Plants*, R. J. Hansen, (ed.), MIT Press, 1970.

2-92 Sabetta, F. and Pugliese, A., "Attenuation of Peak Horizontal Acceleration and Velocity from Italian Strong-Motion Records," *Bull. Seism. Soc. Am.* 77, No. 5, 1491–1513, 1987.

Chapter 3

Dynamic Response of Buildings

*James C. Anderson, Ph.D.**

3.1 INTRODUCTION

The main cause of damage to structures during an earthquake is their response to ground motions which are input at the base. In order to evaluate the behavior of the structure under this type of loading condition, the principles of structural dynamics must be applied to determine the stresses and deflections which are developed in the structure. Structural engineers are familiar with the analysis of structures for static loads in which a load is applied to the structure and a single solution is obtained for the resulting displacements and member forces. When considering the analysis of structures for dynamic motions, the term dynamic simply means "time-varying". Hence the loading and all aspects of the response vary with time. This results in an infinite number of possible solutions at each instant during the time interval under consideration. From an engineering standpoint, the maximum values of the structural response are usually the ones of particular interest, especially in the case of structural design.

The purpose of this chapter is to introduce the principles of structural dynamics with emphasis on earthquake response analysis. Attention will initially be focused on the response of simple structural systems which can be represented in terms of a single degree of freedom. The concepts developed for these systems will then be extended to include generalized single-degree-of-freedom (SDOF) systems using the generalized-coordinate approach. This development in turn leads to the consideration of the response of structures having multiple degrees of freedom. Finally, concepts and techniques used in nonlinear dynamic-response analysis will be introduced.

*Associate Professor of Civil Engineering, University of Southern California, Los Angeles, California.

3.2 DYNAMIC EQUILIBRIUM

The basic equation of static equilibrium used in the displacement method of analysis has the form,

$$p = kv \tag{3-1}$$

where p is the applied force, k is the stiffness or resistance, and v is the resulting displacement. If the statically applied force is now replaced by a dynamic or time-varying force $p(t)$, the equation of static equilibrium becomes one of dynamic equilibrium and has the form

$$p(t) = m\ddot{v}(t) + c\dot{v}(t) + kv(t) \tag{3-2}$$

where a dot represents differentiation with respect to time.

A direct comparison of these two equations indicates that two significant changes, which distinguish the static problem from the dynamic problem, were made to Equation 3-1 in order to obtain Equation 3-2. First, the applied load and the resulting response are now functions of time, and hence Equation 3-2 must be satisfied at each instant of time during the time history under consideration. For this reason it is usually referred to as an equation of motion. Secondly, the time dependence of the displacements gives rise to two additional forces which resist the applied force and have been added to the right-hand side.

The equation of motion represents an expression of Newton's second law of motion, which states that a particle acted on by a force (torque) moves so that the time rate of change of its linear (angular) momentum is equal to the force (torque):

$$p(t) = \frac{d}{dt}\left(m\frac{dv}{dt}\right) \qquad (3\text{-}3)$$

where the rate of change of the displacement with respect to time, dv/dt, is the velocity, and the momentum is given by the product of the mass and the velocity. Recall that the mass is equal to the weight divided by the acceleration of gravity. If the mass is constant, Equation 3-3 becomes

$$p(t) = m\frac{d}{dt}\left(\frac{dv}{dt}\right) = m\ddot{v}(t) \qquad (3\text{-}4)$$

which states that the force is equal to the product of mass and acceleration. According to d'Alembert's principle, mass developes an inertia force which is proportional to its acceleration and opposing it. Hence the first term on the right-hand side of Equation 3-2 is called the inertia force; it resists the acceleration of the mass.

Dissipative or damping forces are inferred from the observed fact that oscillations in a structure tend to diminish with time once the time-dependent applied force is removed. These forces are represented by viscous damping forces which are proportional to the velocity with a constant of proportionality referred to as the damping coefficient. The second term on the right-hand side of Equation 3-2 is called the damping force.

Inertia forces are the more significant of the two and are a primary distinction between static and dynamic analyses.

It must also be recognized that all structures are subjected to gravity loads such as self-weight (dead load) and occupancy load (live load) in addition to dynamic base motions. In an elastic system, the principle of superposition can be applied, so that the responses to static and dynamic loadings can be considered separately and then combined to obtain the total structural response. However, if the structural behavior becomes nonlinear, the response becomes load-path-dependent and the gravity loads must be considered concurrently with the dynamic motions.

Under strong earthquake motions, the structure will most likely display nonlinear behavior, which can be caused by material nonlinearity and/or geometric nonlinearity. Material nonlinearity occurs when stresses at certain critical regions in the structure exceed the elastic limit of the material. The equation of dynamic equilibrium for this case has the general form

$$p(t) = m\ddot{v}(t) + c\dot{v}(t) + k(t)v(t) \qquad (3\text{-}5)$$

in which the stiffness or resistance k is a function of the yield condition in the structure, which in turn is a function of time. Geometric nonlinearity is caused by the gravity loads acting on the deformed position of the structure. If the lateral displacements are small, this effect, which is often referred to as P-delta, can be neglected. However, if the lateral displacements become large, this effect must be considered.

In order to define the inertia forces completely, it would be necessary to consider the accelerations of every mass particle in the structure and the corresponding displacements. Such a solution would be prohibitively time-consuming. The analysis procedure can be greatly simplified if the mass of the structure can be concentrated (lumped) at a finite number of discrete points and the dynamic response of the structure can be represented in terms of this limited number of displacement components. The number of displacement components required to specify the position of the mass points is called the number of dynamic degrees of freedom. The number of degrees of freedom required to obtain an adequate solution will depend upon the complexity of the structural system. For some structures a single degree of freedom may be sufficient, whereas for others several hundred degrees of freedom may be required.

3.3 SINGLE-DEGREE-OF-FREEDOM SYSTEMS

3.3.1 Time-Dependent Force

The simplest structure that can be considered for dynamic analysis is an idealized, one-story structure in which the single degree of freedom is the lateral translation at the roof level as shown in Figure 3-1. In this idealization, three important assumptions are made. First, the mass is assumed to be concentrated (lumped) at the roof level. Second, the roof system is assumed to be rigid, and third, the axial deformation in the columns is neglected. From these assumptions it follows that all lateral resistance is in the resisting elements such as

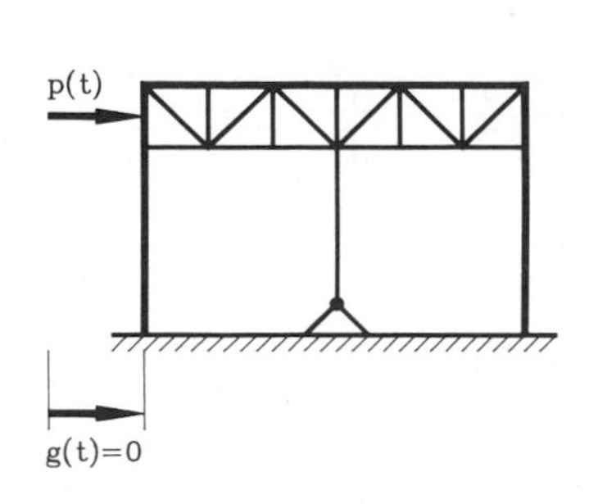

a) SINGLE STORY FRAME

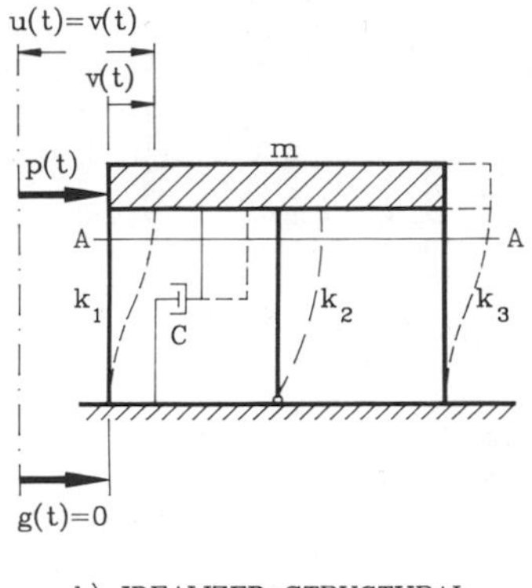

b) IDEALIZED STRUCTURAL SYSTEM

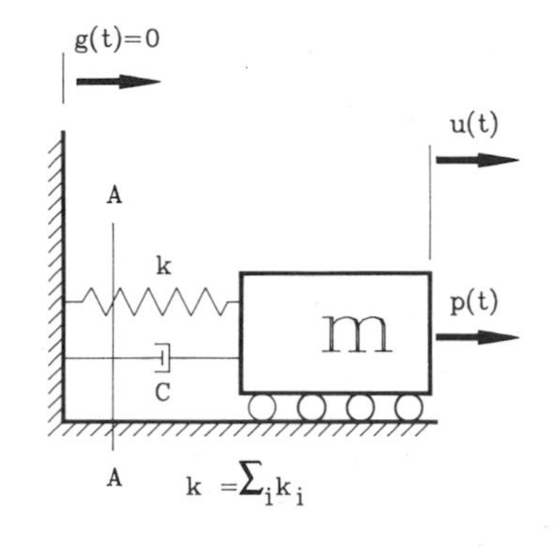

c) EQUIVALENT SPRING–MASS–DAMPER SYSTEM

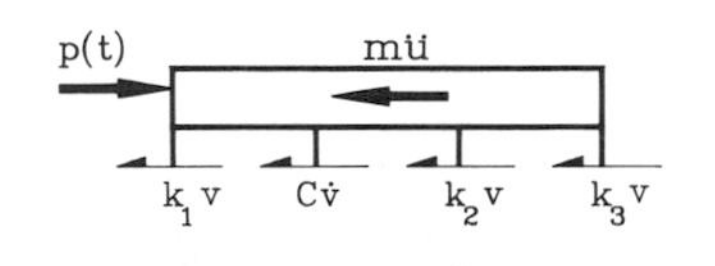

d) FREE BODY DIAGRAMS, SECTION A–A

Figure 3-1 Single-degree-of-freedom system subjected to time-dependent force.

columns, walls, and diagonal braces located between the roof and the base. Application of these assumptions results in a discretized structure that can be represented as shown in either Figure 3-1b or 3-1c with a time-dependent force applied at the roof level. The total stiffness k is simply the sum of the stiffnesses of the resisting elements in the story level.

The forces acting on the mass of the structure are shown in Figure 3-1d. Summing the forces acting on the free body results in the following equation of equilibrium which must be satisfied at each instant of time:

$$f_i + f_d + f_s = p(t) \tag{3-6}$$

where

$$f_i = \text{inertia force} = m\ddot{u}$$
$$f_d = \text{damping (dissipative) force} = c\dot{v}$$
$$f_s = \text{elastic restoring force} = kv$$
$$p(t) = \text{time-dependent applied force}$$

$\ddot{u}$ is the total acceleration of the mass, and $\dot{v}$, v are the velocity and displacement of the mass relative to the base. Writing Equation 3-6 in terms of the physical response parameters results in

$$m\ddot{u} + c\dot{v} + kv = p(t) \tag{3-7}$$

It should be noted that the forces in the damping element and in the resisting elements depend upon the relative velocity and relative displacement, respectively, across the ends of these elements, whereas the inertia force depends upon the total acceleration of the mass. The total acceleration of the mass can be expressed as

$$\ddot{u}(t) = \ddot{g}(t) + \ddot{v}(t) \tag{3-8}$$

where

$$\ddot{v}(t) = \text{acceleration of the mass relative to the base}$$
$$\ddot{g}(t) = \text{acceleration of the base}$$

In this case, the base is assumed to be fixed with no motion, and hence $\ddot{g}(t) = 0$ and $\ddot{u}(t) = \ddot{v}(t)$. Making this substitution for the acceleration, Equation 3-7 for a time-dependent force becomes

$$m\ddot{v} + c\dot{v} + kv = p(t) \tag{3-9}$$

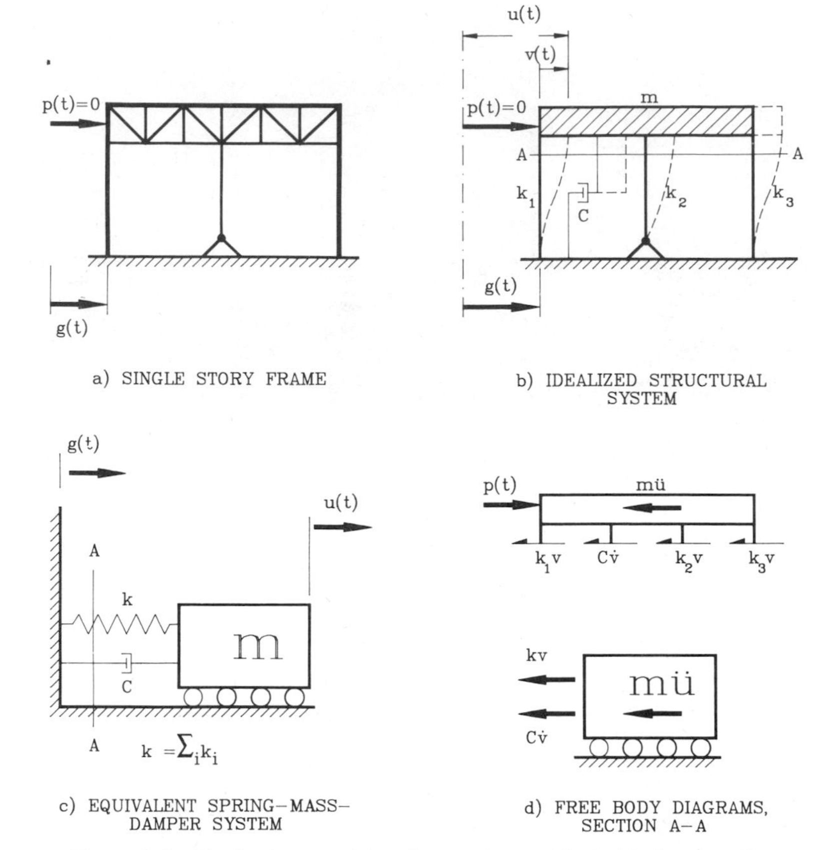

Figure 3-2 Single-degree-of-freedom system subjected to base motion.

3.3.2 Earthquake Ground Motion

When a single-story structure, shown in Figure 3-2a, is subjected to earthquake ground motions, no external dynamic force is applied at the roof level. Instead, the system experiences an acceleration of the base. The effect of this on the idealized structure is shown in Figure 3-2b and c. Summing the forces shown in Figure 3-2d results in the following equation of dynamic equilibrium:

$$f_i + f_d + f_s = 0 \qquad (3\text{-}10)$$

Substituting the physical parameters for f_i, f_d, and f_s in Equation 3-10 results in an equilibrium equation of the form

$$m\ddot{u} + c\dot{v} + kv = 0 \qquad (3\text{-}11)$$

This equation can be written in the form of Equation 3-9 by substituting Equation 3-8 into Equation 3-11 and rearranging terms to obtain

$$m\ddot{v} + c\dot{v} + kv = p_e(t) \qquad (3\text{-}12)$$

where

$$p_e(t) = \text{effective time-dependent force}$$
$$= -m\ddot{g}(t)$$

Hence the equation of motion for a structure subjected to a base motion is similar to that for a structure subjected to a time-dependent force if the base motion is represented as an effective time-dependent force which is equal to the product of the mass and the ground acceleration.

3.3.3 Mass and Stiffness Properties

Most SDOF models consider structures which experience a translational displacement of the roof relative to the base. In this case the translational mass is simply the concentrated weight divided by the acceleration of

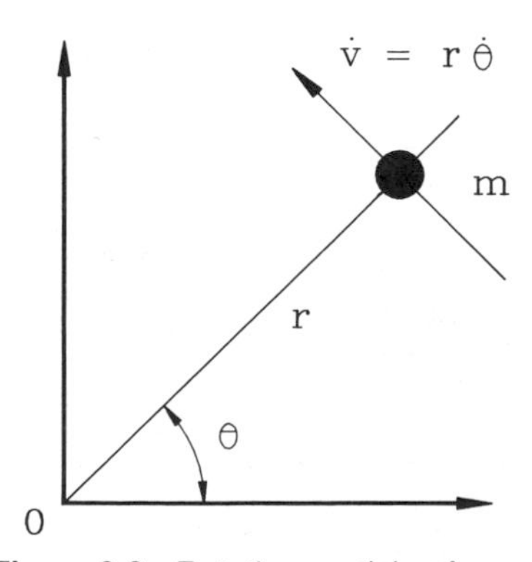

Figure 3-3 Rotating particle of mass.

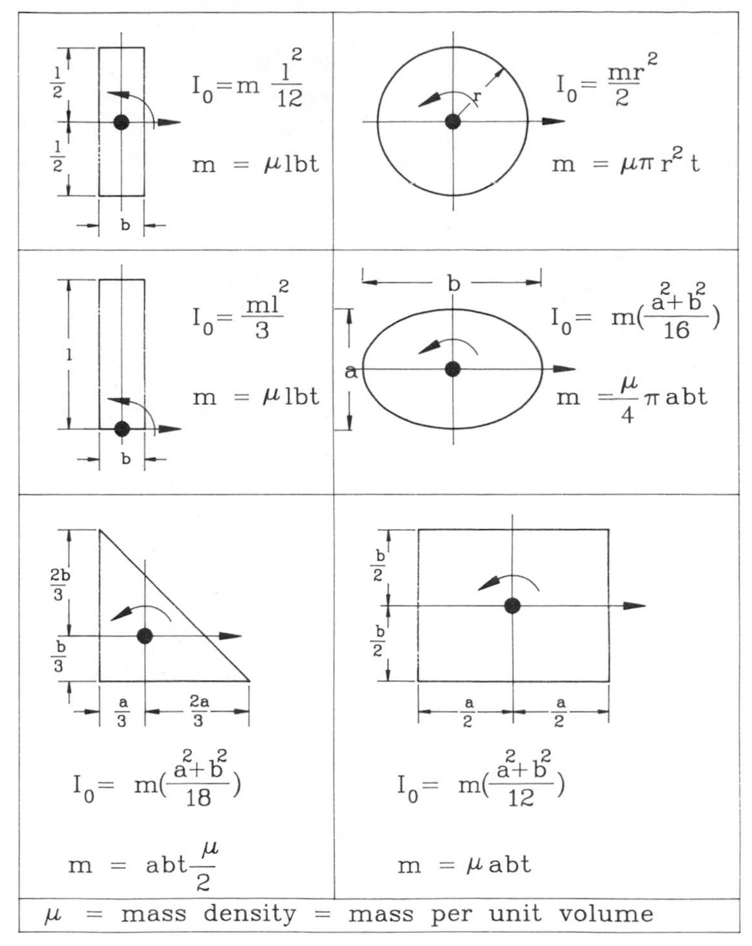

Figure 3-4 Rigid-body mass and mass moment of inertia.

gravity (32.2 ft/sec^2 or 386.4 in./sec^2). However, cases do arise in which the rotational motion of the system is significant. An example of this might be the rotational motion of a roof slab which has unsymmetrical lateral supports. Newton's second law of motion states that the time rate of change of the angular momentum (moment of momentum) equals the torque. Considering a particle of mass rotating about an axis o, as shown in Figure 3-3, the moment of momentum can be expressed as

$$L = rm\dot{v}(t) = mr^2 \frac{d\theta}{dt} \tag{3-13}$$

The torque N is then obtained by taking the time derivative:

$$N = \frac{dL}{dt} = I\ddot{\theta} \tag{3-14}$$

where

$$I = mr^2 = \text{mass moment of inertia}$$

For a rigid body, the mass moment of inertia can be obtained by summing over all the mass particles making up the rigid body. This can be expressed in integral form as

$$I = \int \rho^2 \, dm \tag{3-15}$$

where ρ is the distance from the axis of rotation to the incremental mass dm. For dynamic analysis it is convenient to treat the rigid-body inertia forces as though the translational mass and the mass moment of inertia were concentrated at the center of mass. The mass and mass moment of inertia of several common rigid bodies are summarized in Figure 3-4.

Example 3-1 (Determination of Mass Properties)
Compute the mass and mass moment of inertia for the rectangular plate shown in Figure 3-5.

- Translational mass:

$$m = \mu V = \mu abt$$

where

μ = mass density = mass per unit volume
V = total volume

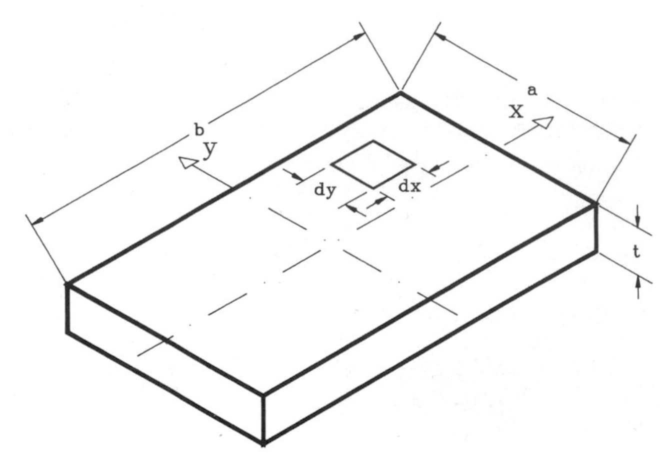

Figure 3-5 Rectangular plate of Example 3-1.

- Rotational mass moment of inertia:

$$I = \int \rho^2 \, dm, \qquad \text{where} \quad \rho^2 = x^2 + y^2$$

$$dm = \mu \, dV = \mu t \, dx \, dy$$

$$I = \int \rho^2 \, dm = 4\mu t \int_0^{a/2} \int_0^{b/2} (x^2 + y^2) \, dx \, dy$$

$$I = 4\mu t \frac{b^3 a + a^3 b}{48} = \mu abt \frac{b^2 + a^2}{12}$$

$$I = m \frac{a^2 + b^2}{12}$$

In order to develop dynamic models of SDOF systems, it is necessary to review the force–displacement (stiffness) relationships of several of the more common lateral force members used in building structures. As indicated previously, the assumptions used in developing the SDOF model restrict lateral resistance to structural members between the roof and base. These might include such members as columns, diagonal braces, and walls. Stiffness properties for these elements are summarized in Figure 3-6.

3.3.4 Free Vibration

Free vibration occurs when a structure oscillates under the action of forces that are inherent in the structure without any externally applied time-dependent loads or ground motions. These inherent forces arise from the initial velocity and displacement the structure has at the beginning of the free-vibration phase.

Undamped Structures The equation of motion for an undamped SDOF system in free vibration has the form

$$m\ddot{v}(t) + kv(t) = 0 \tag{3-16}$$

which can be written as

$$\ddot{v}(t) + \omega^2 v(t) = 0 \tag{3-17}$$

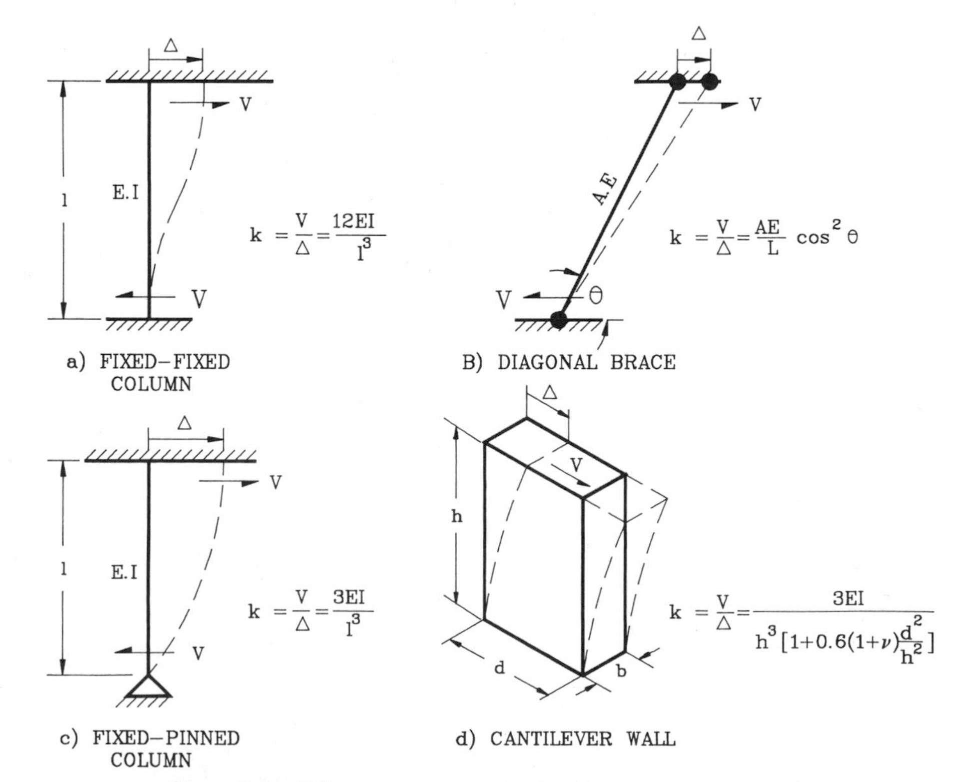

Figure 3-6 Stiffness properties of lateral force resisting elements.

where $\omega^2 = k/m$. This equation has the general solution

$$v(t) = A \sin \omega t + B \cos \omega t \tag{3-18}$$

in which the constants of integration A and B depend upon the initial velocity $\dot{v}(0)$ and initial displacement $v(0)$. Applying the initial conditions, the solution has the form

$$v(t) = \frac{\dot{v}(0)}{\omega} \sin \omega t + v(0) \cos \omega t \tag{3-19}$$

This solution in time is represented graphically in Figure 3-7.

Several important concepts of oscillatory motion can be illustrated with this result. The amplitude of vibration is constant, so that the vibration would, theoretically, continue indefinitely with time. This cannot physically be true, because free oscillations tend to diminish with time, leading to the concept of damping. The time it takes a point on the curve to make one complete cycle and return to its original position is called the period of vibration, T. The quantity ω is the circular frequency of vibration and is measured in radians per second. The cyclic frequency f is defined as the reciprocal of the period and is measured in cycles per second, or *hertz*. These three vibration properties depend only on the mass and stiffness of the structure and are related as follows:

$$T = \frac{2\pi}{\omega} = 2\pi\sqrt{\frac{m}{k}} = \frac{1}{f} \tag{3-20}$$

It can be seen from these expressions that if two structures have the same stiffness, the one having the larger mass will have the longer period of vibration and the lower frequency. On the other hand, if two structures have the same mass, the one having the higher stiffness will have the shorter period of vibration and the higher frequency.

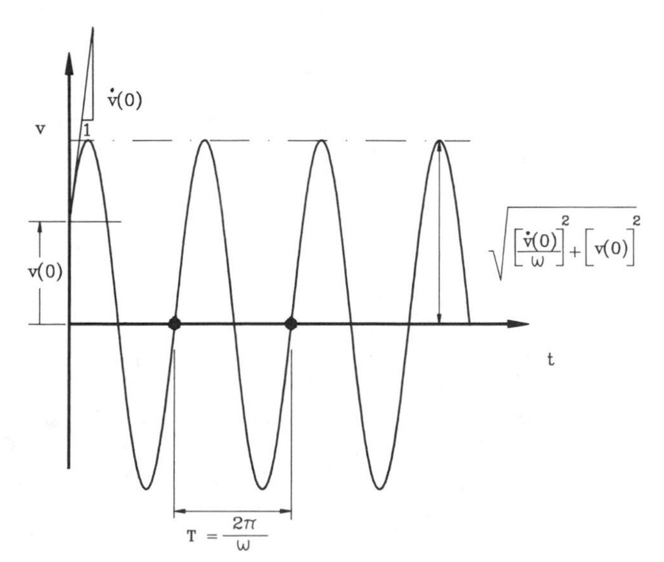

Figure 3-7 Free-vibration response of an undamped SDOF system.

Example 3-2 (Period of Undamped Free Vibration) Construct an idealized SDOF model for the industrial building shown in Figure 3-8, and estimate the period of vibration in the two principal directions. Note that vertical cross bracings are made of 1-inch-diameter rods, horizontal cross bracing is at the bottom chord of trusses, and all columns are W8 × 24.

- Weight determination:

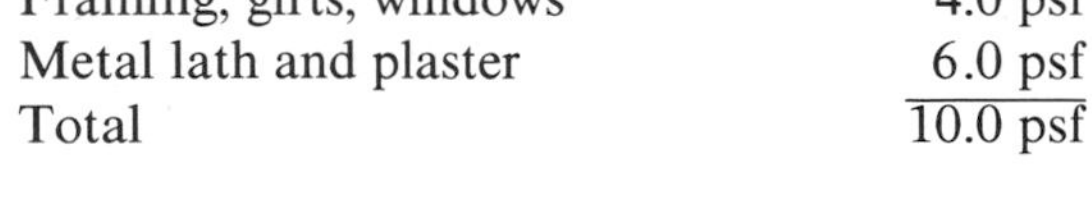

Roof level:	
Composition roof	9.0 psf
Lights, ceiling, mechanical	6.0 psf
Trusses	2.6 psf
Roof purlins, struts	2.0 psf
Bottom chord bracing	2.1 psf
Columns (10 ft, 9 in.)	0.5 psf
Total	22.2 psf
Walls:	
Framing, girts, windows	4.0 psf
Metal lath and plaster	6.0 psf
Total	10.0 psf

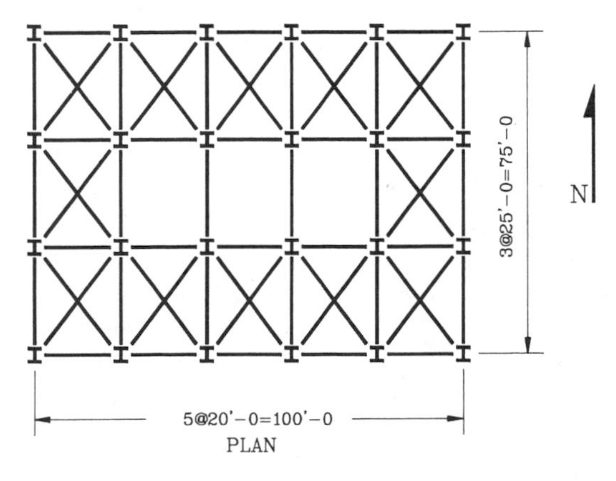

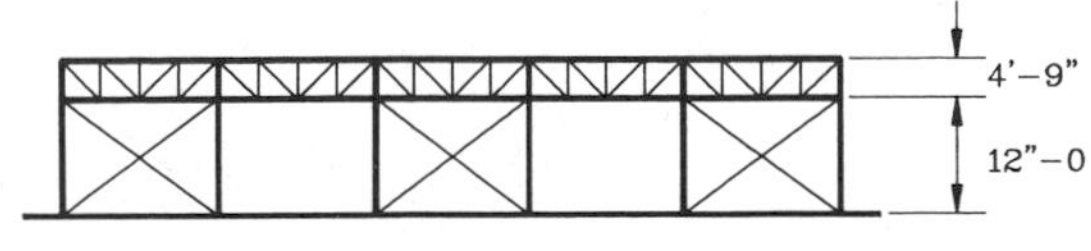

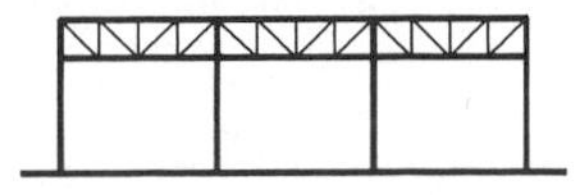

Figure 3-8 Building of Example 3-2.

Total weight and mass:

$$W = (22.2)(100)(75) + (10)(6)(200 + 150)$$

$$W = 187{,}500 \text{ lb} = 187.5 \text{ kips}$$

$$m = \frac{W}{g} = \frac{187.5}{386.4} = 0.485 \text{ kips-sec}^2/\text{in.}$$

- Stiffness determination:

North–south (moment frames):

$$k_i = \frac{12EI}{L^3} = \frac{(12)(29000)(82.8)}{(144)^3}$$

$$k_i = 9.6 \text{ kips/in.}$$

$$k = \sum_{i=1}^{24} k_i = 24(9.6) = 231.6 \text{ kips/in.}$$

East–west (braced frames):

$$k_i = \frac{AE}{L}\cos^2\theta$$

$$A = \pi d^2/4 = 0.785 \text{ in.}$$

$$L = \sqrt{12^2 + 20^2} = 23.3 \text{ ft} = 280 \text{ in.}$$

$$\theta = \tan^{-1}(12/20) = 31°, \qquad \cos(31°) = 0.858$$

$$k_i = \frac{(0.785)(29000)(0.858)^2}{280} = 59.7 \text{ kips/in.}$$

$$k = \sum_{i=1}^{6} k_i = 6(59.7) = 358.7 \text{ kips/in.}$$

- Period determination:

North–south:

$$\omega = \sqrt{\frac{k}{m}} = \sqrt{\frac{236.1}{0.485}} = 21.8 \text{ rad/sec}$$

$$T = 2\pi/\omega = 2\pi/21.8 = 0.287 \text{ sec}$$

$$f = 1/T = 3.48 \text{ Hz}$$

East–west:

$$\omega = \sqrt{\frac{f}{m}} = \sqrt{\frac{358.7}{0.485}} = 27.2 \text{ rad/sec}$$

$$T = \frac{2\pi}{\omega} = \frac{2\pi}{27.2} = 0.23 \text{ sec}$$

$$f = 1/T = 4.3 \text{ Hz}$$

Damped Structures In an actual structure which is in free vibration under the action of internal forces, the amplitude of the vibration tends to diminish with time and eventually the motion will cease. This decrease with time is due to the action of viscous damping forces which are proportional to the velocity. The equation of motion for this condition has the form

$$m\ddot{v}(t) + c\dot{v}(t) + kv(t) = 0 \tag{3-21}$$

This equation has the general solution

$$v(t) = e^{-\lambda\omega t}\left([\dot{v}(0) + v(0)\lambda\omega]\frac{\sin\omega_d t}{\omega_d} + v(0)\cos\omega_d t\right) \tag{3-22}$$

where

$$\lambda = \frac{C}{C_{cr}} = \frac{C}{2\omega m} = \text{percentage of critical damping}$$

$$\omega_d = \omega\sqrt{1 - \lambda^2} = \text{damped circular frequency}$$

The solution to this equation with time is shown in Figure 3-9. The damping in the oscillator is expressed in terms of a percentage of critical damping, where critical damping is defined as $2m\omega$ and is the least amount of damping that will allow a displaced oscillator to return to its original position without oscillation. For most structures, the amount of viscous damping in the system will vary between 3% and 10% of critical. Substituting an upper value of 20% into the above expression for the damped circular frequency gives the result that $\omega_d = 0.98\omega$. Since the two values are approximately the same for values of damping found in structural systems, the undamped circular frequency is used in place of the damped circular frequency.

One of the more useful results of the free-vibration response is the estimation of the damping characteristics of a structure. If a structure is set in motion by some external force which is then removed, the amplitude will decay exponentially with time as shown in Figure 3-9. It can further be shown that the ratio

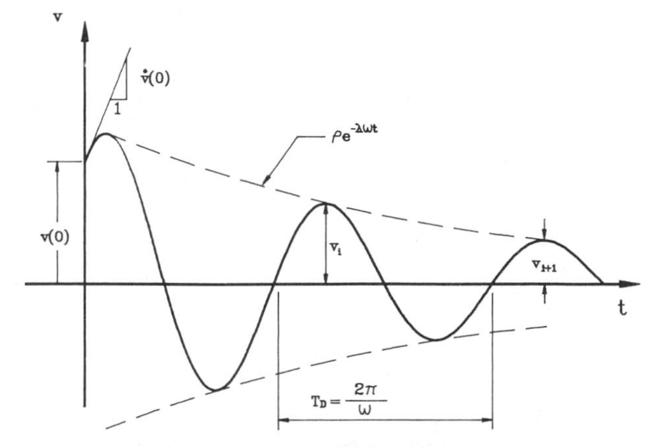

Figure 3-9 Free-vibration response of a damped SDOF system.

between any two successive amplitude peaks can be approximated by the expression

$$\frac{v(i)}{v(i+1)} = e^{2\pi\lambda} \tag{3-23}$$

Taking the natural logarithm of both sides results in

$$\delta = \ln \frac{v(i)}{v(i+1)} = 2\pi\lambda \tag{3-24}$$

where the parameter δ is called the *logarithmic decrement*. Solving for the percentage of critical damping, λ, gives

$$\lambda \approx \delta/2\pi \tag{3-25}$$

The above equation provides one of the more useful means of experimentally estimating the damping characteristics of a structure.

3.3.5 Response to Impulse Loading

In order to develop a method for evaluating the response of a structural system to a general dynamic loading, it is convenient to first consider the response of a structure to a short-duration impulse load as shown in Figure 3-10. If the duration of the applied impulse load, t, is short relative to the fundamental period of vibration of the structure, T, then the effect of the impulse can be considered as an incremental change in velocity. Using the impulse–momentum relationship, which states that the impulse is equal to the change in momentum,

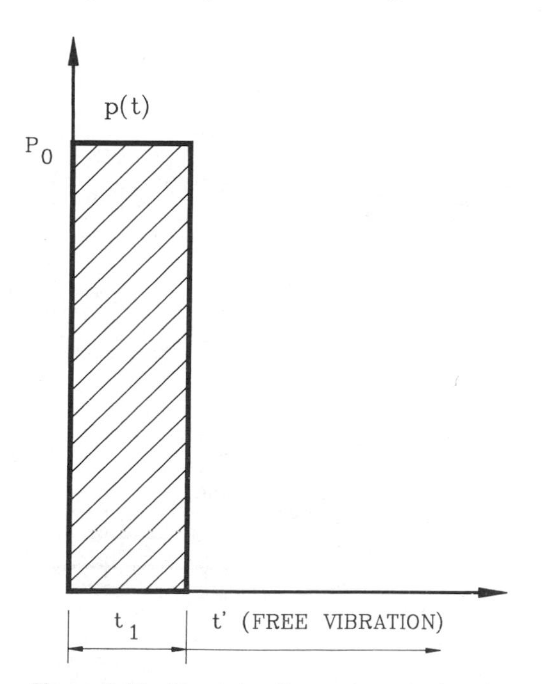

Figure 3-10 Short duration rectangular impulse.

the following equation is obtained:

$$\dot{v}(t) = \frac{1}{m}\int_0^t p(t)\,dt \tag{3-26}$$

Following the application of the short-duration impulse load, the system is in free vibration and the response is given by Equation 3-19. Applying the initial conditions at the beginning of the free-vibration phase,

$$\dot{v}(t_1) = \frac{1}{m}\int_0^{t_1} p(t)\,dt, \qquad v(t_1) \text{ negligible}$$

Equation 3-19 becomes

$$v(t-t_1) = \frac{1}{m\omega}\int_0^{t_1} p(t)\,dt \sin\omega(t-t_1) \tag{3-27}$$

For a damped structural system, the free-vibration response is given by Equation 3-22. Applying the above initial conditions to Equation 3-22 results in the following equation for the damped response:

$$v(t-t_1) = \frac{1}{m\omega_d}\int_0^{t_1} p(t)\,dt\; e^{-\lambda\omega(t-t_1)} \times \sin\omega_d(t-t_1) \tag{3-28}$$

3.3.6 Response to General Dynamic Loading

The above discussion of the dynamic response to a short-duration impulse load can readily be expanded to produce an analysis procedure for systems subjected to an arbitrary loading time history. Any arbitrary time history can be represented by a series of short-duration impulses as shown in Figure 3-11. Consider one of these impulses which ends at time τ after the beginning of the time history and has a duration $d\tau$. The magnitude of this differential impulse is $p(\tau)\,d\tau$, and it produces a differential response which is given as

$$dv(\tau) = \frac{p(\tau)\sin\omega t'\,d\tau}{m\omega} \tag{3-29}$$

The time variable t' represents the free-vibration phase following the differential impulse loading and can be expressed as

$$t' = t - \tau \tag{3-30}$$

Substituting this expression into Equation 3-29 results in

$$dv(\tau) = \frac{p(\tau)\sin\omega(t-\tau)\,d\tau}{m\omega} \tag{3-31}$$

The total response can now be obtained by superimposing the incremental responses of all the differential

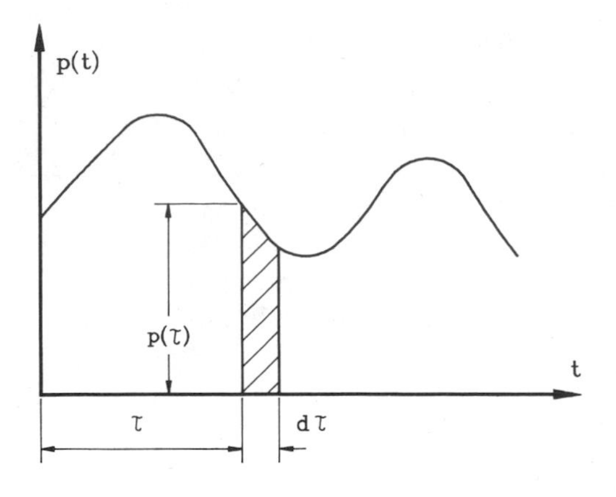

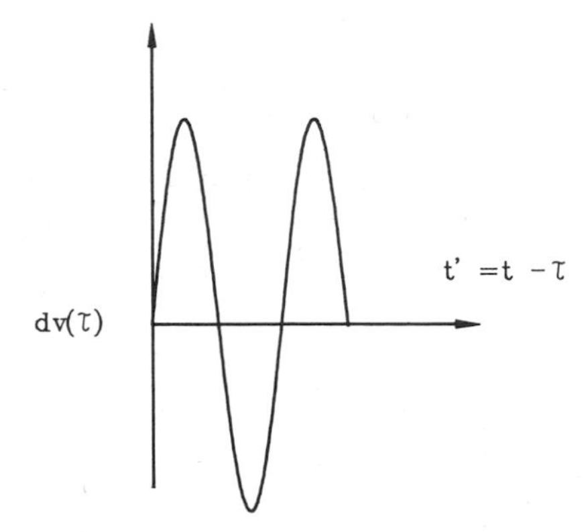

Figure 3-11 Differential impulse response.

impulses making up the time history. Integrating Equation 3-31, the total displacement response becomes

$$v(t) = \frac{1}{m\omega}\int_0^t p(\tau)\sin\omega(t-\tau)\,d\tau \qquad (3\text{-}32)$$

which is known as the Duhamel integral. When considering a damped structural system, the differential response is given by Equation 3-28 and the Duhamel-integral solution becomes

$$v(t) = \frac{p(\tau)e^{-\lambda\omega(t-\tau)}\sin\omega_d(t-\tau)\,d\tau}{m\omega_d} \qquad (3\text{-}33)$$

Since the principle of superposition was used in the derivation of Equations 3-32 and 3-33, the results are only applicable to linear structural systems. Furthermore, evaluation of the integral will require the use of numerical methods. For these two reasons, the use of a direct numerical integration procedure may be preferable for solving for the response of a dynamic system subjected to general dynamic load. This will be addressed in a later section on nonlinear response analysis. However, the Duhamel-integral result can be applied in a convenient and systematic manner to obtain a solution for the linear elastic structural response for earthquake load.

3.3.7 Earthquake Response of Elastic Structures

Time-History Response The response to earthquake loading can be obtained directly from the Duhamel integral if the time-dependent force $p(t)$ is replaced with the effective time-dependent force $p_e(t)$, which is the product of the mass and the ground acceleration. Making this substitution in Equation 3-33 results in the following expression for the displacement:

$$v(t) = \frac{V(t)}{\omega} \qquad (3\text{-}34)$$

where the response parameter $V(t)$ represents the velocity and is defined as

$$V(t) = \int_0^t \ddot{g}(\tau)e^{-\lambda\omega(t-\tau)}\sin\omega_d(t-\tau)\,d\tau \qquad (3\text{-}35)$$

The displacement of the structure at any instant of time during the entire time history of the earthquake under consideration can now be obtained using Equation 3-34. It is convenient to express the forces developed in the structure during the earthquake in terms of the effective inertia forces. The inertia force is the product of the

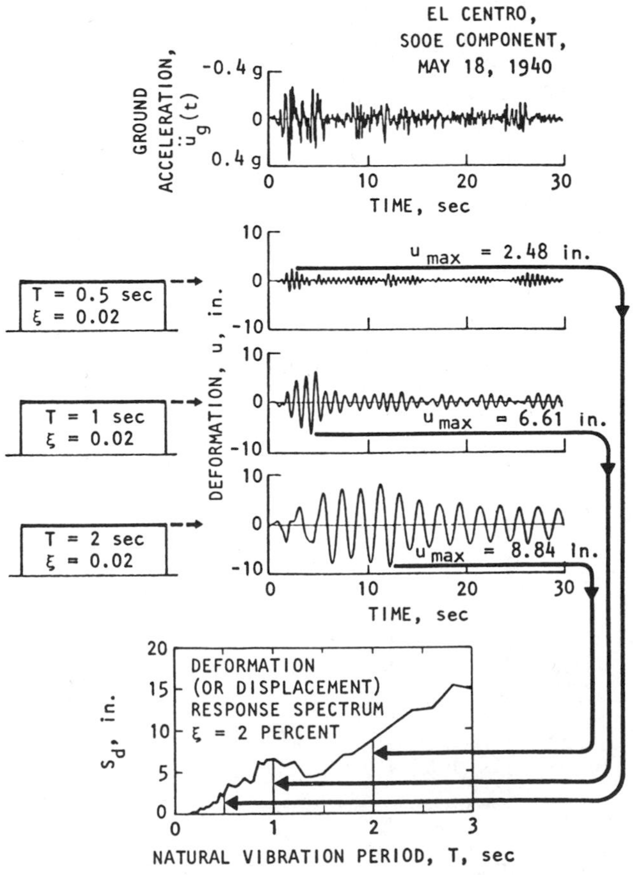

Figure 3-12 Computation of deformation (or displacement) response spectrum. [After Chopra (3-1)].

RESPONSE SPECTRUM

IMPERIAL VALLEY EARTHQUAKE

MAY 18, 1940 — 2037 PST

IIIA001 40.001.0 EL CENTRO SITE

IMPERIAL VALLEY IRRIGATION DISTRICT COMP SOOE

DAMPING VALUES ARE 0, 2, 5, 10, AND 20 PERCENT OF CRITICAL

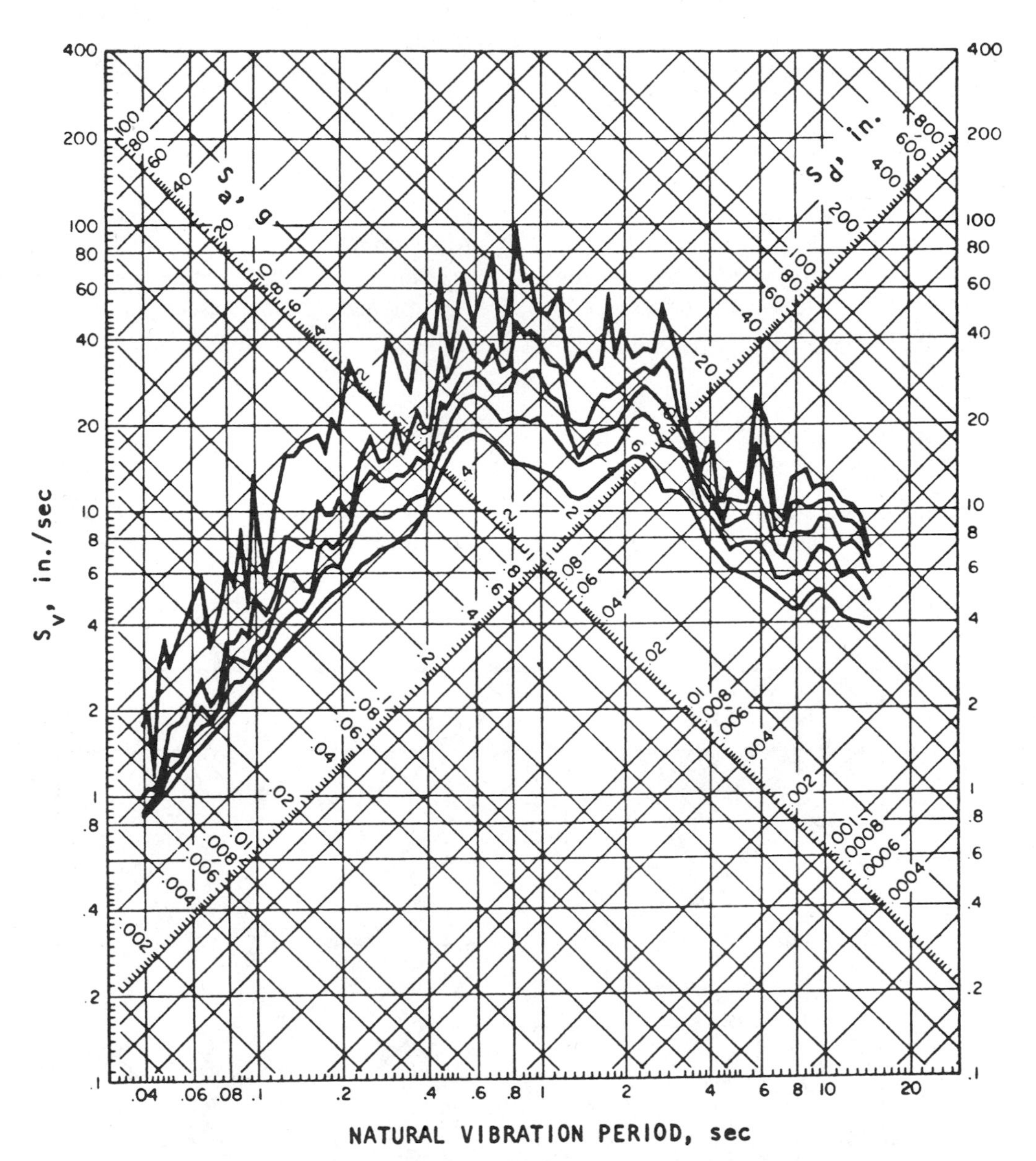

Figure 3-13 Typical tripartite response-spectra curves.

mass and the total acceleration. Using Equation 3-11, the total acceleration can be expressed as

$$\ddot{u}(t) = -\frac{c}{m}\dot{v}(t) - \frac{k}{m}v(t) \tag{3-36}$$

If the damping term can be neglected as contributing little to the equilibrium equation, the total acceleration can be approximated as

$$\ddot{u}(t) = -\omega^2 v(t) \tag{3-37}$$

The effective earthquake force is then given as

$$Q(t) = m\omega^2 v(t) \tag{3-38}$$

The above expression gives the value of the base shear in a single-story structure at every instant of time during the earthquake time history under consideration. The overturning moment acting on the base of the structure can be determined by multiplying the inertia force by the story height:

$$M(t) = hm\omega^2 v(t) \tag{3-39}$$

Response Spectra Consideration of the displacements and forces at every instant of time during an earthquake time history can require considerable computational effort, even for simple structural systems. As mentioned previously, for many practical problems and especially for structural design, only the maximum response quantities are required. The maximum value of the displacement, as determined by Equation 3-34, will be defined as the spectral displacement

$$S_d = v(t)_{max} \tag{3-40}$$

Substituting this result into Equations 3-38 and 3-39 results in the following expressions for the maximum base shear and maximum overturning moment in a SDOF system:

$$Q_{max} = m\omega^2 S_d \tag{3-41}$$

$$M_{max} = hm\omega^2 S_d \tag{3-42}$$

An examination of Equation 3-34 indicates that the maximum velocity response can be approximated by multiplying the spectral displacement by the circular frequency. This response parameter is defined as the *spectral pseudovelocity* and is expressed as

$$S_{pv} = \omega S_d \tag{3-43}$$

In a similar manner, Equation 3-37 indicates that the maximum total acceleration can be approximated as the spectral displacement multiplied by the square of the circular frequency. This product is defined as the *spec-*

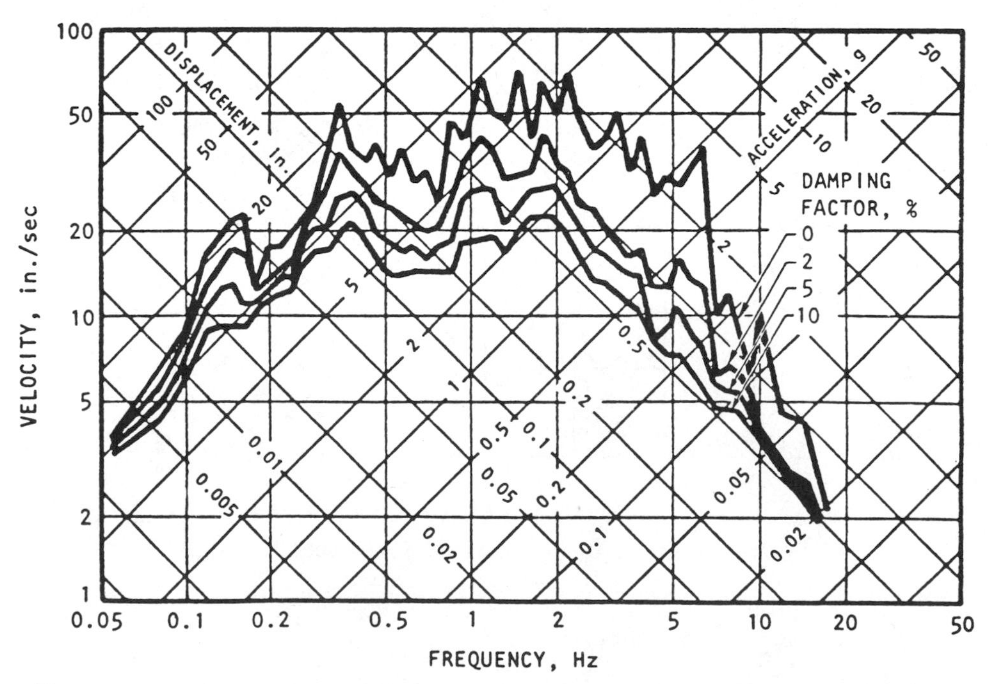

Figure 3-14 Response spectra, El Centro earthquake, May 18, 1940, north–south direction.

tral pseudoacceleration and is expressed as

$$S_{pa} = \omega^2 S_d \tag{3-44}$$

A plot of the spectral response parameter against frequency or period constitutes the response spectrum for that parameter. A schematic representation of the computation of the displacement spectrum for the north–south component of the motion recorded at El Centro on May 18, 1940 has been presented by Chopra[3-1] and is shown in Figure 3-12. Because the three response quantities are related to the circular frequency, it is convenient to plot them on a single graph with log scales on each axis. This special type of plot is called a tripartite log plot. The three response parameters for the El Centro motion are shown plotted in this manner in Figure 3-13. For a SDOF system having a given frequency (period) and given damping, the three spectral response parameters for this earthquake can be read directly from the graph.

Two types of tripartite log paper are used for plotting response spectra. Note that on the horizontal axis at the bottom of the graph in Figure 3-13, the period is increasing from left to right. For this reason, this type of tripartite log paper is often referred to as *period*

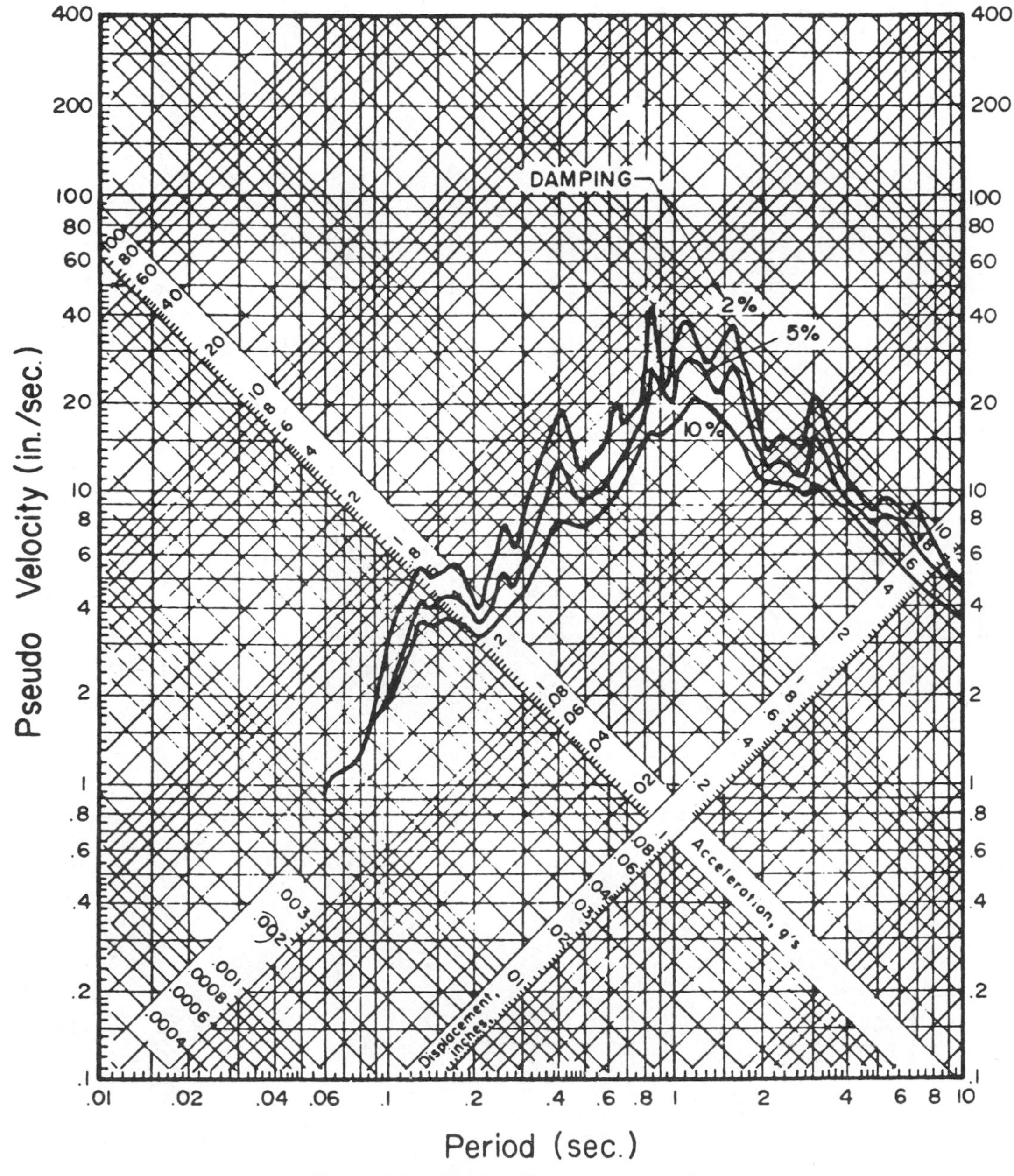

Figure 3-15 Site-specific response spectra.

paper. A similar plot of the response spectra for the El Centro N–S ground motion is shown in Figure 3-14. Here it can be seen that frequency, plotted on the horizontal axis, is increasing from left to right. This type of tripartite paper is referred to as *frequency* paper.

3.3.8 Design Response Spectra

Use of the elastic response spectra for a single component of a single earthquake record (Figure 3-13), while suitable for purposes of analysis, is not suitable for purposes of design. The design response spectra for a particular site should not be developed from a single acceleration time history, but rather should be obtained from the ensemble of possible earthquake motions that could be experienced at the site. This should include the effect of both near and distant earthquakes. Furthermore, a single earthquake record has a particular frequency content which gives rise to the jagged, sawtooth appearance of peaks and valleys shown in Figure 3-13. This feature is also not suitable for design, since for a given period, the structure may fall in a valley of the response spectrum and hence be underdesigned for an earthquake with slightly different response characteristics. Conversely, for a small change in period, the struc-

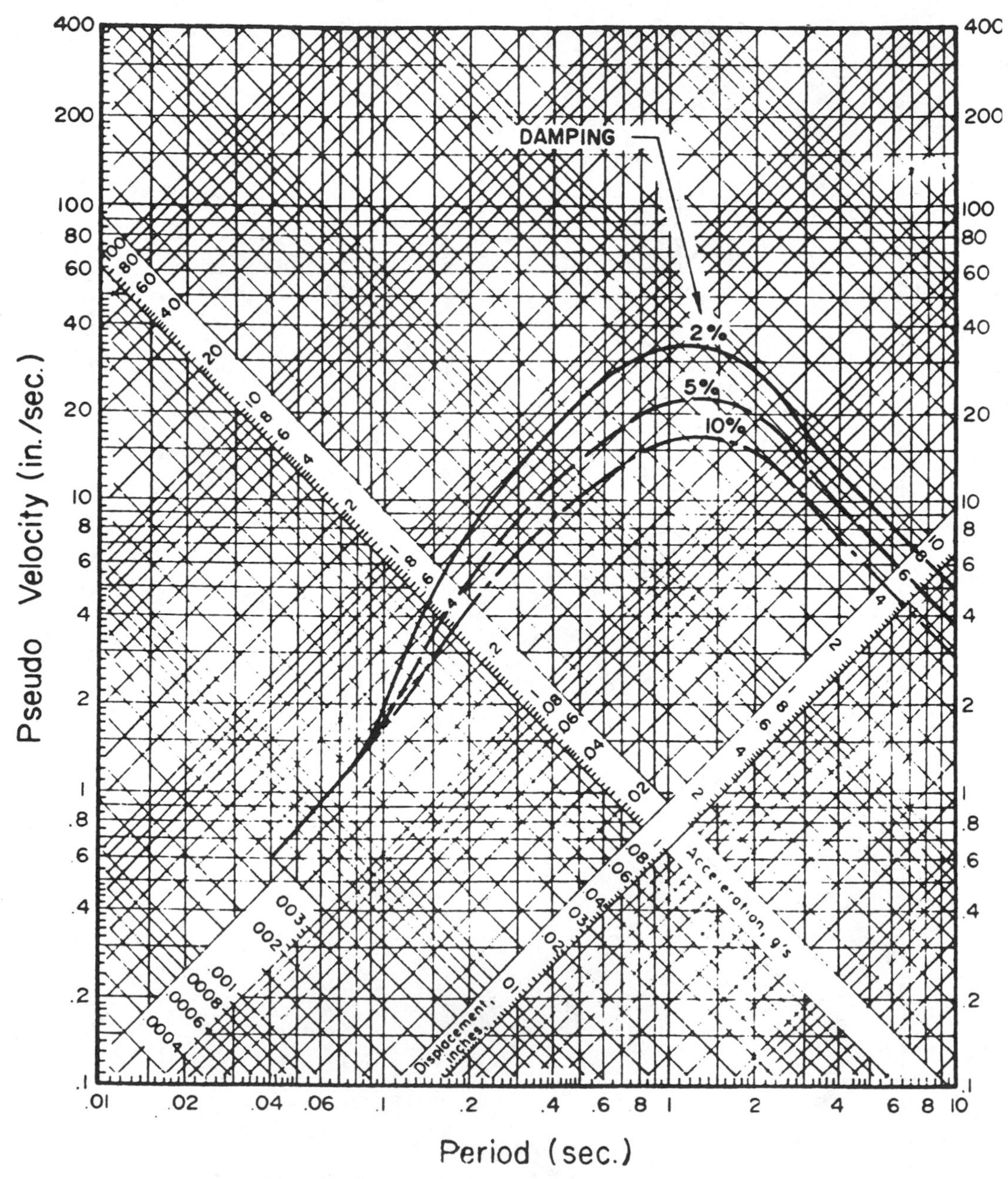

Figure 3-16 Smoothed site-specific design spectra.

ture might fall on a peak and be overdesigned. To alleviate this problem the concept of the smoothed response spectrum has been introduced for design. Statistics are used to create a smoothed spectrum at some suitable design level. The mean value or median spectrum can generally be used for earthquake-resistant design of normal building structures. Use of this spectrum implies there is a 50% probability that the design level will be exceeded. Structures that are particularly sensitive to earthquakes or that have a high risk may be designed to a higher level such as the mean plus one standard deviation, which implies that the probability of exceedance is only 15.9%. Structures having a very high risk are often designed for an enveloping spectrum which envelopes the spectra of the entire ensemble of possible site motions. Response spectra which are representative of a magnitude-6.5 earthquake at a distance of 15 miles, developed by the Applied Technology Council,(3-2) are shown in Figure 3-15. The corresponding smoothed design spectra are shown in Figure 3-16.

Table 3-1 Relative Values of Spectrum Amplification Factors(3-3)

Percentage of Critical Damping	Amplification Factor for		
	Displacement	Velocity	Acceleration
0	2.5	4.0	6.4
0.5	2.2	3.6	5.8
1	2.0	3.2	5.2
2	1.8	2.8	4.3
5	1.4	1.9	2.6
10	1.1	1.3	1.5
20	1.0	1.1	1.2

Newmark and Hall(3-3) have proposed a method for constructing an elastic design response spectrum in

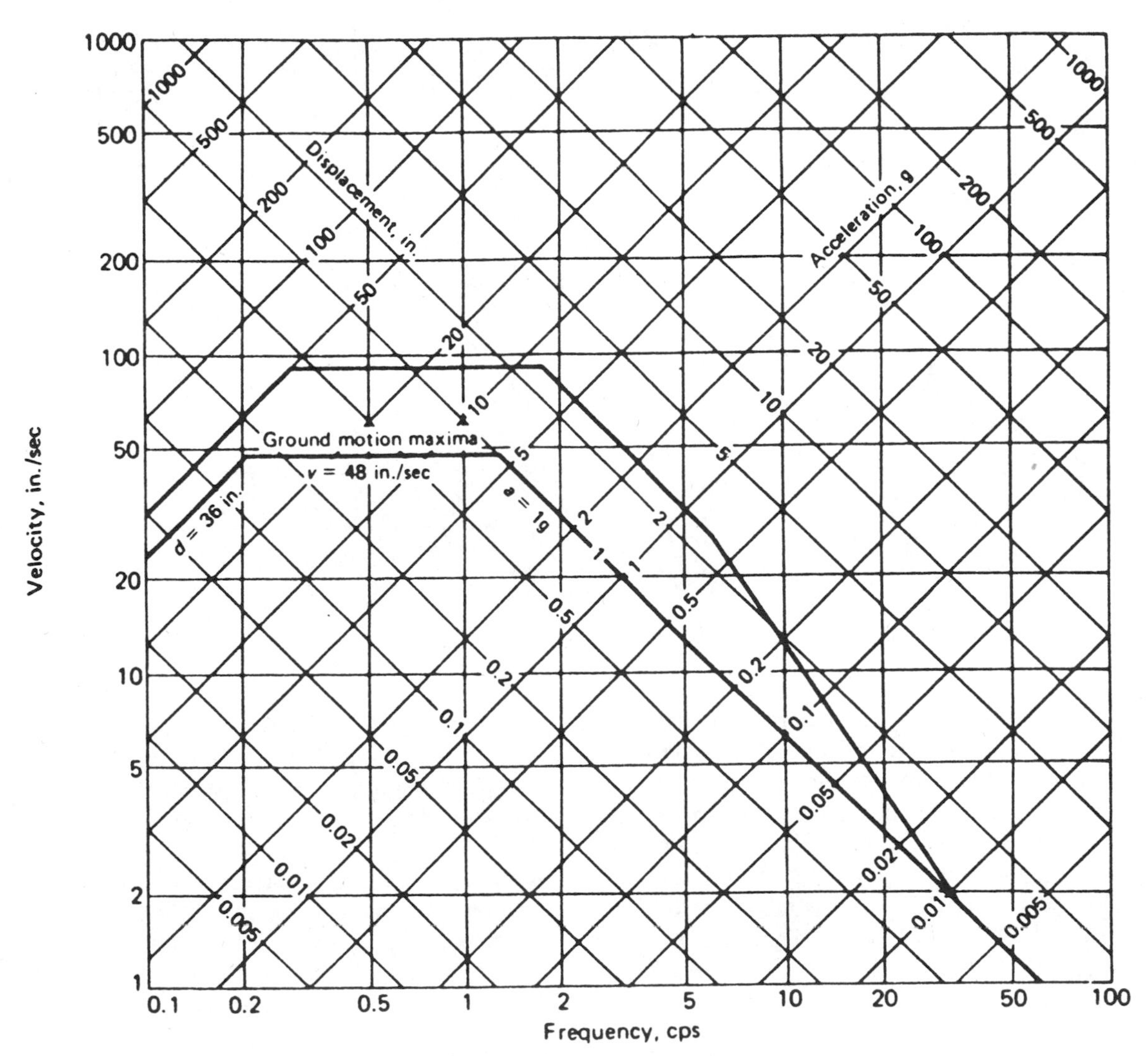

Figure 3-17 Basic Newmark–Hall design spectrum normalized to 1.0*g* for 5% damping (3-3).

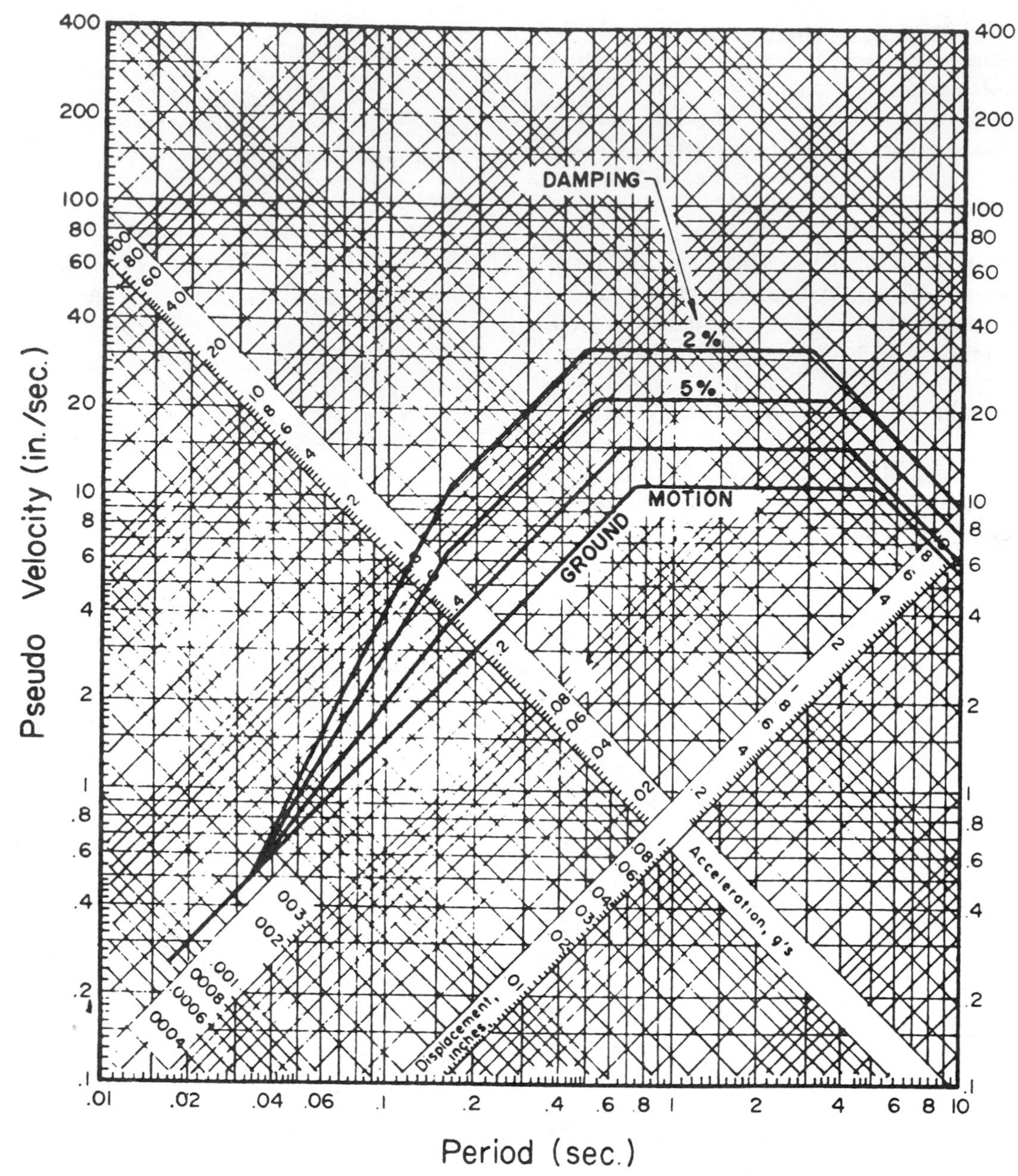

Figure 3-18 A Newmark – Hall design spectra.

which the primary input datum is the anticipated maximum ground acceleration. The corresponding values for the maximum ground velocity and the maximum ground displacement are proportioned relative to the maximum ground acceleration, which is normalized to 1.0g. The maximum ground velocity is taken as 48 in./sec, and the maximum ground displacement is taken as 36 in. It should be noted that these values represent motions which are more intense than those normally considered for earthquake-resistant design; however, they are approximately in the correct proportion for earthquakes occurring on competent soils and can be scaled for earthquakes having lower ground acceleration.

Three principal regions of the response spectrum are identified, in which the structural response is an approximately constant, amplified value. Amplification factors are applied to the ground motions in these three regions to obtain the design spectrum for a SDOF elastic system. Based on a large data base of recorded earthquake motions, amplification factors which give a probability of exceedance of about 10% or less are given in Table 3-1 for various values of the structural damping. The basic shape of the Newmark–Hall design spectrum using the normalized ground motions and the amplification factors given in Table 3-1 for 2% damping is shown in Figure 3-17. The displacement region is the low-

Table 3-2 Recommended Damping Values[(3-3)]

Stress Level	Type and Condition of Structure	Percentage of Critical Damping	Stress Level	Type and Condition of Structure	Percentage of Critical Damping
Working stress, $\lesssim \frac{1}{2}$ yield point	Vital piping	1–2	At or just below yield point	Vital piping	2–3
	Welded steel, prestressed concrete, well-reinforced concrete (only slight cracking)	2–3		Welded steel, prestressed concrete (without complete loss in prestress)	5–7
	Reinforced concrete with considerable cracking	3–5		Prestressed concrete with no prestress left	7–10
	Bolted and / or riveted steel, wood structures with nailed or bolted joints	5–7		Reinforced concrete	7–10
				Bolted and / or riveted steel, wood structures with bolted joints	10–15
				Wood structures with nailed joints	15–20

frequency region with frequencies less that 0.33 Hz (periods greater than 3.0 sec). The maximum displacement of the SDOF system is obtained by multiplying the maximum ground displacement by the displacement amplification factor given in Table 3-1. The velocity region is in the mid-frequency region between 0.33 Hz (3.0 sec) and 2.0 Hz (0.5 sec). Maximum velocities in this region are obtained by multiplying the maximum ground velocity by the amplification factor for the velocity (Table 3-1). An amplified acceleration region lies between 2.0 Hz (0.5 sec) and 6.0 Hz (0.17 sec). The amplified response is obtained in the same manner as in the previous two cases. Structures having a frequency greater than 30 Hz (period less than 0.033 sec) are considered to be rigid and have an acceleration which is equal to the ground acceleration. In the frequency range between 6 Hz (0.17 sec) and 30 Hz (0.033 sec) there is a transition region between the ground acceleration and the amplified acceleration region.

Similar design spectra corresponding to the postulated ground motion presented in Figures 3-15 and 3-16 are shown in Figure 3-18. In order to further define which response spectrum should be used for design, it is necessary to estimate the percentage of critical damping in the structure. A summary of recommended damping values[(3-3)] for different types of structures and different stress conditions is given in Table 3-2 as a guideline.

Example 3-3 (Construction of a Newmark–Hall Design Spectrum)

Construct a Newmark–Hall design spectrum for a maximum ground acceleration of $0.2g$, and use it to estimate the maximum base shear for the industrial building of Example 3-1. Assume the damping is 5 percent of critical.

- Determine ground motion parameters:

$$\text{ground acceleration} = (1.0)(0.2) = 0.2g$$
$$\text{ground velocity} = (48.0)(0.2) = 9.6 \text{ in./sec}$$
$$\text{ground displacement} = (36.0)(0.2) = 7.2 \text{ in.}$$

- Amplified response parameters:

$$\text{acceleration} = (0.2)(2.6) = 0.52g$$
$$\text{velocity} = (9.6)(1.9) = 18.2 \text{ in./sec}$$
$$\text{displacement} = (7.2)(1.4) = 10.0 \text{ in.}$$

The constructed design spectrum is shown in Figure 3-19.

From Example 3-1:

N–S:

$T = 0.287$ sec, $\omega = 21.8$ rad/sec, $f = 3.48$ Hz

From the design spectrum for $f = 3.48$ Hz:

$$S_d = v(t)_{max} = 0.42 \text{ in.}$$

From Equation 3-42:

$$Q_{max} = (0.485)(21.8)^2(0.42) = 96.8 \text{ kips}$$

E–W:

$T = 0.23$ sec, $\omega = 27.2$ rad/sec, $f = 4.3$ Hz

From the design spectrum for $f = 4.3$ Hz:

$$S_d = 0.28 \text{ in.}$$

From Equation 3-42:

$$Q_{max} = (0.485)(21.8)^2(0.28) = 64.5 \text{ kips}$$

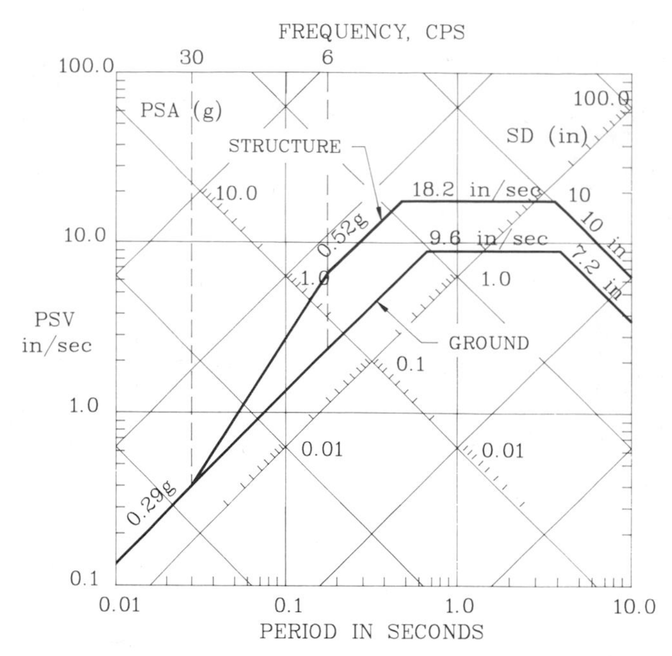

Figure 3-19 Response spectrum of Example 3-3.

3.4 GENERALIZED-COORDINATE APPROACH

Up to this point, the only structures which have been considered are single-story buildings which can be idealized as SDOF systems. The analysis of most structural systems requires a more complicated idealization even if the response can be represented in terms of a single degree of freedom. The generalized-coordinate approach provides a means of representing the response of more complex structural systems in terms of a single, time-dependent coordinate, known as the generalized coordinate.

Displacements in the structure are related to the generalized coordinate as

$$v(x, t) = \phi(x) Y(t) \tag{3-45}$$

where $Y(t)$ is the time-dependent generalized coordinate and $\phi(x)$ is a spatial shape function which relates the structural degrees of freedom, $v(x, t)$, to the generalized coordinate. For a generalized SDOF system, it is necessary to represent the restoring forces in the damping elements and the stiffness elements in terms of the relative velocity and relative displacement between the ends of the element:

$$\Delta \dot{v}(x, t) = \Delta \phi(x) \dot{Y}(t) \tag{3-46}$$

$$\Delta v(x, t) = \Delta \phi(x) Y(t) \tag{3-47}$$

Most structures can be idealized as a vertical cantilever, which limits the number of displacement functions that can be used to represent the horizontal displacement. Once the displacement function is selected, the structure is constrained to deform in that prescribed manner. This implies that the displacement functions must be selected carefully if a good approximation of the dynamic properties and response of the system are to be obtained. This section will develop the equations for determining the generalized response parameters in terms of the spatial displacement function and the physical response parameters. Methods for determining the shape function will be discussed, and techniques for determining the more correct displacement function for a particular structure will be presented.

3.4.1 Displacement Functions and Generalized Properties

Formulation of the equation of motion in terms of a generalized coordinate will be restricted to systems which consist of an assemblage of lumped masses and discrete elements. Lateral resistance is provided by discrete elements whose restoring force is proportional to the relative displacement between the ends of the element. Damping forces are proportional to the relative velocity between the ends of the discrete damping element. Formulation of the equation of motion for systems having distributed elasticity is described by Clough and Penzien.[3-4] The general equation of dynamic equilibrium is given in Equation 3-6, which represents a system of forces which are in equilibrium at any instant of time. The *principle of virtual work* in the form of virtual displacements states that

> If a system of forces which are in equilibrium is given a virtual displacement which is consistent with the boundary conditions, the work done is zero.

Applying this principle to Equation 3-6 results in an equation of virtual work in the form

$$f_{\mathrm{d}}\, \delta v + f_{\mathrm{d}}\, \delta \Delta v + f_{\mathrm{s}}\, \delta \Delta v - p(t)\, \delta v = 0 \tag{3-48}$$

where it is understood that $v = v(x, t)$ and that the virtual displacements applied to the damping force and the elastic restoring force are virtual relative displacements. The virtual displacement can be expressed as

$$\delta v(x, t) = \phi(x)\, \delta Y(t) \tag{3-49}$$

and the virtual relative displacement can be written as

$$\delta \Delta v(x, t) = \Delta \phi(x)\, \delta Y(t) \tag{3-50}$$

where

$$\Delta v(x, t) = \phi(x_i) Y(t) - \phi(x_j) Y(t) = \Delta \phi(x) Y(t).$$

The inertia, damping and elastic restoring forces can be

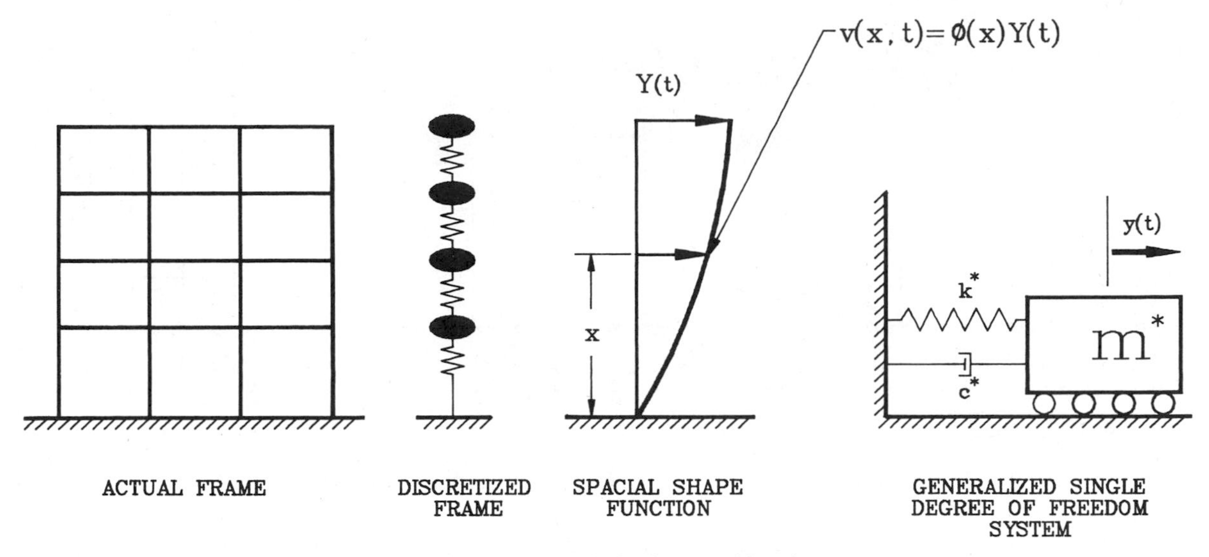

Figure 3-20 Generalized single-degree-of-freedom system.

expressed as

$$f_i = m\ddot{v} = m\phi\ddot{Y}$$
$$f_d = c\,\Delta\dot{v} = c\,\Delta\phi\,\dot{Y} \qquad (3\text{-}51)$$
$$f_s = k\,\Delta v = k\,\Delta\phi\,Y$$

Substituting Equations 3-49, 3-50, and 3-51 into Equation 3-48 results in the following equation of motion in terms of the generalized coordinate:

$$m^*\ddot{Y} + c^*\dot{Y} + k^*Y = p^*(t) \qquad (3\text{-}52)$$

where m^*, c^*, k^*, and p^* are referred to as the *generalized parameters* and are defined as

$$m^* = \sum_i m_i\phi_i^2 = \text{generalized mass}$$
$$c^* = \sum_i c_i\,\Delta\phi_i^2 = \text{generalized damping}$$
$$k^* = \sum_i k_i\,\Delta\phi_i^2 = \text{generalized stiffness} \qquad (3\text{-}53)$$
$$p^* = \sum_i p_i\phi_i = \text{generalized force}$$

For a time-dependent base acceleration the generalized force becomes

$$p^* = \ddot{g}\mathscr{L} \qquad (3\text{-}54)$$

where

$$\mathscr{L} = \sum_i m_i\phi_i$$
$$= \text{earthquake participation factor} \qquad (3\text{-}55)$$

It is also convenient to express the generalized damping in terms of the percent of critical damping in the following manner:

$$c^* = \sum_i c_i\,\Delta\phi(i)^2 = 2\lambda m^*\omega \qquad (3\text{-}56)$$

where ω represents the circular frequency of the generalized system and is given as

$$\omega = \sqrt{k^*/m^*} \qquad (3\text{-}57)$$

The effect of the generalized-coordinate approach is to transform a multiple-degree-of-freedom dynamic system into an equivalent single-degree-of-freedom system in terms of the generalized coordinate. This transformation is shown schematically in Figure 3-20. The degree to which the response of the transformed system represents the actual system will depend upon how well the assumed displacement shape represents the dynamic displacement of the actual structure. The displacement shape depends on the aspect ratio of the structure, which is defined as the ratio of the height to the base dimension. Possible shape functions for high-rise, mid-rise, and low-rise structures are summarized in Figure 3-21. It should be noted that most building codes use the straight-line shape function which is shown for the mid-rise system. Once the dynamic response is obtained in terms of the generalized coordinate, Equation 3-45 must be used to determine the displacements in the structure, and these in turn can be used to determine the forces in the individual structural elements.

In principle, any function which represents the general deflection characteristics of the structure and satisfies the support conditions could be used. However, any shape other than the true vibration shape requires the addition of external constraints to maintain equilib-

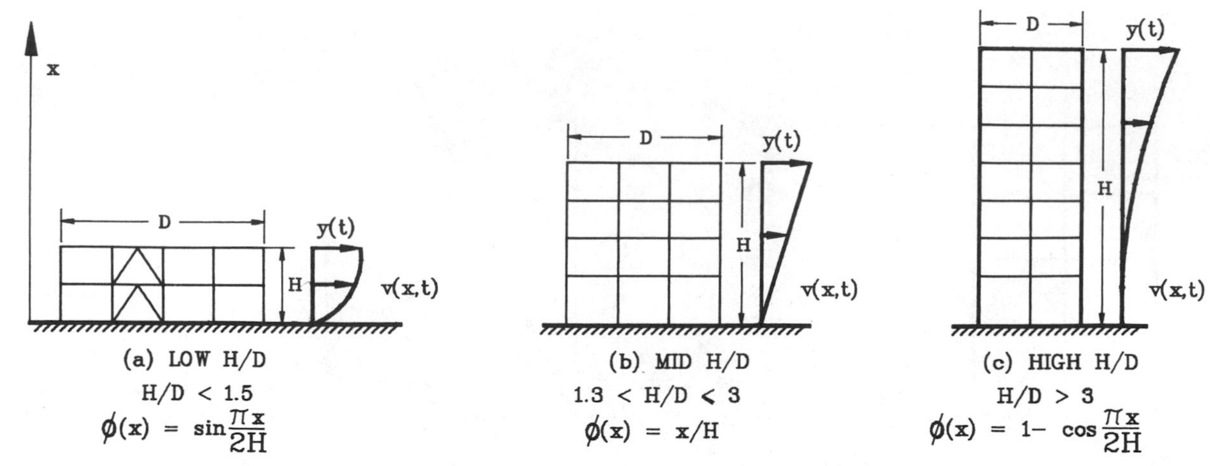

Figure 3-21 Possible shape functions based on aspect ratio.

rium. These extra constraints tend to stiffen the system and thereby increase the computed frequency. The true vibration shape will have no external constraints and therefore will have the lowest frequency of vibration. When choosing between several approximate deflected shapes, the one producing the lowest frequency is always the best approximation. A good approximation to the true vibration shape can be obtained by applying forces representing the inertia forces and letting the static deformation of the structure determine the spatial shape function.

Example 3-4 (Determination of Generalized Parameters)

Considering the four-story, reinforced-concrete moment frame building shown in Figure 3-22, determine the generalized mass, generalized stiffness, and fundamental period of vibration in the transverse direction using the following shape functions: (a) $\phi(x) = \sin(\pi x/2L)$ and (b) $\phi(x) = x/L$. All beams are 12 in. × 20 in., and all columns are 14 in. × 14 in. $f_c' = 4000$ psi, and the modulus of elasticity of concrete is 3.6×10^6 psi. Reinforcing steel is made of grade-60 bars. Floor weights (total dead load) are assumed to be 390 kips at the roof, 445 kips at the fourth and third levels, and 448 kips at the first level. Live loads are 30 psf at the roof and 80 psf per typical floor level.

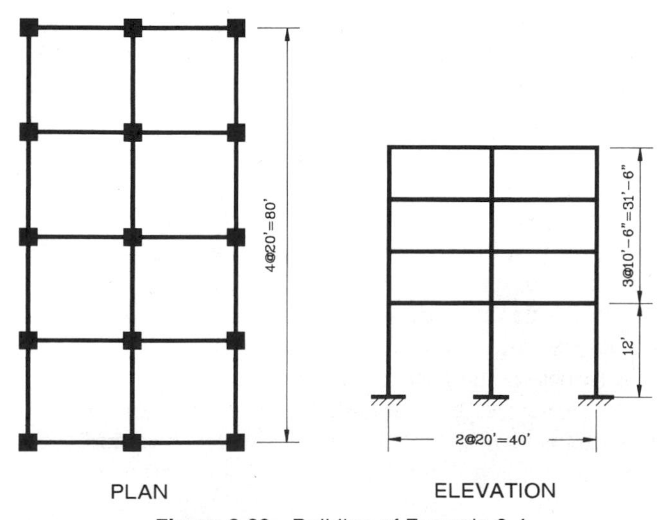

Figure 3-22 Building of Example 3-4.

Assuming beams are rigid relative to columns (Figure 3-23),

$$V = 12EI\Delta/L^3$$

$$K_i = V/\Delta = 12EI/L^3$$

$$I_{col} = \frac{14(14)^3}{12} = 3201 \text{ in.}^4$$

$$I_{beam} = \frac{12(20)^3}{12} = 8000 \text{ in.}^4$$

$$K_{story} = \sum_{i=1}^{3} K_i = 3K_i \text{ (one frame)}$$

$$K_{4,3,2} = \frac{(3)(12)(3.6 \times 10^3)(3201)}{(126)^3} = 209 \text{ kips/in.}$$

$$K_1 = \frac{(3)(12)(3.6 \times 10^3)(3201)}{(144)^3} = 140 \text{ kips/in.}$$

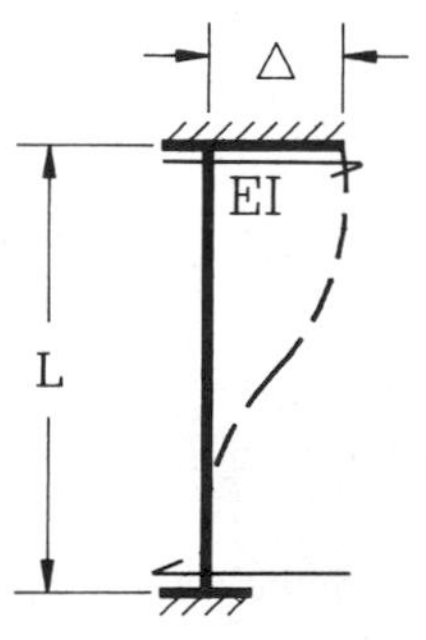

Figure 3-23 Assumed shape of column deformation.

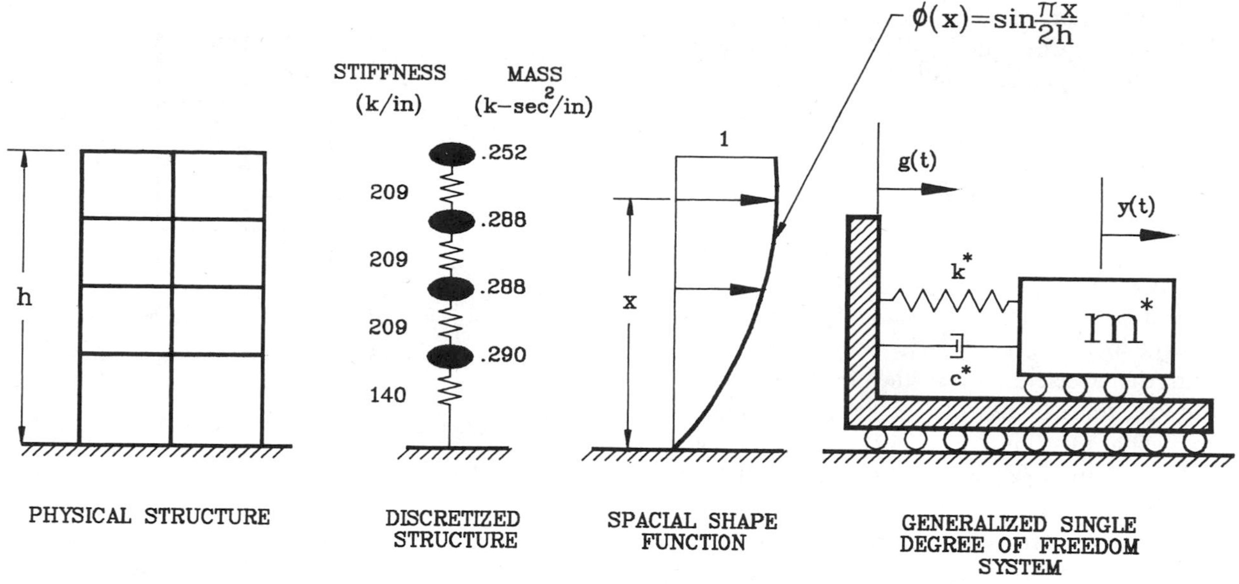

Figure 3-24 Development of a generalized SDOF model for building of Example 3-4.

Calculating generalized properties (see Figure 3-24):
(a) Assuming $\phi(x) = \sin(\pi x/2L)$:

Level	K	M	ϕ_i	$\Delta\phi_i$	$M\phi_i^2$	$K\,\Delta\phi_i^2$
4	209	0.252	1.000	0.071	0.252	1.054
3	209	0.288	0.929	0.203	0.249	8.613
2	209	0.288	0.726	0.306	0.152	19.570
1	140	0.290	0.420	0.420	0.051	24.696
					$M^* = 0.704$	$K^* = 53.933$

$$\omega = \sqrt{\frac{k^*}{m^*}} = \sqrt{\frac{53.93}{0.704}} = 8.75 \text{ rad/sec and } T_a = 0.72 \text{ sec}$$

(b) Assuming $\phi(x) = x/L$:

Level	K	M	ϕ_i	$\Delta\phi_i$	$M\phi_i^2$	$K\,\Delta\phi_i^2$
4	209	0.252	1.000	0.241	0.252	12.139
3	209	0.288	0.759	0.242	0.166	12.240
2	209	0.288	0.517	0.241	0.077	12.139
1	140	0.290	0.276	0.276	0.022	10.665
					$M^* = 0.517$	$K^* = 47.183$

$$\omega = \sqrt{\frac{k^*}{m^*}} = \sqrt{\frac{47.183}{0.517}}$$

$$= 9.55 \text{ rad/sec} \quad \text{and} \quad T_b = 0.66 \text{ sec}$$

Since $T_a > T_b$, $\phi(x) = \sin(\pi x/2L)$ is a better approximation to the deflected shape than $\phi(x) = x/L$.

3.4.2 Rayleigh's Method

Rayleigh's method is a procedure developed by Lord Rayleigh[3-5] for analyzing vibrating systems using the law of conservation of energy. Its principal use is for determining an accurate approximation of the natural frequency of a structure. The success of the technique in accomplishing this has been recognized by most building codes, which have adopted the procedure as an alternative for estimating the fundamental period of vibration. In addition to providing an estimate of the fundamental period, the procedure can also be used to estimate the shape function $\phi(x)$.

In an undamped elastic system, the maximum potential energy can be expressed in terms of the external work done by the applied forces. In terms of a generalized coordinate this expression can be written as

$$(\text{PE})_{\max} = \frac{Y}{2}\sum p_i\phi_i = \frac{p^*Y}{2} \tag{3-58}$$

Similarly, the maximum kinetic energy can be expressed in terms of the generalized coordinate as

$$(\text{KE})_{\max} = \frac{\omega^2 Y^2}{2}\sum_i m_i\phi_i^2 = \frac{\omega^2 Y^2 m^*}{2} \tag{3-59}$$

According to the principle of conservation of energy for an undamped elastic system, these two quantities must be equal to each other and to the total energy of the system. Equating Equation 3-58 to Equation 3-59 results in the following expression for the circular frequency:

$$\omega = \sqrt{p^*/m^*Y} \tag{3-60}$$

Substituting this result into Equation 3-20 for the period results in

$$T = 2\pi\sqrt{m^*Y/p^*} \tag{3-61}$$

Multiplying the numerator and denominator of the radical by Y and using Equation 3-45 results in the expression for the fundamental period:

$$T = 2\pi\sqrt{\frac{\Sigma_i w_i v_i^2}{g\Sigma p_i v_i}} \tag{3-62}$$

which is the expression found in most building codes.

The forces which must be applied laterally to obtain either the shape function $\phi(x)$ or the displacement $v(x)$ represent the inertia forces, which are the product of the mass and the acceleration. If the acceleration is assumed to vary linearly over the height of a building with uniform weight distribution, a distribution of inertia force in the form of an inverted triangle will be obtained, being maximum at the top and zero at the bottom. This is similar to the distribution of base shear used in most building codes and can be a reasonable one to use when applying the Rayleigh method. The resulting deflections can be used directly in Equation 3-62 to estimate the period of vibration or they can be normalized in terms of the generalized coordinate (maximum displacement) to obtain the spatial shape function to be used in the generalized-coordinate method.

Example 3-5 (Application of Rayleigh's Method) Use Rayleigh's method to determine the spatial shape function and estimate the fundamental period of vibration in the transverse direction for the reinforced-concrete building given in Example 3-4.

We want to apply static lateral loads that are representative of the inertial loads on the building. Since the story weights are approximately equal, it is assumed that the accelerations and hence the inertial loads vary linearly from the base to the roof (see Figure 3-25).

Note that the magnitude of loads is irrelevant and is chosen for ease of computation. The following computations are a tabular solution of Equation 3-61:

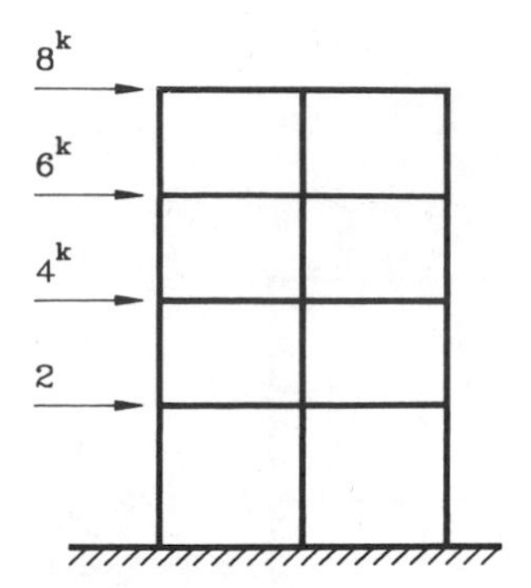

Figure 3-25 Frame of Example 3-5.

3.4.3 Earthquake Response of Elastic Structures

Time-History Analysis Substituting the generalized parameters of Equations 3-53 and 3-54 into the Duhamel-integral solution, Equation 3-33, results in the following solution for the displacement:

$$v(x,t) = \frac{\phi(x)\mathscr{L}V(t)}{m^*\omega} \tag{3-63}$$

Using Equation 3-37, the inertia force at any position x above the base can be obtained from

$$q(x,t) = m(x)\ddot{v}(x,t) = m(x)\omega^2 v(x,t) \tag{3-64}$$

which, using Equation 3-63, becomes

$$q(x,t) = \frac{m(x)\phi(x)\mathscr{L}\omega V(t)}{m^*} \tag{3-65}$$

The base shear is obtained by summing the distributed inertia forces over the height H of the structure:

$$Q(t) = \int q(x,t)\,dx = \frac{\mathscr{L}^2}{m^*}\omega V(t) \tag{3-66}$$

Level	k	m	P	V	$\Delta = V/k$	v	ϕ	$m_i\phi_i^2$	$P_i\phi_i$
4		0.252	8.0			0.3343	1.000	0.252	8.000
	209			8	0.0383				
3		0.288	6.0			0.2960	0.886	0.226	5.316
	209			14	0.0670				
2		0.288	4.0			0.2290	0.685	0.135	2.740
	209			18	0.0861				
1		0.288	2.0			0.1429	0.428	0.053	0.856
	140			20	0.1429				
						0.000	0.000		
								0.666	16.912

$$T = 2\pi\sqrt{m^*Y/p^*}, \quad \text{or}$$

$$T = 2\pi\sqrt{\frac{(0.666)(0.3343)}{16.912}} = 0.712 \text{ sec}$$

Note that since $T = 0.721$ is greater than either of the periods calculated in Example 3-4, the deflected shape given by applying the static loads is a better approximation than either of the two previous deflected shapes.

The above relationships can be used to determine the displacements and forces in a generalized SDOF system at any time during the time history under consideration.

Response-Spectrum Analysis The maximum value of the velocity given by Equation 3-35 is defined as the spectral pseudovelocity (S_{pv}), which is related to the spectral displacement (S_d) by Equation 3-43. Substituting this value into Equation 3-63 results in an expression for the maximum displacement in terms of the spectral displacement:

$$v(x)_{max} = \frac{\phi(x)\mathscr{L}S_d}{m^*} \tag{3-67}$$

The forces in the system can readily be determined from the inertia forces, which can be expressed as

$$q(x)_{max} = m(x)\ddot{v}(x)_{max} = m(x)\omega^2 v(x)_{max} \tag{3-68}$$

Rewriting this result in terms of the spectral pseudo-acceleration (S_{pa}) results in the following:

$$q(x)_{max} = \frac{\phi(x)m(x)\mathscr{L}S_{pa}}{m^*} \tag{3-69}$$

Of considerable interest to structural engineers is the determination of the base shear. This is a key parameter in determining seismic design forces in most building codes. The base shear Q can be obtained from the above expression by simply summing the inertia forces and using Equation 3-55:

$$Q_{max} = \frac{\mathscr{L}^2 S_{pa}}{m^*} \tag{3-70}$$

It is also of interest to express the base shear in terms of the effective weight, which is defined as

$$W^* = \frac{(\Sigma_i w_i \phi_i)^2}{\Sigma_i w_i \phi_i^2} \tag{3-71}$$

The expression for the maximum base shear becomes

$$Q_{max} = W^* S_{pa}/g \tag{3-72}$$

This form is similar to the basic base-shear equation used in the building codes. In the code equation, the effective weight is taken to be equal to the total dead weight W, plus a percentage of the live load for special occupancies. The seismic coefficient C is determined by a formula but is equivalent to the spectral pseudoacceleration in terms of g. The basic code equation for base shear has the form

$$Q_{max} = CW \tag{3-73}$$

The effective earthquake force can also be determined by distributing the base shear over the story height. This distribution depends upon the displacement shape function and has the form

$$q_i = Q_{max}\frac{m_i\phi_i}{\mathscr{L}} \tag{3-74}$$

If the shape function is taken as a straight line, the code force distribution is obtained. The overturning moment at the base of the structure can be determined by multiplying the inertia force by the corresponding story height above the base and summing over all story levels:

$$M_o = \sum_i h_i q_i \tag{3-75}$$

Example 3-6 (Spectrum Analysis of Generalized SDOF System)

Using the design spectrum given in Figure 3-26, the shape function determined in Example 3-5, and the reinforced-concrete moment frame of Example 3-4, determine the base shear in the transverse direction, the corresponding distribution of inertia forces over the height of the structure, and the resulting overturning moment about the base of the structure.

$$T = 0.721 \text{ sec}, \qquad f = 1/T = 1.39 \text{ Hz},$$
$$\omega = 8.715 \text{ rad/sec}$$

From the design spectrum $S_{pa} = 0.185g$.

Level	m_i	ϕ_i	$m_i\phi_i^2$	$m_i\phi_i$	$m_i\phi_i/\mathscr{L}$	q_{max}	V_{max}
4	0.252	1.000	0.252	0.252	0.305	27.10	27.10
3	0.288	0.866	0.226	0.255	0.308	27.36	54.46
2	0.288	0.685	0.135	0.197	0.238	21.14	75.60
1	0.288	0.428	0.053	0.123	0.149	13.24	88.84
			0.666	0.827			

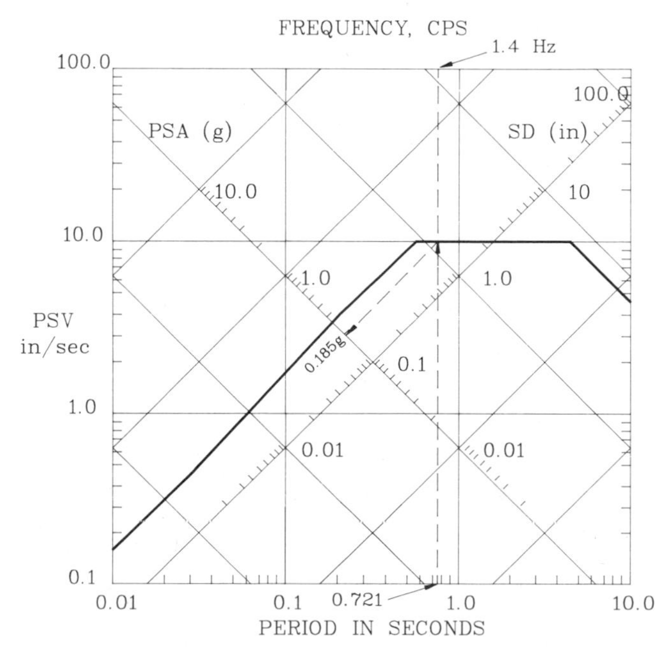

Figure 3-26 Design spectrum for Example 3-6.

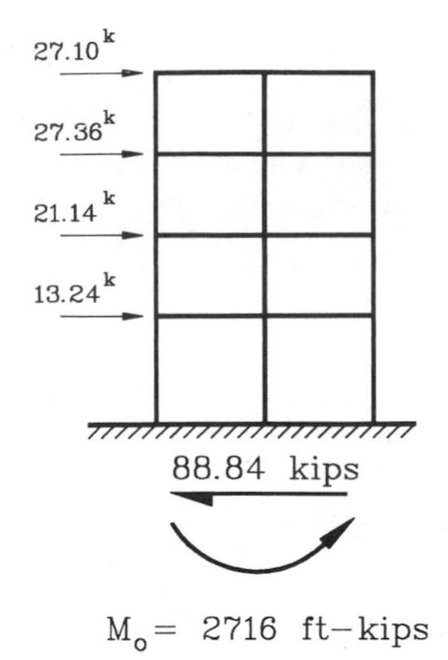

Figure 3-27 Story shears and overturning moment (Example 3-6).

From Equation 3-66,

$$Q_{max} = \frac{(0.827)^2(0.185)(386.4)}{0.666} = 88.84 \text{ kips}$$

The overturning moment is (see Figure 3-27)

$$\begin{aligned} M_o &= 27.10(43.5) + 27.36(33) \\ &\quad + 21.14(22.5) + 13.24(12) \\ &= 2716 \text{ ft-kips} \end{aligned}$$

The displacement is

$$v_{max} = \phi(\mathscr{L}/m^*)S_d = \phi\alpha S_d$$

where

$$S_d = S_{pa}/\omega^2 \quad \text{and} \quad \alpha = \mathscr{L}/m^*$$

$$S_d = \frac{(0.185)(386.4)}{(8.715)^2} = 0.941$$

$$\alpha = \frac{0.827}{0.666} = 1.242$$

$$v_i = (1.242)(0.941)\phi_i = 1.168\phi_i$$

$$v_4 = 1.168 \text{ in.} \qquad v_3 = 1.035 \text{ in.}$$

$$v_2 = 0.80 \text{ in.} \qquad v_1 = 0.50 \text{ in.}$$

3.5 RESPONSE OF NONLINEAR SDOF SYSTEMS

In an earlier section it was shown that the response of a linear structural system could be evaluated using the Duhamel integral. The approach was limited to linear systems because the Duhamel-integral approach makes use of the principle of superposition in developing the method. In addition, evaluation of the Duhamel integral for earthquake input motions will require the use of numerical methods in evaluating the integral. For these reasons it may be more expedient to use numerical integration procedures directly for evaluating the response of linear systems to general dynamic loading. These methods have the additional advantage that with only a slight modification they can be used to evaluate the dynamic response of nonlinear systems. Many structural systems will experience nonlinear response sometime during their life. Any moderate to strong earthquake will drive a structure designed by conventional methods into the inelastic range, particularly in certain critical regions. A very useful numerical integration technique for problems of structural dynamics is the so called step-by-step integration procedure. In this procedure the time history under consideration is divided into a number of small time increments Δt. During a small time step, the behavior of the structure is assumed to be linear. As nonlinear behavior occurs, the incremental stiffness is modified. In this manner, the response of the nonlinear system is approximated by a series of linear systems having a changing stiffness. The velocity and displacement computed at the end of one time interval become the initial conditions for the next time interval, and hence the process may be continued step by step.

3.5.1 Numerical Formulation of Equation of Motion

This section considers SDOF systems with properties m, c, $k(t)$, and $p(t)$, of which the applied force and the stiffness are functions of time. The stiffness is actually a function of the yield condition of the restoring force, and this in turn is a function of time. The damping coefficient may also be considered to be a function of time; however, general practice is to determine the damping characteristics for the elastic system and to keep these constant throughout the complete time history. In the inelastic range the principle mechanism for energy dissipation is through inelastic deformation, and this is taken into account through the hysteretic behavior of the restoring force.

The numerical equation required to evaluate the nonlinear response can be developed by first considering the equation of dynamic equilibrium given previously by Equation 3-6. It has been stated previously that this equation must be satisfied at every increment of time. Considering the time at the end of a short time step, Equation 3-6 can be written as

$$f_i(t + \Delta t) + f_d(t + \Delta t) + f_s(t + \Delta t) = p(t + \Delta t) \tag{3-76}$$

where the forces are defined as

$$
\begin{aligned}
f_{\mathrm{i}} &= m\ddot{v}(t+\Delta t)\\
f_{\mathrm{d}} &= c\dot{v}(t+\Delta t)\\
f_{\mathrm{s}} &= \sum_{i=1}^{n} k_i(t)\,\Delta v_i(t) = r_t + k(t)\,\Delta v(t)\\
\Delta v(t) &= v(t+\Delta t) - v(t)\\
r_t &= \sum_{i=1}^{n-1} k_i(t)\,\Delta v_i(t)
\end{aligned}
\tag{3-77}
$$

and in the case of ground accelerations

$$p(t+\Delta t) = p_{\mathrm{e}}(t+\Delta t) = -m\ddot{g}(t+\Delta t) \tag{3-78}$$

Substituting Equations 3-77 and 3-78 into Equation 3-76 results in an equation of motion of the form

$$m\ddot{v}(t+\Delta t) + c\dot{v}(t+\Delta t) + \sum k_i\,\Delta v_i = -m\ddot{g}(t+\Delta t) \tag{3-79}$$

It should be noted that the incremental stiffness is generally defined by the tangent stiffness at the beginning of the time interval

$$k_i = \frac{df_{\mathrm{s}}}{dv} \tag{3-80}$$

In addition, the dynamic properties given in Equations 3-77 and 3-78 can readily be exchanged for the generalized properties when considering a generalized SDOF system.

3.5.2 Numerical Integration

Many numerical integration schemes are available in the literature. The technique considered here is a step-by-step procedure in which the acceleration during a small time increment is assumed to be constant. A slight variation of this procedure, in which the acceleration is assumed to vary linearly during a small time increment, is described in detail by Clough and Penzien.[3-4] Both procedures have been widely used and have been found to yield good results with minimal computational effort.

If the acceleration is assumed to be constant during the time interval, the equations for the constant variation of the acceleration, the linear variation of the velocity and the quadratic variation of the displacement are indicated in Figure 3-28. Evaluating the expression for velocity and displacement at the end of the time interval leads to the following two expressions for velocity and displacement:

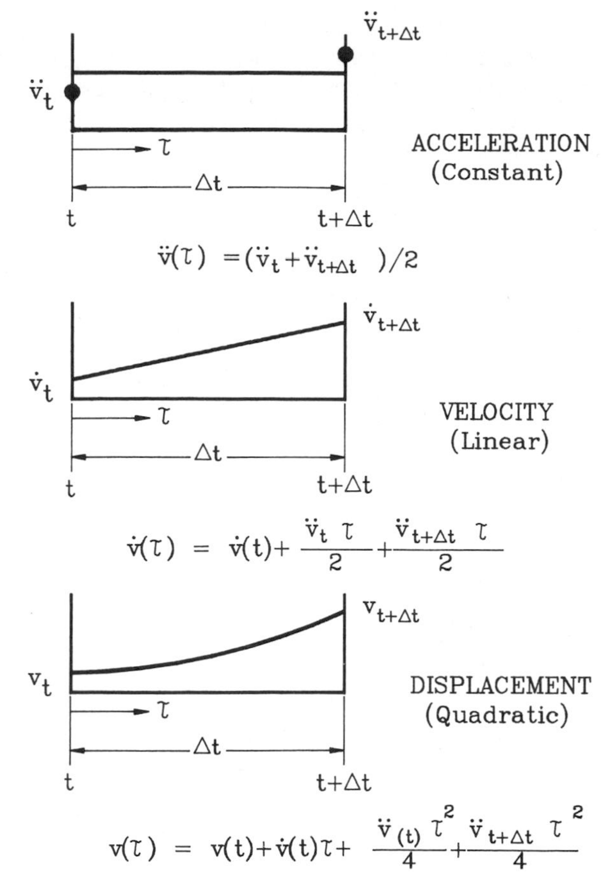

Figure 3-28 Incremental motion (constant acceleration).

$$\dot{v}(t+\Delta t) = \dot{v}(t) + \ddot{v}(t+\Delta t)\frac{\Delta t}{2} + \ddot{v}(t)\frac{\Delta t}{2} \tag{3-81}$$

$$v(t+\Delta t) = v(t) + \dot{v}(t)\,\Delta t + \ddot{v}(t+\Delta t)\frac{\Delta t^2}{4} + \ddot{v}(t)\frac{\Delta t^2}{4} \tag{3-82}$$

Solving Equation 3-82 for the acceleration $\ddot{v}(t+\Delta t)$ gives

$$\ddot{v}(t+\Delta t) = \frac{4}{\Delta t^2}\,\Delta v - \frac{4}{\Delta t}\dot{v}(t) - \ddot{v}(t) \tag{3-83}$$

which can be written as

$$\ddot{v}(t+\Delta t) = \frac{4}{\Delta t^2}\,\Delta v + A(t) \tag{3-84}$$

where

$$\Delta v = v(t+\Delta t) - v(t)$$

$$A(t) = -\frac{4}{\Delta t}\dot{v}(t) - \ddot{v}(t)$$

Note that this equation expresses the acceleration at the end of the time interval as a function of the incremental displacement and the acceleration and velocity at the beginning of the time interval. Substituting Equation 3-83 into Equation 3-81 gives the following expression for the velocity at the end of the time increment:

$$\dot{v}(t+\Delta t) = \frac{2}{\Delta t}\Delta v - \dot{v}(t) \qquad (3\text{-}85)$$

which can be written as

$$\dot{v}(t+\Delta t) = \frac{2}{\Delta t}\Delta v + B(t) \qquad (3\text{-}86)$$

where

$$B(t) = -\dot{v}(t)$$

It is convenient to express the damping as a linear function of the mass:

$$c = \alpha m = \lambda C_{cr} = 2m\omega\lambda \qquad (3\text{-}87)$$

Use of this equation allows the proportionality factor α to be expressed as

$$\alpha = 2\lambda\omega \qquad (3\text{-}88)$$

Substituting Equations 3-85, 3-86, and 3-88 into Equation 3-79 results in the following form of the equation for dynamic equilibrium:

$$m\left[\frac{4}{\Delta t^2}\Delta v + A(t)\right] + \alpha m\left[\frac{2}{\Delta t}\Delta v + B(t)\right] + R(t) + k\,\Delta v = m\ddot{g}(t+\Delta t) \qquad (3\text{-}89)$$

Moving terms containing the response conditions at the beginning of the time interval to the right-hand side of the equation results in the following so-called pseudo-static form of the equation of motion:

$$\bar{k}_t(\Delta v) = \bar{p}(t+\Delta t) \qquad (3\text{-}90)$$

where

$$\bar{k}_t = \frac{4m}{\Delta t^2} + \frac{2\alpha m}{\Delta t} + k_t$$

$$\bar{p}(t+\Delta t) = -m\ddot{g}(t+\Delta t) - R(t) - m[A(t) - \alpha B(t)]$$

The solution procedure for a typical time step is as follows:

1. Given the initial conditions at the beginning of the time interval, calculate the coefficients $A(t)$ and $B(t)$.
2. Calculate the effective stiffness.
3. Determine the effective force.
4. Solve for the incremental displacement

$$v = \bar{p}/\bar{k}_t \qquad (3\text{-}91)$$

5. Determine the displacement, velocity and acceleration at the end of the time interval:

$$\begin{aligned} v(t+\Delta t) &= v(t) + \Delta v \\ \dot{v}(t+\Delta t) &= \frac{2}{\Delta t} + B(t) \\ \ddot{v}(t+\Delta t) &= \frac{4}{\Delta t^2} + A(t) \end{aligned} \qquad (3\text{-}92)$$

6. The values given in Equation 3-92 become the initial conditions for the next time increment, and the procedure is repeated.

The above algorithm can be easily programmed on any microcomputer. If it is combined with a data base of recorded earthquake data such as EQINFOS,[3-6] it can be used to gain considerable insight into the linear and nonlinear response of structures that can be modeled as either a SDOF system or as a generalized SDOF system. It also forms the background material for later developments for multiple-degree-of-freedom systems.

An important response parameter that is unique to nonlinear systems is the ductility ratio. For a SDOF system, this parameter can be defined in terms of the displacement as

$$\mu = \frac{v(\text{max})}{v(\text{yield})} = 1.0 + \frac{v(\text{plastic})}{v(\text{yield})} \qquad (3\text{-}93)$$

As can be seen from the above equation, the ductility ratio is an indication of the amount of inelastic deformation that has occurred in the system. In the case of a SDOF system or generalized SDOF system the ductility obtained from Equation 3-93 usually represents the average ductility in the system. The ductility demand at certain critical regions, such as plastic hinges in critical members, may be considerably higher.

3.6 MULTIPLE-DEGREE-OF-FREEDOM SYSTEMS

In many structural systems it is impossible to model the dynamic response accurately in terms of a single displacement coordinate. These systems require a number of independent displacement coordinates to describe the displacement of the mass of the structure at any instant of time.

3.6.1 Mass and Stiffness Properties

In order to simplify the solution it is usually assumed for building structures that the mass of the structure is lumped at the center of mass of the individual story levels. This results in a diagonal matrix of mass properties in which either the translational mass or the mass moment of inertia is located on the main diagonal.

$$\{f_i\} = \begin{bmatrix} m_1 & & & & \\ & m_2 & & & \\ & & m_3 & & \\ & & & \ddots & \\ & & & & m_n \end{bmatrix} \begin{Bmatrix} v_1 \\ v_2 \\ \vdots \\ v_n \end{Bmatrix} \quad (3\text{-}94)$$

It is also convenient for building structures to develop the structural stiffness matrix in terms of the stiffness matrices of the individual story levels. The simplest idealization for a multistory building is based on the following three assumptions: (i) the floor diaphragm is rigid in its own plane; (ii) the girders are rigid relative to the columns and (iii) the columns are flexible in the horizontal directions but rigid in the vertical. If these assumptions are used, the building structure is idealized as having three dynamic degrees of freedom at each story level: a translational degree of freedom in each of two orthogonal directions, and a rotation about a vertical axis through the center of mass. If the above system is reduced to a plane frame, it will have one horizontal translational degree of freedom at each story level. The stiffness matrix for this type of structure has the tridiagonal form shown below:

$$\{f_s\} = \begin{bmatrix} k_1 & -k_2 & & & & \\ -k_2 & k_1+k_2 & -k_3 & & & \\ & -k_3 & k_2+k_3 & -k_4 & & \\ & & \ddots & \ddots & \ddots & \\ & & & \ddots & \ddots & -k_n \\ & & & & -k_n & k_n \end{bmatrix} \begin{Bmatrix} v_1 \\ v_2 \\ v_3 \\ \vdots \\ v_{n-1} \\ v_n \end{Bmatrix} \quad (3\text{-}95)$$

For the simplest idealization, in which each story level has one translational degree of freedom, the stiffness terms k_i in the above equations represent the translational story stiffness of the ith story level. As the assumptions given above are relaxed to include axial deformations in the columns and flexural deformations in the girders, the stiffness term k_i in Equation 3-95 becomes a submatrix of stiffness terms, and the story displacement v_i becomes a subvector containing the various displacement components in the particular story level. The calculation of the stiffness coefficients for more complex structures is a standard problem of static structural analysis. For the purposes of this chapter it will be assumed that the structural stiffness matrix is known.

3.6.2 Mode Shapes and Frequencies

The equations of motion for undamped free vibration of a multiple-degree-of-freedom (MDOF) system can be written in matrix form as

$$[M]\{\ddot{v}\} + [K]\{v\} = \{0\} \quad (3\text{-}96)$$

Since the motions of a system in free vibration are simple harmonic, the displacement vector can be represented as

$$\{v\} = \{\bar{v}\} \sin \omega t \quad (3\text{-}97)$$

Differentiating twice with respect to time results in

$$\{\ddot{v}\} = -\omega^2\{v\} \quad (3\text{-}98)$$

Substituting Equation 3-98 into Equation 3-96 results in a form of the eigenvalue equation,

$$([K] - \omega^2[M])\{v\} = \{0\} \quad (3\text{-}99)$$

The classical solution to the above equation derives from the fact that in order for a set of homogeneous equilibrium equations to have a nontrivial solution, the determinant of the coefficient matrix must be zero:

$$\det([K] - \omega^2[M]) = \{0\} \quad (3\text{-}100)$$

Expanding the determinant by minors results in a polynomial of degree N, which is called the frequency equation. The N roots of the polynomial represent the frequencies of the N modes of vibration. The mode having the lowest frequency (longest period) is called the first or fundamental mode. Once the frequencies are known, they can be substituted one at a time into the equilibrium Equation 3-99, which can then be solved for the relative amplitudes of motion for each of the displacement components in the particular mode of vibration. It should be noted that since the absolute amplitude of motion is indeterminate, $N - 1$ of the displacement components are determined in terms of one arbitrary component.

This method can be used satisfactorily for systems having a limited number of degrees of freedom. Programmable calculators have programs for solving the polynomial equation and for doing the matrix operations required to determine the mode shapes. However, for problems of any size, digital computer programs which use numerical techniques to solve large eigenvalue systems[3-7] must be used.

Example 3-7 (Mode Shapes and Frequencies)
It is assumed that the response in the transverse direction for the reinforced-concrete moment frame of Example 3-4 can be represented in terms of four displacement degrees of freedom which represent the horizontal displacements of the four story levels. Determine the stiffness matrix and the mass matrix, assuming that the mass is lumped at the story levels. Use these properties to calculate the frequencies and mode shapes of the four-degree-of-freedom system.

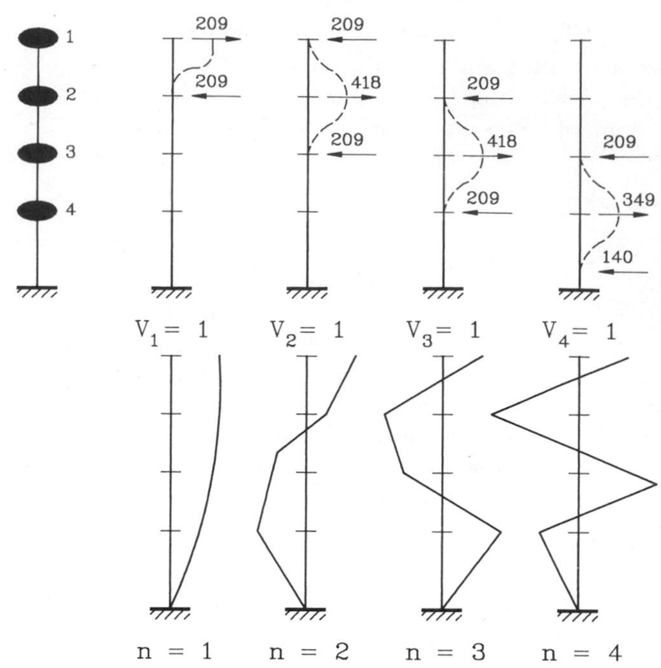

Figure 3-29 Stiffness determination and mode shapes (Example 3-7).

- *Stiffness and mass matrices*: The stiffness coefficient k_{ij} is defined as the force at coordinate i due to a unit displacement at coordinate j, all other displacements being zero (see Figure 3-29):

$$[K] = \begin{bmatrix} 209 & -209 & 0 & 0 \\ -209 & 418 & -209 & 0 \\ 0 & -209 & 418 & -209 \\ 0 & 0 & -209 & 349 \end{bmatrix}$$

$$[M] = \frac{1}{4}\begin{bmatrix} 1.01 & 0 & 0 & 0 \\ 0 & 1.15 & 0 & 0 \\ 0 & 0 & 1.15 & 0 \\ 0 & 0 & 0 & 1.16 \end{bmatrix}$$

$$[K] - \omega^2[M] = 200\begin{bmatrix} 1.05 - 1.01B & -1.05 & 0 & 0 \\ -1.05 & 2.09 - 1.15B & -1.05 & 0 \\ 0 & -1.05 & 2.09 - 1.15B & -1.05 \\ 0 & 0 & -1.05 & 1.74 - 1.16B \end{bmatrix}$$

where $B = \omega^2/800$.

- *Characteristic equation*:

$$|[K] - \omega^2[M]| = 0$$

$$B^4 - 6.183B^3 + 11.476B^2 - 6.430B + 0.486 = 0$$

Solution:

$$B_1 = 0.089 = \frac{\omega_1^2}{800} \quad \omega_1 = 8.438 \quad T_1 = 0.744 \text{ sec}$$

$$B_2 = 0.830 = \frac{\omega_2^2}{800} \quad \omega_2 = 25.768 \quad T_2 = 0.244 \text{ sec}$$

$$B_3 = 2.039 = \frac{\omega_3^2}{800} \quad \omega_3 = 40.388 \quad T_3 = 0.155 \text{ sec}$$

$$B_4 = 0.089 = \frac{\omega_4^2}{800} \quad \omega_4 = 50.800 \quad T_4 = 0.124 \text{ sec}$$

- *Mode shapes* (see Figure 3-29) are obtained by substituting the values of B_i, one at a time, into the equations

$$([K] - \omega^2[M])\{v\} = \{0\}$$

and determining $N - 1$ components of the displacement vector in terms of the first component, which is set equal to unity. This results in the modal matrix

$$[\Phi] = \begin{bmatrix} 1.00 & 1.00 & 1.00 & 1.00 \\ 0.91 & 0.20 & -1.07 & -1.78 \\ 0.74 & -0.78 & -0.75 & 1.75 \\ 0.47 & -1.05 & 1.24 & -0.92 \end{bmatrix}$$

Solution of the above problem using the computer

program ETABS (3-12) gives the following results:

$$\{T\} = \begin{Bmatrix} 0.838 \\ 0.268 \\ 0.152 \\ 0.107 \end{Bmatrix}$$

$$[\Phi] = \begin{bmatrix} 1.00 & 1.00 & 1.00 & 1.00 \\ 0.91 & 0.20 & -1.07 & -1.78 \\ 0.74 & -0.78 & -0.75 & 1.75 \\ 0.47 & -1.05 & 1.24 & -0.92 \end{bmatrix}$$

This program assumes the floor diaphragm is rigid in its own plane but allows axial deformation in the columns and flexural deformations in the beams. Hence, with these added degrees of freedom (fewer constraints) the fundamental period increases. However, comparing the results of this example with those of Example 3-5, it can be seen that for this structure a good approximation for the first-mode response was obtained using the generalized SDOF model and the static deflected shape.

3.6.3 Equations of Motion in Normal Coordinates

Betti's reciprocal work theorem can be used to develop two orthogonality properties of vibration mode shapes which make it possible to greatly simplify the equations of motion. The first of these states that the mode shapes are orthogonal to the mass matrix and is expressed in matrix form as

$$\{\phi_n\}^T[M]\{\phi_m\} = \{0\} \qquad (m \neq n) \quad (3\text{-}101)$$

Using Equations 3-99 and 3-101, the second property can be expressed in terms of the stiffness matrix as

$$\{\phi_n\}^T[K]\{\phi_m\} = \{0\} \qquad (m \neq n) \quad (3\text{-}102)$$

which states that the mode shapes are orthogonal to the stiffness matrix. It is further assumed that the mode shapes are also orthogonal to the damping matrix:

$$\{\phi_n\}^T[C]\{\phi_m\} = \{0\} \qquad (m \neq n) \quad (3\text{-}103)$$

Sufficient conditions for this assumption have been discussed elsewhere.[(3-8)] Since any MDOF system having N degrees of freedom also has N independent vibration mode shapes, it is possible to express the displaced shape of the structure in terms of the amplitudes of these shapes by treating them as generalized coordinates (sometimes called normal coordinates). Hence the displacement at a particular location, v_i, can be obtained by summing the contributions from each mode as

$$v_i = \sum_{n=1}^{N} \phi_{in} Y_n \quad (3\text{-}104)$$

In a similar manner, the complete displacement vector can be expressed as

$$\{v\} = \sum_{n=1}^{N} \{\phi_n\} Y_n = [\Phi]\{Y\} \quad (3\text{-}105)$$

It is convenient to write the equations of motion for a MDOF system in matrix form as

$$[M]\{\ddot{v}\} + [C]\{\dot{v}\} + [K]\{v\} = \{P(t)\} \quad (3\text{-}106)$$

which is similar to the equation for a SDOF system, Equation 3-9. The differences arise because the mass, damping, and stiffness are now represented by matrices of coefficients representing the added degrees of freedom, and the acceleration, velocity, displacement, and applied load are represented by vectors containing the additional degrees of freedom. The equations of motion can be expressed in terms of the normal coordinates by substituting Equation 3-105 and its appropriate derivatives into Equation 3-106 to give

$$[M][\Phi]\{\ddot{Y}\} + [C][\Phi]\{\dot{Y}\} + [K][\Phi]\{Y\} = \{P(t)\} \quad (3\text{-}107)$$

Multiplying the above equation by the transpose of any modal vector $\{\phi_n\}$ results in the following:

$$\{\phi_n\}^T[M][\Phi]\{\ddot{Y}\} + \{\phi_n\}^T[C][\Phi]\{\dot{Y}\} + \{\phi_n\}^T[K][\Phi]\{Y\} = \{\phi_n\}^T\{P(t)\} \quad (3\text{-}108)$$

Using the orthogonality conditions of Equations 3-101, 3-102, and 3-103 reduces this set of equations to the equation of motion for a generalized SDOF system in terms of the generalized properties for the nth mode shape and the normal coordinate Y_n:

$$M_n^* \ddot{Y}_n + C_n^* \dot{Y}n + K_n^* Y = P_n^*(t) \quad (3\text{-}109)$$

where the generalized properties for the nth mode are given as

$$\begin{aligned} M_n^* &= \text{generalized mass} = \{\phi_n\}^T[M]\{\phi_n\} \\ C_n^* &= \text{generalized damping} \\ &= \{\phi_n\}^T[C]\{\phi_n\} = 2\lambda_n \omega_n M_n^* \\ K_n^* &= \text{generalized stiffness} \\ &= \{\phi_n\}^T[K]\{\phi_n\} = \omega_n^2 M_n^* \\ P_n^*(t) &= \text{generalized loading} = \{\phi_n\}^T\{P(t)\} \end{aligned} \quad (3\text{-}110)$$

The above relations can be used to further simplify the equation of motion for the nth mode to the form

$$\ddot{Y}_n + 2\lambda_n \omega_n \dot{Y}_n + \omega_n^2 Y_n = \frac{P_n^*(t)}{M_n^*} \quad (3\text{-}111)$$

The importance of the above transformations to normal coordinates has been summarized by Clough and Penzien,[3-4] who state that

> The use of normal coordinates serves to transform the equations of motion from a set of N simultaneous differential equations which are coupled by off diagonal terms in the mass and stiffness matrices to a set of N independent normal coordinate equations.

It should further be noted that the expressions for the generalized properties of any mode are equivalent to those defined previously for a generalized SDOF system. Hence the use of the normal modes transforms the MDOF system having N degrees of freedom into a system of N independent generalized SDOF systems. The complete solution for the system is then obtained by superimposing the independent modal solutions. For this reason this method is often referred to as the modal-superposition method. Use of this method also leads to a significant saving in computational effort, since in most cases it is not be necessary to use all N modal responses to accurately represent the response of the structure. For most structural systems the lower modes make the primary contribution to the total response. Therefore, the response can usually be represented to sufficient accuracy in terms of a limited number of modal responses in the lower modes.

3.6.4 Earthquake-Response Analysis

Time-History Analysis As in the case of SDOF systems, for earthquake analysis the time-dependent force must be replaced with the effective loads, which are given by the product of the mass at any level, M_i, and the ground acceleration $g(t)$. The vector of effective loads is obtained as the product of the mass matrix and the ground acceleration:

$$P_e(t) = [M]\{\Gamma\}\ddot{g}(t) \qquad (3\text{-}112)$$

where $\{\Gamma\}$ is a vector of influence coefficients of which component i represents the acceleration at displacement coordinate i due to a unit horizontal ground acceleration at the base. For the simple structural model in which the degrees of freedom are represented by the horizontal displacements of the story levels, the vector $\{\Gamma\}$ becomes a unity vector, $\{1\}$, since for a unit ground acceleration in the horizontal direction all degrees of freedom have a unit horizontal acceleration. Using Equation 3-108, the generalized effective load for the nth mode is given as

$$P_{en}^*(t) = \mathscr{L}_n g(t) \qquad (3\text{-}113)$$

where $\mathscr{L}_n = \{\phi_n\}^T[M]\{\Gamma\}$.

Substituting Equation 3-113 into Equation 3-111 results in the following expression for the earthquake response of the nth mode of a MDOF system:

$$\ddot{Y}_n + 2\lambda_n\omega_n\dot{Y}_n + \omega_n^2 Y_n = \mathscr{L}_n\ddot{g}(t)/M_n^* \qquad (3\text{-}114)$$

In a manner similar to that used for the SDOF system, the response of this mode at any time t can be obtained by the Duhamel integral expression

$$Y_n(t) = \frac{\mathscr{L}_n V_n(t)}{M_n^*\omega_n} \qquad (3\text{-}115)$$

where $V_n(t)$ represents the integral

$$V_n(t) = \int_0^t \ddot{g}(\tau)e^{-\lambda_n\omega_n(t-\tau)}\sin\omega_n(t-\tau)\,d\tau \qquad (3\text{-}116)$$

The complete displacement of the structure at any time t is then obtained by superimposing the contributions of the individual modes using Equation 3-105:

$$\{v(t)\} = \sum_{n=1}^{N}\{\phi_n\}Y_n(t) = [\Phi]\{Y(t)\} \qquad (3\text{-}117)$$

The resulting earthquake forces can be determined in terms of the effective accelerations, which for each mode are given by the product of the circular frequency and the displacement amplitude of the generalized coordinate:

$$\ddot{Y}_{ne}(t) = \omega_n^2 Y_n(t) = \frac{\mathscr{L}_n\omega_n V_n(t)}{M_n^*} \qquad (3\text{-}118)$$

The corresponding acceleration in the structure due to the nth mode is given as

$$\{\ddot{v}_{ne}(t)\} = \{\phi_n\}\ddot{Y}_{ne}(t) \qquad (3\text{-}119)$$

and the corresponding effective earthquake force is given as

$$\{q_n(t)\} = [M]\{\ddot{v}_n(t)\} = [M]\{\phi_n\}\omega_n\mathscr{L}_n V_n(t)/M_n^* \qquad (3\text{-}120)$$

The total earthquake force is obtained by superimposing the individual modal forces to obtain

$$q(t) = \sum_{n=1}^{N} q_n(t) = [M][\Phi]\omega^2 Y(t) \qquad (3\text{-}121)$$

The base shear can be obtained by summing the effective earthquake forces over the height of the structure:

$$Q_n(t) = \sum_{i=1}^{H} q_{in}(t) = \{1\}^T\{q_n(t)\} = M_{en}\omega_n V_n(t) \tag{3-122}$$

where $M_{en} = \mathscr{L}_n^2/M_n^*$ is the effective mass for the nth mode.

The sum of the effective masses for all of the modes is equal to the total mass of the structure. This results in a means of determining the number of modal responses necessary to accurately represent the overall structural response. If the total response is to be represented in terms of a finite number of modes and if the sum of the corresponding modal masses is greater than a predefined percentage of the total mass, the number of modes considered in the analysis is adequate. If this is not the case, additional modes need to be considered. The base shear for the nth mode, Equation 3-122, can also be expressed in terms of the effective weight, W_{en}, as

$$Q_n(t) = \frac{W_{en}}{g}\omega_n V_n(t) \tag{3-123}$$

where

$$W_{en} = \frac{\left(\sum_{i=1}^{H} W_i \phi_{in}\right)^2}{\sum_{i=1}^{H} W_i \phi_{in}^2} \tag{3-124}$$

The base shear can be distributed over the height of the building in a manner similar to Equation 3-74, with the modal earthquake forces expressed as

$$\{q_n(t)\} = \frac{[M]\{\phi_n\}Q_n(t)}{\mathscr{L}_n} \tag{3-125}$$

3.6.5 Response-Spectrum Analysis

The above equations for the response of any mode of vibration are exactly equivalent to the expressions developed for the generalized SDOF system. Therefore, the maximum response of any mode can be obtained in a manner similar to that used for the generalized SDOF system. By analogy to Equations 3-34 and 3-43 the maximum modal displacement can be written as

$$Y_n(t)_{\max} = \frac{V_n(t)_{\max}}{\omega_n} = S_{dn} \tag{3-126}$$

Making this substitution in Equation 3-115 results in

$$Y_{n\max} = \mathscr{L}_n S_{dn}/M_n^* \tag{3-127}$$

The distribution of the modal displacements in the structure can be obtained by multiplying this expression by the modal vector

$$\{v_n\}_{\max} = \{\phi_n\}Y_{n\max} = \frac{\{\phi_n\}\mathscr{L}_n S_{dn}}{M_n^*} \tag{3-128}$$

The maximum effective earthquake forces can be obtained from the modal accelerations as given by Equation 3-120:

$$\{q_n\}_{\max} = \frac{[M]\{\phi_n\}\mathscr{L}_n S_{pan}}{M_n^*} \tag{3-129}$$

Summing these forces over the height of the structure gives the following expression for the maximum base shear due to the nth mode:

$$Q_{n\max} = \mathscr{L}_n^2 S_{pan}/M_n^* \tag{3-130}$$

which can also be expressed in terms of the effective weight as

$$Q_{n\max} = W_{en} S_{pan}/g \tag{3-131}$$

where W_{en} is defined by Equation 3-124.

Finally, the overturning moment at the base of the building for the nth mode can be determined as

$$M_o = \langle h\rangle[M]\{\phi_n\}\mathscr{L}_n S_{pan}/M_n^* \tag{3-132}$$

where $\langle h\rangle$ is a row vector of the story heights above the base.

3.6.6 Modal Combinations

Using the response-spectrum method for MDOF systems, the maximum modal response is obtained for each mode of a set of modes which are used to represent the response. The question then arises as to how these modal maxima should be combined in order to get the best estimate of the maximum total response. The modal-response equations such as Equations 3-117 and 3-121 provide accurate results only as long as they are evaluated concurrently in time. In going to the response-spectrum approach, time is taken out of these equations and replaced with the modal maxima. These maximum response values for the individual modes cannot possibly occur at the same time; therefore, a means must be found to combine the modal maxima in such a way as to approximate the maximum total response. One such combination that has been used is to take the sum of the absolute values (SAV) of the modal responses. This combination can be expressed as

$$r \leqslant \sum_{n=1}^{N} |r_n| \tag{3-133}$$

Since this combination assumes that the maxima occur

at the same time and that they also have the same sign, it produces an upper-bound estimate for the response which is too conservative for design application. A more reasonable estimate, which is based on probability theory, can be obtained by using the square-root-of-the-sum-of-the-squares (SRSS) method, which is expressed as

$$r \approx \sqrt{\sum_{n=1}^{N} r_n^2} \tag{3-134}$$

This method of combination has been shown to give a good approximation of the response for two-dimensional structural systems. For three-dimensional systems, it has been shown that the complete-quadratic-combination (CQC) method[3-9] may offer a significant improvement in estimating the response of certain structural systems. The complete quadratic combination is expressed as

$$r \approx \sqrt{\sum_{i=1}^{N} \sum_{j=1}^{N} r_i p_{ij} r_j} \tag{3-135}$$

where for constant modal damping

$$p_{ij} = \frac{8\lambda^2(1+\zeta)\zeta^{3/2}}{(1-\zeta^2)^2 + 4\lambda^2\zeta(1+\zeta)^2} \tag{3-136}$$

and

$$\zeta = \omega_j/\omega_i$$
$$\lambda = c/c_{cr}$$

Using the SRSS method for two-dimensional systems and the CQC method for either two- or three-dimensional systems will give a good approximation to the maximum earthquake response of an elastic system without requiring a complete time-history analysis. This is particularly important for purposes of design.

Example 3-8 (Response Spectrum Analysis)

Use the design response spectrum given in Example 3-6 and the results of Example 3-7 to perform a response-spectrum analysis of the reinforced concrete frame. Determine the modal responses of the four modes of vibration, and estimate the total response using the

Table 3-3 Computation of Response for Model of Example 3-8

Response Quantity	Modal Parameter	Modal Response $n = 1$	2	3	4	Combined Response SAV	SRSS	CQC
	$\omega =$	8.44	25.77	40.39	50.80			
	$\alpha_n =$	1.212	−0.289	0.075	0.010			
	$S_d =$	1.190	0.155	0.062	0.039			
	$\phi =$	1.00	1.00	1.00	1.00			
		0.91	0.20	−1.07	−1.78			
		0.74	−0.78	−0.75	1.75			
		0.47	−1.05	1.24	−0.92			
Displacement	$n = 4$	1.44	−0.045	0.019	−0.002	1.506	1.441	1.441
$v_n = \phi_n \alpha_n S_{dn}$	3	1.31	−0.009	−0.020	0.003	1.342	1.310	1.310
(Eq. 3.125)	2	1.07	0.035	−0.014	−0.003	1.122	1.071	1.071
	1	0.68	0.047	0.023	0.001	0.751	0.682	0.682
Acceleration	4	102.6	−29.9	31.0	−5.1	168.6	111.4	110.7
$\ddot{v}_n = \omega_n^2 v_n$	3	93.3	−6.0	−32.6	7.7	139.6	99.3	98.9
	2	76.2	23.2	−22.8	−7.7	129.9	83.2	83.3
	1	48.4	31.2	37.5	2.6	119.7	68.8	70.0
Inertia force	4	25.91	−7.54	7.83	−1.30	42.6	28.1	27.9
$q_n = M\ddot{v}_n$	3	26.82	−1.72	−9.38	2.23	40.2	28.6	28.4
	2	21.91	6.68	−6.56	−2.23	37.4	23.9	23.9
	1	14.03	9.05	11.35	0.75	35.2	20.2	20.6
Shear	4	25.91	−7.54	7.83	−1.30	42.6	28.1	28.0
$Q_n = \Sigma q_n$	3	52.73	−9.26	−1.55	0.93	64.5	53.6	53.5
	2	74.64	−2.58	−8.11	−1.30	86.6	75.1	75.1
	1	88.67	6.47	3.24	−0.55	98.9	89.0	89.0
Overturning	4	272.1	−79.2	82.2	−13.7	447.2	295.4	293.6
moment	3	825.7	−176.4	65.9	−3.9	1071.9	846.9	845.3
(ft-kips)	2	1609.4	−203.5	−19.2	−17.5	1849.6	1622.4	1621.3
	1	2673.4	−125.9	19.7	−24.1	2843.1	2676.5	2675.7

SAV, SRSS, and CQC methods of modal combination. Present the data in a tabular form suitable for hand calculation. Finally, compare the results with those obtained in Example 3-6 for a generalized SDOF model.

From Example 3-7,

$$[M] = \frac{1}{4}\begin{bmatrix} 1.01 & 0 & 0 & 0 \\ 0 & 1.15 & 0 & 0 \\ 0 & 0 & 1.15 & 0 \\ 0 & 0 & 0 & 1.16 \end{bmatrix}$$

$$\{\omega\} = \begin{Bmatrix} 8.44 \\ 25.77 \\ 40.39 \\ 50.80 \end{Bmatrix}$$

$$[\Phi] = \begin{bmatrix} 1.00 & 1.00 & 1.00 & 1.00 \\ 0.91 & 0.20 & -1.07 & -1.78 \\ 0.74 & -0.78 & -0.75 & 1.75 \\ 0.47 & -1.05 & 1.24 & -0.92 \end{bmatrix}$$

$$\{f\} = \frac{\omega}{2\pi} = \begin{Bmatrix} 1.34 \\ 4.10 \\ 6.43 \\ 8.09 \end{Bmatrix} \text{Hz}$$

$$S_v = \begin{Bmatrix} 10.0 \\ 4.0 \\ 2.5 \\ 2.0 \end{Bmatrix} \text{in./sec}$$

$$S_{dn} = S_{vn}/\omega_n$$

From Equation 3-128,

$$\{v_n\}_{\max} = \{\phi_n\}(\mathscr{L}_n/M_n^*)S_{dn} = \{\phi\}\alpha S_{dn}$$

$$\{q_n\} = [M]\{\ddot{v}_n\} = [M]\omega^2\{v_n\}$$

$$Q_n = \sum_{i=1}^{N} q_{ni}$$

Table 3-4 Comparison of Results Obtained from MDOF and SDOF Models

Response Parameter	MDOF (Example 3-8)	SDOF (Example 3-6)
Period (sec)	0.744	0.721
Displacements (in.)		
Roof	1.44	1.17
3rd	1.31	1.04
2nd	1.07	0.80
1st	0.68	0.50
Inertia force (kips)		
Roof	28.1	27.1
3rd	28.6	27.4
2nd	23.9	21.1
1st	20.2	13.2
Base shear (kips)	89.0	88.8
Overturning moment (ft-kips)	2678	2716

For CQC combination,

$$\lambda = 0.05 = \text{constant for all modes}$$

$$p_{ij} = \begin{bmatrix} 1.0000 & 0.0062 & 0.0025 & 0.0017 \\ 0.0062 & 1.0000 & 0.0452 & 0.0193 \\ 0.0025 & 0.0452 & 1.0000 & 0.1582 \\ 0.0017 & 0.0193 & 0.1582 & 1.0000 \end{bmatrix}$$

The computation of the modal and the combined response is tabulated in Table 3-3. The results are compared with those obtained for the SDOF model in Table 3-4.

3.7 NONLINEAR RESPONSE OF MDOF SYSTEMS

The nonlinear-response analysis of MDOF systems closely parallels the development for SDOF systems presented earlier. The equations of equilibrium for a MDOF system subjected to base excitation can be written in matrix form as

$$[M]\{\ddot{v}\} + [C]\{\dot{v}\} + [K]\{v\} = -[M]\{\Gamma\}\ddot{g}(t) \tag{3-137}$$

This equation is of the same form as that of Equation 3-76 for the SDOF system. The acceleration, velocity, and displacement have been replaced by vectors containing the additional degrees of freedom. The mass has been replaced by the mass matrix, which for a lumped-mass system is a diagonal matrix with the translational-mass and rotational-mass terms on the main diagonal. The incremental stiffness has been replaced by the incremental stiffness matrix, and the damping has been replaced by the damping matrix. This latter term requires some additional discussion. In the mode-superposition method, the damping ratio was defined for each mode of vibration. However, this is not possible for a nonlinear system, because it has no true vibration modes. A useful way to define the damping matrix for a nonlinear system is to assume that it can be represented as a linear combination of the mass and stiffness matrices of the initial elastic system:

$$[C] = \alpha[M] + \beta[K] \tag{3-138}$$

where α and β are scalar multipliers which may be selected so as to provide a given percentage of critical damping in any two modes of vibration of the initial elastic system. These two multipliers can be evaluated from the expression

$$\begin{Bmatrix} \alpha \\ \beta \end{Bmatrix} = 2\begin{bmatrix} \omega_j & -\omega_i \\ -\dfrac{1}{\omega_j} & \dfrac{1}{\omega_i} \end{bmatrix} \frac{\omega_i \omega_j}{\omega_j^2 - \omega_i^2} \begin{Bmatrix} \lambda_i \\ \lambda_j \end{Bmatrix} \tag{3-139}$$

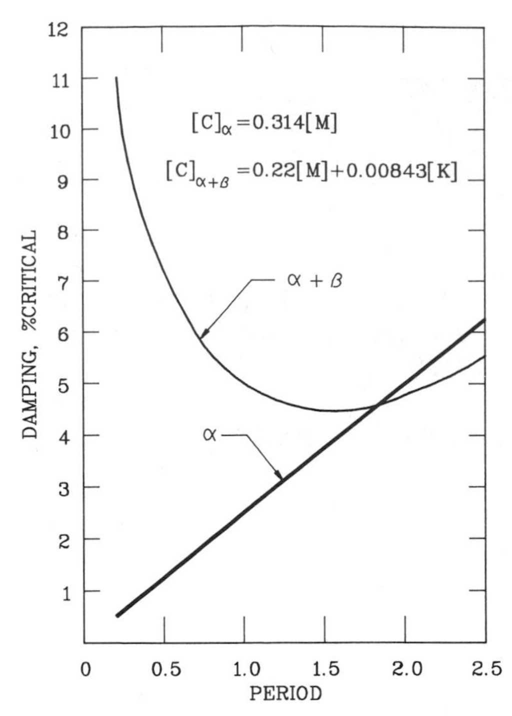

Figure 3-30 Damping functions for a framed tube.

where λ_i and λ_j are the percentages of critical damping in the two specified modes. Once the coefficients α and β are determined, the damping in the other elastic modes is specified by the expression

$$\lambda_k = \frac{\alpha}{2\omega_k} + \frac{\beta\omega_k}{2} \tag{3-140}$$

A typical damping function which was used for the nonlinear analysis of a reinforced concrete frame[3-10] is shown in Figure 3-30. Although the representation for the damping is only approximate, it is justified for these types of analyses in that it gives a good approximation of the damping for a range of modes of vibration and these modes can be selected to be the ones that make the principal contribution to the response. Also, in nonlinear dynamic analyses the dissipation of energy through inelastic deformation tends to overshadow the dissipation of energy through viscous damping. Therefore, an exact representation of damping is not as important in a nonlinear system as it is in a linear system. One should be aware of the characteristics of the damping function to insure that important components of the response are not lost. For instance, if the coefficients are selected to give a desired percentage of critical damping in the lower modes and the response of the higher modes is important, the higher-mode response may be overdamped and its contribution to the total response diminished.

Substituting Equation 3-138 into Equation 3-137 results in

$$[M]\{\ddot{v}\} + \alpha[M]\{\dot{v}\} + \beta[K_i]\{\dot{v}\} + [K_t]\{v\} = -[M]\{\Gamma\}\ddot{g}(t) \tag{3-141}$$

Terms containing the stiffness matrix can be written in matrix form only for the linear behavior that is assumed to occur during a small time increment when the stiffness is represented by the tangent stiffness K_t. Therefore,

$$[K]\{v\} = \sum[K_t]\{\Delta v\} = \{R_t\} + [K_t]\{\Delta v\} \tag{3-142}$$

where

$$\{R_t\} = \sum_{i=1}^{n-1}[K_{ti}]\{\Delta v_i\}$$

Using the step-by-step integration procedure in which the acceleration is assumed to be constant during a time increment, equations similar to Equations 3-84 and 3-86 can be developed for the MDOF system which express the acceleration and velocity vectors at the end of the time increment in terms of the incremental displacement vector and the vectors of initial conditions at the beginning of the time increment:

$$\{\ddot{v}(t)\} = \frac{4}{\Delta t^2}\{\Delta v\} + \{A_t\} \tag{3-143}$$

$$\{\dot{v}(t)\} = \frac{2}{\Delta t}\{\Delta v\} + \{B_t\} \tag{3-144}$$

where

$$\{A_t\} = -\frac{4}{\Delta t}\{\dot{v}(t-\Delta t)\} - \{\ddot{v}(t-\Delta t)\} \tag{3-145}$$

$$\{B_t\} = -\{\dot{v}(t-\Delta t)\} \tag{3-146}$$

Substituting Equations 3-142 through 3-146 into Equation 3-141 and rearranging some terms leads to the pseudostatic form

$$[\tilde{K}]\{\Delta\tilde{v}\} = \{\tilde{P}\} \tag{3-147}$$

where

$$[\tilde{K}] = [C_0[M] + C_1[K_i] + [K_t]]$$

$$\{\tilde{P}\} = \{P(t)\} - \{R_t\} - [M]\{\{A_t\} - \alpha\{B_t\}\} - \beta[K_i]\{B_t\}$$

$$C_0 = \frac{4}{\Delta t^2} + \frac{2\alpha}{\Delta t}$$

$$C_1 = \frac{2\beta}{\Delta t}$$

The displacement vector can be obtained by solving Equation 3-147 for $\{\Delta v\}$. This result can then be used in Equations 3-143 and 3-144 to obtain the acceleration vector and velocity vector at the end of the time interval. These vectors then become the initial conditions for the next time interval, and the process is repeated.

Output from a nonlinear response analysis of a MDOF system may include such response parameters as an envelope of the maximum story displacements, an envelope of the maximum relative story displacements (sometimes referred to as the maximum interstory drift), and the maximum ductility demand of the beam, column, wall, and bracing elements. For MDOF systems, the definition of ductility is not as straightforward as it was for the SDOF systems. Ductility may be expressed in terms of such parameters as displacement, relative displacement, rotation, curvature, or strain. Envelopes of maximum story shear are often plotted for comparison with the design shear specified in most building codes. For some systems, plots of the hysteretic force-vs.-displacement behavior of certain critical members may be of considerable importance in evaluating the ability of these members to withstand the imposed inelastic deformations.

Example 3-9 (Nonlinear-Response Analysis)

The following is a representative nonlinear-response analysis for a moment-resistant steel frame which was done by Anderson and Naeim.(3-11) The frame has ten stories, three 20-ft bays, and a typical story height of 12 ft with a 15-ft story height at the ground level. The frame was designed for combinations of dead load, live load, wind load, and earthquake load. The earthquake load is based on the equivalent lateral forces specified in the Uniform Building Code. The design is done on a computer using a two-step design process. In the first step, an automated, iterative design procedure based on Part 2 of the AISC Specification is used to determine the design forces and to select member sizes from a property table which is included in the program. The frame design obtained from this program is then submitted to another program, which performs a stress check of each member and lists the critical stress ratio due to combined stress. For a proper design the stress ratio will be just less than one; however, this is not always possible, due to the finite number of steel sections which are available. Members which have a stress ratio greater than one are modified, and the stress check is repeated. In this manner the two-step process results

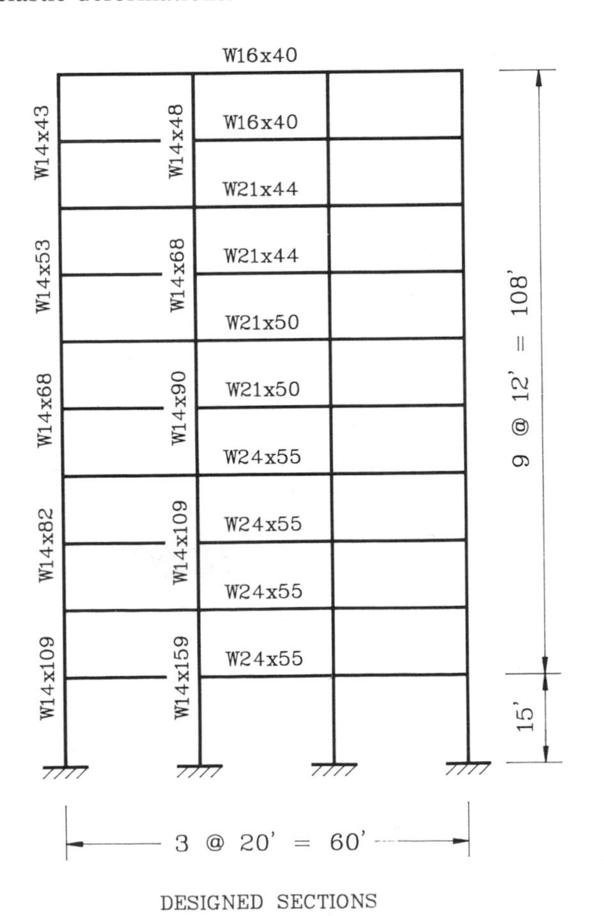

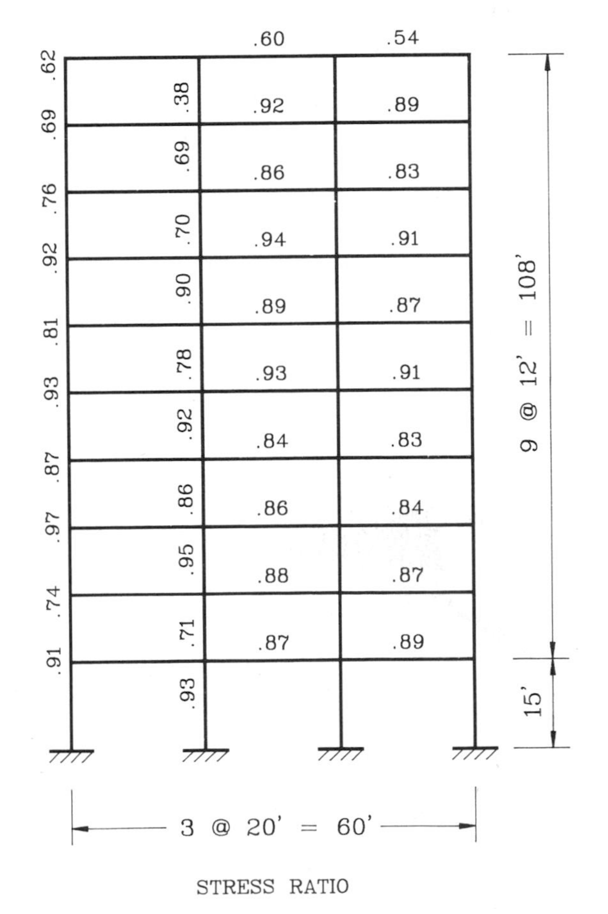

Figure 3-31 Sections and stress ratios for allowable stress frame.

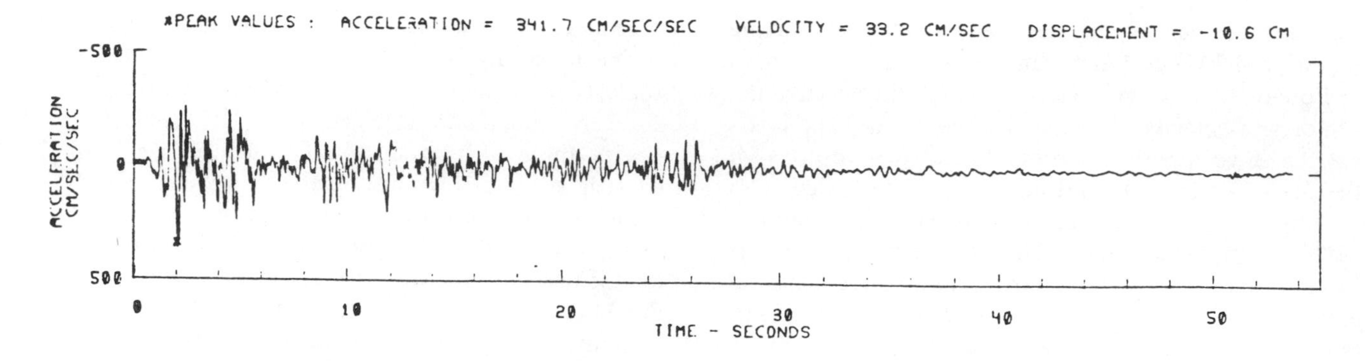

(a) Imperial Valley, 1940, El Centro, N-S.

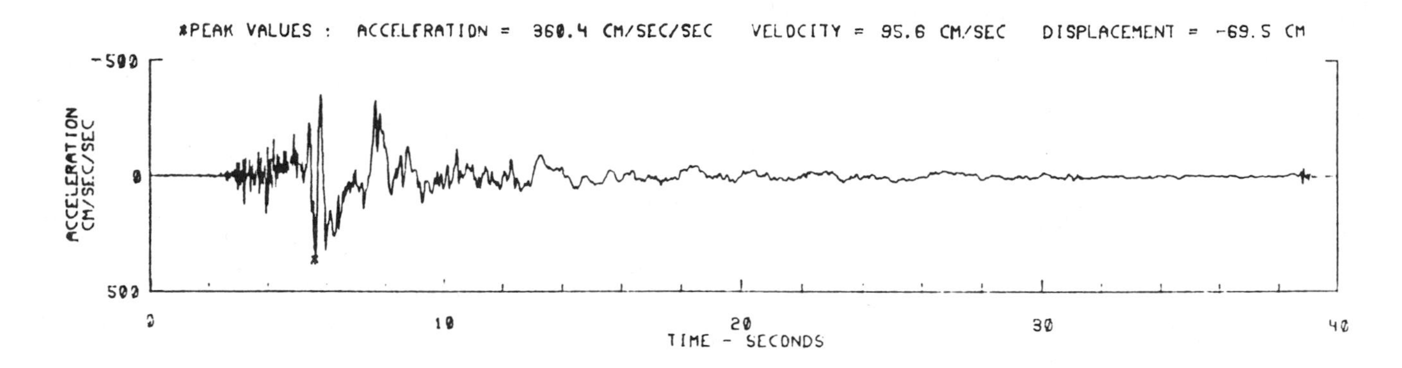

(b) Imperial Valley, 1979, James Road, S50W.

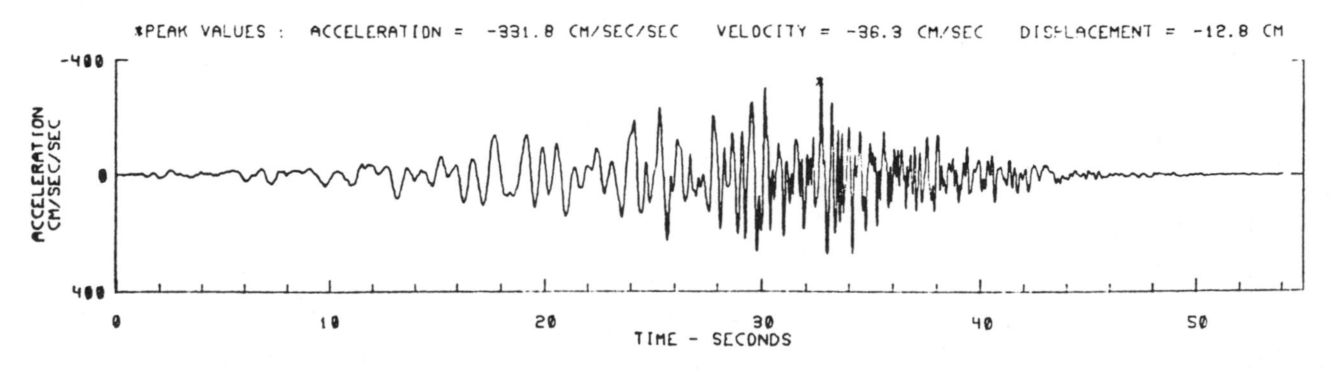

(c) Artificial M8.5 at 40 Km.

Figure 3-32 Acceleration time histories.

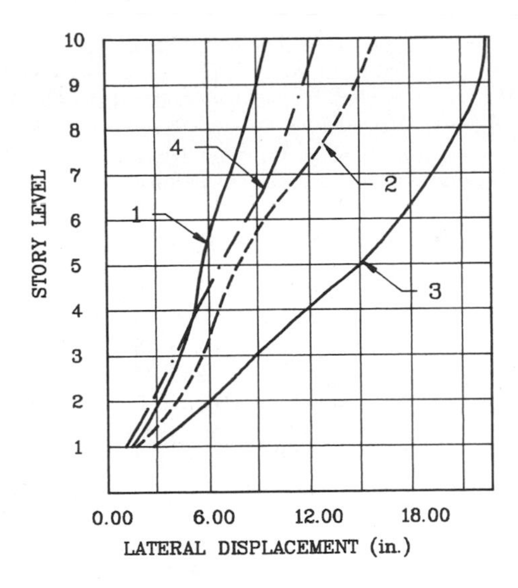

1 – 1.0 X EL CENTRO 1940
2 – 1.5 X EL CENTRO 1940
3 – JAMES ROAD 1979
4 – 8.5 AT 40 KM.

a) MAXIMUM LATERAL DISPLACEMENT

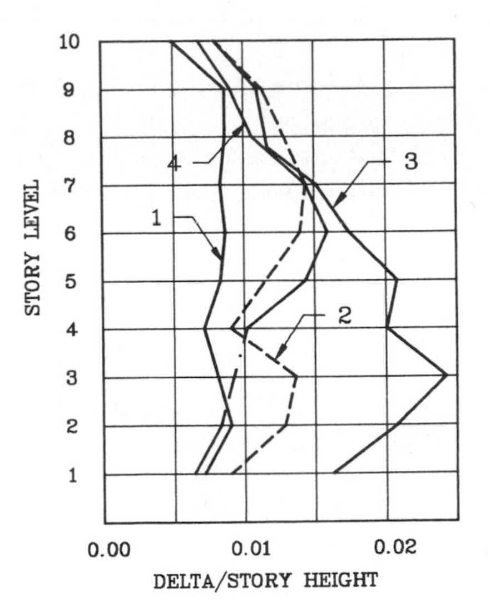

1 – 1.0 X EL CENTRO 1940
2 – 1.5 X EL CENTRO 1940
3 – JAMES ROAD 1979
4 – 8.5 AT 40 KM.

b) MAXIMUM RELATIVE DISPLACEMENT
(Interstory Drift)

Figure 3-33 Nonlinear displacement response.

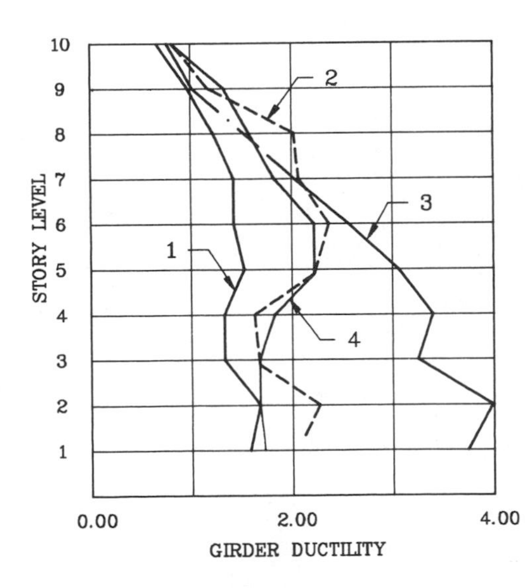

1 – 1.0 X EL CENTRO 1940
2 – 1.5 X EL CENTRO 1940
3 – JAMES ROAD 1979
4 – MAG. 8.5 AT 40 KM.

a) GIRDER DUCTILITY DEMAND

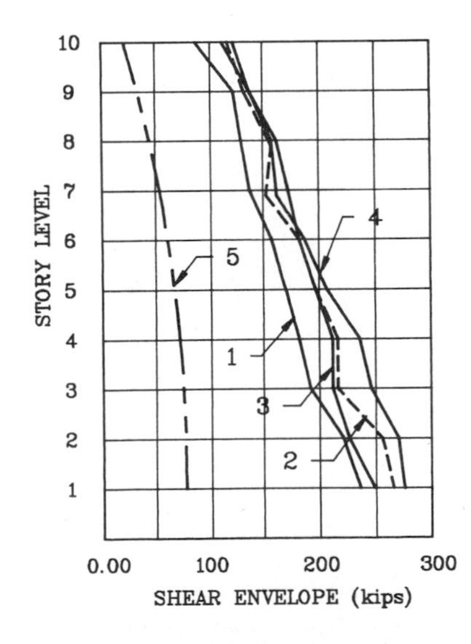

1 – 1.0 X EL CENTRO 1940
2 – 1.5 X EL CENTRO 1940
3 – JAMES ROAD 1979
4 – MAG. 8.5 AT 40 KM.
5 – UBC LATERAL FORCE

b) STORY SHEAR VERSUS DESIGN SHEAR

Figure 3-34 Nonlinear girder ductility and story shear.

in members which meet the design criteria and are economical. The designed frame and the corresponding final stress ratios are summarized in Figure 3-31.

Three carefully selected ground motions having different response characteristics were used for the analysis. This will illustrate the effect different ground motions can have on the nonlinear response of a structure. Acceleration time histories for the three records are shown in Figure 3-32.

The nonlinear response of a structure is often presented in terms of the following response parameters: (i) envelope of maximum total displacement, (ii) envelope of maximum story-to-story displacement divided by the story height (often referred to as interstory drift), (iii) maximum ductility demand for the beams and columns, (iv) envelopes of maximum plastic hinge rotation, and (v) envelopes of maximum story shear. Representative plots of four of these parameters are shown in Figures 3-23 and 3-34.

REFERENCES

3-1 Chopra, A. K., "Dynamics of Structures, a Primer," Earthquake Engineering Research Inst., Berkeley, CA, 1981.

3-2 Applied Technology Council, "An Evaluation of a Response Spectrum Approach to Seismic Design of Buildings," ATC-2, Applied Technology Council, Palo Alto, CA, 1974.

3-3 Newmark, N. M. and Hall, W. J., "Procedures and Criteria for Earthquake Resistant Design," Building Practices for Disaster Mitigation, U.S. Department of Commerce, Building Science Series 46, 1973.

3-4 Clough, R. W. and Penzien, J., *Structural Dynamics*, McGraw-Hill, New York, 1975.

3-5 Rayleigh, Lord, *Theory of Sound*, Dover, New York, 1945.

3-6 Lee, V. W. and Trifunac, M. D., "Strong Earthquake Ground Motion Data in EQINFOS, Part 1," Report No. 87-01, Dept. of Civil Engineering, U.S.C. Los Angeles, 1987.

3-7 Bathe, K. J. and Wilson, E. L., *Numerical Methods in Finite Element Analysis*, Prentice Hall, Englewood Cliffs, NJ, 1976.

3-8 Caughy, T. K., "Classical Normal Modes in Damped Linear Dynamic Systems," *J. Appl. Mech.*, ASME, Paper No. 59-A-62, June 1960.

3-9 Wilson, E. L., Der Kiureghian, A., and Bayo, E. P., "A Replacement for the SRSS Method in Seismic Analysis," Earthquake Engineering and Structural Dynamics, Vol. 9, 1981.

3-10 Anderson, J. C., and Gurfinkel, G., "Seismic Behavior of Framed Tubes," *Int. J. Earthquake Eng. and Structural Dynamics*, 4, No. 2, Oct.–Dec. 1975.

3-11 Anderson, J. C. and Naeim, F., "Design Criteria and Ground Motion Effects on the Seismic Response of Multistory Buildings," Critical Aspects of Earthquake Ground Motion and Building Damage Potential, ATC-10-1, Applied Technology Council, Palo Alto, CA, 1984.

3-12 Wilson, E. L., Hollings, J. P., and Davey, H. H., "Three Dimensional Analysis of Building Systems (Extended Version)," EERC 75-13, University of California, Berkeley, CA, Apr. 1975.

Chapter 4

Static Lateral-Force Procedures

*Roger M. Di Julio Jr., Ph.D., P.E.**

4.1 INTRODUCTION

In order to design a structure to withstand an earthquake, the forces on the structure must be specified. The exact forces that will occur during the life of the structure cannot, of course, be known. A realistic estimate is important, however, since the cost of construction, and therefore the economic viability of the project, depends on a safe and cost-efficient final product.

The seismic forces in a structure depend on a number of factors, including the size and other characteristics of the earthquake, the distance from the fault, the site geology, and the type of lateral-load-resisting system. The importance of the structure may also be of concern in the design. These factors should be included in the specification of the seismic design forces.

There are two commonly used procedures for specifying seismic design forces: The *equivalent-static-force procedure* and *dynamic analysis*. In the equivalent-static-force procedure the inertial forces are specified as static forces using empirical formulas. The empirical formulas do not explicitly account for the dynamic characteristics of the particular structure being designed or analyzed. The formulas were, however, developed to adequately represent the dynamic behavior of what are called *regular* structures, which have a reasonably uniform distribution of mass and stiffness. For such structures the equivalent-static-force procedure is most often adequate.

Structures that do not fit into this category are termed *irregular*. Common irregularities include large floor-to-floor variation in mass or center of mass, and soft stories. Such structures violate the assumptions on which the empirical formulas used in the equivalent-static-force procedure are based. Therefore, its use may lead to erroneous results. In these cases a dynamic analysis should be used to specify and distribute the seismic design forces. A dynamic analysis can take a number of forms, but should take account of the irregularities of the structure by modeling its dynamic characteristics, including natural frequencies, mode shapes, and damping.

The purpose of this chapter is to review and compare the sections of current seismic design provisions which deal with the specification of seismic design forces. Emphasis will be, as in the documents discussed, on the equivalent-static-force procedure.

The following seismic design provisions are included in the discussion which follows:

1. The *Uniform Building Code*, issued by the International Conference of Building Officials, 1985 and 1988 editions, referred to as UBC-85 and UBC-88.
2. The *BOCA National Building Code*, issued by Building Officials and Code Administrators International, Inc., 1987 edition, referred to as BOCA-87.
3. The *American National Standard Minimum Design Loads for Buildings and Other Structures*, issued by the

*Professor of Engineering, California State University, Northridge, California.

American Standards Institute, Inc., 1982 edition, referred to as ANSI-82.

4. *Tentative Provisions for the Development of Seismic Regulations for Buildings*, prepared by the Applied Technology Council, Publication ATC 3-06, issued in 1978, referred to as ATC 3-06.

5. *NEHRP Recommended Provisions for the Development of Seismic Regulations for New Buildings*, prepared by the Building Seismic Safety Council for the Federal Emergency Management Agency, issued in 1985, referred to as NEHRP-85.

6. *Recommended Lateral Force Requirements and Commentary*, issued in draft form by the Seismology Committee of the Structural Engineers Association of California in December 1986, and republished with minor revisions in 1988. Herein referred to as the SEAOC Blue Book.

The Seismic-design-load provisions in UBC-85, BOCA-87, and ANSI-82 are very similar. Hence UBC-85 is discussed in detail in Section 4.3, and the major differences between it and BOCA-87 and ANSI-82 are covered in Section 4.4.

UBC-88, and SEAOC on which it is based, contain significant differences from UBC-85, and are discussed separately in Section 4.5.

NEHRP-85, and ATC 3-06 on which it is based, are significantly different from both UBC-85 and UBC-88, and are discussed in Section 4.6.

Finally, the codes are summarized and compared in Section 4.7.

4.2 CODE PHILOSOPHY

The philosophy of a particular document indicates the general level of protection that it can be expected to provide. Most code documents clearly state that their standards are minimum requirements that are meant to provide for life safety but not to insure against damage.

The code-specified forces are generally lower than those that would occur in a large or moderate earthquake. This is because the structure is designed to carry the specified loads within allowable stresses and deflections which are considerably less than the ultimate or even yield capacities of the materials and system. It is assumed that the larger loads that actually occur will be accounted for by the factors of safety and by the redundancy and ductility of the system. Life safety is thereby insured, but structural damage may be sustained.

4.3 UBC-85 PROVISIONS

4.3.1 Base Shear *V*

UBC-85 specifies the total seismic force, or base shear, acting on a structure by the formula

$$V = ZIKCSW \tag{4-1}$$

The base shear V is seen to be a percentage ($ZIKCS$) of the total dead load W of the structure. This percentage depends on the seismicity of the region, the soil conditions at the site, and the utilization importance, fundamental period of vibration, and type of lateral-load-resisting system of the structure.

4.3.2 Seismic Zone Coefficient *Z*

The seismicity of the region in which the structure is located is represented by the coefficient Z in Equation 4-1. This coefficient is defined by dividing the United States into five seismic zones on the basis of past earthquake activity. The zones are arranged in order of increasing expectation of damage, varying from no damage in zone 0 to a maximum in zone 4. The UBC-85 Seismic Zone Map and a description of the damage expected in each zone are shown in Figure 4-1. The numerical value of the coefficient is:

Zone	Coefficient Z
0	0
1	$\frac{3}{16}$
2	$\frac{3}{8}$
3	$\frac{3}{4}$
4	1

The base shear increases by a factor of more than 5 from zone 1 to zone 4.

4.3.3 Occupancy Importance Factor *I*

The base shear is increased for essential facilities using this factor. The UBC-85 defines essential facilities as "those structures or buildings which must be safe and usable for emergency purposes after an earthquake in order to preserve the health and safety of the general public." For such facilities the base shear is increased by 50 percent ($I = 1.5$). The base shear must also be increased by 25 percent ($I = 1.25$) for "any building where the primary occupancy is for assembly use for more than 300 persons (in one room)." For all other buildings $I = 1.0$.

4.3.4 Horizontal Force Factor *K*

This factor increases or decreases the base shear depending on the ductility assumed to be inherent in the lateral load-resisting system. The values of K and the definitions of the types of framing systems are presented in Table 4-1. The value of K for common building systems varies from 0.67 for a ductile moment-resisting frame to 1.33 for a box system.

The "all building frames except..." classification in Table 4-1 includes braced frames, shear walls, and ordinary moment frames. These systems are assigned a K of 1.0.

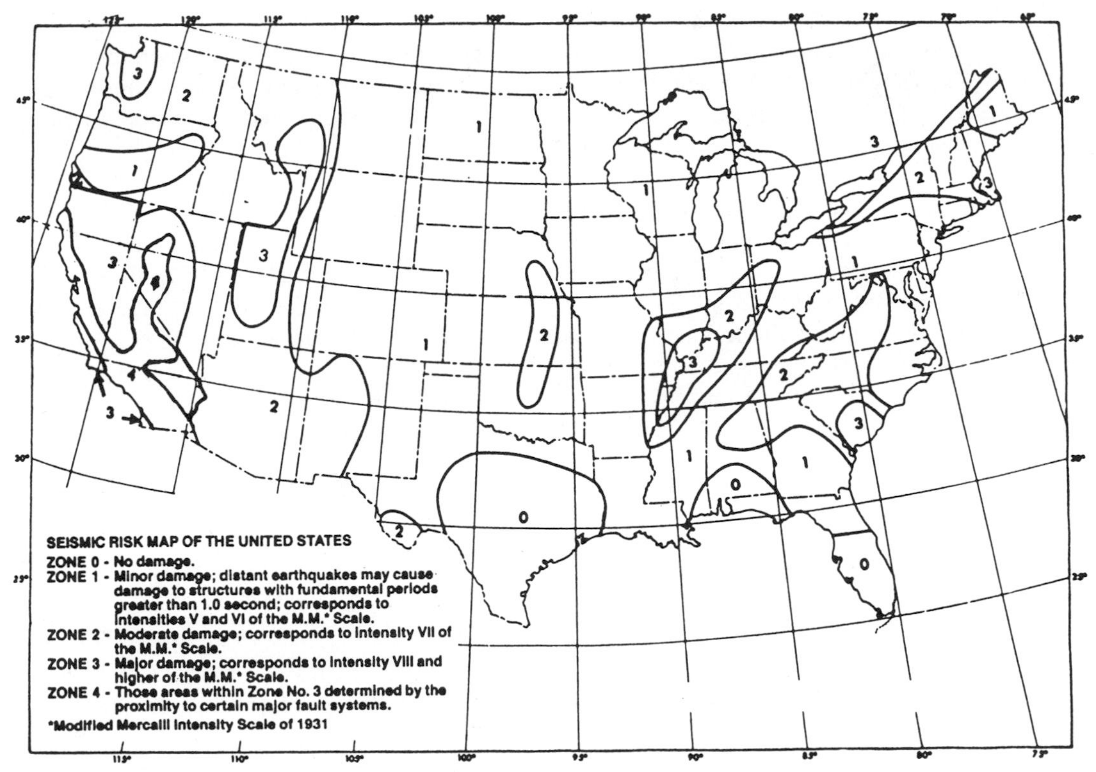

Figure 4-1 Seismic zone map of the United States according to UBC-85. (Reproduced from the Uniform Building Code, 1985 edition. Copyright 1985, with the permission of the publisher, the International Conference of Building Officials.)

Table 4-1 Horizontal force factor K for buildings or other structures according to UBC-85*†

	Type or arrangement of resisting elements	K
1.	All building framing systems except as hereinafter classified	1.00
2.	Buildings with a box system as specified in Section 2312(b) Exception: Buildings not more than three stories in height with stud wall framing and using plywood horizontal diaphragms and plywood vertical shear panels for the lateral force system may use $K = 1.0$.	1.33
3.	Buildings with a dual bracing system consisting of a ductile moment-resisting space frame and shear walls or braced frames using the following design criteria: a. The frames and shear walls or braced frames shall resist the total lateral force in accordance with their relative rigidities considering the interaction of the shear walls and frames b. The shear walls or braced frames acting independently of the ductile moment-resisting portions of the space frame shall resist the total required lateral forces c. The ductile moment-resisting space frame shall have the capacity to resist not less than 25% of the required lateral force	0.80
4.	Buildings with a ductile moment-resisting space frame designed in accordance with the following criteria: The ductile moment-resisting space frame shall have the capacity to resist the total required lateral force	0.67
5.	Elevated tanks plus full contents, on four or more cross-braced legs and not supported by a building	2.5‡
6.	Structures other than buildings and other than those set forth in Table No. 23-J	2.00

*Reproduced from the Uniform Building Code, 1985 edition, Copyright 1985, with the permission of the publisher, the International Conference of Building Officials.
†Where wind load as specified in Section 2311 would produce higher stresses, this load shall be used in lieu of the loads resulting from earthquake forces.
‡The minimum value of KC shall be 0.12 and the maximum value of KC need not exceed 0.25.

Ductile moment frames are detailed so that they are more ductile than ordinary moment frames, braced frames, or shear walls. They are assigned a K of 0.67, which has the effect of reducing the design static lateral forces by 33%. The drift requirement, however, is more strict for such frames.

Ductile moment frames used in combination with shear walls or braced frames are assigned a K of 0.8. The shear walls or braced frames, however, must be designed to resist the total lateral forces independently of the moment frames. In addition, the moment frames must be able to resist 25% of the total lateral load independently.

4.3.5 Period Coefficient *C*

The coefficient C, which represents the influence of the period of the structure on the base shear, is given by the formula

$$C = \frac{1}{15\sqrt{T}} \leqslant 0.12 \tag{4-2}$$

where T is the fundamental period of the building in the direction under consideration. This formula decreases the base shear for longer period buildings.

4.3.6 Building Period *T*

Calculation of the coefficient C requires a determination of the fundamental period of the building. This presents a problem because the period is not known before the building is designed. To circumvent this difficulty, the code provides empirical formulas that may be used to estimate the fundamental period of buildings. The period may be estimated by the formula

$$T = \frac{0.05h_n}{\sqrt{D}} \tag{4-3}$$

where h_n is the height and D is the dimension of the building, in feet, in the direction for which the period is being estimated.

For ductile moment-resisting space frames the period may be estimated by the formula

$$T = 0.1N \tag{4-4}$$

where N is the number of stories above the shear base.

At any point in the design sequence the period can be determined by a "properly substantiated analysis." The code explicitly allows for the use of Rayleigh's formula. Other methods of analysis are acceptable, however. See Chapter 3 for an explanation and an example of the application of Rayleigh's formula.

The period estimated by Equations 4-3 or 4-4 may differ substantially from the actual building period or the period determined by the analysis. For example, for steel buildings with ductile moment-resisting frames the periods predicted by Equation 4-4 are generally shorter than those determined by an analysis using the stiffness method.

An excessively long fundamental period will result in a design base shear which is substantially less than that determined using the empirical Equation 4-3 or 4-4. UBC-85 does not explicitly limit the period that can be used to determine the base shear. In practice, however, it is common to limit the base shear to some percentage of that established using the empirical formulas. A minimum base shear of 80 percent of that determined using the empirical formulas is a typical limitation.

4.3.7 Soil Profile Coefficient *S*

The coefficient S takes account of the soil conditions at the site and the relationship between the soil period and the period of the structure.

If a site period has not been determined, S is defined by a largely qualitative description of the soil. Three soil profiles, types S_1, S_2, and S_3, are defined on the basis of their stiffness. They are arranged in order of increasing flexibility. The value of the coefficient for each soil profile is:

Profile Type	Coefficient S
S_1	1.0
S_2	1.2
S_3	1.5

As can be seen, the base shear increases for more flexible soils. The description of the UBC-85 soil profiles is presented in Table 4-2.

If a site period T_s has been determined on the basis of a soil study, the coefficient S can be determined by formulas which represent the relationship between the period of the structure and the period of the site:

$$\text{for } \frac{T}{T_s} \leqslant 1.0, \quad S = 1.0 + \frac{T}{T_s} - 0.5\left(\frac{T}{T_s}\right)^2 \geqslant 1.0 \tag{4-5}$$

$$\text{for } \frac{T}{T_s} > 1.0, \quad S = 1.2 + 0.6\frac{T}{T_s} - 0.3\left(\frac{T}{T_s}\right)^2 \geqslant 1.0 \tag{4-6}$$

Equations 4-2, 4-5, and 4-6 are subject to the following restrictions:

$$0.5 \leqslant T_s \leqslant 2.5 \tag{4-7}$$

$$T \geqslant 0.3 \tag{4-8}$$

$$CS \leqslant 0.14 \tag{4-9}$$

Table 4-2 Soil Profile Coefficient According to UBC-85*

Soil profile type S_1: Rock of any characteristic, either shale-like or crystalline in nature (such material may be characterized by a shear wave velocity greater than 2500 (ft / sec); or stiff soil conditions where the soil depth is less than 200 ft and the soil types overlying rock are stable deposits of sands, gravels or stiff clays.

Soil profile type S_2: Deep cohesionless or stiff clay soil conditions, including sites where the soil depth exceeds 200 ft and the soil types overlying rock are stable deposits of sands, gravels or stiff clays.

Soil profile type S_3: Soft to medium-stiff clays and sands, characterized by 30 ft or more of soft to medium-stiff clay with or without intervening layers of sand or other cohesionless soils.

In locations where the soil properties are known in sufficient detail to determine the soil profile type or where the profile does not fit any of the three types, soil profile type S_3 shall be used.

*Reproduced from the Uniform Building Code, 1985 edition, Copyright 1985, with the permission of the publisher, the International Conference of Building Officials.

Also, the building period T *must* be determined by an analysis and not by the empirical formulas. If a range of site periods is given, the one closest to the building period must be used in the determination of S. This yields the highest base shear.

Using Equation 4-5 or 4-6, the site coefficient reaches a maximum of 1.5 when the building period and the site period are equal and reaches a minimum of 1.0 when the building period is more than $2\frac{1}{4}$ times longer than the site period.

There is considerable uncertainty in the determination of both T and T_s. Therefore, the calculation of the site coefficient using Equations 4-5 and 4-6 is imprecise. It is partly for this reason that this method of determining S has been removed from subsequent codes.

4.3.8 Dead Load *W*

The dead load W, used to calculate the base shear, includes not only the total dead load but also 25% of the floor live load in storage and warehouse occupancies and the weight of snow when the design snow load is greater than 30 lb/ft^2. The snow load may be reduced by up to 75% if the snow-load duration is short.

The rationale for including a portion of the snow load in heavy-snow areas is the fact that in these areas a significant amount of ice can build up and remain on roofs.

4.3.9 Distribution of Lateral Force F_x

The base shear V is distributed over the height of the structure as a force at each level (F_i) plus an extra force F_t at the top:

$$V = F_t + \sum_{i=1}^{n} F_i \tag{4-10}$$

The extra force at the top is:

$$F_t = 0.07TV \leqslant 0.25V \quad \text{if} \quad T > 0.7 \text{ sec} \tag{4-11a}$$

$$F_t = 0.0 \quad \text{if} \quad T \leqslant 0.7 \text{ sec} \tag{4-11b}$$

F_t takes account of the greater participation of higher modes in the response of longer-period structures.

The remaining portion of the total base shear ($V - F_t$) is distributed over the height, including the top, by the formula

$$F_x = (V - F_t)\frac{w_x h_x}{\sum_{i=1}^{n} w_i h_i} \tag{4-12}$$

where w is the weight at a particular level and h is the height of a particular level above the shear base. At each floor the force is located at the center of mass.

For equal story heights and weights, Equation 4-12 distributes the force linearly, increasing towards the top (see Figure 4-4 in Section 4.3.14 below). Any significant variation from this triangular distribution indicates an irregular structure.

4.3.10 Story Shear and Overturning Moment (V_x, M_x)

The story shear at level x, V_x, is the sum of all the lateral forces at and above that level.

$$V_x = F_t + \sum_{i=x}^{n} F_i \tag{4-13}$$

The overturning moment at a particular level, M_x, is the sum of the moments of the story forces above, about that level (see Figure 4-2). Hence

$$M_x = F_t(h_n - h_x) + \sum_{i=x}^{n} F_i(h_i - h_x) \tag{4-14}$$

The design must be based on the overturning moment as well as the shear at each level.

4.3.11 Torsion and P-Delta Effect

Accidental torsion, due to uncertainties in the mass and stiffness distribution, must be added to the calculated eccentricity. This is done by adding a torsional moment at each floor equal to the story shear multiplied by 5% of the floor dimension, and perpendicular to the direction of the force. This procedure is equivalent to moving the center of mass by 5% of the plan dimension, in a direction perpendicular to the force.

There is no requirement in UBC-85 to include P-delta effects in the design forces. In buildings where this effect is significant, however, it is common practice to include P-delta in the analysis (see Chapter 6).

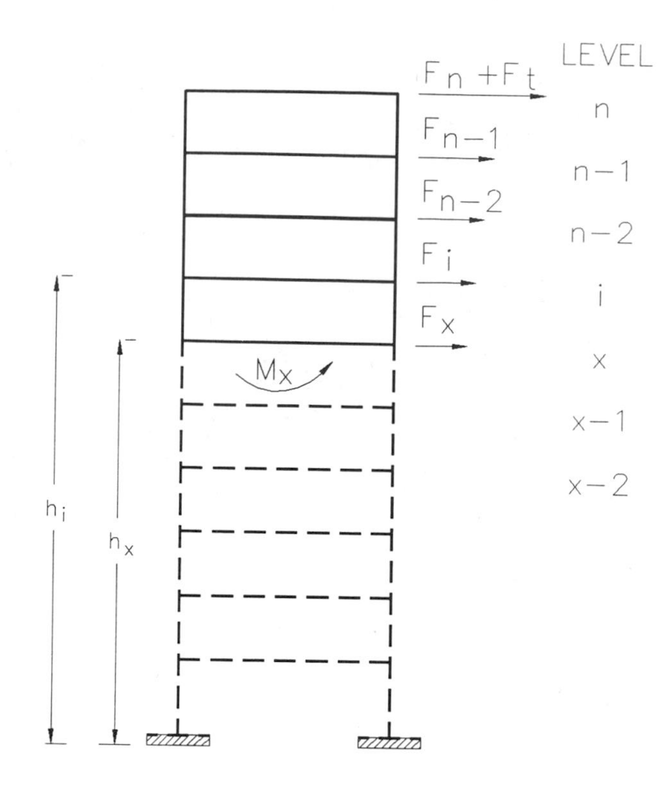

$$M = \sum_{i=x}^{n} F_i(h_i - h_x) + F_t(h_i - h_x)$$

Figure 4-2 Definition of overturning moment.

4.3.12 Drift Limitations

The allowable interstory seismic displacement is given by the formula

$$\Delta \leqslant 0.005Kh \tag{4-15}$$

where h is the story height and K is limited to a value of one or less. The factor K is the same horizontal force factor used in Equation 4-1 and taken from Table 4-1.

There is no requirement to include the P-delta effect in the calculation of lateral displacements.

UBC-85 is not explicit as to whether only average deflections are to be controlled, or if torsional deflections are to be included. If torsional deflections are considered, the deflections in the plane of the lateral load resisting elements must be determined.

4.3.13 Irregular Structures; Dynamic Characteristics

The distribution of force prescribed in the previous paragraphs is based on assumptions that are applicable only to *regular* structures (see Chapter 5 for a detailed discussion of irregularity). Dynamic effects due to unusual mass or stiffness distributions are not included by the equivalent static force procedure.

UBC-85 describes "irregular structures" only qualitatively as

> structures which have highly irregular shapes, large differences in lateral resistance or stiffness between adjacent stories or other unusual features.

For such structures lateral forces must be determined considering the dynamic characteristics of the structure. This code gives no details on how to implement this procedure.

4.3.14 Examples

Example 4-1

Determine the UBC-85 design seismic forces for a nine-story ductile moment-resisting steel-frame office building located in Los Angeles, California on soil profile type 2. The story heights are all 13 ft; the plan area is 100 ft by 170 ft (see Figure 4-3). The total dead load is 100 lb/ft^2 at all levels. Sample calculations are presented below, and a complete tabulation is found in Table 4-3. A plot of the story forces is shown in Figure 4-4.

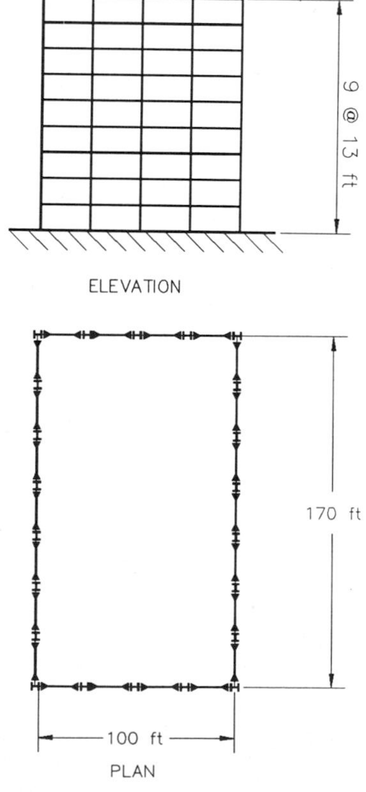

Figure 4-3 Plan and elevation of the building used in Example 4-1.

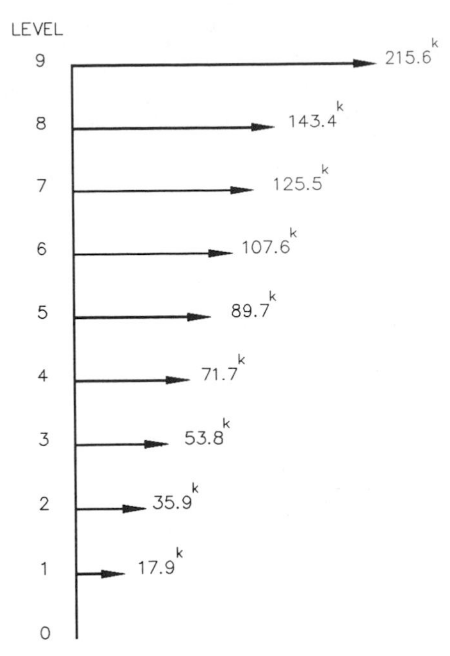

Figure 4-4 Lateral story forces for Example 4-1.

Table 4-3

Level	h_x, ft	w_x, kips	$w_x h_x$, 10^{-3}	$F_i + F_t$, kips	V_x, kips	M_x, ft-kips
9	117	1700	198.9	215.6	215.6	2803
8	104	1700	176.8	143.4	359.0	7469
7	91	1700	154.7	125.5	484.5	13758
6	78	1700	132.6	107.6	592.1	21466
5	65	1700	110.5	89.7	681.8	30329
4	52	1700	88.4	71.7	753.5	40125
3	39	1700	66.3	53.8	807.3	50619
2	26	1700	44.2	35.9	843.2	61581
1	13	1700	22.1	17.9	861.1	72775
Σ		15300	994.5	861.1		

• Base shear:

$$V = ZIKCSW$$

$Z = 1.0$ (zone 4)

$I = 1.0$ (office building)

$K = 0.67$ (ductile moment frame)

$S_2 = 1.2$ (soil type 2)

$T = 0.1N = 0.9$ sec (moment frame)

$$C = \frac{1}{15\sqrt{T}} = \frac{1}{15\sqrt{0.9}} = 0.07 < 0.12.$$

$$CS = 0.07(1.2) = 0.084 < 0.14$$

$$w = 0.1(170)(100) = 1700 \text{ kips/floor}$$

$$W = 9(1700) = 15{,}300 \text{ kips}$$

$$V = (1.0)(1.0)(0.67)(0.07)(1.2)(15{,}300) = 861.1 \text{ kips}$$

• Vertical distribution;

$$T > 0.7 \text{ sec}$$

$$F_t = 0.07\,TV = 0.07(0.9)(861.1) = 54.2$$

$$0.25V = 0.25(861.1) = 215.3 > 54.2$$

$$F_t = 54.2 \text{ kips}$$

$$V - F_t = 861.1 - 54.2 = 806.9$$

$$F_x = (V - F_t)\frac{w_x h_x}{\sum_{i=i}^{n} w_i h_i} \qquad \text{(story force)}$$

$$F_9 + F_t = \frac{806.9(1700)(117)}{994{,}500} + 54.2 = 215.6 \text{ kips}$$

$$F_8 = \frac{806.9(1700)(104)}{994{,}500} = 143.4 \text{ kips}$$

etc.

• Story shear:

$$V_x = F_t + \sum_{i=x}^{n} F_i$$

$$V_9 = 215.6 \text{ kips}$$

$$V_8 = 215.6 + 143.4 = 359.0 \text{ kips}$$

etc.

• Overturning moment:

$$M_x = F_t(h_n - h_x) + \sum_{i=x}^{n} F_i(h_i - h_x)$$

$$M_9 = 215.6(13) = 2803 \text{ ft-kips}$$

$$M_8 = 215.6(26) + 143.4(13) = 7469 \text{ ft-kips}$$

etc.

• Allowable interstory displacement:

$$\Delta \leqslant 0.005Kh = 0.005(0.67)(13 \times 12) = 0.52 \text{ in.}$$

Example 4-2

Determine the UBC-85 design seismic forces for a three-story concrete-shear-wall office building. It is located in Los Angeles on soil type 1. The story heights are 13 ft for the first floor and 11 ft for the second and third floors. The story dead loads are 2200, 2000, and 1700 kips from the bottom up. The plan dimension in the direction under consideration is 120 ft (see Figure 4-5). The shear walls do not carry vertical loads.

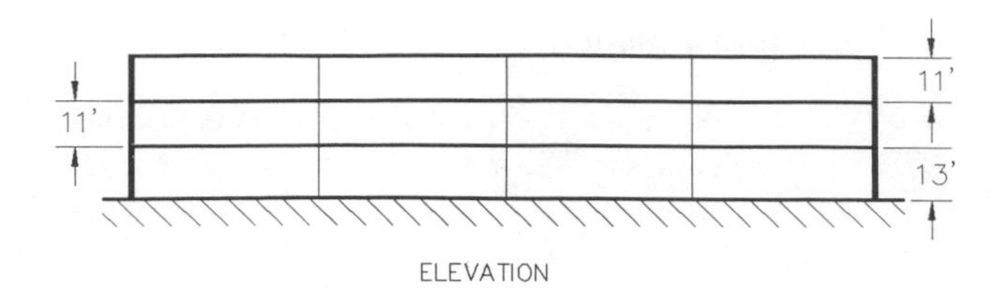

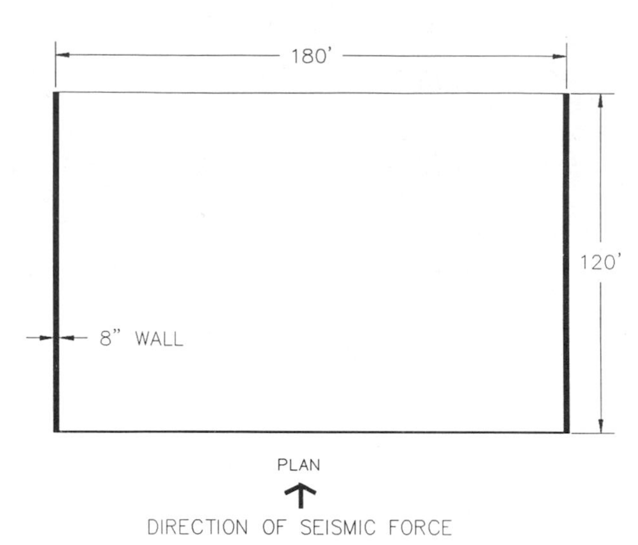

Figure 4-5 Plan and elevation of the building used in Example 4-2.

- Base shear:

$$V = ZIKCSW$$

$Z = 1.0$ (zone 4)

$I = 1.0$ (office building)

$K = 1.0$ (shear-wall system)

$S_1 = 1.0$ (soil type 1)

$$T = \frac{0.05h_n}{\sqrt{D}} = \frac{0.05(35)}{\sqrt{120}} = 0.16 \text{ sec}$$

$$C = \frac{1}{15\sqrt{T}} = \frac{1}{15\sqrt{0.16}} = 0.17 > 0.12$$

$$CS = (0.12)(1.0) = 0.12 < 0.14$$

$$W = 1700 + 2000 + 2200 = 5900 \text{ kips}$$

$$V = (1.0)(1.0)(1.0)(0.12)(1.0)(5900) = 708.0 \text{ kips}$$

- Vertical distribution:

$$T = < 0.7 \text{ sec}$$
$$F_t = 0.0$$

Level	h_x, ft	w_x, kips	$w_x h_x$, 10^{-3}	$F_i + F_t$, kips	V_x, kips	M_x, ft-kips
3	35	1700	59.5	309.5	309.5	3405
2	24	2000	48.0	249.7	559.2	9556
1	13	2200	28.6	148.8	708.0	18760
Σ		5900	136.1	708.0		

- Allowable interstory displacements:
  1st floor:

$$\Delta \leqslant 0.005Kh = 0.005(1.0)(13 \times 12) = 0.78 \text{ in.}$$

2nd and 3rd floors:

$$\Delta \leqslant 0.005(1.0)(11 \times 12) = 0.66 \text{ in.}$$

4.4 BOCA-87 AND ANSI-82 PROVISIONS

The 1987 BOCA National Building Code is a complete revision of the previous edition, and its provisions for seismic loads are essentially equivalent to those of UBC-85. ANSI-82 is virtually identical to BOCA-87 in its major seismic-design-load provisions. Therefore only the differences in major provisions between UBC-85 and BOCA-87 are mentioned here.

4.4.1 Seismic Zone Coefficient *Z*

The seismic zone coefficient Z is given a numerical value of $\frac{1}{8}$ for zone 0. It has a value of 0 in UBC-85.

4.4.2 Building Period *T*

The empirical formula for the period of frame buildings has been revised to

$$T = C_T h_n^{3/4} \tag{4-16}$$

where $C_T = 0.035$ for steel frames, $C_T = 0.030$ for concrete frames, and h_n is the height of the building in feet. This formula is plotted and compared with the UBC-85 formula in Figures 4-9 and 4-10 in Section 4.7.1 below.

The UBC-85 period formula for buildings other than moment-resisting space frames has been clarified. For shear walls and exterior concrete frames with deep beams, or wide piers, or both,

$$T = \frac{0.05h_n}{\sqrt{D}} \tag{4-17}$$

where h_n is the height of the building and D is the plan dimension of the structure, in feet, in the direction parallel to the applied forces. For isolated shear walls not interconnected by frames or for braced frames,

$$T = \frac{0.05h_n}{\sqrt{D_S}} \tag{4-18}$$

where D_S is the largest plan dimension of a shear wall or braced frame in the direction parallel to applied forces.

The period determined by an analysis may be substantially different from that established on the basis of an empirical formula. This problem, which was discussed in the previous section, is specifically addressed in BOCA-87, which specifies that the coefficient C determined using a period established by an analysis must be at least 80 percent of that determined using the appropriate empirical formula. This provision has the effect of establishing a minimum base shear that can be justified on the basis of an analytically determined period.

4.4.3 Soil Profile Coefficient *S*

The coefficient S is specified only on the basis of a qualitative description of the soil. The formulas which relate the site period to the building period (Equations 4-5 and 4-6) have been removed.

The descriptions of the soil types and the values of the coefficient S are identical to those in UBC-85.

4.4.4 Torsion

BOCA-87 explicitly requires that both accidental torsion and torsion resulting from the location of the building center of mass taken into account in computing displacements. The allowable interstory displacements should therefore be checked in the plane of the lateral load-resisting elements.

4.4.5 Overturning Moment M_x

For buildings over ten stories in height the overturning moment may be reduced in the lower floors. The procedure for doing this is identical to that in NEHRP-85 and is explained in Section 4.6.9.

4.4.6 Irregular Structures; Dynamic Characteristics

BOCA-87 provides slightly more guidance than UBC-85 on the treatment of irregular buildings. It says that lateral forces in irregular buildings must be, and in regular buildings may be, determined by an elastic or inelastic dynamic analysis. Such an analysis must consider the dynamic characteristics of the structure.

The ground-motion input can be either a smoothed response spectrum or a number of ground-motion time histories for the site. Most commonly, the dynamic characteristics of the building would be established by its natural frequencies, mode shapes, and damping, the analysis would be linear–elastic, and the ground-motion input would be a smoothed spectrum.

The base shear computed by a dynamic analysis must be at least 90 percent of that determined using the equivalent-lateral-force procedure. In addition, the drift limitations must be satisfied for the equivalent static lateral forces.

4.5 SEAOC AND UBC-88 PROVISIONS

The 1988 SEAOC Blue Book, which has been adopted as UBC-88 with minor differences, provides for the use of the equivalent-static-force procedure or a dynamic analysis for regular structures under 240 ft tall, and requires a dynamic analysis for irregular structures and structures over 240 ft tall.

The equivalent-static-force procedure is similar in concept to that used in UBC-85 but is different in its details. UBC-88 is much more specific in its definition of structural systems, irregularities, and the details of dynamic analysis.

4.5.1 Base Shear *V*

The base shear is specified by the formula

$$V = ZICW/R_w \qquad (4\text{-}19)$$

where Z is a seismic zone factor, I is an importance factor, C is a numerical coefficient dependent on the soil conditions at the site and the period of the structure, W is the dead load of the structure, and R_w is a factor which represents the ductility of the structural system.

4.5.2 Seismic Zone Factor *Z*

Five seismic zones, numbered 1, 2A, 2B, 3, 4, are defined. The zone for a particular site is determined from a seismic-zone map (see Figure 4-6). The numerical values of Z, which are different from those found in UBC-85, are:

Zone	Coefficient Z
1	0.075
2A	0.15
2B	0.2
3	0.3
4	0.4

The value of the coefficient thus normalized can be viewed as the peak ground acceleration, in percentage of gravity, in each zone.

4.5.3 Importance Factor *I*

The importance factor I is used to increase the margin of safety for "essential" and hazardous facilities. For such structures $I = 1.25$. This is less than the maximum value of $I = 1.50$ in UBC-85, but additional quality-assurance requirements are included which should further increase the margin of safety.

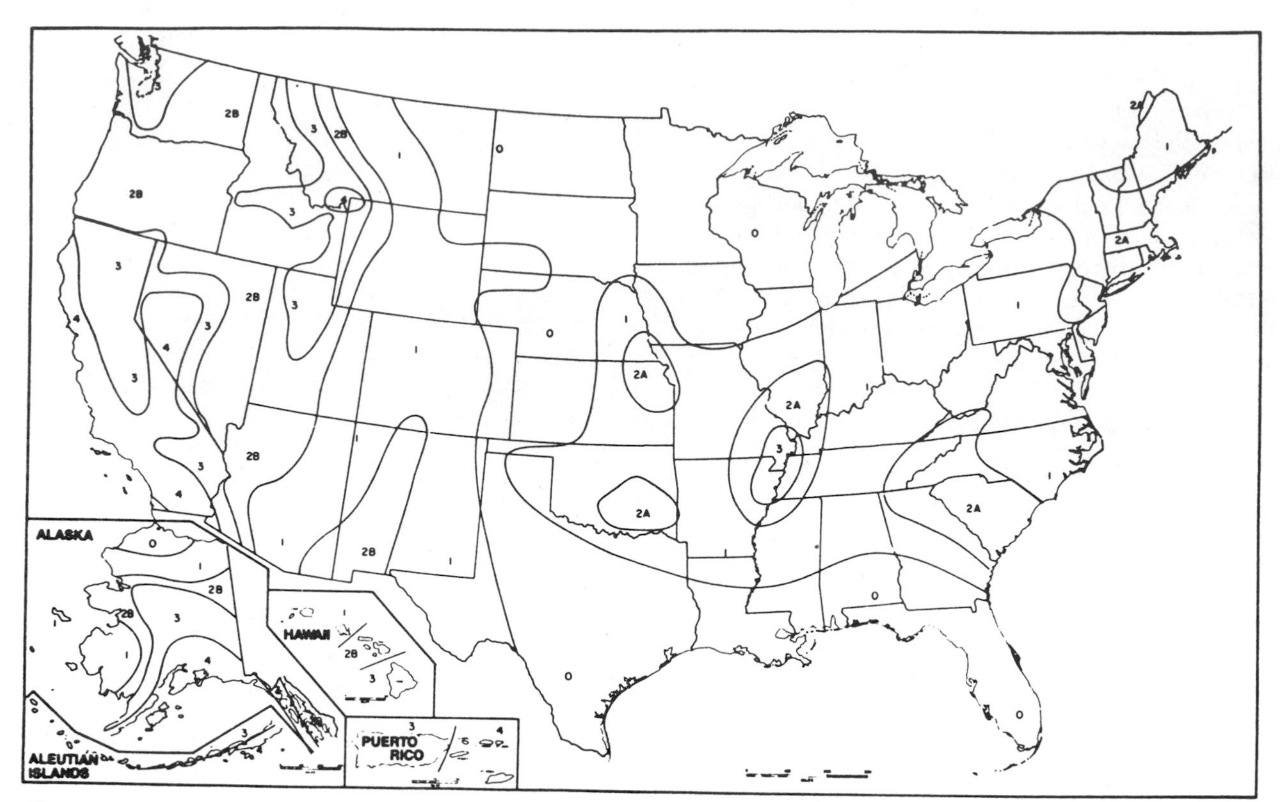

Figure 4-6 Seismic zone map of the United States according to SEAOC-86. (Reprinted from "Lateral Force Requirements, December 1986," with permission of Structural Engineers Association of California.)

"Special occupancy" structures, for which $I = 1.25$ in UBC-85, and "standard occupancy" structures are assigned a value of $I = 1.0$. Both require increased review and quality assurance procedures, however.

4.5.4 Site Coefficient *C*

The coefficient C modifies the zone factor Z to produce a maximum response acceleration CZ. This coefficient, which is dependent on the soil characteristics of the site and the fundamental period of the structure, is specified by the formula

$$C = 1.25S/T^{2/3} \tag{4-20}$$

where S is the site coefficient and T is the fundamental period of the building. S and T are discussed in the following subsections. The value of C need not exceed 2.75.

The minimum value of the ratio C/R_w is 0.075. This has the effect of specifying a minimum ground acceleration for tall moment-frame buildings.

4.5.5 Building Period *T*

The building period may be determined by analysis or using empirical formulas. A single empirical formula, the same one used for frames in BOCA-87, is extended to other framing systems:

$$T = C_t h_n^{3/4} \tag{4-21}$$

where

$$C_t = \begin{cases} 0.035 & \text{for steel moment frames} \\ 0.030 & \text{for reinforced-concrete moment frames and eccentric-braced frames} \\ 0.020 & \text{for all other buildings} \end{cases}$$

h_n = height, in feet, above the base to level n

If the period is determined using Rayleigh's formula or another method of analysis, the value of C must be at least 80% of the value obtained by using the appropriate empirical formula. This provision is included to eliminate the possibility of using an excessively long calculated period to justify an unreasonably low base shear.

4.5.6 Site Coefficient *S*

The site coefficient serves the same function as in UBC-85, that is, to include the effect of the soil conditions at the site on the ground acceleration. As in BOCA-87, the formulas specifying S on the basis of the ratio of the building period to the site period, Equations 4-5 and 4-6, have not been included. The coefficient is specified on the basis of a qualitative description of the soil (see Table 4-4). Four values of S ranging from 1.0 to 2.0 are specified.

Table 4-4 Site Coefficients According to SEAOC*

Type	Description	S Factor†
S_1	A soil profile with either (a) A rock-like material characterized by a shear-wave velocity greater than 2500 ft/sec or by other suitable means of classification, or (b) stiff or dense soil condition where the soil depth is less than 200 ft.	1.0
S_2	A soil profile with dense or stiff soil conditions, where the soil depth exceeds 200 ft or more.	1.2
S_3	A soil profile 40 ft or more in depth and containing more than 200 ft of soft to medium stiff clay but not more than 40 ft of soft clay.	1.5
S_4	A soil profile containing more than 40 ft of soft clay.	2.0

*Reprinted from "Lateral Force Requirements, December 1986," with permission of Structural Engineers Association of California.

†The site factor shall be established from properly substantiated geotechnical data. In locations where the soil properties are not known in sufficient detail to determine the soil profile type, soil profile S_3 will be used unless the Building Official determines that soil profile S_4 may be present at the site, in which case soil profile S_4 will be used.

4.5.7 Structural System Coefficient R_w

The structural system coefficient R_w serves the same function as the coefficient K in UBC-85. It is a measure of the ductility of the structural system, based primarily on the performance of similar systems in past earthquakes. The approximate relationships between the two coefficients is $R_w = 8/K$.

The values of R_w for various structural systems is found in Table 4-5. The definitions of the lateral-load-resisting systems are more detailed and clear than in previous codes.

4.5.8 Dead Load W

The dead load is defined the same as in UBC-85. It is the total dead load plus 25% of floor live load in storage and warehouse occupancies and the weight of snow when the design snow load is greater than 30 lb/ft^2. The snow load may be reduced by up to 75% where the snow-load duration is short.

4.5.9 Distribution of Lateral Force F_x and Overturning Moment M_x

The procedure for distributing the lateral force over the height once the base shear has been determined and the determination of the overturning moment are the same as in UBC-85 (see Section 4.3 and Example 4-1).

4.5.10 Torsion and P-Delta Effect

Accidental torsion must be taken into account. This is done by moving the center of mass in each direction perpendicular to the applied force. As in UBC-85, the center of mass is moved 5 percent of the plan dimension in that direction.

P-delta effects must be included in determining member forces and story displacements where significant.

4.5.11 Drift Limitations

The interstory displacement for buildings which are 65 ft or taller cannot exceed

$$\Delta \leqslant 0.03h/R_w \tag{4-22}$$

nor

$$\Delta \leqslant 0.004h \tag{4-23}$$

For buildings less than 65 ft in height, the interstory displacement cannot exceed

$$\Delta \leqslant 0.04h/R_w \tag{4-22a}$$

nor

$$\Delta \leqslant 0.005h \tag{4-23a}$$

where h is the story height and R_w is the structural system coefficient defined in Table 4-5. The displacement is defined to include both translation and torsion. Hence, the drift must be checked in the plane of the lateral-load-resisting elements. P-delta displacements must be included where significant.

The limitations given by Equations 4-22 and 4-23 are more severe than those in UBC-85 (Equation 4-15), but this is partially mitigated by the fact that the forces specified by UBC-88 are generally lower. In addition, the drift limitations may be satisfied using forces which are not subject to the 80% limit discussed in Section 4.5.5 or the restriction on ratio C/R_w as discussed in Section 4.5.4.

4.5.12 Irregular Structures

UBC-88 quantifies the notion of irregularity, which it breaks into two broad categories: vertical structural and plan structural irregularity. Vertical irregularities include soft or weak stories, large changes in mass from floor to floor, and large discontinuities in the dimensions or in-plane locations of lateral load-resisting elements from floor to floor. Plan irregular buildings include those which undergo substantial torsion when subjected to seismic loads, or which have reentrant corners, discontinuities in floor diaphragms, discontinuity in the lateral force path, or lateral-load-resisting elements which are not parallel to each other or to the axes of the building.

The definitions of these irregularities are found in Tables 4-6 and 4-7. For a more detailed discussion of irregularity see Chapter 5.

Table 4-5 Description of Lateral-Load-Resisting Systems*

Basic structural systems	Lateral-load-resisting system — description	R_w	H†
A. Bearing-wall system	1. Light Framed Walls with Shear Panels		
	a. Plywood walls for structures 3 stories or less	8	65
	b. All other light framed walls	6	65
	2. Shear Walls		
	a. Concrete	6	160
	b. Masonry	6	120
	3. Light Steel Framed Bearing Walls with Tension-Only Bracing	4	65
	4. Braced Frames where Bracing Carries Gravity Loads		
	a. Steel	6	160
	b. Concrete‡	4	—
	c. Heavy Timber	4	65
B. Building-frame system	1. Steel Eccentric Braced Frame (EBF)	10	240
	2. Light Framed Walls with Shear Panels		
	a. Plywood walls for structures 3 stories or less	9	65
	b. All other light framed walls	7	65
	3. Shear Walls		
	a. Concrete	8	240
	b. Masonry	8	160
	4. Concentric Braced Frames		
	a. Steel	8	160
	b. Concrete‡	8	—
	c. Heavy timber	8	5
C. Moment-resisting-frame system	1. Special Moment-Resisting Space Frames (SMRSF)		
	a. Steel	12	N.L.¶
	b. Concrete	12	N.L.
	2. Concrete Intermediate Moment-Resisting Space Frames (IMRSF)	7	—
	3. Ordinary Moment-Resisting Space Frames		
	a. Steel	6	160
	b. Concrete‡	5	—
D. Dual system	1. Shear Walls		
	a. Concrete with SMRSF	12	N.L.
	b. Concrete with concrete IMRSF	9	160
	c. Masonry with SMRSF	8	160
	d. Masonry with concrete IMRSF‡	7	—
	2. Steel EBF with Steel SMRSF	12	N.L.
	3. Concentric Braced Frames		
	a. Steel with Steel SMRSF	10	N.L.
	b. Concrete with concrete SMRSF‡	9	—
	c. Concrete with concrete IMRSF‡	6	—
E. Undefined systems	See Section 1D8c	—	—

*Reprinted from "Lateral Force Requirements, December 1986," with permission of Structural Engineers Association of California.
†H = height limit applicable to Seismic Zones 3 and 4.
‡Prohibited in Seismic Zones 3 and 4.
¶N.L. = no limit

4.5.13 Dynamic Lateral-Force Procedure

A major change from current practice is included. UBC-88 requires that, if the base shear determined by a dynamic analysis is less than that specified by the static lateral-force procedure, it must be scaled to equal that determined by the equivalent-static-force procedure. Similarly, if the base shear obtained from a dynamic analysis is greater than that specified by the static lateral-force procedure, it may be scaled down to the value specified by static lateral-force procedure. In this manner, the dynamic characteristics of the structure are

Table 4-6 Plan Structural Irregularities*

Irregularity type and definition
A. Torsional irregularity, to be considered when diaphragms are not flexible. Torsional irregularity shall be considered to exist when the maximum story drift, computed including accidental torsion, at one end of the structure transverse to an axis is more than 1.2 times the average of the story drifts of the two ends of the structure.
B. Reentrant corners. Plan configurations of a structure and its lateral force resisting system contain reentrant corners, where both projections of the structure beyond a reentrant corner are greater than 15% of the plan dimension of the structure in the given direction.
C. Diaphragm discontinuity. Diaphragms with abrupt discontinuities or variations in stiffness, including those having cutout or open areas greater than 50% of the gross enclosed area of the diaphragm.
D. Out-of-plane offsets. Discontinuities in a lateral force path, such as out-of-plane offsets of the vertical elements.
E. Nonparallel systems. The vertical lateral load resisting elements are not parallel to nor symmetric about the major orthogonal axes of the lateral force resisting system.

*Reprinted from "Lateral Force Requirements, December 1986," with permission of Structural Engineers Association of California.

modeled, and thus the forces are distributed properly, while the code level forces are maintained.

Both the author and the editor feel that this is not a good provision. It is logical to expect a site-specific design spectrum, which has been developed by established procedures, to generate a different base shear (either higher or lower) than the generally applicable spectra provided in the code. It does not seem reasonable that the additional information gathered by detailed study or the specific site be essentially ignored.

4.5.14 Examples

Example 4-3

Rework Example 4-1 using UBC-88.

- Base shear:

$$V = ZICW/R_w$$

$Z = 0.4$ (zone 4)

$I = 1.0$ (office building)

$S_2 = 1.2$ (soil type 2)

$R_w = 12$ (special steel moment frame)

Table 4-7 Vertical Structural Irregularities*

Irregularity type and definition
A. Stiffness irregularity — soft story. A soft story is one in which the lateral stiffness is less than 70% of that in the story immediately above or less than 40% of the combined stiffnesses of the three stories above.
B. Weight (mass) irregularity. Mass irregularity shall be considered to exist where the effective mass of any story is more than 150% of the effective mass of an adjacent story. A roof which is lighter than the floor below need not be considered a mass irregularity.
C. Vertical geometric irregularity. Vertical geometric irregularity shall be considered to exist where the horizontal dimension of the lateral force resisting system in any story is more than 130% of that in an adjacent story. One-story penthouses need not be considered.
D. In-plane discontinuity in vertical lateral force resisting element. An in-plane offset of the lateral load resisting elements greater than the length of those elements.
E. Discontinuity in capacity — weak story. A weak story is one in which the story strength is less than 80% of that in the story above. The story length is the total strength of all seismic resisting elements sharing the story shear for the direction under consideration.

*Reprinted from "Lateral Force Requirements, December 1986," with permission of Structural Engineers Association of California.

$$T = 0.035h_n^{3/4} = 0.035(117)^{3/4} = 1.25 \text{ sec}$$

$$C = 1.25S/T^{2/3} = 1.25(1.2)/1.25^{2/3}$$

$$= 1.30 < 2.75$$

$$C/R_w = 1.30/12 = 0.11 > 0.075$$

$$W = 15{,}300 \text{ kips}$$

$$V = \frac{0.4(1.0)(1.30)(15{,}300)}{12} = 663.0 \text{ kips}$$

- Vertical distribution: The process is the same as for UBC-85. See Table 4-8.

$$T > 0.7 \text{ sec}$$

$$F_t = 0.07TV = 0.07(1.25)(663.0) = 58.0 \text{ kips}$$

$$V - F_t = 663 - 58 = 605 \text{ kips}$$

- Allowable interstory displacements:

$$0.03/R_w = 0.03/12 = 0.0025 < 0.004$$

$$\Delta \leqslant 0.0025h = 0.0025(13)(12) = 0.39 \text{ in.}$$

Table 4-8

Level	h_x, ft	w_x, kips	w_xh_x, 10^{-3}	$F_i + F_t$, kips	V_x, kips	M_x, ft-kips
9	117	1700	198.9	179.0	179.0	2327
8	104	1700	176.8	107.6	286.6	6053
7	91	1700	154.7	94.1	380.7	11002
6	78	1700	132.6	80.7	461.4	17000
5	65	1700	110.5	67.3	528.7	23873
4	52	1700	88.4	53.7	582.4	31444
3	39	1700	66.3	40.3	622.7	39540
2	26	1700	44.2	26.9	649.6	47984
1	13	1700	22.1	13.4	663.0	56603
Σ		15300	994.5	663.0		

Example 4-4

Rework Example 4-2 using UBC-88. The two shear walls are 8 in. thick and run the full 120 ft in the direction under consideration.

- Base shear:

$$V = ZICW/R_w$$

$$Z = 0.4 \quad \text{(zone 4)}$$

$$I = 1.0 \quad \text{(office building)}$$

$$S_1 = 1.0 \quad \text{(soil type 1)}$$

$$R_w = 8 \quad \text{(concrete shear walls)}$$

$$T = 0.02h^{3/4}$$

$$T = 0.02h^{3/4} = 0.02(35)^{3/4} = 0.29 \text{ sec}$$

$$C = \frac{1.25S}{T^{2/3}} = \frac{1.25(1.0)}{0.29^{2/3}} = 2.87 > 2.75$$

$$C/R_w = 2.75/8 = 0.34 > 0.075$$

$$V = \frac{0.4(1.0)(2.75)(5900)}{8} = 811.3 \text{ kips}$$

- Vertical distribution:

$$T > 0.7 \text{ sec}$$

$$F_t = 0$$

Level	h_x, ft	w_x, kips	w_xh_x, 10^{-3}	F_i, kips	V_x, kips	M_x, ft-kips
3	35	1700	59.5	354.7	354.7	3902
2	24	2000	48.0	286.1	640.8	10951
1	13	2200	28.6	170.5	811.3	21497
Σ		5900	136.1	811.3		

Allowable interstory displacements:

$$0.03/R_w = 0.03/8 = 0.00375 < 0.004$$

1st floor:

$$\Delta \leqslant 0.00375h = 0.00375(13)(12) = 0.59 \text{ in.}$$

2nd and 3rd floors:

$$\Delta \leqslant 0.00375h = (0.00375)(11)(12) = 0.50 \text{ in.}$$

4.6 NEHRP-85 AND ATC 3-06 PROVISIONS

ATC 3-06 was released as tentative provisions. After review it was released as a code in NEHRP-85. The differences between the documents in the specification of design lateral loads are minor. Where they do differ the NEHRP-85 provision is cited herein.

4.6.1 Base Shear *V*

NEHRP-85 specifies the base shear acting on a structure by the formula

$$V = C_sW \tag{4-24}$$

Again the total force is specified as a percentage C_s of the total gravity load of the building W.

4.6.2 Gravity Load *W*

The total gravity load consists of the total weight of the structure including permanent attachments plus 25% of the floor load for storage and warehouse structures plus the effective snow load. The effective snow load is defined as 70% of the full snow load. This can be reduced to as low as 20% if the snow load is of short duration.

4.6.3 Seismic Design Coefficient C_s

The seismic design coefficient is determined from the formula

$$C_s = 1.2A_vS/RT^{2/3} \leqslant 2.5A_a/R \tag{4-25}$$

where

A_v = effective peak velocity-related acceleration
A_a = effective peak acceleration
S = coefficient for the soil characteristics of the site
R = response modification factor
T = fundamental period of the building

The seismic design coefficient includes the ground motion through the peak acceleration coefficients A_v and A_a, the soil conditions at the site through the coefficient S, the period of the building (T), and the type of lateral load resisting system through the response modification factor R.

4.6.4 Building Period *T*

The building period can be estimated using one of the following empirical formulas. For moment-resisting frames,

$$T_a = C_T h_n^{3/4} \tag{4-26}$$

where,

$$C_T = \begin{cases} 0.035 & \text{for steel frames} \\ 0.030 & \text{for concrete frames} \end{cases}$$

h_n = the height of the building in feet

For all other buildings,

$$T_a = 0.05h_n/\sqrt{L} \quad (4\text{-}27)$$

where L is the overall length, in feet, of the building at the base in the direction under consideration.

The period may also be determined on the basis of an analysis. The period used to determine the base shear in Equation 4-25 is subject to an upper limit which is based on the effective peak velocity-related acceleration (A_v).

The relationship between A_v and the maximum allowable period used to specify the base shear is:

A_v	T_{max}/T_a
0.4	1.2
0.3	1.3
0.2	1.4
0.15	1.5
0.1	1.7
0.05	1.7

This limitation reflects the position discussed previously, that an excessively long analytically determined period should not be used to justify an unrealistically low design base shear.

4.6.5 Site Coefficient *S*

The coefficient S represents the effect of the site conditions on the building response. The soil profile types are based on a qualitative description of the soil conditions at the site. These descriptions are identical to those in UBC-85 (see Table 4-2). As in UBC-85, the coefficient is given the following values:

Soil Profile Type	Coefficient S
S_1	1.0
S_2	1.2
S_3	1.5

4.6.6 Response Modification Factor *R*

The response modification factor R performs the same function as K in UBC-85 and R_w in UBC-88. That is to represent the ductility and damping inherent in the structural system when it experiences deflections large enough to cause significant yielding or nears its ultimate capacity. The value of R is based on the performance of buildings in past earthquakes. The values of the coefficient are found in Table 4-9.

Table 4-9 Response Modification Coefficients from NEHRP-85*

Type of structural system	Vertical seismic resisting system	Coefficients R	Coefficients C_d
Bearing-wall system: A structural system with bearing walls providing support for all or major portions of the vertical loads.	Light framed walls with shear panels	$6\frac{1}{2}$	4
Seismic-force resistance is provided by shear walls or braced frames.	Reinforced-concrete shear walls	$4\frac{1}{2}$	4
	Reinforced masonry shear walls	$3\frac{1}{2}$	3
	Braced frames	4	$3\frac{1}{2}$
	Unreinforced masonry shear walls	$1\frac{1}{4}$	$1\frac{1}{4}$
Building-frame system: A structural system with essentially complete space frame providing support for vertical loads	Light framed walls with shear panels	7	$4\frac{1}{2}$
Seismic-force resistance is provided by shear walls or braced frames.	Reinforced concrete shear walls	$5\frac{1}{2}$	5
	Reinforced masonry shear walls	$4\frac{1}{2}$	4
	Braced frames	5	$4\frac{1}{2}$
	Unreinforced masonry shear walls	$1\frac{1}{2}$	$1\frac{1}{2}$

Table 4-9 Continued

Type of Structural System	Vertical Seismic Resisting System	Coefficients	
		R	C_d
Moment-resisting frame system: A structural system with an essentially complete space frame providing support for vertical loads.	Special moment frames of steel	8	$5\frac{1}{2}$
	Special moment frames of reinforced concrete	8	$5\frac{1}{2}$
Seismic-force resistance is provided by ordinary or special moment frames capable of resisting the total prescribed forces.	Ordinary moment frames of steel	$4\frac{1}{2}$	4
	Ordinary moment frames of reinforced concrete	2	2
	Intermediate moment frames of reinforced concrete	4	$3\frac{1}{2}$
Dual system: A structural system with an essentially complete space frame providing support for vertical loads.	Reinforced concrete shear walls	8	$6\frac{1}{2}$
	Reinforced masonry shear walls	$6\frac{1}{2}$	$5\frac{1}{2}$
A special moment frame shall be provided that shall be capable of resisting at least 25% of the prescribed seismic forces. The total seismic-force resistance is provided by the combination of the special moment frame and shear walls or braced frames in proportion to their rigidity.	Wood sheathed shear panels	8	5
	Braced frames	6	5
A structural system with an intermediate moment frame of reinforced concrete *or* an ordinary moment frame of steel in combination with the systems shown.	Reinforced concrete shear walls	6	5
	Reinforced masonry shear walls		
	Wood shear panels	7	$4\frac{1}{2}$
	Braced frames	5	$4\frac{1}{2}$
Inverted-pendulum structures: Structures where the framing resisting the total prescribed seismic forces acts essentially as isolated cantilevers and provides support for vertical load.	Special moment frames of structural steel	$2\frac{1}{2}$	$2\frac{1}{2}$
	Special moment frames of reinforced concrete	$2\frac{1}{2}$	$2\frac{1}{2}$
	Ordinary moment frames of structural steel	$1\frac{1}{4}$	$1\frac{1}{4}$

[a]These values are based on best judgment and data available at time of writing and need to be reviewed periodically.

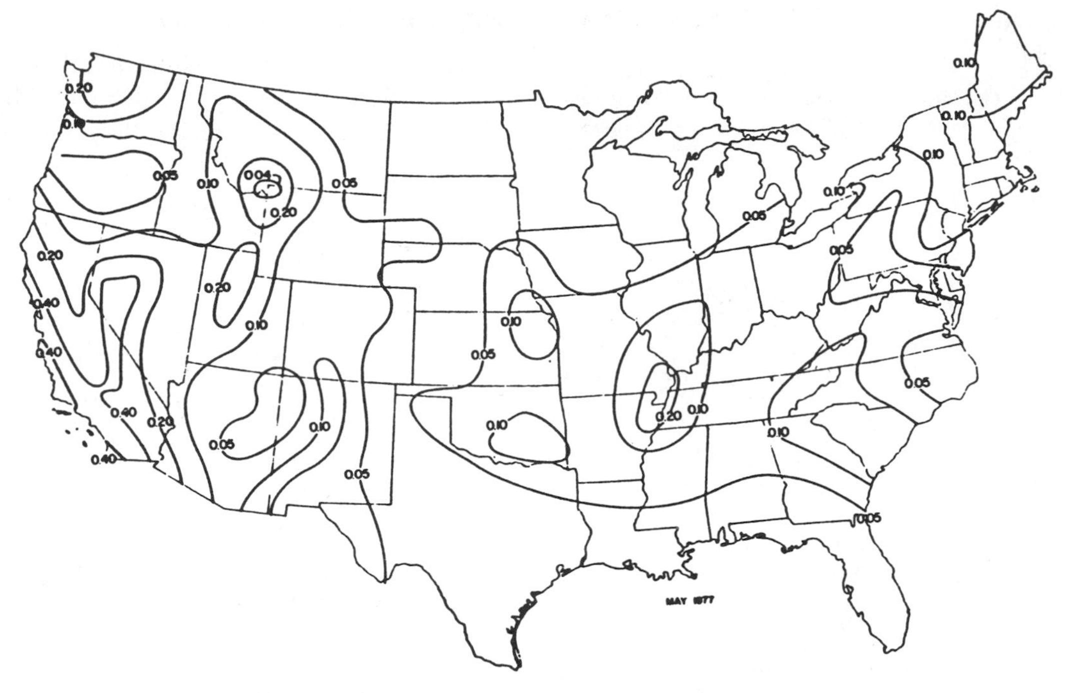

Figure 4-7 Contour map for effective peak velocity-related acceleration coefficient for the continental United States. (Reprinted from NEHRP-85.)

4.6.7 Effective Peak Accelerations A_v, A_a

The design ground motion is defined by two coefficients: the effective peak acceleration A_a, and the effective peak velocity-related acceleration A_v.

The higher frequencies present in ground motion attenuate more rapidly with distance than the lower frequencies. Therefore two ground-motion parameters are incorporated to take account of the distance from anticipated earthquake sources.

Each parameter is specified either by a contour map (Figures 4-7 and 4-8) or by maps assigning A_a and A_v on a county-by-county basis. The value of each coefficient is taken either from Table 4-10 or directly from

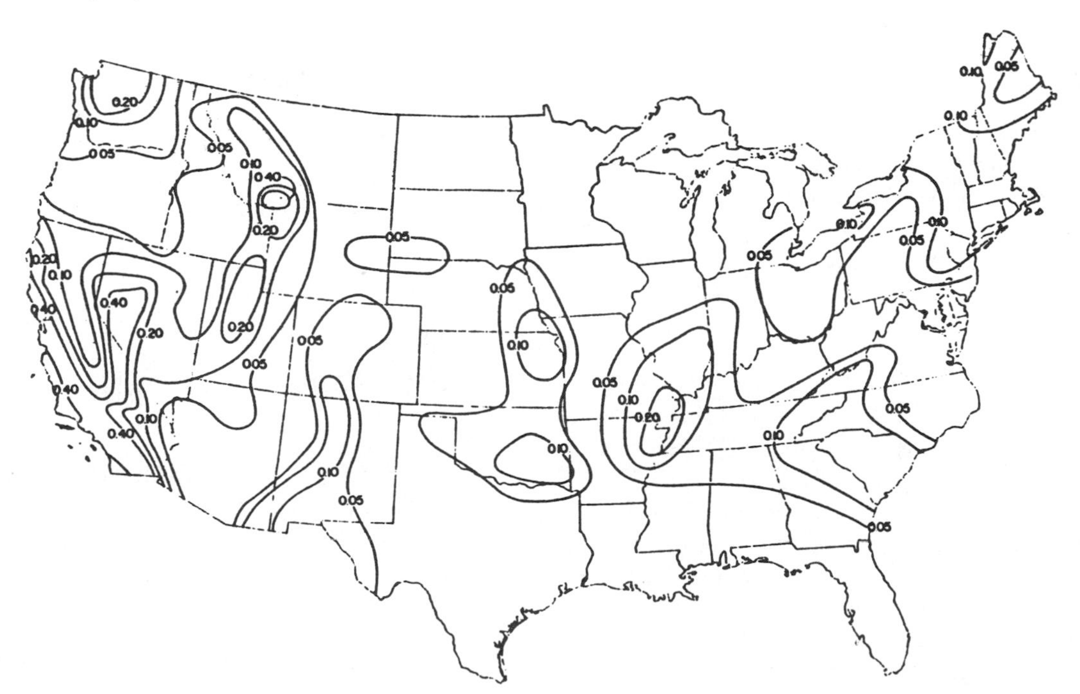

Figure 4-8 Contour map for effective peak acceleration for the continental United States. (Reprinted from NEHRP-85.)

Table 4-10 Coefficients A_a and A_v and Seismicity Index from NEHRP-85*

Coefficient A_a (Figure 1-1)	Map area number	Coefficient A_v (Figure 1-2)	Seismicity index
0.40	7	0.40	4
0.30	6	0.30	4
0.20	5	0.20	4
0.15	4	0.15	3
0.10	3	0.10	2
0.05	2	0.05	2
0.05	1	0.05	1

*The seismicity-index values in this table have been controversial and are of economic concern. The rationale for the numbers of seismicity-index values and the closely affected seismic performance categories are in need of careful review. It has been suggested that the numbers of seismicity-index values and the seismic performance categories should be increased. The BSSC program for updating these provisions will pay particular attention to this subject.

the contour maps, where the design ground acceleration is seen to vary from $0.05g$ to $0.40g$.

4.6.8 Vertical Distribution of the Force F_x

The seismic force at any level is a portion of the total base shear:

$$F_x = C_{vx}V \tag{4-28}$$

where

$$C_{vx} = \frac{w_x h_x^k}{\sum_{i=1}^{n} w_i h_i^k} \tag{4-29}$$

and

w_i, w_x = the portion of W located at or assigned to level i or x,

h_i, h_x = the height above the base to level i or x,

k = an exponent related to the building period as follows: for building periods $\leqslant 0.5$ sec, $k = 1.0$; for building periods $\geqslant 2.5$ sec, $k = 2.0$; for building periods between 0.5 and 2.5 sec, k may be taken as 2.0 or may be determined by linear interpolation between 1.0 and 2.0.

For $k = 1.0$ the distribution is a straight line. This distribution is reasonable for short buildings with a regular distribution of mass and stiffness. Hence, $k = 1.0$ for buildings with a period of 0.5 sec or less.

For $k = 2.0$ the distribution is a parabola with the vertex at the base. This is a reasonable distribution for tall regular buildings where the participation of higher modes is significant. Hence, $k = 2.0$ for buildings with a period of 2.5 sec or more. In UBC-85 and UBC-88 this effect is included through the force F_t placed at the roof.

4.6.9 Overturning Moment M_x

In the previously discussed documents the design overturning moment is given by Equation 4-14. NEHRP-85 allows for a reduction in the design overturning moment:

$$M_x = K \sum_{i=x}^{n} F_i(h_i - h_x) \tag{4-30}$$

where

$$K = \begin{cases} 1.0 & \text{for the top 10 stories} \\ 0.8 & \text{for the 20th story from the top and below} \end{cases}$$

K is the interpolated between 0.8 and 1.0 for stories in between. Part of the reasoning behind this reduction is that the design story forces are an envelope of the maximums at each floor and it is unlikely that they will all reach a maximum simultaneously.

4.6.10 Seismic-Hazard Exposure Groups

The story-drift criterion in NEHRP-85 is dependent on the seismic-hazard exposure group, which is related to the importance factor in other codes. Three groups are defined:

- *Group III:* "Buildings having essential facilities that are necessary for post-earthquake recovery. Essential facilities, and designated systems contained therein, shall have the capacity to function during and immediately after an earthquake."
- *Group II:* "Buildings having a large number of occupants or buildings in which the occupants' movements are restricted or their mobility impaired."
- *Group I:* All other buildings.

4.6.11 Drift Limitations

The allowable interstory-displacement criterion is

$$\Delta \leqslant \Delta a = 0.01 h_{sx} \quad \text{for seismic hazard group III} \tag{4-31}$$

$$\Delta \leqslant \Delta a = 0.015 h_{sx} \quad \text{seismic hazard groups I, II} \tag{4-32}$$

where

Δ = the design interstory displacement

Δa = the allowable story displacement

h_{sx} = the height of the story below level x.

The design interstory displacement Δ is the difference in the deflections δ_x at the top and bottom of the story under consideration. It is based on the calculated deflections and is evaluated by the formula

$$\delta_x = C_d \delta_{xe} \qquad (4\text{-}33)$$

where

C_d = the deflection amplification factor given in Table 4-9,
δ_{xe} = the deflections determined by an elastic analysis.

In determining these deflections the period determined by an analysis may be used to calculate the base shear without considering the limitation on the period discussed in Section 4.6.4. This has the implication that smaller story shears may be used to determine deflections that are used to determine member forces. A similar provision is contained in UBC-88. Where significant, P-delta deflections must be considered in satisfying the drift limitation. This is discussed further in the next section.

The deflection amplification factor C_d is assigned values from 1.25–6.5 and represents the ductility of the system and the properties of the materials from which it is constructed (see Table 4-9).

4.6.12 Torsion and P-Delta Effect

Torsion is treated in same manner as in UBC-85. The torsional moment resulting from the location of the center of mass plus that resulting from an assumed movement of 5% of the plan dimension must be accounted for.

The P-delta effect must be included in the computation of story shears, story drifts, and member forces when the value of the *stability coefficient* θ has a value, for any story, such that

$$\theta = P_x \Delta / V_x h_{sx} C_d > 0.10 \qquad (4\text{-}34)$$

where

Δ = the design story drift,
V_x = the seismic force acting between level x and $x - 1$,
h_{sx} = the story height below level x,
P_x = the total gravity load at and above level x,
C_d = the deflection amplification factor.

The stability coefficient can be visualized as the ratio of the P-delta moment ($P_x \Delta$) to a lateral-force story moment ($V_x h_{sx}$). Hence if the P-delta moment is equal to 10% of the story moment at any floor, the P-delta effect should be considered.

4.6.13 Irregularity

NEHRP-85 defines irregularity in a manner similar to UBC-88 but in less detail. It distinguishes between the two broad categories of plan and vertical irregularity.

- *Plan irregularities* include a nonsymmetric geometric configuration, reentrant corners, significant torsion due to eccentricity between mass and stiffness, and discontinuous diaphragms.
- *Vertical irregularities* include large changes in mass–stiffness ratios between adjacent floors, nonsymmetric geometric configuration about the vertical axis, and significant horizontal offsets.

A more detailed and quantitative description of the various types of irregularity is provided along with examples in the commentary of NEHRP-85.

4.6.14 Dynamic Analysis

Irregular buildings in areas of high seismicity require a dynamic analysis. Design provisions are given for a simplified version of modal analysis which is applicable to buildings with irregular mass or stiffness distribution over the height.

General guidance is given in the commentary on the details of dynamic analysis for more complicated situations.

4.6.15 Examples

Example 4-5
Rework Example 4-1 using NEHRP-85.

- Base shear:

$$V = C_s W$$

$$C_s = 1.2 A_v S / R T^{2/3} \leqslant 2.5 A_a / R$$

$$A_v = 0.4 \qquad \text{(see Figure 4-7)}$$

$$A_a = 0.4 \qquad \text{(see Figure 4-8)}$$

$$S = 1.2 \qquad \text{(soil type 2)}$$

$$R = 8 \qquad \text{(see Table 4-9)}$$

$$T = 0.035 h_n^{3/4} = 0.035(117)^{3/4} = 1.25 \text{ sec}$$

$$\frac{2.5 A_a}{R} = \frac{2.5(0.4)}{8} = 0.125$$

$$C_s = \frac{1.2(0.4)(1.2)}{8(1.25)^{2/3}} = 0.062 < 0.125$$

$$V = C_s W = 0.062(15{,}300) = 948.6 \text{ kips}$$

- Vertical distribution:

$$F_x = C_{vx}V$$

$$C_{vx} = \frac{w_x h_x^k}{\Sigma_{i=1}^n w_i h_i^k}$$

Interpolate to find k:

$$k = 1.0 + 1.0\frac{1.25 - 0.5}{2.5 - 0.5} = 1.375$$

$$h_9^{1.375} = 117^{1.375} = 697.8$$

$$C_{v9} = \frac{1700(697.8)}{1700(3000.9)} = 0.233$$

$$F_9 = 0.233(948.6) = 221.0 \text{ kips}$$

etc. The story shear is determined by the same procedure as in UBC-85.

- Overturning moment:

$$M_x = K\sum_{i=x}^{n} F_i(h_i - h_x)$$

$K = 1.0$ for the top 10 stories. Hence the procedure is the same as in UBC-85:

$$M_9 = 221.0(13) = 2873 \text{ ft-kips}$$

etc. See Table 4-11.

- Allowable interstory displacements: Seismic hazard group I;

$$\Delta_a = 0.015h_{sx} = 0.015(13)(12) = 2.34 \text{ in.}$$

but

$$c_d = 5.5 \quad \text{(see table 4-9)}$$

$$\delta_x = c_d\delta_{xe} = 5.5\delta_{xe}$$

$$\Delta \leqslant 2.34/5.5 = 0.425 \text{ in.}$$

Table 4-11

Level	h_x, ft	w_x, kips	$h_x^{1.375}$	C_{vx}	F_x, kips	V_x, kips	M_x, ft-kips
9	117	1700	697.8	0.233	221.0	221.0	2873
8	104	1700	593.5	0.198	187.8	408.8	8187
7	91	1700	493.9	0.165	156.5	565.3	15536
6	78	1700	399.6	0.133	126.2	691.5	24526
5	65	1700	311.0	0.104	98.7	790.2	34798
4	52	1700	228.8	0.076	72.1	862.3	46008
3	39	1700	154.1	0.051	48.4	910.7	57847
2	26	1700	88.2	0.029	27.5	938.2	70044
1	13	1700	34.0	0.011	10.4	948.6	82376
Σ		15300	3000.9	1.0	948.6		

Example 4-6
Rework Example 4-2 using NEHRP-85.

- Base shear:

$$V = C_s W$$

$$C_s = 1.2A_v S/RT^{2/3} \leqslant 2.5A_a/R$$

$$A_v = 0.4 \quad \text{(see Figure 4-7)}$$

$$A_a = 0.4 \quad \text{(see Figure 4-8)}$$

$$S_1 = 1.0 \quad \text{(soil type 1)}$$

$$T_a = \frac{0.05h_n}{\sqrt{L}} = \frac{0.05(35)}{\sqrt{120}} = 0.16 \text{ sec}$$

$$R = 5.5 \quad \text{(see Table 4-7)}$$

$$\frac{2.5A_a}{R} = \frac{2.5(0.4)}{5.5} = 0.182$$

$$C_s = \frac{1.2(0.4)(1.0)}{5.5(0.16)^{2/3}} = 0.296 > 0.182$$

$$V = 0.182(5900) = 1073.8 \text{ kips}$$

- Vertical distribution:

$$F_x = C_{vx}V$$

$$C_{vx} = \frac{w_x h_x^k}{\Sigma_{i=i}^n w_i h_i^k}$$

$$T < 0.5 \text{ sec} \quad \therefore \quad k = 1.0$$

- Overturning moment:

$$M_x = K\sum_{i=x}^{n} F_i(h_i - h_x)$$

$$K = 1.0 \quad \text{for top 10 stories}$$

Level	h_x, ft	w_x, kips	$w_x h_x$, 10^{-3}	C_{vx}	F_x, kips	V_x, kips	M_x, ft-kips
3	35	1700	59.5	0.44	472.5	472.5	5198
2	24	2000	48.0	0.35	375.8	848.3	14529
1	13	2200	28.6	0.21	225.5	1073.8	28488
Σ		5900	136.1	1.00	1073.8		

- Allowable interstory displacements: Hazard group I;

$$\Delta_a = 0.015h_{sx}$$

$$C_d = 5 \quad \text{(see Table 4-7)}$$

$$\delta_x = 5\delta_{xe}$$

For 1st floor,

$$\Delta = \frac{0.015(13)(12)}{5} = 0.47 \text{ in.}$$

For 2nd and 3rd floors,

$$\Delta = \frac{0.015(11)(12)}{5} = 0.40 \text{ in.}$$

4.7 COMPARISON OF CODE PROVISIONS

A comparison of the design seismic loads specified in UBC-85, NEHRP-85, BOCA-87, and UBC-88 is presented below.

4.7.1 Period *T*

The base shear determined by each of the procedures discussed is dependent on the period used.

Figure 4-9 plots the period vs. height for steel frames with a 13-ft story height using the UBC-85 empirical formula (Equation 4-4), as well as Equation 4-16 used in NEHRP-85, UBC-88, and BOCA-87. The periods determined by dynamic analysis are generally longer than those obtained from the empirical formulas.

In Figure 4-10 the empirical equations used in BOCA-87, UBC-88, and NEHRP-85 (Equation 4-16), and in UBC-85 (Equation 4-4), are compared for concrete frame buildings with an 11-ft story height.

4.7.2 Base Shear *V*

Comparing the base shears generated by each of the procedures is the simplest way of comparing the final results.

UBC-85, BOCA-87, and UBC-88 use working stress design, and NEHRP-85 uses ultimate strength design.

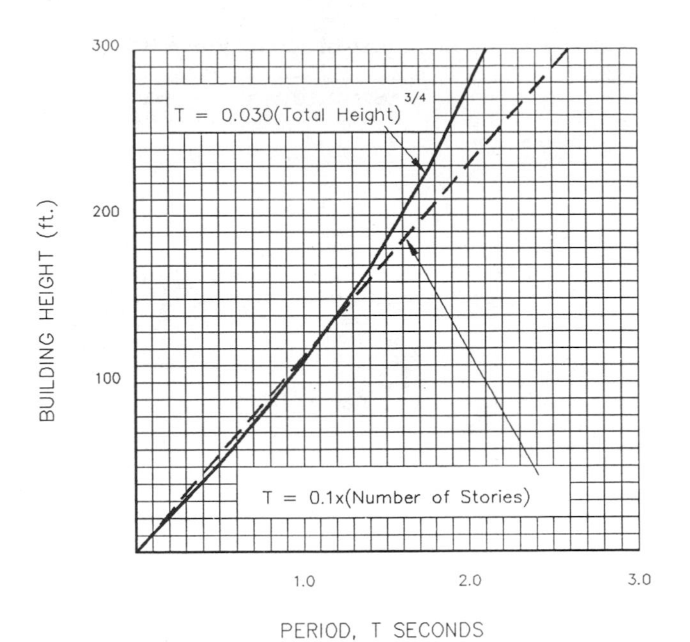

Figure 4-10 Building height versus fundamental period of vibration for concrete frames with a story height of 11 ft.

Therefore, the UBC and BOCA shears must be factored upward for comparison with NEHRP. There is no single factor to use, but 1.4 is a reasonable average.

Figure 4-11 plots base shear coefficient vs. building period for steel special moment-resisting space frames, in the zone of highest seismicity on soil type 1. NEHRP-85, BOCA-87, and UBC-88 use the same formula (Equation 4-16) to specify the building period. Therefore, the base shears specified can be compared directly from the shapes of the curves in Figure 4-11.

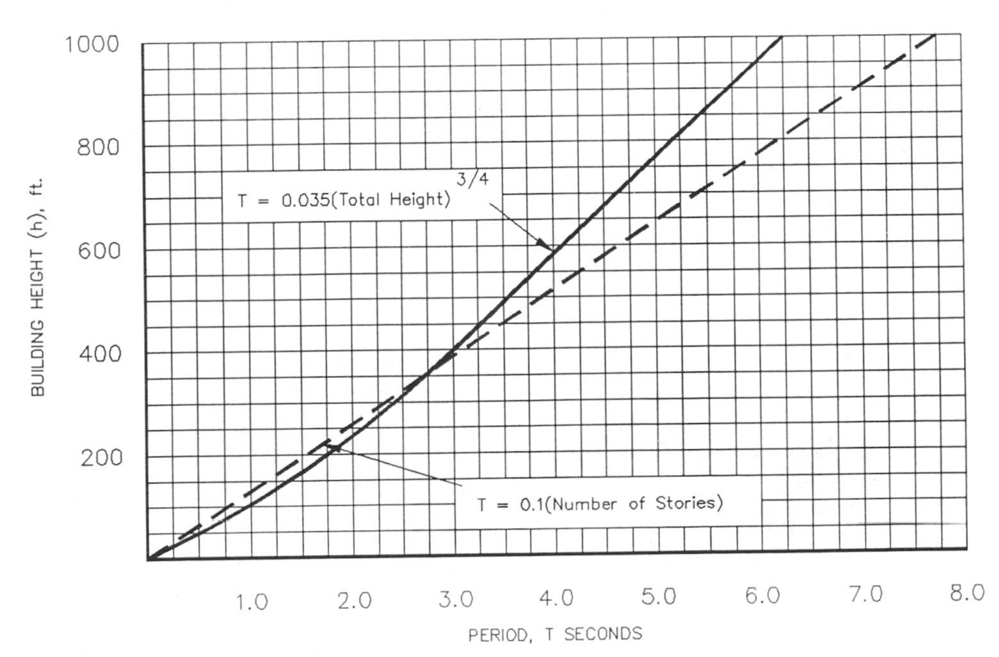

Figure 4-9 Building height versus fundamental period of vibration for steel frames with a story height of 13 ft.

Table 4-12 Comparison of Lateral Force Requirements*

	ANSI-82, BOCA-87	NEHRP-85	SEAOC (UBC-88)	UBC-85
(1) Base shear formula	$V = ZIKCSW$	$V = C_s W$	$V = \frac{ZIC}{R_w} W$	$V = ZIKCSW$
(2) Numerical coefficient C or C_s	$C = \frac{1}{15\sqrt{T}}$	$C_s = \frac{1.2 A_v S}{R T^{2/3}}$	$C = \frac{1.25S}{T^{2/3}}$	$C = \frac{1}{15\sqrt{T}}$
(3) Upper limits	$C \leqslant 0.12$ $CS \leqslant 0.14$ $CS \leqslant 0.11$ for S_3 in zones 3 and 4	$C_s \leqslant \frac{2.5 A_a}{R}$ $C_s \leqslant \frac{2.0 A_a}{R}$ for S_3 when $A_a \geqslant 0.30$	$C \leqslant 2.75$	$C \leqslant 0.12$ $CS \leqslant 0.14$
(4) Lower limits	None	$C_s \geqslant C_s(C_a T_a)$	$C / R_w \geqslant 0.075$ $C \geqslant 0.8C(T_a)$	None
(5) Zone factor	Z	A_a and A_v	Z	Z
(6) Importance factor	I	Seismic-hazard exposure group	I	I
(7) Structural system factor	K	R	R_w	K
(8) Soil factor	$S_1 = 1.0$ $S_2 = 1.2$ $S_3 = 1.5$	$S_1 = 1.0$ $S_2 = 1.2$ $S_3 = 1.5$	$S_1 = 1.0$ $S_2 = 1.2$ $S_3 = 1.5$	$S = \begin{Bmatrix} 1 \\ 1.2 \end{Bmatrix} + \begin{Bmatrix} 1 \\ 0.6 \end{Bmatrix} \frac{T}{T_s} - \begin{Bmatrix} 0.5 \\ 0.3 \end{Bmatrix} \left(\frac{T}{T_s} \right)^2$ for $T / T_s \lessgtr 1.0$ or $S_1 = 1.0$ $S_2 = 1.2$ $S_3 = 1.5$

*Courtesy of Rene W. Luft.

Table 4-13 Comparison of Analysis Provisions*

	ANSI-82, BOCA-87	NEHRP-85	SEAOC (UBC-88)	UBC-85
(a) Direction of base shear	Nonconcurrent in each main axis of structure.	Category B: separate in 2 perpendicular directions. Categories C, D: 100% one direction plus 30% in perpendicular direction.	Nonconcurrent in each principal axis except for certain irregular buildings and for common columns.	Nonconcurrent in each main axis of structure.
(b) Vertical acceleration	Not addressed.	Increases or reduces dead load to increase overall load effect.	Considered for horizontal cantilevers and horizontal prestressed components.	Not addressed.
(c) Period	May use elastic dynamic analysis or empirical formulas. No upper limits.	T less than or equal to $C_a T_a$ for base shear. T as computed for drift.	Upper limit for base shear is indirect by limit on C. No upper limit for drift.	May use elastic dynamic analysis or empirical formulas. No upper limits.
(d) Analysis type	Regular buildings: static equivalent. Irregular buildings: dynamic.	Category A: None. Category B: Static. Categories C, D: regular building, static; irregular building, dynamic.	Regular building < 240 ft high, static. Regular > 240 ft high of irregular building, dynamic.	Regular buildings: static equivalent. Irregular buildings: dynamic.
(e) Modal analysis	Base shear greater than 90 percent of static base shear in each direction.	Three modes for two directions, but all modes with $T > 0.4$ sec	90% of participating mass in each direction.	Not addressed.
(f) Vertical distribution (regular buildings)	Concentrated force at top. Remainder is proportional to $w_x h_x$.	Proportional to $w_x h_x$ with k between 1 and 2 depending on period.	Concentrated force at top. Remainder is proportional to $w_x h_x$.	Concentrated force at top. Remainder is proportional to $w_x h_x$.
(g) Overturning movement	In tall buildings, effect of overturning may be reduced up to 20%.	In tall buildings, overturning moment is reduced up to 20%.	No reduction permitted.	No reduction permitted.
(h) Drift	.005 times story height for $1 / K$ times force.	0.01 – 0.015, depending on SHEG for amplified displacement.	Less than $0.04R_w$ or 0.005. May use T for base shear.	.005 times story height for $1 / K$ times force.
(i) P-delta	None.	Required. Based on elastic drift, if significant.	Not required if drift limits are met. Otherwise required but not specified.	None.

*Courtesy of Rene W. Luft.

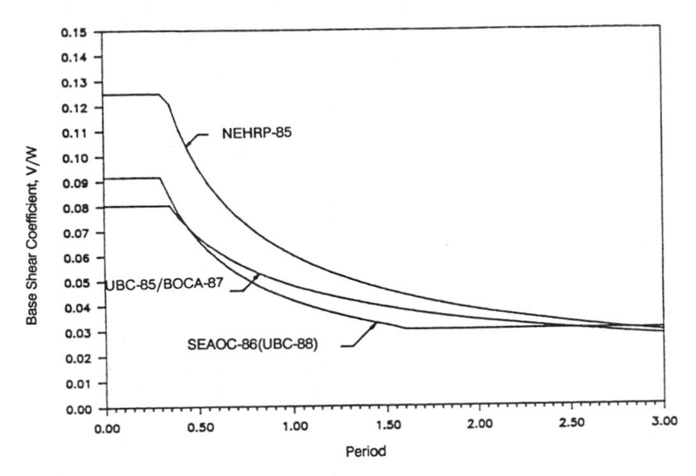

Figure 4-11 Base shear coefficient versus building period for special steel moment resisting space frames. (Courtesy of Rene W. Luft.)

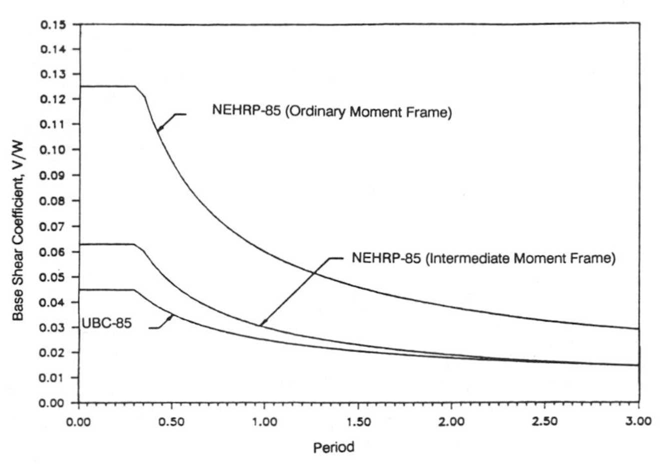

Figure 4-12 Base shear coefficient versus building period for reinforced concrete moment resisting space frames. (Courtesy of Rene W. Luft.)

The base shear specified by NEHRP-85 can be compared with other codes if the factor of 1.4, discussed above, is taken into account.

Since UBC-85 uses a different formula to determine the building period (Equation 4-4), the shape of the curve in Figure 4-11 cannot be compared directly with the NEHRP-85, BOCA-87, and UBC-88 (SEAOC) curves.

Base shear coefficient vs. building period is plotted for reinforced-concrete moment-resisting space frames for areas of moderate seismicity and soil type 1 in Figure 4-12. A plot of reinforced-concrete shear-wall buildings, in areas of highest seismicity, is shown in Figure 4-13. The restriction on comparison mentioned for Figure 4-11 also applies to these figures.

The lateral-force requirements of the codes discussed are presented for comparison in Table 4-12. A summary of the analysis provisions is presented in Table 4-13.

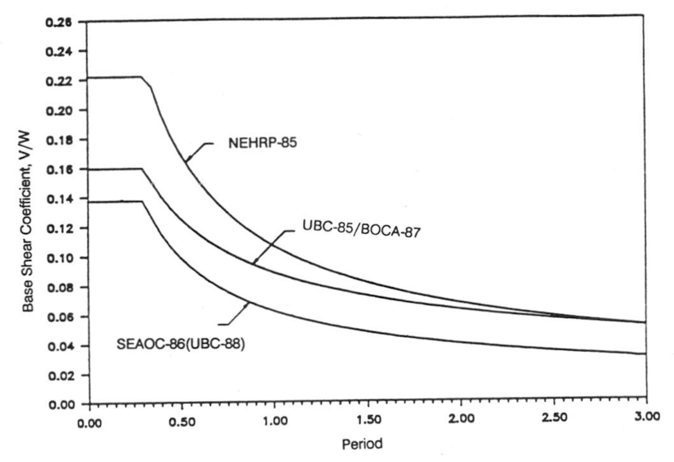

Figure 4-13 Base shear coefficient versus building period for reinforced concrete shear wall systems. (Courtesy of Rene W. Luft.)

Chapter 5

Architectural Considerations

*Christopher Arnold, AIA**

5.1 INTRODUCTION

While the provision of earthquake resistance is accomplished through structural means, the architectural design, and the decisions that create it, play a major role in determining the building's seismic performance. The building architecture must permit as effective a seismic structural design as possible; at the same time the structure must permit the functional and aesthetic aims of the building to be realized.

The architectural design decisions that influence the building's seismic performance can be grouped into three categories. These categories are not exclusive, and each category of decision may influence the others, but it is useful to structure the decisions in this way because it clarifies the influences and their mutual interactions.

The three categories are:

1. *The building configuration*: This means the size, shape, and proportions of the three-dimensional form of the building. In addition, for our purposes, configuration includes the location, shape, and approximate size of the major structural elements, because these are often determined by the architectural concept of the building. Building concept, or conceptual design, is also loosely used to identify the configuration. Strictly speaking, configuration refers only to geometrical properties of the form.

2. *Structurally restrictive detailed architectural design*: This refers to the architectural design of building details, such as columns or walls, that may affect the structural detailing in ways that are detrimental to good seismic design practice.

3. *Hazardous nonstructural components*: The design of nonstructural components is the architect's responsibility, and if inadequately designed against seismic forces, they may themselves present a hazard to life. In addition, they may represent a major cause of property loss, and in the case of essential utilities or other services, their damage may cause loss of building function.

Engineering issues in the design of these components are dealt with in Chapter 14.

This chapter discusses one other issue that bears on the architectural decisions that affect seismic performance. This is the activity of structuring the consideration of mutual architectural and engineering seismic design issues in relation to the typical building design and construction process. This, in turn, leads to some consideration of the architect–engineer relationship as it affects the seismic design problem.

*President, Building Systems Development, Inc., San Mateo, California.

5.2 CONFIGURATION CHARACTERISTICS AND THEIR EFFECTS

5.2.1 Introduction

For our purposes building configuration can be defined as building size and shape: the latter includes the characteristics of proportion. In addition, our definition includes the nature, size, and location of the structural elements. This extended definition of configuration is necessary because of the intricate relationship bearing on seismic performance between all these elements.

The nature, size, and location of nonstructural elements that may affect structural performance, or that are significant items in their own right, may form part of the initial concept of the building or may be added later. In this chapter they are discussed later as separate issues, apart from their relationship to configuration. These include such elements as walls, columns, service cores, and staircases, and also the quantity and type of interior partitions, and the nature of their exterior wall.

There are three major determinants of configuration: building planning and function, urban design and planning requirements, and the need for a distinctive or attractive image.

Building function and planning produce a demand for certain settings and kinds of space division, connected by a circulation pattern for the movement of people, supplies, and equipment. These demands ultimately lead to certain building arrangements, dimensions, and determinants of configuration.

Urban design and planning requirements affect the exterior form of the building. A height limit may set a maximum height; the street pattern may, particularly in a dense urban situation, determine the plan shape of the building, at least for its lower floors. City planning requirements sometimes dictate the need for open first floors, for vertical setbacks, or other characteristics of architectural form.

The need for a distinctive or attractive image is often the most important determinant and, in engineering terms, the most controversial, because the results rarely appear suceptible to the logic or quantitative evaluation that the engineer understands. Nevertheless, it is for his ability to create distinctive forms that the architect in large measure exists, and his images follow their own set of disciplines. The dictates of style and fashion are no less strong for being outside rational analysis. The world's populations, including engineers, have for centuries toured the great architecural centers of Europe and the Far East to enjoy the magic of the unquantifiable.

The final configuration choice is the result of a decision process that balances these varying requirements and influences and, within a budget, resolves conflicts into an architectural concept.

5.2.2 Configuration Influence in General

Configuration largely determines the ways in which seismic forces are distributed throughout the building, and also influences the relative magnitude of those

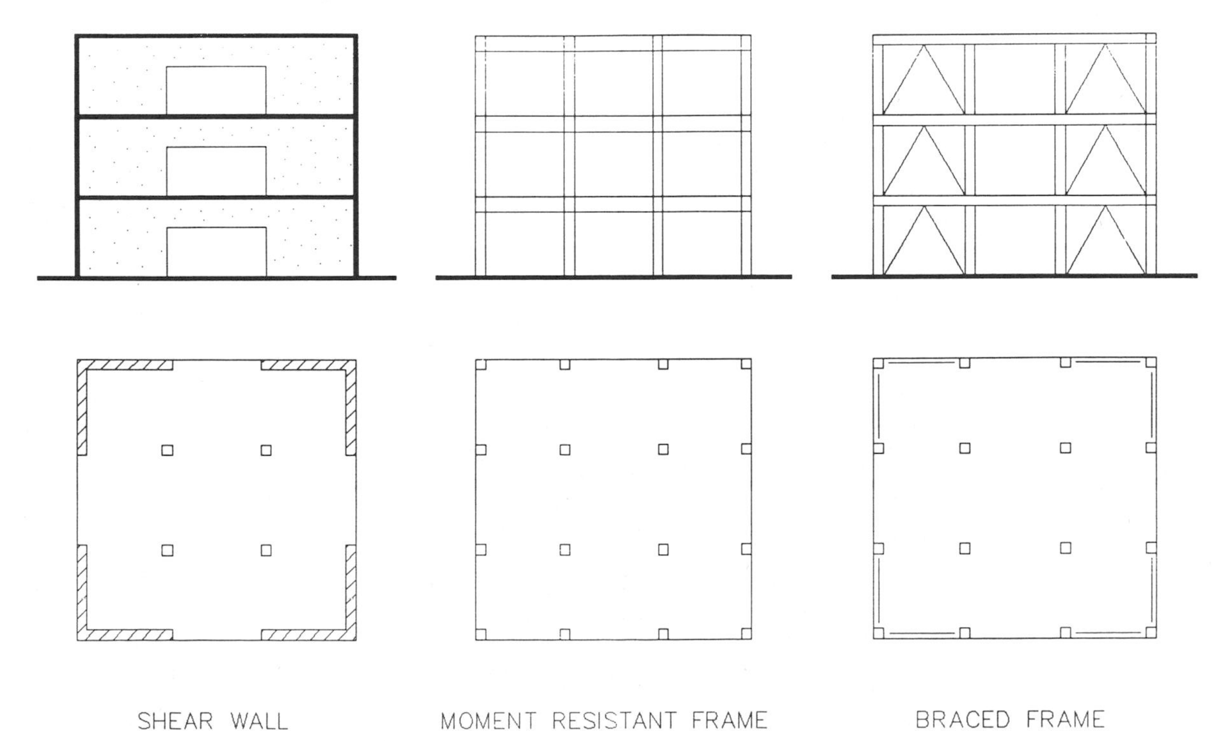

Figure 5-1 The optimal seismic configurations.

forces. For a given ground motion, the major determinant of the total inertial force in the building is the building mass. While the size and shape of the building establish its mass (together with the choice of materials), the building square footage and volume are determined by the building program: the listing of required spaces and the activities and equipment that they contain. But for any given program a variety of configurations can provide a solution, and it is the variables in these configurations that affect the distribution of inertial forces due to ground shaking.

Thus the discussion of configuration influence on seismic performance becomes the identification of configuration variables that affect the distribution of forces. For our purposes, these variables represent irregularities, or deviations from a "regular" configuration that is an optimum, or ideal, with respect to dealing with lateral forces.

5.2.3 Regular Configurations

Figure 5-1 shows three structures, employing three typical alternatives for lateral resistance, that are seismically ideal, while at the same time providing a useful architectural configuration. The designs shown in Figure 5-1 all have the attributes of:

- low height-to-base ratio,
- balanced resistance,
- symmetrical plan,
- uniform section and elevation,
- maximum torsional resistance (due to location of shear walls and bracing),
- short spans,
- direct load paths,
- uniform floor heights.

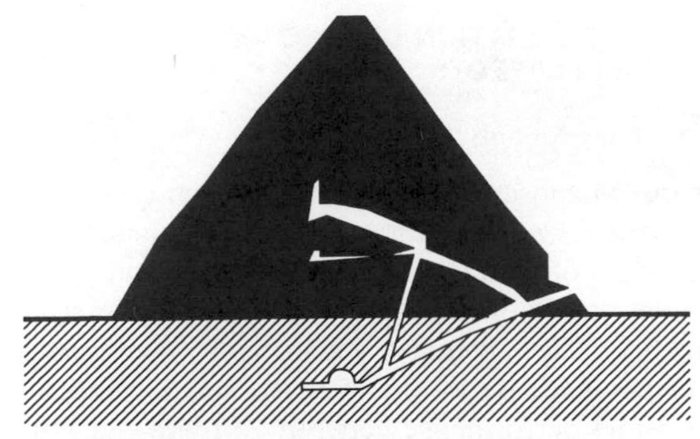

Figure 5-2 The Great Pyramid, Gizeh, Egypt.

The significance of these attributes, and the effects of their absence, will be discussed in the following sections.

A better seismic form, though much less useful architecturally in its provision of usable space, is that of the great Pyramid at Gizeh, Egypt (Figure 5-2). This form has the desirable attributes of:

- symmetrical plan,
- short span,
- direct load paths,
- low unit stress,
- broad base (no overturning),
- symmetrically reducing plan size with height,
- large structural density.

Figure 5-3 The engineer's street.

This list of attributes is so strong, and the configuration so beneficial, that the fact of this structure using unreinforced masonry as a material becomes incidental to its seismic performance.

At this point the engineer may ask why it is not possible for all buildings to follow these models, and why it is necessary to introduce configuration variations that are detrimental to seismic performance. The answer lies in a combination of functional requirements and creativity. Sometimes the functional requirements of the building dictate a less than ideal seismic configuration. The needs for interest and variety in our built environment would make a city composed only of our ideal configurations a dull place indeed (Figure 5-3).

5.2.4 Methods of Analysis and the Uniform Building

An important aspect of a building's response to ground motion is the method of analysis used to establish the seismic forces. The estimate of total forces and their distribution is both a function of and a determinant of the lateral-force-resisting system employed in the building. The great majority of designs estimate lateral forces through use of the equivalent-static-lateral-force method (ELF) established in typical seismic codes, which involves estimating a base shear and then distributing the resulting forces through the structural elements of the building.

It is important to recognize that the equivalent-force method assumes a regular building, comparable to our ideal form described above. This assumption is made clear in the Commentary to the Structural Engineers Association of California (SEAOC) *Recommended Lateral Force Requirements and Commentary* (1980):

> These minimum standards have, in general, been written for uniform buildings and conditions. The subsequent application of these minimum standards to unusual buildings or conditions has, in many instances, led to an unrealistic evaluation.

Hence, to the extent that the building design deviates from the regular code model, the code-derived forces will be inapplicable.

It is safe to say, based on studies of building inventories, that over half the buildings that have been designed in the last few decades do not conform to the simple uniform building configuration upon which the code is based. For new designs, the simple equivalent-static-force method of the code must often be augmented by engineering judgment and experience.

A review of the purpose of seismic codes by Shah et al.[5-1] provides a useful historical context necessary to understand these limitations of the code's provisions:

> In the 1930's, 40's, and 50's the structural engineers of California (with recognition of the experience of Japanese engineers) generated the basic earthquake code and design procedures which are employed throughout the world today. It is most important to recognize that these engineers had developed these provisions for the types of building construction which were prevalent in California at that time, specifically structures in Los Angeles and San Francisco. These buildings typically had strong steel (with concrete fireproofing) framing skeletons, filled with very well constructed brick masonry walls and strong concrete flooring systems. They were usually symmetrical and regular in their configuration, and in most cases they qualified as good tough earthquake resistant structures. It is a most educational experience to walk along Market Street in San Francisco and see some of these structures that survived the motion effects of the disastrous 1906 earthquake without even significant damage. The California engineers, having a knowledge of the good performance record of these structures, formulated the following type of design philosophy:
>
> - relatively low lateral earthquake forces for the design of structural members
> - relatively strict rules governing the types of allowable materials, the methods of member design and tough connections, and an implied need for symmetry and regularity.
>
> For a time up to the 1960's, before which much construction in California did not differ substantially from that of the tough buildings, this philosophy was appropriate to provides seismic resistant structure.
>
> However, architectural configuration along with methods of construction have changed significantly in the past two decades. Frames have become much more open and irregular, and the rugged systems of masonry partition walls and concrete floors have been replaced by largely prefabricated elements with very flexible characteristics. The low seismic design forces which were quite appropriate for the classical old methods of construction were applied without change for the newer structures. The basic error was that the new buildings did not have the regularity, stiffness, and reserve toughness necessary to justify classical low design values.

Progressive evolution of seismic codes has resulted in increasing force levels and the consideration of additional parameters in estimating force levels, but the impact of configuration factors has remained a matter of judgment. More recently attempts have been made to quantify some configuration parameters (as discussed later in this section), but it remains to be seen how effective these approaches will be in practice.

That a situation should prevail in which the basis upon which seismic forces are derived by code is made to some extent unrealistic by the way in which the building is conceived calls not for alarm but for the increased need for conceptual understanding of how seismic forces are derived and how they are distributed, besides relying only on analysis.

Conceptual understanding of structural response and of the problems caused by configuration predates by many decades today's analytical methods. Much of the information is empirical: early observers noted the behavior in earthquakes of buildings of certain types of material, construction, and configuration.

5.2.5 Irregular Configurations: Code Definitions

In the Commentary to the SEAOC *Recommended Lateral Force Requirements and Commentary* (1980), over 20 types of "irregular structures or framing systems" are noted as examples of designs that should involve extra analysis and dynamic consideration rather than use of the normal equivalent-static-force method. These types are illustrated in Figure 5-4, which is a graphical interpretation of the SEAOC list. Scrutiny of these conditions shows that the majority of irregularities are configurational issues within the terms of our definitions.

This list of irregularities defines the conditions, but provides no quantitative basis for establishing the relative significance of a given irregularity. These irregularities vary in the importance of their effects, and their influence also varies in accord with the particular geometry or dimensional basis of the condition. Thus, while in an extreme form the reentrant corner is a serious type of plan irregularity, in a lesser form it may have little significance (Figure 5-5). The determination of the point at which a given irregularity becomes serious is a matter of judgment.

The SEAOC Commentary explains the difficulty of going beyond this basic listing as follows:

> Due to the infinite variation of irregularities (in configuration) that can exist, the impracticality of establishing definite parameters and rational rules for the application of this Section are readily apparent.

However, in the most recent version of the SEAOC *Requirements and Commentary*, and in the 1988 revisions to the Uniform Building Code (which is based on the SEAOC document), an attempt has been made to quantify some critical irregularities, and to define geometrically or by use of dimensional ratios the points at which the specific irregularity becomes an issue of such concern that the equivalent-lateral-force method of analysis cannot be used, and a dynamic-analysis lateral-force procedure must be used. The specific requirements covered in this section of the code should be referred to, but in general the equivalent lateral force method may only be used for irregular structures that are:

- in seismic zone 1,
- in seismic zone 2, if of occupancy III or IV (non hazardous or nonessential),
- not more than five stories or 50 ft in height,
- wood frame structures, having lath and plaster, gypsum, gypsum board, or wood shear walls and wood diaphragms.

However, this code also mandates that structures with a soft story discontinuity in capacity that is less than 65% of that of the story above are not permitted over two stories or 30 ft in height.

The vertical and plan (or horizontal) irregularities are defined in Tables 23-M and 23-N of the new code document. These conditions are interpreted in diagram form in Figure 5-6.

The NEHRP *Recommended Provisions for the Development of Seismic Regulations for New Buildings* defines irregularity (in the Commentary) in an approximately similar way to SEAOC (1980). These definitions are shown in Figures 5-7 and 5-8, which are reproduced from the provisions. The NEHRP provisions make somewhat similar, predominantly nonquantitative comments about the use of the equivalent lateral-force procedure for irregular buildings:

> The ELF procedure is likely to be inadequate in the following cases: buildings with irregular mass and stiffness properties in which case the simple formulas for vertical distribution of lateral forces may lead to erroneous results; buildings (regular or irregular) in which the lateral motions in two orthogonal directions and the torsional motions are strongly coupled; and the buildings with irregular distribution of story strengths leading to possible concentration of ductility demand in a few stories of the building. In such cases, a more rigorous procedure which considers the dynamic behavior of the structure should be employed. Such special consideration is necessary only for irregular buildings which are located in areas with high seis-

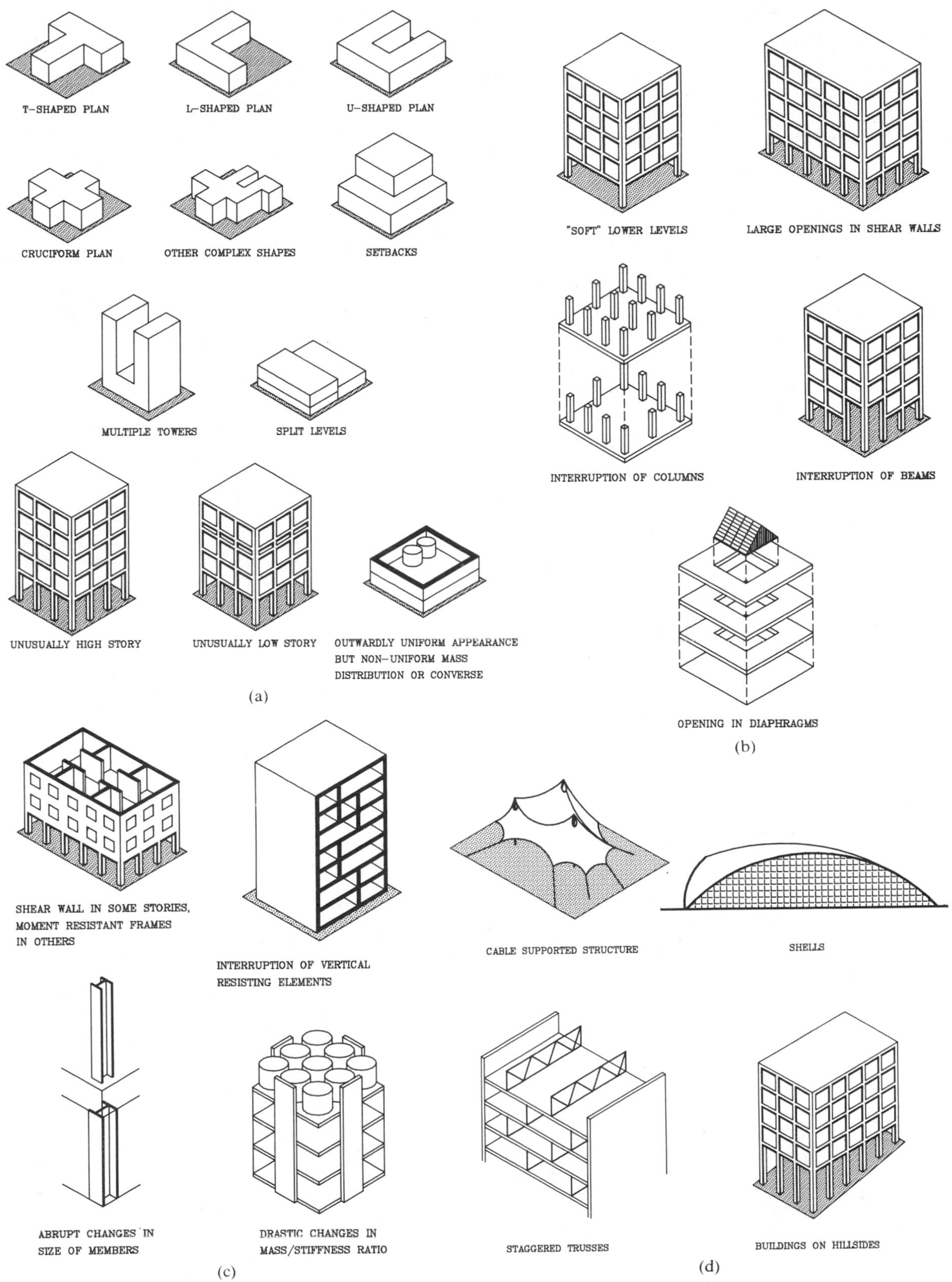

Figure 5-4 Graphic interpretation of "Irregular Structures or Framing Systems" from the commentary to the "SEAOC Recommended Lateral Force Requirements and Commentary." (a) Buildings with irregular configuration. (b) Buildings with abrupt changes in lateral resistance. (c) Buildings with abrupt changes in lateral stiffness. (d) Unusual or novel structural features.

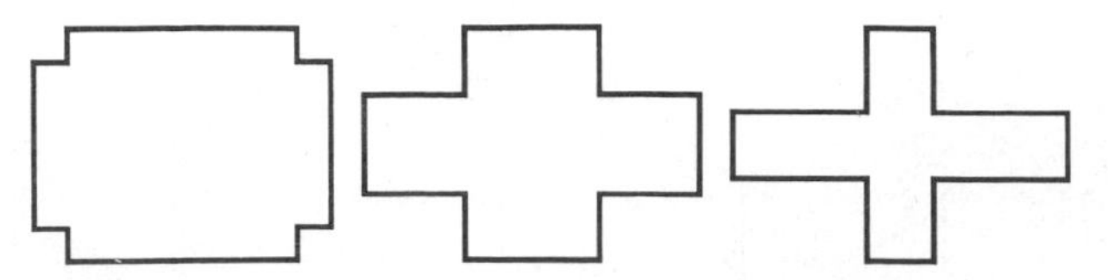

Figure 5-5 The reentrant corner plan: A range of significance.

micity . . . and whose failures would cause significant hazard to the public, those housing [essential facilities and having a large number of occupants or buildings in which occupants' movements are restricted or their mobility is impaired].

Buildings [requiring a high level of seismic performance] with certain types of vertical irregularities may be analyzed as regular buildings in accordance with normal ELF procedures. These buildings are generally referred to as setback buildings. The procedure delineated below may be used.

1. The base and tower portions of a building having a setback vertical configuration may be analyzed as indicated in (2) below if all of the

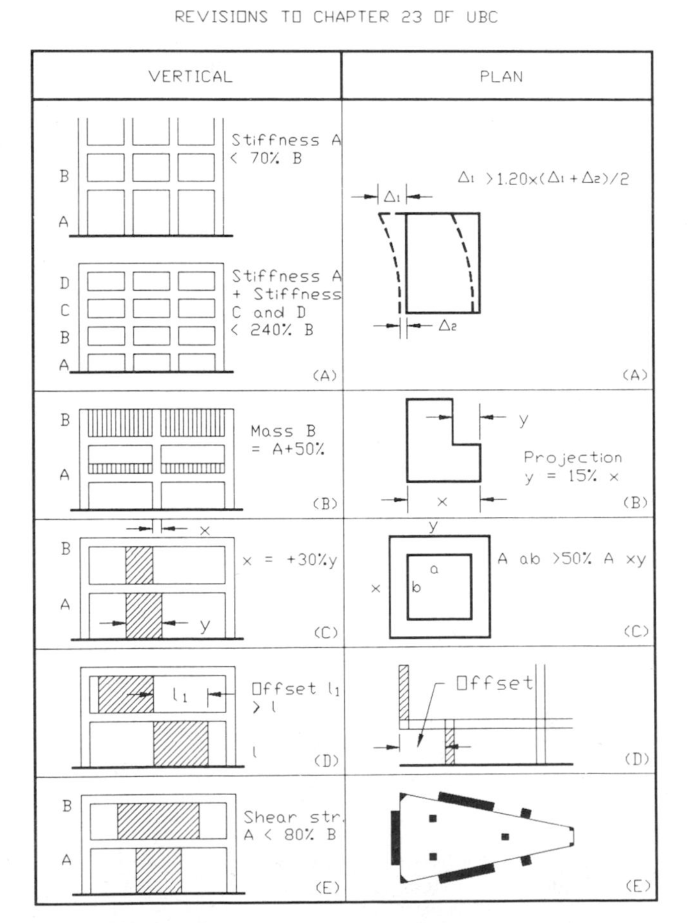

Figure 5-6 Problem configuration definitions interpreted from the 1988 Revisions to the Uniform Building Code.

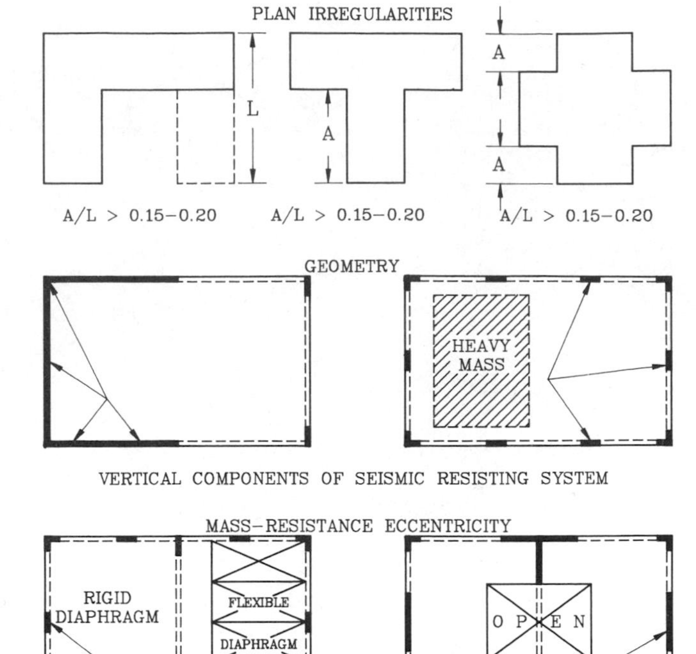

Figure 5-7 Plan irregularities, NEHRP Recommended Provisions (1985).

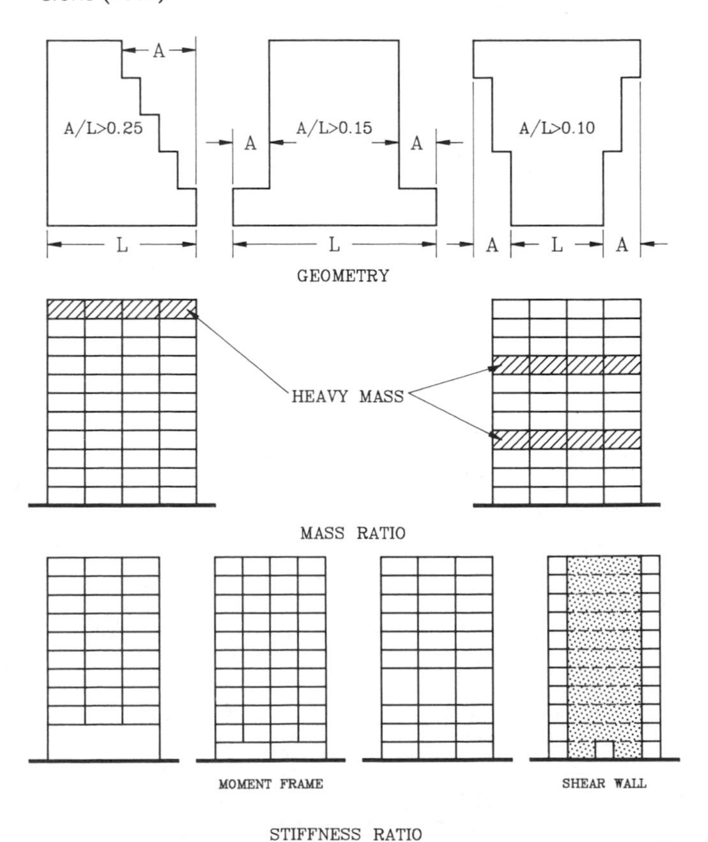

Figure 5-8 Vertical irregularities, NEHRP Recommended Provisions (1985).

following conditions are met:

a. The base portion and the tower portion, considered as separate buildings, can be classified as regular.
b. The stiffness of the top story of the base is at least five times that of the first story of the tower.

Where these conditions are not met, the building shall be analyzed using modal analysis.

2. The base and tower portions may be analyzed as separate buildings in accordance with the following:

a. The tower may be analyzed in accordance with the normal ELF procedure with the base taken at the top of the base portion.
b. The base portion shall then be analyzed in accordance with the ELF procedure using the height of the base portion and with the gravity load and base shear of the tower portion acting at the top level of the base portion.

The procedures defined in the *Provisions* include a simplified modal analysis which takes account of irregularity in mass and stiffness distribution over the height of the building. It would be inadequate, in general, to use the ELF procedure for buildings whose seismic resisting system has the same configuration in all stories and in all floors, and whose floor masses and cross-sectional areas and moments of inertia of structural members do not differ by more than 30 percent in adjacent floors and in adjacent stories. For other buildings, the following criteria should be applied to decide whether modal analysis procedures should be used. The story shears should be computed using the ELF procedure. On this basis, structural members should be approximately dimensioned. The lateral displacements of the floor can then be computed. Using these displacements in the formulae for computing lateral seismic shear at any level, one recomputes lateral forces, and from these, the new story shears are obtained. If at any story the recomputed story shear differs from the corresponding value as obtained from the ELF procedure by more than 30%, the building should be analyzed using the modal analysis procedure. If the difference is less than this value, the building may be designed using the story shear obtained in the application of the present criterion.

5.3 SEISMIC SIGNIFICANCE OF TYPICAL CONFIGURATIONS

5.3.1 Introduction

The code discussions on irregularity outlined above cover the majority of configuration variables that have a significant effect on the seismic performance of a building. Although definitions vary, there is general agreement on those configuration irregularities that are important.

However, the listing is not complete: issues of building proportion and size are not included, nor is the problem of pounding, which combines the issue of drift with that of building adjacency. The discussion of configuration issues that follows incorporates all the code-defined issues but, in going back to our original definition of configuration, categorizes configuration problems in a way more related to their architectural origins as decisions made at the conceptual stages of the design.

5.3.2 Scale, Size, Proportion, and Symmetry

It is possible to introduce configuration irregularities into a wood frame house that would be serious problems in a large building, and yet produce a safe structure with the inclusion of relatively inexpensive and unobtrusive provisions. This is because a small wood structure is light in weight and inertial forces will be low. In addition, spans are small and, relative to the floor area, there will be a large number of walls to distribute the loads.

For a larger building, the violation of basic layout and proportion principles exacts an increasingly severe cost, and as the forces become greater, good performance cannot be relied upon as in an equivalent building of better configuration.

As the absolute size of a structure increases, the number of alternatives for its structure decrease. A bridge span of 300 ft may be built as a beam, arch, truss, or suspension system, but a span of 3000 ft can only be designed as a suspension structure. And as the size increases, the structural discipline becomes more rigorous: architectural flourishes that are perfectly acceptable at the size of a house become physically impossible at the scale of a suspension bridge (Figure 5-9).

In looking at the influence of building size on seismic performance, the influence of both the dynamic environment and the characteristics of ground motion result in more complexity than does the influence of size on vertical forces. Increasing the height of a building may seem equivalent to increasing the span of a cantilever beam, and so it is (all other things being equal). The problem with the analogy is that as a building grows taller its period will tend to increase, and a change in

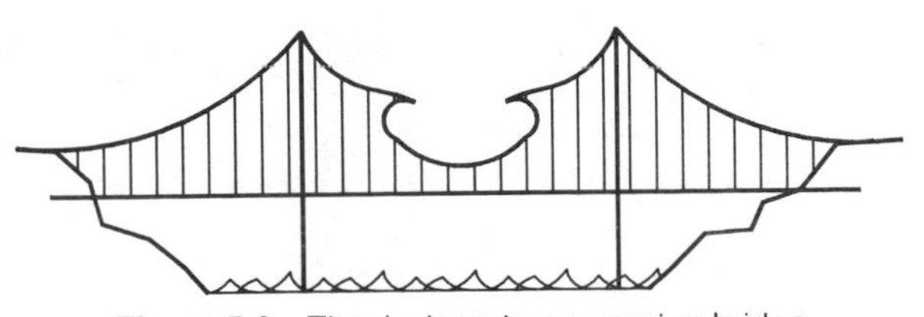

Figure 5-9 The designer's suspension bridge.

period means a change (either upward or downward) of the level of response and size of forces.

The effect of the building period must be considered in relation to the period of ground motion, and if amplification occurs, the effect of an increase in height may be quite disproportionate to the increase itself. Thus the doubling of building height from 5 to 10 stories may, if amplification occurs, result in a four- or fivefold increase in seismic forces. The earthquake in Mexico City in 1985 resulted in major response and amplification in buildings in the 6- to 20-story range, with generally reduced response in well-built buildings below and above these heights.

Although a 100-ft height limit throughout Japan was enforced until 1964, a 150-ft 13-story limit was the maximum in Los Angeles until 1957, and the limit was 80 ft and later 100 ft in San Francisco, height is rarely singled out as a variable to be limited to mitigate the earthquake hazard. The present approach is not to legislate seismic height limits but to enforce more specific seismic design and performance criteria. Generally, urban design, real-estate, or programmatic factors will be more significant, and earthquake performance must be engineered with the height predetermined by these factors.

It is easy to visualize the overturning forces associated with height as a seismic problem, but large plan areas can also be detrimental. When the plan becomes extremely large, even if it is symmetrical and of simple shape, the building can have trouble responding as one unit to the ground motion. Unless there are numerous interior lateral force-resisting elements, large-plan buildings impose unusually severe requirements on their diaphragms, which have large lateral spans, and can build up large forces to be resisted by shear walls or frames. The solution is to add walls or frames to reduce the span of the diaphragm, although it is recognized that this may introduce problems in the use of the building.

An interesting example of a correct "intuitive" response to this problem is that of the design of the Imperial Hotel, Tokyo, by the architect Frank Lloyd Wright in the early 1920s. He subdivided this large, complex building, with long wings and many reentrant corners, into small regular boxes, each about 35 ft by 60 ft in plan. In doing this, he appears to have been concerned about the possibility of differential settlements caused by a traveling surface wave on the site. In the use of this concept, to which is attributed in large measure the success of the building in surviving the 1923 Kanto earthquake, Wright was well ahead of his time. The short-pile foundation system, which Wright claimed as a major invention, probably had much less to do with the building's good performance.[5-2]

In seismic design, the proportions of a building may be more important that its absolute size. For tall buildings the slenderness ratio (height/least depth) of a building, calculated in the same way as for an individual member, is a more important consideration than just height alone. Dowrick[5-3] suggests attempting to limit the height/depth ratio to 3 or 4, explaining:

> The more slender a building the worse the overturning effects of an earthquake and the greater the earthquake stresses in the outer columns, particularly the overturning compressive forces which can be very difficult to deal with.

As urban land becomes more expensive, there is a trend towards designing very slender "sliver" buildings which, although not necessarily very high, may have a large height/depth ratio. Nowhere is this trend more apparent than in downtown Tokyo, where multistory buildings may be built on sites that are of the order of 15 to 20 ft wide (Figure 5-10). However, the same economic forces often dictate that these buildings will be built very close together, so that they will tend to respond as a unit rather than as individual free-standing buildings.

The term symmetry denotes a geometrical property of building plan configuration. Structural symmetry means that the center of mass and center of resistance are located at the same point (unless live loads affect the actual center of mass).

The single admonition that occurs in all codes and in textbooks that discuss configuration is that symmetrical forms are preferred to asymmetrical ones. The two basic reasons are that eccentricity between the centers of mass and resistance will produce torsion and stress concentrations.

However, a building with reentrant corners is not necessarily asymmetrical (a cruciform building may be symmetrical), but it is irregular, as defined, for example in the SEAOC Commentary discussed earlier. Thus symmetry is not sufficient on its own, and only when it is combined with simplicity is it beneficial.

Nevertheless, it is true that as the building becomes more symmetrical, its tendency to suffer torsion and stress concentration will reduce, and performance under seismic loads will tend to be less difficult to analyze. This suggests that when good seismic performance must be achieved with maximum economy of design and construction, the symmetrical, simple shapes are much to be preferred. But these tendencies must not be mistaken for an axiom that a symmetrical building will not suffer torsion.

The effects of symmetry refer not only to the overall building shape, but to its details of design and construction. Study of building performance in past earthquakes indicates that performance is sensitive to quite small variations in symmetry within the overall form. This is

Figure 5-10 Slender buildings, Tokyo, Japan.

particularly true in relation to shear-wall design and where service cores are designed to act as major lateral resistant elements.

Experience in Mexico City shows that many buildings that were symmetrical and simple in overall plan suffered severely because of unsymmetrical location of service cores and escape staircase. Moreover, as soon as a structure begins to suffer damage (cracking in shear walls or columns, for example), its distribution of resistance elements changes, so that even the most symmetrical of structures becomes dynamically asymmetrical and subject to torsional forces.

Finally, it must be recognized that architectural requirements will often make the symmetrical design impossible. In these circumstances, it may be necessary, depending on the size of the building and the type of asymmetry, to subdivide the building into simple elements.

There is a tendency, as noted above, for the very tall building to tend towards symmetry and simplicity. The seismic problems are most apparent in the medium-height building, where considerable choice exists as to planform and the disposition of the major masses of the building.

5.3.3 Plan Density, Perimeter Resistance, and Redundancy

The size and density of structural elements in the buildings of former centuries is strikingly greater than in today's buildings. Structural technology has allowed us to push this trend continually further.

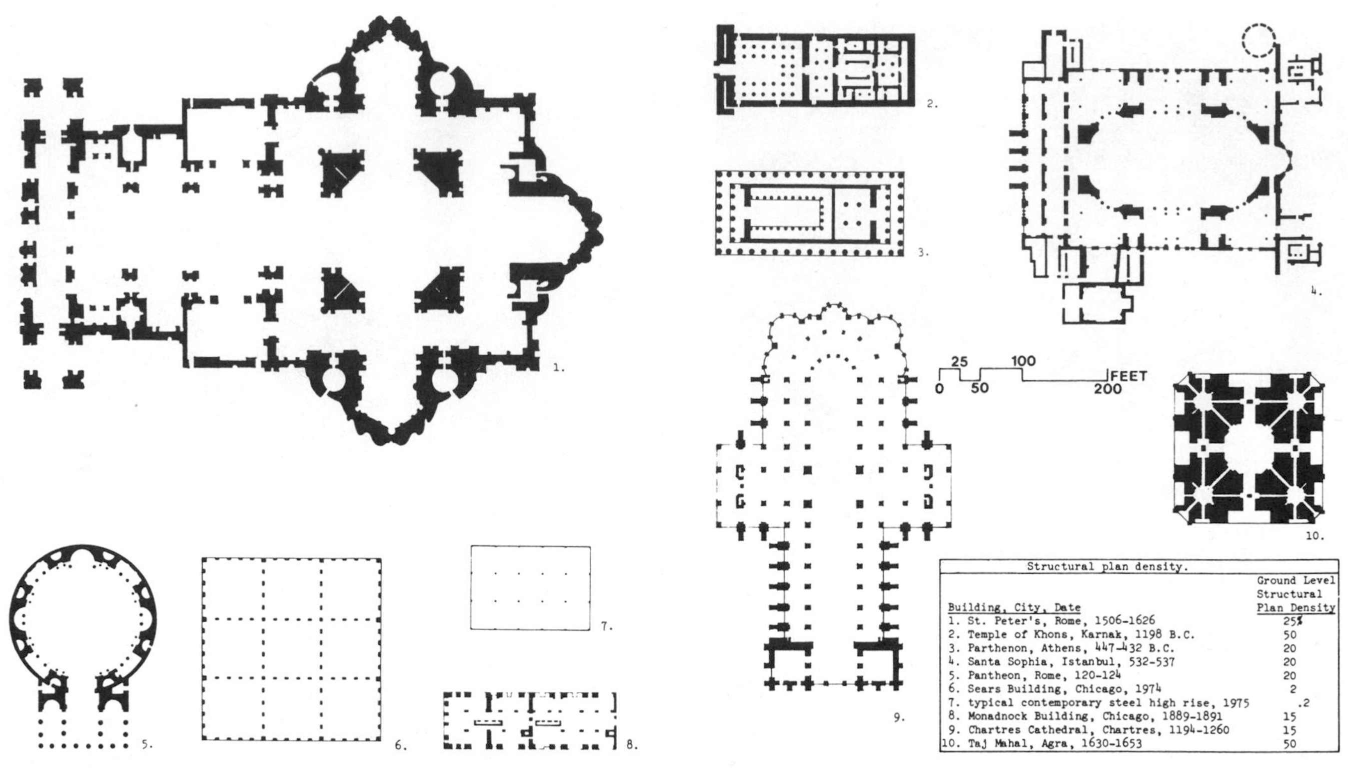

Structural plan density.

Building, City, Date	Ground Level Structural Plan Density
1. St. Peter's, Rome, 1506-1626	25%
2. Temple of Khons, Karnak, 1198 B.C.	50
3. Parthenon, Athens, 447-432 B.C.	20
4. Santa Sophia, Istanbul, 532-537	20
5. Pantheon, Rome, 120-124	20
6. Sears Building, Chicago, 1974	2
7. typical contemporary steel high rise, 1975	.2
8. Monadnock Building, Chicago, 1889-1891	15
9. Chartres Cathedral, Chartres, 1194-1260	15
10. Taj Mahal, Agra, 1630-1653	50

Figure 5-11 Structural plan density.

Earthquake forces are generally greatest at the ground level. The bottom story is required to carry its own lateral load in addition to the shear forces of all the stories above, which is analogous to the downward buildup of vertical gravity loads. At this same lowest level, programmatic and aesthetic criteria are often imposed on the building that demand the removal of as much solid material as possible. This requirement is the opposite of the most efficient seismic configuration, which would provide the greatest intensity of vertical resistant elements at the base, where they are most needed.

An interesting statistical measure in this regard is the ground level *structural plan density*, defined as the total area of all vertical structural elements divided by the gross floor area. The most striking characteristic of the modern framed building is the tremendous reduction of structural plan density compared to historic buildings.

For instance, a typical 10- to 20-story, moment-resisting steel-frame building will touch the ground with its columns over 1% or less of its plan area, and combined frame–shear-wall designs will typically reach structural plan densities of only 2%. The densely filled-in "footprints" of buildings of previous eras present a striking contrast: the structural plan density can go as high as 50%, in the case of the Taj Mahal; the ratio for St. Peter's in Rome is about 25%, and for Chartres Cathedral 15%. The 16-story Monadnock Building in Chicago, which used exterior bearing walls of brick 6 ft thick at the ground level, has a ratio of 15% (Figure 5-11).

Analogous to structural plan density is the measure of the extent of walls in a structure. Surveys of damaged buildings in Japan and Turkey have indicated a clear relationship between the length of walls in a box-system building and the amount of damage. This relationship has been incorporated in the seismic codes of these and other countries to provide prescriptive guidance for the design of simple structures.

In Figure 5-12, although both configurations are symmetrical and contain the same amount of shear wall, the location of the walls is significantly different. The walls on the right form greater lever arms for resisting overturning and torsional moments. In resisting torsion, with the center of twist of a symmetrical building located at or near the geometrical center, the further the

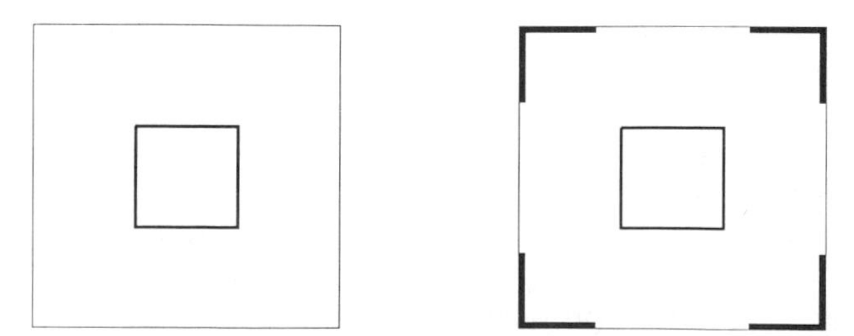

Figure 5-12 Location of lateral resistance systems.

resisting material is placed from the center, the greater the lever arm through which it acts, and hence the greater the resisting moment that can be generated. Placing resisting members on the perimeter whenever possible is always desirable, whether the members are walls, frames, or braced frames, and whether they have to resist direct lateral forces, torsion, or both.

The engineering attribute of redundancy plays an important role in seismic performance. It is significant in several aspects. The detailing of connections is often cited as a key factor in seismic performance, since the more integrated and interconnected a structure is, the more load distribution possibilities there are.

In addition, the redundant design will almost certainly offer direct load paths and many of them, and in this it tends to result in higher plan density as discussed above. Historic buildings tended to be highly redundant, because short spans required many points of support, and thus each supporting member incurs much lower stresses, often within the capability of unreinforced masonry. Thus the limitations of traditional materials forced the designers into good design practices such as redundancy, direct load paths, and high plan density.

5.4 PLAN-CONFIGURATION PROBLEMS

5.4.1 Reentrant Corners

Definition The reentrant corner is the common characteristic of overall building configurations that, in plan, assume the shape of an L, T, H, +, or a combination of these shapes.

Seismic Effects There are two related problems created by these shapes. The first is that they tend to produce variations of rigidity, and hence differential motions, between different parts of the building, resulting in a local stress concentration at the "notch" of the reentrant corner. In Figure 5-13, if the ground motion occurs with a north–south emphasis at the L-shaped building shown, the wing oriented north–south will, purely for geometrical reasons, tend to be stiffer than

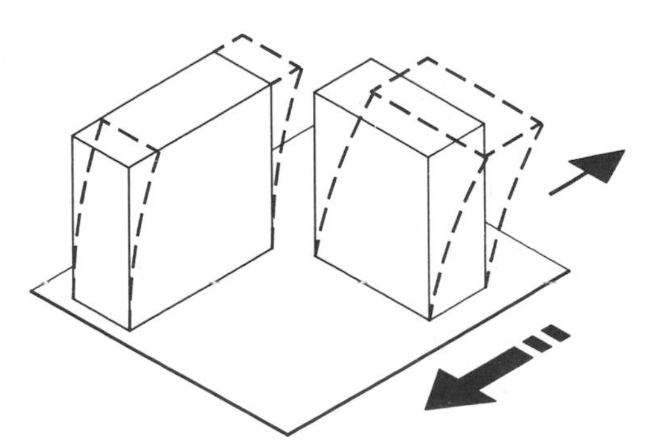

Figure 5-13 Separated buildings.

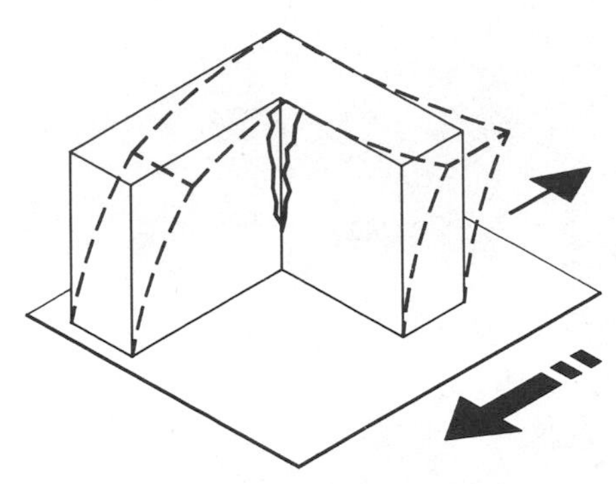

Figure 5-14 The L-shaped building.

the wing oriented east–west. The north–south wing, if it were a separate building, would tend to deflect less than the east–west wing, but the two wings are tied together and attempt to move differentially at their notch, pulling and pushing each other (Figure 5-14). For ground motion along the other axis, the wings reverse roles, but the differential problem remains.

The second problem is torsion. This is because the center of mass and the center of rigidity in this form can not geometrically coincide for all possible earthquake directions. The result is rotation, which tends to distort the form in ways that will vary in nature and magnitude depending on the nature and direction of the ground motion, and result in forces that are very difficult to analyze and predict.

The stress concentration at the notch and the torsional effects are interrelated. The magnitude of the forces and the seriousness of the problem will be dependent on:

- the mass of the building,
- the structural systems,
- the length of the wings and their aspect ratios,
- the height of the wings and their height/depth ratios.

In addition, it is not uncommon for wings of a reentrant-corner building to be of different height, so that the vertical discontinuity of a setback in elevation is combined with the horizontal discontinuity of the reentrant corner, resulting in an even more serious problem.

Architectural Implications Reentrant shapes are a most useful set of building shapes, enabling large plan areas to be accommodated in compact form, while still providing a high percentage of perimeter rooms with access to light and air. Thus such configurations are common for high-density housing and hotel projects.

Current concerns for daylighting and natural ventilation may result in increased use of narrow buildings and the traditional set of reentrant corner configurations. The courtyard form, most appropriate for hotels in tight

Figure 5-15 Damage concentrated at the intersection of two wings of an L-shaped school, Anchorage, Alaska, 1964.

urban sites, has always remained useful. In its contemporary form the courtyard becomes a glass-covered atrium, but the structural form is the same.

Historical Performance Examples of damage to reentrant-corner buildings are common, and this problem was one of the first to be identified by observers. It had been identified before the turn of the century, and by the 1920s was generally acknowledged by the experts of the day. Naito[5-4] attributed significant damage in the 1923 Kanto earthquake to this factor. The same damage phenomena were reported for the 1925 Santa Barbara and the 1964 Alaska earthquakes, and for the 1985 Mexico City earthquake (Figure 5-15).

Solutions There are two basic alternative solutions to this problem: to separate the building structurally into simple shapes, or to tie the building together strongly at lines of stress concentration and locate resistant elements to reduce torsion.

If a decision is made to use separation joints, they must be designed and constructed correctly to achieve the intent. Structurally separated entities of a building must be fully capable of resisting vertical and lateral forces on their own. To design a separation joint, the maximum drift (or some other reasonable criterion) of the two units must be calculated by the structural consultant. The worst case is when the two units would lean towards one another simultaneously, and hence the dimension of the separation space must allow for the sum of the deflections. In a tall building the relative motion between portions of the building will become very large, and create major problems of architectural detailing.

One of these is to preserve integrity against fire and smoke spread. The MGM Grand Hotel in Las Vegas is a T-shaped building in plan, with seismic joints approximately 12 in. in dimension. In the fire of 1983 these joints allowed smoke to propagate to the upper floors, resulting in many deaths.

Several considerations arise if it is decided to dispense with separation joints and tie the building together. Collectors at the intersection can transfer forces across the intersection areas, but only if the design allows for these beamlike members to extend straight across without interruption. Walls in this same location are even more efficient than collectors.

Since the free end of the wing tends to distort most under torsion, it is desirable to place resisting members at this location.

The use of splayed rather than right-angle reentrant corners lessens the stress concentration at the notch, which is analogous to the way a rounded hole in a steel beam creates less stress concentration problems than a

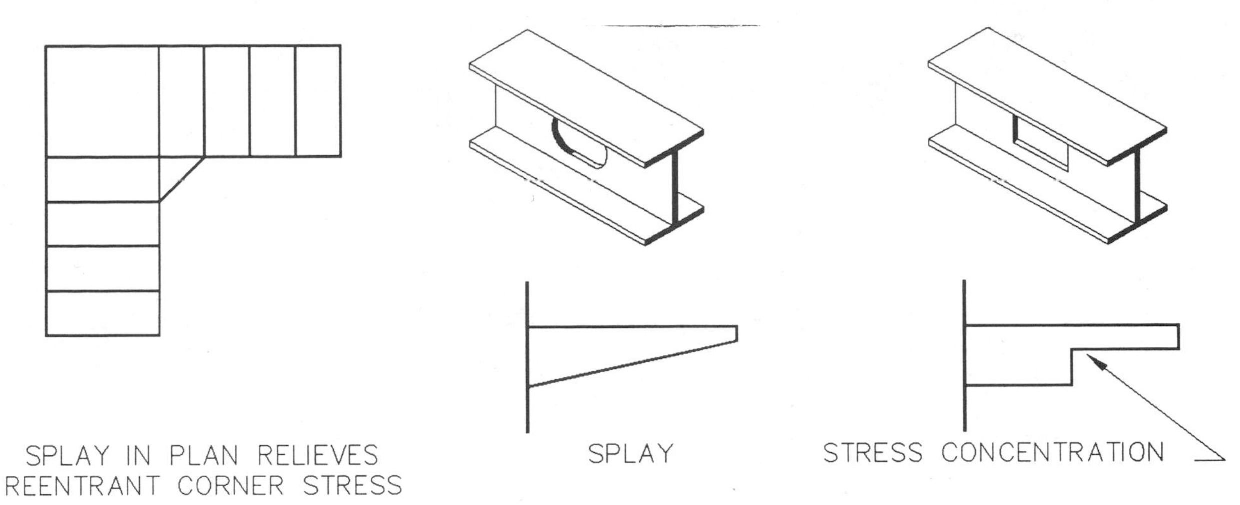

Figure 5-16 Splay in plan relieves reentrant corner stress.

rectangular hole, or the way a tapered cantilever beam is more desirable than one that is abruptly notched (Figure 5-16).

5.4.2 Variations in Perimeter Strength and Stiffness

Definition We here consider the bad effects of wide variations in strength and stiffness around a building perimeter that provides seismic resistance.

Seismic Effects This problem may occur in buildings whose configuration is geometrically symmetrical and simple, but nonetheless irregular for seismic design purposes. If there is wide variation in strength and stiffness around the perimeter, the centers of mass and resistance will not coincide, and torsional forces will tend to cause the building to rotate around the center of resistance. This effect is illustrated in Figure 5-17.

The weaknesses of open-front designs have been discussed by Degenkolb(5-5):

> Figure [5-18] shows the plans of three similar buildings, each with three shear walls so arranged that there is an open end and therefore major torsions in the building. If the buildings are similar, with uniform shear elements (uniform distribution of stiffness) and considering only shear deformations, it can rather simply be proved that the torsional deflection of the open end varies as the *square* of the length of the building.

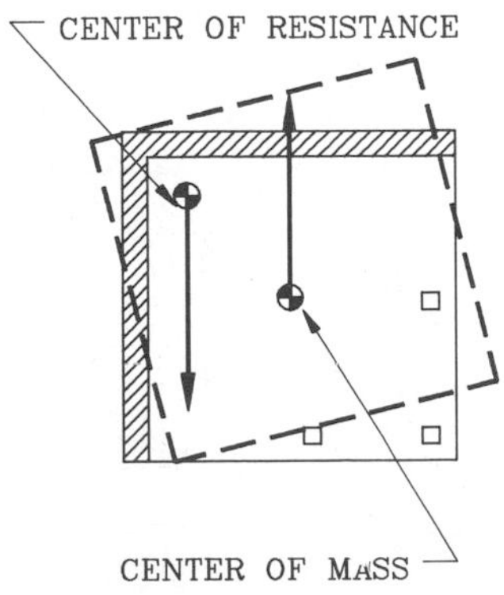

Figure 5-17 Torsional response.

Architectural Implications A common example of this condition occurs in storefront design, particularly on corner lots, and in free-standing department stores with varied openings around the perimeter. A special case is that of fire stations that require large doors for the movement of equipment. In these buildings it is particularly important to avoid major distortion of the front opening, for if the doors jam and cannot be opened, the fire station is out of action at a time when its equipment is most needed.

Industrial and warehouse buildings, particularly if of tiltup concrete construction in which lateral resistance is provided by the perimeter walls, often also require a variety of openings for entrances, loading docks, and office windows, with a consequent variation in seismic resistance around the perimeter.

Historical Performance A classical instance of this problem occurred in the J. C. Penney Department Store in Anchorage, Alaska, in the 1964 earthquake. The building was so badly damaged that it had to be demolished. The store was a five-story building of reinforced-concrete construction. The exterior walls were a combination of poured-in-place concrete, concrete block, and precast concrete nonstructural panels which were heavy, but unable to take large stresses. The first story had shear walls on all four elevations. The upper stories, however, had a structurally open north wall, resulting in U-shaped shear wall bracing system (similar to a typical open-front store) which, when subjected to east–west

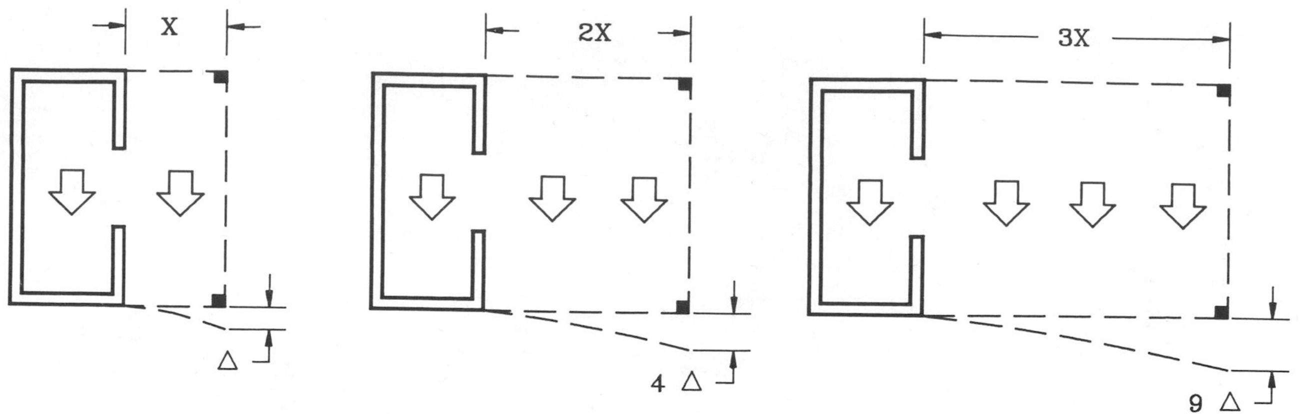

Figure 5-18 Open front design: Torsional deflection varies as the square of the length.

lateral forces, would result in large torsional forces (Figure 5-19).

Solutions The object of any solution to this problem is to reduce the possibility of torsion, and to balance the resistance around the perimeter. Four alternative strategies can be employed, and are diagrammed in Figure 5-20.

The first approach is to design a frame structure with approximately equal strength and stiffness for the entire perimeter. The opaque portions of the perimeter can be constructed of nonstructural cladding that will not affect the seismic performance of the frame. This can be done either by using lightweight cladding, or by insuring that heavy materials (such as concrete or masonry) are isolated from the frame.

A second approach is to increase the stiffness of the open facades by adding shear walls at or near the open face. This solution is, of course, dependent on a design which permits this addition.

A third solution is to use a very strong moment-resisting or braced frame at the open front, which approaches the solid walls in stiffness. The ability to do this will be dependent on the size of the facades: a long steel frame can never approach a long concrete wall in stiffness. This is, however, a good solution for wood frame structures, such as apartment houses with a ground-floor garage space, because even a rather long steel frame can be made to approach plywood walls in stiffness.

Finally, the possibility of torsion may be accepted and the structure designed to resist it. This solution will only apply to small structures with stiff diaphragms, which can be designed to act as a unity.

5.4.3 Nonparallel Systems

Definition The vertical load-resisting elements are not parallel to or symmetric about the major orthogonal axes of the lateral-force-resisting system.

Seismic Effects This condition results in a high probability of torsional forces under ground motion, because the centers of mass and resistance cannot coincide for all directions of ground motion. Moreover, the narrower portions of the building will tend to be more flexible than the wider ones, which will increase the tendency to torsion.

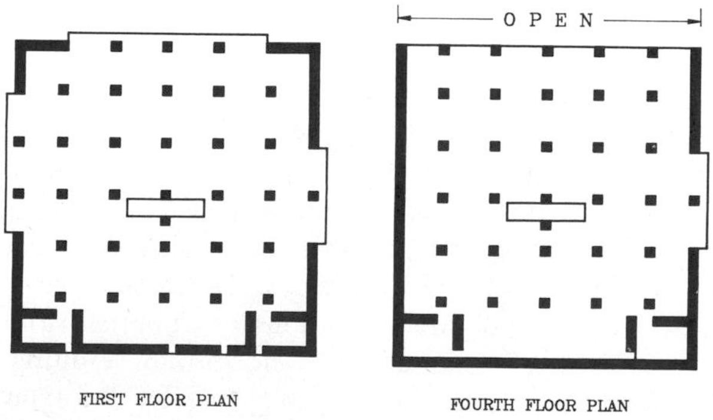

FIRST FLOOR PLAN — FOURTH FLOOR PLAN (OPEN)

Figure 5-19 J.C. Penney Department Store, Anchorage, Alaska.

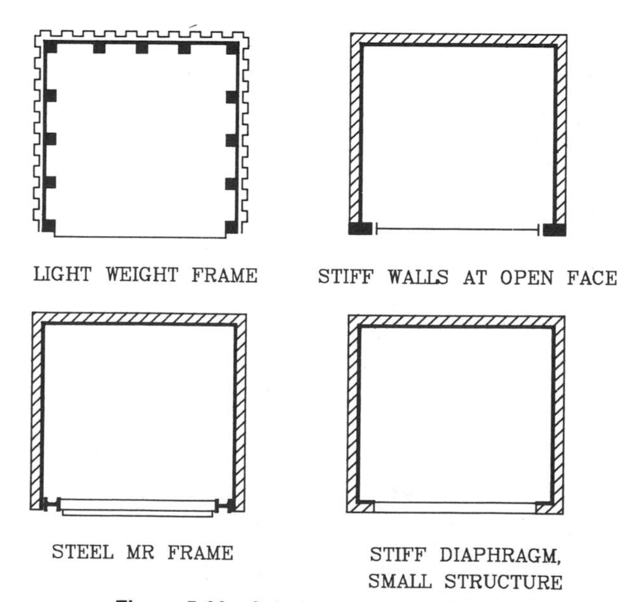

Figure 5-20 Solutions to open front design.

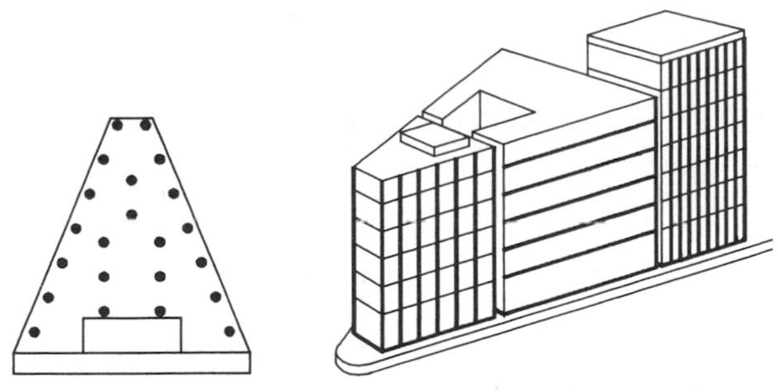

Figure 5-21 Wedge-shaped plan: invitation to torsion.

The problem is often exacerbated by perimeters with variations of strength and stiffness (Figure 5-21). A characteristic form of this condition is the triangular or wedge-shaped building that results from street intersections at an acute angle. These forms often employ a solid, stiff party wall in combination with two more open flexible walls facing the street. The result is a form that is very prone to torsion.

Architectural Implications Nonrectilinear forms have become fashionable forms in the last few years as a reaction against the rectangular "box." Forms that are triangular, polygonal, or curved have become commonplace, even in very large buildings.

The traditional triangular, trapezoidal, or "flatiron" form resulting from street-layout constraints is still common in high-density urban locations.

Historical Performance This form has been identified as a problem configuration for some time, though more on the basis of theory than through observed performance. The form is not identified as irregular in the SEAOC Commentary (1980), but it is identified as irregular in the 1988 UBC.

Many buildings of this type were constructed in Mexico City, resulting from the high density and street layout in the city, and instances of poor performance were observed in the 1985 earthquake. Many buildings suffered severe distortion particularly wedge-shaped buildings with stiff party walls opposite the apex of the triangular form (Figure 5-22). In many cases the condition was exacerbated by other irregularities such as a soft story.

Solutions The 1985 UBC places some special requirements on the design of these types of configuration. Particular care must be exercised to reduce the effects of torsion. In general, opaque walls should be designed as frames clad in lightweight materials, to reduce the stiffness discrepancy between these walls and the rest of the structure.

Figure 5-22 Distortions in wedge-shaped building, Mexico City, 1985. The building was subsequently demolished.

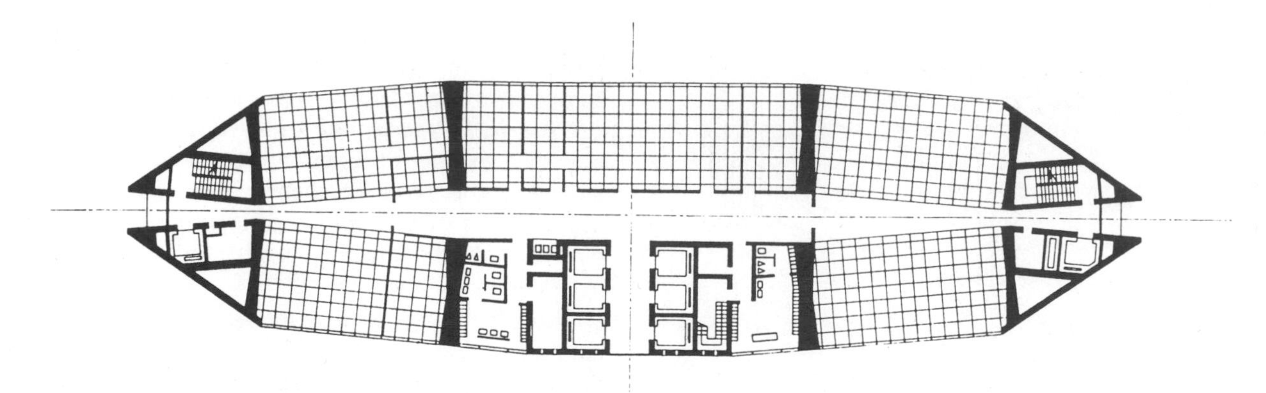

Figure 5-23 Pirelli Building, Milan, Italy (plan at first floor).

Alternatively, special design solutions may be introduced to increase the torsional resistance of the narrow parts of the building, although this may be difficult to achieve while still retaining desired open facades or internal areas.

Although not designed to resist seismic forces, the Pirelli Building in Milan, Italy, designed by the architect Gio Ponti and the engineer Pier Luigi Nervi between 1956 and 1959, is an interesting example of nonrectilinear design. In this instance, the sharply pointed (in plan) ends of the building are designed as four stiff cantilever boxes, to assist in dealing with wind loads at the ends of the long facades of the building (Figure 5-23). Nervi, who has always demonstrated in his structures a strong concern for symmetry and integrity, has pointed out that even these stiff cantilevers could not absorb the entire horizontal wind load. For this reason, the four triangular elements are complemented by four tapering "wall–columns" (walls at the lower floors, columns at the upper) that act with the end elements to resist wind forces. The eight resisting elements are of similar rigidity, which insures their collaboration.

5.4.4 Diaphragm Configuration

Definition The diaphragm configuration is the shape and arrangement of horizontal resistance elements that transfer forces between vertical resistance elements.

Seismic Effects Diaphragms perform a crucial role in distributing forces to the vertical seismic-resisting elements. The diaphragm acts as a horizontal beam: the diaphragm itself forms the web of the beam, and its edges act as flanges. Diaphragm penetration and geometrical irregularities are analogous to such irregularities in other building elements, leading to torsion and stress concentration.

The size and location of these penetrations is critical to the effectiveness of the diaphragm. The reason for this is not hard to see when the diaphragm is visualized as a beam: it is obvious that openings cut in the tension flange of a beam will seriously weaken its load-carrying capacity. In a vertical load system, a penetration in a beam flange would occur in either a tension or a compression area; in a lateral load system, the hole will be in a region of both tension and compression, since the loading alternates in direction.

When diaphragms form part of a resistant system, they may act in either a stiff or a flexible manner. This depends partly on the size of diaphragm (its area between enclosing resistance members or stiffening beams), and also on its material. The flexibility of a diaphragm, relative to the shear walls whose forces it transmits, also has a major influence on the nature and magnitude of those forces.

Architectural Implications Diaphragms are floors or roofs, and so have major architectural functions aside from their seismic role. The shape of the diaphragm is dependent on the overall plan form of the building, and how it can be subdivided by walls or collectors.

In addition, however, architectural requirements such as staircases, elevators and duct shafts, skylights, and atria result in a variety of diaphragm penetrations. In some cases, as in the need for elevators in an L-shaped building, the logical planning location for elevators (at the hinge of the L) is also the area of greatest seismic stress.

Historical Performance Failures specifically due to diaphragm design are difficult to identify, but there is general agreement that poor diaphragm layout is a potential contributor to failure.

Solutions Diaphragm penetrations are a form of irregularity specifically called out in the Commentary to the UBC (1980), requiring engineering judgment. In addition the 1988 revisions to this code specifically define such penetrations, and impose some additional requirements on the diaphragm design in such cases.

The general approach to the design of penetrations in diaphragms is to:

- ensure that penetrations do not interfere with diaphragm attachment to walls or frames;

- ensure that multiple penetrations are spaced sufficiently far from one another to allow reinforcing to develop their required capacity;
- ensure that collectors and drag struts are uninterrupted by openings.

5.5 VERTICAL-CONFIGURATION PROBLEMS

5.5.1 Soft Stories

Definition A soft story is one that shows a significant decrease in lateral stiffness or strength from that immediately below. The term is also sometimes restricted to a decrease in stiffness, and "weak story" used to indicate a decrease in strength.

Seismic Effects The condition is most critical when it occurs at the first story, because the loads are generally greatest at this level.

The essential characteristics of a soft first story consist of a discontinuity of strength or stiffness, which occurs at the second-story connections. This discontinuity is caused because lesser strength, or increased flexibility, in the first story structure results in extreme deflections in the first-story which, in turn, result in a concentration of forces at the second-story connections.

If all stories are approximately equal in strength and stiffness, the entire building deflection under earthquake forces is distributed approximately equally to each story. If the first story is significantly less strong or more flexible, a large portion of the total building deflection tends to concentration there, with consequent concentration of stresses at the second-story connections (Figure 5-24).

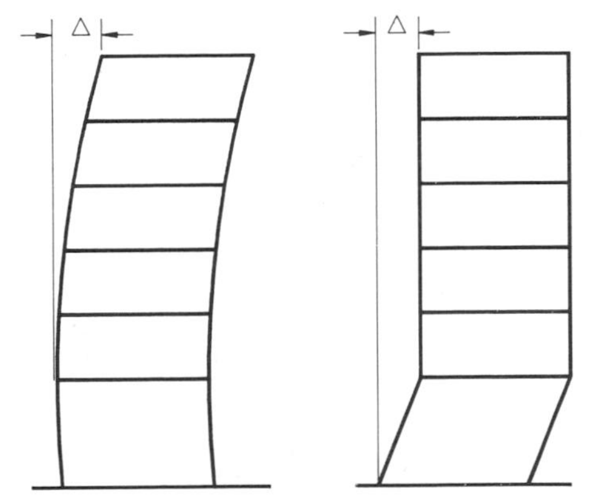

Figure 5-24 The soft story effect.

In more detail, the soft-story problem may result from four basic conditions. These are diagramed in Figure 5-25 and are:

1. A first-story structure significantly taller than upper floors, resulting in less stiffness and more deflection in the first story.
2. An abrupt change of stiffness at the second story, though the story heights remain approximately equal. This is caused primarily by material choice: the use, for example, of heavy precast concrete elements above an open first story.
3. The use of a discontinuous shear wall, in which shear forces are resisted by walls which do not continue to the foundations, but stop at second floor level, thus creating a similar condition to that of item 2 above.
4. Discontinuous load paths, created by a change of vertical and horizontal structure at the second story.

The above characteristics, individually or in combination (high first story combined with discontinuous shear walls, for example, is a particularly bad idea) are readily identifiable.

Architectural Implications A taller first story often has a strong programmatic justification, when large spaces, such as meeting rooms or a banking hall, must

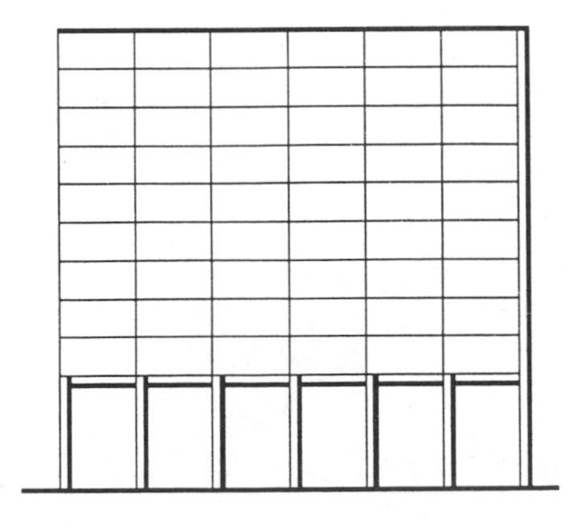

FLEXIBLE FIRST FLOOR

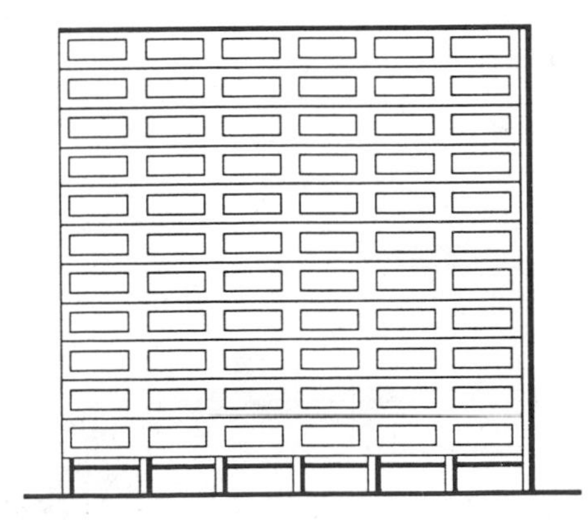

CHANGE OF STIFFNESS ABOVE FIRST FLOOR

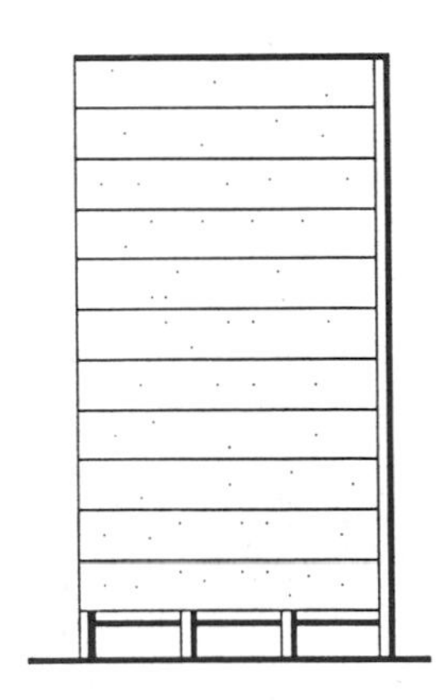

DISCONTINUOUS SHEAR WALL

Figure 5-25 Types of soft story.

be provided at ground level. Similarly, an open ground floor often meets urban design needs by providing both real and symbolic access to a plaza or street. The changes of proportion provided by a high story, or the "floating box" concept (now somewhat outdated), are very real aesthetic tools for the architect, although engineers may find such concepts hard to rationalize in their terms.

Engineers must accept that some form of variation in the first story will remain a desirable architectural characteristic for the foreseeable future; whether it is "soft" in seismic terms is a matter for the architect and engineer to resolve.

Historical Performance The general type of soft first-story configuration was early identified as a problem. Failures in masonry buildings in the 1925 Santa Barbara earthquake were identified by Dewell and Willis[5-6] as soft-first-story failures.

In more recent times, with extensive use of frame structures, damage to reinforced-concrete buildings in Caracas (1967) clearly identified the risk to tall buildings with this condition. In the Mexico City earthquake of 1985, researchers determined that soft first stories were a major contributor to 8% of the serious failures, and the actual percentage is probably greater because many of the total collapses were precipitated by this condition.

The particular case of the discontinuous shear wall has led to clearly diagnosed failures in United States buildings. Olive View hospital, a new structure that was badly damaged in the 1971 San Fernando earthquake, represents a classic case of the problem.

The vertical configuration of the main building was a soft two-story layer of rigid frames on which was supported a four-story shear wall-frame structure (Figure 5-26). The second floor extended out to form a large plaza.

The severe damage occurred in the soft-story portion: the upper floors moved so much as a unit that the columns at ground level could not accommodate such a huge displacement between their bases and tops and failed. The largest amount by which a column was left permanently out of plumb was $2\frac{1}{2}$ ft.

Though not widely understood, the stair towers at Olive View also show a clear and separate example of a discontinuous-shear-wall failure. These seven-story towers were independent structures, and proved incapable of standing up on their own: three stair towers overturned completely, while the fourth leaned outward 10°. The six upper stories were rigid reinforced-concrete walls, but the bottom story was composed of six free-standing reinforced-concrete columns, which failed. The exception was the north tower, whose walls came down to the foundation directly without any discontinuity; this was the only tower to remain standing. Olive View hospital was demolished after the earthquake, and a new hospital built on the same site.

The performance of the Imperial County Services Building, El Centro, in the Imperial Valley Earthquake of 1979, provides another example of the effects of architectural characteristics on seismic resistance. The building was a six-story reinforced-concrete structure built in 1969. In this mild earthquake the building suffered a major structural failure, resulting in column fracture and shortening (by compression) at one end—the east—of the building (Figure 5-27). The origin of this failure lies in the discontinuous shear wall at that end of the building.

The fact that this failure originated in the configuration is made clear by the architectural difference between the east and west ends. The difference in location of the small ground-floor shear walls was sufficient to

Figure 5-26 Olive View Hospital, San Fernando Valley, California. (a) Elevation of staircase towers. (b) Section through main building.

Figure 5-27 Imperial County Services Building, El Centro, California. Failure of end bay at discontinuous shear wall (Imperial Valley Earthquake of 1979).

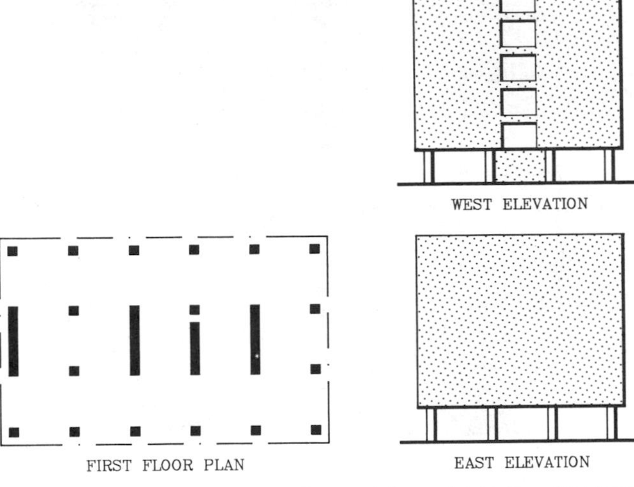

Figure 5-28 Plan and elevations of Imperial Valley Services Building, El Centro, California.

create a major difference in response to the rotational forces on the large end shear walls (Figure 5-28).

Solutions If a high first story is desired, either:

- introduce bracing that stiffens the columns up to a level comparable to the superstructure,
- add columns at the first story to increase stiffness, or
- change the design of the first-story columns to increase stiffness.

If a large opaque wall is required in a location that could create a soft first story:

- Insure that such a wall is not part of the lateral resistance system.
- Reduce the mass of the wall by the use of light materials and hollow construction.
- If a heavy wall is necessary, then insure that the wall is detached in such a way that the superstructure is free to deflect in a comparable way to the first floor.

If the architect insists on such material and design constraints that a major discontinuous shear wall is the only solution, refuse to do it. The liabilities involved in using such a proven failure mechanism are too great.

If the lateral resistance system is based on the use of an interior core (for a high-rise office building, for example), the perimeter columns may be tall, but there is no soft first story, provided the core is brought down to the ground. In such a building it is not difficult, if the core-plan dimensions are sufficient, to insure that the stiffness of a tall first story is adequate to prevent structural discontinuity at the second floor.

It should be noted that in the 1988 UBC, structures with a soft-story discontinuity in capacity that is less than 65% of that of the story above are not permitted over two stories or 30 ft in height.

5.5.2 Columns: Variations in Stiffness, Short Columns, and Weak Column, Strong Beam

Definition Here we consider the use of columns of varying stiffness, by reason of either differences in length or deliberate or inadvertent bracing; the use of columns that are significantly weaker than connecting beams; and the use of columns in one floor that are significantly shorter than those in the other floors.

Seismic Effects Seismic loads are distributed in proportion to the stiffness of the resisting members. Hence, if the stiffness of the supporting columns (or walls) varies, those that are stiffest (usually shorter) will "attract" the most forces. The effect of this phenomenon is explained in Figure 5-29. The important point is that stiffness (and hence force) varies approximately as the cube of the column length.

Similarly, a uniform arrangement of short columns supporting a floor will attract greater forces to that floor, with a corresponding possibility of failure. Typically such an arrangement may also involve deep and stiff spandrel beams, making the columns significantly weaker than the beams.

Such design is in conflict with a basic principle of seismic design, which is to design a structure in such a way that under severe seismic forces, beams will perform plastically before columns. This is based on the reasoning that as beams progress from elastic to inelastic behavior they start to deform permanently. This action will dissipate and absorb some of the seismic

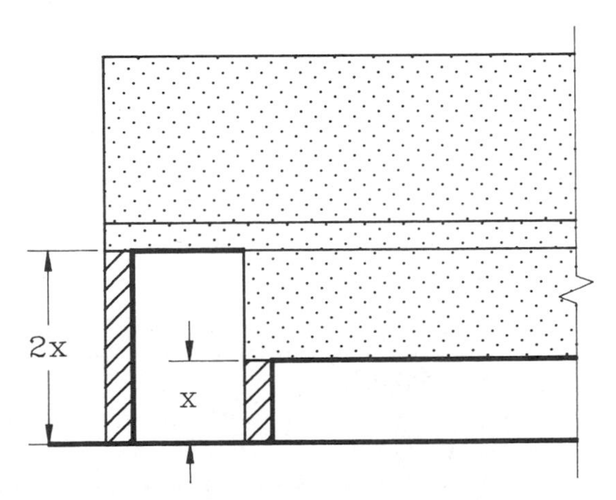

Figure 5-29 Variation of stiffness.

energy. Conversely, if the column fails first and begins to deform and buckle, major vertical compressive loads may quickly lead to total collapse.

Mixing of columns of varying stiffness on different facades may also lead to torsional effects, since the building assumes the attributes of varying perimeter resistance discussed above.

Architectural Implications The origin of variations in column stiffness generally lies in architectural considerations: hillside sites, infilling of portions of frames with nonstructural but stiff material to create high strip windows, desire to raise a portion of the building off the ground on tall pilotis while leaving other areas on shorter columns, or stiffening some columns with a mezzanine or a loft, while leaving others at their full, unbraced height.

The importance of these issues lies in the fact that their effects may be counterintuitive. For example, infilling may be done as a remodel activity later in the building life for which the engineer is not consulted, because intuition may suggest to the designer that he is strengthening it in the act of shortening it rather than introducing a serious stress concentration for which the structure was not designed.

Variations in openings in different facades are often required from a daylighting or energy-conservation requirement. Where openings are created by variations in structural arrangement, rather than by variations in cladding, some of these conditions may well arise.

Historical Performance Significant column failures, sometimes leading to collapse, have been attributed to these conditions in a number of recent earthquakes, particularly in Japan, Latin America, and Algeria.

Many Japanese schools, employing short columns on one side of an elevation, or using a weak-column, strong-beam configuration, suffered severe damage in the Tokaichi-oki earthquake in 1968, and the 1978 Miyagi-ken-oki earthquake.

In Latin America, the problem has frequently been caused by inadvertent stiffening of columns through nonstructural infill, which when combined with high glazing creates short columns.

In the El Asnam (Algeria) earthquake of 1980, many apartment-structure failures were caused by short columns used at ground level to provide a ventilated open area (called a "vide sanitaire") in a semibasement location (Figure 5-30).

Solutions The general solution is to match the detailed seismic design carefully to the architectural requirements. The weak-column, strong-beam condition can be avoided by insuring that deep spandrels are isolated from the columns; in the same way the lengths of columns around a facade can be kept approximately equal.

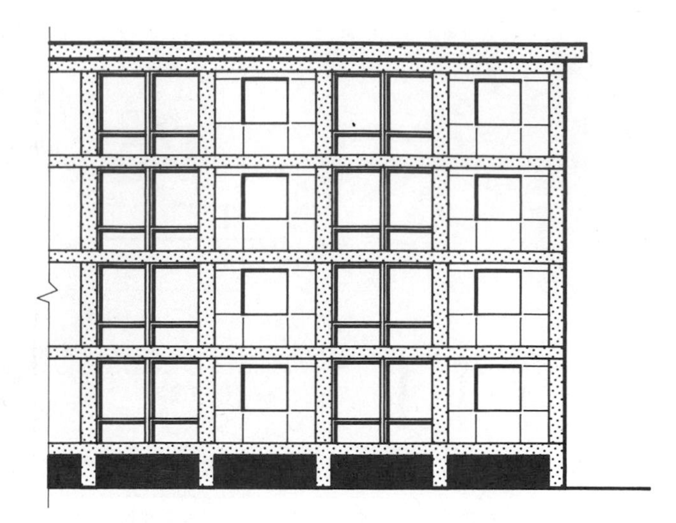

Figure 5-30 Typical apartment house, El Asnam: "Vide Sanitaire" creates short column condition.

Horizontal bracing can be inserted to equalize the stiffness of a set of columns of varying height (Figure 5-31). Heavy nonstructural walls must be isolated from columns to insure that a short-column condition is not created.

5.5.3 Vertical Setbacks

Definition A vertical setback is a horizontal offset in the plane of an exterior wall of a structure.

Seismic Effects The problem of this shape lies in the general problem of discontinuity: the abrupt change of strength and stiffness. In the case of this complex configuration, it is most likely to occur at the line of setback, or *notch*.

The seriousness of the setback effect depends on the relative proportions and absolute size of the separate parts of the building. In addition, the symmetry or asymmetry in plan of the tower and base affect the nature of the forces. If the tower, or base, or both are dynamically asymmetrical, then torsional forces will be introduced into the structure, resulting in great complexity of analysis and behavior.

The setback configuration can also be visualized as a vertical reentrant corner. Stresses must go around a corner, because a notch has been cut out, preventing a more direct route. Hence, the smaller the steps or notches in a setback, the smaller the problem. A smooth taper avoids the notch problem altogether. A tapering beam will not experience stress concentrations, whereas a notched beam will.

Setbacks with shear walls in the tower portion that are not continued to the ground are highly undesirable. Besides the change of stiffness where the shear wall enters the base structure, the shear wall will transmit large forces to the top diaphragm of the base.

Figure 5-31 Horizontal bracing to stiffen an open corner entrance.

Although, typically, setbacks occur in a single building, the condition can also be created by adjoining buildings of different heights which have inadequate or nonexistent seismic separation.

Architectural Implications Setbacks may be introduced for several reasons. The three most common are zoning requirements that require upper floors to be set back to admit light and air to adjoining sites, program requirements that require smaller floors at upper levels, or stylistic requirements relating to the building form.

Setbacks relating to zoning were common a few decades ago when daylighting was a major concern, and resulted in the characteristic shapes of older high-rise buildings in New York City. Stylistic fashions replaced these forms with those of simple rectangular solids, made possible by advances in artificial lighting and air conditioning. Now, there is a renewed interest in setback shapes for stylistic reasons, while at the same time energy conservation requirements have reinstated a functional interest in setbacks for daylighting purposes.

An interesting example of this stylistic trend is that of the new planning code for San Francisco, which specifically mandates setbacks. These represent relatively minor variations in the vertical plane of the facade, rather than the abrupt rising tower on a base, which is of more serious seismic significance. The trend is, however, away from vertical structural continuity at the perimeter and thus introduces complexity and cost into the structural solution.

A type of setback configuration only made possible by modern framed construction is that of the building that grows larger with height. This type is termed *inverted setback*. Its geometrical definition is the same as that of the setback, but, because of the problems of overturning, its extremes of shape are less. Nevertheless, some surprising demonstrations of this shape have appeared, and it appears to be one whose image has a powerful design appeal (Figure 5-32).

Historical Performance Although commonly identified as a configuration problem, severe failures of modern buildings attributed to this condition are few. While traditional towers, primarily churches, have suffered their share of failures, the number of those that have survived severe damage is remarkable.

Figure 5-32 Dallas City Hall: The inverted setback.

While there have been recorded failures of inverted-setback buildings, notably in the Agadir (Morocco) earthquake of 1960, some of the more striking examples have performed well. This is probably because the appearance of instability inherent in this form results in special attention being paid to its structural design. Typically, such buildings devote a much larger percentage of their construction cost to structure than more conventional buildings.

Solutions Setbacks have long been recognized as a problem, and so the Uniform Building Code has attempted to mandate special provisions for them. Currently, the earthquake regulations of the Code refer to setback configurations as follows:

> Buildings having setbacks wherein the plan dimension of the tower in each direction is at least 75% of the corresponding plan dimension of the lower part may be considered as uniform buildings without setbacks, provided other irregularities as defined in this section do not exist.

The SEAOC Commentary (1980) to this section includes:

> Setbacks in many instances constitute a degree of irregularity that requires consideration of the structure's dynamic characteristics in order to achieve a reasonable distribution of lateral forces.

In addition, the Commentary refers the reader to Appendix C, the report of the 1958 Setback Committee, which is reproduced in the Commentary. Thus, in dealing with setbacks, the code essentially leaves the designer to determine the extent of his problem and to devise an appropriate solution.

In general, conceptual solutions to the setback problem are analogous to those for its horizontal counterpart, the reentrant-corner plan. The first type of solution consists of a complete seismic separation in plan, so that portions of the building are free to react independently. For this solution, the guidelines for seismic separation, discussed elsewhere, should be followed.

When the building is not separated, the analysis proposed by the 1958 Setback Committee provides the best guidelines, with some necessary interpretation to fit the particular case. Particular attention should be paid to avoiding vertical column discontinuity, so that setbacks should be arranged to coincide with normal bay sizes (which may result in a series of small bays).

Any large building with major setback conditions should be subject to special analysis, or at least to careful investigation of probable dynamic behavior. Finally, the inverted setback configuration of any extreme form and size should be avoided in seismic areas, unless the owner is willing to assume the considerable additional extra structural costs that will be incurred.

The NEHRP provisions, as noted earlier, permit vertical setback configurations to be analyzed using the simple ELF method if the stiffness on the top story of the base is at least five times that of the first story of the tower. The 1988 UBC permits use of the standard ELF method for a two-stage analysis of tower and base if the average story stiffness of the base is at least 10 times greater than the average story stiffness of the tower.

5.6 STRUCTURALLY RESTRICTIVE ARCHITECTURAL DETAILING

5.6.1 Components and Connections

Definition By *structurally restrictive detailing* we mean detailed architectural design of a component or connection that prevents good seismic design practice in the structural design.

Seismic Effects This problem represents a micro version of typical overall building-configuration problems. Architectural detailing may place dimensional or locational constraints on structural design resulting in weakness or eccentricity of force actions that can lead to stress concentration or local torsion. The problem is most critical at beam–column connections, which are highly stressed, but often represent a critical element in the aesthetic scheme of the building.

Structural detailing ideally provides for direct load transfer and minimum local eccentricity, with forces resolved at a point. Architectural detailing may result in inadequate size and eccentric or discontinuous load paths (Figure 5-33). The problem is particularly critical for reinforced-concrete structures, where constraints may provide inadequate room for proper placing of reinforcing.

Architectural Implications Detailed design is an important element in architectural expression. As an example, the design of the perimeter beam–column connection can provide the building with a predominantly horizontal, vertical, or neutral emphasis (Figure 5-34). But the structural implications of these variations may not be understood by the architect.

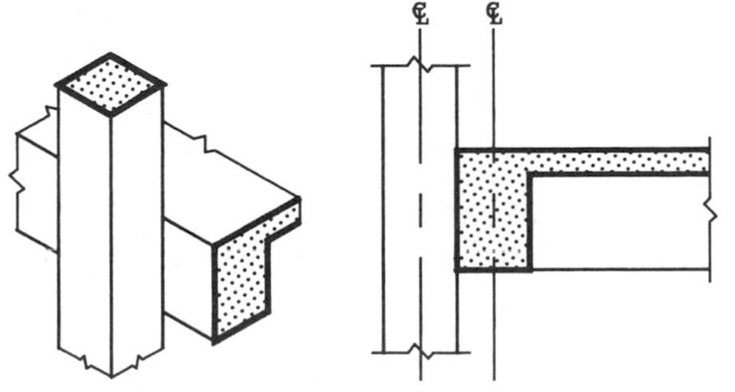

Figure 5-33 Eccentric load paths created by architectural detailing at a joint.

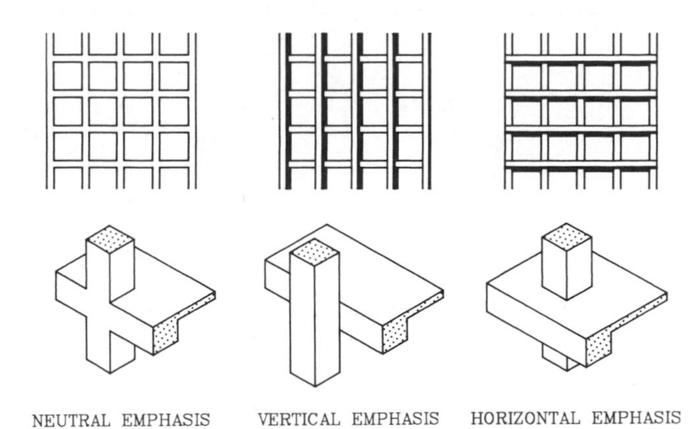

Figure 5-34 Facades: Differences in architectural emphasis.

Another example is the use of taper or the insertion of recesses in columns. Tapered columns may be a correct expression of structural forces, and be easy to accommodate, or they may directly contradict structural action and lead to weakness. Recesses are often designed by architects to accentuate the line at which materials meet one another, particularly when the materials are different or meet at right angles, as in a column–slab junction.

Historical Performance Specific performance attributable to this condition is difficult to document, but the problem is generally recognized by engineers. Two well-documented cases do exist where architectural detailing contributed to failure.

The first is that of the column design of Olive View Hospital, damaged in the 1971 San Fernando earthquake (discussed previously as an example of soft-first-story failure). A significant difference in performance was observed between corner and internal columns in this building. The 12 L-shaped corner columns were completely shattered and their load-carrying capacity reduced almost to zero. The interior columns, of square section, had spiral ties, and although they lost much of their concrete cover, they retained load-carrying capacity and probably saved the building

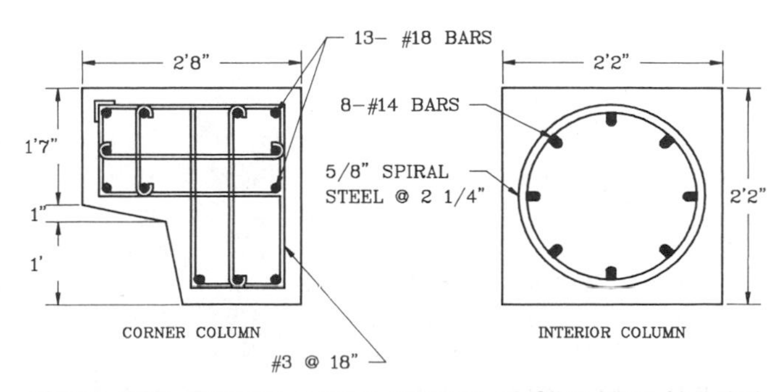

Figure 5-35 Exterior column sections of Olive View Hospital, San Fernando Valley, California. Due to their shapes, corner columns could not use spiral reinforcement.

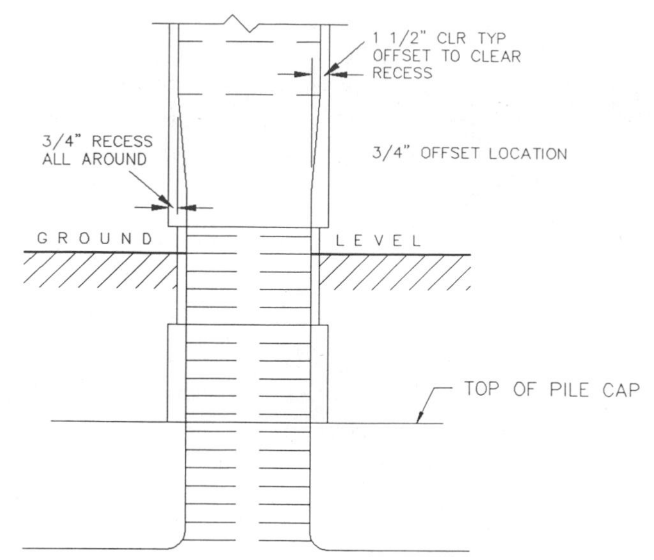

Figure 5-36 Column detail, Imperial County Services Building. Note architectural recess.

from collapse. Because of their architectural form, it was not possible for the corner columns to use spiral ties (Figure 5-35). Higher stress and torsion in the corner columns may also have contributed to their poor performance.

The Imperial County Service Building at El Centro, California, suffered severe damage in the 1979 Imperial Valley Earthquake, and four columns at one end of the building were badly shattered. Detailed study of these columns showed that an architectural recess had been placed at the line where the columns met the ground (Figure 5-36). This recess caused a reduction in sectional area of the column and a reduction in axial load-carrying capacity. Analytical and experimental studies have shown that this change in section accentuated the undesirable performance of these columns.[5-7]

Solutions Close coordination between architect and engineer is necessary to insure that architectural detailing does not result in undesirable structural design constraints.

5.7 PROBLEMS OF ADJACENCY

5.7.1 Pounding

Definition Pounding is damage caused by two buildings, or different parts of a building, hitting one another.

Seismic Effects The possibility of pounding is a function of the vertical deflection or drift of adjoining buildings (or parts of a building). Drift is calculated by applying the code design forces to the building and then deriving the deflections that result. Since these estimated forces will be less than what we know can occur, calculated deflections must be corrected to obtain a more realistic estimate of how much the building may actually move. Alternatively, an accurate estimate of drift may be made that accounts for all foreseeable factors.

Architectural Implications Pounding is included in this discussion of configuration issues because it is a matter of where buildings are located relative to other structures, which is an early architectural decision. The problem has considerable architectural implications for the construction of buildings on constricted urban sites, because to make provision for the worst-case condition could result in large building separations and significant loss of usable space.

While codes place modest limits on drift (for example, 0.005 times story height) based on static analysis, actual experience of drift and calculations of realistic figures provide some startling numbers. Freeman[5-8] calculated the actual drift on flexible buildings up to 20 stories under $0.4g$ as being 0.020–0.055 times the story height. For a 12-story building this translates into 40–110 in. for a 14-ft story height. A separation that could accommodate two such buildings vibrating out of phase would have to be 18 ft 4 in wide.

Clearly, compromise is necessary, but nonetheless, loss of usable space measure in linear feet becomes serious. In addition, the idea of urban buildings with spaces of 2–3 ft between them suggests a very difficult maintenance problem.

Historical Performance Pounding has been noted routinely by earthquake investigators over the past several decades. In the 1972 Managua earthquake, the five-story Grant Hotel suffered a complete collapse of its third floor when battered by the roof level of the adjacent two-story building.

In the 1964 Alaska earthquake, the 14-story Anchorage Westward Hotel pounded against its low rise ballroom and an adjoining six-story wing, although separated by a 4-in. gap. The pounding was severe enough in the high rise to dislocate some of the metal floor decking from its steel beam supports.

More recently the extent of pounding damage in Mexico City in 1985 confirmed this as a major problem. Of the severely damaged buildings 40% suffered a mid-story failure, generally caused by pounding from an adjoining building (Figure 5-37). The situation was probably exacerbated by the long duration of shaking and the relative flexibility of the heavy reinforced-concrete frame buildings that were the major damage victims.

Solutions To assume that the code limits on drift provide an accurate estimate of possible drift is unrealistic, but realistic estimates may provide very large worst case figures. Blume, Corning, and Newmark sug-

Figure 5-37 Pounding Failure, Hotel de Carlo, Mexico City, 1985.

gest an alternative method: [5-9]

> Compute the required separation as the sum of the deflections computed for each building separately on the basis of an increment in deflection for each story equal to the yield-point deflection for that story, arbitrarily increasing the yield deflections of the two lowest stories by multiplying them by a factor of 2.

An earlier edition of the Uniform Building Code contained a rule of thumb intended for the relatively stiff structures of that day [5-10]: separations should be "one inch plus one half-inch for each ten feet of height above twenty feet."

A possible alternative approach is to place an energy-absorbing material between the buildings.

Many buildings in Mexico City were, in fact, protected from collapse because they were erected hard up against adjoining buildings on both sides, so that whole blocks of buildings acted as a unit, and the group was stronger than the individual structures. As evidence of this, Mexican studies show that 42% of severely damaged buildings were corner buildings, lacking the protection of adjoining structures. This finding suggests the need for serious research on the subjects of allowable drift, pounding, and the design and construction of closely spaced buildings.

5.8 SEISMIC ISSUES IN THE DESIGN PROCESS

In this section, a set of issues relating to seismic design is outlined for each of the conventional phases of design and construction. The issues are those that are appropriate for review between the architect and engineer. This set, in turn, is related to a broad list of the structural decisions that would normally be expected for each phase.

5.8.1 Program Phase

The architect should review program requirements and his initial design ideas briefly with the engineer to insure that no needless conflicts will later arise because of programmatic or design assumptions or constraints.

Issues for Review

Building size:

- gross area
- floor area
- probable number of floors

Site characteristics:

- geology
- zoning restrictions
- height limits
- plan area
- orientation
- foundation characteristics

Interior planning requirements:

- Types of spaces:
  - large
  - small
  - column-free
- circulation requirements:
  - vertical
  - horizontal
- special planning requirements

Fire standards:

- code construction type options

Budget:

- general level of quality

Seismic code:

- determination of applicable code
- importance factor

5.8.2 Schematic Design

The architect should review with the engineer before beginning schematic design, or very early in the process, as soon as alternative configurations begin to appear. Complex plans or significant configuration issues should be discussed with the engineer at the earliest opportunity so that their importance can be assessed.

Issues for Review

Configuration:

- size
- shape
- number of floors
- significant problems
- floor to floor heights

Vertical circulation:

- stairs
- elevators
- cores:
  - size
  - location

HVAC:

- general type
- distribution pattern
- space for ducts

Materials:

- code requirements
- cladding

Structural decisions (general structural strategies):

- horizontal framing
- vertical framing
- lateral systems:
  - moment frames
  - braced frames
  - shear walls
  - mixed
- perimeter requirements
- special aesthetic requirements

During this phase, the architect and engineer should confer frequently, particularly on the issues of general structural strategies, vertical circulation, configuration, and their interactions.

5.8.3 Preliminary Design and Design Development

Availability of detailed design information gives the engineer the necessary data to pursue analysis, and on this basis, the design is revised to take into account member sizes or other variations of location and type of resisting elements and connection details. Detailed analysis may be required at earlier stages for complex or unusual designs if the usual approximations cannot be applied.

Issues for Review

Architectural systems

- exterior cladding
- appendages

Preliminary details:

- interior partitions
- ceilings

- depressions in floor slabs
- vertical transportation

Mechanical, Electrical, Plumbing Design:

- HVAC and utilities distribution
- preliminary duct size and location
- openings in floors, walls, beams, girders
- equipment locations: roof, floors, basement
- vertical shafts
- lighting

Structural decisions—structural system design:

- bay size
- horizontal framing: materials, foundation requirements
- vertical lateral framing
- shear wall, braced frame locations

Structural decisions—preliminary structural analysis:

- preliminary member sizing
- preliminary seismic details

5.8.4 Contract Documents

While it is obvious that the architect will want to know the engineering decisions (exact column dimensions, separation-joint detailing, etc.), the engineer also needs to know what the architect is doing. If it is assumed that partitions will carry no lateral load, and will not be deformed by frame deflections, the architectural detailing must insure this. The engineer should review the seismic aspects of nonstructural features designed by others—anchorage of ceilings and shelving, heavy-equipment location, exterior cladding, etc. This task must be called out in contract negotiation between architecture and engineer.

Issues for Review
Architectural systems—final details:

- partitions
- exterior cladding
- ceilings
- vertical shafts
- stairways
- floor slab depressions

Mechanical, electrical, plumbing details:

- responsibility for final seismic safety
- duct size and location
- piping size and location
- size, weight, and location of major equipment
- all penetrations of structural members
- lighting systems

Structural design–final details

- member sizes, location
- connection details

Final structural analysis:

- review of shop drawings

5.8.5 Construction

Current issues of liability are affecting the traditional inspection roles of architect and engineer. However, some mechanism is necessary to insure conformance with drawings and specifications, and with construction quality in the field. The engineer should also review any alternatives proposed by the contractor that may have seismic implications. Contractor proposed cladding details require special attention.

Issues for Review
Conformance to drawings and specifications:

- construction quality, workmanship
- contractor-developed details

5.8.6 Operation

While the architect and engineer may not be consulted after occupancy with regard to equipment purchases or remodeling, larger buildings are often under the authority or facilities managers or building engineers. These should be aware of the seismic importance of certain kinds of nonstructural additions or changes that may affect the seismic performance of the building.

5.9 CONCLUSION

It should be clear from the discussion in this chapter that seismic design is a shared architectural and engineering responsibility, which stems from the physical relationship between architectural forms and structural systems. An understanding of these relationships should be present in the mind of any designer working in a seismic area. Unfortunately, our methods of education and practice have tended to diminish such understanding, because we separate our architects and engineers during their education and, for the most part, in their practice.

The interrelations between issues of engineering and architecture demand that architect and engineer work together from the inception of a project. The idea of engineers participating in early design concepts is not new, yet it often does not happen, for a variety of reasons, economic, cultural, and professional. If they are to work effectively together, the architect and engineer must be able to communicate using a shared language within a common conceptual framework.

One of the traditional problems of communication has been—and continues to be—that designers tend to think visually and express themselves in sketches that are almost a form of shorthand, whereas engineers like the precise but abstract language of mathematics, and their visual language is that of curves and algebraic formulas. Because these two languages are almost completely incompatible, the architect and engineer may be able to say little to one another.

Perhaps the last word may be left to Mete Sozen[5-11]:

> In resistance to gravity loads, architectural and structural decisions may be made independently of each other. But in resistance related to earthquake effects, separating the engineer from the architect is a formula for disaster.

REFERENCES

5-1 Shah, H. C., Zsutty, T. C., and Padilla, L., "The Purpose and Effects of Earthquake Codes," Internal Study Report No. 1, The John A. Blume Earthquake Engineering Center, Stanford Univ., 1977.

5-2 Reitherman, R. K., "Frank Lloyd Wright's Imperial Hotel: A Seismic Re-evaluation," Proc. Seventh World Conference on Earthquake Engineering, Istanbul, 1980.

5-3 Dowrick, D. J., *Earthquake Resistant Design*, John Wiley & Sons, London, 1977.

5-4 Naito, T., "Earthquake-Proof Construction," *Bull. Seism. Soc. Am.* 17, No. 2, June 1977.

5-5 Degenkolb, H., "Seismic Design: Structural Concepts," Summer Seismic Institute for Architectural Faculty, AIA Research Corp., Washington, 1977.

5-6 Dewell, H. and Willis, B., "Earthquake Damage to Buildings," *Bull. Seism. Soc. Am.* 15, No. 4, Dec. 1925.

5-7 Zeris, C. A., Mahin, S. A., and Bertero, V. V., "Analysis of the Seismic Behavior of the Imperial County Services Building," Proc. Third U.S. National Conference on Earthquake Engineering, El Cerrito, CA, Earthquake Engineering Research Inst., 1986.

5-8 Freeman, S. A., "Drift Limits: Are They Realistic," Structural Moments, Structural Engineers Association of Northern California, Berkeley, CA, 1980.

5-9 Blume, J. A., Newmark, N. M., and Corning, L. H., "Design of Multistory Concrete Buildings for Earthquake Motions," Portland Cement Assoc., Skokie, IL, 1961.

5-10 International Conference of Building Officials (ICBO), "1958 Uniform Building Code" (UBC), Whittier, CA, 1958.

5-11 Sozen, M. A., "Earthquake Resistant Design," Designing for Earthquakes, Proceedings from the 1978 Summer Seismic Institute for Architectural Faculty, AIA Research Corp., Washington, 1979.

Chapter 6

Design for Drift and Lateral Stability

*Farzad Naeim, Ph.D., P.E.**

6.1 INTRODUCTION

This chapter deals with the problems of drift and lateral stability of building structures. Design for drift and lateral stability is an issue which should be addressed in the early stages of design development. In many cases, especially in tall buildings or in cases where torsion is a major contributor to structural response, the drift criteria can become a governing factor in the selection of the proper structural system.

In the design of building structures, different engineers attribute various meanings to the term "stability."[6-1] Here, we consider as stability problems only those problems related to the effects of deformation on equilibrium of the structure. Furthermore, we will limit the discussion to the stability of the structure as a whole. Local stability problems, such as the stability of individual columns or walls, are discussed in Chapters 8, 9, and 10 of this handbook.

The concerns that have resulted in code requirements for limiting lateral deformation of structures are explained in Section 6.2. The concept of lateral stability, its relationship to drift and the *P*-delta effect, and factors affecting lateral stability of structures are discussed in Section 6.3.

*Director of Research and Development, John A. Martin and Associates, Los Angeles, California.

Several practical methods for the inclusion of stability effects in structural analysis are presented in Section 6.4. Simplified drift design procedures are presented in Section 6.5. These approximate methods can be valuable in evaluating the potential drift in the early stages of design.

Section 6.6 covers the drift and *P*-delta analysis requirements of major United States seismic design codes.

Several numerical examples are provided to aid in understanding the concepts, and to provide the reader with the "hands-on" experience needed for successful utilization of the material in everyday design practice.

The relative lateral displacement of buildings is sometimes measured by an overall drift ratio or index, which is the ratio of the maximum lateral displacement to the height of the building. More commonly, however, an interstory drift ratio or index is used, which is defined as the ratio of the relative displacement of a particular floor to the story height at that level (see Figure 6-1). In this chapter, unless otherwise noted, the term *drift* means the relative lateral displacement between two adjacent floors, and the *drift index* is defined as the drift divided by the story height.

6.2 THE NEED FOR DRIFT DESIGN

The lateral displacement, or drift, of a structural system under wind or earthquake forces is important from

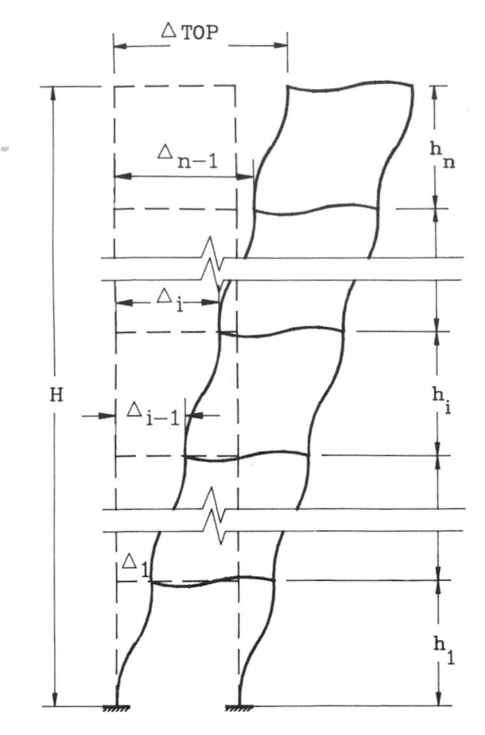

$$\text{OVERALL DRIFT} = \Delta_{TOP}$$

$$\text{INTER-STORY DRIFT} = \Delta_i - \Delta_{i-1}$$

$$\text{OVERALL DRIFT INDEX} = \frac{\Delta_{TOP}}{H}$$

$$\text{INTER-STORY DRIFT INDEX} = \frac{\Delta_i - \Delta_{i-1}}{h_i}$$

Figure 6-1 Definition of drift.

three different perspectives: (i) structural stability, (ii) architectural integrity and potential damage to various nonstructural components, and (iii) human comfort while and after the building experiences these motions.

6.2.1 Structural Stability

Excessive and uncontrolled lateral displacements can create severe structural problems. Empirical observations and theoretical dynamic-response studies have indicated a strong correlation between the magnitude of interstory drift and building damage potential.[6-2] Scholl,[6-3] emphasizing the fact that the potential for drift-related damage is highly variable and is dependent on the structural and nonstructural detailing provided by the designer, has proposed the following generalization of damage potential in relationship to the interstory drift index δ:

1. at $\delta = 0.001$, nonstructural damage is probable;
2. at $\delta = 0.002$, nonstructural damage is likely;
3. at $\delta = 0.007$, nonstructural damage is almost certain and structural damage is likely;
4. at $\delta = 0.015$, nonstructural damage is certain and structural damage is likely.

Drift-control requirements are included in the design provisions of most building codes. However, in most cases, the codes are not specific about the analytical assumptions to be used in the computation of the drifts. Furthermore, most of the codes are not clear about how the magnifying effects of stability-related displacements, such as *P*-delta deformations, are to be incorporated in the evaluation of final displacements and corresponding member forces.

6.2.2 Architectural Integrity

Architectural systems and components, and a variety of other nonstructural items in a building, constitute a large portion of the total investment in the project. In many cases the monetary value of these items exceeds the cost of the structural system by a large margin. In addition, these nonstructural items can be sources of injury, and even loss of life, for building occupants and those who are in the vicinity of the building. Past earthquakes have proven that nonstructural components can also greatly influence the seismic response of the building. Chapter 12 of this handbook is devoted to this important aspect of seismic design.

6.2.3 Human Comfort

Human comfort and motion perceptability, which are of importance in the design of structures for wind-induced motions, are relatively insignificant in seismic design, where the primary objective is to limit damage and prevent loss of life. For very essential structures, where continued operation of facilities is desired during and immediately after an earthquake, a more conservative design or application of special techniques, such as seismic isolation (see Chapter 13), may be considered. However, here again, the primary goal is to keep the system operational, and to prevent damage, rather than to provide for comfort of the occupants during strong ground motion.

Recently some investigators have studied the behavior of building occupants during strong ground motions.[6-4, 6-5, 6-6] Such studies can provide owners, architects, and hazard-mitigation authorities with valuable guidelines for considering these human factors in the planning, design, and operation of building structures.

6.3 DRIFT, *P*-DELTA, AND LATERAL STABILITY

6.3.1 The Concept of Lateral Stability

To illustrate the concept of stability, consider an ideal column without geometrical or material imperfections. Furthermore, assume that there are no lateral loads, and that the column remains elastic regardless of the force magnitude. If the axial force is slowly increased, the column will undergo axial deformation, and no lateral

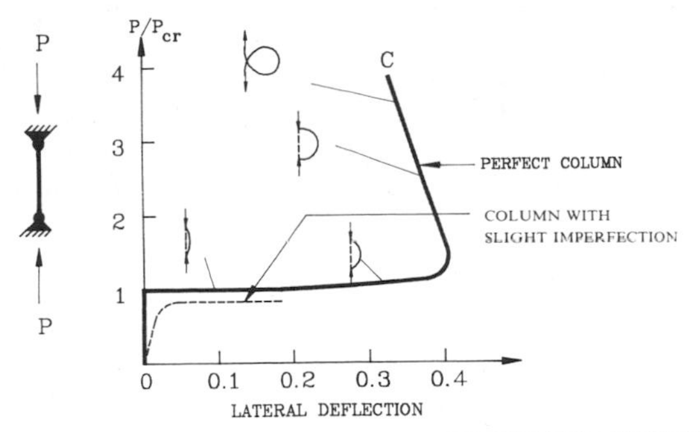

(a) LARGE DISPLACEMENT LOAD–DEFLECTION BEHAVIOR FOR AN IDEAL ELASTIC COLUMN (6-7)

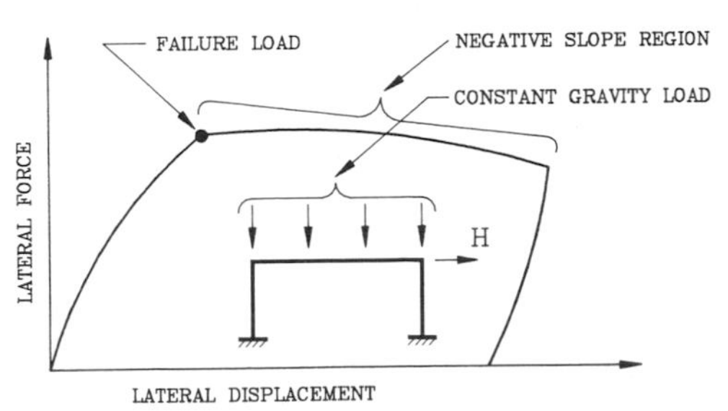

(b) LOAD–DEFLECTION RELATIONSHIP FOR A FRAME SUBJECTED TO COMBINED GRAVITY AND LATERAL LOADS

Figure 6-2 Structural stability of an idealized column and a real frame.

displacements will occur. However, when the applied forces reach a certain magnitude called the critical load (P_{cr}), significant lateral displacements may be observed.

Figure 6-2a shows the load–deflection behavior of this ideal column. It is important to notice that when the magnitude of axial force exceeds P_{cr}, there are two possible paths of equilibrium: the original path, with no lateral displacements, and a path with lateral displacements. However, equilibrium along the original path is not stable, and any slight disturbance can cause a change in the equilibrium position and significant lateral displacements. The force P_{cr} is called the bifurcation load or first critical load of the system. For this ideal column, reaching the bifurcation point does not imply failure, simply because it was assumed that the column will remain elastic regardless of the deflection magnitude. However, in a real column, such large deformations can cause yielding, stiffness reduction, and failure. In a structural system, buckling of critical members and the corresponding large lateral displacements can cause a major redistribution of forces and overall collapse of the system.

It is important to note that the bifurcation point exists only for perfectly symmetric members under pure axial forces. If the same ideal column is simultaneously subjected to lateral loads, or if asymmetry or material or geometric imperfections are present (as they are in any real system), lateral displacements will be observed from very early stages of loading.

When a frame under constant gravity load is subjected to slowly increasing lateral loads, the lateral displacement of the system slowly increases, until it reaches a stage such that in order to maintain static equilibrium a reduction in the gravity or lateral loads is necessary (Figure 6-2b). This corresponds to the region with negative slope on the force–displacement diagram. If the loads are not reduced, the system will fail.

When the same frame is subjected to earthquake ground motion, reaching the negative-slope region of the load–displacement diagram does not necessarily imply failure of the system (see Figure 6-3). In fact, it has been shown that in the case of repeated loads with direction reversals, such as those caused by earthquake ground motion, the load capacity of the system will be significantly larger than the stability load for the same system subjected to unidirectional monotonic loads.[6-1, 6-8] Perhaps this is one reason for the scarcity of stability-caused building failures during earthquakes.

Exact computation of critical loads, for real buildings, is a formidable task. This is true even in a static environment, let alone the added complexities of dynamic loading and inelastic response. Exact buckling analysis is beyond the capacity and resources of a typical design office, and beyond the usual budget and time frame allocated for structural analysis of buildings.

In everyday structural analysis, the stability effects are accounted for either by addressing the problem at the element level (via effective-length factors), or by application of one of the various *P*-delta analysis methods.

The simplest way to minimize lateral-stability problems is to limit the expected lateral displacement, or drift of the structure. In fact several

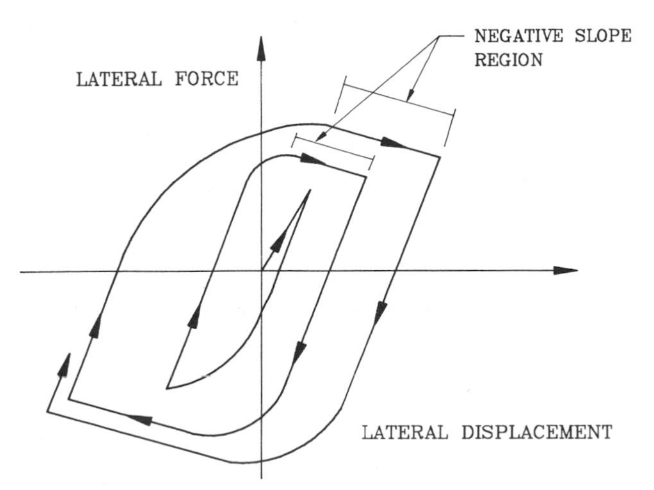

Figure 6-3 A typical load-displacement curve for a frame under constant gravity load and reversing lateral load.

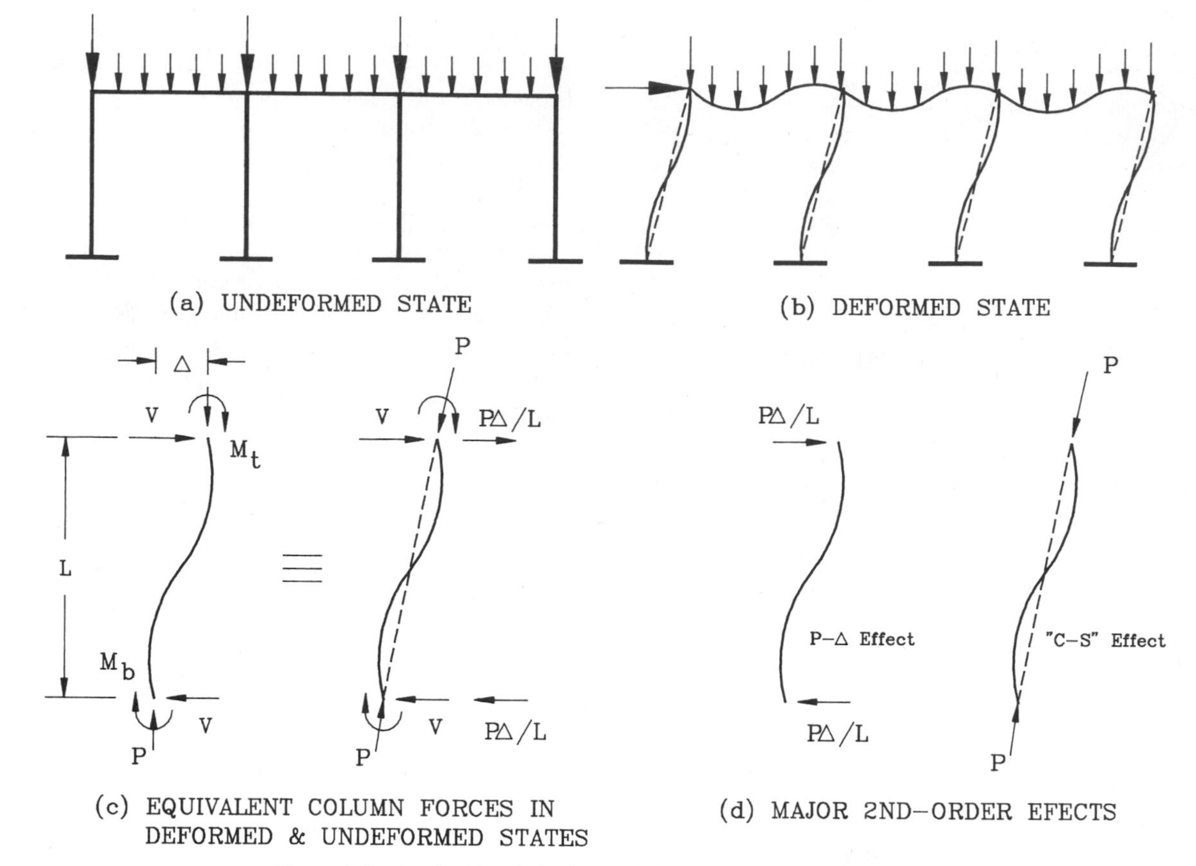

Figure 6-4 Applied loads in the undeformed and deformed states.

studies(6-9, 6-10, 6-11, 6-12) have shown that by increasing the lateral stiffness, the critical load of the building will increase and the chances of stability problems are reduced. Drift limitations are imposed by seismic-design codes primarily to serve this purpose.

6.3.2 *P*-Delta Analysis

For most practical purposes, an accurate estimate of the stability effects may be obtained by what is commonly referred to as *P*-delta analysis.

Overall stability failures of structures have not been common during past earthquakes. However, with the continuing trend towards lighter structural systems, and recent discoveries about the nature of near-field ground motion,(6-13, 6-14, 6-15) the second-order effects are beginning to receive more attention. It is believed that, in most cases, observance of proper drift limitations will provide the necessary safeguards against overall lateral-stability failure of the structure.

In conventional first-order structural analysis, the equilibrium equations are formulated for the undeformed shape of the structure. However, when a structure undergoes deformation, it carries the applied loads into a deformed state along with it (Figure 6-4). The changes in position of the applied forces are cumulative in nature and cause additional second-order forces, moments, and displacements which are not included in a first-order analysis. Studies(6-16) have shown that the single most important second-order effect is the *P*-delta effect. Figure 6-5 illustrates the *P*-delta effect on a simple cantilever column.

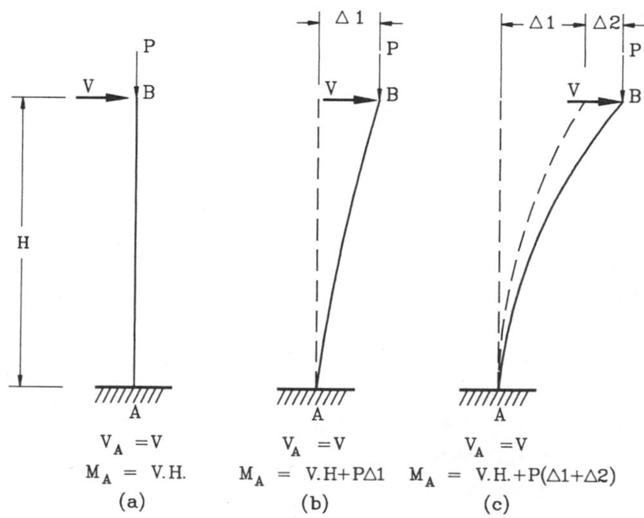

Figure 6-5 The *P*-delta effect. (a) Equilibrium in the undeformed state. (b) Immediate *P*-delta effect. (c) Accumulation of the *P*-delta effect.

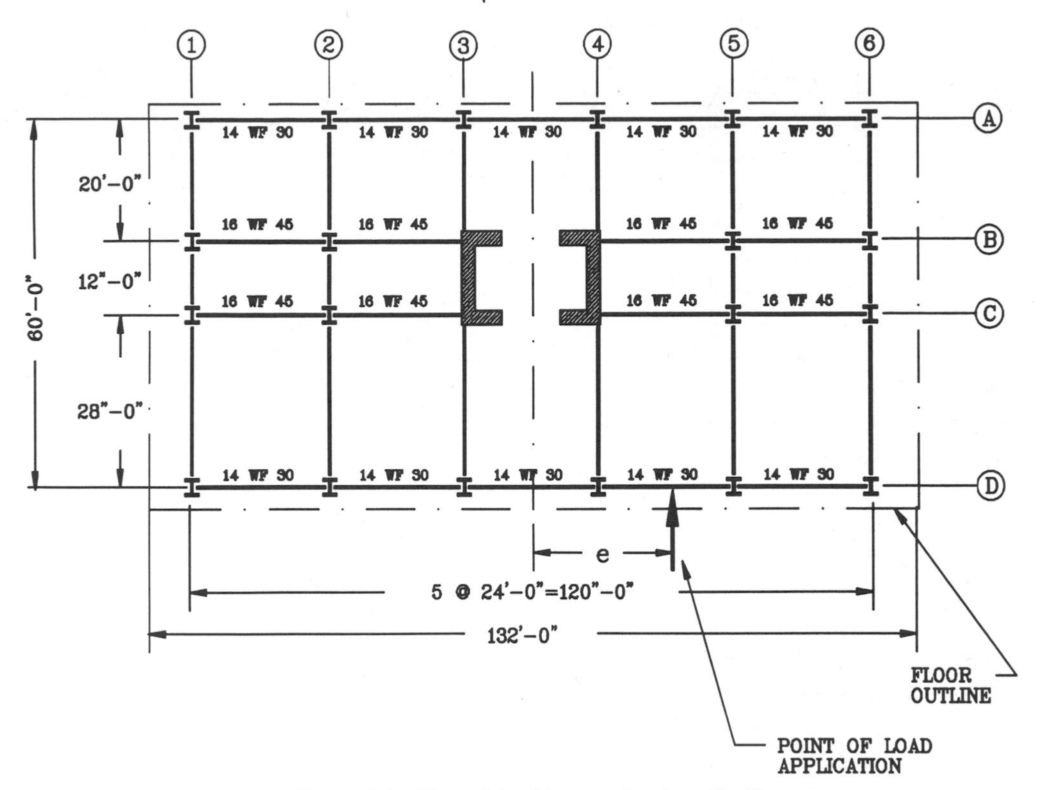

Figure 6-6 Plan of the 24 story structure (6-17).

In some cases, stability or second-order effects are small and can be neglected. However, in many other cases, such as tall buildings, systems under significant gravity loads, soft-story buildings, or systems with significant torsional response, the second-order effects may be quite significant and hence should be considered in the structural analysis.

Although it is true that ignoring second-order effects is not likely to result in overall stability failure of typical buildings subjected to earthquake ground motion, these effects can frequently give rise to a series of premature material failures at levels of forces that would seem safe from a first-order analysis. Strong evidence relating excessive drift to seismic damage during earthquakes supports this point.

6.3.3 Factors Affecting Lateral Stability

In general, the magnitude of the gravity loads and factors that increase lateral displacement, affect the lateral stability of the structure. Chief among these factors are rotation at the base of the structure,[6-12] any significant rotation at any level above the base (as that caused by formation of plastic hinges in the columns or walls), and significant asymmetry or torsion in the structure.

Wynhoven and Adams[6-17] studied the effects of asymmetry and torsion on the ultimate load-carrying capacity of a 24-story frame–shear-wall building with typical plan and elevation layouts as shown in Figures 6-6 and 6-7. The behavior of individual members was idealized as elastic–perfectly plastic. To consider the influence of torsion on the load-carrying capacity of the structure, two asymmetric models were constructed by moving the shear-wall couple from grid lines 3 and 4 to grid lines 4 and 5 in one model, and to grid lines 5 and 6 in another model. Load–displacement diagrams for the three configurations are shown in Figure 6-8, where λ is the ratio of the ultimate lateral loads to the working-stress lateral loads. The gravity loads were not changed. The reduction in the ultimate lateral load-carrying capacity due to induced asymmetry proved to be drastic (51% in one case and 66% in the other case).

6.4 PRACTICAL SECOND-ORDER ANALYSIS TECHNIQUES

6.4.1 The Effective-Length-Factor Method

This method is an attempt to reduce the complex problem of overall frame stability to a relatively simple problem of elastic stability of individual columns with various end conditions. The role of the effective length factor K is to replace an actual column of length L with complex end conditions to an equivalent column of length KL with both ends pinned, so that the classic Euler buckling equation can be used to examine column stability. It is further assumed that if the buckling

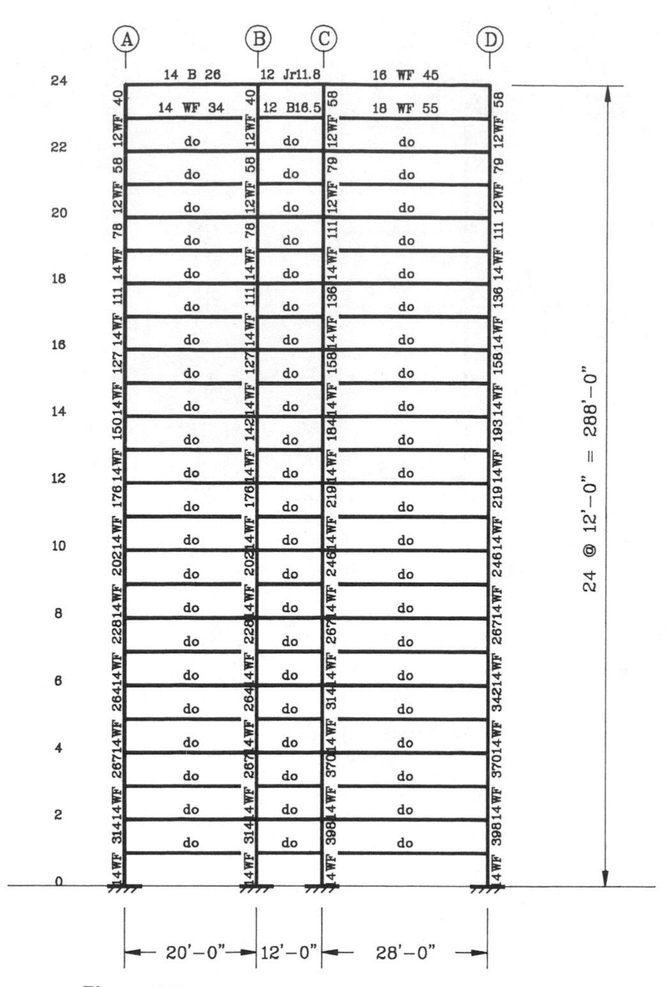

Figure 6-7 Elevation of the 24 story structure (6-17).

stability of each individual column has been verified by this method, then a system instability will not occur.

The general equations for effective-length factors are derived from the elastic-stability analysis of simple beam–column models such as those shown in Figure 6-9. These equations are[6-18]:

- for the sidesway prevented case,

$$\frac{G_A G_B}{4}\left(\frac{\pi^2}{K^2}\right) + \left(\frac{G_A + G_B}{2}\right)\left(1 - \frac{\pi/K}{\tan \pi/K}\right) + \frac{2}{\pi/K}\tan\frac{\pi}{2K} = 1 \tag{6-1}$$

- for the sidesway permitted case,

$$\left[\frac{(\pi/K)^2 G_A G_B}{36} - 1\right]\tan\frac{\pi}{K} - \left(\frac{G_A + G_B}{6}\right)\frac{\pi}{K} = 0 \tag{6-2}$$

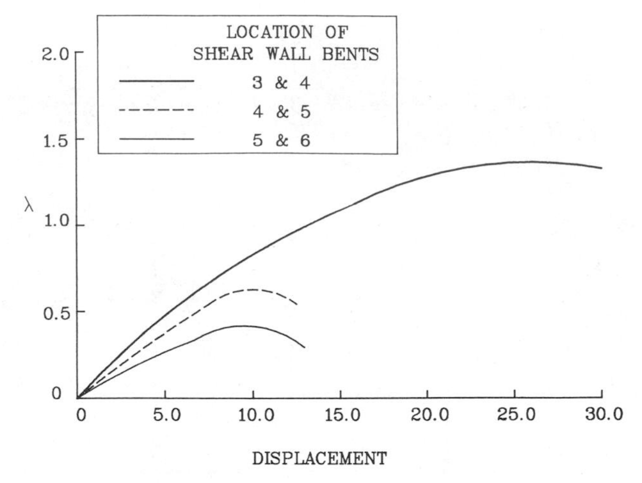

Figure 6-8 Load-displacement relationships for various configurations of the 24 story structure (6-17).

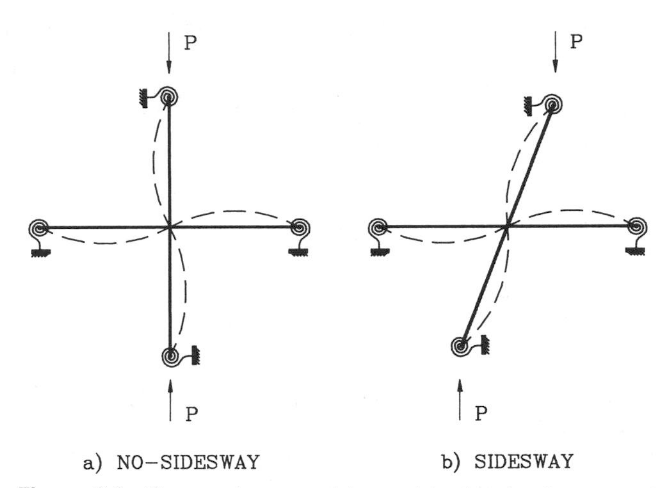

Figure 6-9 Beam-column models used in the development of the effective length factor equations.

where G_A and G_B are the relative rotational stiffness of the beams to the columns, measured at ends A and B of the column under consideration:

$$G = \frac{\Sigma \dfrac{I_c}{L_c}}{\Sigma \dfrac{I_g}{L_g}} \tag{6-3}$$

Graphical solutions to these equations are given by the well-known SSRC alignment charts,[6-19] shown in Figure 6-10. The SSRC Guide[6-19] recommends that for pinned column bases, G be taken as 10, and for column bases rigidly attached to these foundation, G be taken as unity. Furthermore, when certain conditions are known to exist at the far end of a beam, the corresponding beam-stiffness term in Equation 6-3 should be multiplied by a factor. For the sidesway-prevented case, this

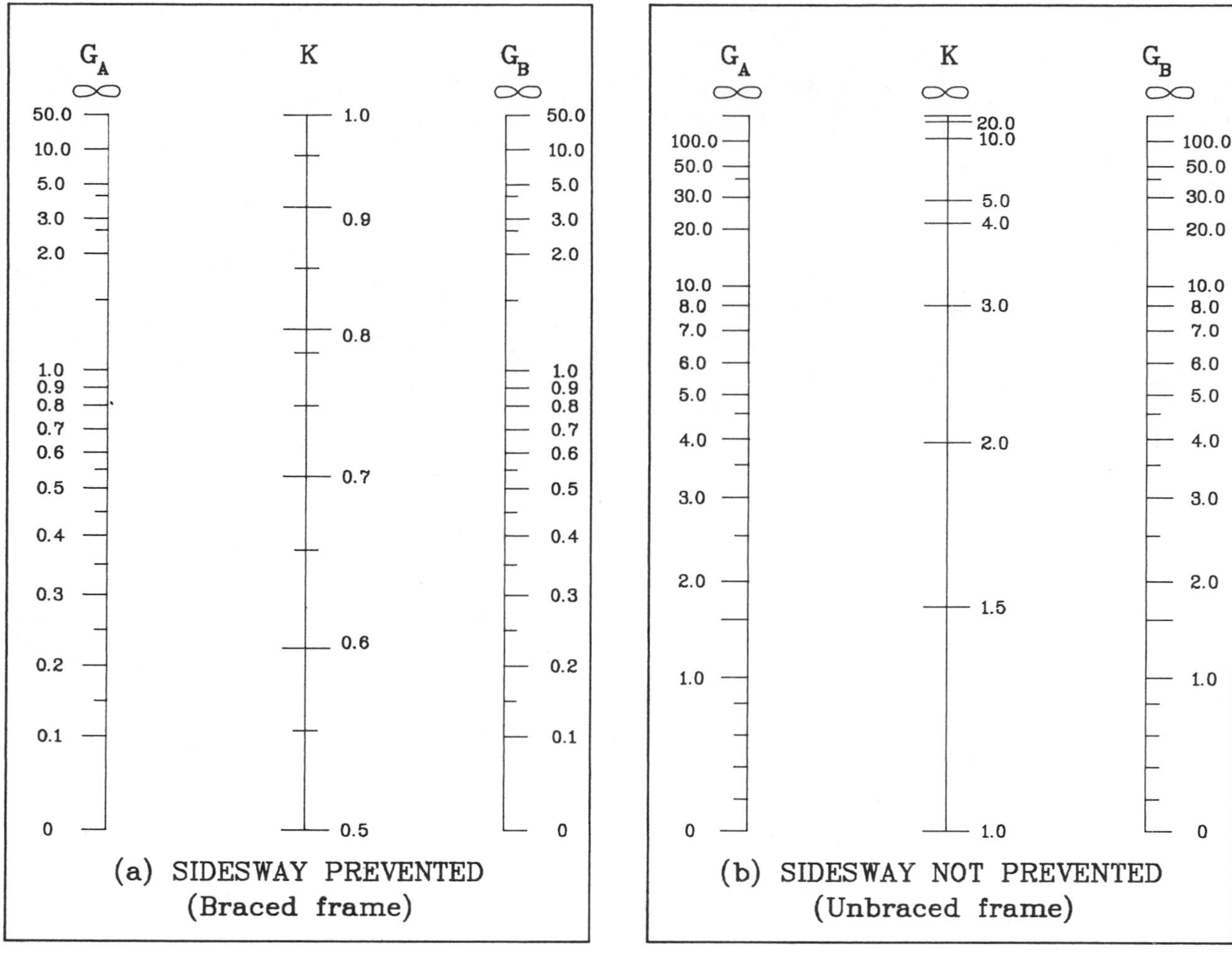

Figure 6-10 Alignment charts for determination of effective length factors (6-19).

factor is 1.5 for the far end hinged and 2.0 for the far end fixed. For the sidesway-permitted case, it is 0.5 for the far end hinged and 0.67 for the far end fixed. Effective-length factors have been incorporated in the column design interaction equations of several building design codes.

The effective-length-factor method has been subjected to serious criticism by various researchers. The main criticism is that this method, which is based on elastic-stability analysis of highly idealized cases, can not be trusted to provide reasonable estimates of the stability behavior of real structural systems. Furthermore, several studies have shown that the lateral stability of a frame, or individual story, is controlled by the collective behavior of all the columns in the story, rather than the behavior of a single column. Hence, if a stability failure is to occur, the entire story must fail as a unit.[6-12]

Examples and evidence of the shortcomings of the effective-length-factor method have been documented, among others, by MacGregor and Hage[6-16] and Choeng-Siat-Moy.[6-20, 6-21] In spite of this evidence, the method has continued to survive as a part of the requirements for many building codes. New editions of some building codes are moving away from this tradition.

6.4.2 Approximate Buckling Analysis[a]

In approximate buckling analysis, the buckling load of a single story, or that of the structure as a whole, is estimated. A magnification factor μ, which is a function of the ratio of the actual gravity load to the buckling load, is defined, and for the design of structural members, all lateral-load effects are multiplied by this magnification factor. Then, member design is performed by assuming an effective-length factor of one.

Several approximate methods have been developed for estimation of critical loads of building structures.[6-10, 6-11, 6-12, 6-22] Among these, a simple method developed by Nair[6-22] is explained here. This method takes advantage of the fact that most multistory buildings have lateral load–displacement characteristics that

[a] Parts of Section 6.4.2 have been extracted from Reference 6-22 with permission from Van Nostrand Reinhold Company.

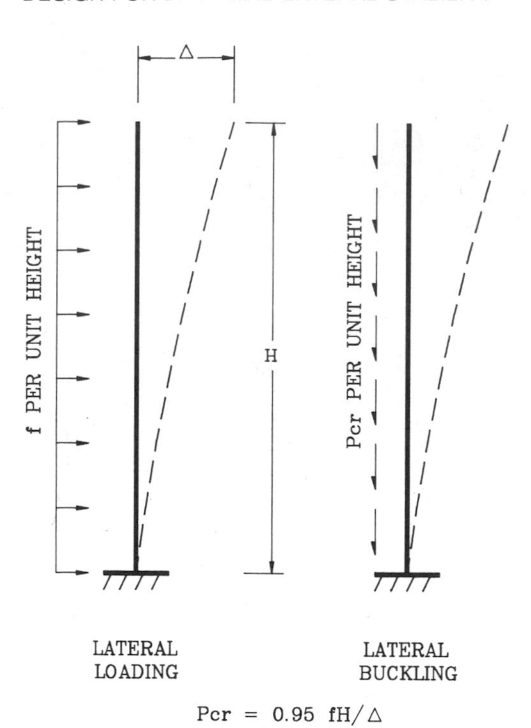

Figure 6-11 Lateral loading and buckling of a flexural cantilever (6-22).

are similar to those of either a flexural cantilever or a shear cantilever. Buildings with braced frames or shear walls, and tall buildings with unbraced frames or tubular frames, usually have lateral load–deformation characteristics that approach those of a flexural cantilever. On the other hand, buildings of low or moderate height with unbraced frames (in which column axial deformations are not significant) usually have characteristics similar to those of a shear cantilever.

The above observations can be extended to the torsional behavior of structures. If in a multistory building, torsional stiffness is provided by braced frames, shear walls, or tall unbraced frames not exhibiting tube action, the torsion–rotation characteristics of the building will be similar to the lateral load–displacement characteristics of a flexural cantilever. If a building's torsional stiffness is provided by low- to mid-rise unbraced frames, or by tubular frames, the building will have torsion–rotation characteristics that are similar to the lateral load–displacement characteristics of a shear cantilever.

Buildings Modeled as Flexural Cantilevers For a flexural cantilever of height H and constant stiffness EI, the uniformly distributed vertical load per unit height (Figure 6-11), p_{cr}, that will cause lateral buckling is given by the equation

$$p_{cr} = 7.84EI/H^3 \tag{6-4}$$

If the stiffness varies according to the equation $EI = (a/H)EI_0$, where EI_0 is the stiffness at the base and a is the distance from the top, the critical load is given by

$$p_{cr} = 5.78EI_0/H^3 \tag{6-5}$$

If the stiffness varies according to the equation $EI = (a/H)^2EI_0$, the critical load is

$$p_{cr} = 3.67EI_0/H^3 \tag{6-6}$$

These solutions can be found in basic texts on elastic stability.

If a uniformly distributed lateral load of f per unit height is applied to a flexural cantilever, the lateral displacement Δ at the top is:

for a constant EI: $$\Delta = 0.125fH^4/EI \tag{6-7}$$

for $EI = (a/H)EI_0$: $$\Delta = 0.167fH^4/EI_0 \tag{6-8}$$

for $EI = (a/H)^2EI_0$: $$\Delta = 0.250fH^4/EI_0 \tag{6-9}$$

If the lateral load is not uniform, an approximate answer may be obtained by defining f as the equivalent uniform lateral load that would produce the same base moment as the lateral load used in the analysis. By combining Equations 6-4, 6-5, and 6-6 with Equations 6-7, 6-8, and 6-9, EI can be eliminated and p_{cr} can be expressed in terms of f/Δ, as follows:

for a constant EI: $$p_{cr} = 0.98fH/\Delta \tag{6-10}$$

for $EI = (a/H)EI_0$: $$p_{cr} = 0.96fH/\Delta \tag{6-11}$$

for $EI = (a/H)^2EI_0$: $$p_{cr} = 0.92fH/\Delta \tag{6-12}$$

From the above equations it is obvious that the relation between p_{cr} and f/Δ is not very sensitive to stiffness variation over the height of the structure. Hence, regardless of the distribution of stiffness, the following equation is sufficiently accurate for design purposes:

$$p_{cr} = 0.95fH/\Delta \tag{6-13}$$

The magnification factor μ, as previously defined, is given by

$$\mu = \frac{1}{1 - \gamma p/\phi p_{cr}} \tag{6-14}$$

where p is the actual average gravity load per unit height of the building, γ is the design load factor, and ϕ is the strength-reduction factor. Note that p must include the load on all vertical members, including those that are not part of the lateral-load-resisting system.

Thus, if the lateral displacement is known from a first-order analysis, the critical load and the corresponding magnification factor can be estimated using Equations 6-13 and 6-14.

For buildings whose torsional behavior approaches that of a flexural cantilever, the following formula may be used to estimate the torsional buckling load of the structure:

$$r^2 p_{cr} = 0.95tH/\theta \qquad (6\text{-}15)$$

where t is an applied torsional load, per unit height of the building, θ is the rotation at the top of the building in radians, p_{cr} is the critical vertical load for torsional buckling per unit height of the building, and r is the polar radius of gyration of the vertical loading about the vertical axis at the center of twist of the building. For a doubly symmetric structure, uniformly distributed gravity loading, and a rectangular floor plan with dimensions a and b,

$$r^2 = \frac{a^2 + b^2}{12} \qquad (6\text{-}16)$$

Buildings Modeled as Shear Cantilevers If a portion of a vertical shear cantilever undergoes lateral deformation δ, over a height h, when subjected to a shear force V, the critical load for lateral buckling of that portion of the cantilever is given by

$$P_{cr} = Vh/\delta \qquad (6\text{-}17)$$

When the above equation is applied to a single story of a building, h is the story height, δ is the story drift caused by the story shear force V, and P_{cr} is the total vertical force that would cause lateral buckling of the story (see Figure 6-12).

The magnification factor μ is given by

$$\mu = \frac{1}{1 - \gamma P/\phi P_{cr}} \qquad (6\text{-}18)$$

where P is the total gravity force in the story, γ is the load factor, and ϕ is the strength-reduction factor.

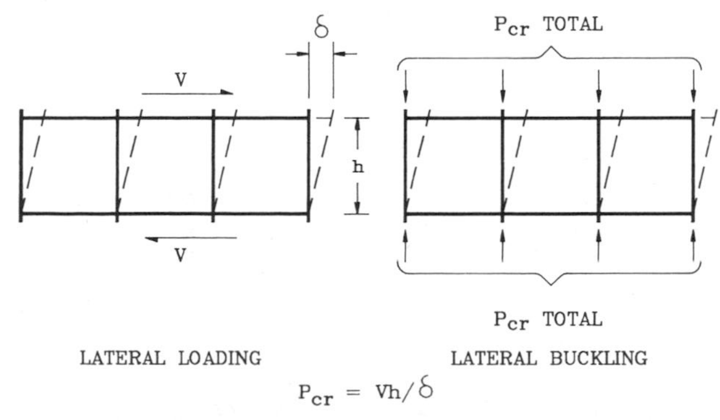

Figure 6-12 Lateral loading and buckling of a story in a shear cantilever type building (6-22).

The accuracy of Equation 6-17, when applied to a single story of a framed structure, depends on the relative stiffness of the beams and columns, and on the manner in which the gravity loads are distributed among the columns. The error is greatest for stiff beams and slender columns and may be as high as 20%.

For buildings whose torsional behavior approaches that of a shear cantilever, the following equation may be used to estimate the torsional buckling load of a particular story of the building:

$$r^2 P_{cr} = Th/\theta \qquad (6\text{-}19)$$

where T is an applied torsional load on the story, θ is the torsional deformation of the story (in radians) due to the torque T, h is the story height, P_{cr} is the critical load for torsional buckling of the story, and r is the polar radius of gyration of the vertical load.

Application Examples Consider the 20-story buildings shown in Figure 6-13. The buildings are analyzed using a linear elastic-analysis program for a constant lateral load of 25 psf applied in the north–south direction. The east–west plan widths are 138 ft. The gravity load is assumed to be 130 psf on each floor.

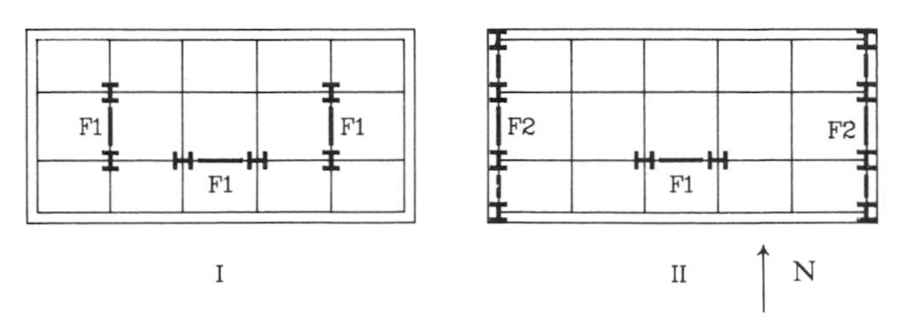

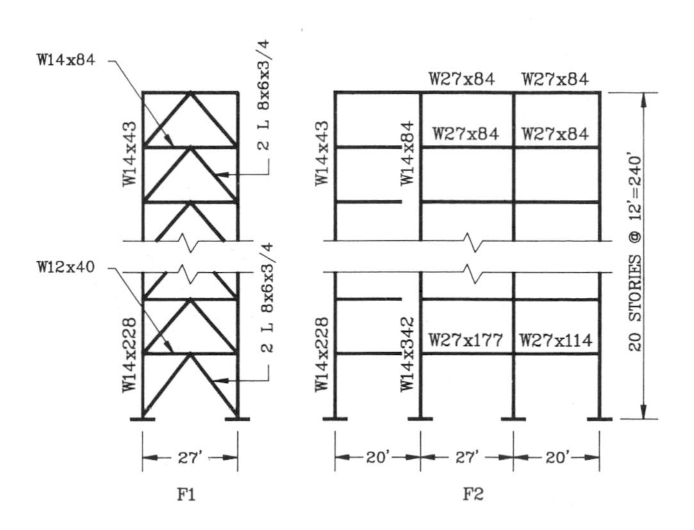

Figure 6-13 Buildings analyzed in references (6-22) and (6-23).

For building I, the first-order displacement at the top is 0.729 ft. Using Equation 6-13,

$$H = 240 \text{ ft}$$
$$f = 0.025(138) = 3.45 \text{ kips/ft}$$
$$\Delta = 0.729 \text{ ft}$$
$$p_{cr} = \frac{0.95(3.45)(240)}{0.729} = 1079 \text{ kips/ft}$$

The estimated critical load of 1079 k/ft corresponds to 12,948 kips or 1,360 psf on each floor. The corresponding magnification factor, assuming $\gamma = \phi = 1.0$, is

$$\mu = \frac{1}{1 - 130/1360} = 1.106$$

and the magnified lateral displacement at the roof is given by

$$\gamma\Delta = 1.106(0.729) = 0.806 \text{ ft}$$

An elastic-stability analysis of this building(6-23) indicates a critical load of 1369 psf for north–south buckling. A large-deformation analysis for combined gravity load and north–south lateral loading indicates a roof displacement of 0.805 ft.

For building II, the computed story drifts for the 15th, 10th, and 5th levels are 0.0522, 0.0609, and 0.0582 ft, respectively. The corresponding story shears at these levels are 228, 435, and 642 kips. Using Equation 6-17,

for the 15th story: $$P_{cr} = \frac{228(12)}{0.0522} = 52{,}414 \text{ kips}$$

for the 10th story: $$P_{cr} = \frac{435(12)}{0.0609} = 85{,}714 \text{ kips}$$

for the 5th story: $$P_{cr} = \frac{642(12)}{0.0582} = 132{,}371 \text{ kips}$$

The corresponding magnification factors assuming $\gamma = \phi = 1.0$ are

for the 15th story: $$\mu = \frac{1}{1 - 7427/52{,}414} = 1.165$$

for the 10th story: $$\mu = \frac{1}{1 - 13{,}616/85{,}714} = 1.189$$

for the 5th story: $$\mu = \frac{1}{1 - 19{,}806/132{,}371} = 1.176$$

and the magnified story drifts are

for the 15th story: $\mu\Delta = 1.165(0.0522) = 0.0608$ ft

for the 10th story: $\mu\Delta = 1.189(0.0609) = 0.0724$ ft

for the 5th story: $\mu\Delta = 1.176(0.0582) = 0.0684$ ft

A large-deformation analysis of this building(6-23) indicates story drifts of 0.0607, 0.0723, and 0.0686 ft for the 15th, 10th, and 5th stories, respectively.

6.4.3 Approximate *P*-Delta Analysis

Three methods for approximate P-delta analysis of building structures are presented in this section: the iterative P-delta method, the direct P-delta method, and the negative-bracing-member method. All three methods are shown to be capable of providing accurate estimates of P-delta effects.

Iterative P-Delta Method The iterative P-delta method(6-16, 6-24, 6-25, 6-26) is based on the simple idea of correcting first-order displacements by adding the P-delta shears to the applied story shears. Since P-delta effects are cumulative in nature, this correction and subsequent reanalysis should be performed iteratively until convergence is achieved. At each cycle of iteration a modified set of story shears are defined as

$$\sum V_i = \sum V_1 + \left(\sum P\right)\Delta_{i-1}/h \qquad (6\text{-}20)$$

where ΣV_i is the modified story shear at the end of ith cycle of iteration, ΣV_1 is the first-order story shear, ΣP is the sum of all gravity forces acting on and above the floor level under consideration, Δ_{i-1} is the story drift as obtained from first-order analysis or the previous cycle of iteration, and h is the story height for the floor level under consideration. Iteration may be terminated when $\Sigma V_i \approx \Sigma V_{i-1}$ or $\Delta_i \approx \Delta_{i-1}$.

Generally for elastic structures of reasonable stiffness, convergence will be achieved within one or two cycles of iteration.(6-16) One should note that since the lateral forces are being modified to approximate the P-delta effect, the column shears obtained will be slightly in error.(6-16) This is true for all approximate methods which use sway forces to approximate the P-delta effect.

Example 6-1

For the 10-story moment-resistant steel frame shown in Figure 6-14, modify the first-order lateral displacements to include the P-delta effects by using the iterative P-delta method. The computed first-order lateral displacements and story drifts for the frame are shown in Table 6-1. The tributary width of the frame is 30 ft. The gravity load is 100 psf on the roof and 120 psf on typical floors. Use center-to-center dimensions.

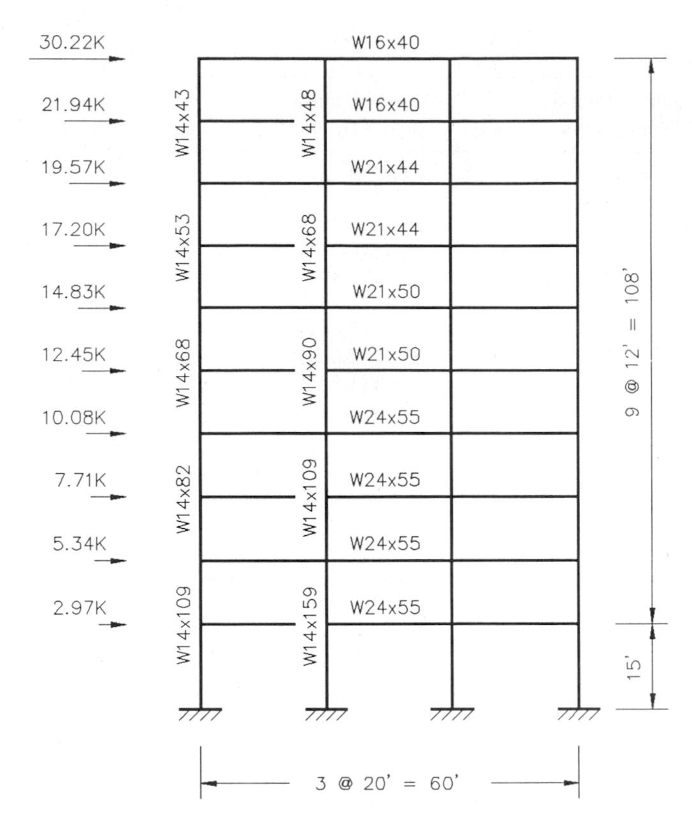

Figure 6-14 Elevation of the 10 story moment frame used in Example 6-1.

The calculations for this example, using the iterative P-delta method, are presented in Tables 6-2 and 6-3. The convergence was achieved in two cycles of iteration. Table 6-3 also shows results obtained by an "exact" P-delta analysis.

To further explain the steps involved in the application of this method, let us consider the bent at the 8th level of the frame. The story height (h) is 12 ft (144 in.), the total gravity force at this level (ΣP) is 612 kips, the story shear (ΣV) is 71.74 kips, and the first-order story drift is 0.785 in. (see Table 6-1).

The P-delta contribution to the story shear is

$$\frac{(\Sigma P)\Delta_1}{h} = \frac{(612)(0.785)}{144} = 3.34 \text{ kips}$$

and the modified story shear is

$$\sum V_2 = \sum V_1 + \left(\sum P\right)\Delta_1/h$$

$$= 71.74 + 3.34 = 75.08 \text{ kips}$$

Repeating this operation for all stories results in a modified set of story shears, from which a modified set of applied lateral forces is obtained (Table 6-2). A new first-order analysis of the frame subjected to these modified lateral forces results in a modified set of lateral

Table 6-1 Applied Forces and Computed First-Order Displacements for the 10-Story Frame

Level	Story height h, in.	Gravity force ΣP, kips	Lateral load V, kips	Story shear ΣV_1, kips	Lateral disp. D_1, in.	Story drift Δ_1, in.
10	144	180	30.22	30.22	7.996	0.517
9	144	396	21.94	52.17	7.479	0.736
8	144	612	19.57	71.74	6.743	0.785
7	144	828	17.20	88.93	5.958	0.907
6	144	1044	14.83	103.76	5.051	0.899
5	144	1260	12.45	116.21	4.152	0.914
4	144	1476	10.08	126.30	3.238	0.838
3	144	1692	7.71	134.01	2.400	0.867
2	144	1908	5.34	139.34	1.533	0.768
1	180	2124	2.97	142.31	0.765	0.765

Table 6-2 Iterative *P*-Delta Method (First Cycle of Iteration)

Level	$(\Sigma P)\Delta_1/h$, kips	$\Sigma V_1 + (\Sigma P)\Delta_1/h$, kips	Modified lateral Force V_2, kips	Modified lateral Disp. D_2, in.	Modified story drift Δ_2, in.
10	0.65	30.87	30.87	8.478	0.533
9	2.02	54.19	23.32	7.945	0.767
8	3.34	75.08	20.89	7.178	0.823
7	5.22	94.15	19.07	6.355	0.959
6	6.52	110.28	16.13	5.396	0.955
5	8.00	124.21	13.93	4.441	0.976
4	8.59	134.89	10.68	3.465	0.897
3	10.19	144.20	9.31	2.568	0.930
2	10.18	149.52	5.32	1.638	0.823
1	9.03	151.34	1.82	0.815	0.815

Table 6-3 Iterative *P*-Delta Method (Second Cycle of Iteration)

Level	$(\Sigma P)\Delta_2/h$, kips	$\Sigma V_2 + (\Sigma P)\Delta_2/h$, kips	Modified lateral Force V_3, kips	Modified lateral Disp.* D_3, in.	Modified story Drift* Δ_3, in.
10	0.67	30.89	30.89	8.508 (8.510)	0.534 (0.534)
9	2.11	54.28	23.39	7.975 (7.976)	0.768 (0.768)
8	3.50	75.24	20.96	7.207 (7.209)	0.825 (0.825)
7	5.51	94.44	19.20	6.382 (6.384)	0.962 (0.963)
6	6.92	110.68	16.24	5.419 (5.421)	0.959 (0.959)
5	8.54	124.75	14.07	4.461 (4.462)	0.980 (0.980)
4	9.19	135.49	10.74	3.480 (3.481)	0.900 (0.901)
3	10.93	144.94	9.45	2.580 (2.581)	0.935 (0.935)
2	10.90	150.24	5.30	1.645 (1.646)	0.827 (0.827)
1	9.62	151.93	1.69	0.818 (0.819)	0.818 (0.819)

*Values in parentheses represent results of an "exact" *P*-delta analysis.

displacements (D_2) and story drifts (Δ_2) as shown in Table 6-2. The maximum displacement obtained from the second analysis was 8.478 in., which is 9% larger than the original first-order displacement. Hence, a second iteration is necessary. Again performing the calculations for the bent at the 8th floor,

$$\left(\sum P\right)\Delta_2/h = \frac{(612)(0.823)}{144} = 3.50 \text{ kips}$$

$$\sum V_3 = \sum V_2 = \left(\sum P\right)\Delta_2/h = 71.74 + 3.50$$
$$= 75.24 \text{ kips}$$

Another first-order analysis for the new set of lateral forces indicates a maximum displacement of 8.508 in., which is less than 1% larger than the displacements obtained in the previous iteration. Hence, the iteration was terminated at this point.

The first-order and second-order lateral displacements and story drifts are shown in Figures 6-15 and 6-16. As indicated by these figures, the results are virtually identical to the exact results.

Direct P-Delta Method The direct *P*-delta method[(6-16)] is a simplification of the iterative method. Using this method, an estimate of final deflections is obtained directly from the first-order deflections.

The simplification is based on the assumption that the story drift at the ith level is proportional only to the applied story shear at that level (ΣV_i). This assumption allows the treatment of each level independent of the others.

If F is the drift caused by a unit lateral load at the ith level, then the first-order drift Δ_1 is

$$\Delta_1 = F\sum V_1 \tag{6-21}$$

After the first cycle of iteration,

$$\Delta_2 = F\sum V_2 = F\left(\sum V_1\right)\left(1 + \left(\sum P\right)\frac{F}{h}\right) \tag{6-22}$$

and after the ith cycle of iteration,

$$\Delta_{i+1} = F\sum V_1\left[1 + \left(\sum P\right)\frac{F}{h}\right) + \left(\left(\sum P\right)\frac{F}{h}\right)^2 + \cdots + \left(\sum P\right)\left(\frac{F}{h}\right)^{i+1}\right] \tag{6-23}$$

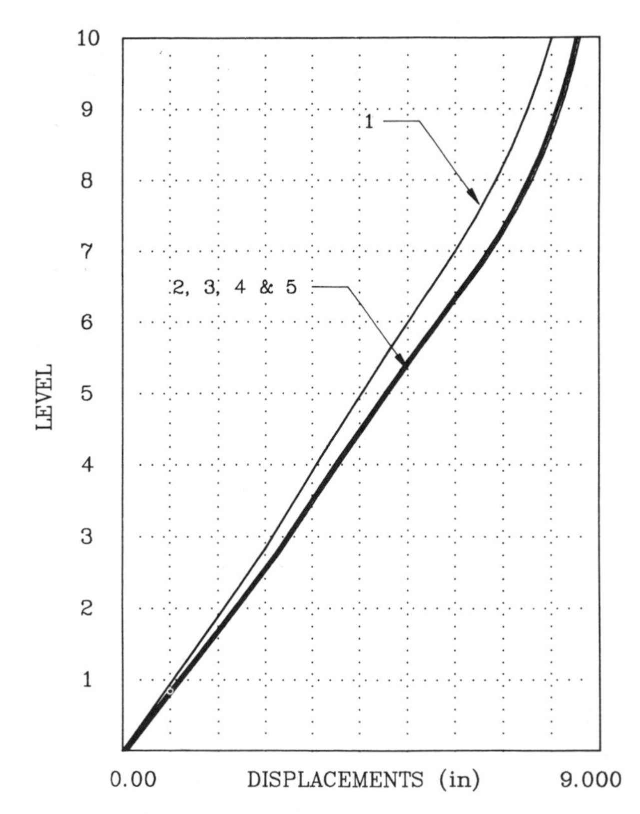

Figure 6-15 Lateral displacement of the 10 story frame as obtained by various *P*-delta methods.

Equation 6-23 is a geometric series that converges if $(\Sigma P)F/h < 1.0$, to

$$\Delta_{\text{Final}} = \frac{F\Sigma V_1}{1 - F_1(\Sigma P)/h} \tag{6-24}$$

But $F\Sigma V_1 = \Delta_1$. Hence, the final second-order deflection is

$$\Delta_{\text{Final}} = \frac{\Delta_1}{1 - (\Sigma P)\Delta_1/(\Sigma V_1)h} \tag{6-25}$$

Equation 6-25 can be expressed as $\Delta_{\text{Final}} = \mu\Delta_1$, where $\mu = 1/[1 - (\Sigma P)\Delta_1/(\Sigma V_1)h]$ is a magnification factor by which the first-order effects should be multiplied to include the second-order effects. All internal forces and moments related to the lateral loads should also be magnified by μ. Member design may be carried out using an effective-length factor of one.

An estimate of the critical load for an individual story, or the entire frame, can be obtained directly from Equation 6-25. Note that if $(\Sigma P)\Delta_1/(\Sigma V_1)h = 1$, the second-order displacement would go to infinity. Hence, $\Sigma P = (\Sigma V_1)h/\Delta_1$ may be considered to be the critical load of the system.

Similarly, $\Sigma Pr^2 = (\Sigma T_1)h/\theta_1$ can be viewed as the torsional critical load of the system. It is interesting to note that the critical loads and the magnification factor obtained here are in essence the same as those obtained in Section 6.4.2 by an approximate buckling analysis.

The term $(\Sigma P)\Delta_1/(\Sigma V_1)h$ is commonly referred to as the *stability index*. Similarly, a *torsional stability index* may be defined as $(\Sigma Pr^2)\theta_1/(\Sigma T_1)h$.

It has been suggested[6-16] that if the stability index is less than 0.0475 for all three axes of the building, the second-order effects can be ignored. For values of the stability index between 0.0475 and 0.20, the direct *P*-delta method can provide accurate estimates of the second-order effects. Designs for which values of the stability index exceed 0.20 should be avoided.

Example 6-2

For the 10-story frame of Example 6-1 compute the second-order displacements and story drifts by the direct *P*-delta method.

The calculations using the direct *P*-delta method are shown in Table 6-4. For example, for the first floor, which has a story height of 15 ft (180 in.), the story shear is 142.31 kips, the total gravity force is 2124 kips, and the first-order drift is 0.765 in. The magnification factor and the second-order displacements are

$$\mu = \frac{1}{1 - (2124)(0.765)/(142.31)(180)} = 1.068$$

$$\Delta_2 = \mu\Delta_1 = (1.068)(0.765) = 0.817 \text{ in.}$$

A comparison with the exact results (Figures 6-15 and 6-16) reveals the remarkable accuracy of this simple technique.

Negative-Bracing-Member Method The negative-bracing-member method,[6-16, 6-26, 6-27] which was first introduced by Nixon, Beaulieu, and Adams,[6-27] provides a direct estimate of the *P*-delta effect via any standard first-order analysis program. Fictitious bracing members with negative areas are inserted (Figure 6-17) to model the stiffness reduction due to the *P*-delta effect.

The cross-sectional area of the negative braces for each floor level can be obtained by a simple analogy to Hooke's law ($F = K\Delta$). The additional shear due to the *P*-delta effect is $(\Sigma P)\Delta/h$, where ΣP is the total gravity force and h is the story height. The factor $\Sigma P/h$ is a stiffness term but it is contributing to lateral displacement instead of resisting it. Hence, it can be considered as a negative stiffness. A brace with cross-sectional area A, length L_{br}, and modulus of elasticity E, making an angle α with the floor, provides a stiffness equal to $(AE\cos^2\alpha)/L_{br}$ against lateral displacement. By equating the brace stiffness to $-\Sigma P/h$, the required area of

Table 6-4 *P*-Delta Analysis by Direct *P*-Delta Method (Example 6-2)

Level	h, in.	ΣV_1, kips	ΣP, kips	Δ_1, in.	μ	$\Delta_2 = \mu\Delta_1$, in.	2nd-Order Disp., in.
10	144	30.22	180	0.517	1.022	0.528	8.505
9	144	52.17	396	0.736	1.040	0.766	7.977
8	144	71.74	612	0.785	1.049	0.823	7.211
7	144	88.93	828	0.907	1.062	0.964	6.388
6	144	103.76	1044	0.899	1.067	0.959	5.424
5	144	116.21	1260	0.914	1.074	0.982	4.465
4	144	126.30	1476	0.838	1.073	0.899	3.483
3	144	134.01	1692	0.867	1.082	0.938	2.584
2	144	139.34	1908	0.768	1.079	0.829	1.646
1	180	142.31	2124	0.765	1.068	0.817	0.817

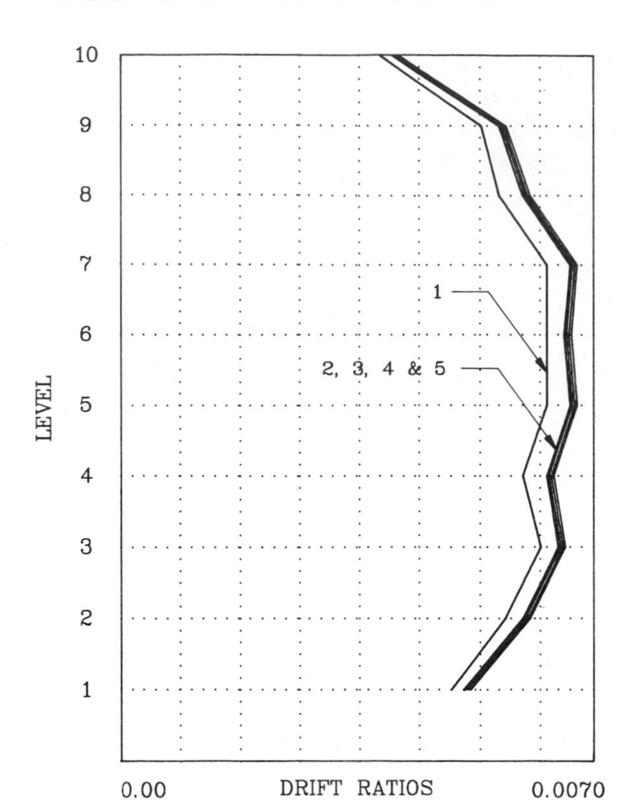

Figure 6-16 Story drift ratios of the 10 story frame as obtained by various *P*-delta methods.

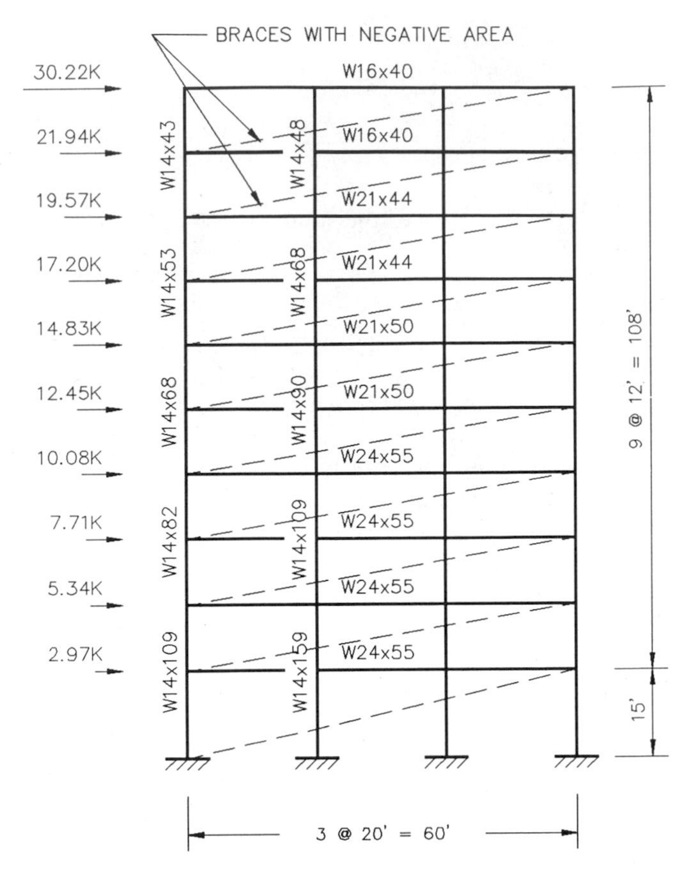

Figure 6-17 Frame modeled with negative braces.

the equivalent negative brace is obtained:

$$A = -\frac{\Sigma P}{h}\frac{L_{br}}{E\cos^2\alpha} \tag{6-26}$$

It is important to note that, due to the horizontal and vertical forces in the braces, the axial forces and shears in the columns will be slightly in error. These errors can be reduced by making the braces as long as possible (see Figure 6-17).

Example 6-3

For the 10-story frame of Example 6-1, compute the second-order displacements and story drifts by the negative-bracing-member method. The modulus of elasticity of the braces is

$$E = 29{,}000 \text{ ksi}$$

For a typical floor

$$L_{br} = \sqrt{(60)^2 + (12)^2} = 61.188 \text{ ft} = 734.26 \text{ in.}$$

$$\cos^2\alpha = (60/61.188)^2 = 0.9615$$

For the first floor,

$$L_{br} = \sqrt{(60)^2 + (15)^2} = 61.847 \text{ ft.} = 742.16 \text{ in}$$

$$\cos^2\alpha = (60/61.847)^2 = 0.9412$$

The negative brace area for each floor level may now be calculated using Equation 6-26. For example, for the fourth floor, where the total gravity force is 1476 kips, the negative brace area is

$$A_4 = -\frac{(1476)(734.26)}{(144)(29000)(0.9615)} = -0.2699 \text{ in.}^2$$

The brace areas, and the displacements obtained using the negative braces, are shown in Table 6-5. The very good agreement with the "exact" results (Table 6-3) is evident.

Modified Versions of Approximate P-Delta Methods The *P*-delta methods presented in this chapter ignore the "C-S" effect (Figure 6-4d). For most practical problems, the C-S effects are much smaller than the *P*-delta effects, and can be ignored. However, if needed, the *P*-delta methods described in previous sections can be simply modified to include this effect.

Table 6-5 ***P*-Delta Analysis by Negative-Bracing-Member Method**

Level	h, in.	ΣP, kips	L_{br}, in.	$E\cos^2\alpha$	A_{br}, in.	2nd-Order Disp., in.
10	144	180	734.26	27,884	−0.0329	8.458
9	144	396	734.26	27,884	−0.0724	7.929
8	144	612	734.26	27,884	−0.1120	7.168
7	144	828	734.26	27,884	−0.1514	6.350
6	144	1044	734.26	27,884	−0.1909	5.394
5	144	1260	734.26	27,884	−0.2341	4.442
4	144	1476	734.26	27,884	−0.2699	3.468
3	144	1692	734.26	27,884	−0.3094	2.572
2	144	1908	734.26	27,884	−0.3489	1.642
1	180	2124	742.16	27,295	−0.3209	0.817

The modification is achieved by multiplying the member axial forces by a flexibility factor γ. For a single column, γ is given by[6-26]

$$\gamma = 1 + 0.22\frac{4(G_A - G_B)^2 + (G_A + 3)(G_B + 2)}{[(G_A + 2)(G_B + 2) - 1]^2} \tag{6-27}$$

where G_A and G_B are the stiffness ratios as defined in Section 6.4.1. The flexibility factor γ has a rather small range of variation (from 1.0 for $G_A = G_B = \infty$, to 1.22 for $G_A = G_B = 0$). For design purposes a conservative average value of γ can be used for the entire frame. Lai and MacGregor[6-26] suggest an average value of $\gamma = 1.15$, while Stevens[6-10] has proposed an average value of $\gamma = 1.11$.

To include the C-S effect in the previously discussed P-delta methods, it is sufficient to use $\gamma\Sigma P$ instead of ΣP whenever the expression ΣP appears.

Example 6-4

For the 10-story frame and Example 6-1, compute the second-order displacements and story drifts at the first, fifth, and roof levels by the modified direct P-delta method. An average value of $\gamma = 1.11$ is assumed for all calculations.

Using the values listed in Table 6-4, we have:

- at the roof:

$$\frac{\gamma(\Sigma P)\Delta_1}{(\Sigma V_1)h} = \frac{(1.11)(180)(0.517)}{(30.22)(144)} = 0.024$$

$$\mu = \frac{1}{1 - 0.024} = 1.025$$

$$\Delta_2 = \mu\Delta_1 = (1.025)(0.517) = 0.530 \text{ in.}$$

- at the fifth level:

$$\frac{\gamma\Sigma P\Delta_1}{\Sigma V_1 h} = \frac{(1.11)(1260)(0.914)}{(116.21)(144)} = 0.076$$

$$\mu = \frac{1}{1 - 0.076} = 1.082$$

$$\Delta_2 = \mu\Delta_1 = (1.082)(0.914) = 0.989 \text{ in.}$$

- and at the first level:

$$\frac{\gamma(\Sigma P)\Delta_1}{(\Sigma V_1)h} = \frac{(1.11)(2124)(0.765)}{(142.31)(180)} = 0.070$$

$$\mu = \frac{1}{1 - 0.070} = 1.075$$

$$\Delta_2 = \mu\Delta_1 = (1.075)(0.765) = 0.822 \text{ in.}$$

Comparison of these results with those obtained by the original method reveals an increase of less than 1% in the story drifts due to this modification.

6.4.4 "Exact" *P*-Delta Analysis

Construction of the geometric stiffness matrix is the backbone of any exact second-order analysis. The same matrix is also essential for any finite-element buckling analysis procedure. In this section, the concept of geometric stiffness matrix is introduced, and a general approach to "exact" second-order structural analysis is discussed.

Consider the deformed column shown in Figure 6-18. For the sake of simplicity, neglect the axial deformation of the member, and the small C-S effect. The slope deflection equations for this column can be written as[6-12]

$$M_t = \frac{EI}{L}\left(4\theta_t + 2\theta_b - \frac{6\Delta_t}{L} + \frac{6\Delta_b}{L}\right) \tag{6-28}$$

$$M_b = \frac{EI}{L}\left(2\theta_t + 4\theta_b - \frac{6\Delta_t}{L} + \frac{6\Delta_b}{L}\right) \tag{6-29}$$

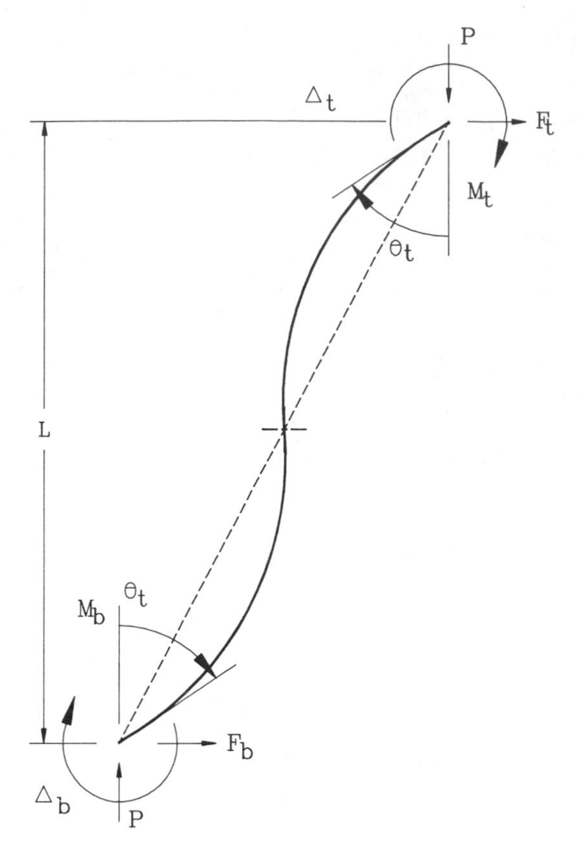

Figure 6-18 Forces acting on a deformed column.

From force equilibrium:

$$F_t = -\frac{M_t + M_b}{L} - \frac{P(\Delta_t - \Delta_b)}{L} \quad (6\text{-}30)$$

$$F_b = -F_t \quad (6\text{-}31)$$

Substituting Equations 6-28 and 6-29 into Equation 6-30,

$$F_t = -\frac{6EI}{L^2}(\theta_t + \theta_b) + 12\left(\frac{EI}{L^3} - \frac{P}{L}\right)(\Delta_t - \Delta_b) \quad (6\text{-}32)$$

Now if we rewrite the above equations in a matrix form, we obtain

$$\begin{bmatrix} M_t \\ M_b \\ F_t \\ F_b \end{bmatrix} = \begin{bmatrix} \frac{4EI}{L} & \frac{2EI}{L} & -\frac{6EI}{L^2} & \frac{6EI}{L^2} \\ \frac{2EI}{L} & \frac{4EI}{L} & -\frac{6EI}{L^2} & \frac{6EI}{L^2} \\ -\frac{6EI}{L^2} & -\frac{6EI}{L^2} & \frac{12EI}{L^3} - \frac{P}{L} & -\frac{12EI}{L^3} + \frac{P}{L} \\ \frac{6EI}{L^2} & \frac{6EI}{L^2} & -\frac{12EI}{L^3} + \frac{P}{L} & \frac{12EI}{L^3} - \frac{P}{L} \end{bmatrix} \begin{bmatrix} \theta_t \\ \theta_b \\ \Delta_t \\ \Delta_b \end{bmatrix} \quad (6\text{-}33)$$

Since we wrote the equilibrium equations for the deformed shape of the member, this is a second-order stiffness matrix. Notice that the only difference between this matrix and a standard first-order beam stiffness matrix is the presence of P/L, or geometric terms. The stiffness matrix given by Equation 6-33 can also be written as

$$[K] = [K_f] - [K_g] \quad (6\text{-}34)$$

where $[K_f]$ is the standard first-order stiffness matrix (material matrix) and $[K_g]$ is the geometric stiffness matrix given by

$$\begin{bmatrix} 0 & 0 & 0 & 0 \\ 0 & 0 & 0 & 0 \\ 0 & 0 & +P/L & -P/L \\ 0 & 0 & -P/L & +P/L \end{bmatrix}$$

Inspection of the simple second-order stiffness matrix given by Equation 6-33 shows why general second-order structural analysis has an iterative nature. The matrix includes P/L terms, but the axial force P is not known before an analysis is performed. For the first analysis cycle, P can be assumed to be zero (standard first-order analysis). In each subsequent analysis cycle, the member forces obtained from the previous cycle are used to form a new geometric stiffness matrix, and the analysis continues until convergence is achieved. If inelastic material behavior is to be considered, then the material stiffness matrix must also be revised at appropriate steps in the analysis.

Substantial research has been performed on the formulation of geometric stiffness matrices and finite-element stability analysis of structures (References 6-28 to 6-36). Most recently, a complete formulation of the three-dimensional geometric stiffness matrix for wide-flange beam–columns has been proposed by Yang and McGuire.[6-36]

The common assumption that floor diaphragms are rigid in their own plane allows condensation of lateral degrees of freedom into three degrees of freedom per floor level: two horizontal translations and a rotation about the vertical axis. This simplification significantly reduces the effort required for an "exact" second-order analysis. A number of schemes have been developed to permit direct and noniterative inclusion of P-delta effects in the analysis of rigid-diaphragm buildings.[6-37, 6-38, 6-39]

The geometric stiffness matrix for a three-dimensional rigid-diaphragm building is given in Figure

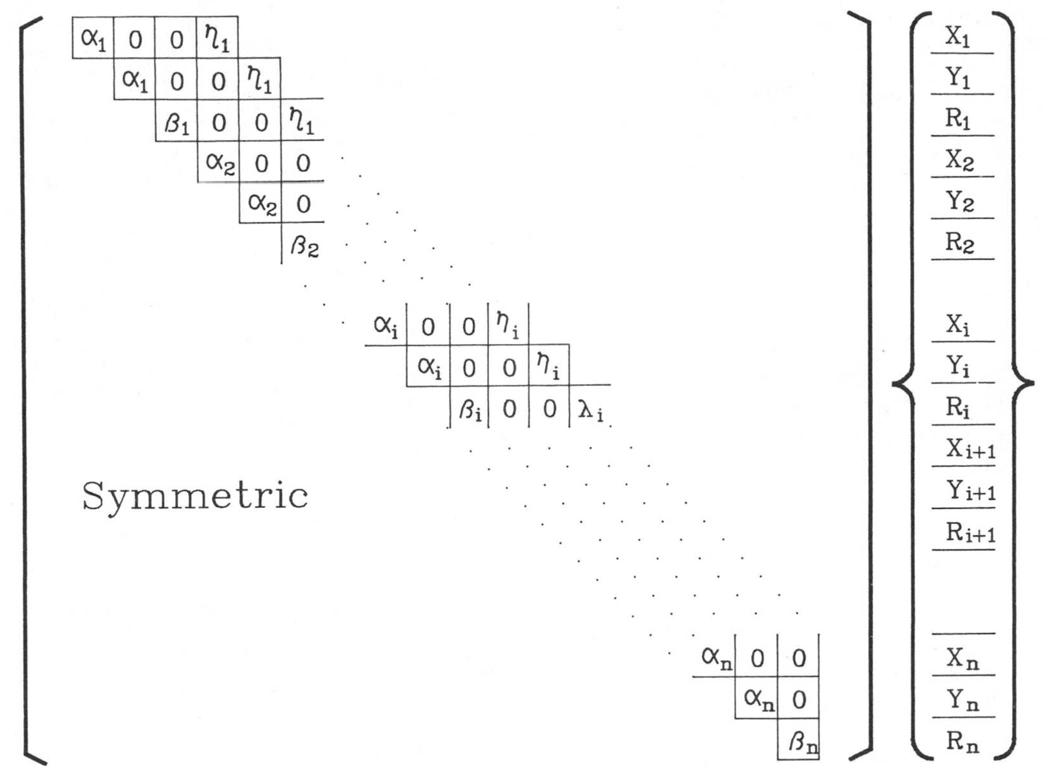

Figure 6-19 Geometric stiffness matrix for three-dimensional rigid diaphragm buildings.

6-19.[6-37, 6-38] For a three-dimensional building with N floor levels, $[K_g]$ is a $3N \times 3N$ matrix. For planar frames, the matrix reduces to a $N \times N$ tridiagonal matrix. The nonzero terms of this matrix are given by

$$\alpha_i = \frac{(\Sigma P)_i}{h_i} + \frac{(\Sigma P)_{i+1}}{h_{i+1}} \tag{6-35}$$

$$\beta_i = \frac{(\Sigma T)_i}{h_i} + \frac{(\Sigma T)_{i+1}}{h_{i+1}} \tag{6-36}$$

$$\eta_i = -\frac{(\Sigma P)_i}{h_i} \tag{6-37}$$

$$\lambda_i = -\frac{(\Sigma T)_i}{h_i} \tag{6-38}$$

where h_i is the floor height for level i, P_i is weight of the ith level, T_i is the second-order story torque, and

$$\left(\sum P\right)_i = \sum_{j=i}^{n} P_j \tag{6-39}$$

$$\left(\sum T\right)_i = \sum_{j=i}^{n} T_j \tag{6-40}$$

$(\Sigma P)_i$ can also be represented in terms of the story mass m_i and gravitational acceleration g as

$$\left(\sum P\right)_i = \left(\sum_{j=i}^{n} m_j\right) \times g \tag{6-41}$$

The story torque T_i is given by[6-38]

$$T_i = \left(\sum_{j=i}^{n} p_j d_j^2\right) \frac{\theta}{h_i} \tag{6-42}$$

where p_j is the vertical force carried by the jth column, d_j is the distance of the jth column from the center of rotation of the floor, and θ is an imposed unit rigid-body rotation of the floor. Assuming that the dead load is evenly distributed over the floor and that a roughly uniform vertical support system is provided over the plan area of the floor, Equation 6-42 can be further simplified to

$$T_i = m_{Ri} \frac{g}{h_i} \tag{6-43}$$

where m_{Ri} is the rotational mass moment of inertia of the ith floor and g is the gravitational acceleration. The approximation involved in the derivation of Equation 6-43 is usually insignificant.[6-39] Hence, for most practical problems, Equation 6-43 can be used instead of Equation 6-42, thereby allowing the direct inclusion of the P-delta effect in a three-dimensional structural analysis.

6.4.5 Choice of Member Stiffnesses for Drift and *P*-Delta Analysis

A common difficulty in seismic analysis of reinforced-concrete structures is the selection of a set of rational stiffness values to be used in force and displacement analyses. Should one use gross concrete-section properties? Should one use some reduced section properties? Or should the gross concrete properties be used for one type of analysis and reduced section properties for another?

The seismic design codes in the United States are not specific about this matter. Hence, the choice of section properties used in lateral analysis in general, and seismic analysis in particular, varies widely.

Contributing to the complexity of this issue are the following factors:

1. Although elastic material behavior is usually assumed for the sake of simplicity, reinforced concrete is not a homogeneous, linearly elastic material.
2. The stiffness and idealized elastic material properties of a reinforced concrete section vary with the state of behavior of the second (e.g. uncracked, cracked, and ultimate states).
3. Not all reinforced-concrete members in a structure, and not all cross sections along a particular member, are in the same state of behavior at the same time.
4. For many beams and other asymmetrically reinforced members, the stiffness properties for positive bending and negative bending are different.
5. The stiffness of reinforced concrete members and structures varies with the time, and with the history of past exposure to wind forces and earthquake ground motions.
6. The stiffness of reinforced concrete members and structures varies with the amplitude of the applied forces.

Analytical and experimental studies[6-40] have indicated that for motions which are within the working-stress design limits of members, the measured fundamental periods of concrete structures are generally slightly less than the periods computed using gross-concrete-section properties. According to Reference 6-40, in the case of large-amplitude motions up to the yield level, the stiffness of the building is usually somewhere between the computed values based on the gross concrete-section properties and the cracked-section properties. Based on this observation, the same reference suggests that for force analysis, the gross concrete-section properties and the clear-span dimensions be used, and the effect of nonseismic structural and nonstructural elements be considered. For drift calculations, either the lateral displacements determined using the above assumptions should be doubled, or the center-to-center dimensions along with the average of the gross section and the cracked-section properties, or one-half of the gross section properties should be used. Furthermore, the nonseismic structural and nonstructural elements should be neglected if they do not create a potential torsional reaction.

Similar sets of assumptions have been proposed by research workers who have been concerned about the choice of member stiffnesses to be used in the *P*-delta analysis of concrete structures. For example, for second-order analysis of concrete structures subjected to combinations of gravity and wind loads, MacGregor and Hage[6-16] recommend using 40% of the gross section moment of inertia for beams and 80% of the gross section moment of inertia for columns.

6.5 DRIFT-DESIGN PROCEDURES

6.5.1 Drift Design of Moment Frames and Framed Tubes

The lateral displacements and story drifts of moment-resistant frames and symmetrical framed tubes are caused by bent action, cantilever action, the shear leak effect, and panel zone distortions. With the simplified methods presented in this section, the contribution of each of these actions to the story drift can be estimated separately. The story drifts so obtained are then added to obtain an estimate of the total story drift. Once an estimate of the drift and the extent of the contribution of each of these actions to the total drift are known, proper corrective measures can be adopted to reduce story drifts to an acceptable level.

Bent Displacements A significant portion of drift in rigid frames and framed tubes is caused by end rotations of beams and columns (Figure 6-20). This phenomenon is commonly referred to as *bent action* (also called frame action, or racking). For most typical low- to mid-rise rigid frames, almost all of the drift is caused by the bent action. However, for taller frames, other actions, such as axial deformation of columns (cantilever or chord action), become more significant. For extremely tall frames, the contribution of cantilever action to drift may be several times larger than that of bent action.

In the design of framed tubes, it is usually desirable to limit the bent-action drifts to 30 to 40% of the total drift. If a framed tube is also braced, the bent-action drifts are usually limited to about 20 to 25% of the total drift.[6-1] The bent-action Δ_{bi} for any level i of a frame may be estimated by[6-41]

$$\Delta_{bi} = \frac{(\Sigma V)_i h_i^2}{12E}\left(\frac{1}{(\Sigma K_g)_i} + \frac{1}{(\Sigma K_c)_i}\right) \qquad (6\text{-}44)$$

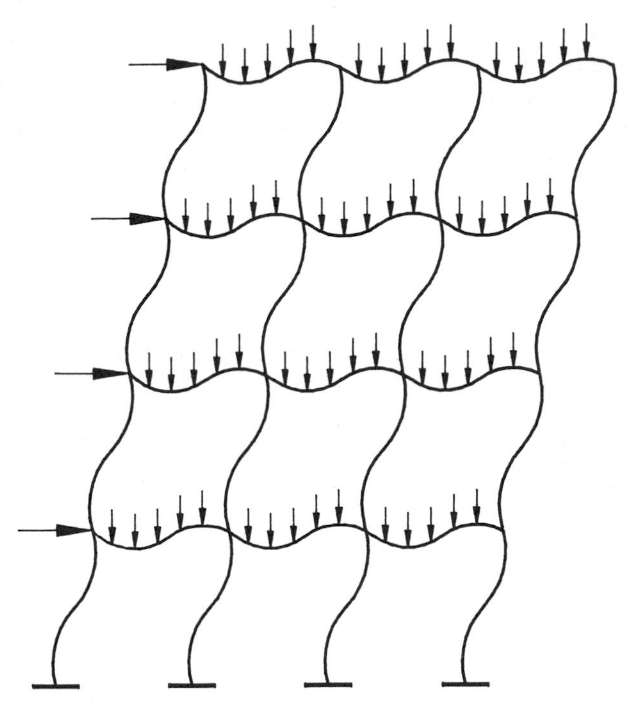

Figure 6-20 Frame deformation caused by the bent action.

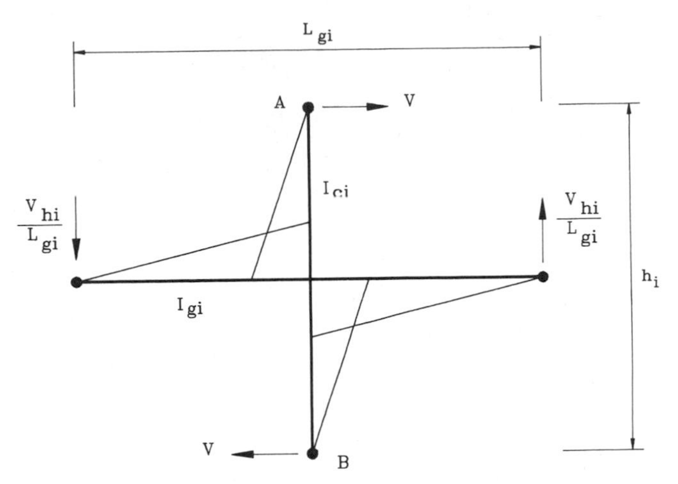

Figure 6-21 Typical subassemblage used in derivation of the bent action drift equation (6-41).

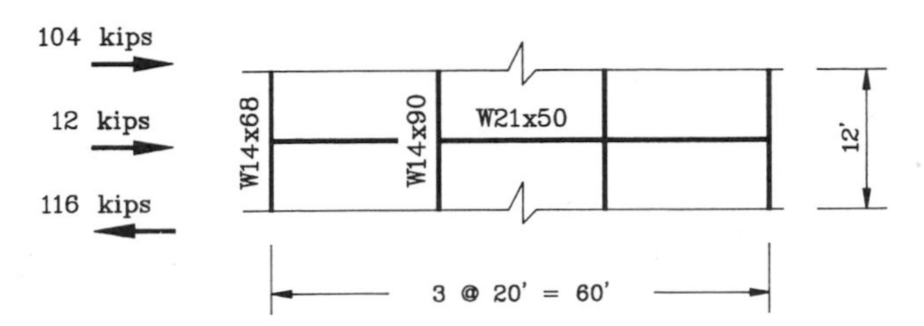

Figure 6-22 The bent at the 5th floor (Example 6-5).

where $(\Sigma V)_i$ is the story shear, h_i is the story height,[b] and

$(\Sigma K_g)_i$ = summation of I_{gi}/L_{gi} for all girders

$(\Sigma K_c)_i$ = summation of I_{ci}/h_i for all columns

I_{gi} = individual girder moment of inertia

L_{gi} = individual bay length[b]

I_{ci} = individual column moment of inertia

Equation 6-44 can be derived by applying the slope–deflection equations to the typical subassemblage shown in Figure 6-21. In the derivation of Equation 6-44, it is assumed that the points of the contraflexure are at the mid-span of beams and columns.

Other, but similar, relationships for bent-drift design have been proposed.(6-42, 6-43) Equation 6-44 can also be used to modify existing beam and column sizes to satisfy a given drift limit. Example 6-5 illustrates such an application.

Example 6-5

For the bent at the 5th floor of the 10-story frame of Example 6-1 (Figure 6-22), estimate the story drift caused by bent action. Modify member sizes, if necessary, to limit the bent drift ratio to 0.0030. Neglect the P-delta effect.

We have

W14 × 68 $\quad I_{c1} = 723 \text{ in.}^4$

W14 × 90 $\quad I_{c2} = 999 \text{ in.}^4$

W21 × 50 $\quad I_g = 984 \text{ in.}^4$

$$\Sigma\left(\frac{I_g}{L_g}\right) = \frac{(3)(984)}{(12)(20)} = 12.30 \text{ in.}^3$$

$$\Sigma\left(\frac{I_c}{h}\right) = \frac{(2)(723+999)}{(144)} = 23.92 \text{ in.}^3$$

$$\Delta_{bi} = \frac{116(144)^2}{(12)(29000)}\left(\frac{1}{12.3} + \frac{1}{23.92}\right)$$

$$= 0.85 \text{ in.}$$

$$\delta_{bi} = \frac{0.85}{144} = 0.0059 > 0.0030 \qquad \text{N.G.}$$

1. Increasing both beam and column sizes:

$$\Delta_{\text{Limit}} = (0.0030)(144) = 0.432 \text{ in.}$$

$$\Delta_{\text{Limit}} = \frac{\Delta_{bi}}{\Phi} \text{ or } 0.432 = \frac{0.85}{\Phi} \quad \rightarrow \quad \Phi = 1.97$$

[b]Depending on the modeling assumption, center-to-center length or clear length may be used.

Select new beam and column sizes:

$$I_{c1} = (1.97)(723) = 1424 \text{ in.}^4 \quad \rightarrow \quad \text{use W14} \times 120\text{: } I = 1380 \text{ in}^4.$$

$$I_{c2} = (1.97)(999) = 1968 \text{ in.}^4 \quad \rightarrow \quad \text{use W14} \times 176\text{: } I = 2140 \text{ in.}^4$$

$$I_g = (1.97)(984) = 1938 \text{ in.}^4 \quad \rightarrow \quad \text{use W24} \times 76\text{: } I = 2100 \text{ in.}^4$$

Check the new bent drift:

$$\sum\left(\frac{I_g}{L_g}\right) = \frac{(3)(2100)}{240} = 26.25 \text{ in.}^3$$

$$\sum\left(\frac{I_c}{h}\right) = \frac{(2)(1380 + 2140)}{144} = 48.89 \text{ in.}^3$$

$$\Delta_{bi} = 6.912\left(\frac{1}{26.25} + \frac{1}{48.89}\right) = 0.405 \text{ in.} < 0.432 \text{ in.} \qquad \text{O.K.}$$

Additional member weight required for drift control:

$$W = 3(76 - 50)(20) + 2(176 + 120 - 68 - 90)(12) = 4872 \text{ lb}$$

2. Increasing beam sizes only:

$$0.432 = 6.912\left(\frac{1}{12.3\Phi_g} + \frac{1}{23.92}\right) \quad \rightarrow \quad \Phi_g = 3.93$$

$$I_g = (3.93)(984) = 3867 \text{ in.}^4 \quad \rightarrow \quad \text{use W30X99: } I = 3990 \text{ in.}^4$$

Check the new bent drift:

$$\sum\left(\frac{I_g}{L_g}\right) = \frac{(3)(3990)}{240} = 49.9 \text{ in.}^3$$

$$\Delta_{bi} = 6.912\left(\frac{1}{49.9} + \frac{1}{23.92}\right) = 0.427 \text{ in.} < 0.432 \text{ in.} \qquad \text{O.K.}$$

Additional member weight required for drift control:

$$W = 3(99 - 50)(20) = 2940 \text{ lb}$$

3. Increasing column sizes only:

$$0.432 = 6.912\left(\frac{1}{12.3} + \frac{1}{23.92\Phi_c}\right) \quad \rightarrow \quad \Phi_c < 0.$$

Hence bent-drift control by increasing column sizes only is not feasible.

In this case, drift control by increasing beam sizes alone requires less material. However, in general one should be careful about increasing beam sizes alone, since it can jeopardize the desirable strong-column–weak-girder behavior.

Cantilever Displacements In tall frames and tubes, there is significant axial deformation in the columns caused by the overturning moments. The distribution of axial forces among the columns due to the overturning moments is very similar to distribution of flexural stresses in a cantilever beam. The overturning moments cause larger axial forces and deformations on the columns which are farther from the center line of the frame. This action, which causes a lateral deformation that closely resembles the deformation of a cantilever beam (Figure 6-23) is called the cantilever or chord action. In a properly proportioned framed tube, the cantilever deflections are significantly smaller than in a similar rigid frame. As shown in Figure 6-24, this is due to the participation of some of the columns in the flange frames in resistance to cantilever deformations. The taller the framed tube, the closer the column spacings, and the stronger the spandrel girders, the more significant the tube action becomes.

Cantilever displacements may be estimated by simple application of the moment-area method. The moment of inertia for an equivalent cantilever beam is computed as:

$$I_{0i} = \sum\left(A_{ci} d_i^2\right) \tag{6-45}$$

where A_{ci} is cross sectional area of an individual column and d_i is its distance from the center-line of the frame. The summation is carried over all the columns of the web frames, and those columns of the flange frames which are believed to participate in resistance to can-

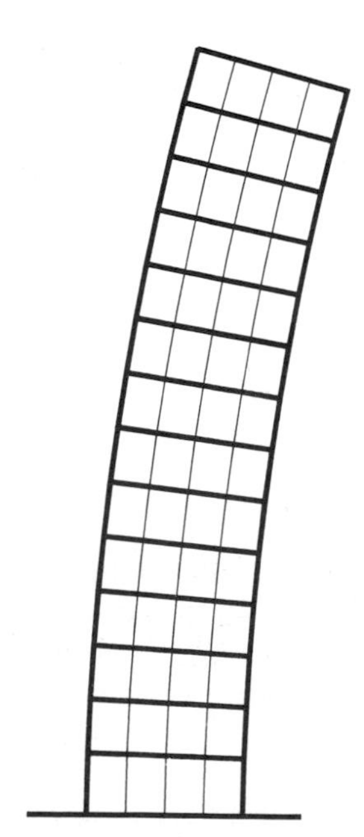

Figure 6-23 Cantilever or chord deformation.

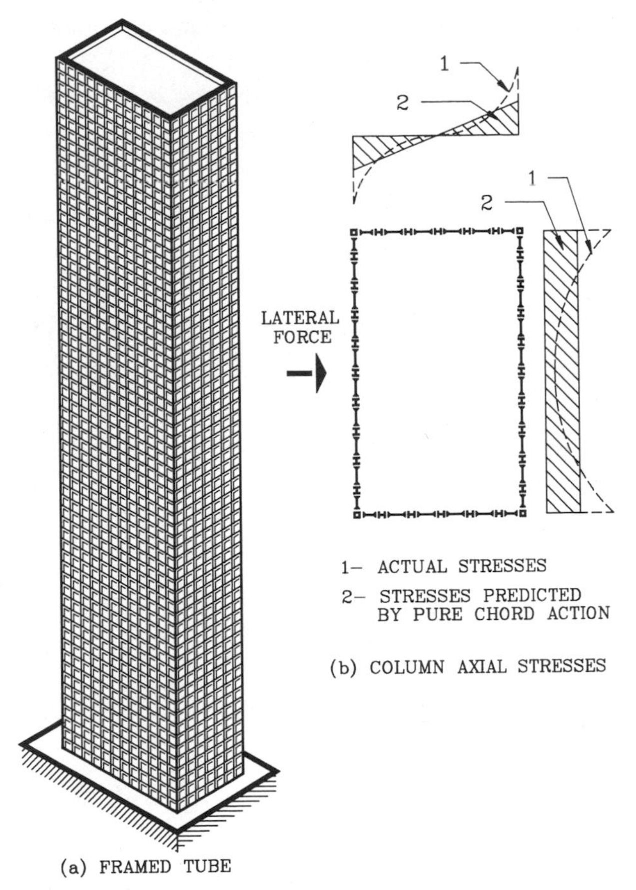

Figure 6-24 Tube action in response to lateral loads.

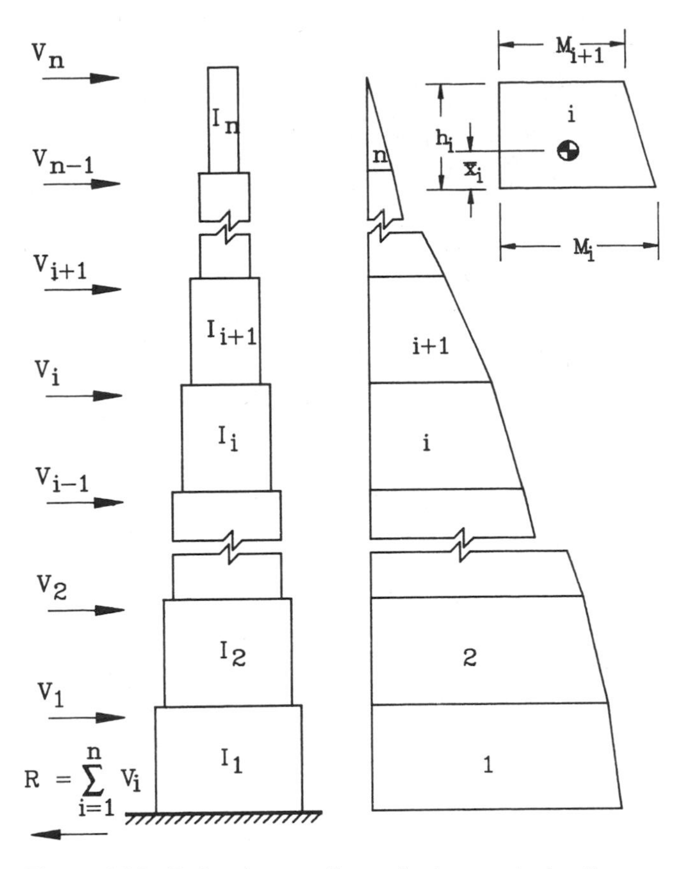

Figure 6-25 Estimating cantilever displacements by the moment area method.

tilever deflections. The computation of cantilever displacements for each floor level can be summarized in the following steps:

Step 1. Compute story moment of inertia I_{0i} using Equation 6-45.

Step 2. Compute overturning moments M_i.

Step 3. Compute area under the M/EI_{0i} from

$$A_i = \frac{(M_{i-1} + M_i)h_i}{2EI_{0i}} \tag{6-46}$$

Step 4. Compute $\bar{x}_i$ (see Figure 6-25) from

$$\bar{x}_i = \frac{h_i}{3}\,\frac{M_{i-1} + 2M_i}{M_{i-1} + M_i} \tag{6-47}$$

Step 5. Compute story displacement from

$$\Delta_{ci} = A_i(h_i - \bar{x}_i) + \sum_{j=1}^{i-1} A_j(H_i - \bar{x}_j) \tag{6-48}$$

where H_i is the total height of the ith floor measured from the base of the structure.

Example 6-6

Use the moment-area method and the procedure explained in this section to compute displacements at points 1, 2, and 3 of the simple cantilever column shown in Figure 6-26. Assume $EI = 58 \times 10^6$, kips-in^2.

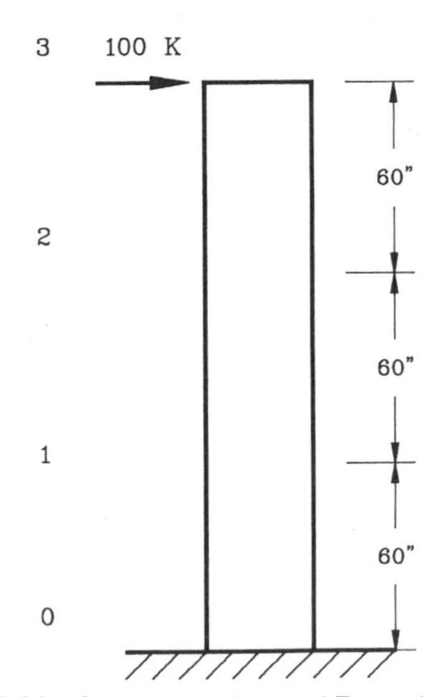

Figure 6-26 Cantilever column of Example 6-6.

Overturning moments:

$$M_3 = 0$$
$$M_2 = (100)(60) = 6000 \text{ in.-kips}$$
$$M_1 = (100)(120) = 12{,}000 \text{ in.-kips}$$
$$M_0 = (100)(180) = 18{,}000 \text{ in.-kips}$$

Area under M/EI curve:

$$A_0 = 0.$$
$$A_1 = \frac{(18{,}000 + 12{,}000)(60)}{(2)(58 \times 10^6)} = 0.01552$$
$$A_2 = \frac{(12{,}000 + 6000)(60)}{(2)(58 \times 10^6)} = 0.00931$$
$$A_3 = \frac{(6000 + 0)(60)}{(2)(58 \times 10^6)} = 0.00310$$

Distances $\bar{x}_i$:

$$\bar{x}_0 = 0$$
$$\bar{x}_1 = \frac{(20)(18{,}000 + 24{,}000)}{18{,}000 + 12{,}000} = 28.00 \text{ in.}$$
$$\bar{x}_2 = \frac{(20)(12{,}000 + 12{,}000)}{12{,}000 + 6000} = 26.67 \text{ in.}$$
$$\bar{x}_3 = \frac{(20)(6000 + 0)}{6000 + 0} = 20.00 \text{ in.}$$

Displacements:

$$\Delta_1 = 0.01552(60 - 28) = 0.497 \text{ in.}$$
$$\Delta_2 = 0.01552(120 - 28) + 0.00931(60 - 26.67) = 1.738 \text{ in.}$$
$$\Delta_3 = 0.01552(180 - 28) + 0.00931(120 - 26.67) + 0.00310(60 - 20) = 3.352 \text{ in.}$$

Shear Leak Displacements In buildings with closely spaced columns and deep girders, such as framed tubes, the contribution of shearing deformations to the lateral displacements (called the shear leak effect) may be significant. Story drifts due to the shear leak effect at level i, Δ_{shi}, may be estimated as[6-41]

$$\Delta_{shi} = \frac{\Sigma V_i h_i^2}{G}\left(\frac{1}{\Sigma A'_{gi} L_{gi}} + \frac{1}{\Sigma A'_{ci} h_i}\right) \quad (6\text{-}49)$$

where G is the shear modulus and A'_{gi} and A'_{ci} are the shear areas of individual girders and columns at level i.

In order to simplify the design process, an effective moment of inertia, I_{eff}, can be defined where the contributions of both flexural and shearing deformations are considered:

$$I_{eff} = \frac{A'L^2 I}{24(1 + \upsilon)I + A'L^2} \quad (6\text{-}50)$$

where A' is the shear area, L is span length, I is the moment of inertia of the section, and υ is Poisson's ratio.

Example 6-7

For the bent of Example 6-5, estimate the additional story drift caused by the shear leak effect.

We have

$$\text{W14} \times 68: \; A' = dt_w = (14.04)(0.415) = 5.83 \text{ in.}^2$$
$$\text{W14} \times 90: \; A' = dt_w = (14.02)(0.440) = 6.17 \text{ in.}^2$$
$$\text{W21} \times 50: \; A' = dt_w = (20.83)(0.380) = 7.92 \text{ in.}^2$$
$$\sum A'_{gi} L_{gi} = (3)(7.92)(240) = 5702.4$$
$$\sum A'_{ci} h_i = (2)(6.17 + 5.83)(144) = 3456.0$$

Using Equation 6-49,

$$\Delta_{shi} = \frac{116(144)^2}{11{,}200}\left(\frac{1}{5702.4} + \frac{1}{3456.0}\right) = 0.10 \text{ in.}$$

Panel-Zone Distortions When joint shear forces are high, and the beam–column panel zones are not adequately stiffened, panel-zone distortions can have a measurable impact on the story drift. The panel-zone force–deformation behavior is complex and nonlinear. Currently, there are no standard procedures or code recommendations for calculating additional story drifts caused by the panel-zone distortions.[a]

Cheong-Siat-Moy[6-44] has recommended a simple method based on elastic theory to estimate this effect. The method assumes a linear relationship between the shearing forces and the panel-zone distortions. It also assumes a uniform distribution of shear stress throughout the panel zone.

A simple beam–column subassemblage and the corresponding force and displacement diagrams, as assumed by this method, are shown in Figure 6-27. It can be shown that the deformation angle γ and the additional lateral story drift due to panel-zone distortion, Δ_p, are

$$\gamma = \frac{2(M_c/d_g) - V}{Gtd_c} \quad (6\text{-}51)$$

$$\Delta_p = \frac{\gamma(h - d_g)}{2} \quad (6\text{-}52)$$

[a] Designers usually consider this effect by using a value between the center-to-center and clear-span lengths in their analysis.

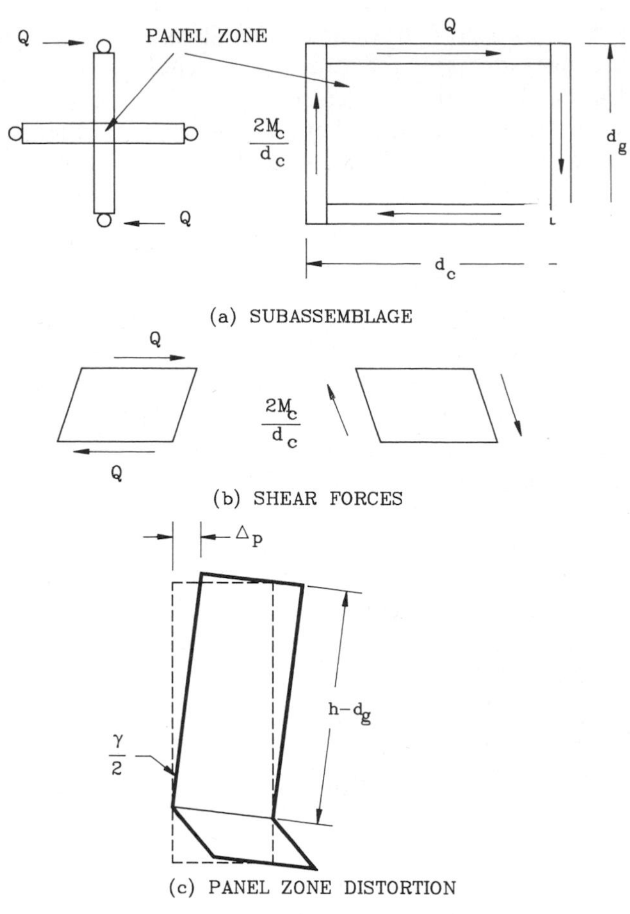

Figure 6-27 Effect of panel zone deformation (6-44).

where M_c is the moment from one column, d_g is the girder depth, V is the column shear, G is the shear modulus, t is the panel-zone thickness, d_c is the column depth, and h is the story height. Hence, $h - d_g$ is the clear column height.

If the points of contraflexure are assumed to be at mid-span of the beams and columns, Equation 6-51 can be further simplified to

$$\gamma = V\frac{(h/d_g) - 2}{Gtd_c} \tag{6-53}$$

Considering the approximate nature of the above formula, it is not necessary to apply it to each individual column. Instead, it can be used in an average sense (see Example 6-8).

A series of experimental and analytical studies on the behavior of steel beam–column panel zones have been conducted by various research institutions (references 6-45 to 6-48). In one of these studies,[(6-48)] conducted at Lehigh University, several beam–column subassemblage specimens were subjected to cyclic loads far beyond their elastic limits. Based on these tests, a formula, similar to Equation 6-53, for estimation of panel-zone distortions was recommended:

$$\gamma = \frac{V}{Gd_ct}\left(\frac{L_c}{d_g} - \frac{L}{h}\right) \tag{6-54}$$

where L is the beam span length, L_c is the clear column length, G is the shear modulus, which is taken as 11,000 ks$i$, and γ is the panel-zone distortion in radians.

There is a need for further research on the behavior of beam–column panel zones for different types of construction.

Example 6-8

For the bent of Example 6-5, estimate the contribution of panel zone distortion to story drift assuming two conditions: (i) no doubler plates, and (ii) $\frac{1}{4}$-in. doubler plates.

We have

W14 × 68: $d = 14.04$ in., $t = 0.450$ in.

W14 × 90: $d = 14.02$ in., $t = 0.440$ in.

W21 × 50: $d_g = 20.83$ in.,

Using Cheong-Siat-Moy's method (Equations 6-52 and 6-53), we have

$$\gamma = V\frac{h/d_g - 2}{Gtd_c}$$

Without doubler plates,

$$\text{Average } t = 2\frac{0.450 + 0.440}{4} = 0.445 \text{ in.}$$

$$\text{Average } V = 116/4 = 29 \text{ kips}$$

$$\gamma = 29\frac{144/20.83 - 2}{(11200)(0.445)(14.03)} = 0.0020$$

$$\Delta_p = \gamma\frac{h - d_g}{2} = 0.0010(144 - 20.83)$$

$$= \underline{0.123} \text{ in.}$$

With $\frac{1}{4}$-in. doubler plates,

$$\text{Average } t = 0.445 + 0.25 = 0.695 \text{ in.}$$

$$\Delta_p = 0.0013\frac{144 - 20.83}{2} = \underline{0.080} \text{ in.}$$

Using Lehigh's formula (Equation 6-54), we have

$$\gamma = \frac{V}{Gd_ct}\left(\frac{L_c}{d_g} - \frac{L}{h}\right)$$

$$L_c = 144 - 20.83 = 123.17 \text{ in.}$$

$$L = 12(20) = 240 \text{ in.}$$

Without doubler plates,

$$t = 0.445 \text{ in.}$$

$$\gamma = \frac{(29)(123.17/20.83 - 240/144)}{(11000)(14.03)(0.445)} = 0.00179 \text{ rad}$$

$$\Delta_p = (0.00179)(144 - 20.83)/2 = \underline{0.110} \text{ in.}$$

With doubler plates,

$$t = 0.695 \text{ in.}$$

$$\gamma = \frac{(0.00179)(0.445)}{0.695} = 0.00115 \text{ rad}$$

$$\Delta_p = \frac{(0.110)(0.00115)}{0.00179} = \underline{0.071} \text{ in.}$$

Drift Design of a 10-Story Moment-Resistant Frame In this subsection the approximate methods for drift and P-delta analysis, which were explained previously, are put into practice by performing a complete drift design for the 10-story moment-resistant steel frame introduced in Example 6-1. The goal is to achieve an economical design that meets the story drift-index limitation of 0.0033.

The first step is to estimate the lateral displacements and story drifts of the structure. Calculations of story drifts and lateral displacements due to bent action, the shear leak effect, and chord action are presented in Tables 6-6, 6-7, and 6-8 respectively. Center-to-center dimensions are used and panel-zone deformations are ignored in subsequent analyses.

The total displacements and story drifts are magnified using the direct P-delta method. These calculations are shown in Table 6-9. Notice that in sizing the members for strength, all lateral load-related forces and moments should also be multiplied by the corresponding story magnification factors (see μ in Table 6-9). Once the internal forces are thus magnified, it is rational to design the members using an equivalent-length factor of one.

Figures 6-28 and 6-29 depict the contribution of each action to the total lateral displacement and story drift. The dominance of bent action in the lateral response of this frame can be clearly seen in these figures. As explained previously, if the frame were significantly taller, bent action would be replaced by chord action as the dominant contributor to lateral displacement.

The results of this approximate analysis are compared with the results of an exact elastic analysis in Figures

Table 6-6 Calculation of Bent-Action Story Drifts and Lateral Displacements for the 10-Story Unbraced Frame

Level	h, in.	ΣV, kips	$\Sigma(I_g/L_g)$, in.	$\Sigma(I_c/h)$, in.	Δ_{bi}, in. (Eq. 7-44)	Bent Disp., in.
10	144	30.22	6.475	12.68	0.420	6.802
9	144	52.17	6.475	12.68	0.725	6.382
8	144	71.74	10.538	17.56	0.649	5.657
7	144	88.93	10.538	17.56	0.805	5.001
6	144	103.76	12.300	23.92	0.761	4.203
5	144	116.21	12.300	23.92	0.856	3.442
4	144	126.30	16.875	29.47	0.701	2.588
3	144	134.01	16.875	29.47	0.744	1.877
2	144	139.34	16.875	43.61	0.682	1.143
1	180	142.31	16.875	52.33*	0.461	0.461

*Two-thirds of the first-story height was used in calculation of the bent-action drift.

Table 6-7 Calculation of Shear-Leak Story Drifts and Lateral Displacements for the 10-Story Unbraced Frame

Level	h, in.	ΣP, kips	$\Sigma(A'_g L_g)$, in.	$\Sigma(A'_c h)$, in.	Δ_{shi}, in. (Eq. 7-49)	Shear-leak Disp., in.
10	144	30.22	3516	2550	0.0379	0.8377
9	144	52.17	3516	2550	0.0653	0.7998
8	144	71.74	5206	3161	0.0675	0.7345
7	144	88.93	5206	3161	0.0837	0.6670
6	144	103.76	5999	3455	0.0893	0.5833
5	144	116.21	5999	3455	0.1000	0.4940
4	144	126.30	6703	4267	0.0897	0.3939
3	144	134.01	6703	4267	0.0951	0.3042
2	144	139.34	6703	5379	0.0864	0.2091
1	180	142.31	6703	5379	0.1226	0.1226

Table 6-8 Calculation of Chord-Action Story Drifts and Lateral Displacements for the 10-Story Unbraced Frame

Level	h, in.	ΣV, kips	M_{0v},[a] in.-kips	I_{0i}, in.	A, in.	$\bar{x}$, in.	Chord disp., in.	Chord drift, in.
10	144	30.22	4,352	3,672,000	0.294×10^5	48.00	0.3936	0.0592
9	144	52.17	11,864	3,672,000	1.096×10^5	60.88	0.3344	0.0582
8	144	71.74	22,194	4,619,520	1.830×10^5	64.72	0.2762	0.0562
7	144	88.93	35,001	4,619,520	3.074×10^5	66.63	0.2200	0.0528
6	144	103.76	49,942	5,947,200	3.546×10^5	67.78	0.1674	0.0479
5	144	116.21	66,677	5,947,200	4.868×10^5	68.56	0.1195	0.0418
4	144	126.30	84,864	7,168,320	5.249×10^5	69.12	0.0777	0.0346
3	144	134.01	104,161	7,168,320	6.547×10^5	69.55	0.0431	0.0260
2	144	139.34	124,226	9,639,360	5.882×10^5	69.89	0.0171	0.0171
1	180	142.31	149,841	9,639,360	8.824×10^5	87.20	0.0000	0.0000

[a]Overturning moment.

Table 6-9 Calculation of Total First- and Second-Order Story Drifts and Lateral Displacements for the 10-Story Unbraced Frame

Level	h, in.	ΣV, kips	ΣP, kips	Δ_1, in.	μ	$\Delta_2 = \mu\Delta_1$, in.	2nd-Order Disp., in.
10	144	30.22	180	0.517	1.022	0.528	8.547
9	144	52.17	396	0.849	1.047	0.889	8.019
8	144	71.74	612	0.773	1.048	0.810	7.130
7	144	88.93	828	0.941	1.065	1.002	6.320
6	144	103.76	1044	0.898	1.067	0.958	5.318
5	144	116.21	1260	0.987	1.080	1.066	4.360
4	144	126.30	1476	0.833	1.073	0.894	3.294
3	144	134.01	1692	0.865	1.082	0.936	2.400
2	144	139.34	1908	0.786	1.081	0.850	1.464
1	180	142.31	2124	0.584	1.051	0.614	0.614

6-30 and 6-31, where the good agreement between the two sets of results may be observed.

Given the dominance of bent action in this case, a simple drift-design strategy based on reducing the bent drift is adopted. The maximum bent drift is about 80% of the maximum total drift. Hence, it would be rational to reduce the bent-drift ratios to 80% of the maximum allowable value of 0.0033 ($\approx$ 0.0026). It should be noted that increasing member sizes would further reduce the contribution of chord and shear-leak actions to the drift. Assuming that the drift control is to be achieved by increasing both beam and column sizes, the average magnification factors Φ by which the moments of inertia of beams and columns should be multiplied can be calculated as described in part 1 of Example 6-5. Based on the average values of Φ, new member sizes for beams and columns are selected. These member sizes are shown in Figure 6-32, where the computed values of Φ are shown in parenthesis.

At this stage, another round of displacement analysis, similar to that performed in Tables 6-6 to 6-9, is necessary to make sure that the new design satisfies the drift-design criteria. Results of this analysis are shown in Figures 6-33 and 6-34, which indicate that the new design satisfies the drift-design criteria. This was also confirmed by performing an exact structural analysis (Figures 6-35 and 6-36).

The last item on the agenda is to check the satisfaction of the strength criteria by the new design. Codified equivalent static lateral forces, which are based on a predetermined fundamental period for the structure, do not change with variation of stiffness. In reality, however, increasing member sizes for drift control increases the stiffness of the structure and reduces its natural periods. In multistory buildings, reduction of natural periods usually implies an increase in the inertial forces exerted on the structure. Hence, the adequacy of the modified design to withstand increased inertial forces should be examined.

Let us assume that the design ground motion for this example is represented by the response spectrum shown in Figure 6-37. Application of the Rayleigh method, or a simple dynamic analysis, reveals that the fundamental period of the original design (Figure 6-14) is about 2.7 sec. The fundamental period of vibration of the structure after drift design (Figure 6-32) is about 1.9 sec. Given the design response spectrum of Figure 6-37, the spectral acceleration corresponding to the first mode of

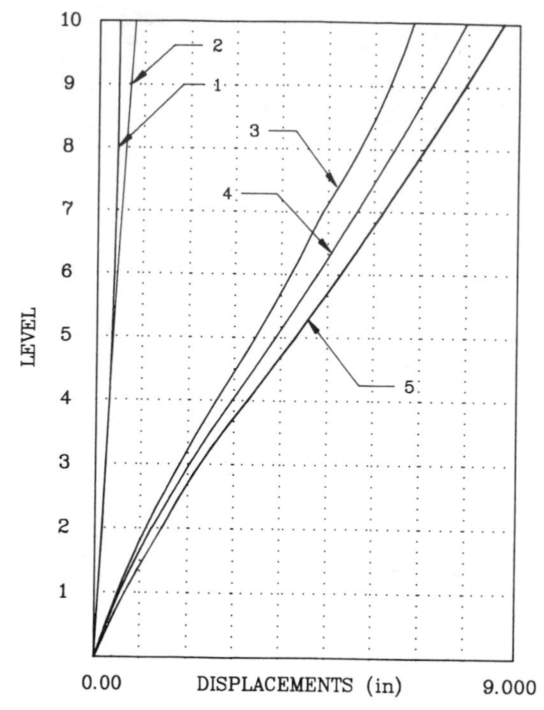

1 – SHEAR LEAK
2 – CHORD ACTION
3 – BENT ACTION
4 – TOTAL
5 – TOTAL + P Δ

Figure 6-28 Contribution of various actions to the total lateral displacement of the 10 story frame.

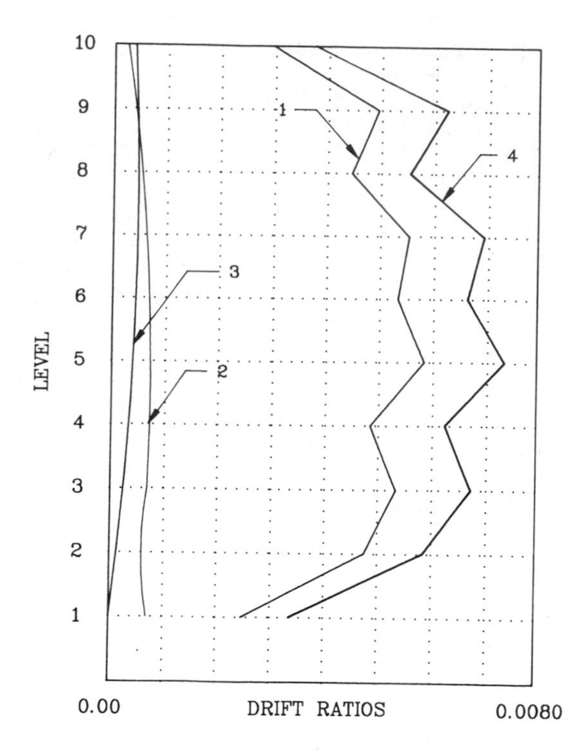

1 – BENT DRIFT
2 – SHEAR LEAK DRIFT
3 – CHORD DRIFT
4 – TOTAL + P –Δ DRIFT

Figure 6-29 Contribution of various actions to the total interstory drift ratios of the 10 story frame.

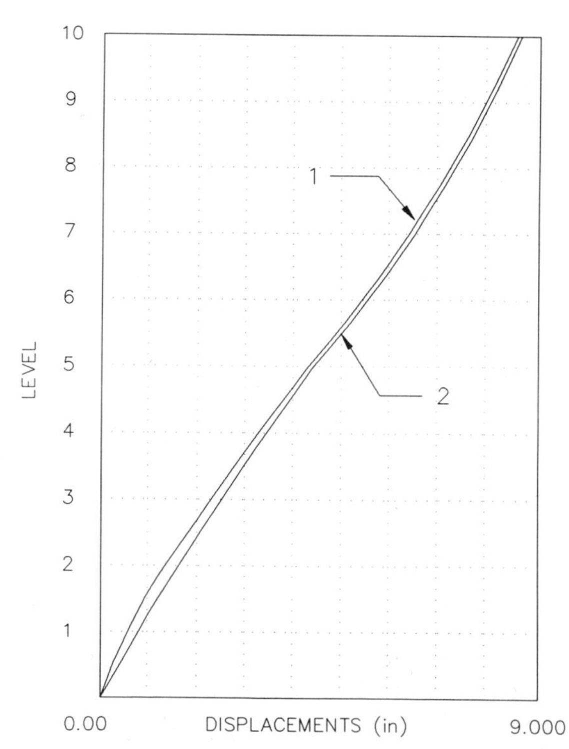

1 – APPROXIMATE
2 – EXACT

Figure 6-30 Comparison of approximate and "exact" second-order displacements.

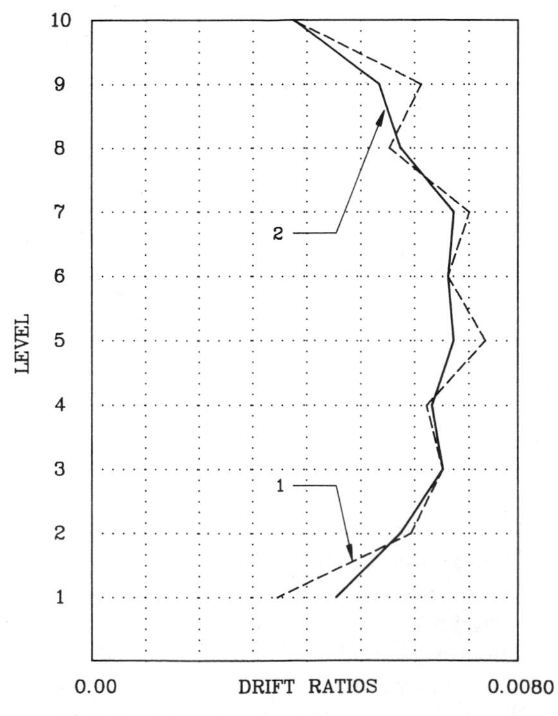

1 – APPROXIMATE
2 – EXACT

Figure 6-31 Comparison of approximate and "exact" second-order interstory drift ratios.

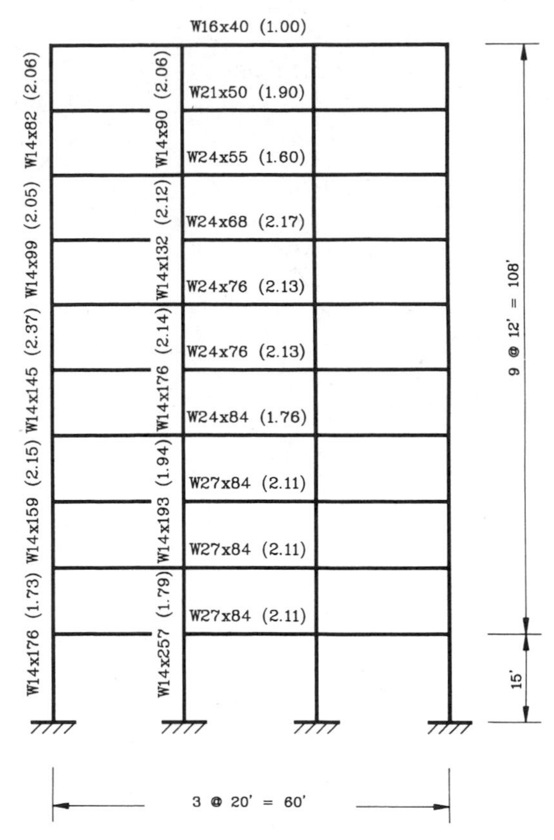

Figure 6-32 Member sections after drift design.

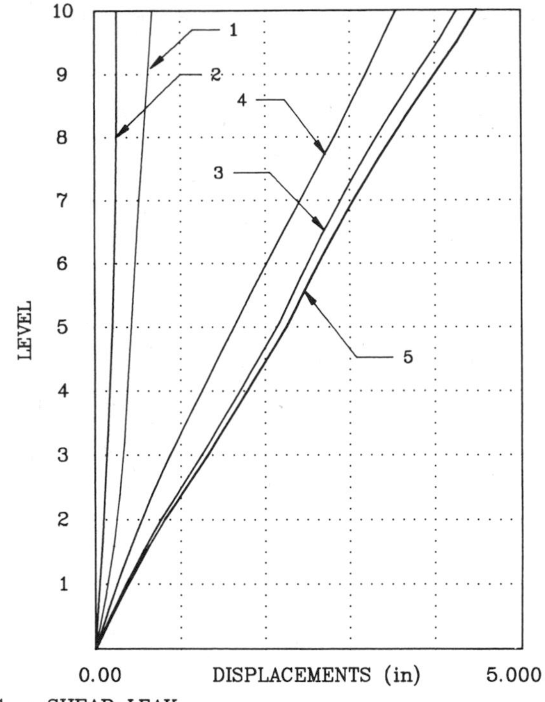

Figure 6-33 Approximate lateral displacements for the 10 story frame after drift design.

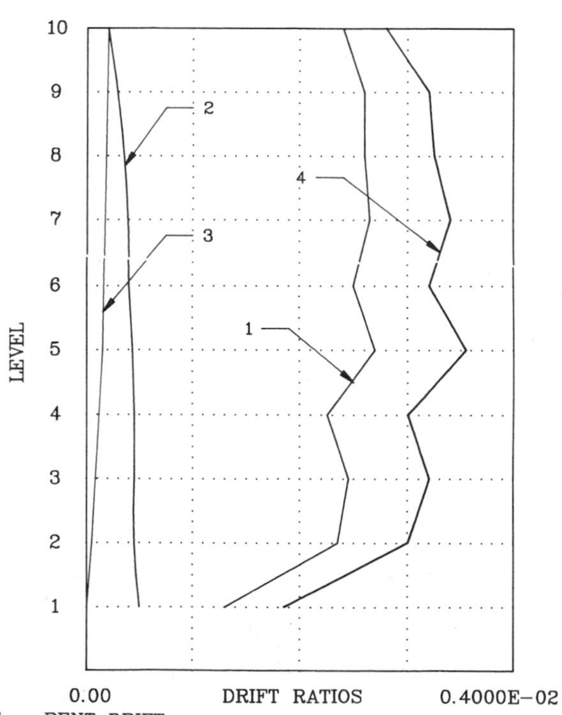

Figure 6-34 Approximate interstory drift ratios for the 10 story frame after drift design.

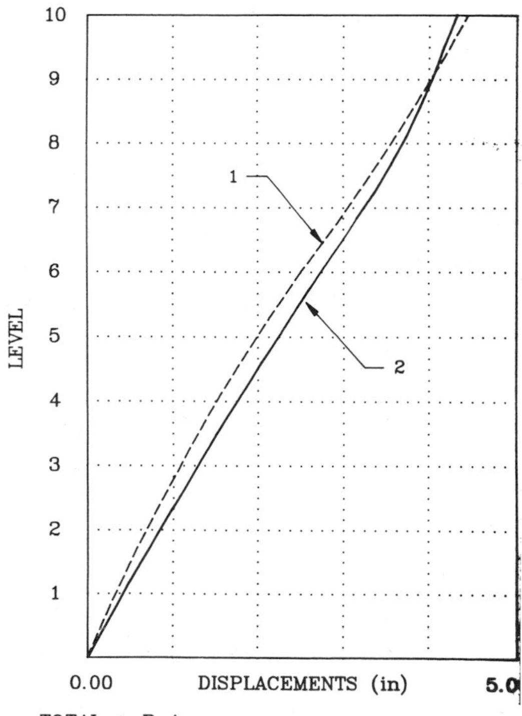

Figure 6-35 "Exact" versus approximate displacements for the 10 story frame after drift design.

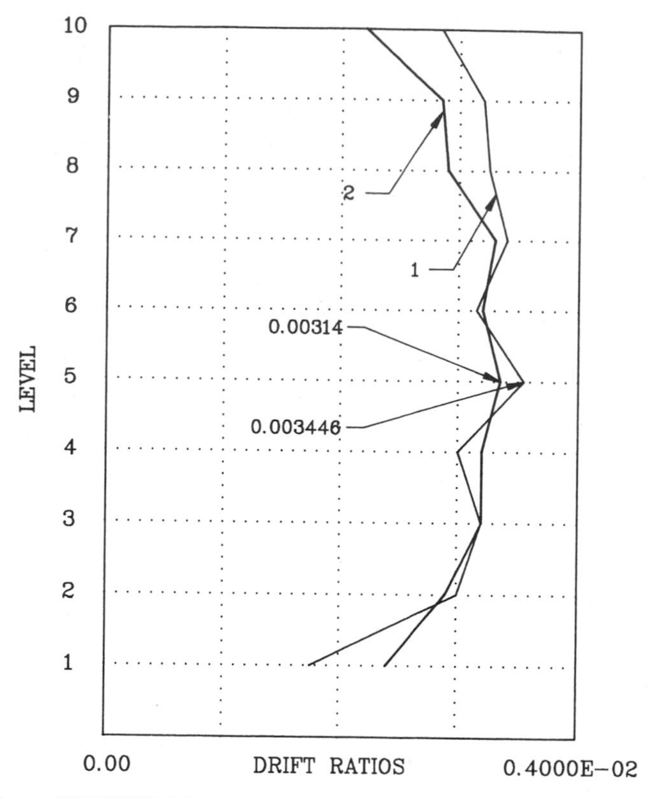

Figure 6-36 "Exact" versus approximate interstory drift ratios for the 10 story frame after drift design.

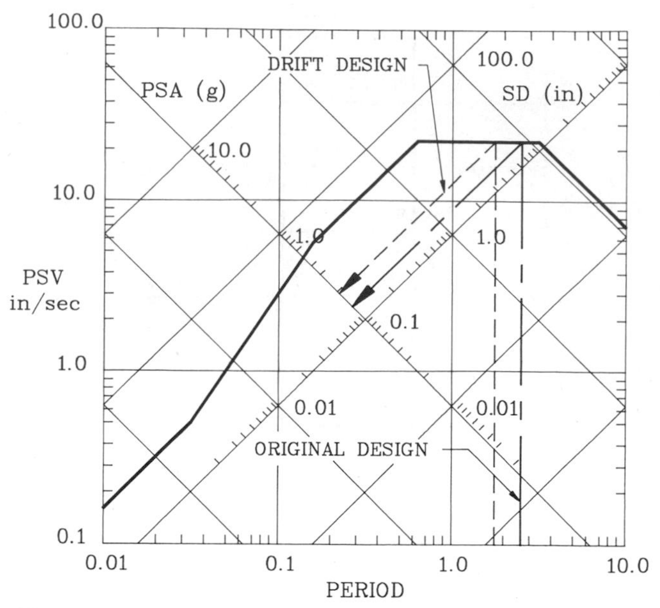

Figure 6-37 Influence of drift design on imposed inertial forces.

vibration of the structure is about $0.15g$ for the original design and $0.20g$ for the modified design. Hence, the modified design will be expected to withstand about 33% more inertial forces than the original one.

6.5.2 Drift Design of Braced Frames

Lateral displacements of braced frames are primarily caused by two actions: deformation of the braces, and axial deformation of the columns (chord action). Several methods are available for estimation of braced-frame displacements.(6-44, 6-49, 6-50) The contribution of brace deformations to story drift may be estimated by(6-44)

$$S_{br} = \sum \frac{A_{br} E \cos^2 \alpha}{L_{br}} \tag{6-55}$$

$$\Delta_{br} = \frac{\Sigma V}{S_{br}} \tag{6-56}$$

where Δ_{br} is story drift due to brace deformations, ΣV is the story shear, S_{br} is the sum of stiffnesses of the braces at the level under consideration, E is the modulus of elasticity of brace, A_{br} and L_{br} are the cross-sectional area and the length of each brace, and α is the angle that a brace makes with the horizontal axis. The summation is carried out over all braces at the level under consideration. Equation 6-55 is valid as long as the braces do not yield or buckle.

For ordinary braced frames, the bent story stiffness is negligible in comparison with the brace stiffness. However, in cases where rigid beam–column connections are utilized, the bent stiffness can be significant. In these situations, the bent story stiffness (see Section 6.5.1, "Bent Displacements") should be added to the brace stiffness.

The cantilever drifts may be computed via the moment-area method as explained in Section 6.5.1, "Cantilever Displacements." Note that in ordinary braced frames, where beams and columns are not joined by moment connections, only some of the columns (those in the vicinity of braces) provide significant resistance to cantilever deflections.

Example 6-9

Estimate the first- and second-order lateral displacements and story drifts for the 10-story braced steel frame shown in Figure 6-38. All beam-to-column connections are simple. The tributary width of the frame is 30 ft. The gravity load is 100 lb/ft² on the roof level and 120 lb/ft² on typical floors. Assume that the braces are so proportioned that none of them either yield or buckle under the given loads.

We have

W8 × 35: $A = 10.3 \text{ in.}^2$

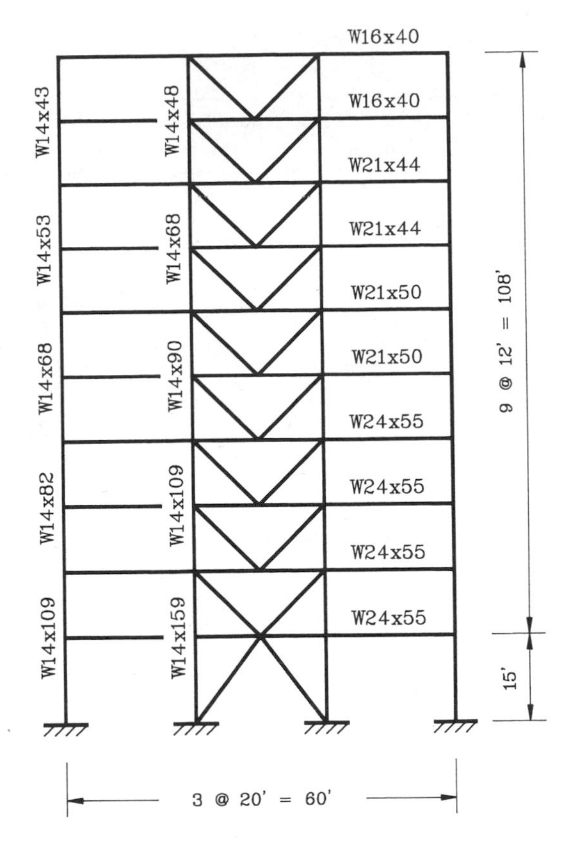

ALL BRACES ARE W8x35
ALL BEAM-COLUMN CONNECTIONS ARE SIMPLE

Figure 6-38 Braced frame elevation (Example 6-9).

For braces at typical floors,

$$L_{br} = \sqrt{(10)^2 + (12)^2} = 15.62 \text{ ft.} = 187.44 \text{ in.}$$

$$\cos\alpha = 10/15.62 = 0.6402$$

$$S_{br} = \sum E \frac{A_{br}}{L_{br}} \cos^2 \alpha$$

$$= 2(29000)(10.3)(0.6402)^2/187.44$$

$$= 1306.27 \text{ kips/in.}$$

For braces at the first floor,

$$L_{br} = \sqrt{(10)^2 + (15)^2} = 18.03 \text{ ft} = 216.33 \text{ in.}$$

$$\cos(\alpha) = 10/18.03 = 0.5547$$

$$S_{br} = \frac{2(29000)(10.3)(0.5547)^2}{216.33} = 849.67 \text{ kips/in.}$$

The brace-action story drifts and lateral displacements are calculated in Table 6-10. To show the accu-

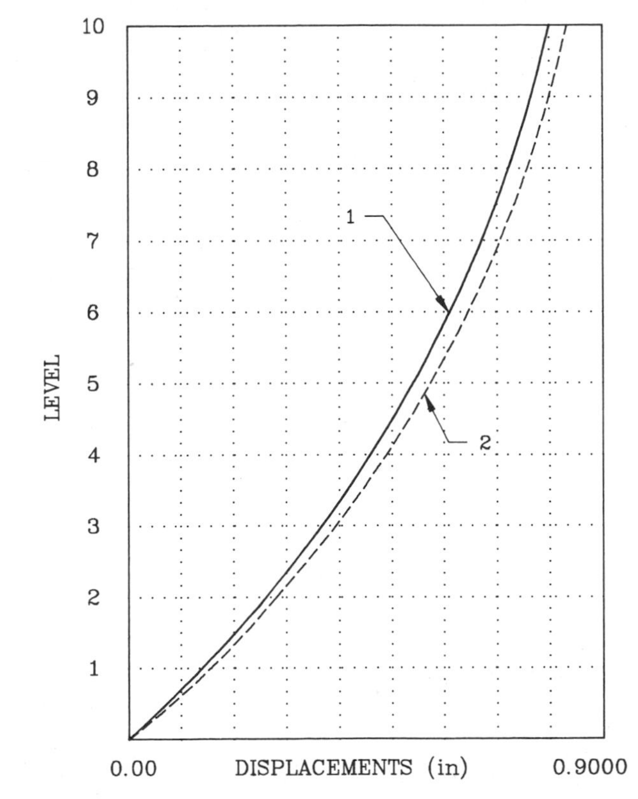

1 – EXACT (COLUMN DEFORMATION IGNORED)
2 – APPROXIMATE (BRACE ACTION ONLY)

Figure 6-39 Lateral displacements caused by brace deformations.

Table 6-10 Calculation of Brace-Action Story Drifts and Lateral Displacements for the 10-Story Braced Frame of Example 6-9

Level	h, in.	ΣV, kips	S_{br}, kips/in.	Δ_{br}, in.	Lat. disp., in.
10	144	30.22	1306	0.0231	0.8279
9	144	52.17	1306	0.0399	0.8048
8	144	71.74	1306	0.0549	0.7649
7	144	88.93	1306	0.0681	0.7100
6	144	103.76	1306	0.0794	0.6419
5	144	116.21	1306	0.0890	0.5625
4	144	126.30	1306	0.0967	0.4735
3	144	134.01	1306	0.1026	0.3768
2	144	139.34	1306	0.1067	0.2742
1	180	142.31	850	0.1675	0.1675

Table 6-11 Calculation of Chord-Action Story Drifts and Lateral Displacements for the Braced Frame of Example 6-9

Level	h, in.	ΣV, kips	M_{ov}, in.-kips	I_{oi}, in.^{4}	A, in.^{2}	$\bar{x}$, in.	Chord Disp., in.	Chord Drift, in.
10	144	30.22	4,352	406,080	2.66×10^5	48.00	2.958	0.452
9	144	52.17	11,864	406,080	9.92×10^5	60.88	2.506	0.443
8	144	71.74	22,194	576,000	14.7×10^5	64.72	2.063	0.426
7	144	88.93	35,001	576,000	24.6×10^5	66.63	1.637	0.397
6	144	103.76	49,942	763,200	27.6×10^5	67.78	1.240	0.360
5	144	116.21	66,677	763,200	37.9×10^5	68.56	0.880	0.312
4	144	126.30	84,864	921,600	40.8×10^5	69.12	0.568	0.256
3	144	134.01	104,161	921,600	50.9×10^5	69.55	0.312	0.190
2	144	139.34	124,226	1,344,960	42.2×10^5	69.89	0.122	0.122
1	180	142.31	149,841	1,344,960	63.2×10^5	87.20	0.000	0.000

Table 6-12 Calculation of Total First- and Second-Order Story Drifts and Lateral Displacements for the Braced Frame of Example 6-9

Level	h, in.	ΣV, kips	ΣP, kips	Δ_1, in.	μ	$\Delta_2 = \mu\Delta_1$, in.	2nd-Order Disp., in.
10	144	30.22	180	0.475	1.020	0.485	3.897
9	144	52.17	396	0.483	1.026	0.496	3.412
8	144	71.74	612	0.481	1.029	0.495	2.916
7	144	88.93	828	0.465	1.031	0.479	2.421
6	144	103.76	1044	0.439	1.032	0.453	1.942
5	144	116.21	1260	0.401	1.031	0.413	1.489
4	144	126.30	1476	0.353	1.029	0.363	1.076
3	144	134.01	1692	0.301	1.027	0.309	0.713
2	144	139.34	1908	0.229	1.022	0.234	0.404
1	180	142.31	2124	0.168	1.014	0.170	0.170

racy of the above simple procedure, an exact first-order elastic analysis was also performed, in which large column areas were used to eliminate axial deformation of the columns. Results of the exact and approximate analyses are compared in Figure 6-39, where good agreement can be observed.

The chord-action story drifts and lateral displacements are calculated in Table 6-11. The total drifts are magnified using the direct *P*-delta method in Table 6-12. The extent of contribution of each action to the lateral response of the frame is shown in Figure 6-40, where the dominance of chord action is evident. The results obtained by the above simple procedure are compared with those obtained by an exact second-order analysis in Figures 6-41 and 6-42.

6.5.3 Drift Design of Frame — Shear-Wall Systems

Estimates of the lateral displacements of frame–shear-wall systems may be obtained using the charts developed by Khan and Sbarounis.[6-51] Some of these charts, for the case of constant stiffness over the height, are reproduced in Figures 6-43 to 6-45. A sample application of the charts is presented in Example 6-10. In order to utilize the charts, the sum of stiffnesses of beams (S_b), columns (S_c), and shear walls (S_s) should be computed by adding the corresponding EI/L terms.

The charts provide the ratio of the lateral deflection of the frame–shear-wall system to the free deflection (at the top) of the shear wall alone. Note that the ratio S_s/S_c should be normalized by multiplying it by $(10/N)^2$, where N is the number of stories in the structure.

Another method for estimating drift and natural periods of frame–shear-wall systems has been developed by Stafford Smith et al.[6-52, 6-53] The method has been shown to provide accurate estimates of lateral displacements for a variety of structural systems. It can be easily adapted to programmable calculators. It is rather tedious, however, for hand calculations.

Example 6-10

Use the Khan–Sbarounis charts to estimate the east-west lateral displacement at the top of the 30-story frame–shear-wall building shown in Figure 6-46. Assume a uniform lateral pressure of 30 lb/ft^2. Story heights are 12.5 ft. Use gross-concrete-section properties and $E = 4000$ ksi.

Column stiffnesses:

Col. Type	b, in.	h, in.	I, ft^4	$\frac{I}{L}$, ft^3
C1	24	24	1.333	0.1067
C2	28	28	2.470	0.1976
C3	32	32	4.214	0.3371
C4	36	36	6.750	0.5400

$$\text{Total } I/L = 4(0.1067) + 6(0.1976) + 4(0.3371) + 2(0.5400) = 4.041\text{ft}^3$$

Beams:

B1: $$I = \frac{(14)(36)^3}{(12)^5} = 2.625 \text{ ft}^4$$

B2: $$I = \frac{(18)(24)^3}{(12)^5} = 1.000 \text{ ft}^4$$

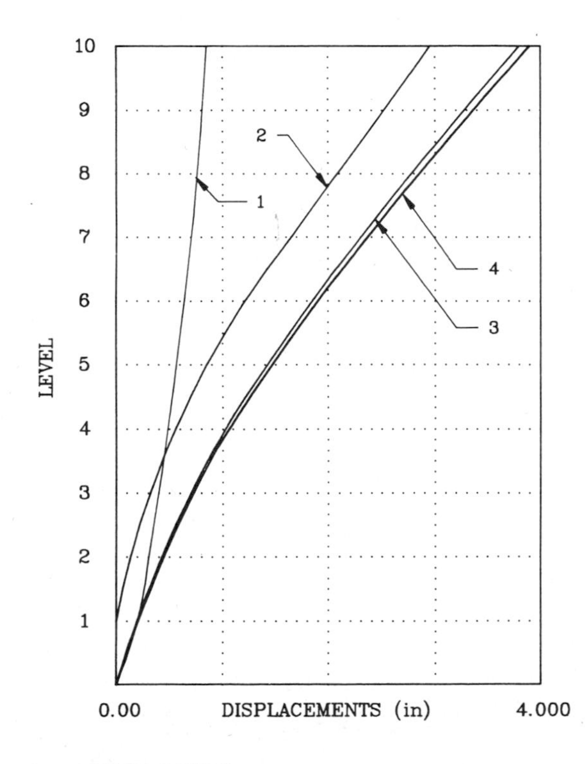

Figure 6-40 Contribution of various actions to the total lateral displacement of the braced frame of Example 6-9.

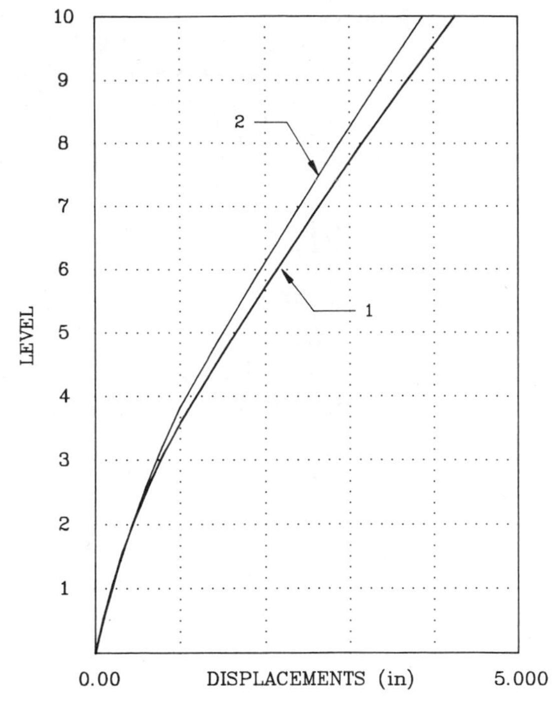

Figure 6-41 "Exact" versus approximate lateral displacements for the braced frame of Example 6-9.

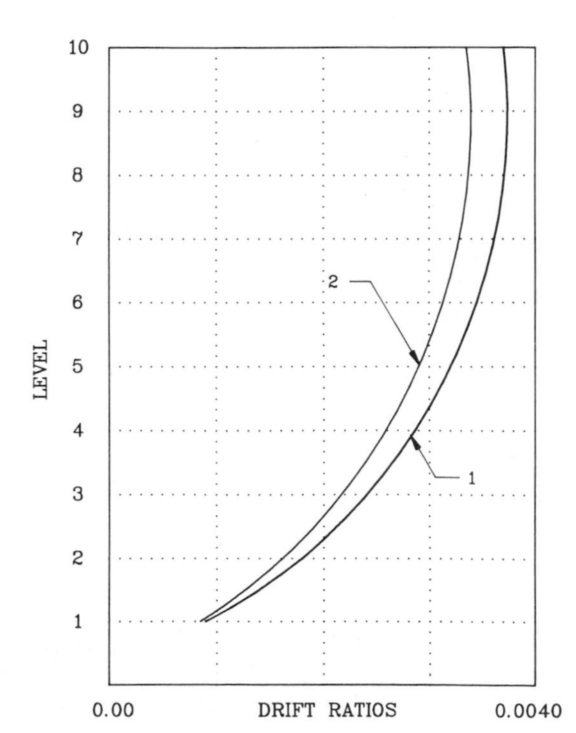

Figure 6-42 "Exact" versus approximate interstory drift ratios for the braced frame of Example 6-9.

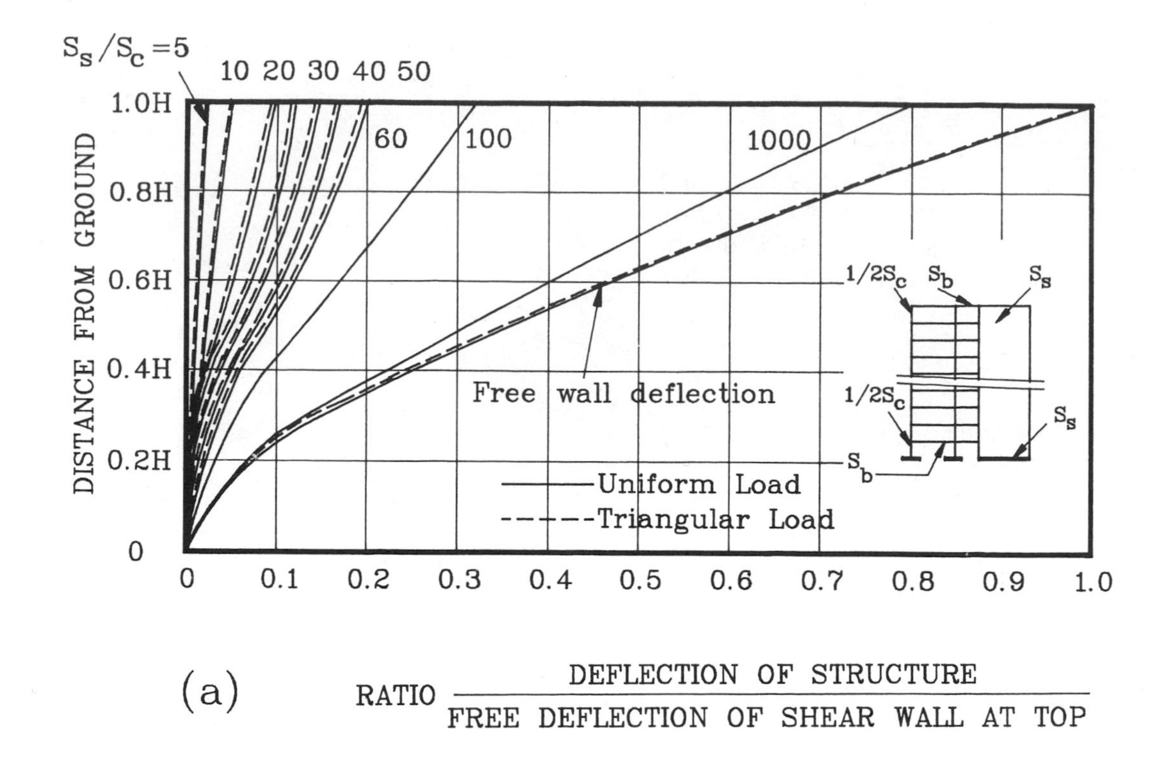

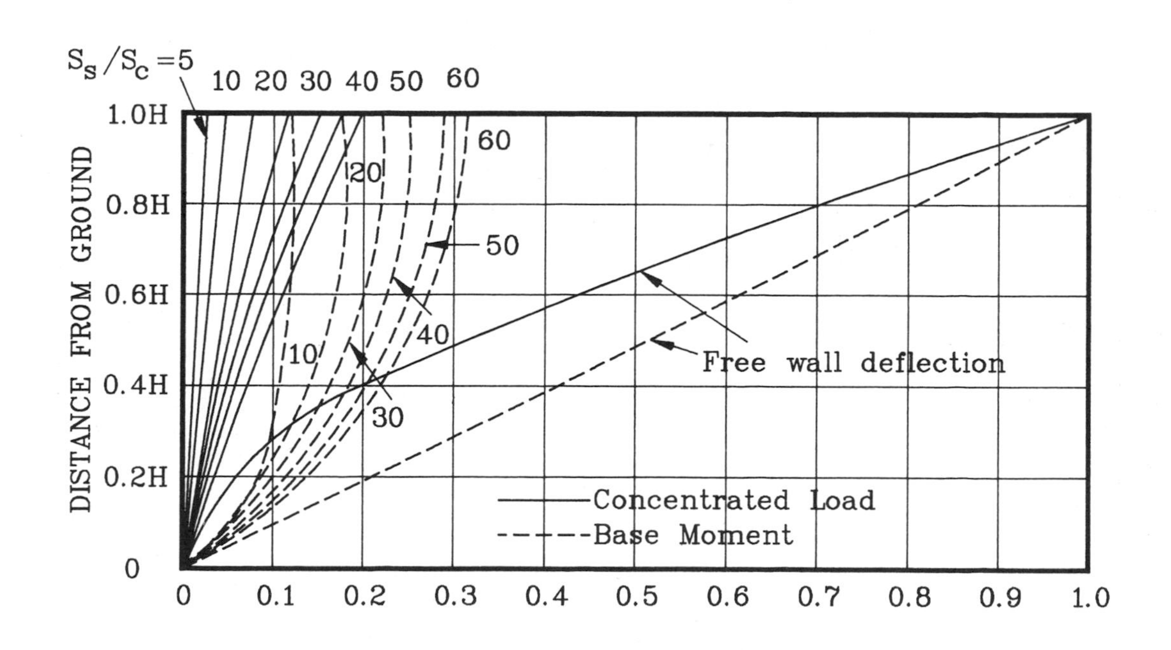

Figure 6-43 Design aid for drift design of frame-shear wall systems (6-51) (S_c / S_b = 1).

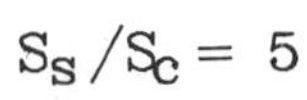

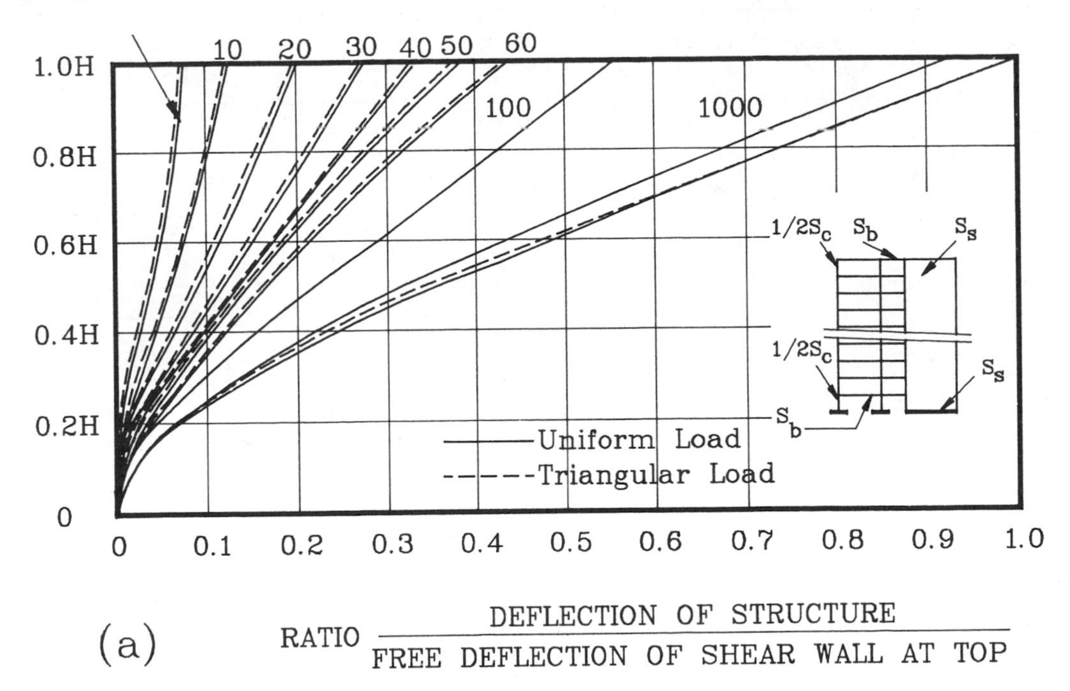

(a) RATIO $\frac{\text{DEFLECTION OF STRUCTURE}}{\text{FREE DEFLECTION OF SHEAR WALL AT TOP}}$

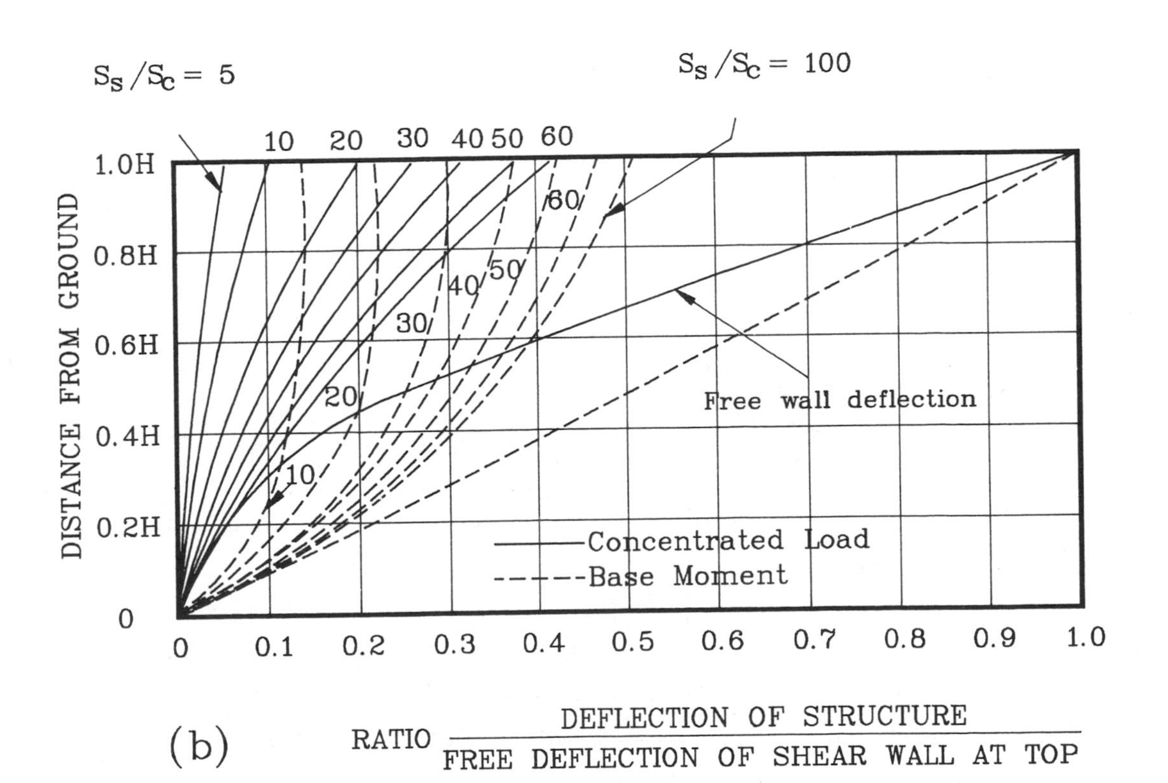

(b) RATIO $\frac{\text{DEFLECTION OF STRUCTURE}}{\text{FREE DEFLECTION OF SHEAR WALL AT TOP}}$

Figure 6-44 Design aid for drift design of frame-shear wall systems (6-51) (S_c / S_b = 5).

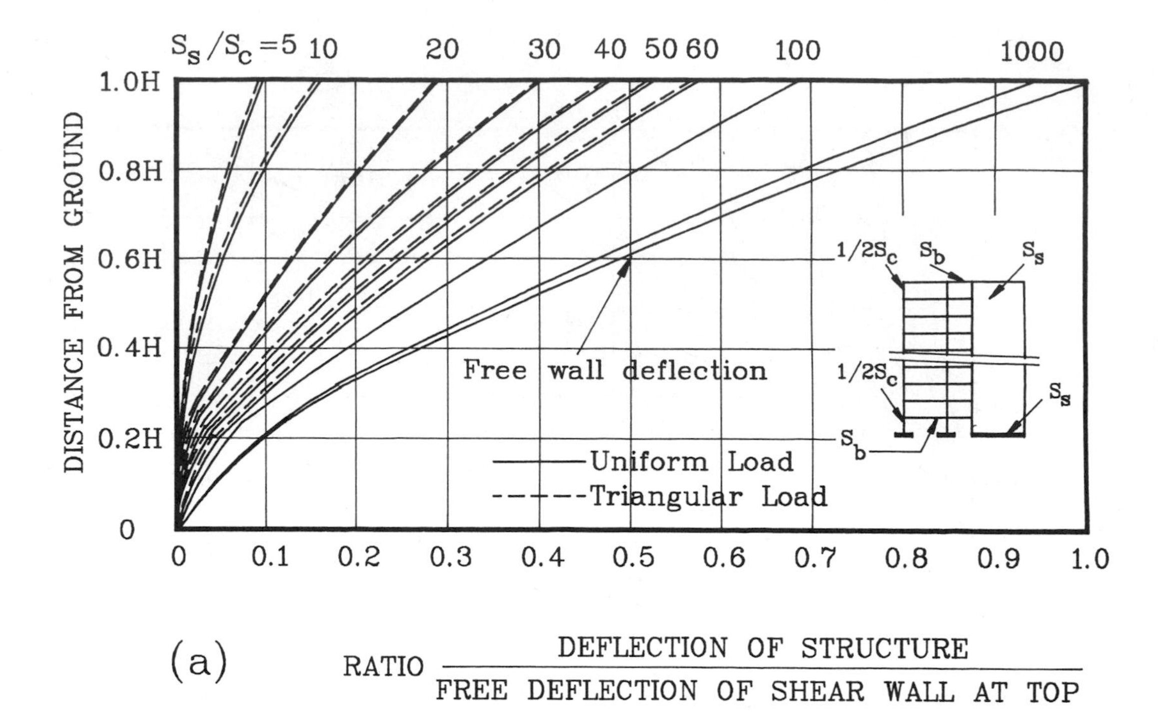

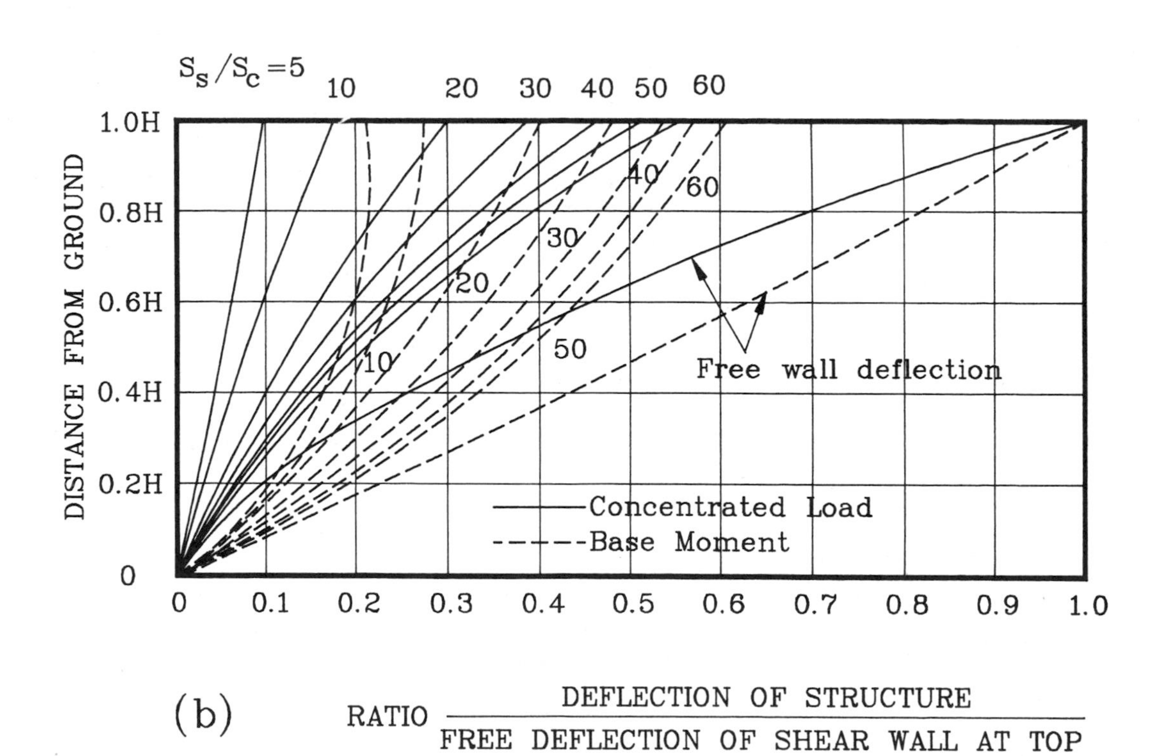

Figure 6-45 Design aid for drift design of frame-shear wall systems (6-51) (S_c / S_b = 10).

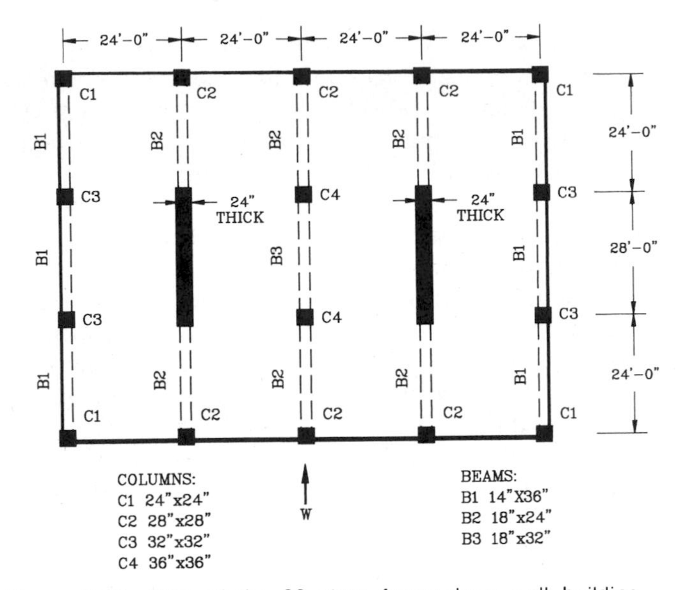

Figure 6-46 Plan of the 30 story frame-shear wall building (6-52).

B3:

$$I = \frac{(18)(32)^3}{(12)^5} = 2.370 \text{ ft}^4$$

$$\text{Total } I/L = \frac{(4)(2.625)}{24} + \frac{(2)(2.625)}{28} + \frac{(6)(1.00)}{28} + \frac{(1)(2.37)}{28} = 0.924 \text{ ft}^3$$

Walls:

$$I = \frac{(2)(28)^3}{12} = 3658.67 \text{ ft}^4$$

$$\text{Total } I/L = \frac{(2)(3658.67)}{12.5} = 585.39$$

$$\frac{S_s}{S_c} = \frac{585.39}{4.041}\left(\frac{10}{30}\right)^2 = 16.10$$

$$\frac{S_c}{S_b} = \frac{4.041}{0.924} = 4.37$$

Free deflection of the wall:

$$w = \frac{30(4)(24)}{1000} = 2.88 \text{ kips/ft}$$

$$\Delta = \frac{wl^4}{8EI} = \frac{(2.88)(375)^4}{(8)(576000)(3658.67)(2)} = 1.69 \text{ ft} = 20.28 \text{ in.}$$

Using the curve corresponding to $S_s/S_c = 20$ from chart (a) of Figure 6-44, we have $D_{top} = (0.20)(20.28) =$ 4.06 in., which compares very well with the computed exact displacement of 4.23 in. (see Figure 6-47).

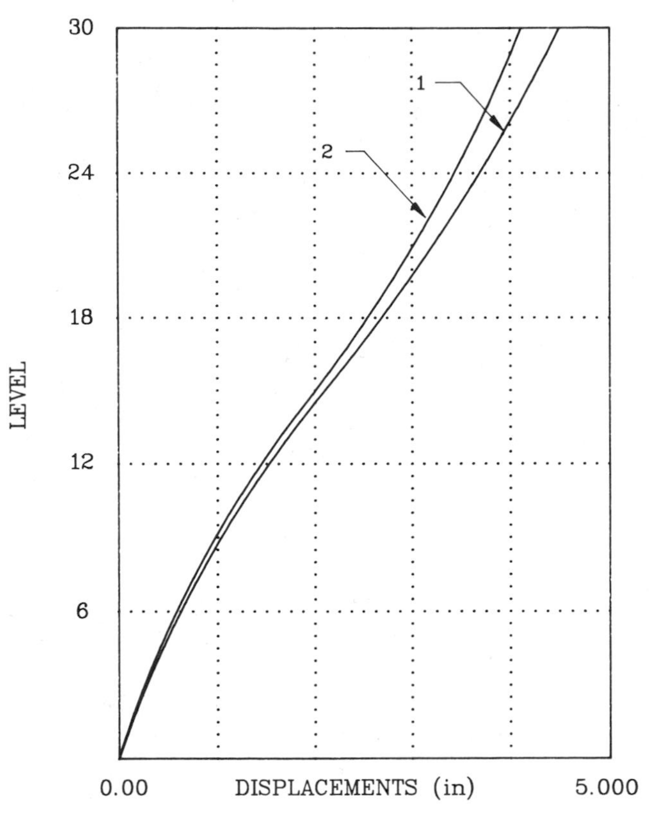

Figure 6-47 Lateral displacement of the 30 story frame-shear wall building.

6.5.4 Torsional Effects

One of the most important tasks in the selection and subsequent proportioning of a structural system is the minimization of torsional response. In general, this is a rather difficult task, and its success is strongly dependent on the intuition and experience of the designer.

For buildings in which the locations and relative stiffnesses of the lateral load-resisting subsystems (e.g. frames and walls) do not vary significantly along the height, the torsional displacements may be estimated as follows:

1. For buildings which are composed of only one type of lateral load-resisting system (moment frames, braced frames, or walls), the torsional rotation at the ith floor, θ_i, and the corresponding torsional drift of the jth frame at this floor, Δ_j, may be estimated as

$$\theta_i = \frac{(\Sigma V_i)e_i^2}{J} \tag{6-57}$$

$$\Delta_j = R_j\theta_i \tag{6-58}$$

where $(\Sigma V)_i$ is the story shear, e_i is the eccentricity of the "center of rigidity" from the center of mass, R_j is the closest distance from the jth frame to the center of rigidity, and J is the torsional story stiffness given by

$$J = \sum K_j R_j^2 \tag{6-59}$$

2. For combination systems (frame–shear-wall systems, moment-frame and braced-frame combinations), the process is more complex:

- The direct lateral displacements and story drifts of the structure are obtained via the Khan–Sbarounis charts or any other appropriate method.
- The total direct story shear carried by the frames subjected to the above displacements, V_{fi}, are calculated (see Section 6.5.1, "Bent Displacements").
- The shear V_{fi} is distributed among the various frames according to their relative stiffness in the direction of applied load.
- The rest of the story shear $(\Sigma V_i - V_{fi})$ is distributed among the various walls (braced frames) according to their relative stiffness in the direction of applied loads.
- The shear in each frame or wall, as calculated in the two preceding steps, is used as a measure of rigidity, and the center of rigidity of the entire system is located.
- The torsional rotation and the corresponding torsional drift of individual frames and walls are calculated using Equations 6-57 and 6-58.

It may be noticed that the concept of the "center of rigidity" is of significant use in the preliminary evaluation of the torsional response. However, the physical limitations of such a concept when applied to the seismic response of general, three-dimensional, multistory structures should be clearly understood. In a three-dimensional, multistory structure, if it exhibits significant plan and elevation irregularities, the lateral resistance is provided by a combination of strongly interdependent actions, both within a single story, and among various stories. In general, for such a complex system, centers of rigidity (points of application of forces for a torsion-free response) do not exist. Furthermore, if and when they exist, they must all lie on a single vertical line.[6-54]

6.6 SEISMIC-CODE REQUIREMENTS FOR DRIFT AND *P*-DELTA ANALYSIS

6.6.1 ANSI-82, BOCA-87, and UBC-85 Requirements

ANSI-82,[6-55] BOCA-87,[6-56] and UBC-85[6-57] address design for drift and lateral stiffness within the framework of working-stress design. Among the three, only BOCA-87 explicitly requires that the contribution of torsion should be considered in drift calculations. All three codes have provisions that quantify the drift limitations. These codes limit the maximum interstory drift index, under code seismic lateral forces, to 0.005 for buildings with a structural system factor K greater than one, and to $0.005K$ for buildings with K less than one. Structural system factors, as defined by these two codes, are shown in Table 4-1. The K factors are smaller for more ductile buildings, and initially it might seem as if the codes were prescribing stricter stiffness requirements for more ductile structures. However, as indicated by Berg,[6-58] the lateral-force requirements for buildings with smaller K factors are more relaxed, and the drift limits are essentially the same for all types of structures, independent of the K factor.

ANSI-82 and UBC-85 codes have escape clauses relieving the designer from strict adherence to the above limits (by permitting greater drifts "if it can be demonstrated that greater drifts can be tolerated"). They also lack specific requirements for consideration of *P*-delta displacements in the evaluation of interstory drifts.

6.6.2 ATC 3-06 and NEHRP-85 Requirements

The Applied Technology Council's ATC 3-06[6-59] document, published in 1978, and its adaptation by the Federal Emergency Management Agency, referred to as NEHRP-85,[6-60] address design for drift and lateral stiffness at the ultimate limit state of building behavior. This more rational approach is expected to be followed by other model codes in the future.

According to these provisions, the design story drift Δ is computed as the difference of the deflections δ_x at the top and bottom of the story under consideration. The deflection is evaluated in accordance with the following formula:

$$\delta_x = C_d \, \delta_{xe} \tag{6-60}$$

where

C_d = the deflection amplification factor as given in Table 4-9

δ_{xe} = the deflection determined by an elastic analysis for the seismic forces prescribed by these codes considering the building fixed at the base

The maximum interstory drift index is 0.015 for buildings in Seismic Hazard Exposure Groups I and II, and 0.010 for buildings in Seismic Hazard Exposure Group III. Seismic Hazard Exposure Groups were defined in Chapter 4. ATC 3-06 permits a one-third increase in drift limitation for buildings in Seismic Hazard Exposure Group I with three stories or less, if there are no brittle-type finishes. NEHRP-85 has adopted this one-third increase rule. NEHRP-85 provisions permit exceeding specified drift limits for one-story buildings in Seismic Hazard Group I if there are no brittle finishes and if the seismic-resisting system is not attached to equipment or processes. Furthermore, for one-story steel buildings in Seismic Hazard Exposure Group II, NEHRP-85 permits exceeding the drift limitation if there are no brittle finishes and if it can be demonstrated that greater drifts can be tolerated by all components, including nonstructural ones.

To determine if a P-delta analysis is required, a stability coefficient is used. This is in fact the same as the stability index introduced in Section 6.4.3, "Direct P-Delta Method." P-delta effects need not be considered when the stability coefficient θ is less than 0.10, where one uses

$$\theta = \frac{P_x \Delta}{V_x h_{sx} C_d} \tag{6-61}$$

where

Δ = the design-story drift

V_x = the seismic shear force acting between level x and $x - 1$

h_{sx} = the story height below level x

P_x = the total unfactored vertical design load at and above level x

When θ is greater than 0.10, the incremental factor related to p-delta effects is to be determined by rational analysis. The codes do not explicitly specify what is a rational analysis. However, in the commentary to the codes, the direct P-delta method, which was explained in Section 6.4.3, is described as an option. In accordance with this method, the commentary to ATC 3-06 recommends multiplication of the design story drifts by a factor of $1/(1 - \theta) > 1$. NEHRP-85 requires multiplication of design story drifts by the factor $0.9/(1 - \theta) > 1$.

6.6.3 SEAOC-1986 (UBC-1988) Requirements

Historically, recommendations of the Seismological Committee of the Structural Engineers Association of California (SEAOC) have served as the backbone of the seismic-design requirements of the Uniform Building Code (UBC).

In 1985, SEAOC published *Tentative Lateral Force Requirements*, which was reviewed and modified several times before it was adopted by the International Conference of Building Officials for inclusion in the 1988 edition of the UBC Code.[6-61]

The new UBC provisions exhibit a transition from the previous editions to the ideas embodied in ATC 3-06 and NEHRP documents. While many of the provisions of the latter documents have been adopted in the new UBC requirements, they have been converted to fit in the context of working-stress design.

According to these requirements, the calculated story-drift index, which should include both translational and torsional deflections, should not exceed either $0.03/R_w$ or 0.004, where R_w is the structural system factor defined in Table 4-5. These drift limits may be exceeded, however, when it is demonstrated that greater drift can be tolerated by both structural elements and nonstructural elements that could affect life safety. The design lateral forces used to determine the calculated drift may be derived from the value of C based on the period determined from Equation 4-20, neglecting the lower-bound ratio for C/R_w of 0.075 and the 80% limitation as discussed in Section 4.5. Generally, the P-delta effects should be considered in the evaluation of overall structural frame stability. However, they may be neglected if story-drift indices do not exceed $0.02/R_w$.

REFERENCES

6-1 Council on Tall Buildings, Committee 16, "Stability," Chapter SB-4, Vol. SB of *Monograph on Planning and Design of Tall Buildings*, ASCE, New York, 1979.

6-2 Scholl, R. E. (ed.), "Effects Prediction Guidelines for Structures Subjected to Ground Motion," Report No. JAB-99-115, URS/Blume Engineers, San Francisco, 1975.

6-3 Scholl, R. E., "Brace Dampers: An Alternative Structural System for Improving the Earthquake Performance of Buildings," *Proceedings of the 8th World Conference on Earthquake Engineering*, San Francisco, Vol. 5, Prentice-Hall, 1984.

6-4 Ohta, Y., and Omote, S., "An Investigation into Human Psychology and Behavior During an Earthquake," *Proceedings of the 6th World Conference on Earthquake Engineering*, India, 1977.

6-5 Mileti, D. S., and Nigg, J. M., "Earthquakes and Human Behavior," *Earthquake Spectra EERI*, 1, No. 1, Feb. 1984.

6-6 Durkin, M. E., "Behavior of Building Occupants in Earthquakes," *Earthquake Spectra EERI* 1, No. 2, Feb. 1985.

6-7 Chen, W. F., and Lui, E. M., *Structural Stability*, Elsevier Science Publishing Company, New York, 1987.

6-8 Wakabayashi, M., *Design of Earthquake Resistant Buildings*, McGraw-Hill, New York, 1986.

6-9 Rosenblueth, E., "Slenderness Effects in Buildings," Proc. Paper 4235, *J. Structural Div. ASCE* 91, No. ST1, 229–252, Jan. 1967.

6-10 Stevens, L. K., "Elastic Stability of Practical Multi-story Frames," *Proc. Inst. Civil Engineers* 36, 99–117, 1967.

6-11 Goldberg, J. E., "Approximate Methods for Stability and Frequency Analysis of Tall Buildings," *Proceedings, Regional Conference on Tall Buildings*, Madrid, Spain, Sept. 1973, pp. 123–146.

6-12 Council on Tall Buildings, Committee 23, "Stability," Chapter CB-8, Vol. CB, *Monograph on Planning and Design of Tall Buildings*, ASCE, New York, 1978.

6-13 Singh, J. P., "Earthquake Ground Motions: Implications for Designing Structures and Reconciling Structural Damage," *Earthquake Spectra EERI* 1, No. 2, Feb. 1985.

6-14 Anderson, J. C. and Naeim, F., "Design Criteria and Ground Motion Effects on the Seismic Response of Multistory Buildings," *Critical Aspects of Earthquake Ground Motion and*

Building Damage Potential, ATC-10-1, Applied Technology Council, 1984.

6-15 Anderson, J. C. and Bertero, V. V., "Uncertainties in Establishing Design Earthquakes," *Proceedings of a Two-Day Course from EERI on Strong Ground Motion*, San Francisco and Los Angeles, Apr. 1987.

6-16 MacGregor, J. G. and Hage, S. E., "Stability Analysis and Design of Concrete Frames," *J. Structural Div. ASCE* 103, No. ST10, 1953–1970, Oct. 1977.

6-17 Wynhoven, J. H. and Adams, P. F., "Behavior of Structures Under Loads Causing Torsion," *J. Structural Div. ASCE* 98, No. ST7, 1361–1376, July 1972.

6-18 Salmon, C. G. and Johnson, J. E., *Steel Structures, Design and Behavior*, 2nd Edition, Harper and Row, 1980, pp. 843–851.

6-19 Johnston, B. G. (ed.), *Structural Stability Research Council, Guide to Stability Design Criteria for Metal Structures*, 3rd Edition, John Wiley and Sons, New York, 1976.

6-20 Cheong-Siat-Moy, F., "Frame Design without Using Effective Column Length," *J. Structural Div. ASCE* 104, No. ST1, 23–33, Jan. 1978.

6-21 Cheong-Siat-Moy, F., "*K*-Factor Paradox," *J. Structural Eng. ASCE* 112, No. 8, 1747–1760, Aug. 1986.

6-22 Nair, R. S., "A Simple Method of Overall Stability Analysis for Multistory Buildings," *Developments in Tall Buildings-1983*, Council on Tall Buildings and Urban Habitat, Lynn S. Beedle (editor-in chief), Van Nostrand Reinhold, New York, 1983.

6-23 Nair, R. S., "Overall Elastic Stability of Multistory Buildings," *J. Structural Div. ASCE*, 101, No. ST12, 2487–2503, Dec. 1975.

6-24 Springfield, J. and Adams, P. F., "Aspects of Column Design in Tall Steel Buildings," *J. Structural Div. ASCE* 98, No. ST5, 1069–1083, May 1972.

6-25 Wood, B. R., Beaulieu, D., and Adams, P. F., "Column Design by *P*-Delta Method," *J. Structural Div. ASCE* 102, No. ST2, 411–427, Feb. 1976.

6-26 Lai, S. A. and MacGregor, J. G., "Geometric Nonlinearities in Unbraced Multistory Frames," *J. Structural Eng. ASCE* 109, No. 11, 2528–2545, Nov. 1983.

6-27 Nixon, D., Beaulieu, D., and Adams, P. F., "Simplified Second-Order Frame Analysis," *Can. J. Civil Eng.* 2, No. 4, 602–605, Dec. 1975.

6-28 Renton, J. D., "Stability of Space Frames by Computer Analysis," *J. Structural Div. ASCE* 88, No. ST4, 81–103, Aug. 1962.

6-29 Chu, K. H., and Rampetsreiter, R. H., "Large Deflection Buckling of Space Frames," *J. Structural Div. ASCE* 98, No. ST12, 2701–2722, Dec. 1972.

6-30 Connor, J. J., Jr., Logcher, R. D., and Chen, S. C., "Nonlinear Analysis of Elastic Frame Structures," *J. Structural Div. ASCE* 94, No. ST6, 1525–1547, June 1968.

6-31 Zarghammee, M. S. and Shah, J. M., "Stability of Space Frames," *J. Eng. Mech. Div. ASCE* 94, No. EM2, 371–384, Apr. 1968.

6-32 Krajcinovic, D., "A Consistent Discrete Element Technique for Thin-Walled Assemblages," *Int. J. Solids and Structures* 5, 639–662, 1969.

6-33 Barsoum, R. S. and Gallagher, R. H., "Finite Element Analysis of Torsional Flexural Stability Problems," *Int. J. Numer. Methods Eng.* 2, 335–352, 1970.

6-34 Bazant, Z. P. and El Nimeiri, M., "Large-Deflection Buckling of Thin Walled Beams and Frames," *J. Eng. Mech. Div. ASCE* 79, No. EM6, 1259–1281, Dec. 1973.

6-35 Yoo, C. H., "Bimoment Contribution to Stability of Thin-Walled Assemblages," *Computers and Structures* 11, 465–471, 1980.

6-36 Yang, Y. and McGuire, W., "Stiffness Matrix for Geometric Nonlinear Analysis," *J. Structural Eng. ASCE* 112, No. 4, 853–877, Apr. 1986.

6-37 Naeim, F., "An Automated Design Study of the Economics of Earthquake Resistant Structures," Ph.D. Dissertation, Department of Civil Engineering, University of Southern California, Aug. 1982.

6-38 Neuss, C. F., Maison, B. F., and Bouwkamp, J. G., *A Study of Computer Modeling Formulation and Special Analytical Procedures for Earthquake Response of Multistory Buildings, A Report to National Science Foundation*, J. G. Bouwkamp, Inc., Berkeley, CA, Jan. 1983, pp. 335–362.

6-39 Wilson, E. L. and Habibullah, A., "Static and Dynamic Analysis of Multi-Story Buildings Including the *P*-Delta Effects," *Earthquake Spectra EERI* 3, No. 2, May 1987.

6-40 Freeman, S. A., Czarncki, R. M., and Honda, K. K., "Significance of Stiffness Assumptions on Lateral Force Criteria," *Reinforced Concrete Structures Subjected to Wind and Earthquake Forces*, Publication SP-63, American Concrete Institute, Detroit, MI, 1980.

6-41 Wong, C. H., El Nimeiri, M. M., and Tang, J. W., "Preliminary Analysis and Member Sizing of Tall Tubular Steel Buildings," *AISC Eng. J.* (American Institute of Steel Construction), Second Quarter, 1981, pp. 33–47.

6-42 Council on Tall Buildings, Committee 14, "Elastic Analysis and Design," Chapter SB-2, Vol. SB of *Monograph on Planning and Design of Tall Buildings*, ASCE, New York, 1979.

6-43 Cheong-Siat-Moy, F., "Multistory Frame Design Using Story Stiffness Concept," *J. Structural Div. ASCE* 102, No. ST6, 1197–1212, June 1976.

6-44 Cheong-Siat-Moy, F., "Consideration of Secondary Effects in Frame Design," *J. Structural Div. ASCE* 103, No. ST10, 2005–2019, Oct. 1977.

6-45 Krawinkler, H., Bertero, V. V., and Popov, E. P., "Inelastic Behavior of Steel Beam-to-Column Subassemblages," Earthquake Engineering Research Center, University of California, Berkeley, Report No. EERC 71-7, Oct. 1971.

6-46 Becker, E. R., "Panel Zone Effect on the Strength and Stiffness of Rigid Steel Frames," Structural Mechanics Laboratory Report, University of Southern California, June 1971.

6-47 Richards, R. M. and Pettijohn, D. R., "Analytical Study of Panel Zone Behavior in Beam-Column Connections," University of Arizona, Nov. 1981.

6-48 Slutter, R. G., "Tests of Panel Zone Behavior in Beam–Column Connections," Fritz Engineering Laboratory, Lehigh University, Report No. 200-81-403-1, 1981.

6-49 Teal, E. J., *Practical Design of Eccentric Braced Frames to Resist Seismic Forces*, Structural Steel Education Council, Los Angeles, CA, 1987.

6-50 White, R. N. and Salmon, C. G. (eds.), *Building Structural Design Handbook*, John Wiley and Sons, New York, 1987.

6-51 Khan, F. R. and Sbarounis, J. A., "Interaction of Shear Walls and Frames," *J. Structural Div. ASCE* 90, No. ST3, 285–335, June 1964.

6-52 Stafford Smith, B., Kuster, M., and HoenderKamp, J. C. D., "Generalized Method for Estimating Drift in High-Rise Structures," *J. Structural Eng., ASCE* 110, No. 7, 1549–1562, July 1984.

6-53 Stafford Smith, B. and Crowe, E., "Estimating Periods of Vibration of Tall Buildings," *J. Structural Eng., ASCE* 112, No. 5, 1005–1018, May 1986.

6-54 Riddell, R., and Vasques, J., "Existence of Centers of Resistance and Torsional Uncoupling of Earthquake Response of Buildings," *Proceedings of the 8th World Conference on Earthquake Engineering*, San Francisco, Vol. 5, Prentice-Hall, 1984.

6-55 Buildings Officials and Code Administrators International, *The BOCA Basic Building Code*, Homewood, IL, 1987.

6-56 American National Standards Institute, *American National Standards Building Code Requirements for Minimum Design Loads in Buildings and Other Structures*, ANSI A58.1-1982, New York, 1982.

6-57 International Conference on Building Officials, *Uniform Building Code—1985*, Whittier, CA, 1985.

6-58 Berg, G. V., *Seismic Design Codes and Procedures*, EERI Monograph, Earthquake Engineering Research Institute, 1983.

6-59 Applied Technology Council, "Tentative Provisions for the Development of Seismic Regulations for Buildings," Publication ATC-3-06, 1978.

6-60 Federal Emergency Management Agency, *1985 Edition of NEHRP Recommended Provisions for the Development of Seismic Regulations for New Buildings,* FEMA-96, Feb. 1986.

6-61 Seismology Committee of Structural Engineers Association of California, "Tentative Lateral Force Requirements," Nov.–Dec. 1986.

Chapter 7

Seismic Design of Floor Diaphragms

*Farzad Naeim, Ph.D., P.E.** and *Rao Boppana, Ph.D., S.E.*†

7.1 INTRODUCTION

The primary function of floor and roof framing systems is to support gravity loads and to transfer these loads to the vertical structural members such as columns and walls. Furthermore, they play a central role in the distribution of wind and seismic forces to the vertical elements of the lateral load-resisting system (such as frames and structural walls). The behavior of the floor and roof systems under the influence of gravity loads is well established, and guidelines for structural design have been adopted.[7-1, 7-2]

In the earthquake-resistant design of building structures, the building is designed and detailed to act as a single unit under the action of seismic forces. Design of a building as a single unit helps to increase the redundancy and the integrity of the building. The horizontal forces generated by earthquake excitations are transferred to the ground by the vertical systems of the building, which are designed for lateral load resistance (e.g. frames, bracings, and walls). These vertical systems are generally tied together as a unit by means of the building floors and roof. In this sense, the floor and roof structural systems, used primarily to create enclosures and resist gravity (or out of plane) loads, are also designed as horizontal diaphragms to resist and to transfer horizontal (or in-plane) loads to the appropriate vertical elements.

The analysis and design of a floor or roof deck under the influence of horizontal loads is performed assuming that the floor or roof deck behaves as a horizontal continuous beam supported by the vertical lateral-load-resisting elements (hereafter referred to as VLLR elements). The floor deck is assumed to act as the web of the continuous beam, and the members at the floor periphery are assumed to act as the flanges of the continuous beam (see Figure 7-1).

Accurate determination of the in-plane shears and bending moments acting on a floor diaphragm, and the corresponding horizontal force distribution among various VLLR elements, requires a three-dimensional analysis that takes account of the relative rigidity of the various elements. In general, however, simplifying assumptions are made on the horizontal-diaphragm rigidity, and a fairly simple analysis is performed to determine distribution of lateral forces. Obviously, the accuracy of the results obtained depends on the validity of the assumptions made. In addition, the behavior of certain floor systems, such as plywood, metal deck, and precast concrete diaphragms, is difficult to model analytically due to their various attachments. In some cases

*Director of Research and Development, John A. Martin and Associates, Los Angeles, California.
† President, Boppana and Associates, Los Angeles, California.

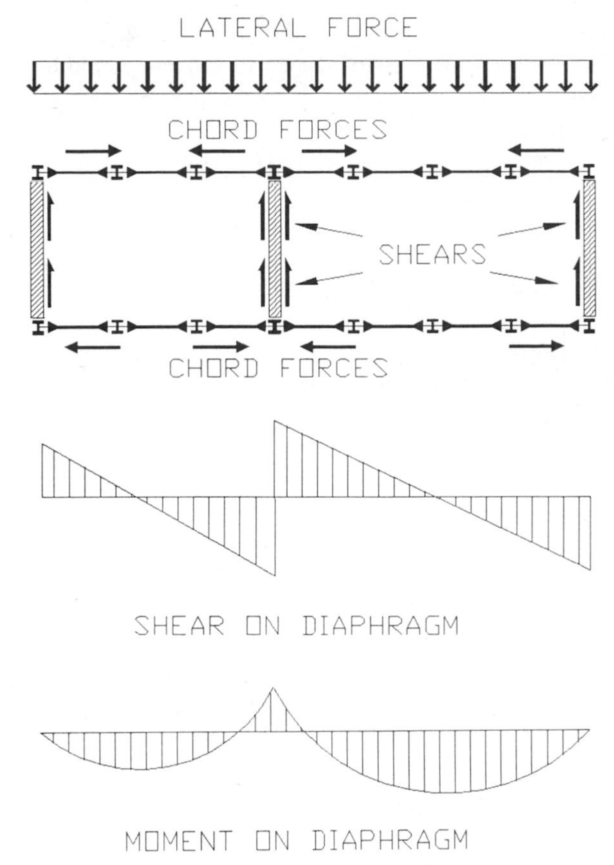

Figure 7-1 Design forces on a diaphragm.

testing may be required to establish the strength and stiffness properties of such systems.

While for the great majority of structures, simplified analysis procedures result in a safe design, studies indicate that neglecting the true behavior of floor diaphragms can sometimes lead to serious errors in accessing the required lateral load-resistance capacities of the VLLR elements.[7-3, 7-4, 7-5]

This chapter addresses the major issues of seismic behavior and design of diaphragms. It starts by classification of diaphragm behavior in Section 7.2, and a discussion on the determination of diaphragm rigidity in Section 7.3. Potential diaphragm problems are explained in Section 7.4, where examples are provided to clarify the subject. Provisions of major United States building codes for seismic design of diaphragms are summarized in Section 7.5. Finally, in Section 7.6, the current standard procedures for design of diaphragms are presented via their application in a number of realistic design examples.

7.2 CLASSIFICATION OF DIAPHRAGM BEHAVIOR

The distribution of horizontal forces by the horizontal diaphragm to the various VLLR elements depends on the relative rigidity of the horizontal diaphragm and the VLLR elements. For analysis purposes diaphragms are classified as *rigid*, *flexible*, and *semirigid* based on this relative rigidity.

A diaphragm is classified as rigid if it can distribute the horizontal forces to the VLLR elements in direct proportion to their relative stiffness. In the case of rigid diaphragms, the diaphragm deflection will be insignificant compared to that of the VLLR elements. A diaphragm is called flexible if the distribution of horizontal forces to the VLLR elements is independent of their relative stiffness. In the case of a flexible diaphragm, the diaphragm deflection will be significantly large compared to that of the VLLR elements. A flexible diaphragm distributes lateral loads to the VLLR elements as a series of simple beams spanning between these elements.

In reality no diaphragm is perfectly rigid or perfectly flexible. Reasonable assumptions, however, can be made as to a diaphragm's rigidity or flexibility in order to simplify the analysis. In some cases the diaphragm deflection and the deflection of the VLLR elements are of the same order of magnitude, hence the diaphragm cannot reasonably be assumed as either rigid or flexible. Such a diaphragm is classified as semirigid.

Exact analysis of structural systems containing semirigid diaphragms is complex, since any such analysis should take account of the relative rigidity of all structural elements including the diaphragm. The horizontal load distribution of a semirigid diaphragm may be best approximated as that of a continuous beam supported on elastic supports. In most cases consisting of semirigid diaphragms, assumptions can be made to bound the exact solution without resorting to a complex analysis.

The absolute size and stiffness of a diaphragm, while important, are not the final determining factors whether or not a diaphragm will behave as rigid, flexible, or semirigid.[7-3] Consider the one-story concrete shear wall building shown in Figure 7-2a. Keeping the width and the thickness of walls and slabs constant, it is possible to simulate rigid (Figure 7-2b), flexible (Figure 7-2c), and semirigid (Figure 7-2d) diaphragms as the wall heights and diaphragm spans are varied. The wall stiffness decreases with an increase in the floor height H. Similarly, the diaphragm stiffness decreases with an increase in the span L.

The dashed line in Figure 7-2b indicates the deflection of the system under the influence of horizontal forces when the diaphragm is rigid. This can be accomplished by increasing H and decreasing L so that the stiffness of the diaphragm relative to the wall is significantly larger. In such a situation, the deflection of the diaphragm under horizontal loads is insignificant when

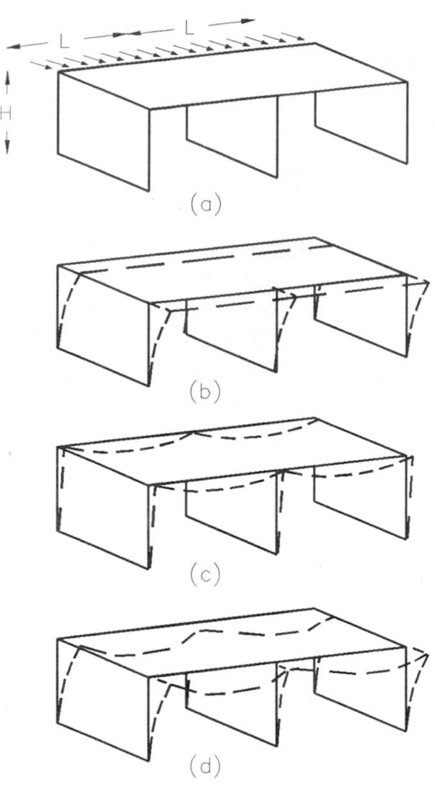

Figure 7-2 Diaphragm behavior. (a) Loading and building proportions. (b) Rigid diaphragm behavior. (c) Flexible diaphragm behavior. (d) Semirigid diaphragm behavior.

compared to the deflections of the walls. The diaphragm will move as a rigid body and will force the walls to move together accordingly. The force distribution among the walls will depend only on the relative stiffnesses of the walls. In Figure 7-2b it is assumed that the applied load and the wall stiffness are symmetric about the center wall. If this is not the case, in addition to the rigid-body translation, the diaphragm will experience rigid-body rotation.

Figure 7-c shows the deflection of the system under the influence of horizontal forces when the diaphragm is flexible. This can be accomplished by decreasing H and increasing L so that the stiffness of the diaphragm compared to the walls is small. In such a situation, the diaphragm segments between the walls act as a series of simply supported beams and the load distribution to the walls can be determined from the tributary area of the diaphragm to the wall. Obviously a flexible diaphragm cannot experience the rigid-body rotation that a rigid diaphragm can undergo.

The dashed line in Figure 7-2d indicates the deflection pattern of a semirigid diaphragm under the influence of lateral forces. Here the stiffnesses of the walls and the diaphragm are of the same order. Both wall deflections and diaphragm deflections do contribute to the total system deflection. Determination of the exact load distribution among the walls requires a three-dimensional analysis of the entire system (including the diaphragm).

7.3 DETERMINATION OF DIAPHRAGM RIGIDITY

In order to estimate the diaphragm rigidity, it is necessary to predict the deflection of the diaphragm under the influence of lateral loads. The various floor and roof systems that have evolved primarily for the purpose of supporting gravity loads do not lend themselves easily to analytical calculation of lateral deflections. Some of the more common floor systems in use today are: (i) cast-in-place concrete; (ii) precast planks or tees with or without concrete topping; (iii) metal deck with or without concrete fill, and (iv) wood framing with plywood sheathing.

With the single exception of a cast-in-place concrete floor system, which is a monolithic construction, all the floor systems mentioned above consist of discrete units that are connected. In precast-concrete construction, adjacent units are generally connected together by field welding embedded plates or rebars. This will help the units to deflect vertically without separation while providing some diaphragm action. The strength and rigidity of such a diaphragm depends to a great extent on the type and spacing of connections. Analytical computation of the deflections and stiffness of such a diaphragm is complex. In floor systems consisting of metal decks, each unit of deck is welded to the supports below at regular intervals. Adjacent units of the deck are connected together by means of button punching or seam welding. Here again, the diaphragm stiffness is directly related to the spacing and type of connections. In the wood construction, the plywood sheathing is nailed directly to the framing members. Again, strength and stiffness depend on the spacing of the nails and whether or not the separate plywood panels are connected by blocking.

It is general practice to consider diaphragms made of cast-in-place concrete, precast concrete with concrete topping, and metal deck with concrete fill as rigid, but diaphragms made of precast planks without concrete topping, metal deck without concrete fill, and plywood sheathing as flexible. This classification is valid for most cases. Gross errors in force distribution, however, can be made if the above assumption is used without paying attention to the relative rigidity of the VLLR elements and the diaphragm.[7-3, 7-4, 7-5]

To aid engineers in evaluating relative rigidities, metal-deck manufacturers have established test programs to provide strength and deflection characteristics of various metal decks and various connection patterns.[7-6, 7-7] Similarly, the Uniform Building Code provides an empirical formula to compute deflections of plywood diaphragms and tables to establish the strength of such diaphragms.

7.4 SIGNIFICANT FACTORS AFFECTING DIAPHRAGM BEHAVIOR

Identifying every situation where special attention should be given to the design and detailing of floor diaphragms requires substantial experience and a good amount of engineering judgment. Certain cases, however, more often than not require special attention, and in this section we try to provide some guidelines for identifying such cases.

In general, low-rise buildings and buildings with very stiff vertical elements such as shear walls are more susceptible to floor-diaphragm flexibility problems than taller structures.

In buildings with long and narrow plans, if seismic resistance is provided either by the end walls alone, or if the shear walls are spaced far away from each other, floor diaphragms may exhibit the so-called bow action (see Figure 7-3). The bow action subjects the end walls to torsional deformation and stresses. If sufficient bond is not provided between the walls and the diaphragm, the two may be separated from each other, starting at the wall ends. This separation results in a dramatic increase in the wall torsion and can lead to collapse.

The Arvin High School Administrative Building in California, which suffered extensive damage during the Kern County earthquake of July 21, 1952, is a good example in this regard. Schematic plans and elevations of this building are shown in Figure 7-4. An analytical study of building by Jain[7-8] indicated that the lowest two natural frequencies of the building were close to the fundamental frequencies of the floor and roof diaphragms modeled as simply supported beams. When an analytical model of the building was subjected to a 0.20g constant spectral acceleration, with four translational modes considered, the two diaphragm modes represented 74% of the sum of the modal base shears.

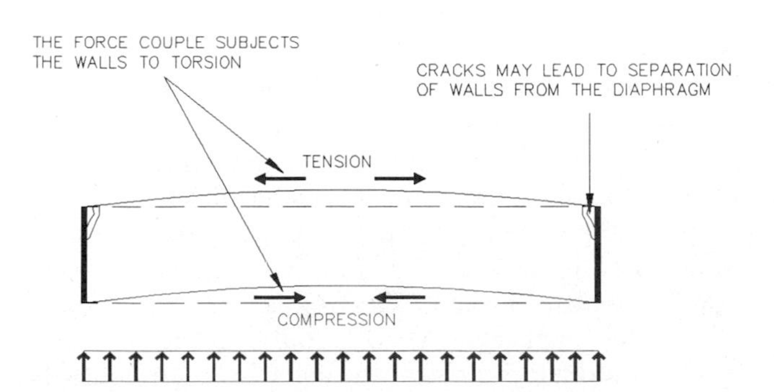

Figure 7-3 A plan showing how the so-called bow action subjects the end walls to torsion.

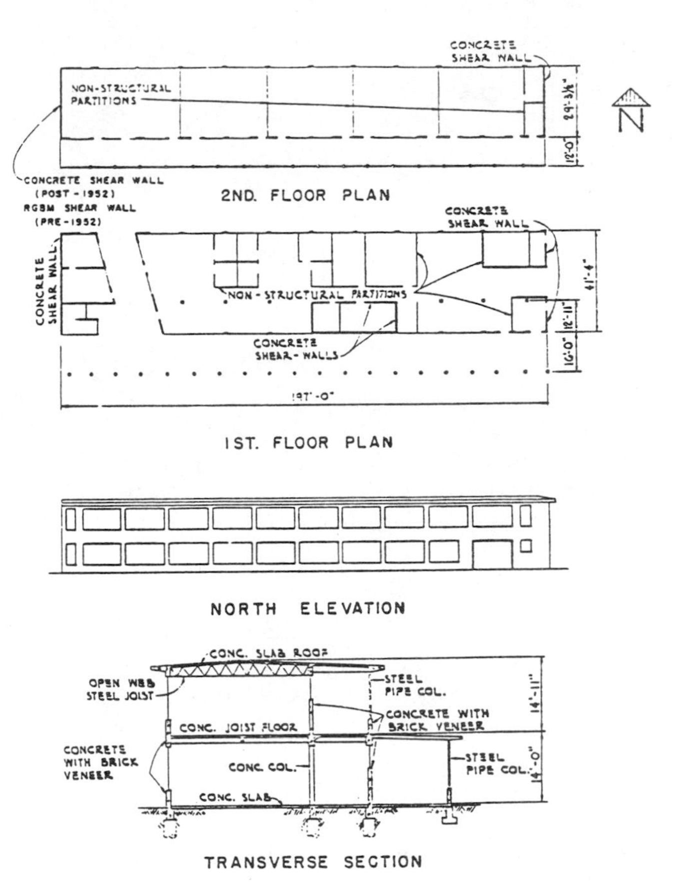

Figure 7-4 Plan and elevation of the Arvin High School Administrative Building.[7-8]

As documented by Steinburgge et al.,[7-9] diaphragm deflections caused a separation between the roof diaphragm and the wall ends at the second-story wall located at the west end of the building. This action subjected the wall to significant torsional stresses beyond its capacity.

Any abrupt and significant changes in a wall stiffness below and above a diaphragm level, or any such changes in the relative stiffness of adjacent walls in passing through one floor level to another (Figure 7-5), can cause high shear stresses in the floor diaphragm and/or a redistribution of shear forces among the walls. As an example consider the three-story concrete shear wall building shown in Figure 7-6. The concrete floor diaphragms are 8 in. thick. A set of static lateral forces of 24, 48, and 73 kips are applied at the center of mass of the first, second, and third levels, respectively. The base of the building is assumed to be fixed, and the reported results are based on an elastic analysis. An analysis based on a rigid-diaphragm assumption, and a finite-element analysis considering the uncracked diaphragm stiffness, yield very close results. However, if we make a simple change in the elevation of the building by moving the opening at the second level from the wall on line A to the wall on line B (Figure 7-7), the results of the

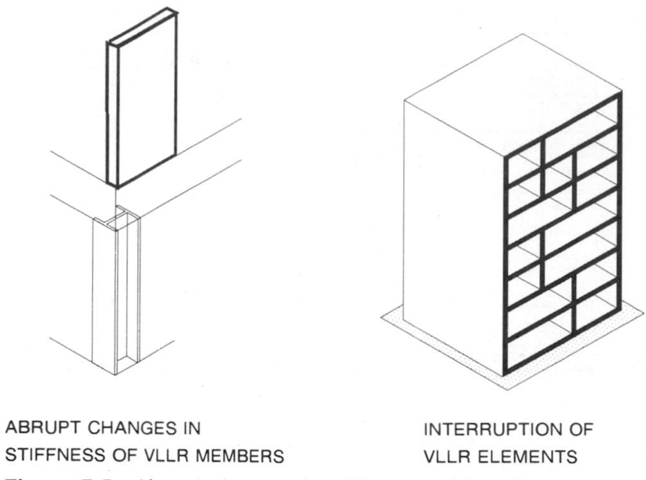

Figure 7-5 Abrupt changes in stiffness and location of VLLR elements can cause drastic redistribution of forces.

two methods will be markedly different (see Figure 7-8). For example, the rigid-diaphragm assumption suggests that the shear force in wall A is reduced from 94.3 kips above the first floor diaphragm to 26 kips below this level, while the finite-element model of the building shows that no such large portion of the shear force is transferred away from this wall by the floor diaphragm.

In buildings with significant plan irregularities, such as multiwing, *L*-shape, *H*-shape, and *V*-shape plans (Figure 7-9), particular attention should be paid to the accurate accessment of the in-plan diaphragm stress at the joints of the wings and to design for them. In this type of building, the fanlike deformations in the wings of diaphragm can lead to a stress concentration at the junction of the diaphragms (see Figure 7-10). If these stress concentrations are not considered, serious problems can arise. Sometimes the diaphragm stresses at the junctions may be so excessive that a feasible diaphragm thickness and reinforcement cannot be accommodated. In these cases the wings should be separated by seismic joints. One example for this type of problem was provided by the West Anchorage High School Building in Anchorage, Alaska, which suffered severe damage during the Alaskan earthquake of March 27, 1964 (see Figure 5-15).

Other classes of buildings deserving special attention to diaphragm design include those with relatively large openings in one or more of the floor decks (Figure 7-11) and tall buildings resting on a significantly larger low-rise part (Figure 7-12). In the latter case, the action of the low-rise portion as the shear base and the corresponding redistribution of shear forces (kickbacks) may subject the diaphragm located at the junction of the low-rise and high-rise parts (and sometimes a number of floor diaphragms above and below the junction) to some significant in-plane shear deformations.

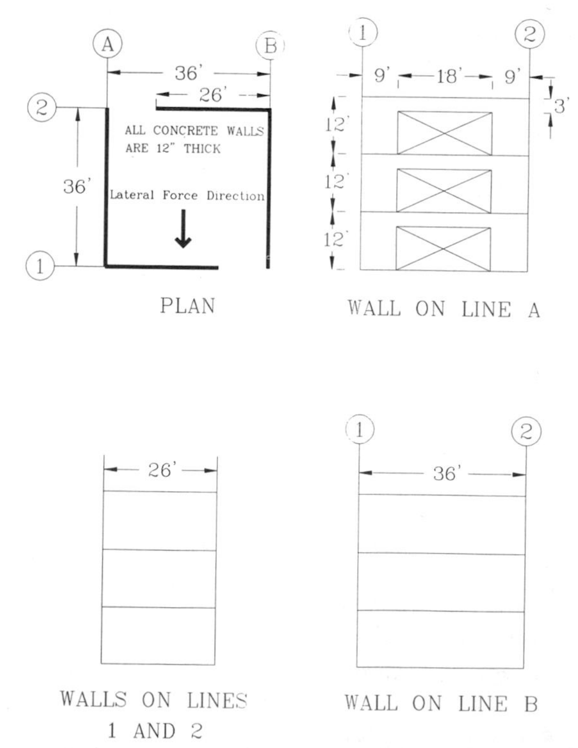

Figure 7-6 Plan and elevation of a simple three story shear wall building. (Note the uniform stiffness along the height of walls on lines *A* and *B*.)

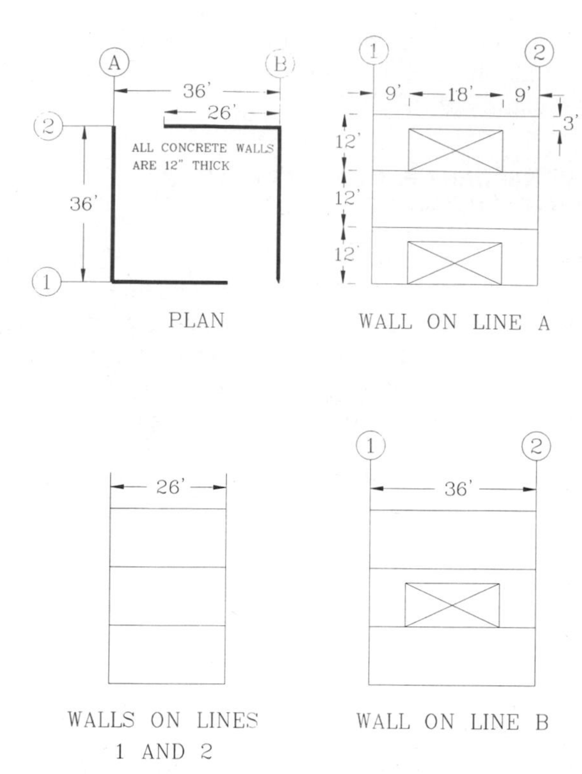

Figure 7-7 Altered plan and elevation of the three story shear wall building. (Note the abrupt change of stiffness along the height of walls on lines *A* and *B*.)

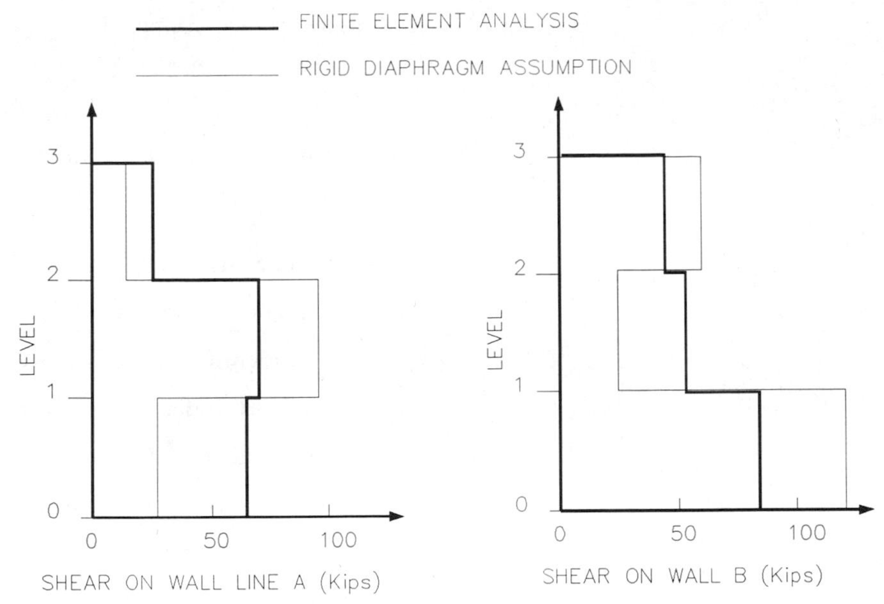

Figure 7-8 Computed shears of walls on lines *A* and *B*.

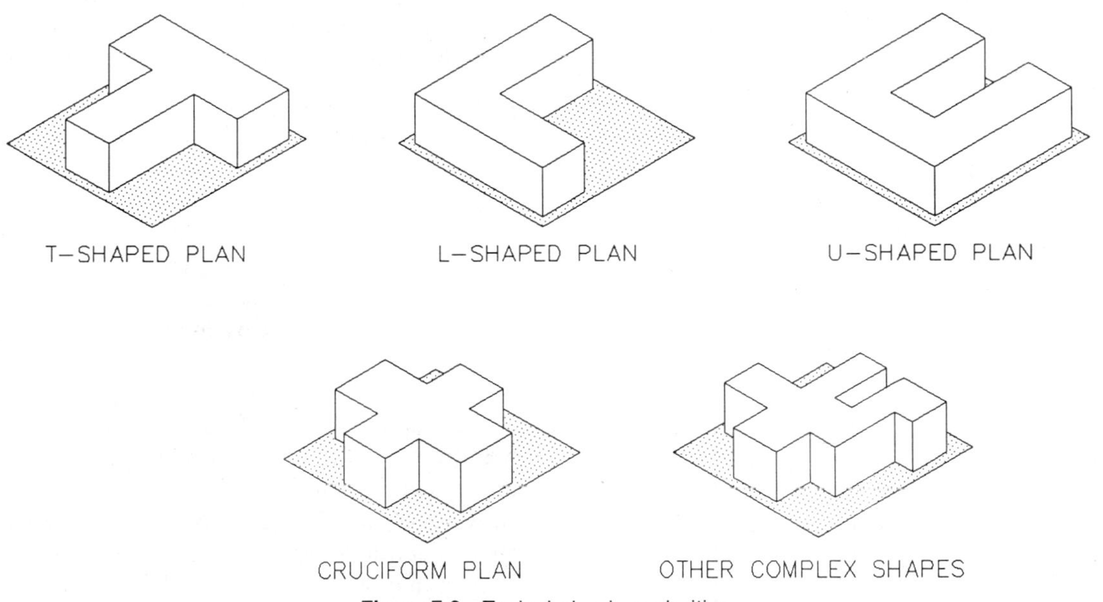

Figure 7-9 Typical plan irregularities.

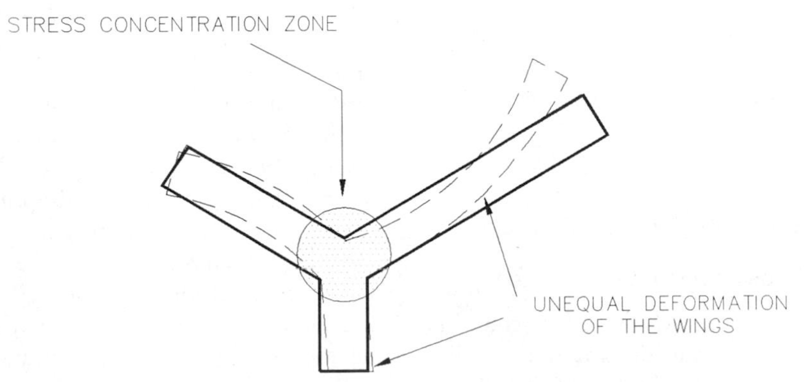

Figure 7-10 Fan-like deformation of wings causes stress concentration at the junction.

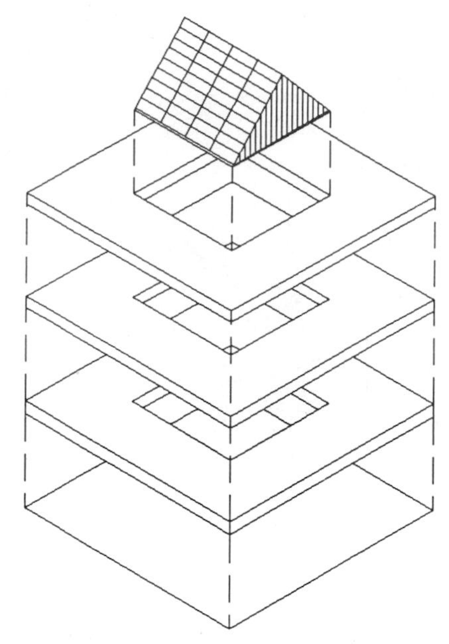

Figure 7-11 Significant floor openings are cause for concern.

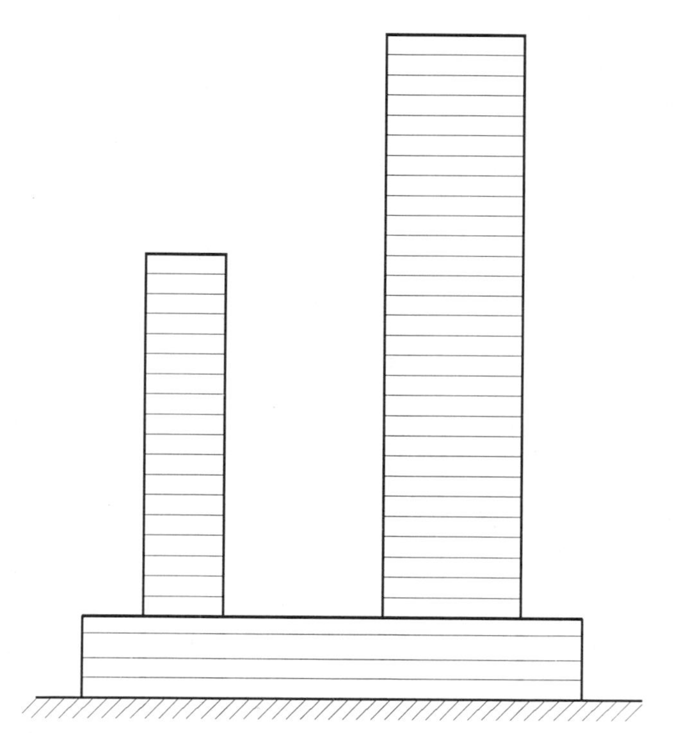

Figure 7-12 Elevation of towers on an expanded low-rise base.

7.5 CODE PROVISIONS FOR DIAPHRAGM DESIGN

7.5.1 UBC-85, BOCA-87, and ANSI-82 Provisions

UBC-85,[7-1] BOCA-87,[7-10] and ANSI-82[7-11] contain very similar provisions for seismic design of floor diaphragms. These documents require the roof and floor diaphragm collectors to be designed to resist the forces determined in accordance with

$$F_{px} = \frac{\sum_{i=x}^{n} F_i}{\sum_{i=x}^{n} w_i} w_{px} \tag{7-1}$$

The minimum value of F_{px} to be used in analysis is $0.14ZIw_{px}$. However, F_{px} need not exceed $0.30ZIw_{px}$, where

Z = seismic Zone Coefficient (see Section 4.3)

I = occupancy Importance Factor (see Section 4.3)

i = index identifying the ith level above the base

x = floor level under design consideration

n = total number of floor levels above the base

W = total weight of the building

F_i = lateral force applied to level i

w_i = portion of W at level i

w_{px} = weight of the diaphragm and the elements tributary thereto at level x, including 25% of the floor live load in storage and warehouse occupancies

Diaphragms supporting concrete or masonry walls should be designed and detailed such that the anchorage forces are distributed into the diaphragm. Diaphragm deformations should also be considered in the design of walls braced by the diaphragm. In design of wood diaphragms providing lateral support for concrete or masonry walls in seismic zones 2, 3, and 4, anchorage must be accomplished by metal straps anchored in the wall and connected to wood framing members by nails in shear or by through bolts. Anchorage may not be accomplished by use of toenails or nails subjected to withdrawal. In addition, wood ledgers must not be used in cross-grain bending or tension.

Per Equation 7-1, UBC-85 requires that the floor or roof diaphragm at every level be designed to span horizontally between the VLLR elements and to transfer the force F_{px} to these elements (see Figure 7-13a). In addition, all diaphragms must be designed to transfer lateral forces from the VLLR elements above the diaphragm to the other VLLR elements below the diaphragm due to offsets in the placement of VLLR elements or due to changes in stiffness of these elements. For example, in Figure 7-13b, the force P_1 has to be transferred by the diaphragm to the VLLR elements below the diaphragm, since the VLLR element above the diaphragm has been discontinued at this level. In addition, the force P_2 from the other VLLR elements above has to be redistributed among the VLLR elements below the diaphragm. Obviously, the diaphragm must be designed to transfer these additional loads.

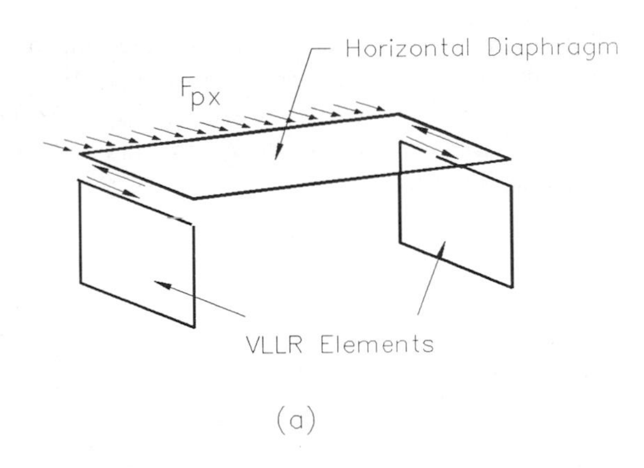

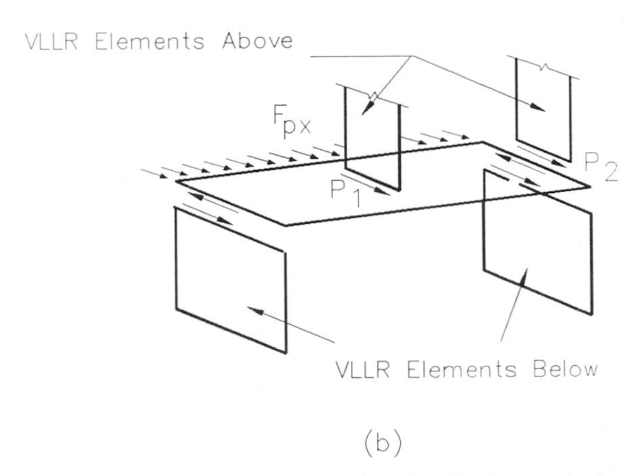

Figure 7-13 Code provisions for diaphragm design.

7.5.2 NEHRP-85 Provisions

In contrast to the NEHRP-85's other provisions,[(7-12)] which are rather comprehensive, the provisions regarding seismic design of floor diaphragms are rather short and minimal.

NEHRP-85 contains a clause limiting the in-plane deflection of the floor diaphragms:

> The deflection in the plane of the diaphragm, *as determined by engineering analysis*, shall not exceed the permissible deflection of attached elements. Permissible deflection shall be that deflection which will permit the attached element to maintain its structural integrity under the individual loading and continue to support the prescribed loads *without endangering the occupants of the building*. [Italics supplied.]

In addition, NEHRP-85 requires that the floor nad roof diaphragms be designed to resist a minimum seismic force of $0.5A_v$ times the weight of the diaphragm and other elements of the building attached thereto plus the portion of the seismic shear force at that level, V_x, required to be transferred to the VLLR elements because of offset or changes in stiffness of VLLR elements above and below the diaphragm. A_v is the velocity-related ground acceleration as defined in Section 4.6.

Diaphragms should be designed to resist both shear and bending stresses resulting from these forces. Ties or struts should be provided to distribute the wall anchorage forces, as determined from Equation 12-13, into the diaphragm.

7.5.3 SEAOC-86 (UBC-88) Provisions

The SEAOC-86 document,[(7-13)] which has been adopted as the seismic design provisions of the UBC-88 code, has expanded the requirements embodied in the previous SEAOC documents and reflected in the old versions of the UBC code.

SEAOC-86 has incorporated the diaphragm-deflection clause from the NEHRP-85 code. However, it has dropped the two statements that we italicized in quoting the NEHRP-85 requirement.

The concentrated force at the top of the structure, F_t, as discussed in Section 4.3.9, has been added to Equation 7-1. Hence the SEAOC-86 (UBC-88) formula for determination of seismic force acting on a floor diaphragm is

$$F_{px} = \left(F_t + \frac{\sum_{i=x}^{n} F_i}{\sum_{i=x}^{n} w_i} \right) w_{px} \qquad (7\text{-}2)$$

The expressions for the minimum and maximum values of the force F_{px} have been changed to $0.35ZIw_{px}$ and $0.75ZIw_{px}$ respectively. The values obtained, however, are essentially the same as for UBC-85, since the seismic-zone factors are reduced in SEAOC-86 (UBC-88). For example, for seismic zone 4, SEAOC-86 (UBC-88) specifies $Z = 0.4$ as opposed to $Z = 1$ in the UBC-85 code.

SEAOC-86 (UBC-88) contains special provisions for seismic design of floor diaphragms in buildings with plan irregularities located in seismic zones 3 and 4. For structures having a plan irregularity of type A, B, C, or D as described in Table 4-6 and shown graphically in Figure 5-6, connections of drag members to the VLLR elements should be designed without considering the common one-third increase in allowable stresses. In addition, for structures having a plan irregularity of type B, diaphragm chords and drag members should be designed considering independent movement of the projecting wings of the structure. Each of these diaphragm elements should be designed for the more severe of the following two conditions:

1. motion of the projecting wings in the same direction, and
2. motion of the projecting wings in opposing directions.

This requirement is considered satisfied if a three-dimensional dynamic analysis according to the code provisions is performed.

7.6 DESIGN EXAMPLES

As discussed in Chapter 5, it is desirable from the structural point of view to have regular buildings with minimal offset in the location of VLLR elements and without sudden changes in stiffness from floor to floor. Quite often, however, other requirements of the project (such as architectural considerations) control these parameters, and the structural engineer is faced with buildings that are considered irregular in terms of seismic behavior and design.

Diaphragm design consists primarily of the following tasks:

1. Determining the lateral-force distribution on the diaphragm and computing diaphragm shears and moments at different locations.
2. Providing adequate in-plane shear capacity in the diaphragm to transfer lateral forces to the VLLR elements.
3. Providing suitable connection between the diaphragm and the VLLR elements.
4. Design of boundary members or reinforcement to develop chord forces.
5. Computing diaphragm deflections, when necessary, to ascertain that the diaphragm is stiff enough to support the nonstructural elements such as curtain walls, etc. without excessive deflections.

In addition, the diaphragm must be designed and detailed for local effects caused by various openings such as those caused by the elevator shafts. Parking-structure diaphragms with ramps are a special case of diaphragms with openings. The effect of the ramp attachment to floors above and below the ramp should be considered in lateral-force distribution, especially for non-shear-wall buildings.

Slab-on-grade in a building is often assumed to redistribute the shear from the VLLR elements to other foundation locations that offer additional resistance to sliding. In such cases, the slab-on-grade must be properly designed and detailed as a diaphragm. Any detail specified to control shrinkage cracking in the slab must also be verified for capacity to transfer the diaphragm forces.

In this section, procedures for seismic design of floor diaphragms are demonstrated by means of four design examples which are worked out in detail. In the first example, a concrete floor diaphragm at the top of a parking level under a two-story wood-framed apartment building is designed. The second example explains diaphragm design for a four-story concrete parking structure, which has setbacks in elevation of the building and the shear walls. In the third example, the metal-deck diaphragm of a three-story steel-framed office building is designed. Finally, the fourth example explains the wood-diaphragm design for a typical one-story neighborhood shopping center.

Example 7-1

It is proposed to build a two-story wood-framed apartment building on top of one-story concrete parking. The building will be located in UBC seismic zone 4. The concrete floor supporting the wood construction (see Figure 7-14) will be a 14-in.-thick, hard-rock-concrete, flat plate ($f_c' = 4000$ lb/in.^{2}). The lateral-force-resisting system for the concrete parking structure consists of concrete-block masonry walls ($f_m' = 3000$ lb/in.^{2}). Given that the superimposed dead load from the two-story wood framing above is 65 lb/ft^2, design the concrete diaphragm per requirements of the UBC-85 Code.[a] The floor-to-floor height is 10 ft.

- Dead loads and seismic shears:

superimposed dead load from wood framing above $= 65$ lb/ft^2

concrete slab at 150 lb/ft^3 $= (14/12)(150) = 175$ lb/ft^2

miscellaneous ($M + E + \frac{1}{2}$ column weights) $= 10$ lb/ft^2

total floor weight $= (175)(89.66)(65 + 175 + 10) = 3922.6$ k

N–S walls:

12-in walls at 124 lb/ft^2 $= 4(5)(17.33)(0.124) = 43$ kips

E–W walls:

8-in. wall at 78 lb/ft^2 $= (5)(175)(0.078) = 68.25$ kips

12-in. walls at 124 lb/ft^2 $= (5)(17.33 + 35.33)(0.124) = 32.65$ kips

The weight of the walls parallel to the applied seismic force does not contribute to the diaphragm shears. However, in general, they are included conservatively in the design of concrete floor diaphragms. In this example, the weight of the walls parallel to the applied seismic force is

[a] Except for the base shear calculation formulas the UBC-88 procedure is essentially the same as the UBC-85 procedure for all of the design examples presented in this chapter.

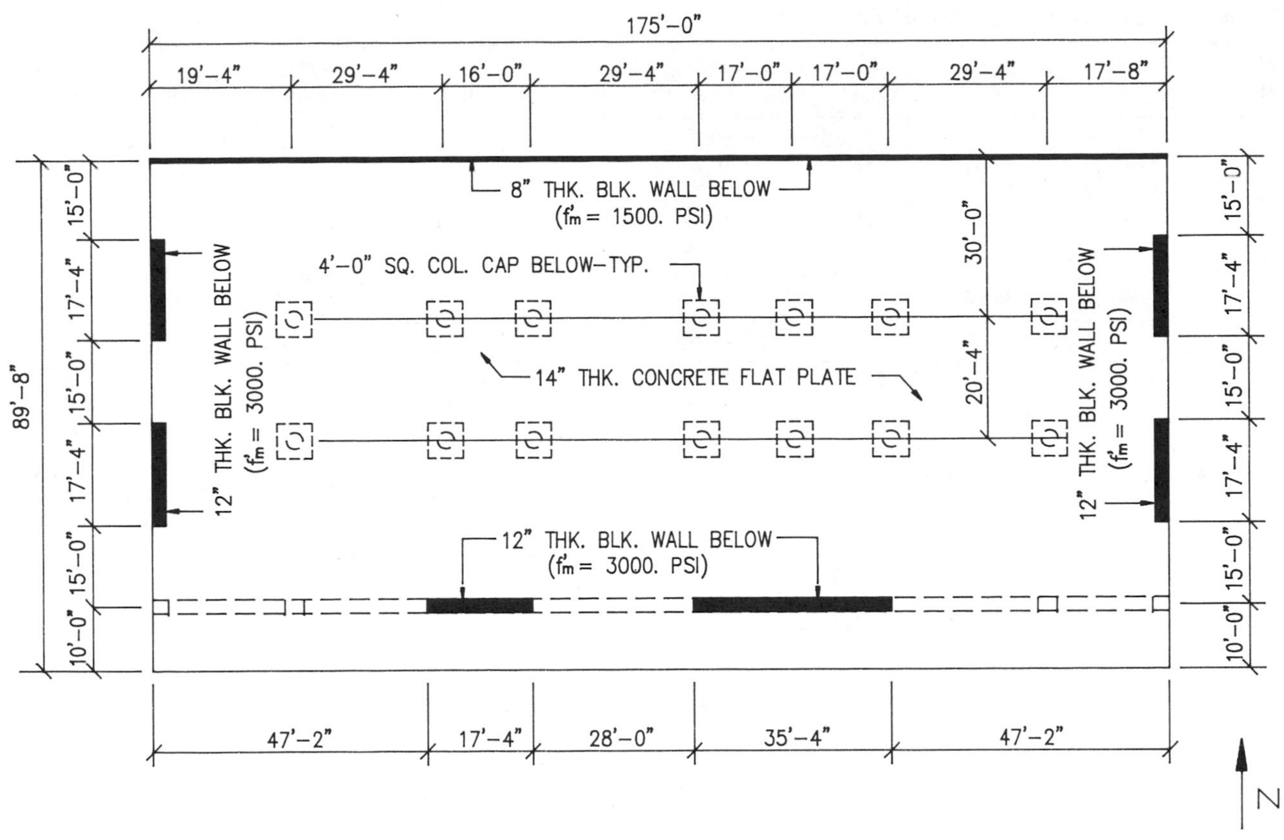

Figure 7-14 Second floor framing plan (Example 7-1).

not included in calculating diaphragm shears:

$$\text{E–W weight} = W_x = 3922.6 + 43 = 3965.6 \text{ kips}$$

$$\text{N–S weight} = W_y = 3922.6 + 100.9 = 4023.5 \text{ kips}$$

Base shears:

$$V = ZIKCSW = (1)(1)(1.33)(0.14)W = 0.186W$$

$$V_x = 0.186(3965.6) = 737.6 \text{ kips}$$

$$V_y = 0.186(4023.5) = 748.4 \text{ kips}$$

- Center of mass (see Figure 7-15): In computing the location of the center of mass of the walls it is generally assumed that one-half of the height of a wall above and below the diaphragm will contribute to the mass of each floor. The parameters needed for determination of the center of mass of the walls are calculated in Table 7-1. Therefore, the center of mass of the walls is located at

$$x_1 = \frac{\Sigma xW}{\Sigma W} = \frac{12{,}703.0}{143.85} = 88.31 \text{ ft}$$

$$y_1 = \frac{\Sigma yW}{\Sigma W} = \frac{8564.1}{143.85} = 59.53 \text{ ft}$$

Since the slab is of uniform thickness, the center of mass of the floor coincides with its geometric centroid:

$$x_2 = 87.50 \text{ ft}$$

$$y_2 = 44.83 \text{ ft}$$

Location of the combined center of mass:

$$x_m = \frac{143.9(88.31) + 3922.6(87.5)}{143.9 + 3922.6} = 87.53 \text{ ft}$$

$$y_m = \frac{143.9(59.53) + 3922.6(87.5)}{143.9 + 3922.6} = 45.35 \text{ ft}$$

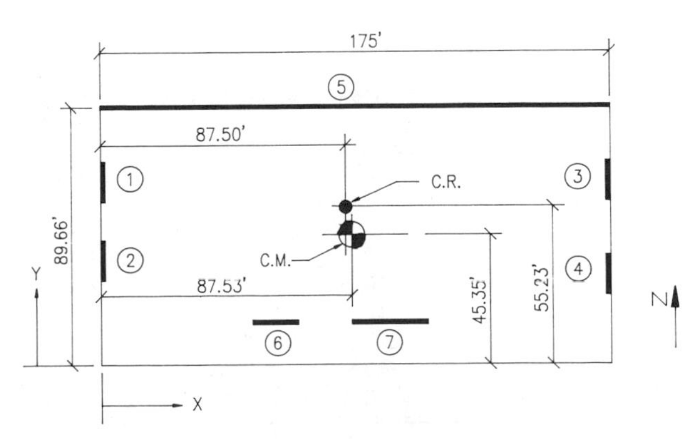

Figure 7-15 Locations of centers of mass and rigidity.

Table 7-1 Center-of-Mass Calculations for Example 7-1

Wall No.	Weight, lb / ft²	Length, ft	Area, ft²	Weight, kips	Dir.	x, ft	xW ft-kips	y, ft	yW ft-kips
1	124	17.33	86.65	10.74	y	0.50	5.37	66.00	708.84
2	124	17.33	86.65	10.74	y	0.50	5.37	33.67	361.62
3	124	17.33	86.65	10.74	y	174.50	1,874.10	66.00	708.84
4	124	17.33	86.65	10.74	y	174.50	1,874.10	33.67	361.62
5	78	175.00	875.00	68.25	x	87.50	5,971.88	89.33	6096.78
6	124	17.33	86.65	10.74	x	55.84	559.72	10.00	107.40
7	124	35.33	176.70	21.90	x	110.16	2,412.50	10.00	219.00
Σ				143.85			12,703.00		8564.10

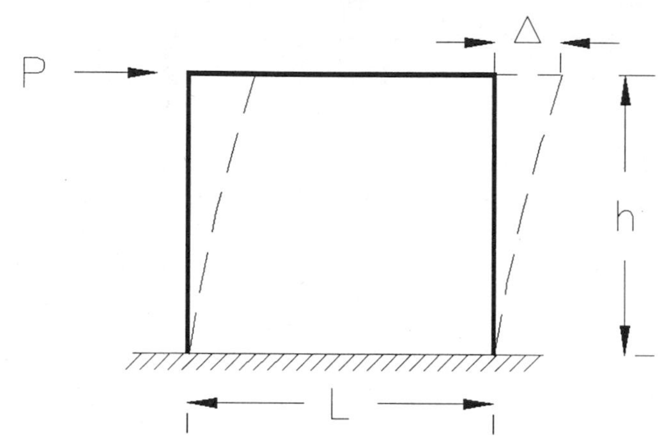

Figure 7-16 Deformation of a free-standing wall panel.

- Center of rigidity: For a free-standing wall (see Figure 7-16),

$$\Delta = \frac{Ph^3}{3EI} + \frac{1.2Ph}{AG}$$

Denoting the wall thickness by t and assuming $G = 0.40E$ for masonry this relation may be rewritten as

$$\Delta = \frac{4P(h/L)^3}{Et} + \frac{3P(h/L)}{Et}$$

The relative wall rigidities, $R = 1/\Delta$, may be computed assuming a constant value of P, say $P = 1{,}000{,}000$ lb. Using the parameters generated in Tables 7-2 and 7-3, the location of the center of rigidity is established as

$$x_r = \frac{\Sigma x R_y}{\Sigma R_y} = \frac{4886.0}{55.84} = 87.50 \text{ ft}$$

$$y_r = \frac{\Sigma y R_x}{\Sigma R_x} = \frac{6506.93}{117.8} = 55.23 \text{ ft}$$

- Torsional eccentricity:

$$e_x = x_r - x_m = 87.5 - 87.53 \approx 0 \text{ ft}$$

$$e_y = y_r - y_m = 55.23 - 45.35 = 9.88 \text{ ft}$$

UBC code requires a minimum of 5% eccentricity on either side of the center of rigidity. Therefore,

$$e_{min} = 0.05(175) = 8.75 \text{ ft} > e_x$$

$$\therefore \quad \text{use} \quad e_x = \pm 8.75 \text{ ft}$$

Torsional moments:

$$T_y = V_y e_x = 748.4(\pm 8.75) = \pm 6548.5 \text{ ft-k}$$

$$T_x = V_x e_y = 737.6(-9.88) = -7287.5 \text{ ft-k}$$

In-plane forces in the walls due to direct shear

Table 7-2 Relative Rigidity of the Walls*

Wall No.	Height, ft	Length, ft	H / L	E, lb / in.²	t in.	Δ	R = 1 / Δ
1	10	17.33	0.5770	3,000,000	11.625	0.0716	13.96
2	10	17.33	0.5770	3,000,000	11.625	0.0716	13.96
3	10	17.33	0.5770	3,000,000	11.625	0.0716	13.96
4	10	17.33	0.5770	3,000,000	11.625	0.0716	13.96
5	10	175.00	0.0571	1,500,000	7.625	0.0150	66.67
6	10	17.33	0.5770	3,000,000	11.625	0.0716	13.96
7	10	35.33	0.2830	3,000,000	11.625	0.0269	37.17

*Computed assuming a constant force of $P = 1{,}000{,}000$ lb.

Table 7-3 Center-of-Rigidity Calculations for Example 7-1

Wall No.	Dir.	*x*	*y*	R_x	R_y	xR_y	yR_x
1	*y*	0.50	—	—	13.96	6.98	—
2	*y*	0.50	—	—	13.96	6.98	—
3	*y*	174.50	—	—	13.96	2436.02	—
4	*y*	174.50	—	—	13.96	2436.02	—
5	*x*	—	89.33	66.67	—	—	5995.63
6	*x*	—	10.00	13.96	—	—	139.60
7	*x*	—	10.00	37.17	—	—	371.70
Σ				117.80	55.84	4886.00	6506.93

Table 7-4 Wall Shears for Seismic Force in the N – S Direction

Wall No.	R_x	R_y	d_x, ft	d_y, ft	*Rd*	Rd^2	F_v, kips	F_t, kips	F_{total}, kips
1	0	13.96	−87.00	—	−1214.52	105,663	187.1	−13.13	187.10
2	0	13.96	−87.00	—	−1214.52	105,663	187.1	−13.13	187.10
3	0	13.96	87.00	—	1214.52	105,663	187.1	13.13	200.20
4	0	13.96	87.00	—	1214.52	105,663	187.1	13.13	200.20
5	66.67	0	—	34.10	2273.45	77,524	0	24.62	24.62
6	13.96	0	—	−45.23	−631.41	28,559	0	−6.84	6.84
7	37.17	0	—	−45.23	−1681.20	76,041	0	−18.20	18.20
Σ	117.80	55.84				604,776			

Table 7-5 Wall Shears for Seismic Force in the E – W Direction

Wall No.	R_x	R_y	d_x, ft	d_y, ft	*Rd*	Rd^2	F_v, kips	F_t, kips	F_{total}, kips
1	0	13.96	−87.00	—	−1214.52	105,663	0	14.63	14.63
2	0	13.96	−87.00	—	−1214.52	105,663	0	14.63	14.63
3	0	13.96	87.00	—	1214.52	105,663	0	−14.63	14.63
4	0	13.96	87.00	—	1214.52	105,663	0	−14.63	14.63
5	66.67	0	—	34.10	2273.45	77,524	417.45	−27.39	417.45
6	13.96	0	—	−45.23	−631.41	28,559	87.40	7.60	95.00
7	37.17	0	—	−45.23	−1681.20	76,041	232.70	20.26	252.96
Σ	117.80	55.84				604,776			

are computed from

$$F_{vx} = V_x \frac{R_x}{\Sigma R_x}$$

$$F_{vy} = V_y \frac{R_y}{\Sigma R_y}$$

while the in-plane wall forces due to torsion are computed from

$$F_{tx} = T_x \frac{Rd}{\Sigma Rd^2}$$

$$F_{ty} = T_y \frac{Rd}{\Sigma Rd^2}$$

Table 7-6 Design Shear Forces

Wall No.	Wall Length, ft	Design Shear Forces, kips: For Wall Design	For Diaphragm Design*: E – W Seismic	N – S Seismic
1	17.33	200.20	−14.63	174.00
2	17.33	200.20	−14.63	174.00
3	17.33	200.20	14.63	200.20
4	17.33	200.20	14.63	200.20
5	175.00	417.45	390.00	24.62
6	17.33	95.00	95.00	−6.84
7	35.33	252.96	252.96	−18.20

*Eccentricity was assumed.

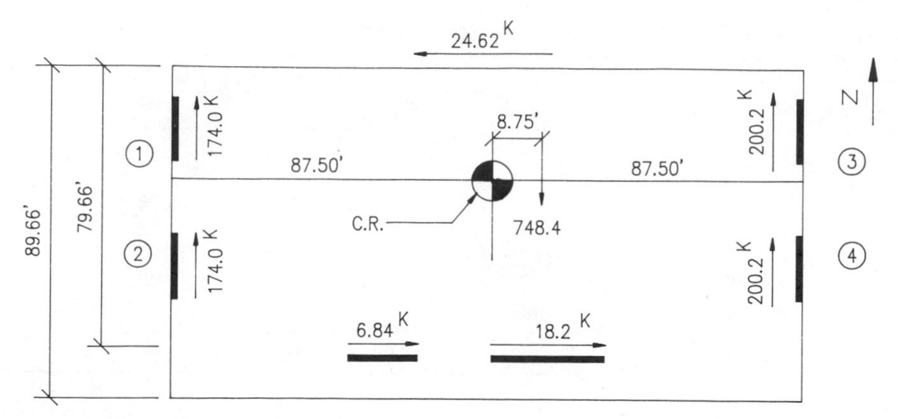

Figure 7-17 Design wall forces for seismic load in the N–S direction.

where d is the distance of each wall from the center of rigidity. Using these formulas, the wall forces for seismic forces acting in the N–S and E–W directions are calculated and reported in Tables 7-4 and 7-5, respectively. Note that the contribution of torsion, if it reduces the magnitude of the design wall shears, is ignored. The design shear forces are summarized in Table 7-6.

- Diaphragm design for seismic force in the N–S direction: The wall forces and the assumed direction of torque due to the eccentricity are shown in Figure 7-17. Using this information, the distribution of the applied force on the diaphragm may be calculated. Denoting the left and right diaphragm reactions by V_L and V_R, from force equilibrium (see Figure 7-18) we get

$$V_L \frac{175}{2} + V_R \frac{175}{2} = 748.4 \text{ kips}$$

or

$$V_L + V_R = 8.55 \qquad \text{(I)}$$

From moment equilibrium,

$$\left(\frac{175}{3}\right)\frac{175}{2}V_L + \left(2x\frac{175}{3}\right)\left(\frac{175}{2}\right)V_R = 748.4(96.25)$$

or

$$V_L + 2V_R = 14.11 \qquad \text{(II)}$$

solving Equations I and II for V_L and V_R yields

$$V_L = 2.99 \text{ kips/ft} \quad \text{and} \quad V_R = 5.56 \text{ kips/ft}$$

The mid-span diaphragm moment[b] (Figure 7-18) is

$$M = 348(87.5) - 12.31(79.66) - \frac{2.99(87.5)(58.33)}{2} - \frac{4.28(87.5)(29.17)}{2} = 16{,}377 \text{ ft-kips}$$

Check slab shear stress along walls 1 and 2:

$$L = 17.33 \text{ ft}, \qquad t = 14 \text{ in.}$$

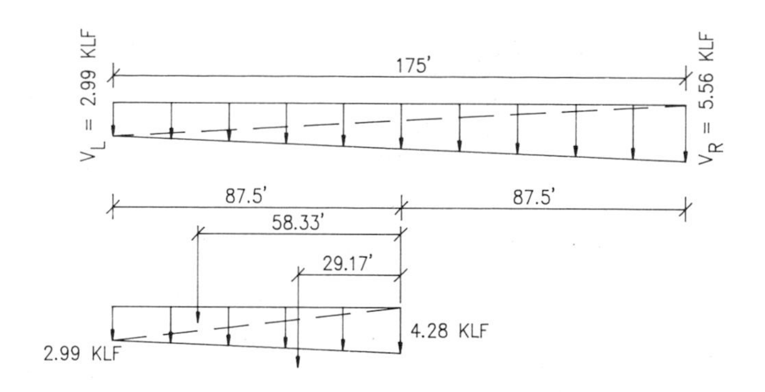

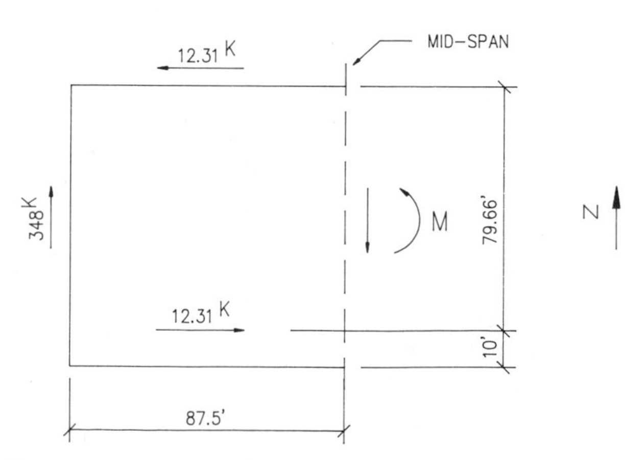

Figure 7-18 Force distribution and diaphragm moments for seismic load in the N–S direction.

[b]The mid-span moment has been used in this example to demonstrate the chord design procedures. This moment, however, is not necessarily the maximum moment. In a real design situation the maximum moment should be calculated and used for chord design.

The slab capacity without shear reinforcement is

$$\phi\, 2\sqrt{f_c'}\, bt = \frac{0.85(2)\sqrt{4000}\,(14)(17.33)(12)}{1000}$$
$$= 313 \text{ kips}$$
$$> 1.4(200.2) = 280.3 \qquad \text{O.K.}$$

Chord design:

$$T_u = \frac{1.4M}{d} = \frac{1.4(16{,}377)}{89.66 - 1.0} = 258 \text{ kips}$$

$$A_s = \frac{T_u}{\phi f_y} = 258/0.90(60) = 4.77 \text{ in.}^2$$

Provide four #10 chord bars ($A_s = 5.08$ in.^{2}) along the slab edges at the north and south sides of the building.

- Diaphragm design for seismic force in the E–W direction: A sketch of the wall forces indicating the assumed direction of the torque due to eccentricity is shown in Figure 7-19. Similar to the N–S direction, the force and moment equilibrium equations may be used to obtain the distribution of lateral force on the diaphragm:

$$V_L \frac{89.66}{2} + V_R \frac{89.66}{2} = 737.6 \text{ kips}$$

or

$$V_L + V_R = 16.45 \qquad \text{(III)}$$

and

$$29.89 \frac{89.66}{2} V_L + 59.77 \frac{89.66}{2} V_R = 737.6(45.35)$$

or

$$V_L + 2V_R = 24.96 \qquad \text{(IV)}$$

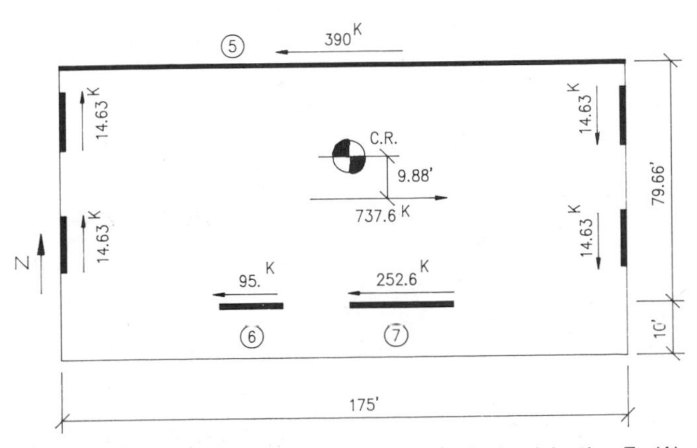

Figure 7-19 Design wall forces for seismic load in the E–W direction.

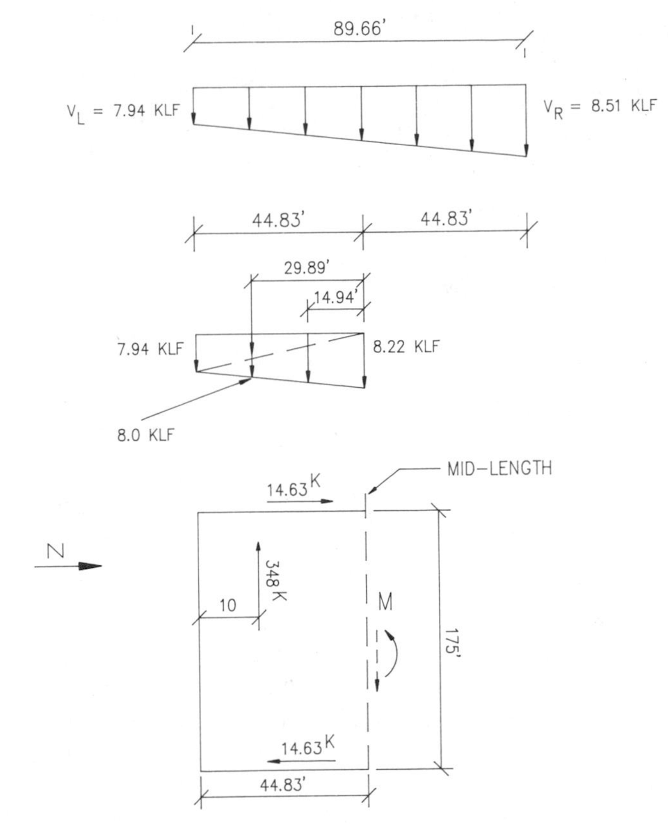

Figure 7-20 Force distribution and diaphragm moments for seismic load in the E–W direction.

Solving Equations III and IV for V_L and V_R,

$$V_L = 7.94 \text{ kips/ft} \quad \text{and} \quad V_R = 8.51 \text{ kips/ft}$$

The mid-span diaphragm moment (Figure 7-20) is

$$M = 348(34.83) + 14.63(175) - \frac{7.94(44.83)(29.83)}{2} - \frac{8.22(44.83)(14.94)}{2}$$
$$= 6609 \text{ ft-kips}$$

Similarly, diaphragm moments at other locations, including the cantilever portion of the diaphragm, can be calculated. Check diaphragm shear capacity: Along wall 5,

$$L = 175 \text{ ft}, \qquad t = 14 \text{ in.}$$

$$V_c = \frac{0.85(2)\sqrt{4000}\,(14)(175)(12)}{1000}$$
$$= 3161 \text{ kips}$$
$$> 1.4(444.84) = 623 \qquad \text{O.K.}$$

Along wall 6,

$$L = 17.33 \text{ ft}, \qquad t = 14 \text{ in.}$$

$$V_c = \frac{0.85(2)\sqrt{4000}\,(14)(17.33)(12)}{1000} = 313 \text{ kips}$$
$$> 1.4(95) = 133 \qquad \text{O.K.}$$

Along wall 7,

$$L = 35.33 \text{ ft}, \qquad t = 14 \text{ in.}$$

$$V_c = \frac{0.85(2)\sqrt{4000}\,(14)(35.33)(12)}{1000} = 638 \text{ kips}$$

$$> 1.4(252.96) = 354 \qquad \text{O.K}$$

Chord design:

$$T_u = \frac{1.4M}{d} = \frac{1.4(6609)}{175.0 - 1.0} = 53 \text{ kips}$$

$$A_s = \frac{T_u}{\phi f_y} = \frac{53}{0.90(60)} = 0.98 \text{ in.}^2$$

Provide two #7 chord bars ($A_s = 1.20$ in.^{2}) along the slab edges at the east and west sides of the building.

Example 7-2

Design the third-floor diaphragm of the four-story parking structure shown in Figures 7-21 through 7-25. The building is to be located in UBC seismic zone 4. Access to each floor will be provided from an adjacent parking structure that will be separated by a seismic joint. Typical floor and roof framing consists of 5-in.-thick posttensioned slabs spanning to 36-in.-deep posttensioned beams. A typical floor dead load for purposes of seismic design is estimated at 150 lb/ft^2. This includes contributing wall and column weights. The typical floor-to-floor height is 10 ft. UBC-85 code requirements are to be used. $f_c' = 4000$ lb/in.^{2}, $f_y = 60{,}000$ psi.

- Weight computations:

$$\text{roof weight} = (68 \text{ ft})(185 \text{ ft})(0.15 \text{ kips/ft}^2) = 1887 \text{ kips}$$

$$\text{4th-floor weight} = (85 \text{ ft})(185 \text{ ft})(0.15 \text{ kips/ft}^2) = 2359 \text{ kips}$$

$$\text{3rd-floor weight} = (104 \text{ ft})(185 \text{ ft})(0.15 \text{ kips/ft}^2) = 2886 \text{ kips}$$

$$\text{2nd-floor weight} = (104 \text{ ft})(185 \text{ ft})(0.15 \text{ kips/ft}^2) = 2886 \text{ kips}$$

$$\text{total weight} = 1887 + 2359 + 2(2886) = 10018 \text{ kips}$$

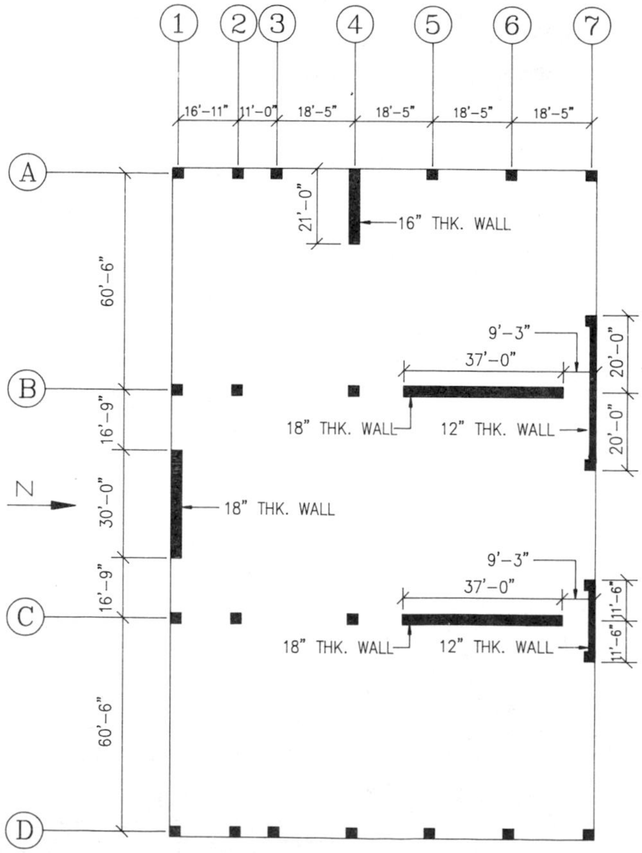

Figure 7-21 Ground floor framing plan (Example 7-2).

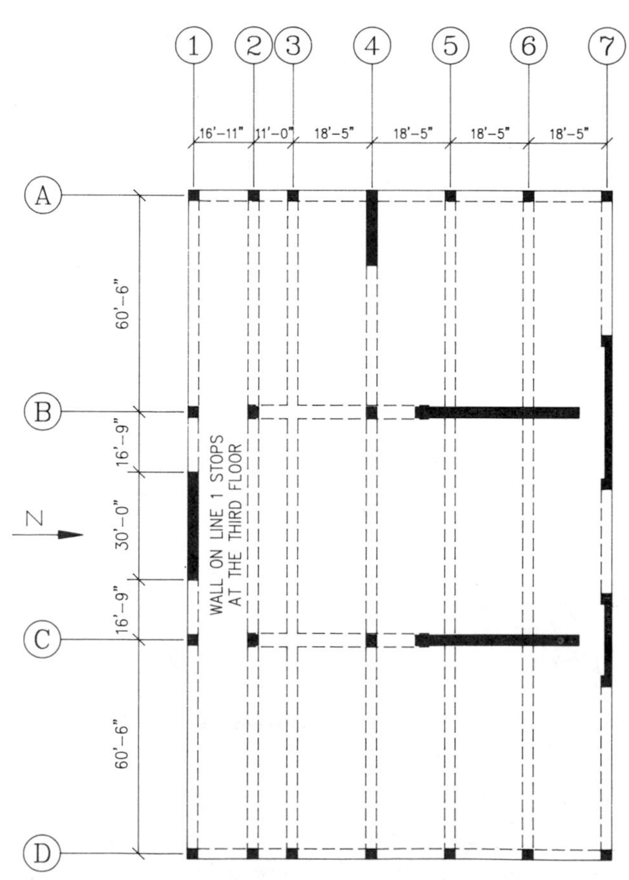

Figure 7-22 Second and third floor framing plan.

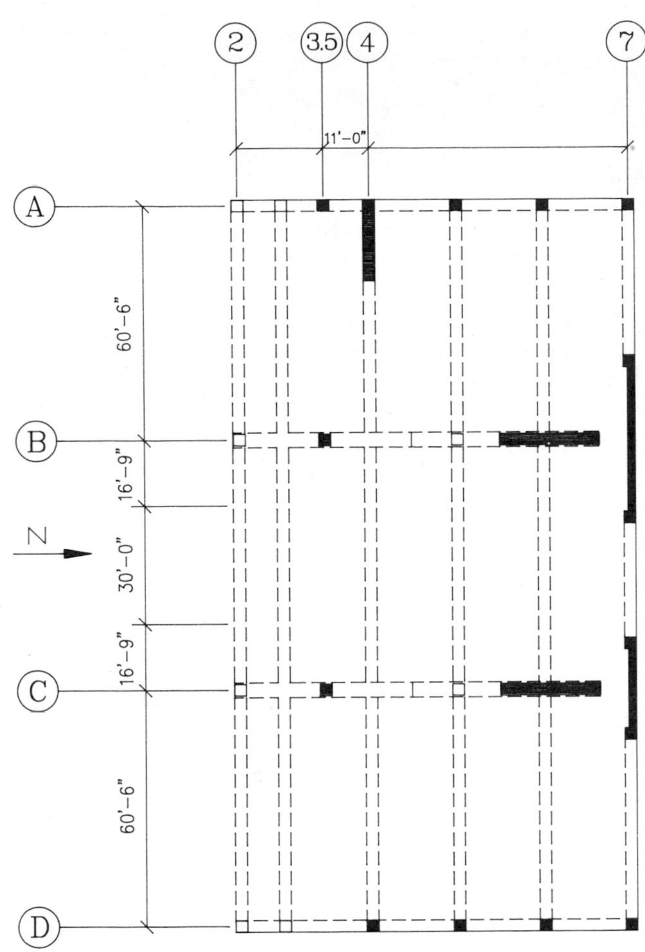

Figure 7-23 Fourth floor framing plan.

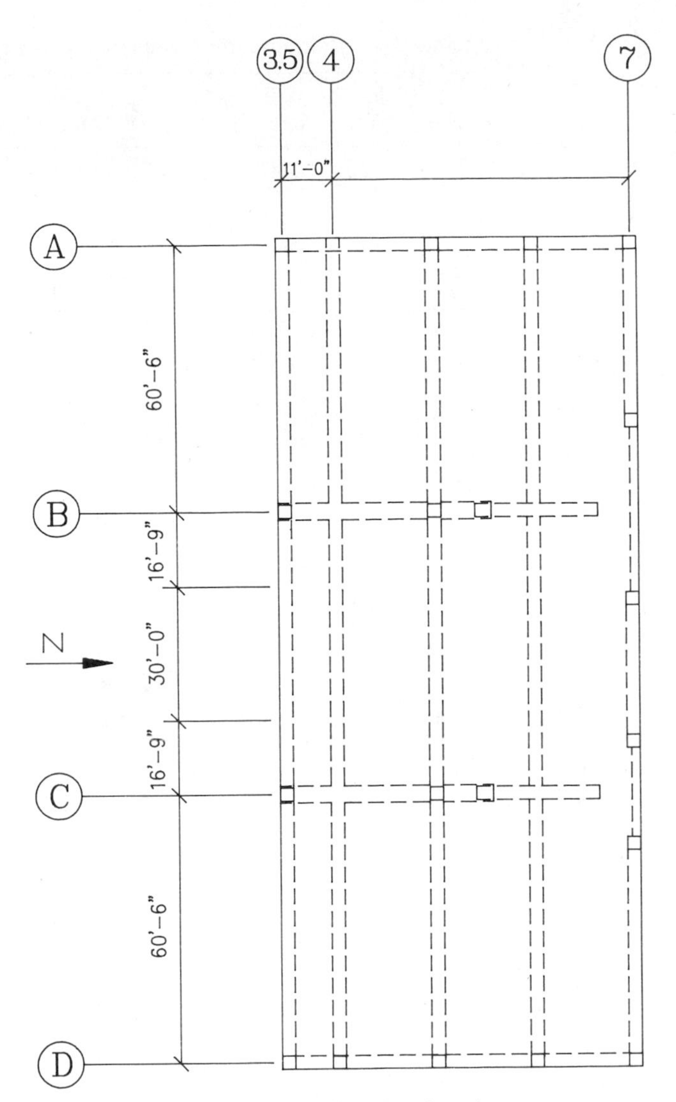

Figure 7-24 Roof framing plan.

- Design lateral forces:

$$\text{base shear } V = ZIKCSW$$
$$V = (1)(1)(1.33)(0.14)W$$
$$= 0.186W = 1863 \text{ kips}$$

lateral force applied to level x,

$$F_x = (V - F_t)\frac{W_x h_x}{\sum_{i=1}^{n} W_i h_i}$$

lateral force on diaphragm at level x,

$$F_{px} = \frac{W_{px}\sum_{l=x}^{n} F_l}{\sum_{l=x}^{n} W_l}$$

$$F_{px}(\text{minimum}) = 0.14ZIW_{px}$$

Values of F_{px} for various floors are calculated in Table 7-7. The concrete diaphragm is assumed to be rigid. The seismic shear forces acting on the walls were obtained by a computer analysis and are shown in Figures 7-26 and 7-27.

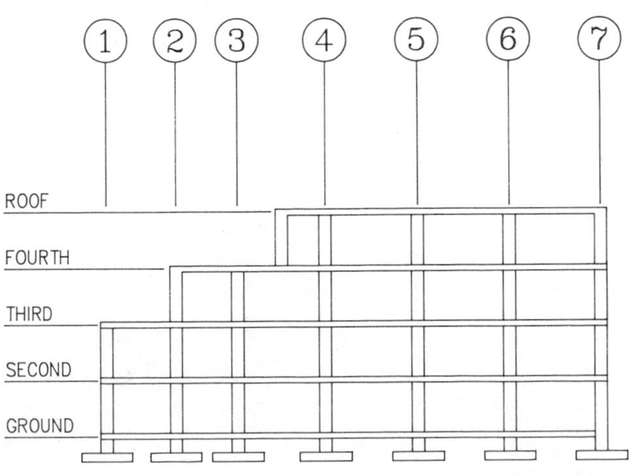

Figure 7-25 A section through the building (Example 7-2).

Table 7-7 Calculation of Diaphragm Design Forces for Example 7-2

Level	h_x, ft	W_x, kips	$W_x h_x$	$W_x h_x / \Sigma W_i h_i$	F_x, kips	ΣF_i, kips	ΣW_i, kips	$\Sigma F_i / \Sigma W_i$	F_{px}, kips
Roof	40	1887	75,480	0.324	604	604	1887	0.320	603
4th	30	2359	70,770	0.304	566	1170	4246	0.276	651
3rd	20	2886	57,720	0.248	462	1632	7132	0.229	661
2nd	10	2886	28,860	0.124	231	1863	10,018	0.186	537
Σ		10,018	232,830	1.0					

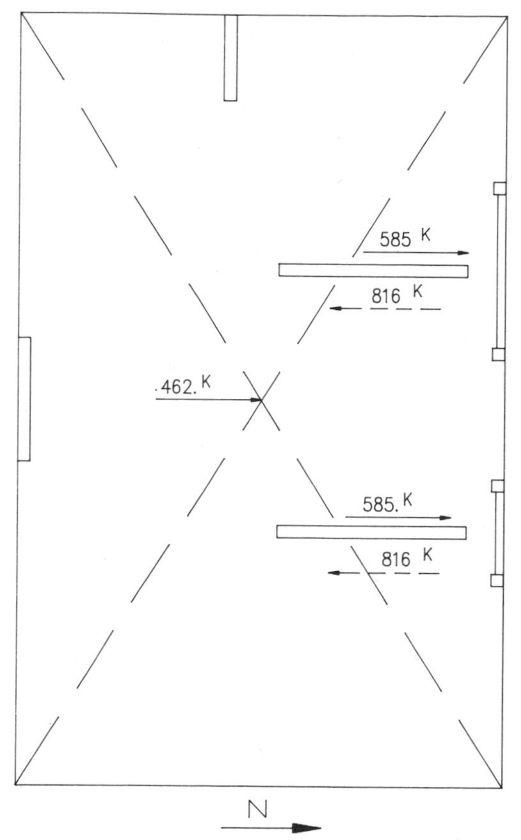

Figure 7-26 Forces on the third floor diaphragm due to N–S seismic loading. (Wall shears above the diaphragm are shown with solid arrows while wall shears below the diaphragm are indicated by dashed lines.)

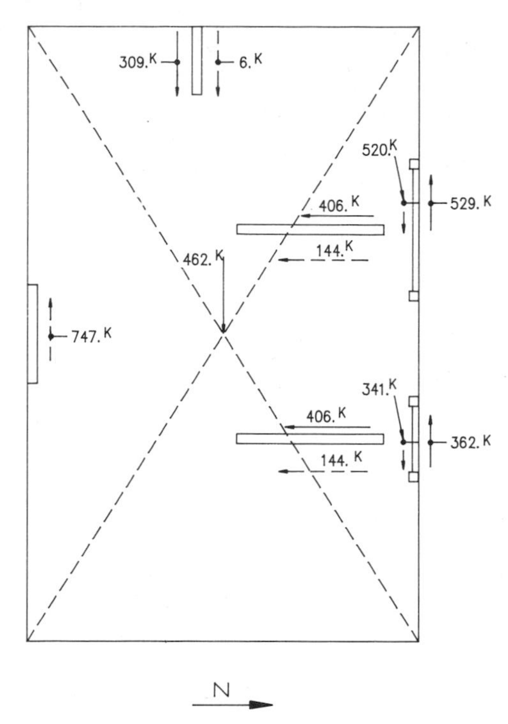

Figure 7-27 Forces on the third floor diaphragm due to E–W seismic loading. (Wall shears above the diaphragm are shown with solid arrows while wall shears below the diaphragm are indicated by dashed lines.)

- Diaphragm design in the N–S direction: Net shear forces acting on the walls and the corresponding diaphragm load, shear, and moment diagrams are shown in Figure 7-28. Check shear capacity of 5 in.-thick slab along the walls on grid lines B and C:

  Maximum slab shear = 151.3 kips

  Slab capacity without shear reinforcement =

$$\phi 2\sqrt{f_c'}\,bt = \frac{(0.85)(2)\sqrt{4000}\,(5)(37)(12)}{1000}$$

$$= 239 \text{ kips}$$

$$> 1.4 \times 151.3 = 212 \text{ kips}$$

Therefore no shear reinforcement is needed. Chord design:

$$T_u = \frac{1.4M}{d} = \frac{(1.4)(4577)}{101.58 - 1.0}$$

$$= 63.7 \text{ kips}$$

$$A_s = \frac{T_u}{\phi f_y} = \frac{63.7}{(0.90)(60)} = 1.17 \text{ in.}^2$$

Therefore provide three #6 chord bars (A_s = 1.325 in.^{2}) along slab edges on the north and south sides of the building.

- Diaphragm design in the E–W direction: The net shear forces acting on the walls and the corresponding diaphragm load, shear, and moment diagrams are shown in Figure 7-29. Moment

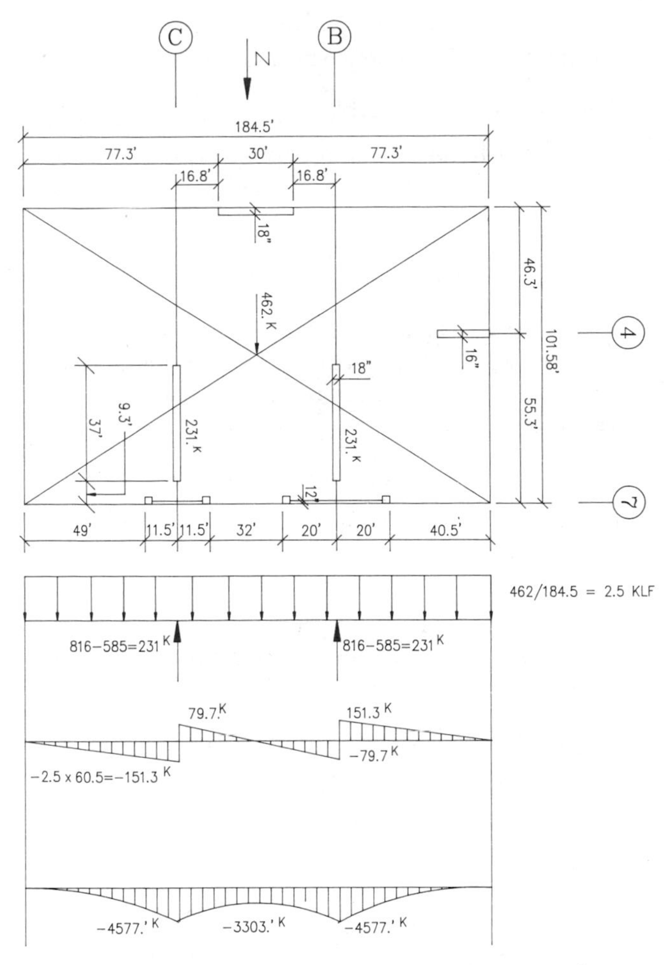

Figure 7-28 Diaphragm loading, shear, and moment diagrams for seismic load in the N–S direction.

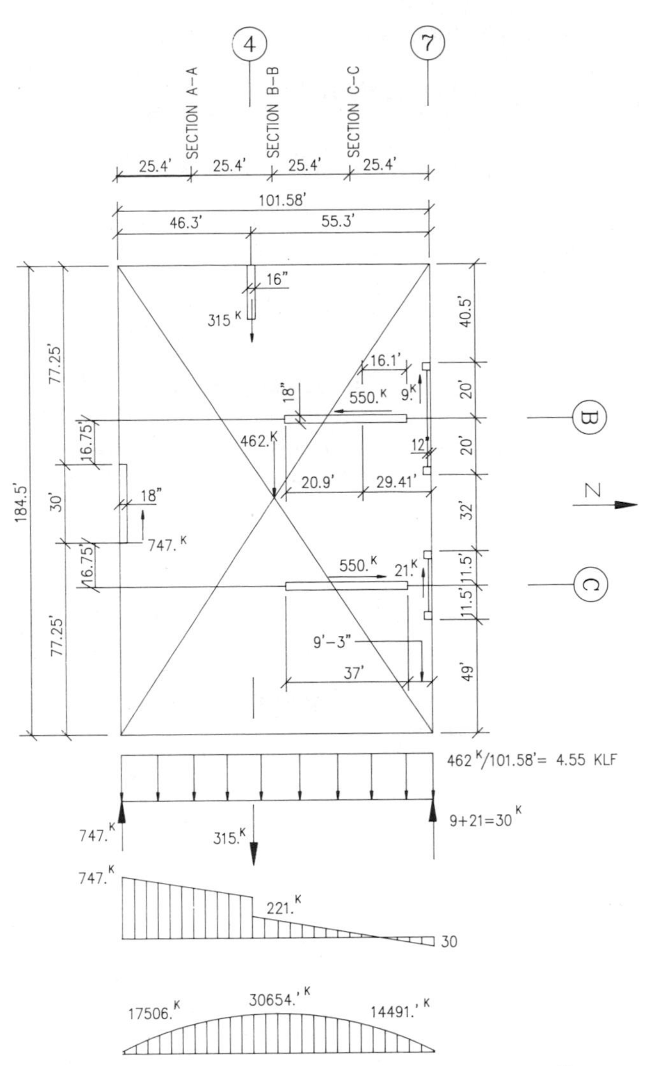

Figure 7-29 Diaphragm loading, shear, and moment diagrams for seismic load in the E–W direction.

calculations: at section A–A,

$$M_{A-A} = 747(25.4) - \frac{4.55(25.4)^2}{2}$$
$$= 17{,}506 \text{ ft-kips}$$

at section B–B,

$$M_{B-B} = 747(50.8) - 315(4.5) - \frac{4.55(50.8)^2}{2}$$
$$= 30{,}659 \text{ ft-kips}$$

at section C–C,

$$M_{C-C} = 30(25.4) - \frac{4.55(25.4)^2}{2} + \frac{16.1}{37}(550)(63.5)$$
$$= 14{,}491 \text{ ft-kips}$$

$\therefore$ [c]Estimated maximum moment = 30,659 ft-kips

[c]A more accurate value of the maximum moment may be obtained by reading the moment diagram plotted to a larger scale.

Chord design:

$$T_u = \frac{1.4M}{d} = \frac{1.4(30659)}{184.5 - 1.0}$$
$$= 233.9 \text{ kips}$$

$$A_s = \frac{T_u}{\phi f_y} = \frac{233.9}{(0.90)(60)} = 4.33 \text{ in.}^2$$

Therefore provide six #8 chord bars (A_s = 4.74 in.^{2}) along slab edges on the east and west sides of the building

$$C_u = T_u$$

Compression C_u to be resisted by edge beam and slab concrete. Check 5-in.-thick slab shear capacity along the wall on line 1: For L = 30 ft, slab capacity without shear reinforcement is

$$\frac{(0.85)(2)\sqrt{4000}\,(5)(30)(12)}{1000} = 194 \text{ kips}$$
$$< 1.4(744) = 1042 \text{ kips} \qquad \text{N.G.}$$

For $L = 184.5$ ft, slab capacity without shear reinforcement is

$$\frac{(0.85)(2)\sqrt{4000}\,(5)(184.5)(12)}{(1000)} = 1190 \text{ kips}$$
$$> 1.4(744) = 1042 \text{ kips} \qquad \text{O.K.}$$

Check the capacity of a 30-ft-long slab with #4 bars at 18 in., at the top and bottom of the slab, to carry 1042 kips of force:

$$\phi V_c = 194 \text{ kips}$$
$$A_s = 0.13 \text{ in.}^2/\text{ft}$$
$$\phi V_s = (0.85)(2 \times 0.13)(60)(30 \text{ ft}) = 398 \text{ kips}$$
$$\phi V_n = 398 + 194 = 592 \text{ kips} < 1042 \text{ kips}$$

Drag struts are needed to transfer the difference ($1042 - 592 = 450$ kips).

- Design of drag struts (see Figure 7-30): The two beams along the grid line 1 may be designed to transfer the slab shear into the walls:

$$A_s = \frac{450/2}{(0.90)(60)} = 4.2 \text{ in.}^2$$

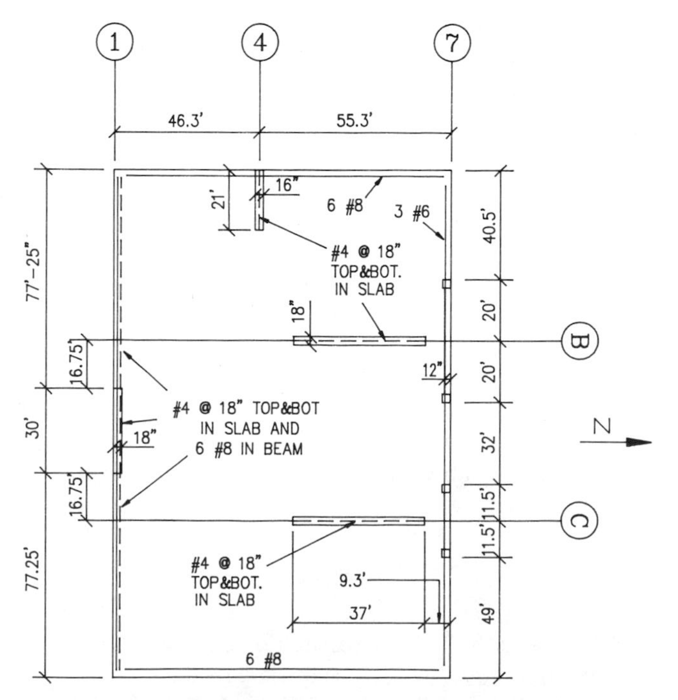

Figure 7-30 Diaphragm chord, drag, and shear reinforcement.

Therefore provide six #8 bars ($A_s = 4.74$ in.^{2}) in the beams for seismic shear transfer.

$$\text{Drag strut length provided} = 2(77.3) = 154.6 \text{ ft}$$

Capacity of slab along drag strut

$$= \frac{(0.85)(2)\sqrt{4000}\,(5)(154.6)(12)}{1000}$$
$$= 997 \text{ kips} > 450 \text{ kips} \qquad \text{O.K.}$$

Check shear capacity of 5-in.-thick slab at the wall on grid line 4 to carry $315/2 = 157.5$ kips of shear (notice that slab occurs on both sides of the wall):

$$\phi V_c = \frac{(0.85)(2)\sqrt{4000}\,(5)(21)(12)}{1000} = 135 \text{ kips}$$
$$< 1.4(157.5) = 220.5 \text{ kips}$$

Therefore shear reinforcement is required. Using #4 bars at 18 in. at the top and bottom of the slab,

$$\phi V_s = (0.85)(2 \times 0.13)(60)(21) = 278 \text{ kips}$$
$$\phi V_n = 135 + 278 = 413 \text{ kips}$$
$$> (1.4)(157.5) = 220.5 \text{ kips}$$

Therefore drag struts are not required. It can be seen that the slab shear capacity along the walls on the grid line 7 is sufficient. Check the shear capacity of the slab along the cross walls on grid lines *B* and *C*. Here again, slab occurs on both sides of the wall:

$$\phi V_c = \frac{(0.85)(2)\sqrt{4000}\,(5)(37)(12)}{1000} = 239 \text{ kips}$$
$$< (1.4)\frac{550}{2} = 385 \text{ kips}$$

Therefore shear reinforcement is required. Try #4 bars at 18 in. at the top and bottom of the slab:

$$\phi V_s = (0.85)(0.13 \times 2)(60)(37) = 490 \text{ kips}$$
$$\phi V_n = 239 + 490 = 729 \text{ kips}$$
$$> (1.4)(275) = 385 \text{ kips}$$

Therefore drag struts are not required.

Example 7-3

Design the roof diaphragm of the three-story steel-framed building shown in Figure 7-31. The building is supported on the top of a one-story subterranean concrete parking structure. The parking-structure deck may

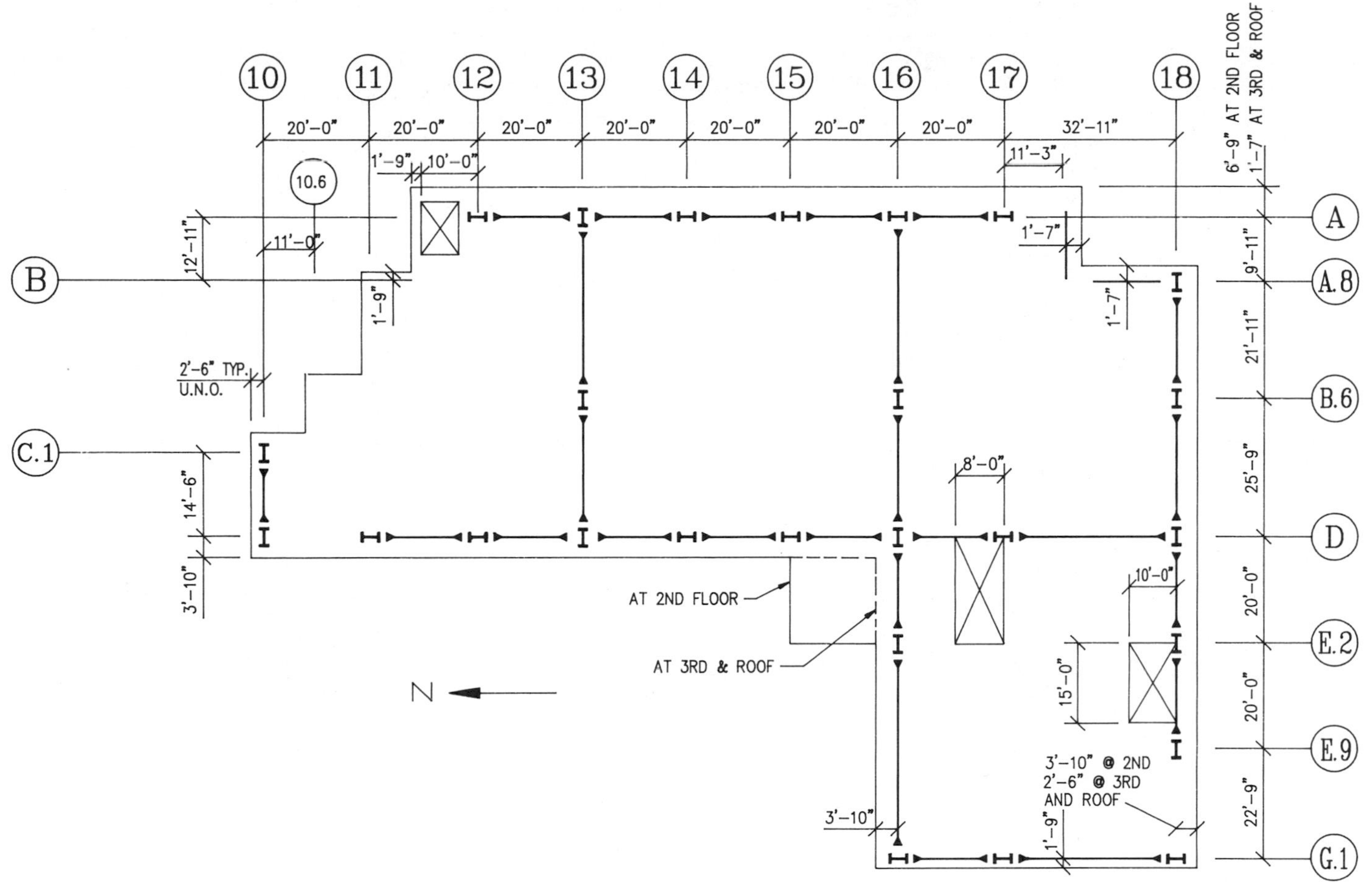

Figure 7-31 Typical floor framing plan for building of Example 7-3 (Openings shown exist on second and third floors only).

be considered as the shear base for the steel structure. The lateral-load-resisting system for the steel building consists of moment-resisting frames in both directions. Beams and columns which are not part of the lateral system are not shown in Figure 7-31. The floor construction consists of $3\frac{1}{4}$ in. of light weight concrete on the top of a 3-in.-deep, 20-gauge metal deck. The maximum spacing of floor purlins is 10 ft. Mechanical equipment is located on the roof, west of grid line *D*. The roof construction west of grid line *D* consists of $4\frac{1}{2}$ in. of hard-rock concrete on the top of a 3-in.-deep, 18-gauge metal deck. The maximum spacing of the roof purlins is 8 ft. The roof construction east of grid line *D* is similar to the typical floor construction.

The estimated total dead loads for seismic design are 100 lb/ft^2 at the typical floors, 200 lb/ft^2 at the mechanical areas of the roof, and 70 lb/ft^2 elsewhere on the roof. The building is located in UBC seismic zone 4. Use UBC-85 static lateral loads. Story height is 15 ft between the first and second floors and 13 ft elsewhere. The fundamental period of the building in both directions is estimated to be 0.87 sec.

- Weight computations:

$$\begin{aligned}
\text{roof weight} &= (3593\ \text{ft}^2)(0.20\ \text{kips/ft}^2) \\
&\quad + (10170\ \text{ft}^2)(0.07\ \text{kips/ft}^2) \\
&= 1431\ \text{kips} \\
\text{3rd-floor weight} &= (13764\ \text{ft}^2)(0.10\ \text{kips/ft}^2) \\
&= 1376\ \text{kips} \\
\text{2nd-floor weight} &= (14751\ \text{ft}^2)(0.10\ \text{kips/ft}^2) \\
&= 1475\ \text{kips} \\
\text{total weight} &= 1431 + 1376 + 1475 \\
&= 4282\ \text{kips}
\end{aligned}$$

- Design lateral forces: Base shear, $V = ZIKCSW = (1)(1)(1)(1/15\sqrt{0.87})(1.5)W = 458$ kips. The lateral-force distribution and minimum diaphragm forces are shown in Table 7-8. The floor and roof diaphragms are assumed to be rigid compared to the moment frames. The seismic shear forces acting on the frames were obtained

Table 7-8 Calculation of Diaphragm Design Forces for Example 7-3

Level	h_x, ft	W_x, kips	$W_x h_x$	$W_x h_x / \Sigma W_i h_i$	F_x, kips	ΣF_i, kips	ΣW_i, kips	$\frac{\Sigma F_i}{\Sigma W_i}$	F_{px}, kips
Roof	41	1431	58671	0.492	241*	241	1431	0.168	240.7
3rd	28	1376	38528	0.323	138	379	2807	0.135†	192.6
2nd	15	1475	22125	0.185	79	458	4282	0.107†	206.5
Σ		4282	119,324	1.0					

*Includes $F_t = 0.07TV = (0.07)(0.87)V = 0.061V = 28$ kips.
† The minimum value of 0.14 is used.

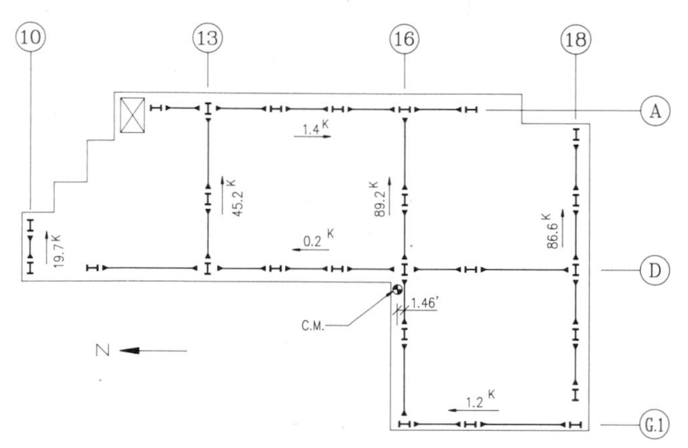

Figure 7-32 Frame shears for E – W seismic loading.

Figure 7-33 Frame shears for N – S seismic loading.

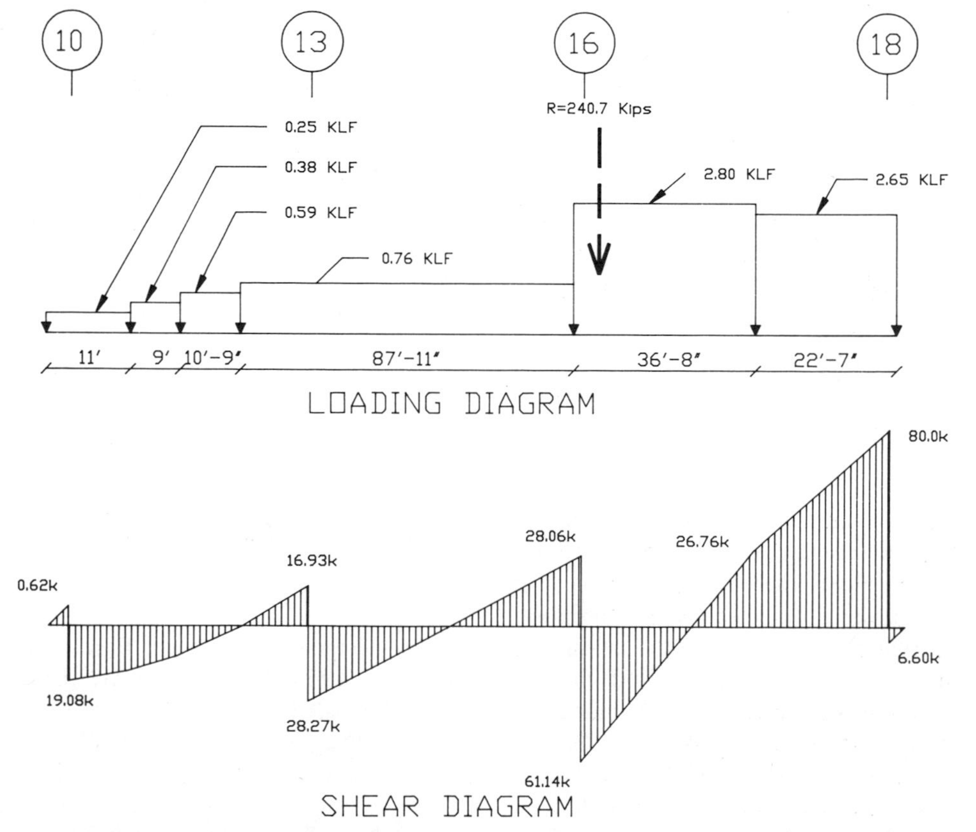

Figure 7-34 Diaphragm loading and shear diagrams for the E – W seismic loading.

W3 FORMLOK

GALVANIZED

LIGHT WEIGHT CONCRETE (110 pcf)

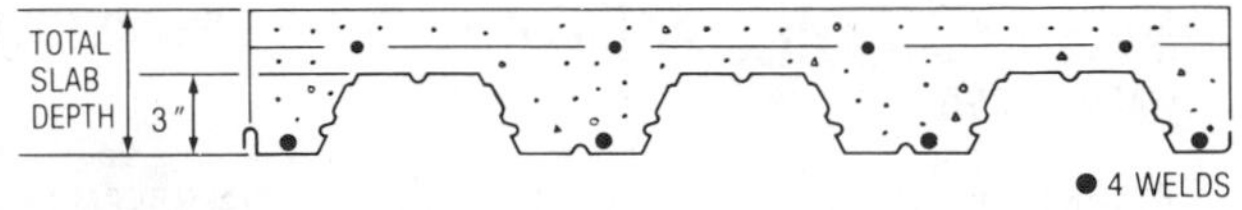

ALLOWABLE SUPERIMPOSED LOADS (Lbs./Sq. Ft.), DIAPHRAGM SHEAR VALUES (q) (Lbs./L.F.) AND FLEXIBILITY FACTORS (F)

TOTAL SLAB DEPTH & CONCRETE WT. psf	DECK GAGE & WT. psf	NUMBER OF SPANS / q / F	SPAN 8′-0″	8′-6″	9′-0″	9′-6″	10′-0″	10′-6″	11′-0″	11′-6″	12′-0″	12′-6″	13′-0″	13′-6″	14′-0″	14′-6″	15′-0″
6¼″	22	1	309	238	213	191	172	155	141	128	116	106	97	88	81	74	68
43.5		2	309	279	254	232	172	155	141	128	116	106	97	88	81	74	68
2 HOUR FIRE RATING		3	309	279	254	232	213	155	141	128	116	106	97	88	81	74	68
		q	1780	1760	1740	1725	1710	1695	1680	1670	1660	1650	1640	1630	1620	1615	1610
	1.9	F	.47	.48	.48	.49	.49	.50	.50	.50	.50	.51	.51	.52	.52	.52	.52
	21	1	341	309	281	215	194	176	160	145	133	121	111	102	93	86	79
		2	341	309	281	256	235	176	160	145	133	121	111	102	93	86	79
		3	341	309	281	256	235	217	201	145	133	121	111	102	93	86	79
		q	1810	1790	1770	1750	1730	1715	1700	1685	1670	1660	1650	1640	1630	1620	1610
	2.1	F	.43	.44	.44	.45	.45	.46	.46	.47	.47	.48	.48	.48	.48	.49	.49
	20	1	357	323	294	268	205	186	169	154	141	129	118	108	100	92	84
		2	357	323	294	268	246	227	169	154	141	129	118	108	100	92	84
		3	357	323	294	268	246	227	210	195	141	129	118	108	100	92	84
		q	1830	1805	1780	1760	1740	1720	1700	1690	1680	1665	1650	1640	1630	1620	1610
	2.3	F	.42	.43	.43	.44	.44	.45	.45	.45	.45	.46	.46	.47	.47	.47	.47
	19	1	400	372	338	309	284	262	201	183	168	154	142	131	121	111	103
		2	400	372	338	309	284	262	242	225	168	154	142	131	121	111	103
		3	400	372	338	309	284	262	242	225	209	196	142	131	121	111	103
		q	1890	1860	1830	1805	1780	1760	1740	1725	1710	1695	1680	1670	1660	1645	1630
	2.7	F	.37	.38	.38	.39	.39	.40	.41	.41	.41	.41	.42	.42	.42	.43	.43
	18	1	400	400	369	338	310	286	264	241	187	172	158	146	135	125	116
		2	400	400	369	338	310	286	264	241	229	214	158	146	135	125	116
		3	400	400	369	338	310	286	264	241	229	214	200	188	176	125	116
		q	1940	1905	1870	1845	1820	1795	1770	1750	1730	1715	1700	1690	1680	1665	1650
	2.9	F	.34	.35	.35	.36	.36	.37	.37	.38	.38	.39	.39	.39	.39	.40	.40
	16	1	400	400	400	400	370	341	316	293	273	255	197	182	169	157	146
		2	400	400	400	400	370	341	316	293	273	255	239	224	211	157	146
		3	400	400	400	400	370	341	316	293	273	255	239	224	211	199	188
		q	2070	2025	1980	1945	1910	1880	1850	1830	1810	1790	1770	1750	1730	1715	1700
	3.5	F	.28	.29	.30	.31	.31	.32	.32	.32	.32	.33	.33	.34	.34	.34	.34
7¼″	22	1	315	280	250	224	202	182	165	150	136	124	113	103	94	86	79
52.7		2	364	329	299	224	202	182	165	150	136	124	113	103	94	86	79
3 HOUR FIRE RATING		3	364	329	299	273	202	182	165	150	136	124	113	103	94	86	79
		q	2100	2080	2060	2040	2020	2010	2000	1985	1970	1960	1950	1945	1940	1930	1920
	1.9	F	.40	.41	.41	.41	.41	.42	.42	.42	.42	.43	.43	.43	.43	.44	.44
	21	1	400	314	281	253	228	206	187	170	155	142	130	119	109	100	92
		2	400	363	330	302	228	206	187	170	155	142	130	119	109	100	92
		3	400	363	330	302	277	255	187	170	155	142	130	119	109	100	92
		q	2130	2125	2080	2060	2040	2025	2010	1995	1980	1970	1960	1950	1940	1935	1930
	2.1	F	.37	.38	.38	.38	.38	.39	.39	.40	.40	.40	.40	.40	.40	.41	.41
	20	1	400	380	345	266	240	218	198	180	165	150	138	126	116	107	98
		2	400	380	345	316	290	218	198	180	165	150	138	126	116	107	98
		3	400	380	345	316	290	267	247	180	165	150	138	126	116	107	98
		q	2150	2120	2090	2070	2050	2035	2020	2005	1990	1980	1970	1960	1950	1940	1930
	2.3	F	.35	.36	.36	.37	.37	.38	.38	.38	.38	.39	.39	.39	.39	.39	.39
	19	1	400	400	397	363	333	258	235	214	196	180	166	153	141	130	120
		2	400	400	397	363	333	307	285	214	196	180	166	153	141	130	120
		3	400	400	397	363	333	307	285	264	246	180	166	153	141	130	120
		q	2210	2180	2150	2125	2100	2080	2060	2040	2020	2010	2000	1985	1970	1960	1950
	2.7	F	.31	.32	.32	.33	.33	.34	.34	.34	.34	.35	.35	.35	.35	.36	.36
	18	1	400	400	400	396	364	335	260	238	218	201	185	171	158	146	135
		2	400	400	400	396	364	335	310	288	218	201	185	171	158	146	135
		3	400	400	400	396	364	335	310	288	268	251	235	171	158	146	135
		q	2260	2225	2190	2160	2130	2110	2090	2070	2050	2035	2020	2005	1990	1980	1970
	2.9	F	.29	.30	.30	.31	.31	.31	.31	.32	.32	.33	.33	.33	.33	.33	.33
	16	1	400	400	400	400	400	399	370	343	320	248	229	212	197	183	170
		2	400	400	400	400	400	399	370	343	320	299	280	212	197	183	170
		3	400	400	400	400	400	399	370	343	320	299	280	263	247	183	170
		q	2380	2340	2300	2265	2230	2200	2170	2145	2120	2100	2080	2065	2050	2035	2020
	3.5	F	.25	.26	.26	.26	.26	.27	.27	.28	.28	.28	.28	.29	.29	.29	.29

Figure 7-35 A Verco FormLok diaphragm design table. (Reproduced with permission of Verco Manufacturing Company, Benicia, California.)

W3 FORMLOK NORMAL WEIGHT CONCRETE (145 pcf)

GALVANIZED

TOTAL SLAB DEPTH 3″

ALLOWABLE SUPERIMPOSED LOADS (Lbs./Sq. Ft.), DIAPHRAGM SHEAR VALUES (q) (Lbs./L.F.) AND FLEXIBILITY FACTORS (F)

● 4 WELDS

TOTAL SLAB DEPTH & CONCRETE WT. psf	DECK GAGE & WT. psf	NUMBER OF SPANS / q / F	SPAN														
			8′-0″	8′-6″	9′-0″	9′-6″	10′-0″	10′-6″	11′-0″	11′-6″	12′-0″	12′-6″	13′-0″	13′-6″	14′-0″	14′-6″	15′-0″
6½″ 60.4 1 HOUR FIRE RATING	22 (1.9)	1	266	235	209	186	166	149	134	120	108	97	87	79	71	64	57
		2	322	291	209	186	166	149	134	120	108	97	87	79	71	64	57
		3	322	291	265	242	166	149	134	120	108	97	87	79	71	64	57
		q	2430	2410	2390	2375	2360	2345	2330	2315	2300	2290	2280	2265	2270	2260	2250
		F	.34	.35	.35	.36	.36	.36	.36	.36	.36	.37	.37	.37	.37	.37	.37
	21 (2.1)	1	356	266	236	211	189	170	153	138	125	113	102	93	84	76	69
		2	356	322	293	211	189	170	153	138	125	113	102	93	84	76	69
		3	356	322	293	267	246	170	153	138	125	113	102	93	84	76	69
		q	2460	2435	2410	2390	2370	2355	2340	2330	2320	2305	2290	2280	2270	2265	2260
		F	.32	.33	.33	.33	.33	.33	.33	.34	.34	.34	.34	.34	.34	.35	.35
	20 (2.3)	1	373	337	250	224	201	181	163	147	133	121	110	100	90	82	74
		2	373	337	306	280	201	181	163	147	133	121	110	100	90	82	74
		3	373	337	306	280	257	181	163	147	133	121	110	100	90	82	74
		q	2480	2455	2430	2405	2380	2365	2350	2335	2320	2310	2300	2290	2280	2270	2260
		F	.31	.31	.31	.32	.32	.32	.32	.33	.33	.33	.33	.33	.33	.34	.34
	19 (2.7)	1	400	388	353	322	239	216	196	178	162	147	134	123	112	103	94
		2	400	388	353	322	296	216	196	178	162	147	134	123	112	103	94
		3	400	388	353	322	296	273	252	178	162	147	134	123	112	103	94
		q	2540	2510	2480	2455	2430	2410	2390	2370	2350	2340	2330	2315	2300	2290	2280
		F	.27	.28	.28	.29	.29	.29	.29	.30	.30	.30	.30	.30	.30	.30	.30
	18 (2.9)	1	400	400	385	352	323	241	219	199	181	166	152	139	127	117	107
		2	400	400	385	352	323	298	276	199	181	166	152	139	127	117	107
		3	400	400	385	352	323	298	276	256	238	223	152	139	127	117	107
		q	2590	2555	2520	2490	2460	2440	2420	2400	2380	2365	2350	2335	2320	2310	2300
		F	.25	.26	.26	.27	.27	.27	.27	.28	.28	.28	.28	.28	.28	.29	.29
	16 (3.5)	1	400	400	400	400	385	355	329	248	227	208	191	176	162	150	139
		2	400	400	400	400	385	355	329	305	284	266	191	176	162	150	139
		3	400	400	400	400	385	355	329	305	284	266	249	234	213	150	139
		q	2710	2670	2630	2595	2560	2530	2500	2475	2450	2430	2410	2395	2380	2365	2350
		F	.22	.22	.22	.23	.23	.23	.23	.24	.24	.24	.24	.25	.25	.25	.25
7½″ 72.5 2 HOUR FIRE RATING	22 (1.9)	1	311	275	244	217	194	173	155	140	125	113	101	91	82	73	66
		2	378	275	244	217	194	173	155	140	125	113	101	91	82	73	66
		3	378	342	311	217	194	173	155	140	125	113	101	91	82	73	66
		q	2910	2890	2870	2850	2830	2820	2810	2795	2780	2770	2760	2755	2750	2740	2730
		F	.29	.29	.29	.30	.30	.30	.30	.30	.30	.30	.30	.30	.30	.31	.31
	21 (2.1)	1	350	310	276	246	221	198	178	161	145	131	119	107	97	88	79
		2	400	377	276	246	221	198	178	161	145	131	119	107	97	88	79
		3	400	377	343	313	221	198	178	161	145	131	119	107	97	88	79
		q	2940	2915	2890	2870	2850	2835	2820	2805	2790	2780	2770	2760	2750	2745	2740
		F	.27	.27	.27	.27	.27	.28	.28	.28	.28	.28	.28	.28	.28	.29	.29
	20 (2.3)	1	400	328	292	261	234	210	190	171	155	140	127	115	105	95	86
		2	400	395	359	261	234	210	190	171	155	140	127	115	105	95	86
		3	400	395	359	328	301	210	190	171	155	140	127	115	105	95	86
		q	2950	2925	2900	2880	2860	2845	2830	2815	2800	2790	2780	2770	2760	2750	2740
		F	.26	.26	.26	.27	.27	.27	.27	.27	.27	.27	.27	.27	.27	.28	.28
	19 (2.7)	1	400	400	400	310	279	252	228	207	188	171	156	142	130	119	109
		2	400	400	400	377	346	252	228	207	188	171	156	142	130	119	109
		3	400	400	400	377	346	319	295	207	188	171	156	142	130	119	109
		q	3020	2990	2960	2935	2910	2890	2870	2850	2830	2815	2800	2790	2780	2770	2760
		F	.23	.24	.24	.24	.24	.24	.24	.24	.25	.25	.25	.25	.25	.25	.25
	18 (2.9)	1	400	400	400	400	352	280	254	231	211	193	176	161	148	135	124
		2	400	400	400	400	378	348	254	231	211	193	176	161	148	135	124
		3	400	400	400	400	378	348	322	299	211	193	176	161	148	135	124
		q	3070	3035	3000	2970	2940	2920	2900	2880	2860	2845	2830	2815	2800	2790	2780
		F	.21	.22	.22	.22	.22	.23	.23	.23	.23	.23	.23	.23	.23	.24	.24
	16 (3.5)	1	400	400	400	400	400	400	363	288	263	242	222	204	188	174	160
		2	400	400	400	400	400	400	384	356	263	242	222	204	188	174	160
		3	400	400	400	400	400	400	384	356	332	310	290	204	188	174	160
		q	3190	3150	3110	3075	3040	3010	2980	2955	2930	2910	2890	2875	2860	2845	2830
		F	.18	.19	.19	.19	.19	.20	.20	.20	.20	.20	.20	.21	.21	.21	.21

Figure 7-35 *(Continued)*

by a computer analysis and are shown in Figures 7-32 and 7-33.

- Diaphragm design in the E–W direction: The applied lateral force of 240.7 kips is distributed along the roof in the same proportion as the mass distribution at this level. The loading pattern and the corresponding diaphragm shear diagram are shown in Figure 7-34. The maximum diaphragm shear per linear foot occurs at grid line 10 and is equal to

$$v = \frac{19.08 \text{ kips}}{3.8 + 14.5 + 2.5 \text{ ft}} = 0.92 \text{ kips/ft}$$

This value has to be compared with the allowable shear values supplied by the metal-deck manufacturer. For example, if a Verco 20-gauge, W3 Formlok deck with $3\frac{1}{4}$ light-weight concrete fill and puddle welds in every flute is used, the allowable shear will be 1.74 kips/ft, compared to the required value of 0.92 kips/ft (see Figure 7-35). Check diaphragm chord requirements: As mentioned earlier in this chapter, the frame beams at the perimeter of the building will act as chord members or flanges of the diaphragm. To get a handle on the magnitude of the chord forces, diaphragm moments are computed at various sections. The transverse shear forces (in the N–S frames) are small and hence are ignored in this analysis:

moment at grid line 13

$$= 19.7(60) - 0.25(11)(57) - 0.38(9)(47)$$

$$-0.59(10.75)(37.125) - \frac{0.76(31.75)^2}{2}$$

$$= 246 \text{ kips-ft}$$

chord force at grid line 13

$$= \frac{246}{57.58} = 4.27 \text{ kips}$$

moment at grid line 16

$$= 19.7(120) - 0.25(11)(137) - 0.38(9)(107)$$

$$-0.59(10.75)(97.125) - 0.76(87.92)(47.76)$$

$$- \frac{2.8(3.8)^2}{2} + 45.2(60) = 506 \text{ kips-ft}$$

$$\text{chord force at grid line 16} = \frac{506}{57.58} = 8.8 \text{ kips}$$

Similarly, diaphragm moments and chord forces can be computed at other locations. In design of beams and the beam–column connections, these chord forces must be considered. The metal-deck to beam welds must be verified for the development of the chord forces in addition to the diaphragm shear transfer capability.

- Diaphragm design in the N–S direction: Here again, the applied lateral force of 240.7 kips is distributed in proportion to the mass distribution (see Figure 7-36). Diaphragm shears and moments at any location can be computed similarly to the east–west seismic analysis. For example,

diaphragm shear at grid line $G.1$

$$= \frac{65.5 - 1.99(1.75)}{59.25} = \frac{62}{59.25}$$

$$= 1.05 \text{ kips/ft}$$

diaphragm shear at grid line D

$$= \frac{1.99(60.67) - 65.5}{59.25} = \frac{55.2}{59.25}$$

$$= 0.93 \text{ kips/ft}$$

Both of the above computed diaphragm shears are less than the allowable shear value of 3.07 kips per linear foot for a Verco 18-gauge, W3 Formlok deck with puddle welds in all flutes. As an example of diaphragm moment calculations, we compute the diaphragm moment at grid line D:

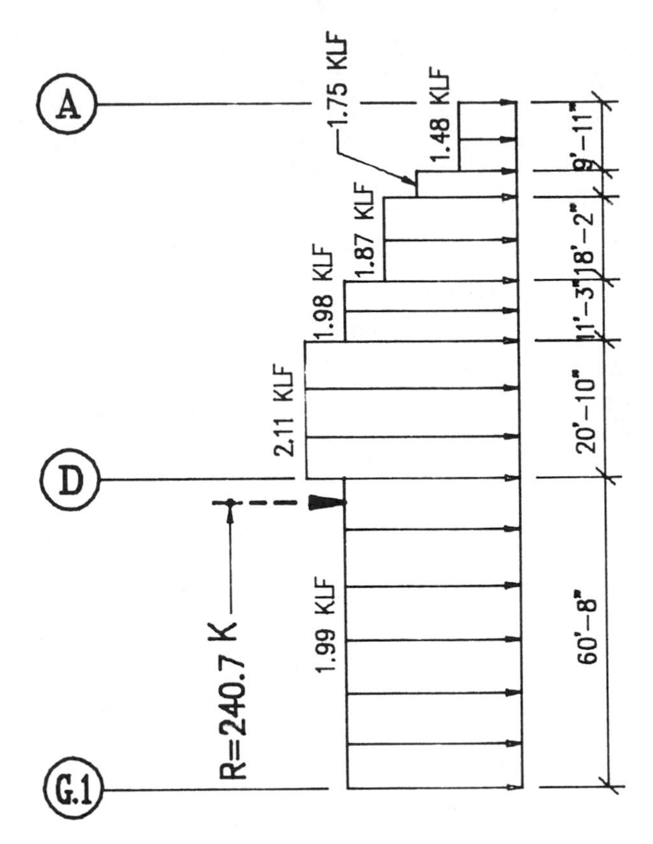

Figure 7-36 Diaphragm loading diagrams for the N–S seismic loading.

diaphragm moment at grid line D

$$= 65.5(58.92) - \frac{1.99(60.67)^2}{2}$$

$$= 3859.3 - 3663 = 197.3 \text{ ft-kips}$$

chord force at grid line D

$$= 197.3/52.92 = 3.7 \text{ kips}$$

The beams along grids 16 and 18, near grid line D, may be designed to carry these chord forces. To complete this design, diaphragm moments should be computed at a few other locations on the diaphragm, in order to establish the maximum moment and the corresponding maximum chord force.

Example 7-4

The ground floor and roof plans of a one-story neighborhood shopping center which is being planned for a city in UBC seismic zone 4 are shown in Figure 7-37. The roof framing consists of plywood panelized roof with glued-laminated beams and purlins. The roof dead load for the purposes of seismic-design calculations is estimated to be 16 lb/ft^2. In addition to the framing

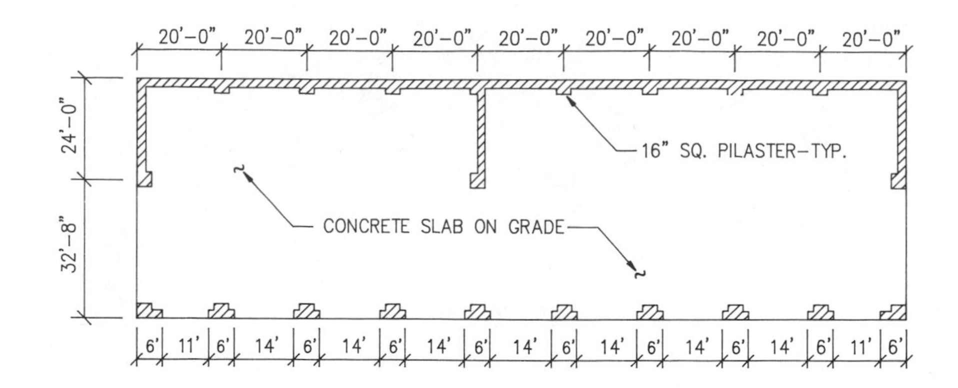

N

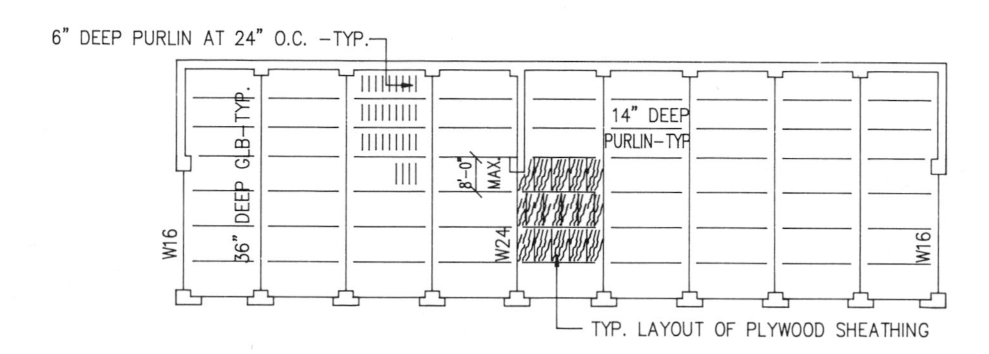

Figure 7-37 Floor plans for building of Example 7-4.

weight, this includes allowances for composition roof, insulation, acoustic tile ceiling, and a miscellaneous load of 1.5 lb/ft^2. Story height is 14 ft and the parapet height is 2 ft. Design the roof diaphragm in accordance with the UBC-85 requirements.

- Dead load and base shear in the N–S direction:

$$\text{north wall at } 75 \text{ lb/ft}^2 = 75\left(\frac{14}{2} + 2\right)(180) = 121{,}500 \text{ lb}$$

$$\text{pilasters in north wall} = 75\frac{14}{2}(1.33 \times 8) = 5600 \text{ lb}$$

$$\text{south piers at } 75 \text{ lb/ft}^2 = 75\left(\frac{14}{2} + 2\right)(10 \times 6) = 40{,}500 \text{ lb}$$

$$\text{pilasters in south piers} = 73\left(\frac{14}{2}\right)(1.33 \times 10) = 7000 \text{ lb}$$

$$\text{glass window at } 15 \text{ lb/ft}^2 = 15\left(\frac{14}{2} + 2\right) \times (7 \times 14 + 2 \times 11) = 16{,}200 \text{ lb}$$

$$\text{roof at } 16 \text{ lb/ft}^2 = 16(180)(56.67) = 163{,}210 \text{ lb}$$

$$\begin{aligned}\text{total dead load} &= 121{,}500 + 5600 + 40{,}500 + 7000 + 16{,}200 + 163{,}210\\ &= 354{,}010 \text{ lb}\end{aligned}$$

$$\begin{aligned}\text{base shear} &= V = ZIKCSW\\ &= (1)(1)(1.33)(0.14)W = 0.186W\\ &= 0.186(354{,}010) = 65{,}846 \text{ lb}\end{aligned}$$

- Diaphragm design in the N–S direction (see Figure 7-38): The diaphragm is assumed to be flexible. Therefore, in both directions, the wall loads will be based on the tributary diaphragm areas:

$$\text{N–S diaphragm load} = \frac{65{,}846 \text{ lb}}{180 \text{ ft}} = 365.8 \text{ lb/ft}$$

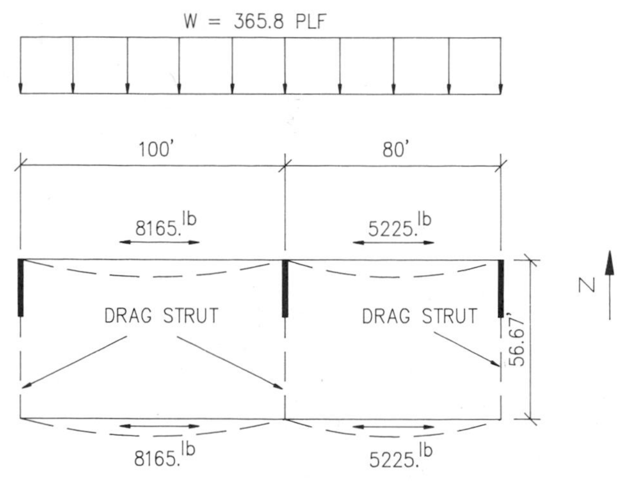

Figure 7-38 Chord forces for N–S seismic loading.

East wall:

$$\text{wall shear} = 365.8(80/2) = 14{,}632 \text{ lb}$$
$$\text{diaphragm shear} = 14{,}632/56.67 = 258.2 \text{ lb/ft}$$
$$\text{force in the drag strut} = 258.2(32.67) = 8435 \text{ lb}$$

Center wall:

$$\text{east side shear} = 365.8(80/2) = 14{,}632 \text{ lb}$$
$$\text{diaphragm shear} = 14{,}632/56.67 = 258.2 \text{ lb/ft}$$
$$\text{west side shear} = 365.8(100/2) = 18{,}290 \text{ lb}$$
$$\text{diaphragm shear} = 18{,}290/56.67 = 322.7 \text{ lb/ft}$$
$$\text{force in the drag strut} = (258.2 + 322.7)(32.67) = 18{,}978 \text{ lb}$$

West wall:

$$\text{wall shear} = 365.8(100/2) = 18{,}290 \text{ lb}$$
$$\text{diaphragm shear} = 18{,}290/56.67 = 322.7 \text{ lb/ft}$$
$$\text{force in the drag strut} = 322.7(32.67) = 10{,}543 \text{ lb}$$

Diaphragm plywood requirements: Per UBC-85 Table 25-J, per $\frac{1}{2}$-in. CDX plywood diaphragm, Struct II, blocked, 8d nails at 4 in. on center at the boundaries and continuous panel edges, 8d nails at 6 in. on center at other panel edges, and 12 in. on center on intermediate framing members. Allowable diaphragm shear is 360 lb/ft, which is more than the required value of 322.7 lb/ft.

Chord design (see Figure 7-38): For the 100-ft span,

$$M = \frac{365.8(100)^2}{8} = 457{,}250 \text{ ft-lb}$$
$$d = 56.67 - \frac{8}{12} = 56.0 \text{ ft}$$
$$C \text{ or } T = \frac{457{,}250}{56.0} = 8165 \text{ lb}$$

For the 80-ft span,

$$M = \frac{365.8(80)^2}{8} = 292{,}640 \text{ ft-lb}$$
$$d = 56.67 - \frac{8}{12} = 56.0 \text{ ft}$$
$$C \text{ or } T = 292{,}640/56.0 = 5225 \text{ lb}$$

Provide horizontal bond beam with chord reinforcement in the north wall at the roof level. The required area of steel is

$$A_s = \frac{8165}{1.33 \times 24000} = 0.26 \text{ in.}^2$$

Therefore use one #5 continuous horizontal bar. A chord member is also required on the south side of the diaphragm. Assuming a timber chord member, the required area is

$$A_{\text{chord}} = \frac{8165}{1.33 \times 950} = 6.5 \text{ in.}^2$$

Since the required chord area is small, one can design the edge purlin to act as a chord. Bolt the purlin to the piers and provide metal strap across the beams for continuity of the chord.

Design of drag struts: The steel beams may be designed to act as drag struts. Diaphragm shear is transferred from plywood to the drag strut by means of the nailer as shown in Figure 7-39. The

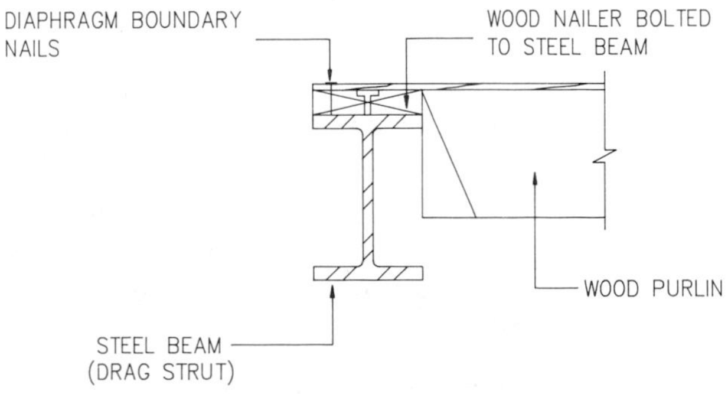

Figure 7-39 Typical detail for transfer of shear from plywood to the drag strut.

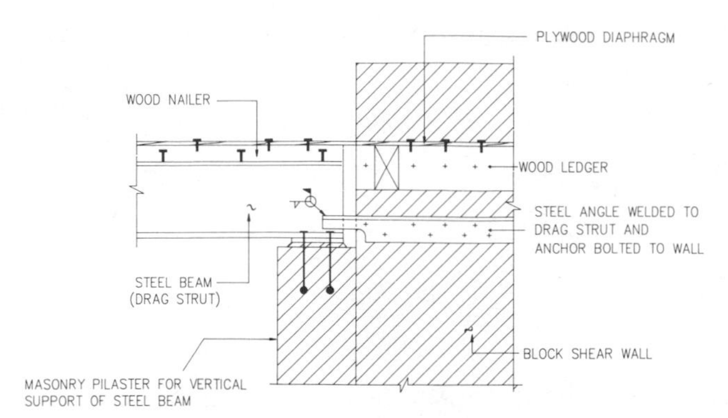

Figure 7-40 Typical detail for transfer of force from drag struts to a block shear wall.

nailer is bolted to the drag strut and the plywood sheathing is nailed to the nailer. The drag-strut force is transferred to the block shear wall by means of the steel angle as shown in Figure 7-40. The steel angle is welded to the steel beam and bolted to the wall. A wood ledger is used to transfer the diaphragm shear from the plywood to the wall, and to attach purlins to the wall.

- Dead load and base shear in the E–W direction: East and west walls at 75 lb/ft.^{2}:

$$75(14/2 + 2)(2)(24) + 75\frac{14}{2}(24) = 45{,}000 \text{ lb}$$

Pilasters at 75 lb/ft.^{2}:

$$75\frac{14}{2}(16/12)(3) = 2100 \text{ lb}$$

Glass windows at 15 lb/ft.^{2}:

$$15(14/2 + 2)(2)(32.67) = 8821 \text{ lb}$$

Roof at 16 lb/ft.^{2}:

$$16(180)(56.67) = 163{,}210 \text{ lb}$$

$$\text{total dead load} = 45{,}000 + 2100 + 8821 + 163{,}210 = 219{,}131 \text{ lb}$$

Then

$$\begin{aligned}\text{base shear} &= V = ZIKCSW \\ &= (1)(1)(1.33)(0.14)W = 0.186W \\ &= 0.186(219{,}131) = 40{,}758 \text{ lb}\end{aligned}$$

- Diaphragm design in the E–W direction (see Figure 7-41):

$$\text{base shear per unit length} = 40{,}758/56.67 = 719.2 \text{ lb/ft}$$

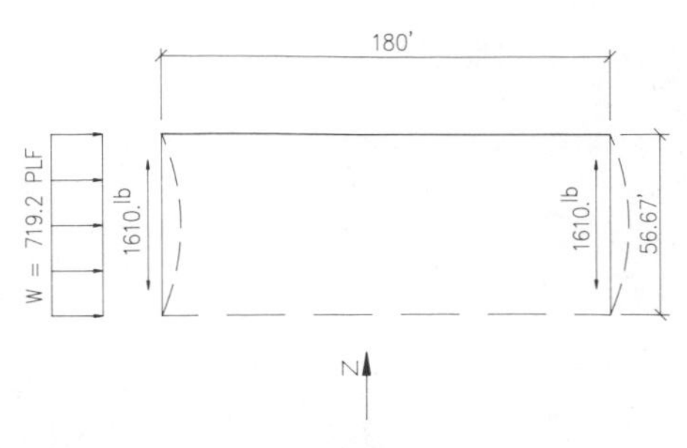

Figure 7-41 Chord forces for E – W seismic loading.

North wall:

$$\begin{aligned}&\text{wall shear} = 719.2(56.67/2) = 20{,}378 \text{ lb} \\ &\text{effective length of diaphragm} = 180 \text{ ft} \\ &\text{diaphragm shear} = 20{,}378/180 = 113 < 360 \text{ lb/ft}\end{aligned}$$

Therefore plywood requirements specified for N–S seismic are adequate along this wall.
South wall:

$$\text{wall shear} = 719.2\frac{56.67}{2} = 20{,}378 \text{ lb}$$

Length of diaphragm in direct contact with the wall is

$$10 \times 6 \text{ ft} = 60 \text{ ft}$$

However, the south-side edge purlins, which were also designed and detailed as the chord for N–S seismic, will act as drag members along the south wall. Therefore diaphragm shear = 20,378/180 = 113 < 360 lb/ft. Hence, previously specified plywood detailing will be adequate. Push or pull at the wall in a typical drag strut is

$$T = (113 \text{ lb/ft})(14/2 \text{ ft}) = 791 \text{ lb}$$

The edge purlin and its bolting to the wall must be verified for the above force.
Chord design:

$$\text{diaphragm span} = 56.67 \text{ ft}$$

$$\begin{aligned}M &= 719.2(56.67)^2/8 = 288{,}712 \text{ ft-lb} \\ d &= 180 - 8/12 = 179.33 \text{ ft} \\ C \text{ or } T &= 288{,}712/179.33 = 1610 \text{ lb}\end{aligned}$$

The chord force is small. It can be easily verified that the steel beam and the horizontal bond beam reinforcement in the block wall will be adequate as chord members.

- Diaphragm deflections: The span-to-width ratio of the diaphragm in both directions is less than 4. Therefore, deflection is not expected to be a problem. However, if a deflection check is necessary, a simple procedure described in the *Timber Construction Manual*[7-14] may be used to estimate diaphragm deflections.

REFERENCES

7-1 International Conference of Building Officials, *Uniform Building Code—1985*, Whittier, CA, 1985.

7-2 International Conference of Building Officials, *Uniform Building Code—1988*, Whittier, CA, 1988.

7-3 Boppana, R. R. and Naeim, F., "Modeling of Floor Diaphragms in Concrete Shear Wall Buildings," *Concrete International, Design & Construction*, ACI, July 1985.

7-4 Roper, S. C. and Iding, R. H., "Appropriateness of the Rigid Floor Assumption for Buildings with Irregular Features," *Proceedings of 8th World Conference on Earthquake Engineering*, San Francisco, CA, 1984.

7-5 Mendes, S., "Wood Diaphragms: Rigid Versus Flexible Inappropriate Assumptions Can Cause Shear-Wall Failures," *Proceedings of the 56th Annual Convention*, Structural Engineers Association of California, San Diego, CA, 1987.

7-6 S. B. Barnes and Associates, "Report on Use of H.H. Robertson Steel Roof and Floor Decks as Horizontal Diaphragms," prepared for H. H. Robertson Company by S. B. Barnes and Associates, Los Angeles, CA, July 1963.

7-7 American Iron and Steel Institute, *Design of Light Gage Steel Diaphragms*, American Iron and Steel Institute, New York, 1982.

7-8 Jain, S. K., "Analytical Models for the Dynamics of Buildings," Earthquake Engineering Research Laboratory, Report No. 83-02, California Institute of Technology, Pasadena, May 1983.

7-9 Steinburgge, K. V., Manning, J. H., and Dagenkolb, H. J., "Building Damage in Anchorage," *The Prince Williams Sound, Alaska, Earthquake of 1964, and Aftershocks*, F. J. Wood (Editor-in-Chief), U.S. Department of Commerce, Washington, 1967.

7-10 Buildings Officials and Code Administrators International, *The BOCA Basic Building Code*, Homewood, IL, 1987.

7-11 American National Standard Institute, *American National Standards Building Code Requirements for Minimum Design Loads in Buildings and Other Structures*, ANSI A58.1-1982, New York, 1982.

7-12 Federal Emergency Management Agency, *1985 Edition of NEHRP Recommended Provisions for the Development of Seismic Regulations for New Buildings,* FEMA-86, Feb. 1986.

7-13 Seismology Committee of Structural Engineers Association of California, *Tentative Lateral Force Requirements*, Nov.–Dec. 1986.

7-14 American Institute of Timber Construction, *Timber Construction Manual*, 2nd Edition, American Institute of Timber Construction, Englewood, CO, 1974.

Chapter 8

Seismic Design of Steel Structures

*Joseph P. Nicoletti, S.E.**

8.1 INTRODUCTION

Structural steel has many physical properties that make it a desirable building material for structures designed to resist earthquake ground motions. Among these properties are:

Isotropy Steel is equally strong in tension or compression. This is a very desirable property because of the cyclic nature of earthquake forces.

Ductility Steel has the ability to sustain large strains, beyond the initial yield limit, without degradation of load-carrying capacity. This property allows steel structures, if they are designed elastically for code forces, to absorb energy with inelastic deformation and avoid collapse for ground motion that is much more severe than that represented by the code forces.

Versatility Structural steel members can be rolled or fabricated to any desired configuration, and modern welding procedures provide great versatility in joining these members into efficient framing systems.

High Strength-to-Weight Ratio Structural steel has a high strength-to-weight ratio and also exhibits a high strength-to-volume ratio as compared with other construction materials. These properties allow steel structures to be efficient in terms of gross floor area and story height and to minimize the building dead loads and thus the earthquake inertia forces.

8.1.1 Scope

This chapter will discuss the design of structural steel lateral-load-resisting systems. It is assumed that the reader is familiar with the basic fundamentals of structural mechanics and structural analysis as well as design procedures for gravity (i.e. dead and live loads). Some of these fundamentals will be repeated in the following sections, but only in so far as they pertain to the design for lateral forces. The following sections will deal with the design of structural steel beams and columns and the necessary connections to assemble these members into moment-resisting and braced frame systems.

8.1.2 Definitions, Symbols, and Notations

Definitions

- *Base shear*, V, is the total design lateral force or shear at the base of a structure.
- A *braced frame* is an essentially vertical truss system of the concentric or eccentric type which is provided to resist lateral forces.
- A *building frame system* is an essentially complete space frame which provides support for gravity loads.

*Senior Vice President, URS/John A. Blume & Associates, San Francisco, California.

- *Chevron bracing* is that form of bracing where a pair of braces located either above or below a beam terminates at a single point within the clear beam span.
- A *compact section* is a structural steel shape meeting certain requirements for the width-to-thickness ratio of the compression flange and the depth-to-thickness ratio of the web.
- *Composite beams* are structural steel shapes that are designed and constructed to resist loads integrally with a reinforced-concrete slab or steel decking with concrete fill.
- A *concentric braced frame* is a braced frame in which the members are subjected primarily to axial forces.
- The *connection* is the group of elements that connect the member to the joint.
- *Diagonal bracing* is that form of bracing that diagonally connects joints at different levels.
- A *diaphragm* is a horizontal or nearly horizontal system acting to transmit lateral forces to the vertical resisting elements. When the term "diaphragm" is used, it includes horizontal bracing systems.
- A *diaphragm strut* (drag strut, tie, collector) is the element of a diaphragm parallel to the applied load which collects and transfers diaphragm shear to vertical resisting elements or distributes loads within the diaphragm. Such members may take axial tension or compression.
- *Drift:* See Story drift.
- A *dual system* consists of moment-resisting space frames and shear walls or braced frames designed in accordance with the prescribed code provisions.
- A *ductile moment-resisting space frame* (DMRSF) is a moment-resisting space frame complying with the requirements for a ductile moment-resisting space frame as prescribed in the 1985 UBC.
- An *eccentric braced frame* (EBF) is that form of braced frame where at least one end of each brace intersects a beam at a point away from a joint.
- A *girder* is a horizontal member in a seismic frame. The words beam and girder may be used interchangeably.
- A *horizontal bracing system* is a horizontal truss system that serves the same function as a diaphragm.
- A *joint* is the entire assemblage at the intersection of the members.
- *K bracing* is that form of bracing where a pair of braces located on one side of a column terminates at a single point within the clear column height.
- The *lateral-force-resisting system* is that part of the structural system assigned to resist lateral forces.
- A *link beam* is that part of a beam in an eccentric braced frame which is designed to yield in shear and/or bending so that buckling of the bracing members is prevented.
- A *moment-resisting space frame* is a space frame in which the members and joints are capable of resisting forces primarily by flexure.
- *Noncompact sections* are structural steel shapes that do not meet the requirements prescribed for compact sections. Allowable stresses for these shapes are reduced on the basis of the width-to-thickness ratio of the compensation flange ($b_f/2t_f$) and/or the depth-to-thickness ratio of the web (d/t_w).
- An *ordinary moment-resisting space frame* is a moment-resisting space frame not meeting special detailing requirements for ductile behavior.
- *Orthogonal effects* are the effects on elements of the structure which are part of the resisting systems along two orthogonal axes due to earthquake forces acting in a direction other than those axes.
- The *P-delta effect* is the secondary effect on shears, axial forces, and moments of frame members induced by the vertical loads acting on the laterally displaced building frame.
- A *space frame* is a three-dimensional structural system without bearing walls composed of interconnected members laterally supported so as to function as a complete self-contained unit with or without the aid of horizontal diaphragms or floor-bracing systems.
- A *special moment-resisting space frame* (SMRSF) is a moment-resisting space frame specially detailed to provide ductile behavior and comply with the requirements given in the SEAOC-86 or UBC-88.
- A *story* is the space between levels. Story x is the story below level x.
- The *story drift* is the displacement of one level relative to the level above or below.
- The *story drift ratio* is the story drift divided by the story height.
- The *story shear*, V_x, is the summation of design lateral forces above the story under consideration.
- A *structure* is an assemblage of framing members designed to support gravity loads and resist lateral forces. Structures may be categorized as building structures or nonbuilding structures.
- V *bracing* is that form of chevron bracing that intersects a beam from above, and *inverted* V

bracing is that form of chevron bracing that intersects a beam from below.

- A *vertical-load-carrying space frame* is a space frame designed to carry all vertical loads.
- X *bracing* is that form of bracing where a pair of diagonal braces cross near midlength of the bracing members.

Symbols and Notation

A_a = the seismic coefficient representing the effective peak acceleration as determined in ATC 3-06

A_e = the effective net area of axially loaded tension members, where the load is transmitted by bolts or rivets through some but not all of the cross-sectional elements of the member

A_f = the area of the compression flange

A_n = the net area of the member

A_{st} = the cross-sectional area of stiffener or pair of stiffeners

A_v = the seismic coefficient representing the effective peak velocity-related acceleration as determined in ATC 3-06

C = the numerical seismic coefficient as specified in UBC-85

C_c = the column slenderness ratio separating elastic and inelastic buckling

C_m = the coefficient applied to the bending term in the interaction formula for prismatic members and dependent upon column curvature caused by applied moments

C_t = the reduction coefficient in computing the effective net area of an axially loaded tension member

E = the modulus of elasticity of steel

F_a = the axial compressive stress permitted in a prismatic members in the absence of bending moment

F_b = the bending stress permitted in a prismatic member in the absence of axial force

F_e' = the Euler stress for a prismatic member divided by a factor of safety

F_u = the specified minimum tensile strength for a specific type of steel

F_y = the specified minimum yield stress for a specific type of steel

I = the importance factor for UBC-85 or SEAOC-86 (UBC-88)

K = the effective length factor for a prismatic member

L = (for beams) the distance between cross-sections braced against twist or lateral displacement of the compression flange; (for columns) actual unbraced length of member

M_p = plastic moment

M_{rs} = the reduced flexural strength of link beams in eccentric braced frames resulting from the presence of axial forces

P_{bf} = the factored beam-flange or connection-plate force in a restrained connection

$P - \text{delta}$ = the moment in a structural member or system resulting from the product of the vertical axial loads and the lateral displacement of the member from its unloaded position

P_e = Euler buckling load

R = seismic response modification coefficient as set forth in ATC 3-06

R_w = seismic response modification coefficient as set forth in the SEAOC-86 (UBC-88)

S = section modulus

V = total design lateral force or shear at the base of the structure

V_p = the yield shear force in a beam or column web

V_s = the shear strength for link beams in eccentric braced frames

Z = the plastic section modulus for a steel member

b_f = the flange width of a rolled beam or plate girder

f_a = computed axial stress

f_b = computed bending stress

f_v = computed shear stress

r = governing radius of gyration

t_f = flange thickness

t_w = web thickness

β = the stress reduction factor for allowable axial stress in bracing members as set forth in the SEAOC-86 (UBC-88)

γ = the rotation of a link beam segment relative to the rest of the beam

8.1.3 Structural Steel Specifications

The following material specifications define structural steel shapes, fasteners, and welding electrodes commonly used in building construction:

American Society for Testing and Materials
1916 Race Street
Philadelphia, PA 19103

- Structural Steel, ASTM A36
- Welded and Seamless Steel Pipe, ASTM A53, Grade B
- High-Strength Low-Alloy Structural Manganese Vandadium Steel, ASTM A441
- Cold Formed Welded and Seamless Carbon Steel Structural Tubing, ASTM A501
- Structural Steel with 42,000 psi Minimum Yield Point, ASTM A529
- High-Strength Low-Alloy Columbium–Vanadium Steels of Structural Quality, ASTM A572
- High-Strength Low-Alloy Structural Steel with 50,000 psi Minimum Yield Point to 4 in. Thick, ASTM A588

American Welding Society
2501 Northwest Seventh Street
Miami, FL 33125

- Specification for Mild Steel Covered Arc-Welding Electrodes, AWS A5.1
- Specification for Bare Mild Steel Electrodes and Fluxes for Submerged-Arc Welding, AWS A5.17
- Specification for Mild Steel Electrodes for Gas Metal–Arc Welding, AWS A5.18
- Specification for Mild Steel Electrodes for Flux-Cored Arc Welding, AWS A5.20

8.1.4 Seismic Design Criteria for Steel Structures

Uniform Building Code (UBC-85) Seismic design of buildings in accordance with model codes, such as the Uniform Building Code,[8-1] prescribes an elastic design based on allowable stresses with equivalent static lateral forces that are substantially less than the dynamic forces associated with a severe earthquake. The UBC-85 design provisions are discussed generally in Section 4.3, and specific provisions applicable to steel lateral-load-resisting systems will be discussed in the following sections of this chapter.

ATC 3-06 Provisions The *Tentative Provisions for the Development of Seismic Regulations for Buildings*, ATC 3-06,[8-2] published by the Applied Technology Council (ATC) in June 1978, were a significant departure from the established format of the UBC seismic provisions. The ATC provisions (see Section 4.5) established seven zones for the definition of the site seismicity, A_a or A_v, corresponding to the five zones for the Z factor in the UBC. The accelerations given for each zone are all based on the same risk level (10% probability of being exceeded in a 50-year period). The earthquake demand forces on a building subjected to these ground accelerations (see Chapter 3) are reduced for design purposes by a reduction factor R. The ATC 3-06 provisions are based on structural design at yield stress levels, so that the above R factors are an indication of the anticipated inelastic response of the structure to the effective peak acceleration for each seismic zone.

NEHRP Provisions The ATC 3-06 provisions have been extensively reviewed and tested by trial designs since their publication. The Building Seismic Safety Council (BSSC) in 1985 published a revised version of these provisions[8-3] as part of the National Earthquake Hazards Reduction Program (NEHRP). These NEHRP provisions, which are being promulgated by the Federal Emergency Management Agency as a national standard for seismic provisions, are currently (1987) undergoing a final evaluation by BSSC for reissue in a further revised edition scheduled in 1988.

SEAOC Recommendations (UBC-88) The seismic provisions in the UBC have traditionally been based on recommendations from the Seismology Committee of the Structural Engineers Association of California (SEAOC). The current recommendations that have been adopted as a part of the 1988 UBC are the result of a five-year effort by SEAOC to adapt many of the features of the ATC 3-06 provisions within the basic framework of the UBC format. The SEAOC recommendations[8-4] retain the five seismic zones (i.e. 0, 1, 2, 3, and 4) and the importance factor I. The C factor has been retained with some modifications that incorporate the current soil factor S. The principal change is the replacement of the K factor with an R_w. The R_w factor is similar in concept to R factor in ATC 3-06, except that R_w is based on working stress levels for all materials, except concrete.

Design vs. Realistic Displacements In the design of steel structures, it is important to recognize the fact that design to code-level forces implies that the structural displacements under severe ground motion may be substantially greater than the calculated static displacement with the design forces. This consideration not only influences the selection of the structural steel system and its design and detailing, but for moment frames, as will be discussed in Section 8.5.3, it may have a serious influence on overall structural stability due to the P-delta effect.

8.2 DESIGN OF STEEL BEAMS

Structural steel beams in buildings are usually attached to or embedded in the floor construction, so that axial loads on these beams are negligible (see Section 8.6 for connections for special cases of beams designed for axial loads). Thus most beams in buildings need only be designed for shears and moments in their vertical plane (i.e. the plane of the beam web).

The allowable bending stresses of rolled or fabricated steel sections used as beams may be limited by the potential for (i) local buckling of the beam web or an unstiffened compression flange or (ii) lateral–torsional buckling between points of lateral support for the compression flange.

8.2.1 Compact Sections

The AISC Specification(8-5) stipulates that tension and compression on the extreme fibers of compact hot-rolled or built-up steel members, symmetrical about, and loaded in, the plane of their minor axis (plane of the web), shall not exceed

$$F_b = 0.66F_y \tag{8-1}$$

where

F_b = bending stress permitted in a prismatic member in the absence of axial force (ksi)
F_y = specified minimum yield stress (ksi)

In order to qualify as a compact section, a member must meet the following requirements:

1. The flanges shall be continuously connected to the web or webs.
2. The width-to-thickness ratio of unstiffened projecting elements of the compression flange shall not exceed

$$\frac{65}{\sqrt{F_y}} \tag{8-2}$$

3. The width-to-thickness ratio of stiffened elements of the compression flange shall not exceed

$$\frac{190}{\sqrt{F_y}} \tag{8-3}$$

4. The depth-to-thickness ratio of the web or webs shall not exceed the value

$$\frac{d}{t} = \frac{640}{\sqrt{F_y}}\left(1 - 3.74\frac{f_a}{F_y}\right) \quad \text{when} \quad \frac{f_a}{F_y} \leqslant 0.16 \tag{8-4}$$

or

$$\frac{d}{t} = \frac{257}{\sqrt{F_y}} \quad \text{when} \quad \frac{f_a}{F_y} > 0.16 \tag{8-5}$$

5. The laterally unsupported length of the compression flange of members other than circular or box members shall not exceed the value

$$\frac{76b_f}{\sqrt{F_y}} \quad \text{or} \quad \frac{20{,}000}{(d/A_f)F_y} \tag{8-6}$$

where

b_f = flange width of rolled beam or plate girder (in.)

d = depth of beam or girders (in.)

A_f = area of compression flange (in.^{2})

6. The laterally unsupported length of the compression flange of a box-shaped member of rectangular cross section whose depth is not more than 6 times the width and whose flange thickness is not more than 2 times the web thickness shall not exceed the value

$$\left(1950 + 1200\frac{M_1}{M_2}\right)\frac{b}{F_y} \tag{8-7}$$

except that it need not be less than $1200(b/F_y)$, where

M_1 = smaller moment at end of unbraced length of beam or column (kip-ft)

M_2 = larger moment at end of unbraced length of beam or column (kip-ft)

b = actual width of stiffened and unstiffened compression elements (in.)

7. The diameter-to-thickness ratio of hollow circular sections shall not exceed

$$3300/F_y. \tag{8-8}$$

The AISC Manual(8-6) lists the properties and dimensions of the standard rolled shapes. These tables also list the limiting yield stresses F_y, below which the section may be considered to be compact in accordance with the above criteria. The limiting yield stresses are given for both the flange-width–thickness criterion and the web-depth–thickness criterion. From these tabulations, it will be noted that most of the heavier *W* sections commonly used in building construction meet the compact criteria for yield stresses in the range of 36 to 50 ksi, but many of the lighter sections will have reduced allowable bending stress f_b for 36-ksi yield stress (A36 steel).

8.2.2 Noncompact Sections

1. Members which meet the requirements in Section 8.2.1, except that $b_f/2t_f$ exceeds $65/\sqrt{F_y}$ but is less than $95/\sqrt{F_y}$, may be designed on the basis of an allowable

bending stress:

$$F_b = F_y\left[0.79 - 0.002\left(\frac{b_f}{2t_f}\right)\sqrt{F_y}\right] \quad (8\text{-}9)$$

where

t_f = flange thickness (in.)

2. Tension and compression on extreme fibers of doubly symmetrical I- and H-shape members meeting the requirements of Section 8.2.1, paragraphs 1 and 2, and bent about their minor axis; solid round and square bars; solid rectangular sections bent about their weaker axis:

$$F_b = 0.75F_y \quad (8\text{-}10)$$

Doubly symmetrical I- and H-shape members bent about their minor axis meeting the requirements of Section 8.2.1, paragraph 1, except where $b_f/2t_f$ exceeds $65/\sqrt{F_y}$, may be designed on the basis of an allowable bending stress

$$F_b = F_y\left[1.075 - 0.005\left(\frac{b_f}{2t_f}\right)\sqrt{F_y}\right] \quad (8\text{-}11)$$

Rectangular tubular sections meeting the requirements of Section 8.2.1, paragraphs 1, 3, and 4, and bent about their minor axis, may be designed on the basis of an allowable bending stress:

$$F_b = 0.66F_y \quad (8\text{-}12)$$

3. Tension and compression on extreme fibers of box-type flexural members whose compression flange or web width-to-thickness ratio does not meet the requirements of Section 8.2.1, but is less than $238/\sqrt{F_y}$:

$$F_b = 0.60F_y \quad (8\text{-}13)$$

Lateral torsional buckling need not be investigated for a box section whose depth is less than 6 times its width. Lateral support requirements for box sections or larger depth-to-width ratios must be determined by special analysis.

4. For allowable tension and compression stresses in noncompact members not covered in the above paragraphs (e.g. angles, channels, tees, etc.) reference is made to Section 1.5.1.4.5 of Reference 8-5.

The provisions of this section and the preceding section apply equally to the design of steel beams for gravity loads as well as lateral loads. Example 8-1 illustrates the compact section check for a light W section and the derivation of reduced allowable flexural stresses. The following sections will discuss additional considerations that may govern the flexural load capacity of beams that are expected to form yield hinges as they respond inelastically to severe ground motion.

Example 8-1 (Check for AISC Compact Section Criteria)
For W6 × 15, from the AISC Handbook:

$t_w = 0.230$ in., $t_f = 0.260$ in., $d = 5.99$ in.
$b_f = 5.99$ in., $d/A_f = 3.85$

Equation 8-2:

$$\frac{b_f}{2t_f} \leqslant \frac{65}{\sqrt{F_y}} \qquad \text{for } F_y = 36 \text{ ksi}$$

$$\frac{5.99}{2} \times 0.260 \leqslant \frac{65}{\sqrt{36}}$$

$$11.52 \geqslant 10.83 \qquad \text{N.G.}$$

Equation 8-4:

$$\frac{d}{t_w} \leqslant \frac{640}{\sqrt{F_y}}\left(1 - 3.74\frac{f_a}{F_y}\right)$$

$$\leqslant \frac{640}{\sqrt{F_y}} \qquad \text{for } f_a = 0 \quad (\text{no axial load})$$

$$\frac{5.99}{0.230} \leqslant \frac{640}{\sqrt{36}}$$

$$26.04 \leqslant 106.67 \qquad \text{O.K.}$$

Allowable F_b (major axis)

$$= F_y\left(0.79 - 0.002\frac{b_f}{2t_f}\sqrt{F_y}\right) \qquad (\text{from Eq. 8-9})$$

$$= F_y(0.79 - 0.002 \times 11.52 \times 6)$$

$$= 0.652F_y$$

Allowable F_b (minor axis)

$$= F_y\left(1.075 - 0.005\frac{b_f}{2t_f}\sqrt{F_y}\right) \qquad (\text{from Eq. 8-11})$$

$$= F_y(1.075 - 0.005 \times 11.52 \times 6)$$

$$= 0.729F_y$$

8.2.3 Lateral–Torsional Buckling

Secondary framing members in building construction (e.g. floor and roof beams and purlins) are usually designed as simple beams. Under this design assumption, the upper flange of the beam is in compression for its entire length and, since it is restrained by the floor system (e.g. concrete slab or steel decking), has continuous lateral support. The possibility of lateral–torsional buckling need not be considered for these beams. How-

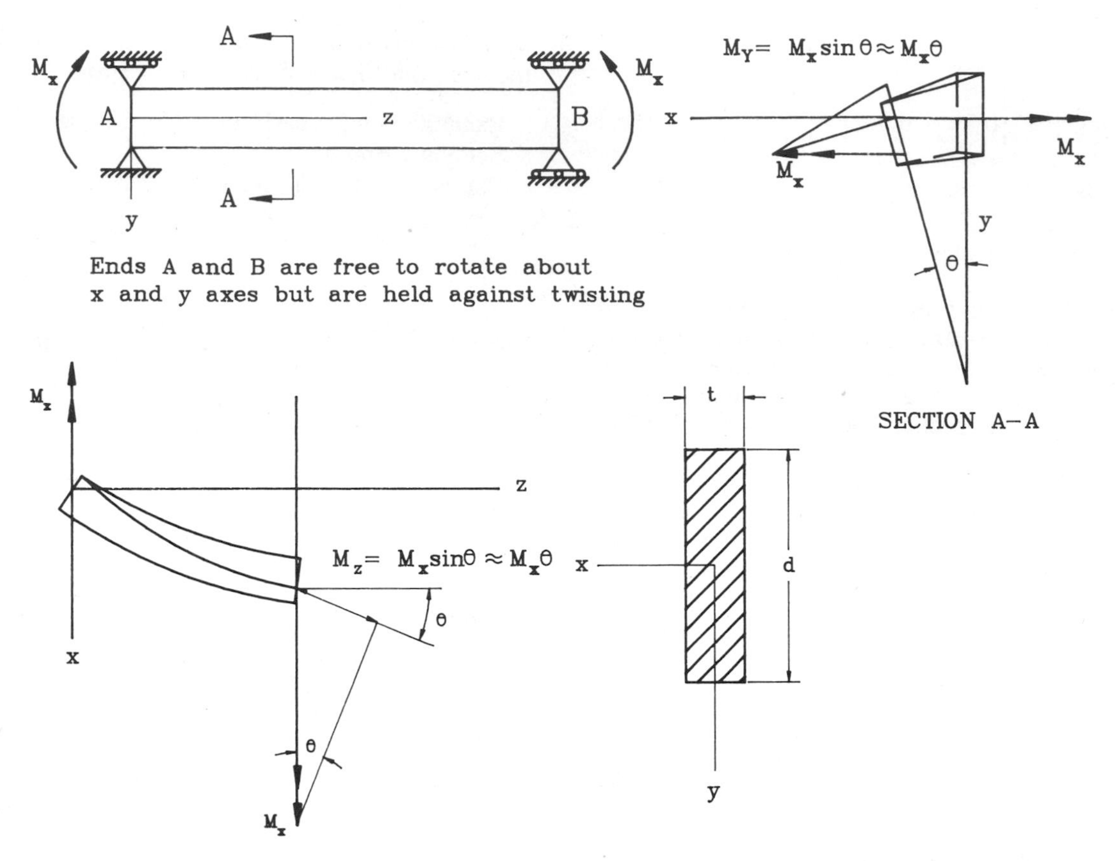

Figure 8-1 Torsion (M_z) and lateral bending moment (M_y) induced during lateral-torsional buckling by uniform bending moment (M_z). (8-7)

ever, for beams continuous over one or more supports and for beams in moment frames, the potential for lateral–torsional buckling of the laterally unsupported lower flange, where it is in compression, must be investigated.

Lateral–torsional buckling may be initiated by geometrical imperfections (e.g. rolling tolerances, or out-of-straightness) in the beam that will cause the normal bending moment to induce torsional and lateral moments as indicated in Figure 8-1. This figure is reproduced from the Structural Stability Research Council *Guide to Stability Design Criteria for Metal Structures*.[8-7] Additional discussion regarding lateral–torsional buckling will be found in that publication.

Lateral Bracing of Compression Flange Framing considerations may dictate situations where it is not feasible to provide lateral support for a steel beam. An example of this situation is illustrated in Figure 8-2, where the beams between elevator-shaft openings do not have a floor slab or other lateral support for either flange. The following alternatives could be considered:

1. Use of a box girder section. This section has excellent torsional and lateral bending resistance.
2. Use of boxed flanges (see Figure 8-3). This solution improves the lateral bending resistance, and testing has demonstrated that it increases the resistance to lateral–torsional buckling.
3. Concrete encasement. This has been shown to be particularly effective for increasing the torsional resistance of open (I-beam and W-shape) sections.

Floor beams are usually adequate to provide lateral support for the compression flange of a continuous girder or moment-frame beam when the depths of the two beams are not too disparate (see Figure 8-4). When the floor beams are of significantly less depth than the beam to be braced, then a direct inclined brace between the compression flange and the floor slab should be considered. Exterior spandrel beams in frames that are parallel to the floor beams may need special bracing as indicated in Figure 8-5.

The sizing of lateral support members does not lend itself to analytical solution. Since there is theoretically no force in these members if there is no twist or misalignment, it is obvious that rigidity is necessary as well as strength. The common rule of thumb is to design the brace for a normal force equal to 2% of the compression force in the flange to be braced. This is considered to provide the necessary rigidity as well as strength. The application of this procedure is shown in Example 8-2.

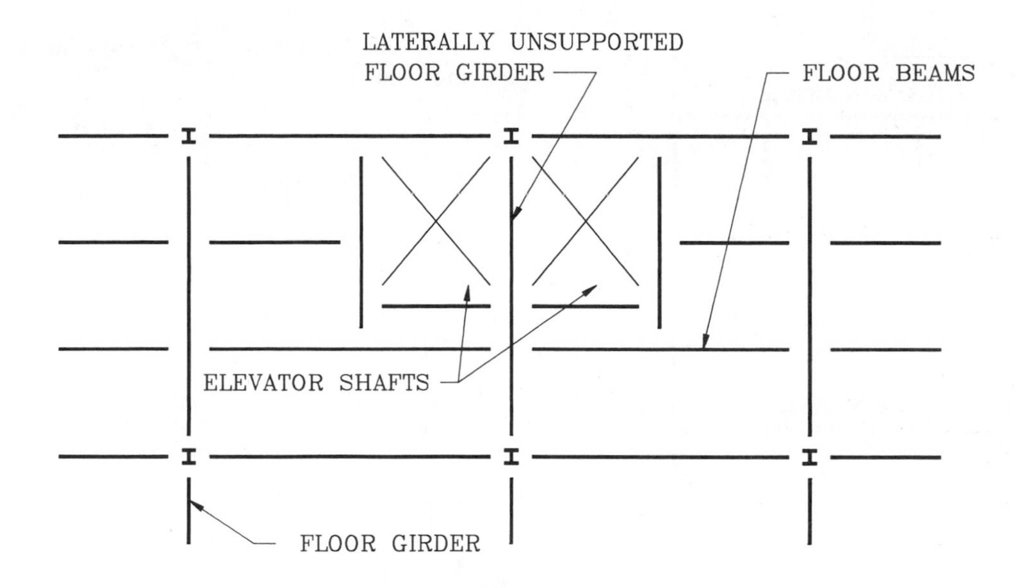

PARTIAL FLOOR FRAMING PLAN

Figure 8-2 Example of laterally unsupported girder in elevator shaft framing.

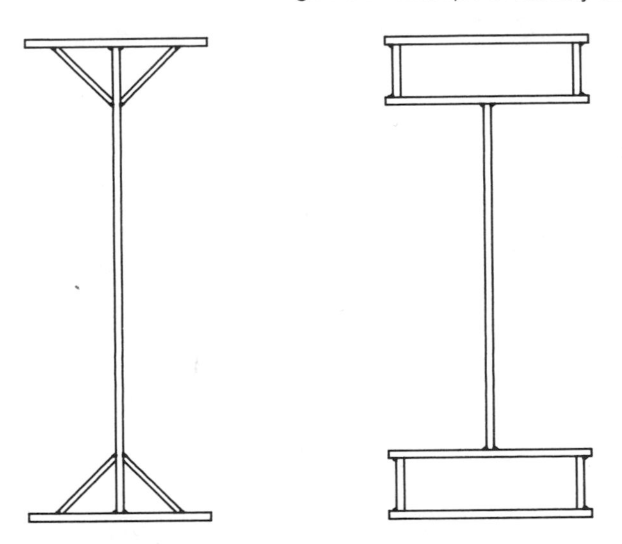

Figure 8-3 Girder sections with boxed flanges.

Figure 8-4 Floor beam connection to provide lateral support for lower flange of girder.

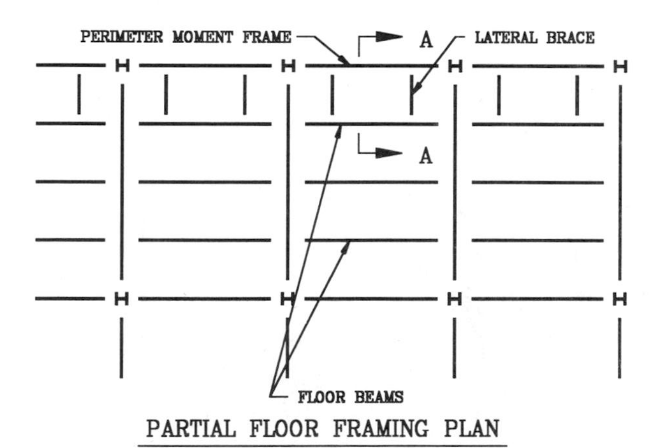

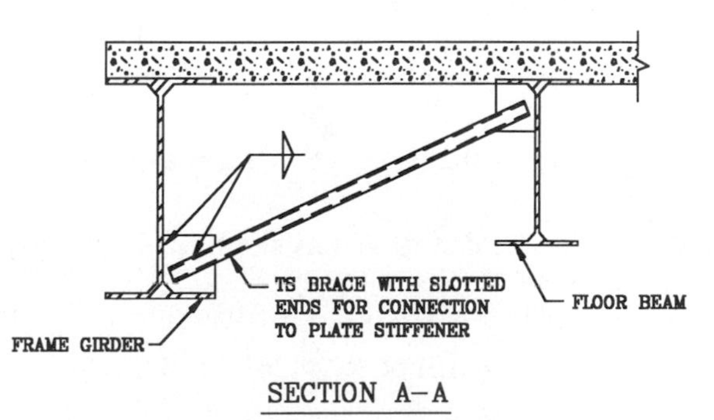

Figure 8-5 Typical detail for lateral bracing of compression flange of perimeter frame girders.

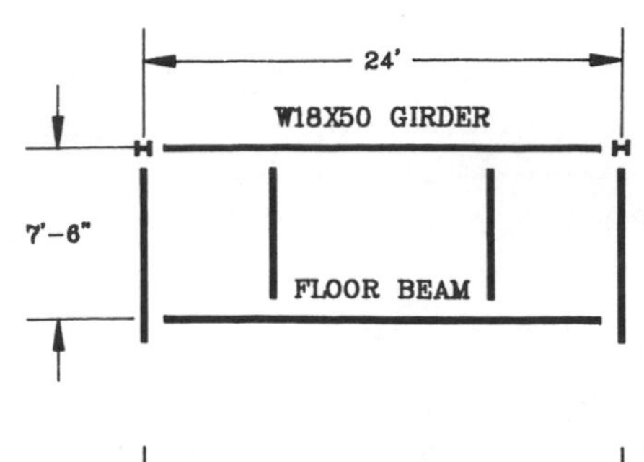

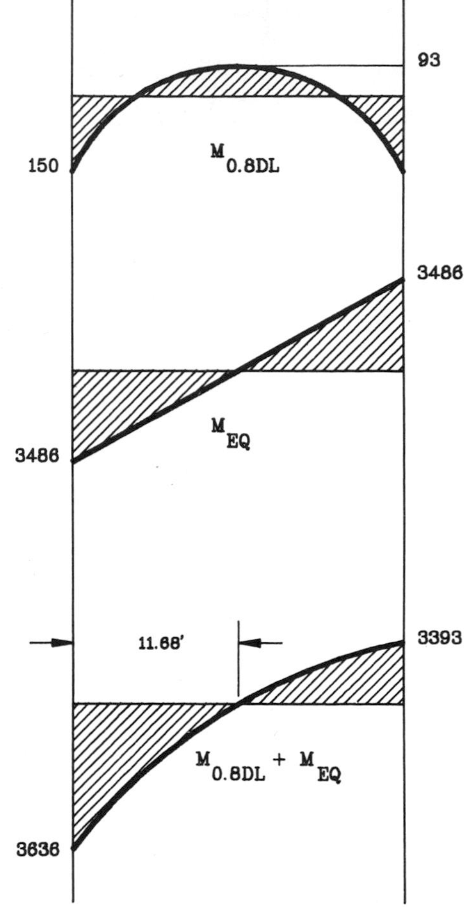

Figure 8-6 Design of lateral bracing (Example 8-2).

Example 8-2 (Design of Lateral Bracing) Refer to Figure 8-6. Assume

$$W_{0.8\mathrm{DL}} = 0.075 \times 7.5/2 = 0.281 \text{ kip/ft}$$

$$\sum M_{0.8\mathrm{DL}} = 0.281 \times \frac{24^2}{8} \times 12 = 243 \text{ in.-kips}$$

Assume moment distributed as shown in Figure 8-6:

$$M_\mathrm{p} = ZF_\mathrm{y} \qquad (Z = 101 \text{ in.}^3)$$
$$= 101 \times 36 = 3636 \text{ in.-kip}$$

Beam capacity available for seismic moment:

$$3636 - 150 = 3486 \text{ in.-kip}$$

Unbraced length of compression flange under ultimate moment conditions:

$$L_\mathrm{u} = 11.68 \text{ ft}$$

Equation 8-6 requires

$$L_\mathrm{u} \leqslant \frac{76b_\mathrm{f}}{\sqrt{F_\mathrm{y}}} \quad \text{and} \quad L_\mathrm{u} \leqslant \frac{20{,}000}{(d/A_\mathrm{f})F_\mathrm{y}}$$

Thus

$$L_\mathrm{u} \leqslant \frac{76 \times 7.495}{\sqrt{36}} = 94.9 \text{ in.}$$

and

$$L_\mathrm{u} \leqslant \frac{20{,}000}{4.21 \times 36} = 132.0 \text{ in.}$$

Use lateral bracing at 6 ft from each end, $L_\mathrm{u} = 72$ in. Design brace for compression force from average moment between brace and end of beam:

$$M = 3636 - \frac{3}{11.68} \times 3636 = 2702 \text{ in.-kips}$$

$$\text{force in beam compression flange} = 2702/17.99$$
$$= 150 \text{ kips}$$

design brace for 2% $\times$ 150 = 3 kips

Brace $L/r \approx 120$ for rigidity

Try $2 \times 2 \times 1/4$ tube. $A = 1.59$ in.^{2}, $r = 0.694$ in.:

$$L = \sqrt{18^2 + 90^2} = 91.8 \text{ in.} \quad \text{(see Figure 8-5)}$$

Clear length = 91.8 − 8 = 83.8 in.

$$\frac{L}{r} = \frac{83.8}{0.694} = 121 \qquad \text{O.K.}$$

$$F_\mathrm{a} = 10.14 \text{ ksi}$$

The brace is overdesigned for the calculated load, but provides good rigidity and simple connection detail.

8.2.4 Composite Beams

Composite beams in building construction are usually composed of a steel shape with a reinforced-concrete slab or with steel decking and concrete fill. Although the shear transfer between the steel beam and the concrete slab or the decking may be enhanced by embedding the top flange in the concrete slab or welding the deck to the beam, usually shear studs are provided as indicated in Figure 8-7 and the studs are designed for the entire shear to be transferred.

Since the composite action is primarily effective in increasing the flexural section modulus when the con-

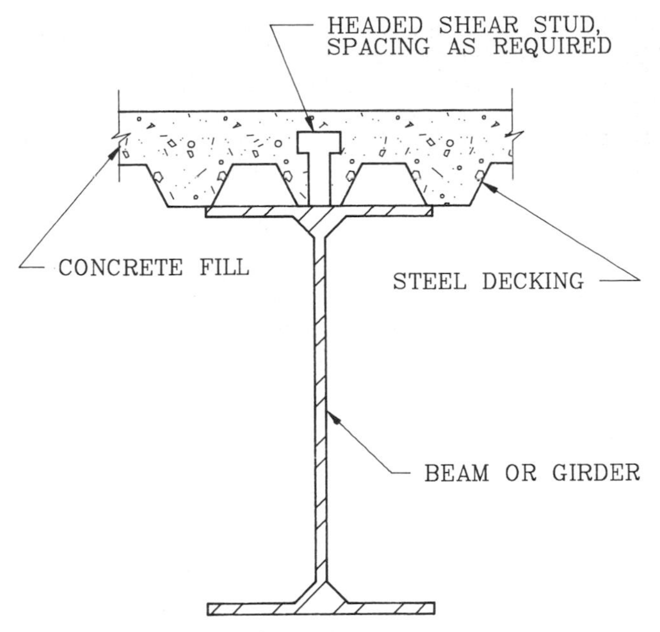

Figure 8-7 Shear stud detail for development of composite action.

crete is in compression, common practice in the design of steel-framed buildings is to design only the secondary beams as simple beams with composite action. The frame beams are designed without composite action. However, it is generally recognized that even though normal floor construction may not furnish dependable additional strength at critical sections of the frame beams, it increases the moment of inertia and reduces drift. This consideration, as well as the use of centerline-to-centerline dimensions instead of clear span and clear height, tends to make calculated drifts larger than actual frame drifts. Because of the difficulty in quantifying these drift adjustments, some designers in the higher seismic zones provide nominal shear studs for composite action over the frame girders. The composite action is neglected for member sizing and drift control under code requirements, but a reduction factor (of 20–25%) will be applied to the calculated drifts in the P–Δ calculations for stability under severe earthquake loading (see Section 8.5.3 under "The P–Delta Effect.")

8.3 DESIGN OF STEEL COLUMNS

Columns in structural-steel lateral-force-resisting systems are usually required to resist bending (due to shears and moments) as well as axial loads. In cases where moment frames are used in both orthogonal directions, the columns may be subjected to biaxial bending as well as a resultant axial load from the two orthogonal frames. Although it may be feasible to design braced frames with simple beam connections to the columns, even these connections will input some moment into the column, and unless the connections coincide with the column neutral axes, additional moments will result from the eccentricity of the shear reaction.

The following subsections will discuss the fundamental considerations for allowable stresses in columns due to axial loads and for the combined effect of axial loads and bending.

8.3.1 Centrally Loaded Columns

Over 200 years ago, Euler[8-8] developed a formula to calculate the critical load for a slender, elastic, and initially straight centrally loaded column. For a pin-ended column, the critical load is given by the expression

$$P_e = \frac{\pi^2 EI}{L^2} \tag{8-14}$$

where

P_e = Euler load as defined above
E = modulus of elasticity
I = moment of inertia of the column
L = length of the column between pins

The more common version of this equation is in terms of the allowable compression stress:

$$F_e = \frac{\pi^2 E}{(KL/r)^2} \tag{8-15}$$

where

F_e = Euler stress = P_e/A

r = radius of gyration of the column

K = effective-length factor for other than pin-ended columns

Early tests of steel columns of practical proportions indicated that failure actually occurred at loads less than the Euler load. Because of residual stresses and initial out-of-straightness, inelastic action was found to begin at an average stress level of $0.5F_y$. The value of KL/r corresponding to this stress level is designated as C_c, and it defines the boundary between elastic and inelastic action. The AISC[8-5] allowable axial column stress, F_a, for columns with KL/r less than C_c is

$$F_a = \frac{\left[1 - \dfrac{(KL/r)^2}{2C_c^2}\right] F_y}{\dfrac{5}{3} + \dfrac{3(KL/r)}{8C_c} - \dfrac{(KL/r)^3}{8C_c^3}} \tag{8-16}$$

where

$$C_c = \frac{2\pi^2 E}{F_y}$$

The allowable compression stress when KL/r exceeds C_c is

$$F_a = \frac{12\pi^2 E}{23(KL/r)^2} \qquad (8\text{-}17)$$

The above allowable stresses provide a variable factor of safety from $\frac{5}{3}$ at $KL/r = 1$ to $\frac{21}{12}$ at $KL/r = C_c$. For KL/r values greater than C_c, the factor of safety against buckling is uniformly maintained at $\frac{23}{12}$. For A36 steel, $C_c = 126$, which indicates that most columns in conventional steel buildings would be expected to fail by inelastic yielding rather than elastic buckling.

Effective-Length Factor The effective-length factor K, when applied to columns with other than pinned ends, is intended to modify the actual column length to make it equivalent to a pinned-end column (for which $K = 1.0$). The effective-length factor is related to the degree of restraint against rotation and translation provided at each end of the column. For a column with ends that are fully fixed and restrained against sideway, the theoretical value of K is 0.5.

For braced frames, or other frames where joint translation is restrained by shear walls or other structures having adequate lateral stability, the effective length factor K for compression members may be taken as unity, unless analysis demonstrates that a smaller value may be used.

For unbraced frames where the lateral stability is dependent upon the bending stiffness of rigidly connected beams and columns, the AISC Specification[8-5] requires determination of the effective length by prescribed procedures based on the restraint provided at each end of the column, but not less than the actual unbraced length (see Figure 6-10).

Several researchers have demonstrated that the choice of the effective-length coefficient K should be based on consideration of overall frame instability (or overall story instability for buildings with rigid diaphragms). After a series of comprehensive studies of 17 frames with various design parameters (number of stories, story height, bay width, dead and live loads, wind, etc.), the Structural Stability Research Council[8-7] has proposed the following tentative recommendations for the design of frames without consideration of frame instability:

1. All columns are to be proportioned to satisfy the interaction formulas with K equal to unity in the calculations of F_a and F_e'. The coefficient C_m is computed from the formula $C_m = 0.6 - 0.4M_1/M_2 \geqslant 0.4$.

2. The maximum column axial-load ratios f_a/F_a and $f_a/0.6F_y$ are not to exceed 0.75.

3. The maximum in-plane column-slenderness ratio L/r is not to exceed 35.

4. The bare-frame first-order story-drift index, Δ/H, at working load shall be controlled so that $\Delta/H < \frac{1}{7}(\Sigma V/\Sigma P)$, where

H = story height

Δ = drift of story due to ΣV

ΣV = total story shear due to lateral loads

ΣP = total gravity load on story

If the drift index is not satisfied, a design based on a second-order P-delta analysis must be carried out.

The limitations in items 2, 3, and 4 above, according to this reference, were based on the limiting values used in the studies and may be increased in the future, if justified by additional studies. Example 8-3 illustrates the design procedures for beams and columns in ductile moment frames. The example initially follows the 1985 UBC procedures for the column design and indicates the effect the above recommendations would have on the selection of the column. See also Chapter 6 for further discussion of column effective-length factors.

8.3.2 Beam – Columns

A beam–column is defined as a structural member that is subjected concurrently to axial force and bending moment. The bending moment may be caused by applied forces between the ends of the column, by beams connecting to the ends of the column, or by the eccentricity of the axial force at either end of the column. It should be noted that as the moments in a beam–column approach zero, it becomes a centrally loaded column, and as the axial load approaches zero, it becomes a beam.

From the above discussion, it follows that beam–columns may be subject to the limitations associated with beams (see Section 8.2) as well as those associated with centrally loaded columns (Section 8.3.1). The design procedure of Reference 8-4 uses empirical interaction formulas, based both on test results and analytical studies. These empirical formulas provide allowable stress combinations for the design loads. These allowable stresses compensate for the potential effects of residual stresses, initial imperfection, and lateral–torsional buckling. For members subject to both axial compression and biaxial bending stresses, Reference 8-5 requires

$$\frac{f_a}{F_a} + \frac{C_{mx} f_{bx}}{\left(1 - \dfrac{f_a}{F'_{ex}}\right) F_{bx}} + \frac{C_{my} f_{by}}{\left(1 - \dfrac{f_a}{F'_{ey}}\right) F_{by}} \leqslant 1.0 \qquad (8\text{-}18)$$

and

$$\frac{f_a}{0.60F_y} + \frac{f_{bx}}{F_{bx}} + \frac{f_{by}}{F_{by}} \leqslant 1.0 \qquad (8\text{-}19)$$

When $f_a/F_a \leqslant 0.15$, Equation 8-20 may be used in lieu of Equations 8-18 and 8-19:

$$\frac{f_a}{F_a} + \frac{f_{bx}}{F_{bx}} + \frac{f_{by}}{F_{by}} \leqslant 1.0 \qquad (8\text{-}20)$$

In Equations 8-18, 8-19, and 8-20, the subscripts x and y, combined with subscripts b, m, and e, indicate the axis of bending about which a particular stress of design property applies, and

F_a = axial compressive stress that would be permitted if axial force alone existed, kips/in.^{2}

F_b = compressive bending stress that would be permitted if bending moment alone existed, kips/in.^{2}

$$F_e' = \frac{12\pi^2 E}{23(KL_b/r_b)^2} \qquad (8\text{-}21)$$

= Euler stress divided by a factor of safety, kips/in.^{2}, where L_b is the actual unbraced length in the plane of bending, r_b is the corresponding radius of gyration, and K is the effective length factor in the plane of bending

f_a = computed axial stress, kips/in.^{2}

f_b = computed compressive bending stress at the point under consideration, kips/in.^{2}

C_m = a coefficient whose value shall be taken as follows:

a for compression members in frames subject to joint translation (sideways), $C_m = 0.85$

b for restrained compression members in frames braced against joint translation and not subject to transverse loading between their supports in the plane of bending.

$$C_m = 0.6 - 0.4\frac{M_1}{M_2}, \text{ but not less than } 0.4 \qquad (8\text{-}22)$$

where M_1/M_2 is the ratio of the smaller to the larger moment at the ends of that portion of the member unbraced in the plane of bending under consideration (positive when the member is bent in reverse curvature, negative when bent in single curvature)

c for compression members in frames braced against joint translation in the plane of loading and subjected to transverse loading between their supports, the value of C_m may be determined by rational analysis, or else the following values may be used:

1 for members whose ends are restrained, $C_m = 0.85$.

2 for members whose ends are unrestrained, $C_m = 1.0$.

Additional discussion regarding the selection of appropriate values for K and C_m for moment frames is given in Section 8.5.4. Example 8-3 provides a numerical example of the beam–column design for biaxial bending and axial compression.

Example 8-3 (Design of a Ductile Moment Frame) Assume:

story height = 12 ft, bay width = 24 ft

Loads and moments as shown in Figure 8-8.

- Beam design:

$$M_{DL+LL+EQ} = 65 + 80 + 120 = 265 \text{ ft-kip}$$

$$F_b = 0.66F_y \quad \text{(compact section)}$$

$$= 24 \text{ ksi} \times 1.33 = 32 \text{ ksi}$$

$$S_{reqd} = 265 \times 12/32 = 99.4 \text{ in.}^3$$

Use W18 × 55:

$$S = 98.3 \quad \text{O.K.}$$

(compact section per AISC table).

- Column design: Try W14 × 145:

$$A = 42.7$$

$$S_{x-x} = 232, \quad r_{x-x} = 6.33, \quad I_{x-x} = 1710$$

$$S_{y-y} = 87.3, \quad r_{y-y} = 3.98, \quad I_{y-y} = 677$$

Assume W18 × 55 beams (x direction) and W18 × 35 (y direction) at top and bottom of column: $I_{g(x-x)} = 890$, $I_{g(y-y)} = 510$. Then

$$G_A = G_B = \frac{\Sigma I_c/L_c}{\Sigma I_g/L_g}$$

$$= \frac{2 \times 1710/12}{2 \times 890/24} = 3.84$$

$$= \frac{2 \times 1710/12}{2 \times 510/24} = 2.65$$

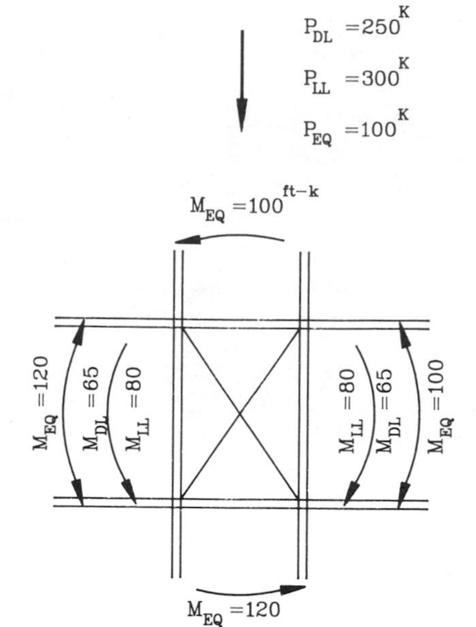

Figure 8-8 Loads and moments on the ductile moment frame connection of Example 8-3.

From alignment chart (Figure 6-10): $K_{x-x} = 2.00$, $K_{y-y} = 1.76$,

$$KL/r = \frac{1.76 \times 12 \times 12}{3.98} = 63.7$$

$$F_a = 17.07 \text{ ksi}, \; f_a = 650/42.7 = 15.22 \text{ ksi}$$

$$F_{bx} = 24 \text{ ksi}, \; f_{bx} = 120 \times 12/232$$

$$= 6.21 \text{ ksi}$$

Assume $M_{EQ\,y-y} = 30$ ft-kip. Then

$$F_{by} = 0.75F_y = 27 \text{ ksi},$$

$$f_{by} = \frac{30 \times 12}{87.3} = 4.12 \text{ ksi}$$

AISC Equation 1.6-1b:

$$\frac{f_a}{0.60F_y} + \frac{f_{bx}}{F_{bx}} + \frac{f_{by}}{F_{by}} \leqslant 1.33$$

$$\frac{15.22}{0.60 \times 36} + \frac{6.21}{24} + \frac{4.12}{27}$$

$$= 0.70 + 0.26 + 0.15$$

$$= 1.11 < 1.33 \quad \text{O.K.}$$

Equation 1.6-1a:

$$\frac{f_a}{F_a} + \frac{C_{mx} f_{bx}}{(1 - f_a/F'_{ex}) F_{bx}} + \frac{C_{my} f_{by}}{(1 - f_b/F'_{ey}) F_{by}} \leqslant 1.33$$

$$\frac{K_{x-x}L}{r_{x-x}} = \frac{2.0 \times 12 \times 12}{6.33} = 45.5, \quad F'_{ex} = 72.16 \text{ ksi}$$

$$\frac{K_{y-y}L}{r_{y-y}} = 63.7, \quad F'_{ey} = 36.81 \text{ ksi}$$

$$\frac{15.22}{17.07} + \frac{0.85 \times 6.21}{\left(1 - \frac{15.22}{72.16}\right)24} + \frac{0.85 \times 4.12}{\left(1 - \frac{15.22}{36.81}\right)27}$$

$$= 0.89 + 0.28 + 0.22 = 1.39 > 1.33$$

Try with clear length of column

$$L = 144 - 18 = 126 \text{ in.}$$

Then

$$\frac{K_{x-x}L}{r_{x-x}} = \frac{2.00 \times 126}{6.33} = 39.8$$

$$\frac{K_{y-y}L}{r_{y-y}} = \frac{1.76 \times 126}{3.98} = 55.7$$

$$F_a = 17.83, \quad F'_{ey} = 48.15, \quad F'_{ex} = 94.24$$

$$\frac{15.22}{17.83} + \frac{0.85 \times 6.21}{\left(1 - \frac{15.22}{94.24}\right)24} + \frac{0.85 \times 4.12}{\left(1 - \frac{15.22}{48.15}\right)27}$$

$$= 0.85 + 0.15 + 0.19 = 1.19 < 1.33 \quad \text{O.K.}$$

- Design of beam web connection: Assume $V_{DL+LL} = 20$ kips. For W18 × 55, $Z = 112$,

$$M_p = ZF_y = 112 \times 36 = 4032 \text{ in.-kips}$$

$$\frac{2M_p}{L} = \frac{2 \times 4032}{24 \times 12} = 28 \text{ kips}$$

$$\Sigma V = 20 + 28 = 48 \text{ kips} \quad \text{[UBC-85 2722(d)]}$$

For $\frac{3}{4}$-in.-diameter A325 bolts,

single shear = (7.7 kips/bolt) × 1.33

bearing = (24.5 kips/bolt) × 1.33

Use three $\frac{3}{4}$-in.-diameter bolts. Note: This provides friction capacity for DL + LL + moderate EQ and bearing capacity in excess of beam moment capacity.

- Check need for continuity, plates:

$$A_{st} \geqslant \frac{P_{bf} - F_{yc}t(t_b + 5k)}{F_{yst}} \quad \text{[UBC-85 2712(e)]}$$

where

A_{st} = area of pair of column web stiffeners
t = thickness of column web
k = distance from face of column flange to toe of web fillet
t_b = thickness of beam flange or moment connection plate
t_f = thickness of column flange
$P_{bf} = A_f F_{yb}$ for ductile moment-resisting frames
F_{yb} = beam yield stress
F_{yc} = column yield stress
F_{yst} = stiffener yield stress

For this example

$$P_{bf} = 7.53 \times 0.63 \times 36 = 170.8 \text{ kips}$$

$$F_{yc} = 36 \text{ ksi}, t = 0.68 \text{ in.}, k = 1.75 \text{ in.},$$

$$F_{yb} = 36 \text{ ksi}, t_b = 0.63 \text{ in.}, F_{yst} = 36 \text{ ksi}$$

$$A_{st} \geqslant \frac{170.8 - 36 \times 0.68(0.63 + 5 \times 1.75)}{36}$$

$$\geqslant \frac{170.8 - 229.6}{36}$$

Stiffener is not required: however, provide $\frac{1}{2}$-in. plate stiffeners as connection plates for beams in the y–y direction. Assume beams framing into web of column are W18 × 35. Then

$$b = 6.0 \text{ in.}, \qquad t = 0.425 \text{ in.}$$

$$P_{bf} = 6.0 \times 0.425 \times 36 = 91.8 \text{ kips}$$

Use $\frac{1}{4}$-in. fillet weld:

$$V = 3.71 \times 1.33 = 4.93 \text{ kips/in.}$$

$$91.8/4.93 = 18.6 \text{ in.}$$

$\frac{1}{4}$-in. fillet both sides:

$$4 \times 7 = 28 \text{ in.} \qquad \text{O.K.}$$

See Figure 8-26 in Section 8.7.2 for typical connection details.

- Check need for web doubler plate:

$$V = 2 \times 170.8 - 16.7 \qquad \text{(see Figure 8-9)}$$

$$= 324.9 \text{ kips}$$

$$A_w = 14.78 \times 0.68 = 10.05 \text{ in.}^2$$

$$V_{allow} = 0.55 f_y t_d$$

$$= 0.55 \times 36 \times 10.05$$

$$= 199.0 \text{ kips}$$

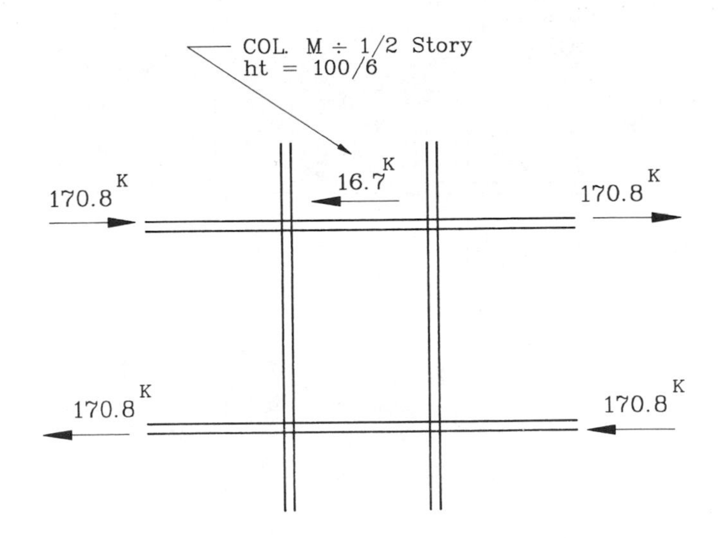

Figure 8-9 Checking the need for doubler plates (Example 8-3).

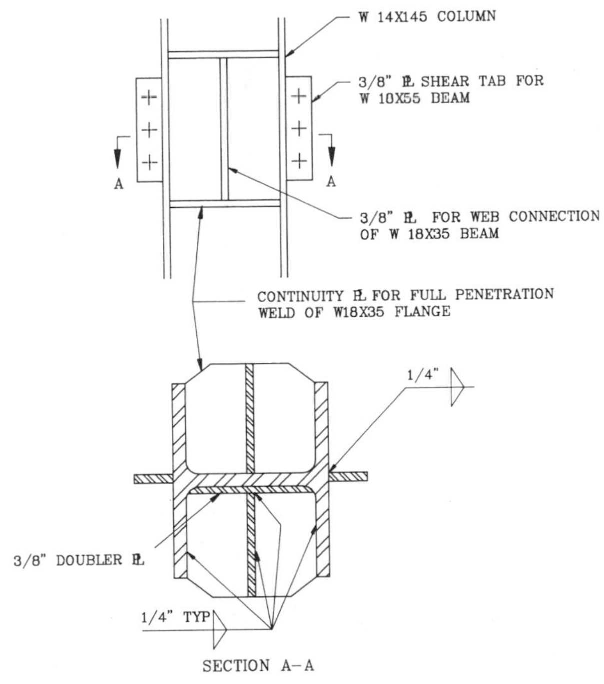

Figure 8-10 Connection design details (Example 8-3).

For $\frac{3}{8}$-in. doubler plate (see Figure 8-10)

$$A_d = 0.375 \times (14.78 - 2 \times 1.75) = 4.23 \text{ in.}^2$$

$$(d = 14.75 \text{ in.}, \quad k = 1.75 \text{ in.})$$

$$V_d = 4.23 \times 36 = 152.3 \text{ kips}$$

$$V_c + V_d = 199.0 + 152.3$$

$$= 351.3 > 324.9 \text{ kips} \qquad \text{O.K.}$$

See Figure 8-11 for typical welding details in panel zone.

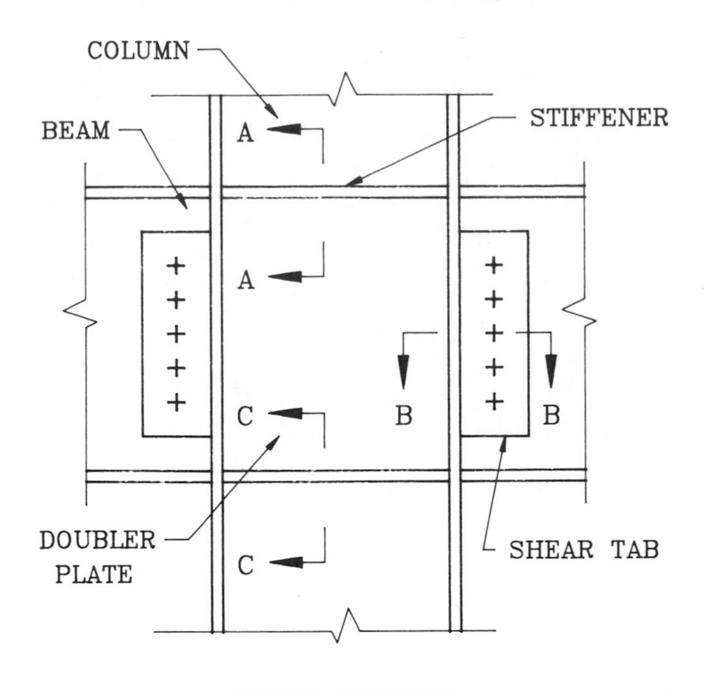

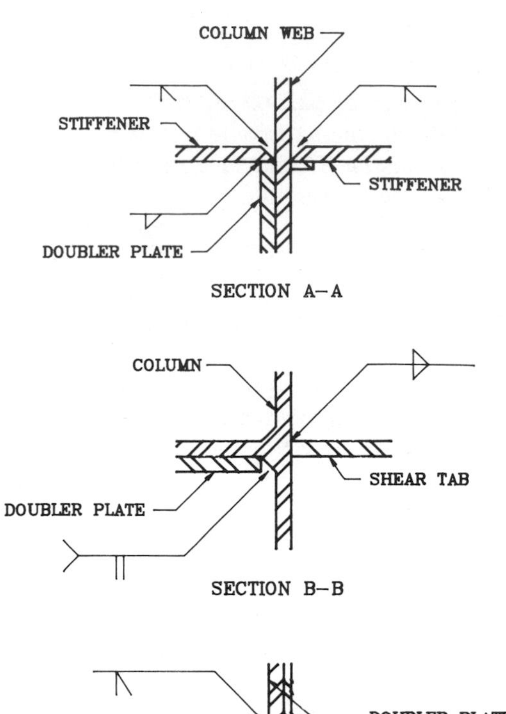

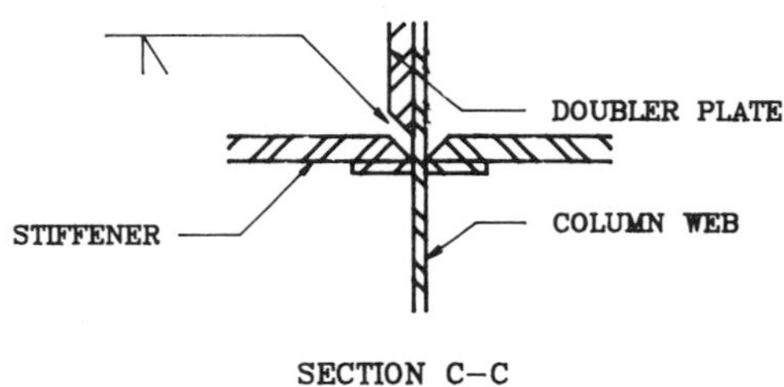

Figure 8-11 Typical panel zone details for moment frames.

- Check for strong column, weak beam, x–x direction:

$$2M_p(\text{beam}) = 2 \times 112 \times 36 = 8064 \text{ in.-kips}$$

$$2M_p(\text{col}) = 2 \times 260 \times 36 = 18{,}720 \text{ in.-kips}$$

y–y direction:

$$2M_p(\text{beam}) = 2 \times 66.5 \times 36 = 4788 \text{ in.-kips}$$

$$2M_p(\text{col}) = 2 \times 133 \times 36 = 9576 \text{ in.-kips}$$

- Redesign column for SSRC recommendations:

$$\frac{L}{r} \leqslant 35; \qquad \text{for W14} \times 135\text{:} \quad \frac{126}{3.98} = 31.7 < 35$$

$$F_a = 19.82 \text{ ksi}$$

$$\frac{f_a}{F_a} = \frac{15.22}{19.82} = 0.77 > 0.75 \quad \text{N.G., but marginal}$$

$$\frac{f_a}{0.60F_y} = \frac{15.22}{0.60 \times 36} = 0.70 < 0.75 \qquad \text{O.K.}$$

$$\frac{L_{x-x}}{r_{x-x}} = \frac{126}{6.33} = 19.9$$

$$F'_{ex-x} = 377.09, \qquad F'_{ey-y} = 14{,}870,$$

$$C_{mx-x} = C_{my-y} = 0.4$$

$$\frac{15.22}{19.82} + \frac{0.4 \times 6.21}{\left(1 - \dfrac{15.22}{377.09}\right)24} + \frac{0.4 \times 4.12}{\left(1 - \dfrac{15.22}{148.70}\right)27}$$

$$= 0.77 + 0.11 + 0.07 = 0.95 < 1.33 \quad \text{O.K.}$$

Assuming that the P-delta restrictions are met, this column would be allowed to resist significantly more moment under the SSRC criteria than under the 1985 UBC criteria; however, this column would be marginally inadequate for compliance with the axial-load provisions of the SSRC recommendations.

8.4 DESIGN OF STEEL BRACED FRAMES

The use of braced frames in steel-framed buildings provides an effective system for the resistance of lateral loads. The use of steel bracing in industrial buildings to provide stability and to resist wind loads was common practice for many years before seismic provisions were introduced into the building codes. Steel bracing may

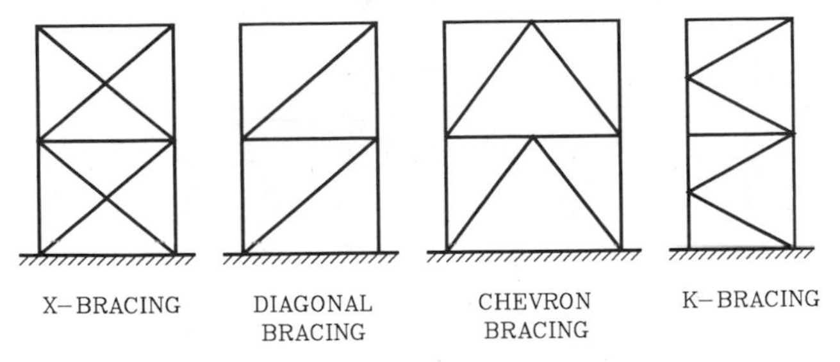

Figure 8-12 Typical concentric bracing configurations.

be used as the only lateral-force-resisting system, or it may be used together with moment frames (Section 8.6 discusses seismic design provisions for such dual systems).

8.4.1 Bracing Configurations

The configuration of bracing systems will be described generically as either concentric or eccentric. In concentric bracing, it is assumed that the neutral axes of the various members (i.e. columns, beams, and bracing members) intersect at a common point at each joint. In actuality, some eccentricity may occur due to the physical limitations of the connections, but this is usually ignored in the design of the members. However, any eccentricities in the connection must be considered in the design of the connecting bolts or weldments.

Concentric Bracing Configurations Figure 8-12 shows typical configurations for concentric bracing. In the design of concentric bracing systems, it is usually assumed that the bracing members are pin-ended (i.e. they are designed only for axial tension or compression). The columns and beams in the bracing system may be assumed to be either pin-ended (i.e. simple beam connections) or to be components of a moment frame (See Section 8.4.4, "1985 Uniform Building Code," paragraph 1, for definition of dual systems).

Eccentric Bracing Configurations Typical eccentric bracing configurations are indicated in Figure 8-13. As is the case with concentric bracing, the bracing member in eccentric-braced frames is usually assumed to be pin-ended, but the beam–column joints are usually welded for full continuity. The eccentric-braced frame has several desirable features as a system for resisting seismic forces:

1. The system can be designed for stiffness, or drift control, comparable to that obtained with concentric bracing.
2. The link beam is designed to yield in shear or flexure prior to initiation of buckling or yielding of the bracing member.
3. Yielding of the link beam is a ductile phenomenon (as opposed to the brittle compression failure of a concentric brace) that absorbs seismic energy and reduces the dynamic structural response by increasing the damping and the natural period of vibration of the building.

8.4.2 Effects of Gravity Loads on Braced Frames

In the case of concentric X bracing or diagonal bracing, the effects of gravity (i.e. vertical dead and live) loads on the bracing members are usually ignored. However, in the case of the concentric chevron or inverted chevron bracing, the bracing acts as a vertical support for the beam to which it is framed, and the tributary gravity loads must be considered. The same analogy also applies to eccentric bracing, where the gravity loads tributary to the framing point on the beam must be considered in the design of the bracing member.

Elastic shortening of the building columns under axial compression will tend to impose additional compressive stress in the bracing members. However, this secondary stress is usually neglected except that it may be included when the analysis is performed using a structural-analysis computer program with the bracing members modeled as inclined columns.

8.4.3 Effects of Lateral Loads on Braced Frames

If all of the members of a braced frame are assumed to be pin-ended, the shear in each story can be distributed to the bracing members in each direction in accordance with their relative rigidities (proportional to the size, length, and angle of inclination of each member). When the bracing is part of a dual system together with moment frames, the distribution becomes more complex (see Section 8.6). For one- and two-story buildings the above simplified distribution provides acceptable design results.

The analysis of concentric bracing systems can be performed, as in truss analysis, by considering static equilibrium at each joint. Usually the story shear is determined and then distributed to each brace in each direction as indicated above. With the known shear (horizontal component) in a bracing member, the vertical component in the member can be calculated, and these two components are then used to find the forces in the other members framing into each joint.

In the analysis of eccentric bracing systems, the simplifying assumption may be made that the entire story shear is resisted by the bracing. This assumption may be conservative for low-rise buildings, but it is not valid for estimating the shear distribution between braced frames and moment frames in a high-rise building (see Section 8.6).

8.4.4 Code Requirements for Braced Frames

1985 Uniform Building Code (UBC-85) Under the provisions of the current edition of this code, the coefficient K for braced frames, in the base-shear formula

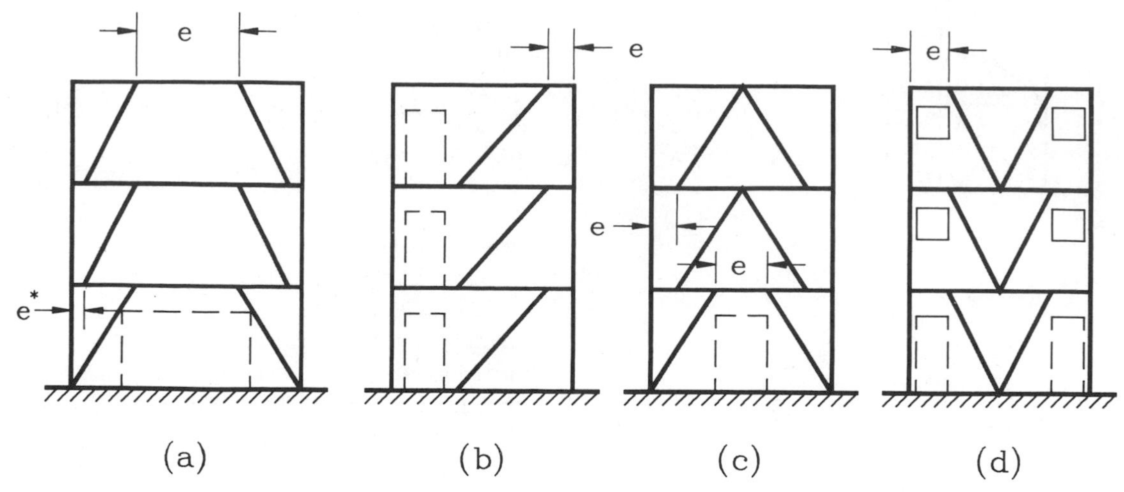

Figure 8-13 Alternative configurations of eccentric bracing showing possible locations of architectural openings.

(see Section 4.3), may be any one of the following:

1. $K = 0.8$ for buildings with dual systems consisting of braced frames and ductile-moment-resisting space frames (DMRSF) provided that:

- The braced frames and the DMRSF are designed to resist the total lateral force in accordance with their relative rigidities.
- The braced frames, acting independently of the DMRSF, are designed to resist the total lateral force.
- The DMRSF shall have the capacity to resist at least 25% of the total lateral force.

2. $K = 1.0$ for buildings with an essentially complete space-frame system designed to resist the vertical gravity loads. For this condition, the braced frames may be designed to resist the entire lateral force or to share the force, in accordance with their relative rigidities, with ordinary moment frames.

3. $K = 1.33$ for buildings without a complete vertical-load-carrying space frame. For this condition the braced frames are required to be designed for the total lateral load.

In seismic zones 3 and 4, buildings more than 160 ft in height with braced frames must comply with the $K = 0.8$ requirements above. This restriction does not apply in seismic zones 1 and 2.

ATC 3-06 Provisions Design under the ATC provisions is to *strength* level rather than allowable stress level. For structural steel the strength level for resisting seismic forces is defined as 1.7 times the allowable stresses given in Reference 8-5. Formulas are provided for approximating the dynamic response of the building to the realistic ground motion (10% probability of exceedance in 50 years) represented by A_a or A_v. The forces derived from those formulas are reduced by a reduction factor R for structural design. The R factors for braced frames corresponding to the UBC-85 K factors of 0.80, 1.00, and 1.33 are 6, 5, and 4 respectively.

SEAOC Recommendations (UBC-88) As previously indicated, the principal change in these recommendations is the replacement of the K factor with a reduction factor R_w that is similar in concept to R factor in ATC 3-06 except that R_w is based on working-stress levels for all materials, except concrete. The R_w values for braced frames corresponding to the ATC 3-06 R values of 6, 5, and 4 are 10, 8, and 6 respectively.

Neither the UBC-85 nor the ATC 3-06 provisions recognize the eccentric braced frame as an entity separate from concentric braced frames. However, the above SEAOC recommendations and the 1988 UBC introduce special provisions for eccentric braced frames, and their greater ductility is acknowledged by larger R_w factors than for the concentric braced frames for dual systems (12 vs. 10) and building frame systems (10 vs. 8). Where braced frames constitute the primary system for gravity as well as lateral loads, an R_w factor of 6 is assigned for these frames, corresponding to a UBC-85 system with $K = 1.33$. Tentative provisions for eccentric braced frames, similar to those in the SEAOC-86 (UBC-88), are currently included as an appendix in the NEHRP provisions and will be incorporated as an integral part of the 1988 edition of the provisions.

8.4.5 Design of Bracing Members

General Provisions Bracing members are designed as columns with allowable stresses determined from the appropriate equations in Reference 8-4. The effective-length factors for X bracing are generally taken as the distance from the end of the brace to the intersection of the two braces (i.e. $K = 0.5$) for the in-plane direction.

For the out-of-plane direction, it has been shown that the tension brace provides an elastic lateral support for the compression brace so that the effective-length factor is greater than 0.5, but less than 1.0. A value of $K = 0.67$ has been used by some designers as a reasonable compromise. Example 8-4 illustrates the use of these assumptions in the design of X bracing. Recent investigations[8-10] have shown that for light bracing members (e.g. single and double angles), the typical connections provide significant end restraint in the in-plane direction so that the K values approach those for fixed end members (i.e. one-half of the length between connections). For other than X bracing, K is usually assumed to be 1.0, although there may be justification for the use of smaller values for the in-plane direction.

1985 UBC Provisions In seismic zones 3 and 4 and for buildings having an importance factor greater than 1.0 located in seismic zone 2, all members in braced frames are required to be designed for 1.25 times the forces obtained from the general seismic provisions. Also the provisions require that the connections for these members are to be designed to develop the full capacity of the member or to be based on the above increased forces without the one-third increase normally permitted for stresses resulting from earthquake forces.

Example 8-4 (Design of Concentric Bracing: 1985 UBC)

Assume a two-story braced bay as shown in Figure 8-14. Shear in 1st story: $V = 200$ kips, assumed equally shared by tension and compression braces. Length of brace,

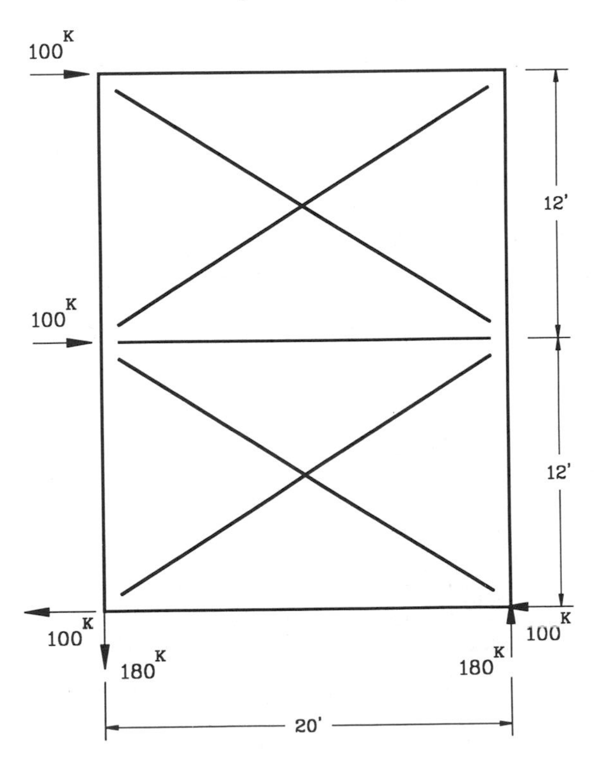

Figure 8-14 Braced frame of Example 8-4.

$$L = \sqrt{12^2 + 20^2} = 23.3 \text{ ft}$$

$$\text{Force in brace} = \frac{23.3}{20} \times 100 = 116.6 \text{ kips}$$

Assume: Unbraced length in plane of brace,

$$L_{x-x} = L/2$$

Unbraced length out of plane of brace,

$$L_{y-y} = 2L/3$$

For W12 columns and W18 beam,

$$L = \sqrt{(240 - 12)^2 + (144 - 9)^2} = 265 \text{ in.}$$

Try two L's $5 \times 5 \times \frac{3}{8}$:

$$A = 7.22 \text{ in.}^2 \qquad r_{x-x} = 1.56, \qquad r_{y-y} = 2.22$$

$$\frac{KL_{x-x}}{r_{x-x}} = \frac{1}{2} \times \frac{265}{1.56} = 84.9$$

$$\frac{KL_{y-y}}{r_{y-y}} = \frac{2}{3} \times \frac{265}{2.22} = 80.0$$

F_a (for $KL/r = 84.9$)

$= 14.80$

$\times 1.33$ ($\frac{1}{3}$ allowable increase)

19.68 ksi

$f_a = 116.6/7.22$

$= 16.15$

$\times 1.25$ (load factor for bracing systems)

20.19 ksi

19.68 < 20.19, but assume O.K., since connection plate reduces clear length in x–x direction.

- Check tension capacity:

$$A_{net} = 7.22 - 2 \times \tfrac{13}{16} \times \tfrac{3}{8} = 6.61,$$

$$A_e = 0.85A_n = 0.85 \times 6.61 = 5.62 \text{ in.}^2$$

$$f_a = 116.6/5.62 = 20.75 \times 1.25 = 25.94 \text{ ksi}$$

$$F_t = 0.5 \times F_u \times 1.33 = 0.5 \times 58 \times 1.33$$

$$= 38.6 \text{ ksi} > 25.94 \text{ ksi} \quad \text{O.K.}$$

- Design of connections: Use $\frac{3}{4}$-in.-diameter A325 high-strength bolts. From AISC Manual for stan-

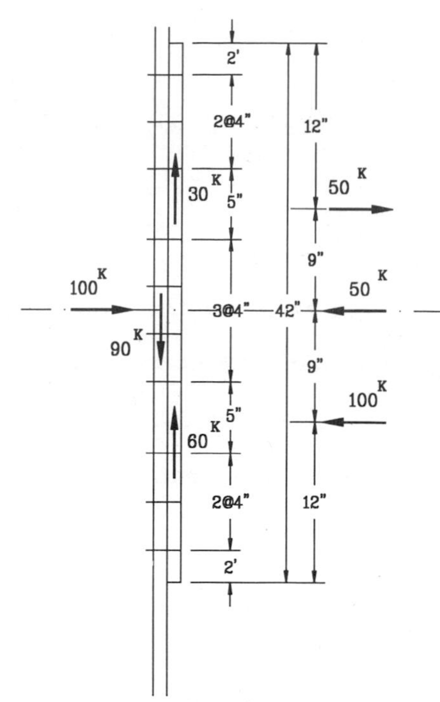

Figure 8-15 Bolt design for Example 8-4.

dard holes and clean mill scale,

$$V_{\text{allow}} \text{ (double shear)} = 15.5 \text{ kips}$$

(no allow. increase)

$$V_{\text{allow}} \text{ (bearing)} = 24.5 \text{ kips } (\tfrac{3}{8}\text{-in. plate})$$

(no allow. increase)

$$\text{No. of bolts} = 116.6 \times 1.25/15.5 = 9.4 \quad \text{(friction)}$$

$$= 116.6 \times 1.25/24.5 = 5.9 \quad \text{(bearing)}$$

Use eight $\frac{3}{4}$-in.-diameter bolts (some bolt slippage permitted).

- Plate connection to column: Assume plate $8 \times 42 \times \frac{1}{2}$ in. Then

$$V = 90 \text{ kips} \times 1.25 = 112.5 \text{ kips}$$

Use $\frac{3}{4}$-in.-diameter A325 bolt (single shear):

$$V_{\text{allow}} = 7.7 \text{ kips/bolt}$$

$$112.5/7.7 = 14.6 \text{ bolts}$$

Use 20 bolts as shown in Figure 8-15. Find I of connection. Assume no net tension:

$$I = \frac{8 \times 42^3}{12} = 49392 \text{ in.}^4$$

$$S = 49{,}392/21 = 2342 \text{ in.}^3$$

$$\begin{aligned} M = 100 \times 9 &= 900 \\ +50 \times 9 &= \underline{450} \\ & \quad 1350 \text{ in.-kips} \end{aligned}$$

$$f = \frac{P}{A} \pm \frac{M_c}{I}$$

$$= \frac{100}{42 \times 8} \pm \frac{1350}{2352}$$

$$= 0.298 \pm 0.574 \qquad \text{(slight tension)}$$

Try neutral axis at 27 in. from bottom of plate:

$$\begin{aligned} I = 8 \times 27^3/12 &= 13{,}122 \\ 8 \times 27 \times 27^2/4 &= 39{,}366 \\ 0.88 \times 5^2 &= 22 \\ 0.88 \times 9^2 &= 72 \\ 0.88 \times 13^2 &= \underline{149} \\ & \quad 52{,}730 \text{ in.}^4 \end{aligned}$$

$$\begin{aligned} M = 50 \times 3 &= 150 \\ 50 \times 6 &= 300 \\ 100 \times 15 &= \underline{1500} \\ & \quad 1950 \text{ in.-kips} \end{aligned}$$

$$f_c = \frac{1950}{52730} \times 27 = 0.998 \text{ ksi}$$

$$f_t = \frac{1950}{52730} \times 13 = 0.481 \text{ ksi}$$

$$C = 0.998 \times \tfrac{1}{2} \times 8 \times 27 = 107.8 \text{ kips}$$

$$T = 0.481 \times \tfrac{1}{2} \times 0.88 \times 3 = \underline{0.6 \text{ kips}}$$

$$107.2 \text{ kips} > 100 \text{ kips}$$

Try neutral axis at 23 in.:

$$\begin{aligned} I = 8 \times 23^3/12 &= 8{,}111 \\ 8 \times 23 \times 23^2/4 &= 24{,}334 \\ 0.88 \times 4^2 &= 14 \\ 0.88 \times 9^2 &= 71 \\ 0.88 \times 13^2 &= 149 \\ 0.88 \times 17^2 &= \underline{254} \\ & \quad 32{,}934 \text{ in.} \end{aligned}$$

$$\begin{aligned} M = 50 \times 7 &= 350 \\ 50 \times 2 &= 100 \\ 100 \times 11 &= \underline{1100} \\ & \quad 1550 \text{ in.-kips} \end{aligned}$$

$$f_c = \frac{1550}{32934} \times 23 = 1.082 \text{ ksi}$$

$$f_t = \frac{1550}{32934} \times 17 = 0.800 \text{ ksi}$$

$$C = 1.082 \times \tfrac{1}{2} \times 8 \times 23 = 99.5 \text{ kips}$$

$$T = 0.800 \times \tfrac{1}{2} \times 0.88 \times 4 = \underline{-1.4 \text{ kips}}$$

$$98.1 \text{ kips} < 100 \text{ kips}$$

- Check for reverse loading: Try all bolts in tension:

$$\begin{aligned} I = 0.88 \times 2^2 &= 4 \\ 0.88 \times 6^2 &= 32 \\ 0.88 \times 11^2 &= 106 \\ 0.88 \times 15^2 &= 198 \\ 0.88 \times 19^2 &= \underline{318} \\ 657 \times 2 &= 1315 \text{ in.}^4 \end{aligned}$$

$$\begin{aligned} M = 50 \times 9 &= 450 \\ 100 \times 9 &= \underline{900} \\ & \quad 1350 \text{ in.-kips} \end{aligned}$$

$$f_b = \frac{1350}{1315} \times 19 = \pm 19.51 \text{ ksi}$$

$$\frac{P}{A} = \frac{100}{10 \times 0.88} = 11.36 \text{ ksi} \quad \text{(some compression)}$$

Try neutral axis at 2 in. from top of plate:

$$\begin{aligned} I = 8 \times 2^3/12 &= 5 \\ 8 \times 2 \times 2^2/4 &= 16 \\ 0.88 \times 4^2 &= 14 \\ 0.88 \times 8^2 &= 56 \\ 0.88 \times 13^2 &= 149 \\ 0.88 \times 17^2 &= 254 \\ 0.88 \times 21^2 &= 388 \\ 0.88 \times 25^2 &= 550 \\ 0.88 \times 30^2 &= 792 \\ 0.88 \times 34^2 &= 1017 \\ 0.88 \times 38^2 &= \underline{1271} \\ & \quad 4513 \text{in.}^4 \end{aligned}$$

$$\begin{aligned} M = -50 \times 10 &= -500 \\ +50 \times 19 &= +950 \\ +100 \times 28 &= \underline{+2800} \\ & \quad 3250 \text{ in.-kips} \end{aligned}$$

$$f_c = \frac{3250}{4513} \times 2 = 1.44 \text{ ksi}$$

$$f_t = \frac{3250}{4513} \times 38 = 27.37 \text{ ksi}$$

$$\begin{aligned} T &= 27.37 \times \tfrac{1}{2} \times 0.88 \times 10 = 120.4 \text{ kips} \\ C &= 1.44 \times \tfrac{1}{2} \times 8 \times 2 = \underline{11.5 \text{ kips}} \\ & \qquad 108.9 \text{ kips} < 100 \text{ kips} \end{aligned}$$

Allowable bolt tension:

$$f_t = 55 - 1.8 f_v \leqslant 44 \qquad \text{(AISC Spec., Table 1.6.3)}$$

$$f_v = \frac{90}{20 \times 0.44} = 10.22 \text{ ksi}$$

$$F_t = 55 - 1.8 \times 10.22 = 36.6 \text{ ksi}$$

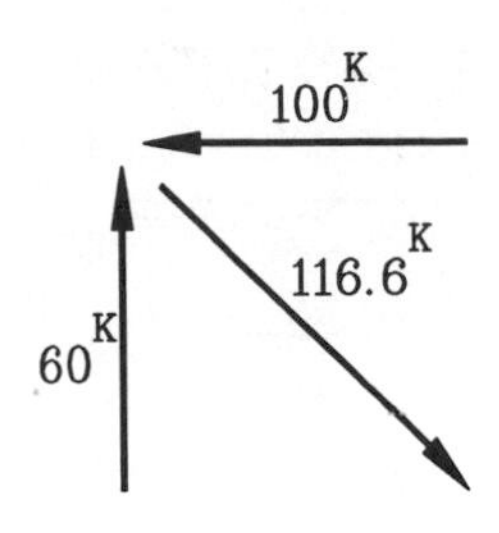

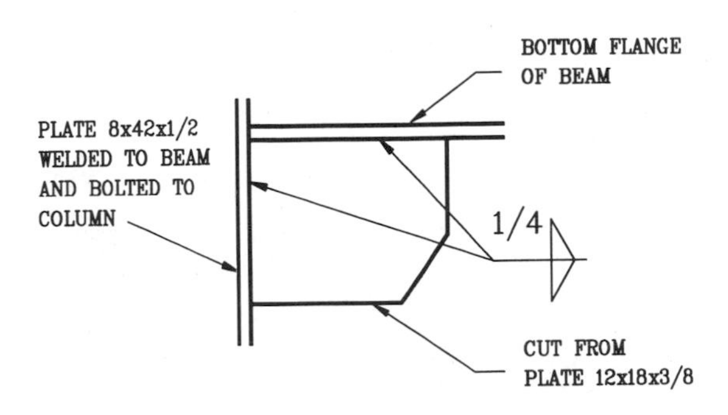

Figure 8-16 Gusset plate connection design (Example 8-4).

- Design of gusset-plate connection for $18 \times 12 \times \frac{3}{8}$-in. plate (see Figure 8-16): Try $\frac{1}{4}$-in. fillet welds. Then

$$A_{eff} = 0.25 \times 0.707 = 0.177 \text{ in./in.}$$

For E70 electrodes, $V_{allow} = 70 \times 0.30 = 21$ ksi. Then

$$V = 21 \times 0.177 = 3.71 \text{ kips/in.}$$

For the base metal,

$$V = 0.25 \times 36 \times 0.4 = 3.6 \text{ kips/in.}$$

The base metal governs. Thus

$$L = 100/3.60 = 27.8 \text{ in.}$$

(use $\frac{1}{4}$ in. fillet on each side: $L = 36$ in.), and

$$L = 60/3.60 = 16.7 \text{ in.}$$

(use $\frac{1}{4}$ in. fillet on each side: $L = 24$ in.). Shear on $\frac{3}{8}$-in. plate:

$$\frac{100}{18 \times 0.375} = 14.8 \text{ ksi} > 14.4 \qquad \text{O.K.}$$

- Design of bracing intersection: Connecting plate:

$$(F_t)_{allow} = 0.50 F_u \qquad \text{[UBC-85 2702(b)]}$$

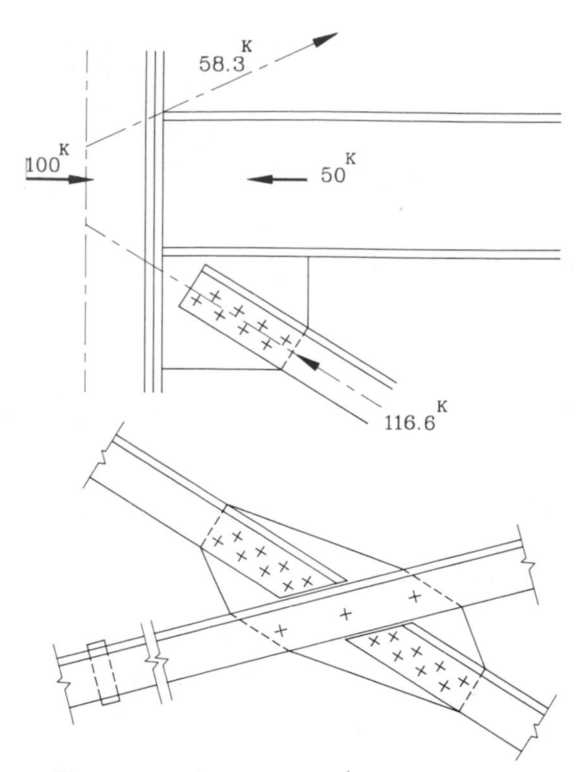

Figure 8-17 Bracing connection (Example 8-4).

For $F_u = 58$ ksi,

$$(F_t)_{allow} = 29 \text{ ksi}$$

$$A_{eff} = 116.6/29 = 4.02 \text{ in.}^2 = C_t A_n$$

For $C_t = 0.85$ [UBC-85 2711(b)3],

$$A_n = 4.02/0.85 = 4.73 \text{ in.}^2$$

For $t = 0.375$,

$$w = \frac{4.73}{0.375} + \frac{13}{16} \times 3 = 15.0 \text{ in.}$$

(The width of three bolt holes is conservatively deducted in the above calculation. See Figure 8-17). Cut connection plate from plate 15 × 32 × $\frac{3}{8}$ in. See Figure 8-17 for bracing connection.

SEAOC Recommendations (1988 UBC) In the SEAOC-86 (UBC-88), there are several new requirements for the design of bracing members, including the following:

1. *Slenderness.* In seismic zones 3 and 4, the L/r ratio for bracing members shall not exceed $720/\sqrt{F_y}$.

2. *Stress reduction.* The allowable stress, F_{as}, for bracing members resisting seismic forces in compression shall be determined from the following formula:

$$F_{as} = \beta F_2 \tag{8-23}$$

where F_2 is the allowable axial compressive stress for columns, and β is the stress-reduction factor determined from

$$\beta = \frac{1}{1 + \dfrac{KL/r}{2C_c}} \tag{8-24}$$

(*Exception*: The value of β is seismic zone 2 need not be less than 0.8.)

3. *Lateral-force distribution.* The seismic lateral force along any line of bracing shall be distributed to the various members so that neither the sum of the horizontal components of the forces in members acting in tension nor the sum of the horizontal components of the forces in members acting in compression exceeds 70% of the total force. (Exception: Where compression bracing acting alone has the strength, neglecting the strength-reduction factor ϕ, to resist $3R_w/8$ times the prescribed seismic force, the above distribution is not required.)

4. *Chevron bracing.* Bracing members in the chevron bracing configurations shall be designed for 1.5 times the otherwise prescribed forces.

5. *K bracing.* In seismic zones 3 and 4, K bracing is prohibited, except as provided in item 6 below. In seismic zone 2, K bracing is permitted when complying with item 4 above.

6. *One- and two-story buildings.* Braced frames need not meet the requirements of items 1 to 5 above in buildings not over two stories in height and in roof structures if they are designed for $3R_w/8$ times the code-equivalent static forces at code-allowable seismic stresses.

7. *Bracing members in eccentric braced frames.* Each brace shall have a compressive strength at least 1.5 times the axial force corresponding to the controlling link beam strength, defined as the lesser of the shear strength or the reduced flexural strength.

8.4.6 Design of Other Members in Braced Frames

General Provisions UBC-85 does not include special provisions for these members in braced frames, except for the multiplier 1.25 on the design forces and the restriction on the connection design indicated under "1985 UBC Provisions" in Section 8.4.5 above. The design of columns in braced frames follows the same provisions as for moment frames, and the allowable stresses given in Reference 8-5. The design of beams in concentric braced frames follows the provisions for beam–columns as prescribed in Reference 8-5, because of the axial tension or compression that these members resist as part of the braced-frame system in addition to the shears and moments from the superimposed gravity loads.

Link Beams in Eccentric Braced Frames In eccentric braced frames, the portion of the beam that is designed to yield in shear or flexure is designated as the link beam. Reference 8-9 recommends the following provisions for these link beams:

1. The flange width–thickness ratio, $b_f/2t_f$, shall not exceed $52/\sqrt{F_y}$.

2. Where the link beam strength is governed by the shear capacity of the web, the flexural and axial capacities within the link shall be calculated using the beam flanges only.

3. A reduced flexural strength, M_{rs}, for flexural yielding is defined as

$$M_{rs} = Z(F_y - F_a) \tag{8-25}$$

If F_a is less than $0.15F_y$, then the reduction for axial load may be neglected.

4. The rotation of the link beam relative to the rest of the beam, γ (see Figure 8-18), at a total frame drift of $3R_w/8$ times the drift calculated for the prescribed seismic forces, shall not exceed the following:

- 0.060 rad for link segments having a clear length of $1.6M_s/V_s$ or less.
- 0.015 rad for link segments having clear length of $2.6M_s/V_s$ or greater.
- The value of γ may be interpolated for values of M_s/V_s intermediate to the above values.

5. The web of the link beam shall be single thickness without doubler plates or other reinforcement. Penetrations shall not be permitted in the web, and the web shear shall not exceed $0.8V_s$ under the prescribed lateral forces.

6. Brace-to-beam connections shall develop the compression strength of the brace and shall transfer this forces to the link beam web, but no part of this connection shall extend into the web area of the link beam.

7. Link beams shall have full-depth web stiffeners on both sides of the beam web at the brace end of the link. In addition, for link beams with clear lengths that exceed $1.6M/V_s$, full-depth stiffeners shall also be provided at a distance of b_f from each end of the link. Stiffeners shall have a thickness not less than $0.75t_w$ or $\frac{3}{8}$ in.

8. Intermediate full-depth web stiffeners shall be provided when:

- The link beam strength is controlled by V_s.
- The link beam strength is controlled by flexure and the shear associated with the reduced moment capacity, M_{rs} exceeds $0.45F_y\,dt$.

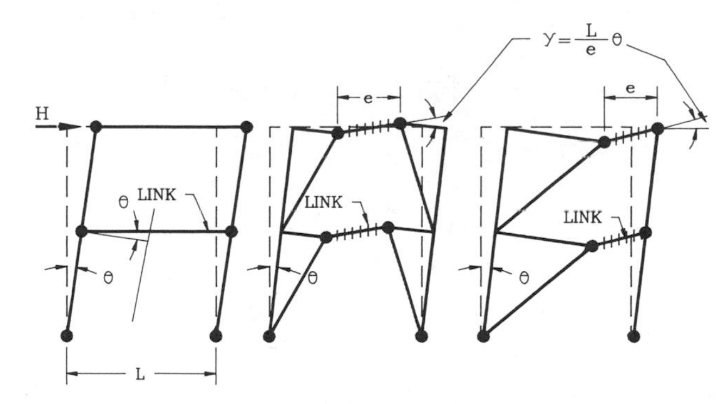

Figure 8-18 Relationship between frame drift θ and link beam rotation γ.

9. The spacing of intermediate web stiffeners shall be as follows:

- For link beams with a rotation angle γ of 0.06 rad, the spacing shall not exceed $38t_w - d/5$.
- For link beams with a rotation angle of 0.03 rad or less, the spacing shall not exceed $56t_w - d/5$.
- For intermediate values of γ the spacing may be interpolated.
- For link beams with a depth of 24 in. or more, the intermediate full-depth web stiffeners, if required, shall be provided on both sides of the web. For beams less than 24 in. in depth, the intermediate stiffeners are required only on one side of the web.

Example 8-5 illustrates the design of a link beam and web stiffness.

Example 8-5 (Design of an Eccentric Braced Frame) Assume a frame configuration as shown in Figure 8-19. Length of brace: $\sqrt{12^2 + 16^2} = 20$ ft. Then

$$P = \tfrac{5}{4} \times 150 = 187.5 \text{ kips.}$$

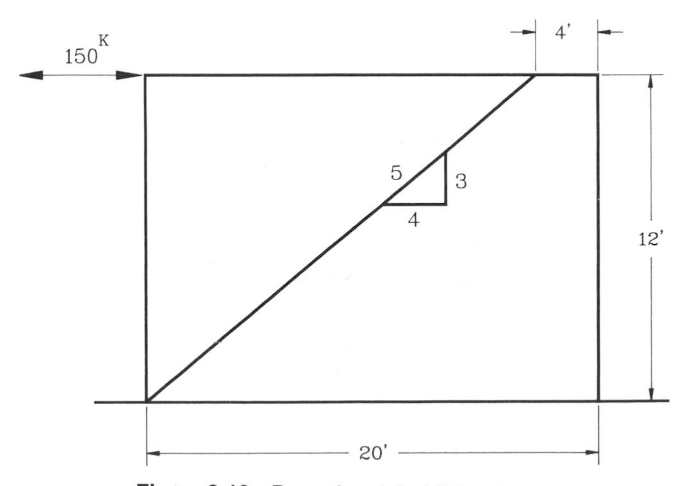

Figure 8-19 Braced panel of Example 8-5.

- Design of link beam:

$$V = 150 \times \tfrac{3}{4} = 112.5 \text{ kips}$$

Try W16 × 45:

$$A = 13.3 \text{ in.}^2, \quad b_f = 7.035 \text{ in.}, \quad t_f = 0.565 \text{ in.}$$

$$Z = 82.3 \text{ in.}^3, \quad d = 16.13 \text{ in.}, \quad t_w = 0.345 \text{ in.}$$

$$M_p = ZF_y{}^a = 82.3 \times 36 = 2962.8 \text{ in.-kips}$$

$$\frac{b_f}{2t_f} = \frac{7.025}{2 \times 0.565} = 6.2 < \frac{52}{\sqrt{F_y}} \quad \text{O.K.}$$

$$V_p = 0.55 t_w d \times 36 = 0.55 \times 0.345 \times 16.13 \times 36$$
$$= 110.2 \text{ kips} \approx 112.5 \text{ kips}$$

$$\frac{1.6M_p}{V_p} = \frac{1.6 \times 2962.8}{110.2} = 43.0 \text{ in.}$$

For W14 columns,

$$e = 48 - 7 = 41 \text{ in.} < 43 \text{ in.} \quad \text{O.K.}$$

- Design of brace: Design brace for 1.5 × 187.5 = 281.3 kips (1988 UBC). Try 8 × 8 × $\frac{5}{8}$ tube:

$$A = 17.4 \text{ in.}^2, \quad r = 2.96 \text{ in.}$$
$$\frac{L}{r} = \frac{20 \times 12}{2.96} = 81.1,$$
$$F_a = 15.3 \times 1.7 = 25.9 \text{ ksi}$$
$$f_a = \frac{281.3}{17.4} = 16.2 \text{ ksi}$$

Try 8 × 8 × $\frac{3}{8}$ tube:

$$A = 11.1, \quad r = 3.09$$
$$\frac{L}{r} = \frac{20 \times 12}{3.09} = 77.7,$$
$$F_a = 15.61 \times 1.7 = 26.5 \text{ ksi}$$
$$f_a = 281.3/11.1 = 25.3$$

(brace is stronger than link beam)

- Brace–beam connection: Assume $\frac{5}{16}$-in. fillet welds.

$$V_{allow} = \tfrac{5}{16} \times 0.707 \times 0.30 \times 70 \times 1.7$$
$$= 7.89 \text{ kips/in.}^2 \quad \text{(ultimate strength)}$$

Length of weld required:

$$281.3/7.89 = 36.7 \text{ in.}$$

[a]Assume that 150 kips force is distributed by the floor diaphragm, so that no reduction of M_p is required.

Use 10 in. of weld each side, each face:

$$4 \times 10 = 40 \text{ in.}$$

- Stiffener spacing:

$$\gamma = L/e\theta$$

Assume $\theta = \delta/H$, and δ is caused by brace deformation only. For $f_a = 25.3$ ksi,

$$\Delta L = 25.3 \times \frac{20 \times 12}{29{,}000} = 0.21 \text{ in.}$$

$$\delta = \tfrac{4}{5} \times 0.21 = 0.17 \text{ in.}$$

$$\theta = \frac{\delta}{H} = \frac{0.17 \text{ in.}}{12 \times 12} = 0.0012 \text{ rad}$$

$$\gamma = \frac{L\theta}{e} = \frac{20 \times 12 \times 0.0012}{41} = 0.007 \text{ rad}$$

Assume the eccentric bracing system was designed under 1988 UBC with $R_w = 10$:

$$\frac{\gamma \times 3 \times R_w}{8} = \frac{0.007 \times 3 \times 10}{8}$$
$$= 0.026 \text{ rad} < 0.060$$

Use full-depth stiffeners at each end of link beam and additional stiffeners at 7 in. (b_f) from end stiffeners [1988 UBC 2722(h)6]. All stiffeners are to be $\frac{3}{8}$-in. plate. Intermediate-stiffener spacing:

$$a = 56t_w - \frac{d}{5} \quad \text{for} \quad \gamma \leqslant 0.03 \text{ rad}$$

$$= 56 \times 0.345 - \frac{16.13}{5} = 16.09 \text{ in.}$$

$$41 - 2 \times 7 = 27 \text{ in.}$$

Use one intermediate stiffener at center of link beam (one side only).

- Stiffener welds: Use $\frac{1}{4}$-in. fillet welds on each side:

$$V_{cap} = 2 \times \tfrac{1}{4} \times 0.707 \times 0.30 \times 70$$
$$\times 1.7 \times (16.13 - 2 \times 0.565)$$
$$= 196 \text{ kips}$$
$$V_{reqd} = A_{st} \times F_y = 3.5 \times 0.375 \times 36$$
$$= 47.3 \text{ kips}$$

See Figure 8-20 for details of the brace connection and stiffeners.

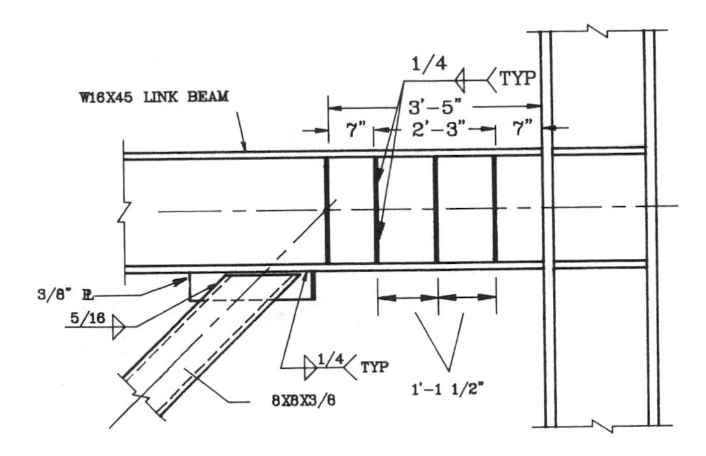

Figure 8-20 Details of brace connection and stiffeners (Example 8-5).

8.5 DESIGN OF STEEL MOMENT FRAMES

Prior to about 1935, design for seismic forces was not required by the various applicable building codes, and was generally neglected in the design of building frames. Design for wind loads was accomplished by the use of concentric bracing or, in some cases, by using knee braces or deep stiffened triangular gusset plates, above and below the beam at each end connection to the columns. Classical analytical procedures for indeterminate structures were known for over a hundred years prior to 1935, and approximate adaptations of these procedures had been in use for steel bridges and buildings since before the turn of the century. Many steel-frame buildings constructed during this period were designed for gravity loads under simple-beam assumptions, with the fixity provided by the above kneebraces or gusset plates analyzed separately by some approximate method.

About 1935, California structural engineers began to accept the need to design for earthquake forces. Fortuitously, at about the same time, Professor Hardy Cross developed a simplified procedure for the analysis of indeterminate structures, and also dependable welding equipment and electrodes became available to simplify frame joints. However, until about 1950, welding for building frames was generally limited to shop fabrication with riveted or bolted details for field erection. After this date, field-riveted connections were generally replaced with high-strength bolts, which are still in use for secondary framing and temporary or permanent shear connections for welded frames. These three developments, followed a few years later by the availability of digital computer analysis, were instrumental in spurring the popular use of moment frames in building construction.

8.5.1 Moment-Frame Configuration

Typical moment-frame configurations are shown in Figure 8-21. The rectilinear frame, shown in Figure 8-21a is, of course, the most common application of moment frames for buildings. Although the most effective and economical use of this configuration is retained when repetition occurs in the width of the column bays and in the story height, moment frames are easily adaptable for variations in these dimensions. Excessive variations, however, may adversely effect the horizontal and/or vertical response to the dynamic earthquake forces (see Section 8.5.3).

The frame configurations shown in Figure 8-21b and c are common variations for one-story buildings or for the top story of a multistory building. These configurations may be used to enhance the exposure to sunlight or to an attractive view, to provide good roof drainage for rain water, or to add esthetic interest to the top of the building.

8.5.2 Effects of Gravity Loads

In the design of moment frames, it is customary in the analysis to treat dead loads and live loads as separate load cases so that they may be combined in the various load combinations with appropriate load factors or reductions. Most building codes permit a reduction of

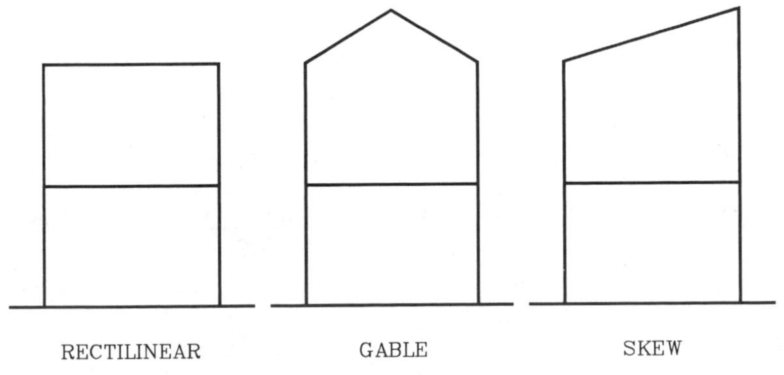

Figure 8-21 Typical moment frame configurations.

the design floor live load for all structural framing members in recognition of the fact that it is unlikely that the entire floor area will be fully loaded at any one time. This reduction is proportional to the area of the floor or roof supported by the members and will have some upper limits (e.g., UBC allows up to 40% reduction in live loads for beams and girders, and 60% for columns supporting more than one level). Most building codes also require that the effects of unbalanced live loading be investigated (e.g. live loads on all spans, live loads on alternate spans, etc.), so that there may be three or more live-load cases.

It should be noted that unbalanced gravity loading in moment frames induces sideways moments and shears that must be considered in the frame design. An additional secondary effect is that due to elastic column shortening. This effect is even more pronounced for the overturning forces due to wind or seismic forces. Column shortening can be neglected for low-rise buildings (less than about six stories), but may be significant for higher buildings. This effect is very difficult and laborious to analyze by hand calculations, but fortunately, there are many computer programs, for both static and dynamic analyses, that include this effect in the analysis.

8.5.3 Effects of Lateral Loads

Moment frames resist lateral loads by means of joint rotations that induce shears and moments in the frame members. Axial forces are also induced in the frame columns due to the overturning moments caused by the lateral forces. Axial forces in the horizontal beams (i.e. strut action) are typically neglected for buildings with floor or roof diaphragms.

The P-Delta Effect Because moment frames resist lateral loads primarily by joint rotations, displacements or drift of these frames may be significant, particularly in the event of a severe earthquake. As indicated in Chapter 4, the forces developed from the seismic design provisions for steel frames in building codes are only a small fraction of those associated with the elastic response of these frames in a severe earthquake. Although steel frames possess excellent ductility that will limit these forces, the displacements may be as great as if the structure had remained elastic. These excessive displacements can lead to overall frame instability due to the P-delta effect. This effect is illustrated in Figure 8-22. It can be seen that the equivalent additional story shear due to P-delta is $P\Delta/h$, where Δ/h is the story drift index. The allowable story drift index, in accordance with the UBC-85, is 0.005 under the code-prescribed seismic forces. UBC-85 requires that the displacement calculated from the specified lateral forces be divided by the K factor to obtain the drift (for $K \leqslant 1.0$). If it is

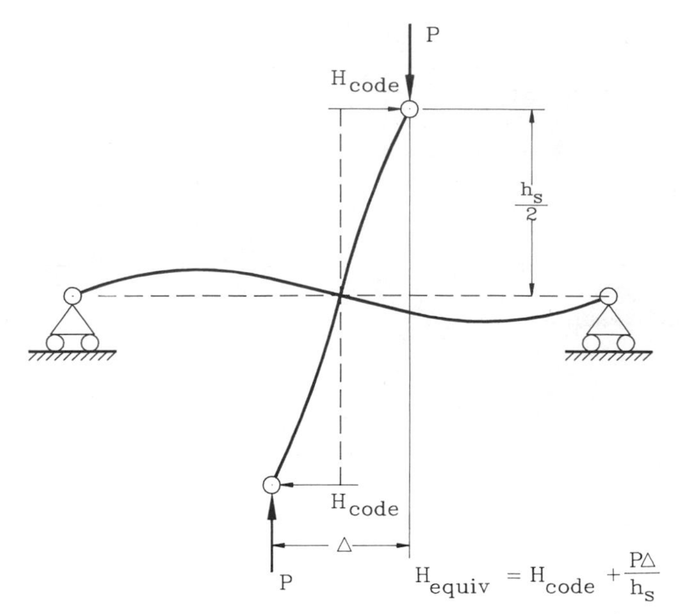

Figure 8-22 Additional equivalent story shear due to P-Δ.

assumed that the frame design for a moment-frame building is governed by drift, but if the drift under the "real" earthquake forces is $3/K$ times the calculated drift, then the story drift index is 0.015 (i.e. 0.005×3.0). If it is further assumed that the dead and live loads are equal, then $P = 2W$, where P is the total load above the story in question and W is the total dead load above the story. Under those assumptions, the equivalent additional story shear due to P-delta becomes $2W \times 0.015 = 0.030W$. It should be noted that this shear is equivalent to the design base shear for a 15-story building with Z, I, and S equal to 1.0, K equal to 0.67, and T equal to 1.5 sec (i.e. $0.10N$) in the equation $C = 1/(15\sqrt{T})$. This would indicate that this building would be subjected to an additional story shear equal to the design story shear, or a doubling of the seismic design forces and the calculated displacements from these forces. Note also that this additional increment of drift would lead to yet another increment, and so on, until either stability is attained or instability and collapse occur. It is therefore very important that the engineer be aware of the fact that the severe earthquake loads and displacements may be significantly larger than the calculated design values and that the P-delta effect be investigated for these larger displacements.

As indicated in Section 4.3, the 1985 UBC seismic provisions have evolved from a combination of analytical studies and historical experience with the response of various structural systems to actual earthquakes. As discussed in the above paragraph, the design forces that are developed from these provisions are significantly less than the real forces from a severe earthquake. The strength-to-weight and strength-to-volume ratios that make structural steel a desirable building material are sometimes detrimental in moment frames when inelastic drift must be controlled to preclude frame or story instability. The UBC in its various editions, up to and including 1985, has not required that the *P*-delta effect be investigated, because of lack of historical evidence of this type of failure, although it has been definitely established by analytical studies and laboratory testing. The SEAOC recommendations[8-4] and UBC-88 attempt to address this apparent shortcoming in the UBC-85. The proposed reduction factors R_w are 6 for ordinary moment frames, and 12 for special moment-resisting space frames (SMRSF). These force reduction factors for use with allowable stresses are intended to be comparable to the ATC 3-06 reduction factor R,[8-2, 8-3] which is $4\frac{1}{2}$ for ordinary frames and 8 for SMRSF. This would imply that the real forces for the SMRSF may be as much as 8 times the design forces, if the structure remains elastic. The allowable drift under the SEAOC recommendations (UBC-88) is given by $0.03/R_w$, but not more than 0.004 times the story height. For SMRSF this becomes 0.0025 (i.e. 0.03/12). It is also stipulated that the resulting member forces, moments, and story shears induced by *P*-delta effects need to be considered in the evaluation of overall frame stability. However, *P*-delta may be ignored when the story drift does not exceed $0.02/R_w$ times the story height. For the SMRSF, this becomes 0.0017 times the story height. If we use the factor of 4.5 for the severe earthquake deflections (they could be as much as 8 times greater), then Δ/h becomes 0.0077 and for $P = 2W$, the additional story shear that is neglected is $0.015W$. This is an improvement over the example in the preceding paragraph, but there is still cause for concern. A more reasonable approach might be to use a multiplier of $3R_w/8$ (i.e. 4.5 for $R_w = 12$) on the calculated drift and then compare the additional story shear with the design shear. If the additional story shear exceeds 10% of the design shear, the *P*-delta effects should be included in the design responses.

Deformation Compatibility The effect of potentially larger than calculated frame displacements on other framing members is recognized in the 1985 UBC. In seismic zones 2, 3, and 4 all framing elements not required by design to be part of the lateral force resisting system are required to be investigated and shown to be adequate for a vertical-load-carrying capacity and induced moment due to $3/K$ times the displacements due to the code required lateral forces. For buildings with ductile moment-resisting spaces frames designed to resist the total seismic lateral force, K equals 0.67 and the above requirement becomes 4.5 (i.e. 3/0.67) times the calculated displacement.

For interior columns that are not part of the lateral force system, the induced moment due to this displacement will depend on the restraint imposed on the interior columns by the floor framing members at the top and bottom of the column. If these columns are pinned at both ends, there is no moment. If pinned at the bottom and fixed at the top, the induced moment is the product of the vertical load and the displacement. If the column is fixed at both ends, the moment is reduced by one-half. The most probable restraint condition is less than fixed at both ends (i.e., assumes some rotation of the base–column joints). Example 8-6 illustrates the extreme case (i.e., column fixed at the top and pinned at the bottom), simulating a slender interior column in a one-story structure with deep rigid roof beams.

Example 8-6 (Check for Deformation Compatibility) Assume an interior column that is supporting floor framing in a building, but is not part of the lateral-load-resisting system.

$$\text{Story height} = 12 \text{ ft}$$

$$P_{DL+LL} = 280 \text{ kips}$$

$$K = 0.67$$

$$\Delta/Kh = 0.005 \qquad (\text{in } y\text{–}y \text{ direction of column})$$

$$\Delta = 0.005 \times 0.67 \times 144 = 0.48 \text{ in.}$$

$$\frac{3}{K} \times \Delta = \frac{3}{0.67} \times 0.48 = 2.16 \text{ in.}$$

Try W12 × 58:

$$A = 17.0 \text{ in.}^2 \qquad r_{y-y} = 2.10 \text{ in.,} \qquad Z_{y-y} = 32.5 \text{ in.}^3$$

$$\frac{L}{r} = \frac{144}{2.10} = 68.6, \qquad F_a = 16.57 \text{ ksi,}$$

$$f_a = \frac{280}{17.0} = 16.47 \text{ ksi}$$

Use Equation 2.4-2 (AISC specification) for check:

$$\frac{P}{P_{cr}} + \frac{C_m M}{(1 - P/P_e)} M_m \leqslant 1.0$$

$$P_{cr} = 1.7AF_a = 1.7 \times 17 \times 16.57 = 478.9 \text{ kips}$$

$$C_m = 0.85$$

$$M = 280 \times 2.16 = 604.8 \text{ in.-kips}$$

$$P_e = \tfrac{23}{12} A F'_{ey},$$

$$F'_{ey} = 31.74 \qquad (\text{AISC, Table 9})$$

$$P_e = \tfrac{23}{12} \times 17.0 \times 31.74 = 1034.2 \text{ kips}$$

$$M_m = \left[1.07 - \frac{(L/r_{y-y})\sqrt{F_y}}{3160}\right] M_p$$

$$M_p = Z_{y-y} F_y = 32.5 \times 36 \times 1170.0 \text{ in.-kips}$$

$$M_m = \left[1.07 - \frac{68.6\sqrt{36}}{3160}\right]1170.0 = 1099.5 \text{ in.-kips}$$

$$\frac{280}{478.9} + \frac{0.85 \times 604.8}{(1 - 280/1034.2)1099.5}$$

$$= 0.584 + 0.641 = 1.23 > 1.0 \qquad \text{N.G.}$$

Try W12 × 65:

$$A = 19.1 \text{ in.}^2, \qquad r_{y-y} = 3.02 \text{ in.}, \qquad Z_{y-y} = 44.1 \text{ in.}^3$$

$$\frac{L}{r_{y-y}} = \frac{144}{3.02} = 47.7,$$

$$F_a = 18.55 \text{ ksi}, \qquad f_a = \frac{280}{19.1} = 14.66 \text{ ksi}$$

$$P_{cr} = 1.7 \times 19.1 \times 18.55 = 602.3 \text{ kips}$$

$$F'_{ey} = 65.65 \text{ ksi},$$

$$P_e = \tfrac{23}{12} \times 19.1 \times 65.65 = 2403.2 \text{ kips}$$

$$M_p = 44.1 \times 36 = 1587.6 \text{ in.-kips}$$

$$M_m = \left[1.07 - \frac{47.7\sqrt{36}}{3160}\right]1587.7 = 1555.2 \text{ in.-kips}$$

$$\frac{280}{602.3} + \frac{0.85 \times 604.8}{(1 - 280/2403.2)1555.2}$$

$$= 0.465 + 0.374 = 0.84 < 1.0 \qquad \text{O.K.}$$

This example demonstrates the need to investigate structural framing, that is not part of the lateral load resisting system, for stability under the imposed deformations.

The SEAOC Recommendations (UBC-88) contain a similar provision that requires these framing elements to be investigated for $3R_w/8$ times the calculated deflections. It will be noted, that for R_w equal to 12, this requirement is also equal to 4.5 (i.e. 3 × 12/8) times the calculated displacement. The recommendations further prescribe that this requirement may be satisfied at the yield stress level or 1.7 times the allowable stresses.

8.5.4 Design of Beams and Columns in Moment Frames

1985 UBC Provisions Under the seismic provisions of the 1985 UBC, ordinary moment frames may be used to resist seismic forces provided:

1. The building is less than 160 ft in height.
2. A K factor equal to 1.0 is used for the base-shear determination.

Ordinary moment frames are required to comply with the seismic drift limitation (i.e. 0.005 times the story height), but otherwise are designed in accordance with the same provisions as for gravity loads. Allowable stresses in these frames when resisting earthquake forces, acting alone or in combination with dead or live loads, may be increased by one-third.

Ductile moment-resisting frames are required for:

1. Buildings designed with a force factor K equal to 0.67 or 0.80.
2. Buildings more than 160 ft in height. Ductile moment-resisting frames in these buildings shall be capable of resisting at least 25% of the total seismic force prescribed for the building as a whole.

In seismic zones 1 and 2, ordinary moment frames are considered as meeting the UBC requirements for ductile moment-resisting frames. For seismic zones 3 and 4, the code contains several special provisions, including:

1. *Materials.* Structural steel shall conform to A36, A441, A500 (grades B and C), A501, A572 (grades 42 and 50), or A588.
2. *Connections.* Each beam or girder moment connection to a column shall be capable of developing in the beam the full plastic capacity of the beam or girder. (*Exception*: The connection need not develop the full plastic capacity of the beam or girder if it can be shown that adequately ductile joint displacement capacity is provided with a lesser connection.) For steel whose specified ultimate strength is less than 1.5 times the specified yield strength, plastic hinges in beams formed during inelastic deformations of the frame shall not occur at locations in which the beam flange area has been reduced, as by holes for bolts.
3. *Local buckling.* Members in which hinges will form during inelastic displacement of the frames shall comply with the requirement for plastic design sections.
4. *Nondestructive testing.* Welded connections between the primary members of ductile moment-resisting space frames shall be tested by nondestructive methods.

Although not required by the above code provisions, many designers in seismic zones 3 and 4 apply the weak-beam, strong-column philosophy to the design of ductile steel frames. This philosophy was initiated in the design of ductile concrete frames to preclude shear failures in the columns, but it is considered prudent, even in steel frames, to force the yielding to occur initially in the beams rather than the columns. Trial designs, with and without enforcement of the weak-beam, strong-column requirement, have been subjected to inelastic dynamic analyses with four carefully selected earthquake accelerograms, adjusted to have about the same peak ground acceleration. The results of this analytical study[8-10] indicate improved response characteristics, particularly in the significant reduction of the ductility demand in the interior columns, when the initial yielding occurs in the girders rather than the columns. Example 8-3 illustrates the design of the principal members of a ductile moment resisting space frame utilizing the weak-beam, strong-column philosophy.

In the design of important or essential facilities special seismic design criteria may be specified (e.g. the Veterans Administration code for hospitals, Tri-Service manuals for essential military facilities, and the Department of Energy manual for hazardous facilities) or proposed by the engineer. These criteria may include specifications for the determination of the site-specific seismic hazard and for the selection of an appropriate level of risk. Dynamic analyses procedures are usually employed using site response spectra or representative time histories. Performance objectives required of the structures to be designed govern the limits of the permitted inelastic response (e.g. allowable ductility). The inelastic response may be computed directly by a suitable nonlinear analytical program or may be inferred from various types of linear elastic analyses. Since the seismic provisions of these criteria are generally significantly beyond code requirements and are supplementary to the basic code requirements for life safety, yield-capacity values are used for the various structural members in the seismic-load cases. The *P*-delta effect may also be investigated using the drifts from the dynamic analyses.

SEAOC Recommendations (1988 UBC) These provisions recognize ordinary moment frames and special moment-resisting space frames (SMRSF). The force factor K has been removed and replaced by the R_w factors described in previous sections of this chapter.

The proposed requirements for both ordinary moment frames and SMRSF include the following:

1. *Column strength.* Columns in frames shall have the strength to resist, in addition to all other applicable load combinations, the axial force resulting from the following load combinations:

- Axial compression:

$$1.0P_{DL} + 0.8P_{LL} + 3(R_w/8)P_E \quad (8\text{-}26)$$

- Axial tension:

$$0.85P_{DL} + 3(R_w/8)P_E \quad (8\text{-}27)$$

[*Exception*: The axial force combination as outlined above need not exceed either of the following: (i) the maximum force that can be transferred to the column, using 125% of the strength of the elements of the structure, or (ii) the limit as determined by the overturning uplift which the foundation is capable of resisting.]

2. *Column splices.* Column splices shall have sufficient strength to develop the column forces determined from item 1 above. Column splices subject to net tension forces shall comply with the more critical of the following:

- Partial penetration welds shall be designed to resist 150% of the above axial tension force.
- Welding shall develop not less than 50% of the flange-area strength of the smaller column.

Splices employing partial-penetration welds shall be located at least 3 ft from girder flanges.

3. *Slenderness evaluation.* This paragraph is applicable to the effective-length determination for columns of moment frames resisting earthquake forces in seismic zones 3 and 4. In the plane of the earthquake forces the factor K may be taken as unity when all of the following conditions are met:

- The column either is continuous or is fixed at each joint.
- The maximum axial compressive stress, f_a, does not exceed $0.4F_y$ under design loads.
- The calculated story drift ratios are less than $0.03/R_w$ or 0.004 times the story height.

Additional requirements pertaining only to the SMRSF include:

1. *Flange detail limitations.* For steel whose specified ultimate strength is less than 1.5 times the specified yield strength, plastic hinges shall not form at locations at which the beam flange area has been reduced, as for bolt holes. Bolted connections of flange plates of beam–column joints shall have net-to-gross area ratio A_e/A_g equal to or greater than $1.2F_y/F_u$, where F_u is the minimum tensile strength.

2. *Girder flanges.* Girders shall have a flange width-to-thickness ratio, $b_f/2t_f$, not to exceed $52\sqrt{F_y}$.

3. *Strength ratio.* At any moment-frame joint, the following relationship shall be satisfied:

$$\frac{\Sigma Z_c(F_{yc} - f_a)}{\Sigma Z_b F_{yb}} > 1.0 \quad (8\text{-}28)$$

where $f_a \geqslant 0$. (*Exception*: This requirement need not apply in any of the following cases, provided the compactness limitations for beams apply to columns as well:

- For columns with $f_a < 0.4F_y$ for all load combinations.
- For columns in any story which have lateral shear strength 50% greater than that of the story above.

- For any column whose lateral shear strength is not included in the design to resist code-required seismic shears.

4. *Girder–column joint restraint.*

- Restrained joint: Where it can be shown that the columns of SMRSF remain elastic, the flanges of the columns need be laterally supported only at the level of the girder top flange. Columns may be assumed to remain elastic if one of the following conditions is satisfied: (i) the ratio in Equation 8-28 is greater than 1.25; (ii) the flexural strength of the column is at least 1.25 times the moment that corresponds to the panel-zone shear strength; (iii) girder flexural strength or panel-zone strength will limit column stress ($f_a + f_{bx} + f_{by}$) to F_y of the column; (iv) the column will remain elastic under gravity loads plus $3(R_w/8)$ times the prescribed seismic forces. Where the column cannot be shown to remain elastic, the column flanges shall be laterally supported at the levels of the girder top and bottom flanges. The column flange lateral support shall be capable of resisting a force equal to 1% of the girder flange capacity at allowable stresses and a limiting displacement perpendicular to the frame of 0.2 in. Required bracing members may brace the column flanges directly or indirectly through the column web or the girder flanges.
- Unrestrained joint: Columns without lateral support transverse to a joint shall be designed to satisfy the following:

$$\frac{f_a}{F_a} + \frac{f_{bx}}{F_{bx}} + \frac{f_{by}}{F_{by}} < 1.0 \tag{8-29}$$

(a) The column stresses in Equation 8-29 shall be determined from gravity loads plus the lesser of the following: (i) $3(R_w/8)$ times code seismic forces, (ii) the forces corresponding to either 125% of the beam flexural strength or the strength of the panel zone.
(b) The allowable column axial stress F_a shall be determined assuming a pin-ended column with length L taken as the height between supports.
(c) L/r for such columns shall not exceed 60.
(d) The stress f_{by} shall include the effects of P-delta and the bracing force specified for restrained joints above.

5. *Beam bracing.* Both flanges of beams shall be braced directly or indirectly. The beam bracing between column center lines shall not exceed $96r_y$. In addition, braces shall be placed at concentrated loads where a hinge may form.

6. *Changes in beam flange area.* Abrupt changes in beam flange area are not permitted within possible plastic-hinge regions of special moment-resistant frames.

7. *Moment-frame drift calculations.* Moment-frame drift calculations shall include bending and shear contributions from the clear girder and column spans, column axial deformation, and the rotation and distortion of the panel zone.

Exceptions:

1. Drift calculations may be based on column and girder center lines where either of the following conditions is met: (i) it can be demonstrated that the drift so computed for frames of similar configuration is typically within 15% of that determined above; (ii) The column panel-zone strength can develop $0.8\Sigma M$ of girders framing to the column flanges at the joint.
2. Column axial deformations may be neglected if they contribute less than 10% to the total drift.

8.6 DESIGN OF DUAL SYSTEMS

Dual systems for resisting seismic forces are defined as structural systems, consisting of a combination of moment frames and shear walls or braced frames, designed to act concurrently in the resistance of lateral forces. Although, as will be discussed in the following paragraphs, the various building codes may have arbitrary requirements for the distribution of the design forces, these codes also stipulate that the design forces for each subsystem (i.e. moment frames, shear walls, or braced frame) shall not be less than calculated in accordance with its relative rigidity considering the interaction of the frames and shear walls. For one- or two-story buildings, simple calculations for relative rigidity are adequate for this distribution. In high-rise structures, this distribution becomes more complex as flexural (i.e. cantilever-beam) deflection of the shear walls becomes more predominant, and the relative rigidity analysis must consider the building as a whole rather than the simple relative rigidity of the elements within a given story. The results of these analyses will sometimes indicate that the moment frames will be restraining the shear walls, or braced frames, in the upper stories while the reverse may be true in the bottom stories. It is not unusual for the shear in the restraining element, at these locations, to be greater than the story shear with compensating negative shears in the restrained element.

8.6.1 1985 UBC Provisions

As indicated in Section 8.4.4, these provisions permit the design of dual systems for all buildings subject to the requirements associated with a K value of 0.8. In seismic zones 3 and 4, only buildings under 160 ft in height may be designed with a K of 1.0, which permits ordinary moment frames, braced frames, or dual systems without the restrictions of the $K = 0.8$ category. In zones 1 and 2, the 160-ft restriction does not apply, and additionally, the ductile moment-resisting space requirement of the $K = 0.8$ category may be satisfied with ordinary moment frames.

8.6.2 SEAOC Recommendations (UBC-88)

The requirements for dual systems proposed by these recommendations are generally similar to those in the 1985 UBC with equivalent R_w factors in lieu of the K factors. An R_w factor of 12 is assigned to a dual system, comprising reinforced-concrete shear walls and special moment-resisting space frames (SMRSF), comparable to the $K = 0.8$ in the current code. Concentric braced frames and SMRSF are assigned an R_w factor of 10, but eccentric braced frames with SMRSF qualify for an R_w of 12. The lateral-force distribution requirements for these systems are the same as for the $K = 0.8$ category in the current code. Ordinary moment frames with an R_w of 6 are permitted in buildings up to 160 ft in height in zones 3 and 4, and without height limit in zones 1 and 2; however, the combination of ordinary moment frames and shear walls or braced frames (i.e. available options for $K = 1.0$ in current code) are not provided in the recommendations.

8.7 DESIGN OF STRUCTURAL STEEL CONNECTIONS

8.7.1 Design of Bracing Connections

Bracing Connections in Concentric Braced Frames Bracing members in concentric braced frames are usually designed for pinned-end conditions. The connections may be designed either to be welded or to be bolted using A325 (high-strength) bolts. Light double angle or double channel bracing members are often field-bolted because of the ease and symmetry of the connection, while tubular members are often field-welded for the same reasons. Figure 8-23 shows typical bracing connections.

1. *The 1985 UBC* requires, for seismic zones 3 and 4 and for buildings in zone 2 with an importance factor I greater than 1.0, that all members in braced frames be designed for 1.25 times the prescribed basic seismic force and that all connections for those members be designed to develop the full capacity of the members or be based on the increased forces without the one-third stress increase permitted for earthquake stresses. See Example 8-4 for the design of typical concentric bracing connections.

2. *The SEAOC Recommendations (UBC-88)* require concentric bracing with chevron configurations to be designed for 1.5 times the otherwise prescribed seismic forces. K bracing is prohibited in seismic zones 3 and 4 except for one- and two-story buildings. In seismic zone 2, K bracing is permitted provided it meets the force requirements specified above for chevron bracing. All

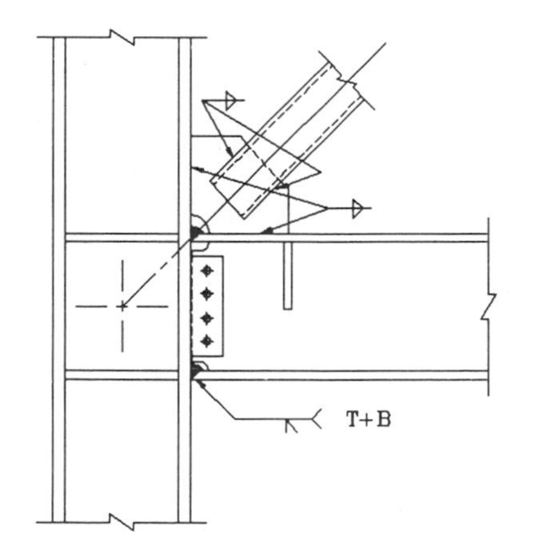

a) BEAM–COLUMN MOMENT CONNECTION

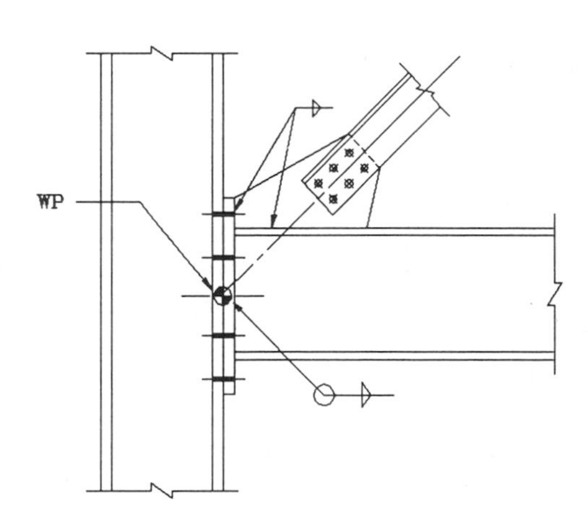

b) BEAM–COLUMN NON–MOMENT CONNECTION

NOTE: CONNECTION SHOWN IS CONCENTRIC WITH RESPECT TO BEAM, BUT SLIGHTLY ECCENTRIC WITH RESPECT TO THE COLUMN.

Figure 8-23 Concentric bracing connections.

bracing members resisting seismic forces in compression are designed using the allowable compressive stress F_a for compressive members with a one-third increase multiplied by a stress-reduction factor β where

$$\beta = \frac{1}{1 + \dfrac{KL/r}{2C_c}} \tag{8-30}$$

Bracing connections are required to be designed for the lesser of the following:

- the tensile strength of the bracing;
- $3R_w/8$ times the force in the brace due to the prescribed seismic forces;
- the maximum force that can be transferred to the brace by the structural system.

In bolted-brace connections, the ratio of the effective net section area to the gross area must satisfy the following formula:

$$\frac{A_e}{A_g} = \frac{1.2\alpha F^*}{F_u} \tag{8-31}$$

where

A_e = effective net area (varies from 0.75 to 0.90 times the net area A_n)

A_g = gross area of bracing members

α = fraction of member force that is transferred across the particular net section

F^* = stress in brace from bracing force in above paragraph

F_u = minimum tensile strength of brace

Bracing Connections in Eccentric Braced Frames The 1985 UBC, as previously indicated, does not contain any specific provisions for eccentric braced frames, and the design of these frames prior to the SEAOC Recommendations (UBC-88), was generally based on adaptation of the concentric braced frame requirements and ad hoc approvals of the cognizant building departments.

Requirements for bracing connections proposed by the SEAOC Recommendations (UBC-88) for eccentric braced frames prescribe that the bracing be designed for a compressive strength at least 1.5 times the axial force corresponding to the controlling strength of the link beam. The brace-to-beam connection is required to develop the compressive strength of the brace and transfer this force to the beam web. The brace-to-beam connection is not permitted to extend into the link beam area. See Figure 8-24 for a typical eccentric bracing connection, and refer to Example 8-5 for the design of the brace member, link-beam connections, and web stiffeners.

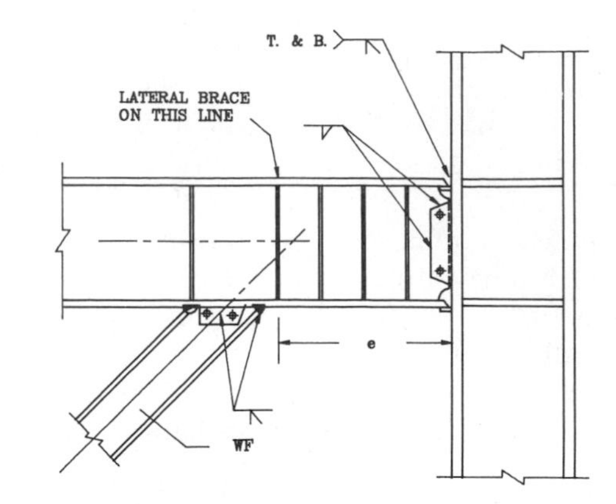

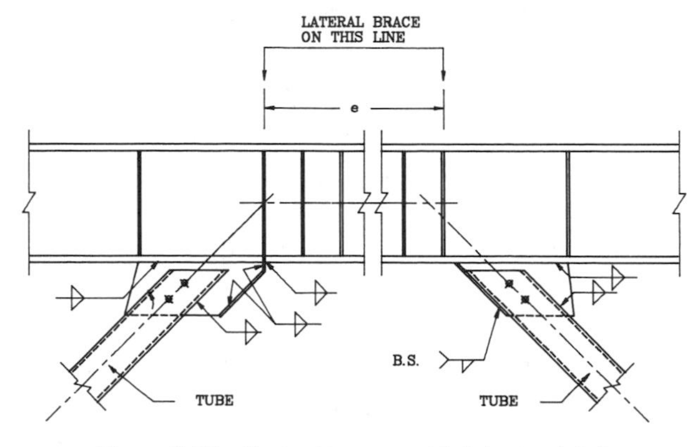

Figure 8-24 Typical brace and link beam details.

8.7.2 Design of Beam Connections

Beam Connections in Concentric Braced Frames The general requirements for these connections is that they shall have the capacity to transfer the prescribed member forces. The 1985 UBC requirements are the same as for the bracing connections (i.e. designed to develop the member, or for the prescribed member force without a one-third stress increase). Because these members are required to transmit axial torsion and compression in addition to moment and shear from gravity loads, semi-rigid web connections (e.g. bolted angles subject to flexural deformation) should be avoided. Typical connections for pinned-end and moment-resisting conditions are shown in Figures 8-25 and 8-26.

Beams in Eccentric Braced Frames For these connections, in the SEAOC Recommendations (UBC-88), a distinction is made between the *link* portion of the beam, which is designed to yield in shear and/or moment, and the remainder of the beam. The nonlink end of the beam may be connected to a column by pinned-end or moment-resisting connections. The connection must be capable of transmitting the design axial tensile

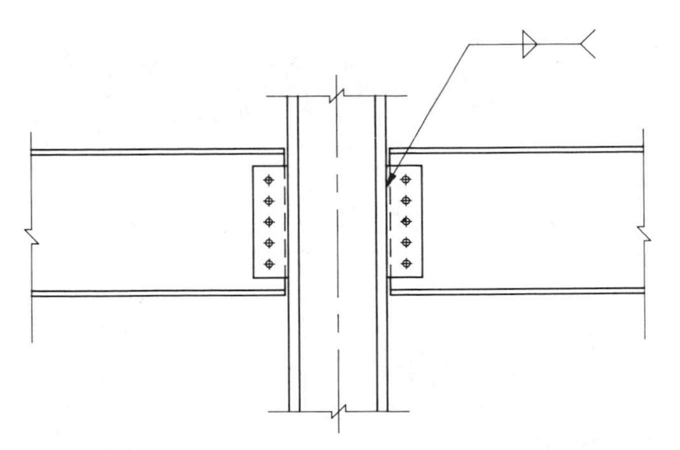

Figure 8-25 Typical beam to column flange nonmoment connection.

or compressive force and is required to be able to resist a torsional moment equal to $0.01F_y b_f t_f d$.

The link beam–column connection, under the proposed recommendations, must comply with the following:

1. *Connection to column flange.*

- The beam flanges shall have full penetration welds to the column.
- Where the link beam strength is controlled by shear, the web connection shall be welded to develop the full link-beam-web shear strength.

2. *Connection to column web.*

- Beam flanges shall have full-penetration welds to the connection plates.

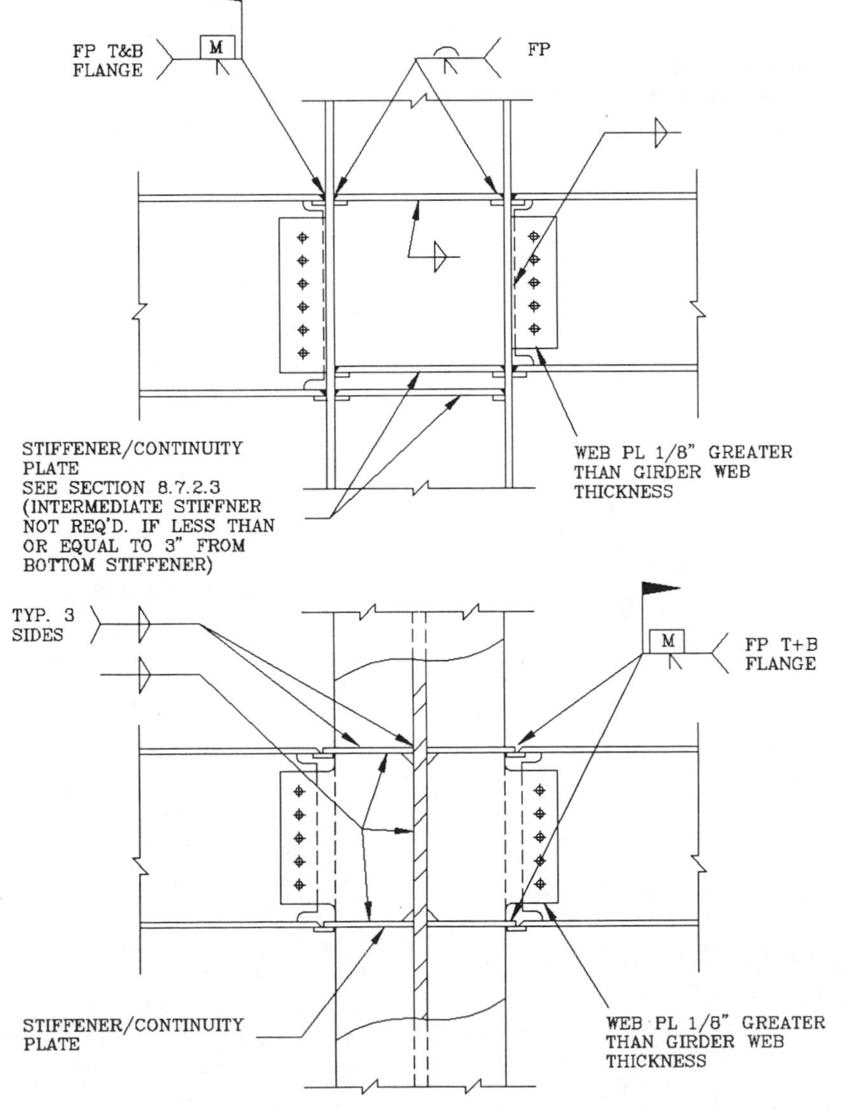

Figure 8-26 Typical beam-column moment connections.

- The web connection shall be welded to develop the link-beam-web shear strength.

Link beams shall be provided with web stiffeners as follows:

1. *At ends of link beam.*

- Full-depth web stiffeners shall be provided on both sides of the beam web at the brace end of the link beam.
- Brace beams with clear lengths between $1.6M_s/V_s$ and $2.6M_s/V_s$ shall have full-depth stiffeners on each side of the beam web at a distance b_f from the end of the link.
- Stiffeners shall have a combined width not less than $b_f - 2t_w$ and a thickness not less than $0.75t_w$ nor $\frac{3}{8}$ in.

2. *Intermediate stiffeners.*

- Full-depth web stiffeners are required where the link-beam strength is controlled by V_s or when the link beam strength is controlled by flexure and the shear associated with the reduced flexural strength, M_{rs}, is greater than $0.45F_y\, dt_w$.
- Spacing for intermediate web stiffeners shall be as follows: (i) for link beams with rotation angles of 0.06 rad, the spacing shall not exceed $38t_w - d/5$; (ii) for link beams with rotation angles of 0.03 rad or less, the spacing shall not exceed $56t_w - d/5$; (iii) spacing may be interpolated for intermediate rotation angles.
- Intermediate stiffeners are required only on one side of the web for beams less than 24 in. in depth and on both sides of the web for deeper beams.
- Each intermediate stiffener shall be not less than $b_f/2 - t_w$ in width and not less than $\frac{3}{8}$ in. in thickness.

3. *Stiffener welding.* Fillet welds connecting the stiffeners to the beam web shall develop a stiffener force of $A_{st}F_y$. Fillet welds connecting the stiffeners to the beam flanges shall develop a force of $A_{st}F_u/4$, where A_{st} is the cross-sectional area of the stiffener.

Typical link-beam details are shown in Figures 8-20 and 8-24.

Beam Connections in Moment-Resisting Frames

1. *Ordinary moment-resisting frames.* The connections need only be designed for the prescribed seismic loads, with a one-third increase on the allowable stresses.

2. *Ductile moment-resisting space frames (DMRSF).* Beam-to-column connections for these frames are required to be able to develop the full plastic moment capacity of the beam. This capacity is represented by ZF_y, where Z is the plastic-section modulus. An acceptable detail for the connection is shown in Figure 8-26, and the design procedures are contained in Example 8-3.

- Flange welding: Full penetration welds are required to develop the beam flanges. A single bevel weld with a backup bar allows downhand welding and is the typical procedure for this connection.
- Web connection: A single plate shear lug is desirable, as it facilitates temporary erection bolting to hold the beam in position for alignment and welding. The permanent shear connection must be designed[8-9] to resist the gravity loads and the design seismic loads. It should be noted that the beam shear associated with development of the plastic moment capacity at each end of the beam under combined gravity and seismic loads will be larger than the above design shear. However, research investigations[8-11] on these connections have demonstrated that the full-penetration flange welds transmit a significant portion of the joint shear as well as the flexural stresses. The SEAOC Recommendations (UBC-88) stipulate that when the flexural strength of the beam flanges is greater than 70% of the flexural strength of the entire section [i.e. $bt_f(d - t_f)F_y > 0.7ZF_y$], the beam web-to-column connection may be made by welding or high-strength bolting, or both. For girders not meeting this criterion, the beam web-to-column connection shall be made by welding the web, directly or through shear tabs, to the column, and that welding shall have a strength capable of developing at least 20% of the flexural strength of the girder web. The use of shear tabs is the preferable option, as they can be used with erection bolts to hold the beam in place for the flange welding. The web welds should not be made until after the flange welds are completed, to minimize the restraint and thus the possibility of lamellar tearing of the column flange due to the beam-flange welds.
- Continuity plates: These plates are usually welded to the column flanges and web and thus serve the dual role of transferring the beam-flange force to the column and also stiffening the column web against buckling. The 1985 UBC requires that pairs of column-web stiffeners having an area given by Equation 8-31 shall be provided whenever the calculated value of A_{st} is positive:

$$A_{st} \geqslant \frac{P_{bf} - F_{yc}t(t_b + 5k)}{F_{yst}} \tag{8-32}$$

In addition, stiffeners are required when

$$d_c > \frac{4100t^3\sqrt{F_{yc}}}{P_{bf}} \quad (8\text{-}33)$$

or

$$t_f < 0.4\frac{\sqrt{P_{bf}}}{F_{yc}} \quad (8\text{-}34)$$

where

A_{st} = area of pair of column-web stiffeners

A_s = area of flange delivering concentrated force

t = thickness of column web

k = distance between outer face of column flange and web toe of its fillet, if column is rolled shape, or equivalent distance if column is welded shape

t_b = thickness of flange or moment connection plate delivering concentrated force

t_f = thickness of column flange

d_c = column-web depth clear of fillets

P_{bf} = computed force delivered by the flange or moment connection plate times $\frac{5}{3}$ when the computed force is due to live and dead load only, or $\frac{4}{3}$ when the computed force is due to live and dead load in conjunction with wind or earthquake forces in seismic zones 1 and 2, except that for ductile moment-resisting frames in seismic zone 2 and all moment-resisting frames in seismic zones 3 and 4, $P_{bf} = A_f F_{yb}$

F_{yb} = beam yield stress

F_{yc} = column yield stress

F_{yst} = stiffener yield stress

When required, the combined width of the pair of stiffeners and the thickness of the column web shall be not less than two-thirds the width of the flange or moment connection plate delivering the concentrated force. The thickness of stiffeners shall be not less than $t_b/2$. When the concentrated force delivered occurs on only one column flange, the stiffener need not exceed one-half the column depth. The weld joining stiffeners to the column web shall be sized to carry the force in the stiffener caused by unbalanced moments on opposite sides of the column.

The SEAOC Recommendations (UBC-88) contain similar requirements except that, in Equation 8-34, the value of P_{bf} is to be taken as $1.8bt_fF_{yb}$.

In actual practice, many designers use continuity plates for all beam-column joints as a matter of good practice. These plates are generally of approximately the same thickness as the connecting beam flange and are full-penetration welded to the column flange and double fillet welded to the column web.

- Web doubler plates: The 1985 UBC stipulates that the allowable shear, V_a, on a column web should not exceed $0.40F_y\, dt$. Reference 8-4 for plastic design, with a load factor 1.3 for combined gravity and seismic loads, stipulates an allowable shear of $0.55F_y\, dt$. The SEAOC Recommendations (UBC-88) prescribe that the panel zone of the joint shall be capable of resisting the shear induced by beam bending moments due to gravity loads plus 1.85 times the prescribed seismic forces, but the shear strength need not exceed that required to develop $0.8\Sigma M$ of the girders framing into the column flanges at the joint. The joint panel-zone shear strength may be obtained from the following formula:

$$V = 0.55F_y d_c t\left(1 + \frac{3b_c t_{cf}^2}{d_b d_c t}\right) \quad (8\text{-}35)$$

where

t = total thickness of the joint panel zone including doubler plates

d_b = depth of the beam

d_c = column depth

b_c = width of the column flange

t_{cf} = thickness of the column flange

In addition, the recommendations propose that the panel-zone thickness t_z comply with the following:

$$t_z \geqslant \frac{d_z + w_z}{90} \quad (8\text{-}36)$$

where

d_z = panel-zone depth between continuity plates

w_z = panel-zone width between column flanges

For this purpose, t_z shall not include any doubler-plate thickness unless the doubler plate is connected to the column web with plug welds adequate to prevent local buckling of the plate.

Doubler plates provided to reduce panel-zone shear stress or to reduce the web depth-to-thickness ratio shall be placed not more than $\frac{1}{16}$ in. from the column web and shall be welded across the plate width top and bottom with at least a $\frac{1}{16}$-in. fillet weld. They shall be either butt- or fillet-welded to the column flanges to develop the shear strength of the doubler plate. Typical details are shown in Figure 8-11.

Although doubler plates are effective in reducing the panel-zone contribution to story drift as well as reducing the panel-zone shear stress, they are expensive and awkward to install, particularly when moment connections must be made to the column web as well as the flanges. Alternatives that should be evaluated are to use steel with a higher yield stress (e.g. 50 ksi in place of 36 ksi) or to use a heavier column section with a thicker web. Rolled sections with $F_y = 50$ ksi are readily available at a relatively small premium over 36 ksi, and this will usually be more economical than heavier sections or doubler plates.

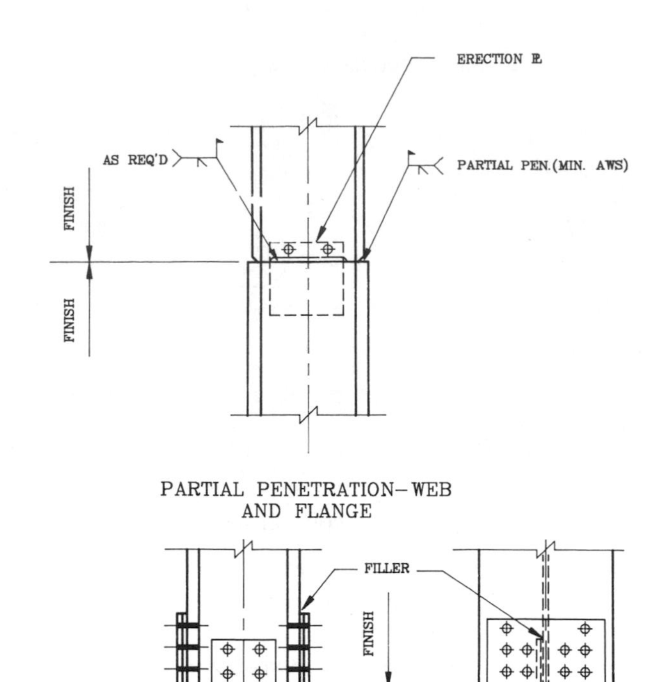

Figure 8-27 Typical flange and web column splices.

8.7.3 Design of Column Connections

Column Splices Splices in steel columns for buildings are generally made with milled ends in bearing for the compressive loads and bolted or welded connections to resist shear, axial tension, and flexural stresses. A common detail may include milled ends, bolted web splice plates for erection, and partial-penetration welds at the flanges. The web plates may be high-strength bolted or welded in the final assembly. Direct welding of the column webs is usually avoided because of the difficulty in achieving good alignment of the two relatively thin plates restrained by the heavy column flanges. Full-penetration welds are preferred for column splices that may be subject to high axial tension or flexural stresses. However, for the very heavy column sections in high-rise buildings, full-penetration welds are not economically justified and may not be necessary if dead load on the column can offset or adequately reduce the tensile stresses. In this case, partial penetration welds may be acceptable. Even though these welds may be symmetrical with respect to a column cross-section, tests have indicated[8-12] that under heavy tensile loads there is a tendency toward "prying" or tearing of the weld because of its eccentricity with respect to the center of gravity of the connected element (e.g. each individual flange). This tendency is recognized by most building codes in the lower allowable stresses permitted by these codes for tension normal to the effective area of the partial penetration weld. A typical splice detail is shown in Figure 8-27.

The 1985 UBC, in addition to the general requirements that all connections be designed to resist the shear forces and moments resulting from all of the applicable load cases, requires that compression members with bearing joints (i.e. columns with milled ends) be designed to resist tension due to lateral loads acting in conjunction with only 75% of the design dead load.

The SEAOC Recommendations (UBC-88) require that column splices shall conform to the following requirements:

1. *Strength*. Column splices shall have sufficient strength to develop the axial forces prescribed in Section 8.5.4, "SEAOC Recommendations for 1988 UBC," item 1, in addition to other applicable loads.
2. *Tension splices*. Column splices that are subject to net tension forces shall comply with the more critical of the following:

- Partial-penetration welds shall be designed to resist 150% of the force determined from Section 8.5.4, "SEAOC Recommendations for 1988

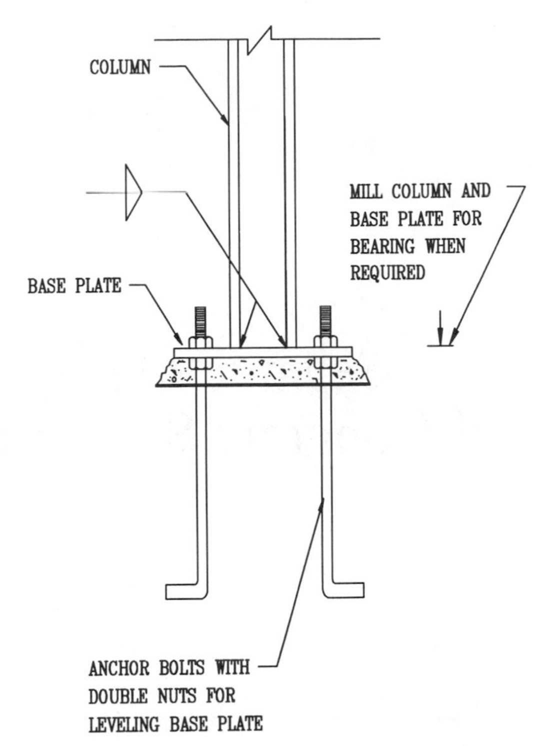

Figure 8-28 Typical detail for column base connection.

UBC," item 1, in addition to other applicable loads.

- The welding shall develop not less than 50% of the flange-area strength of the smaller column.

3. *Location*. Splices employing partial-penetration welds shall be located at least 3 ft from girder flanges.

Column Bases The column-base connection in commercial building construction is generally assumed to be pinned. Fixed-end column bases are sometimes used in industrial buildings, but are very seldom used in commercial construction because of the difficulty of restraining large column sections; the bulkiness of the connections that, when enclosed, may intrude on the usable floor area; and the offsetting effect of elastic footing rotation in average soils.

A typical column-base detail, as indicated in Figure 8-28, may include milling of the base plate and the column end for bearing and welding of the column to the base plate to transfer uplift forces, shear, and the nominal moment that results from the fixity provided by the bearing of the base plate. Anchor bolts are sometimes provided with double nuts, as shown, to level and support the column base in preparation for "dry packing" with nonshrink grout. The size of the base plate is determined so as not to exceed the allowable bearing pressure on the grout. The thickness of the base plate is selected to resist the moments from the bearing pressures due to compression loads or from the pullout forces on the anchor bolts due to uplift loads. Lateral shear forces at the column base may be transferred to the footing by the anchor bolts and/or by friction between the base plate and the grout. For large shear forces shear keys (i.e. steel bars welded to the underside of the base plate) may be used for positive shear transfer. In many cases, the column base will be below a ground floor or basement slab and will be enclosed in concrete, which facilitates the shear transfer.

REFERENCES

8-1 *Uniform Building Code*, 1985 Edition, International Conference of Building Officials, Whittier, CA, 1985.

8-2 "Tentative Provisions for the Development of Seismic Regulations for Building," Applied Technology Council Publication ATC 3-06, National Bureau of Standards, Washington, 1978.

8-3 *National Earthquake Hazards Reduction Program (NEHRP) Recommended Provisions for the Development of Seismic Regulations for New Buildings*, 1985 Edition, Building Seismic Safety Council, Washington, 1985.

8-4 *Recommended Lateral Force Requirements and Commentary*, Seismology Committee, Structural Engineers Association of California, Sacramento, CA, 1986.

8-5 *Specification for the Design, Fabrication, and Erection of Structural Steel for Buildings, with Commentary*, American Institute of Steel Construction, Chicago, 1978.

8-6 *Manual of Steel Construction*, 8th Edition, American Institute of Steel Construction, Chicago, 1980.

8-7 *Guide to Stability Design Criteria for Metal Structures*, 3rd Edition, John Wiley & Sons, New York, 1976.

8-8 Euler, L., "Sur la Forces des Colonnes," English translation by J. A. Van den Broek, *Amer. J. Phys.*, 15, pp. 309–318, New York, 1947.

8-9 Popov, E. P. and Engelhardt, M. D., "Seismic Eccentrically Braced Frames," *J. Constructional Steel Res.*, London, 1987.

8-10 Anderson, J. C. and Naeim, F., "Design Criteria and Ground Motion Effects on the Seismic Response of Multistory Buildings," ATC 10-1, Critical Aspects of Earthquake Ground Motion and Building Damage Potential, Applied Technology Council, Palo Alto, CA, 1984.

8-11 Popov, E. P. and Pinkney, R. B., "Behavior of Steel Building Connections Subjected to Inelastic Strain Reversals," Bulletin No. 14, American Iron and Steel Institute, New York, 1968.

8-12 Popov, E. P. and Stephen, R. M., "Tensile Capacity of Partial Penetration Welds," Report No. EERC 76-28, University of California, Berkeley, 1976.

Chapter 9

Seismic Design of Reinforced Concrete Structures

*Arnaldo T. Derecho, Ph.D.**

9.1 INTRODUCTION

9.1.1 The Basic Problem

The problem of designing earthquake-resistant reinforced concrete buildings, like the design of structures (whether of concrete, steel, or other material) for other loading conditions, is basically one of defining the anticipated forces and/or deformations in a preliminary design and providing for these by proper proportioning and detailing of members and their connections. Designing a structure to resist the expected loading(s) is generally aimed at satisfying established or prescribed safety and serviceability criteria. This is the general approach to engineering design. The process thus consists of determining the expected *demands* and providing the necessary *capacity* to meet these demands for a specific structure. Adjustments to the preliminary design may likely be indicated on the basis of results of the analysis–design–evaluation sequence characterizing the iterative process that eventually converges to the final design. Successful experience with similar structures should increase the efficiency of the design process.

In earthquake-resistant design, the problem is complicated somewhat by the greater uncertainty surrounding the estimation of the appropriate design loadings as well as the capacities of structural elements and connections. However, information accumulated during the last three decades from analytical and experimental studies, as well as evaluations of structural behavior during recent earthquakes, has provided a strong basis for dealing with this particular problem in a more rational manner. As with other developing fields of knowledge, refinements in design approach can be expected as more information is accumulated on earthquakes and on the response of particular structural configurations to earthquake-type loadings.

As in design for other loading conditions, attention in design is generally focused on those areas in a structure which analysis and experience indicate are or will likely be subjected to the most severe demands. Special emphasis is placed on those regions whose failure can affect the integrity and stability of a significant portion of the structure.

9.1.2 Design for Inertial Effects

Earthquake-resistant design of buildings is intended primarily to provide for the inertial effects associated with the waves of distortion that characterize dynamic response to ground shaking. These effects account for most of the damage resulting from earthquakes. In a few cases, significant damage has resulted from condi-

*Senior Consultant, Wiss, Janney, Elstner Associates, Inc., Northbrook, Illinois.

tions where inertial effects in the structure were negligible. Examples of these latter cases occurred in the excessive tilting of several multistory buildings in Niigata, Japan, during the earthquake of June 16, 1964, as a result of the liquefaction of the sand on which the buildings were founded, and the loss of a number of residences due to large landslides in the Turnagain Heights area in Anchorage, Alaska, during the March 28, 1964 earthquake. Both of the above effects, which result from ground motions due to the passage of seismic waves, are usually referred to as secondary effects. They are distinguished from so-called primary effects, which are due directly to the causative process, such as faulting (or volcanic action, in the case of earthquakes of volcanic origin).

9.1.3 Estimates of Demand

Estimates of force and deformation demands in critical regions of structures have been based on dynamic analyses—first, of simple systems, and more recently, with the availability of larger and faster digital computers, on inelastic analyses of more complex structural configurations. The latter approach has allowed estimation of force and deformation demands in local regions of specific structural models. Dynamic inelastic analyses of models of representative structures have been used to generate information on the variation of demand with major structural as well as ground-motion parameters. Such an effort involves consideration of the practical range of values of the principal structural parameters as well as the expected range of variation of the ground-motion parameters. Structural parameters include the structure fundamental period, principal member yield levels, and force–displacement characteristics; input motions of reasonable duration and varying intensity and frequency characteristics normally have to be considered.

A major source of uncertainty in the process of estimating demands is the characterization of the design earthquake in terms of intensity, frequency characteristics, and duration of large-amplitude pulses. Estimates of the intensity of ground shaking that can be expected at particular sites have generally been based on historical records. Variations in frequency characteristics and duration can be included in an analysis by considering an ensemble of representative input motions.

Useful information on demands has also been obtained from tests on specimens subjected to simulated earthquake motions using shaking tables and, more recently, with what has been referred to as the pseudodynamic method of testing. The latter method is a combination of the so-called quasistatic, or slowly reversed, loading test and the dynamic shaking-table test. In this method, the specimen is subjected to essentially statically applied increments of deformation at discrete points, the magnitudes of which are calculated on the basis of a predetermined earthquake input and the measured stiffness and estimated damping of the structure. Each increment of load after the initial increment is based on the measured stiffness of the structure during its response to the imposed loading of the preceding increment.

9.1.4 Estimates of Capacity

Proportioning and detailing of critical regions in earthquake-resistant structures have mainly been based on results of tests on laboratory specimens tested by the quasistatic method, i.e., under slowly reversed cycles of loading. Data from shaking-table tests and, more recently, from pseudodynamic tests have also contributed to the general understanding of structural behavior under earthquake-type loading. Design and detailing practice, as it has evolved over the last two or three decades, has also benefited from observations of the performance of structures subjected to actual destructive earthquakes.

Earthquake-resistant design has tended to be viewed as a special field of study, not only because many engineers do not have to be concerned with it, but also because it involves additional requirements not normally dealt with in designing for wind. Thus, while it is generally sufficient to provide adequate stiffness and strength in designing buildings for wind, in the case of earthquake-resistant design, a third basic requirement, that of ductility or inelastic deformation capacity, must be considered. This third requirement arises because it is generally uneconomical to design most buildings to respond elastically to moderate-to-strong earthquakes. To survive such earthquakes, codes require that structures possess adequate ductility to allow them to dissipate most of the energy from the ground motions through inelastic deformations. In the case of wind, structures are generally expected to respond to the

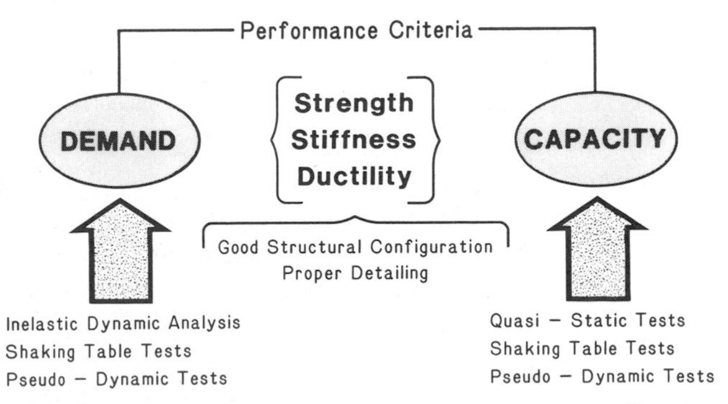

Figure 9-1 Components of and considerations in earthquake-resistant building design.

design wind within their "elastic" range of stresses. Figure 9-1 attempts to depict the interrelationships between the various considerations involved in earthquake-resistant design.

9.1.5 The Need for a Good Design Concept and Proper Detailing

Because of the appreciable forces and deformations that can be expected in critical regions of structures subjected to strong ground motions and a basic uncertainty concerning the intensity and character of the ground motions at a particular site, a good design concept is essential at the start. A good design concept implies a structure with a configuration that behaves well under earthquake excitation and designed in a manner that allows it to respond to strong ground motions according to a predetermined pattern or sequence of yielding. The need to start with a sound structural configuration that minimizes "incidental" and often substantial increases in member forces resulting from torsion due to asymmetry or force concentrations associated with discontinuities cannot be overemphasized. Although the precept may not be met with favor by some architects, clear (mainly economic) benefits can be derived from structural configurations emphasizing symmetry, regularity, and the avoidance of severe discontinuities in mass, geometry, stiffness, or strength. A direct path for the lateral (inertial) forces from the superstructure to an appropriately designed foundation would be desirable. On numerous occasions, failure to take account of the increase in forces and deformations in certain elements due to torsion or to discontinuities has led to severe structural distress and even collapse. The provision of relative strengths in the various types of elements making up a structure with the aim of controlling the sequence of yielding in such elements has been recognized as desirable from the standpoint of structural safety as well as minimizing postearthquake repair work.

An important characteristic of a good design concept and one intimately tied to the idea of ductility is structural redundancy. Since yielding at critically stressed regions and subsequent redistribution of forces to less stressed regions is central to the ductile performance of a structure, good practice suggests providing as much redundancy as possible in a structure. In monolithically cast reinforced concrete structures, redundancy is normally achieved by continuity between moment-resisting elements. In addition to continuity, redundancy or the provision of multiple load paths may also be accomplished by using several types of lateral-load-resisting systems in a building so that a "backup system" can absorb some of the load from a primary lateral-load-resisting system in the event of a partial loss of capacity in the latter.

Just as important as a good design concept is the proper detailing of members and their connections to achieve the requisite strength and ductility. Such detailing should aim at preventing nonductile failures, such as those associated with shear and with bond/anchorage. In addition, a deliberate effort should be made to securely tie all parts of a structure that are intended to act as a unit together. Because dynamic response to strong earthquakes, characterized by repeated and reversed cycles of large-amplitude deformations in critical elements, tends to concentrate deformation demands in highly stressed portions of yielding members, the importance of proper detailing of potential hinging regions should command as much attention as the development of a good design concept. As with most designs—but more so in design for earthquake resistance, where the relatively large repeated deformations tend to "seek and expose," in a manner of speaking, weaknesses in a structure—the proper field implementation of engineering drawings ultimately determines how well a structure performs under the design loading.

Experience and observation have shown that properly designed, detailed, and constructed reinforced-concrete buildings can provide the necessary strength, stiffness, and inelastic deformation capacity to perform satisfactorily under severe earthquake loading.

9.1.6 Accent on Design for Strong Earthquakes

The focus in the following discussion will be on the design of buildings for moderate-to-strong earthquake motions. These cases correspond roughly to buildings located in seismic zones 3 and 4 as defined in the Uniform Building Code.[9-1] By emphasizing design for strong ground motions, it is hoped that the reader will gain an appreciation of the special considerations involved in this most important loading case. Adjustments for buildings located in regions of lesser seismic risk will generally involve relaxation of some of the requirements associated with highly seismic areas.

Because the requirement for greater ductility in earthquake-resistant buildings represents the principal departure from the conventional design for gravity and wind loading, the major part of the discussion in this chapter will be devoted to considerations associated with providing ductility in members and structures.

The discussion in this chapter will be confined to monolithically cast reinforced-concrete buildings.

9.2 DUCTILITY IN EARTHQUAKE-RESISTANT DESIGN

9.2.1 Design Objective

In general, the design of economical earthquake-resistant structures should aim at providing the appropriate dynamic and structural characteristics so that acceptable levels of response result under the design earthquake. The magnitude of the maximum acceptable deformation will vary depending upon the type of structure and/or its function.

In some structures, such as slender, free-standing towers or smokestacks or suspension-type buildings consisting of a centrally located corewall from which floor slabs are suspended by means of peripheral hangers, the stability of the structure is dependent on the stiffness and integrity of the single major element making up the structure. For such cases, significant yielding in the principal element cannot be tolerated and the design has to be based on an essentially elastic response.

For most buildings, however, and particularly those consisting of a rigidly connected frame members and other multiply redundant structures, economy is achieved by allowing yielding to take place in some critically stressed elements under moderate-to-strong earthquakes. This means designing a building for force levels significantly lower than would be required to ensure a linearly elastic response. Analysis and experience have shown that structures having adequate structural redundancy can be designed safely to withstand strong ground motions even if yielding is allowed to take place in some elements. As a consequence of allowing inelastic deformations to take place under strong earthquakes in structures designed to such reduced force levels, an additional requirement has resulted and this is the need to insure that yielding elements be capable of sustaining adequate inelastic deformations without significant loss of strength, i.e., they must possess sufficient ductility. Thus, where the strength (or yield level) of a structure is less than that which would insure a linearly elastic response, sufficient ductility has to be built in.

9.2.2 Ductility vs. Yield Level

As a general observation, it can be stated that for a given earthquake intensity and structure period, the ductility demand increases as the strength or yield level of a structure decreases. To illustrate this point, consider two vertical cantilever walls having the same initial fundamental period. For the same mass and mass distribution, this would imply the same stiffness properties. This is shown in Figure 9-2, where idealized force–deformation curves for the two structures are marked (1) and (2). Analyses[9-2, 9-3] have shown that the maximum lateral displacements of structures with the same initial fundamental period and reasonable properties are approximately the same when subjected to the same input motion. This phenomenon is largely attributable to the reduction in local accelerations, and

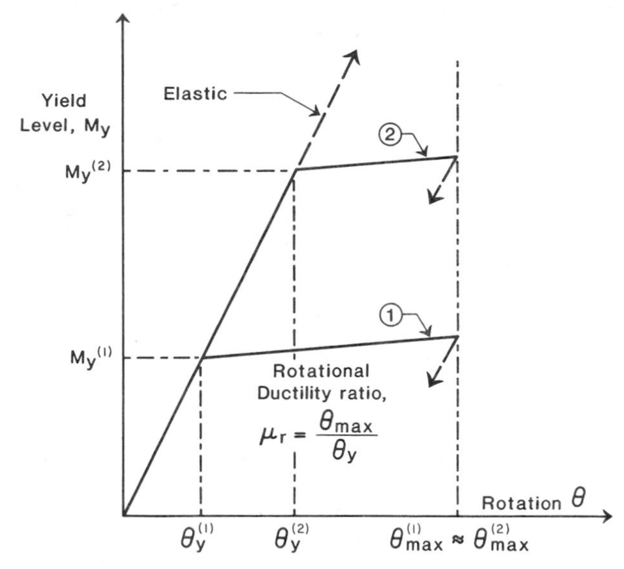

Figure 9-2 Decrease in ductility ratio demand with increase in yield level or strength of a structure.

hence displacements, associated with reductions in stiffness due to yielding in critically stressed portions of a structure. Since in a vertical cantilever the rotation at the base determines to a large extent the displacements of points above the base, the same observation concerning approximate equality of maximum lateral displacements can be made with respect to maximum rotations in the hinging region at the bases of the walls. This can be seen in Figure 9-3, from Reference 9-3, which shows results of dynamic analysis of isolated structural walls having the same fundamental period (T_1 = 1.4 sec) but different yield levels M_y. The structures were subjected to the first 10 sec of the east–west component of the 1940 El Centro record with intensity normalized to 1.5 times that of the north–south component of the same record. It is seen in Figure 9-3a that, except for the structure with a very low yield level (M_y = 500,000 in.-kips), the maximum displacements for the different structures are about the same. The corresponding ductility demands, expressed as the ratio of the maximum hinge rotations, θ_{max}, to the corresponding rotations at first yield, θ_y, are shown in Figure 9-3b. The increase in ductility demand with decreasing yield level is apparent in the figure.

A plot showing the variation of rotational ductility demand at the base of an isolated structural wall with both the flexural yield level and the initial fundamental period is shown in Figure 9-4.[9-4] The results shown in Figure 9-4 were obtained from dynamic inelastic analysis of models representing 20-story isolated structural

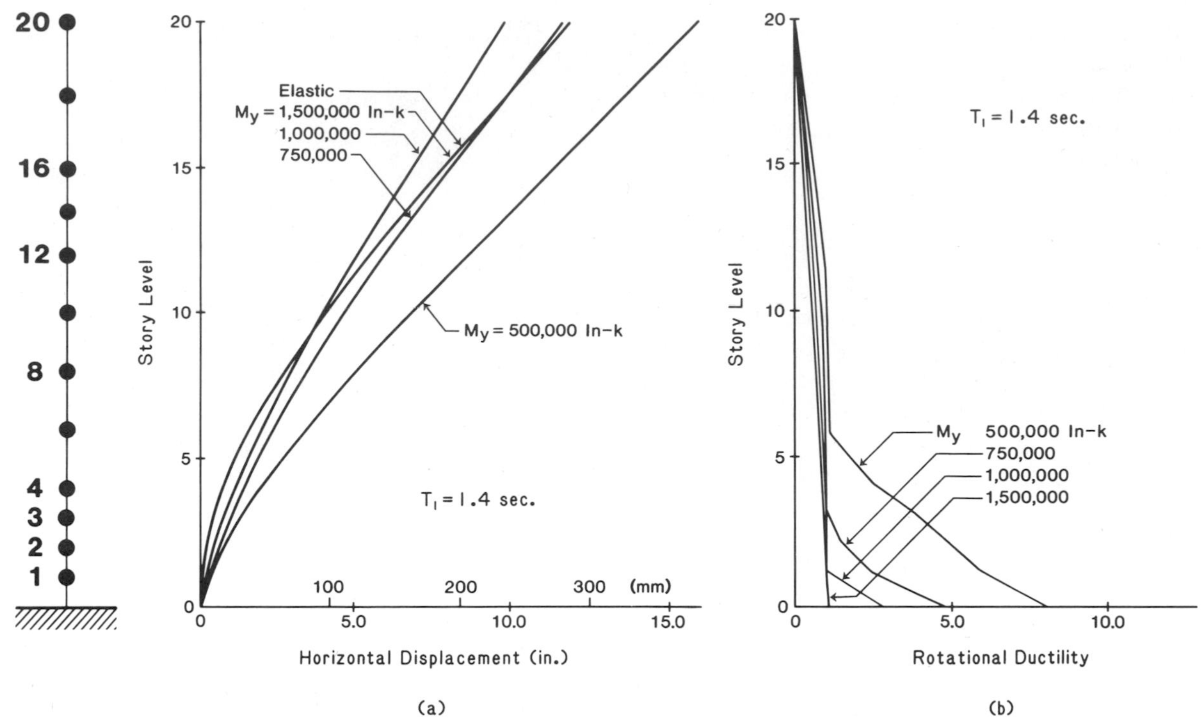

Figure 9-3 Effect of yield level on ductility demand. Note approximately equal maximum displacements for structures with reasonable yield levels. (From Ref. 9-3.)

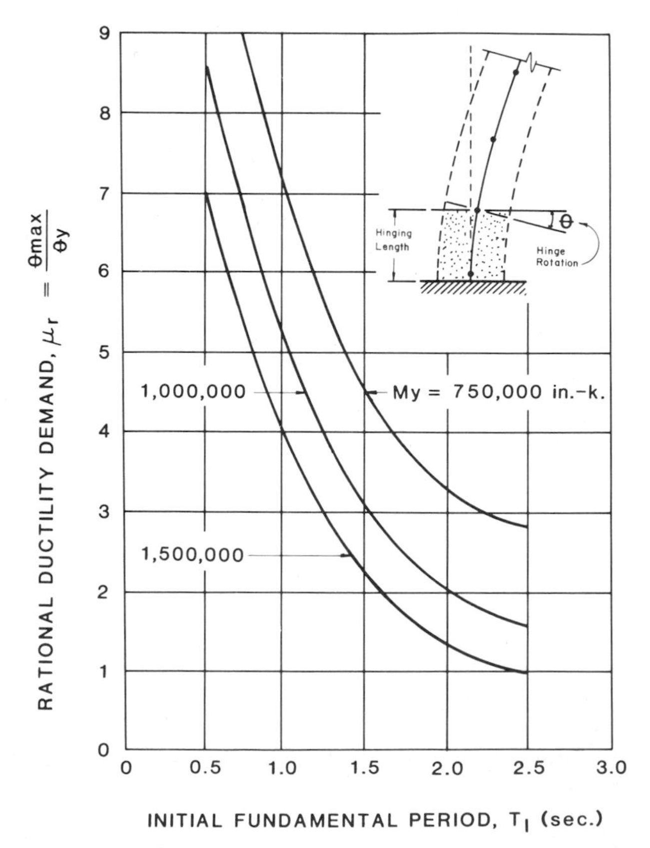

Figure 9-4 Rotational ductility demand as a function of initial fundamental period and yield level of 20-story structural walls. (From Ref. 9-4.)

walls subjected to six input motions of 10-sec duration having different frequency characteristics and an intensity normalized to 1.5 times that of the north–south component of the 1940 El Centro record. Again, note the increase in ductility demand with decreasing yield level; also the decrease in ductility demand with increasing fundamental period of the structure.

The above-noted relationship between strength or yield level and ductility is the basis for code provisions requiring greater strength (by specifying higher design lateral forces) for materials or systems that are deemed to have less available ductility.

9.2.3 Some Remarks about Ductility

One should note the distinction between inelastic deformation demand expressed as a ductility *ratio*, μ (as it usually is) on one hand, and in terms of absolute rotation on the other. An observation made with respect to one quantity may not apply to the other. As an example, Figure 9-5, from Reference 9-3, shows results of dynamic analysis of two isolated structural walls having the same yield level (M_y = 500,000 in.-kips) but different stiffnesses, as reflected in the lower initial fundamental period T_1 of the stiffer structure. Both structures were subjected to the E–W component of the 1940 El Centro record. Even though the maximum rotation for the flexible structure (with T_1 = 2.0 sec) is 3.3 times that of the stiff structure, the ductility ratio for

M_y = 500,000 in-kips

	Yield Rotation θ_y(rad.)	Max. Rotation $\theta_{max.}$(rad.)
1	.00014	.00135
2	.00070	.00448

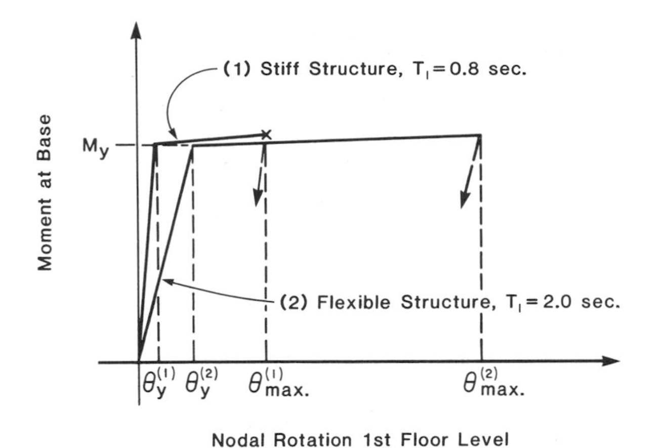

MEASURES OF INELASTIC DEFORMATION

Ductility Ratio vs. Absolute Rotation

$$\mu_r^{(1)} = \frac{\theta^{(1)}_{max.}}{\theta_y^{(1)}} = 9.6$$

but note that:

$$\mu_r^{(2)} = \frac{\theta^{(2)}_{max.}}{\theta^{(2)}} = 6.4 \qquad \theta^{(2)}_{max.} = 3.3\,\theta^{(1)}_{max.}$$

Figure 9-5 Rotational ductility ratio versus maximum absolute rotation as measures of inelastic deformation.

the stiff structure is 1.5 times that of the flexible structure. The latter result is, of course, partly due to the lower yield rotation of the stiffer structure.

Another important distinction worth noting with respect to ductility is the difference between displacement ductility and rotational ductility. The term *displacement ductility* refers to the ratio of the maximum horizontal (or transverse) displacement of a structure to the corresponding displacement at first yield. In a rigid frame or even a single cantilever structure responding inelastically to earthquake excitation, the lateral displacement of the structure is achieved by flexural yielding at local critically stressed regions. Because of this, it is reasonable to expect—and results of analyses bear this out[9-2, 9-3, 9-5]—that rotational ductilities at these critical regions are generally higher than the associated displacement ductility. Thus, overall displacement ductility ratios of 3 to 6 may imply local rotational ductility demands of 6 to 12 or more in the critically stressed regions of a structure.

9.3 BEHAVIOR OF CONCRETE MEMBERS UNDER EARTHQUAKE-TYPE LOADING

9.3.1 General Objectives of Member Design

A general objective in the design of reinforced concrete members is to so proportion such elements that they not only possess adequate stiffness and strength but so that the strength is, to the extent possible, governed by flexure rather than by shear or bond/anchorage. Code design requirements are framed with the intent of allowing members to develop their flexural or axial load capacity before shear or bond/anchorage failure occurs. This desirable feature in conventional reinforced concrete design becomes imperative in design for earthquake motions where significant ductility is required.

In certain members, such a conventionally reinforced short walls—with height-to-width ratios of 1.0 or less—the very nature of the principal resisting mechanism would make a shear-type failure difficult to avoid. Diagonal reinforcement, in conjunction with horizontal and vertical reinforcement, has been shown to improve the performance of such members.[9-6]

9.3.2 Types of Loading Used in Experiments

The bulk of information on behavior of reinforced-concrete members under load has generally been obtained from tests of full-size or near-full-size specimens. The loadings used in these tests fall under four broad categories, namely:

1. *Static monotonic loading*—where load in one direction only is applied in increments until failure or excessive deformation occurs. Data which form the basis for the design of reinforced concrete members under gravity and wind loading have been obtained mainly from this type of test. Results of this test can serve as bases for comparison with results obtained from other types of test that are more representative of earthquake loading.

2. *Slowly reversed cyclic ("quasistatic") loading*—where the specimen is subjected to (force or deformation) loading cycles of predetermined amplitude. In most cases, the load amplitude is progressively increased until failure occurs. This is shown schematically in Figure 9-6a. As mentioned earlier, much of the data upon which current design procedures for earthquake resistance are based have been obtained from tests of this type. In a few cases, a loading program patterned after analytically determined dynamic response[9-7] has been used. The latter, which is depicted in Figure 9-6b, is usually characterized by large-amplitude load cycles early in the test, which can produce early deterioration of the strength of a specimen.[9-8] In both of the above cases, the load application points are fixed so that the moments and shears are always in phase—a condition, incidentally, that does not always occur in dynamic response.

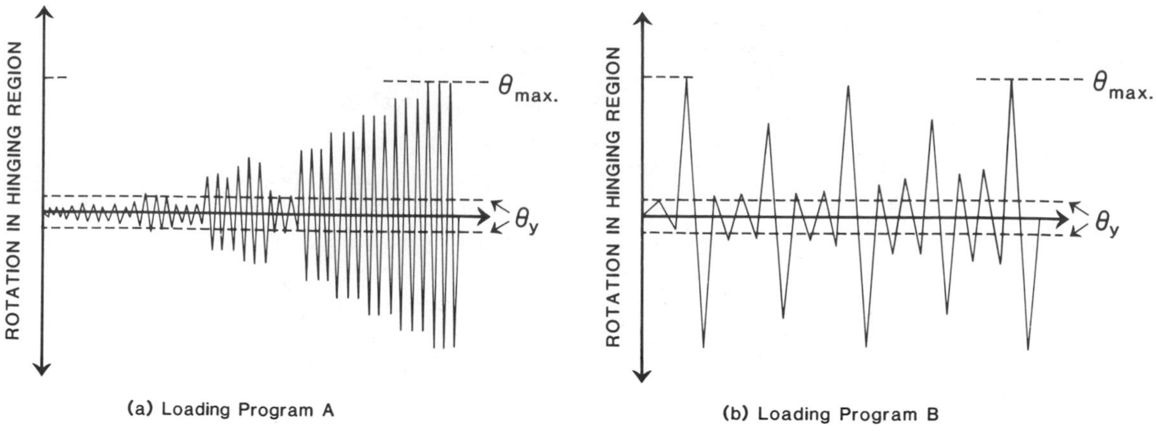

Figure 9-6 Two types of loading programs used in quasistatic tests.

This type of test provides the reversing character of the loading that distinguishes dynamic response from response to unidirectional static loading. In addition, the relatively slow application of the load allows close observation of the specimen as the test progresses. However, questions concerning the effects of the sequence of loading as well as the phase relationship between moment and shear associated with this type of test as it is normally conducted need to be explored further.

3. *Pseudodynamic tests*. In this type of test, which has been developed only during the last few years, the specimen base is fixed to the test floor while time-varying displacements determined by an on-line computer are applied to selected points on the structure. By coupling loading rams with a computer that carries out an incremental dynamic analysis of the specimen response to a preselected input motion, using measured stiffness data from the preceding loading increment and prescribed data on specimen mass and damping, a more realistic distribution of horizontal displacements in the test structure is achieved. The relatively slow rate at which the loading is imposed allows convenient inspection of the condition of the structure during the progress of the test.

This type of test, which has been used mainly for testing structures, rather than members or structural elements, requires a fairly large reaction block to take the thrust from the many loading rams normally used.

4. *Dynamic tests using shaking tables (earthquake simulators)*. The most realistic test conditions are achieved in this setup, where a specimen is subjected to a properly scaled input motion while fastened to a test bed impelled by computer-controlled actuators. Most current earthquake simulators are capable of imparting controlled motions in one horizontal direction and in the vertical direction.

The relatively rapid rate at which the loading is imposed in a typical dynamic test generally does not allow close inspection of the specimen while the test is in progress, although photographic records can be viewed after the test. Most currently available earthquake simulators are limited in their capacity to small-scale models of multistory structures or near-full-scale models of segments of a structure of two or three stories. The difficulty of viewing the progress of damage in a specimen as the loading is applied and the limited capacity of available (and costly) earthquake simulators has tended to favor the recently developed pseudodynamic test as a basic research tool for testing structural systems.

The effect of progressively increasing lateral displacements on actual structures has been studied in a few isolated cases by means of forced-vibration testing. These tests have usually been carried out on buildings or portions of buildings intended for demolition.

9.3.3 Effects of Different Variables on the Ductility of Reinforced Concrete Members

Figure 9-7 shows typical stress–strain curves of concrete having different compressive strengths. The steeper downward slope beyond the point of maximum stress of curves corresponding to the higher strength concretes is worth noting. The greater ductility of the lower-strength concretes is apparent in the figure. Typical stress–strain curves for the commonly available grades of reinforcing steel, with nominal yield strengths of 60 ksi and 40 ksi, are shown in Figure 9-8. Note in the figure that the ultimate stress is significantly higher than the yield stress. Since strains well into the strain-hardening range can occur in hinging regions of flexural members, stresses in excess of the nominal yield stress (normally used in conventional design as the limiting stress in steel) can develop in the reinforcement at these locations.

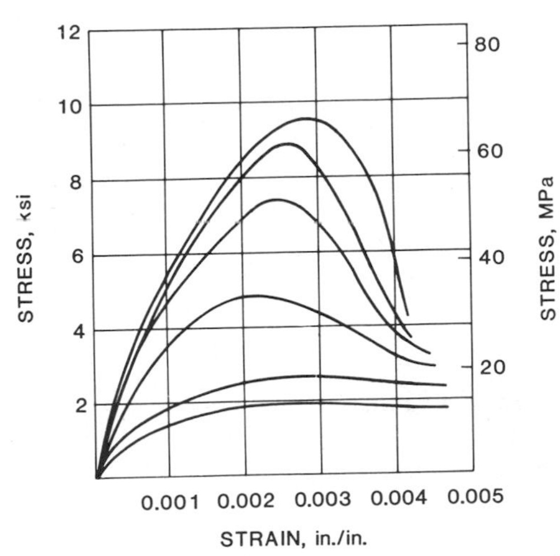

Figure 9-7 Typical stress–strain curves for concrete of varying compressive strengths.

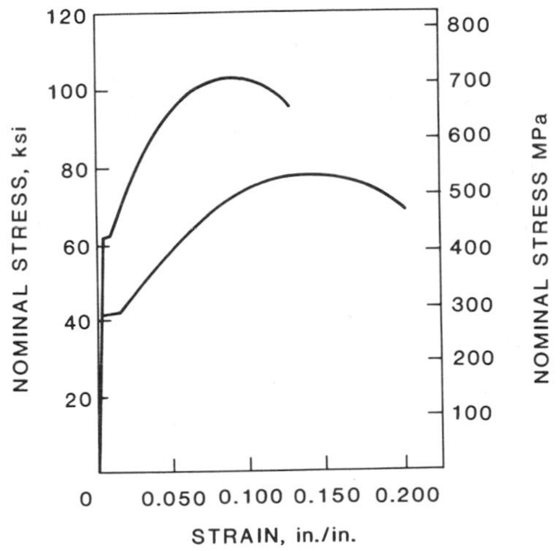

Figure 9-8 Typical stress–strain curves for ordinary reinforcing steel.

Rate of Loading An increase in the strain rate of loading is generally accompanied by an increase in the strength of concrete or the yield stress of steel. The greater rate of loading associated with earthquake response, as compared with static loading, results in a slight increase in the strength of reinforced concrete members, due primarily to the increase in the yield strength of the reinforcement. The calculation of the strength of reinforced concrete members in earthquake-resistant structures on the basis of material properties obtained by static tests (i.e., normal strain rates of loading) is thus reasonable and conservative.

Confinement Reinforcement The American Concrete Institute *Building Code Requirements for Reinforced Concrete*, ACI 318–83[9-9] (hereafter referred to as the ACI Code), specifies a maximum usable compressive strain in concrete, ϵ_{cu}, of 0.003. Lateral confinement, whether from active forces such as transverse compressive loads, or passive restraints from other framing members or lateral reinforcement, tends to increase the value of ϵ_{cu}. Tests have shown that ϵ_{cu} can range from 0.0025 for unconfined concrete to about 0.01 for concrete confined by lateral reinforcement subjected to predominantly axial (concentric) load. Under eccentric loading, values of ϵ_{cu} for confined concrete of 0.04 and more have been observed.[9-10, 9-11]

Effective lateral confinement of concrete increases its compressive strength and deformation capacity in the longitudinal direction, whether such longitudinal stress represents a purely axial load or the compressive component of a bending couple.

In reinforced concrete members, the confinement commonly takes the form of lateral ties or spiral reinforcement covered by a thin shell of concrete. The passive confining effect of the lateral reinforcement is not mobilized until the concrete undergoes sufficient lateral expansion under the action of compressive forces in the longitudinal direction. At this stage, the outer shell of concrete usually has reached its useful load limit and starts to spall. Because of this, the net increase in strength of the section due to the confined core may not amount to much in view of the loss in capacity of the spalled concrete cover. In many cases, the total strength of the confined core may be slightly less than that of the original section. The increase in ductility due to effective confining reinforcement, however, is significant.

The confining action of rectangular hoops mainly involves reactive forces at the corners, with only minor restraint provided along the straight unsupported sides. Because of this, rectangular hoops are generally not as effective as circular spiral reinforcement in confining the concrete core of members subjected to compressive loads. Tests have shown that circular-spiral reinforcement can be twice as effective as rectangular hoops as confinement reinforcement. Square spirals, because of their continuity, are slightly better than separate rectangular hoops.

The stress–strain characteristics of concrete, as represented by the maximum usable compressive strain ϵ_{cu}, is important in designing for ductility of reinforced concrete members. However, other factors also influence the ductility of a section: factors which may increase or diminish the effect of confinement on the ductility of concrete. Note the distinction between the ductility of concrete as affected by confinement and the ductility of a reinforced concrete sectional (i.e., sectional ductility) as influenced by the ductility of the concrete as well as other factors.

Sectional Ductility A convenient measure of the ductility of a section subjected to flexure or combined flexure and axial load is the ratio μ of the ultimate curvature attainable without significant loss of strength,

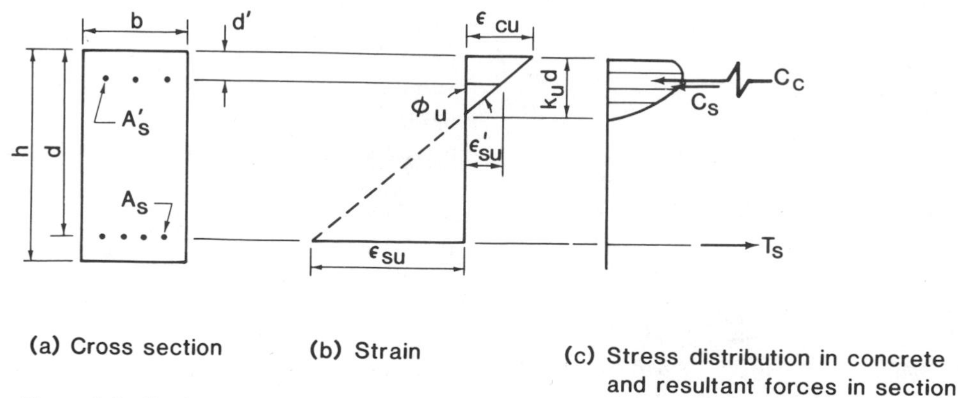

Figure 9-9 Strains and stresses in a typical reinforced concrete section under flexure at ultimate condition.

ϕ_u, to the curvature corresponding to first yield of the tension reinforcement, ϕ_y. Thus

$$\text{sectional ductility, } \mu = \frac{\phi_u}{\phi_y}$$

Figure 9-9, which shows the strains and resultant forces on a typical reinforced concrete section under flexure, corresponds to the condition when the maximum usable compressive strain in concrete, ϵ_{cu}, is reached. The corresponding curvature is denoted as the ultimate curvature, ϕ_u. It will be seen in the figure that

$$\phi_u = \frac{\epsilon_{cu}}{k_u d}$$

where $k_u d$ is the distance from the extreme compression fiber to the neutral axis.

The variables affecting sectional ductility may be classified under three groups, namely: (i) material variables, such as the maximum usable compressive strain in concrete, particularly as this is affected by confinement, and grade of reinforcement; (ii) geometric variables, such as the amount of tension and compression reinforcement, and the shape of the section; (iii) loading variables, such as the level of the axial load and accompanying shear.

As is apparent from the above expression for ultimate curvature, factors that tend to increase ϵ_{cu} or decrease $k_u d$ tend to increase sectional ductility. As mentioned earlier, a major factor affecting the value of ϵ_{cu} is lateral confinement. Tests have also indicated that ϵ_{cu} increases as the distance to the neutral axis decreases, that is, as the strain gradient across the section increases[9-12, 9-13] and as the moment gradient along the span of the member increases or as the shear span decreases.[9-14, 9-15] (For a given maximum moment, the moment gradient increases as the distance from the point of zero moment to the section considered decreases.)

The presence of compressive reinforcement and the use of concrete with a high compressive strength,[a] as well as the use of flanged sections, tend to reduce the required depth of the compressive block, $k_u d$, and hence to increase the ultimate curvature ϕ_u. In addition, the compressive reinforcement also helps confine the concrete compression zone and, in combination with adequate transverse reinforcement, allows the spread of the inelastic action in a hinging region over a longer length than would otherwise occur, thus improving the ductility of the member.[9-17] On the other hand, compressive axial loads and large amounts of tensile reinforcement, especially tensile reinforcement with a high yield stress, tend to increase the required $k_u d$ and thus decrease the ultimate curvature ϕ_u.

Figure 9-10 shows axial-load–moment-strength interaction curves for a reinforced-concrete section subjected to a compressive axial load and bending about the horizontal axis. Both confined and unconfined conditions are assumed. The interaction curve provides a convenient way of displaying the combinations of bending moment M and axial load P which a given section can carry. A point on the interaction curve is obtained by calculating the forces M and P associated with an assumed linear strain distribution across the section, account being taken of the appropriate stress–strain relationships for concrete and steel. For an ultimate load curve, the concrete strain at the extreme compressive fiber, ϵ_c, is assumed to be at the maximum usable strain, ϵ_{cu}, while the strain in the tensile reinforcement, ϵ_s, varies. A loading combination represented by a point on or inside the interaction curve can be safely resisted by the section. The balance point in the interaction curve corresponds to the condition in which the tensile reinforcement is stressed to its yield point at the same

[a] The lower ductility of the higher-strength concretes ($f_c' > 5000$ lb/in.^{2}), however, has been shown to result in a decrease in sectional ductility, particularly for sections with low reinforcement indexes.[9-16]

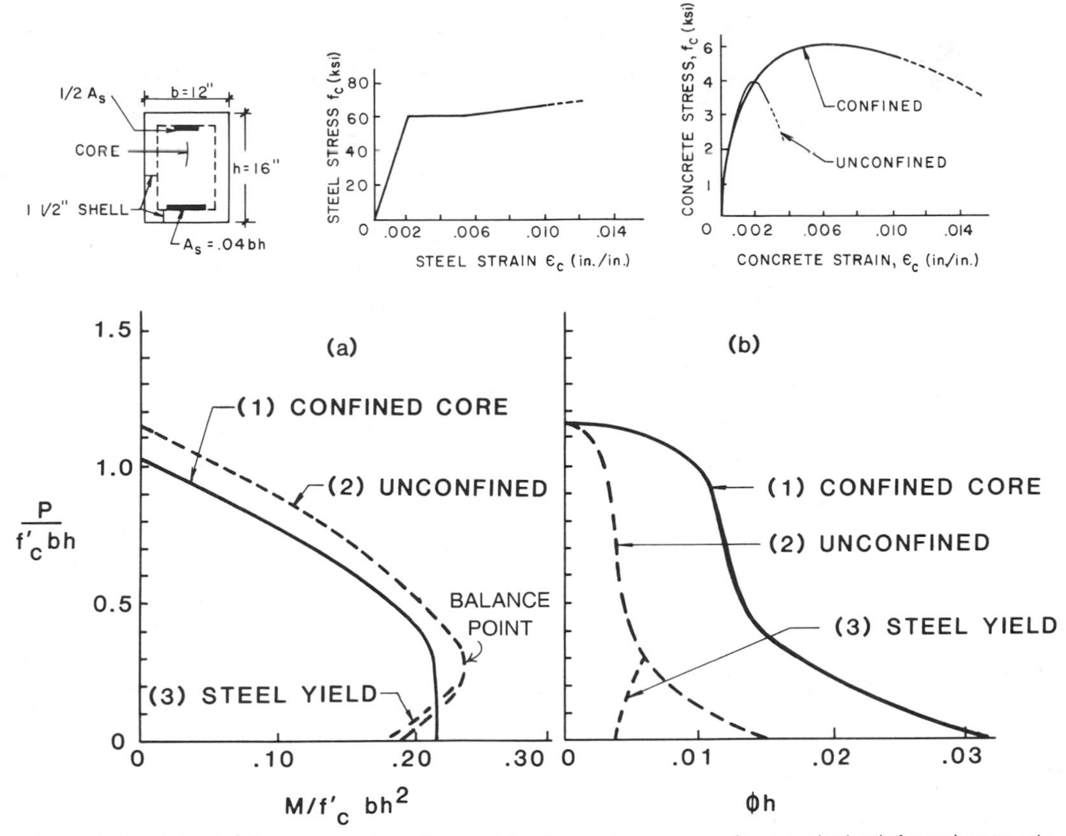

Figure 9-10 Axial load-moment interaction and load-curvature curves for a typical reinforced concrete section with unconfined and confined cores.

time that the extreme concrete fiber reaches its useful limit of compressive strain. Points on the interaction curve above the balance point represent conditions in which the strain in the tensile reinforcement is less than its yield strain ϵ_y, so that the strength of the section in this range is governed by failure of the concrete compressive zone. For those points on the curve below the balance point, $\epsilon_s > \epsilon_y$. Hence, the strength of the section in this range is governed by rupture of the tensile reinforcement.

Figure 9-10 also shows the variation of the ultimate curvature ϕ_u (in units of $1/h$) with the axial load P. It is important to note the greater ultimate curvature (being a measure of sectional ductility) associated with values of P less than that corresponding to the balance condition, for both unconfined and confined cases. The significant increase in ultimate curvature resulting from confinement is also worth noting in Figure 9-10b.

In the preceding, the flexural deformation capacity of the hinging region in members was examined in terms of the curvature at a section, ϕ, and hence the sectional or curvature ductility. Using this simple model, it was possible to arrive at important conclusions concerning the effects of various parameters on the ductility of reinforced concrete members. In the hinging region of members, however, the curvature can vary widely in value over the length of the "plastic hinge." Because of this, the total rotation over the plastic hinge, θ, provides a more meaningful measure of the inelastic flexural deformation in the hinging regions of members and one that can be related directly to experimental measurements. (One can, of course, speak of average curvature over the hinging region, i.e., total rotation divided by length of the plastic hinge.)

Shear The level of shear present can have a major effect on the ductility of flexural hinging regions. To study the effect of this variable, controlled tests of laboratory specimens are required. This will be discussed further in the following section.

9.3.4 Some Results of Experimental and Analytical Studies on the Behavior of Reinforced Concrete Members under Earthquake-Type Loading and Related Code Provisions

Experimental studies of the behavior of structural elements under earthquake-type loading have been concerned mainly with identifying and/or quantifying the effects of variables that influence the ability of critically stressed regions in such specimens to perform properly. Proper performance means primarily possessing adequate ductility. In terms of the quasistatic test that has been the most widely used for this purpose, proper performance would logically require that these critical regions be capable of sustaining a minimum number of deformation cycles of specified amplitude without significant loss of strength.

In the United States, there is at present no standard set of performance requirements corresponding to designated areas of seismic risk that can be used in connection with the quasistatic test. Such requirements would have to specify not only the minimum amplitude (i.e., ductility ratio) and number of deformation cycles, but also the sequence of application of the large-amplitude cycles in relation to any small-amplitude cycles and the permissible reduction in strength at the end of the loading. In New Zealand,[9-18] a building is considered to have adequate ductility if it can sustain a lateral deflection at the top equal to four times the deflection at first yield through eight load reversals (four complete cycles) without losing more than 20% of its original lateral-load-resisting capacity. The top horizontal deflection at first yield can be taken as that when first yield occurs in any main structural element or the deflection under the static design load calculated on the assumption of elastic behavior, whichever is greater.

As mentioned earlier, the bulk of experimental information on the behavior of elements under earthquake-type loading has been obtained by quasistatic tests using loading cycles of progressively increasing amplitude, such as is shown schematically in Figure 9-6a. Adequacy with respect to ductility for regions of high seismicity has usually been inferred when displacement ductility ratios of anywhere from 4 to 6 or greater were achieved without appreciable loss of strength.

In the following, some results of tests and analyses of typical reinforced-concrete members will be briefly reviewed. Where appropriate, related code provisions, mainly those in Appendix A of the ACI Code[9-9] (ACI Appendix A) are also discussed.

Beams Under earthquake loading, beams will generally be most critically stressed at and near their intersections with the supporting columns. An exception may be where a heavy concentrated load is carried at some intermediate point on the span. As a result, the focus of attention in the design of beams is on these critical regions where plastic hinging can take place.

At potential hinging regions, the need to develop and maintain the strength and ductility of the member through a number of cycles of reversed inelastic deformation calls for special attention in design. This special attention relates mainly to the lateral reinforcement, which takes the form of closed hoops or spirals. As might be expected, the requirements governing the design of lateral reinforcement for potential hinging regions are more stringent than those for members designed for gravity and wind loads, or the less critically stressed parts of members in earthquake-resistant structures. The lateral reinforcement in hinging regions of beams is designed to provide (i) confinement of the concrete core, (ii) support for the longitudinal compressive reinforcement against inelastic buckling, and (iii) resistance, in conjunction with the confined concrete, against transverse shear.

In addition to confirming the results of sectional analyses regarding the influence of such variables as concrete strength, confinement of concrete, and amounts and yield strengths of tensile and compressive reinforcement and compression flanges mentioned earlier, tests, both monotonic and reversed cyclic, have shown that the flexural ductility of hinging regions in beams is significantly affected by the level of shear present. A review of test results by Bertero[9-19] indicates that when the nominal shear stress exceeds about $3\sqrt{f_c'}$, members designed according to the present seismic codes can expect to suffer some reduction in ductility as well as stiffness when subjected to loading associated with strong earthquake response. When the shear accompanying flexural hinging is of the order of $5.5\sqrt{f_c'}$ or higher, very significant strength and stiffness degradation has been observed to occur under cyclic reversed loading.

The behavior of a segment at the support region of a typical reinforced-concrete beam subjected to reversed cycles of inelastic deformation in the presence of high shear[9-20, 9-21] is shown schematically in Figure 9-11. In Figure 9-11a, yielding of the top longitudinal steel under a downward movement of the beam end causes flexure–shear cracks to form at the top. A reversal of the load and subsequent yielding of the bottom longitudinal steel is also accompanied by cracking at the bottom of the beam (see Figure 9-11b). If the area of the bottom steel is at least equal to that of the top steel, the top cracks remain open during the early stages of the load reversal until the top steel yields in compression, allowing the top crack to close and the concrete to carry some compression. Otherwise, as in the more typical case where the top steel has greater area than the bottom steel, the top steel does not yield in compression (and we assume it does not buckle), so that the top crack remains open during the reversal of the load (directed upward). Even in the former case, complete closure of the crack at the top may be prevented by loose particles of concrete that may fall into the open cracks. With a crack traversing the entire depth of the beam, the resisting flexural couple consists of the forces in the tensile and compressive steel areas, while the shear along the through-depth crack is resisted primarily by dowel action of the longitudinal steel. With

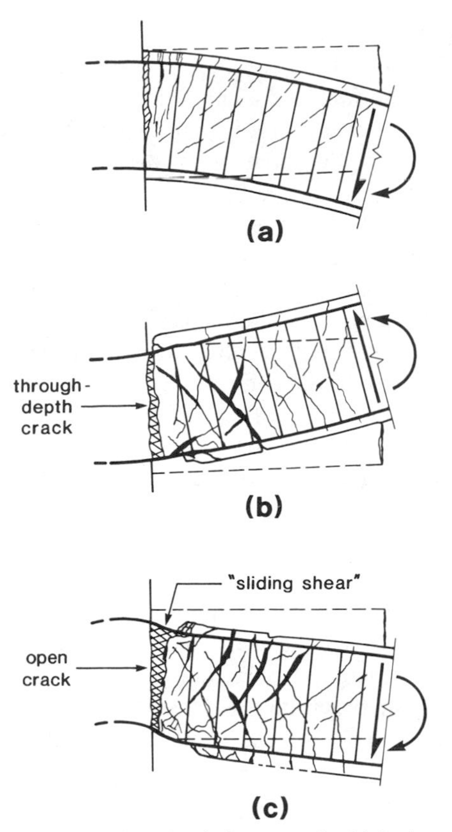

Figure 9-11 Plastic hinging in beam under high shear. (Adapted from Ref. 9-30.)

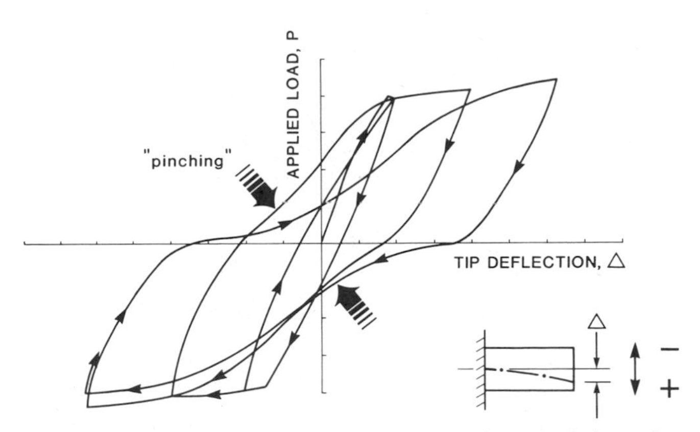

Figure 9-12 Pinching in load-displacement hysteresis loop due mainly to sliding shear.

subsequent reversals of the load and progressive deterioration of the concrete in the hinging region (Figure 9-11c), the through-depth crack widens. The resulting increase in total length of the member due to the opening of through-depth cracks under repeated load reversals is sometimes referred to as *growth* of the member.

Where the shear accompanying the moment is high, sliding along the through-depth crack(s) can occur. This sliding shear displacement, which is resisted mainly by dowel action of the longitudinal reinforcement, is reflected in a *pinching* of the associated load–deflection curve near the origin, as indicated in Figure 9-12. Since the area under the load–deflection curve is a measure of the energy-dissipation capacity of the member, the pinching in this curve due to sliding shear represents a degradation not only of the strength but also the energy-dissipation capacity of the hinging region. Where the longitudinal steel is not adequately restrained by lateral reinforcement, inelastic buckling of the compressive reinforcement followed by a rapid loss of flexural strength can occur.

Because of the significant effect that shear can have on the ductility of hinging regions, it has been suggested(9-22) that when two or more load reversals at a displacement ductility of 4 or more are expected, the nominal shear stress in critical regions reinforced according to normal U.S. code requriements for earthquake-resistant design should be limited to $6\sqrt{f_c'}$. Results of tests reported in Reference 9-22 have shown that the use of crossing diagonal or inclined web reinforcement, in combination with vertical ties, as shown in Figure 9-13, can effectively minimize the degradation of stiffness associated with sliding shear. Relatively stable hysteretic force–displacement loops, with minimal or no pinching, were observed. Tests reported in Reference 9-23 also indicate the effectiveness of intermediate longitudinal shear reinforcement, shown in Figure 9-14, in reducing pinching of the force–displacement loops of specimens subjected to moderate levels of shear stresses, i.e., between $3\sqrt{f_c'}$ and $6\sqrt{f_c'}$.

As mentioned earlier, a major objective in the design of reinforced concrete members is to have the strength controlled by flexure rather than shear or other less ductile failure mechanisms. To insure that beams develop their full strength in flexure before failing in shear, ACI Appendix A requires that the design for shear in beams be based not on the factored shears obtained from a lateral-load analysis but rather on the shears corresponding to the maximum *probable flexural strength* that can be developed at the beam ends. Such a

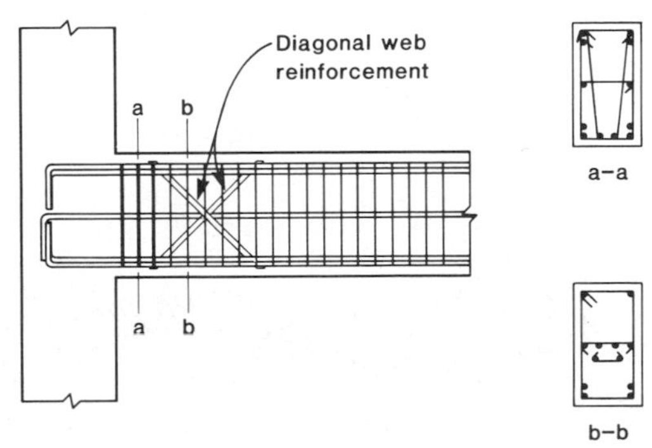

Figure 9-13 Crossing diagonal web reinforcement in combination with vertical web steel for hinging regions under high shear. (Adapted from Ref. 9-22.)

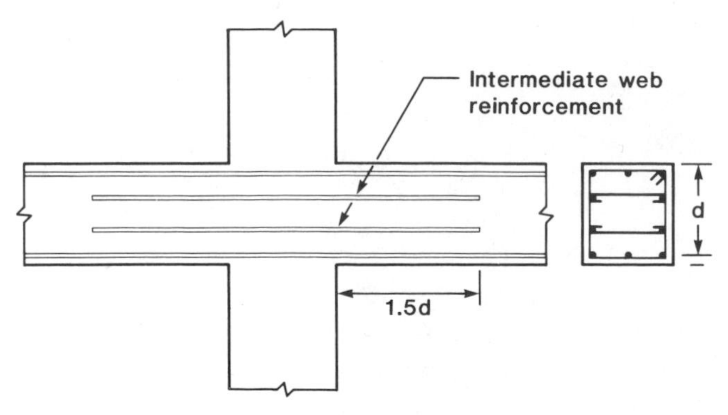

Figure 9-14 Intermediate longitudinal web reinforcement for hinging regions under moderate levels of shear.

probable flexural strength is calculated by assuming the stress in the tensile reinforcement to be equal to $1.25f_y$ and using a strength reduction factor ϕ equal to 1.0 (instead of 0.9). This is illustrated in Figure 9-15. The use of the factor 1.25 to be applied to f_y is intended to take account of the likelihood of the actual yield stress in the steel being greater (tests indicate it to be commonly 10 to 25% greater) than the specified nominal yield stress, and also in recognition of the strong possibility of strain hardening developing in the reinforcement when plastic hinging occurs at the beam ends.

ACI Appendix A requires that when the earthquake-induced shear force calculated on the basis of the maximum probable flexural strength at the beam ends is equal to or more than one-half the total design shear, the contribution of the concrete in resisting shear, V_c, be neglected if the factored axial compressive force including earthquake effects is less than $A_g f_c'/20$, where A_g is the gross area of the member cross-section. In the 1982 New Zealand Code,[9-24] the concrete contribution is to be entirely neglected and web reinforcement provided to carry the total shear force in plastic-hinging regions. It should be pointed out that the New Zealand seismic design code appears to be generally more conservative than comparable U.S. codes. This will be discussed further in subsequent sections.

$$V_a = \frac{M_{p1} + M_{p2}}{\ell} + 0.75\left(\frac{w_T \ell}{2}\right)$$

$$V_b = \frac{M_{p1} + M_{p2}}{\ell} - 0.75\left(\frac{w_T \ell}{2}\right)$$

M_p based on $f_s = 1.25\, f_y$ and $\phi = 1.0$

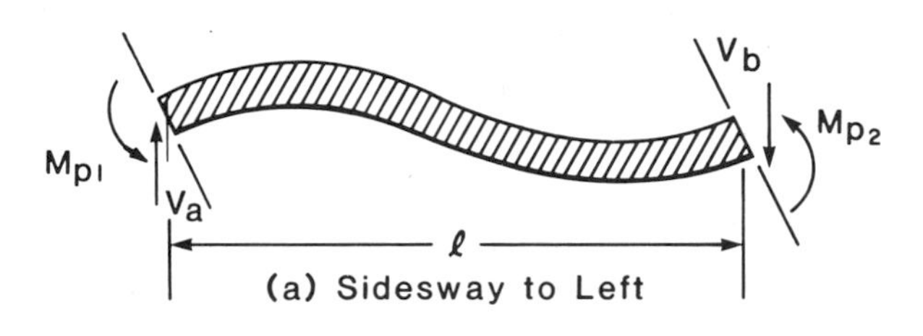

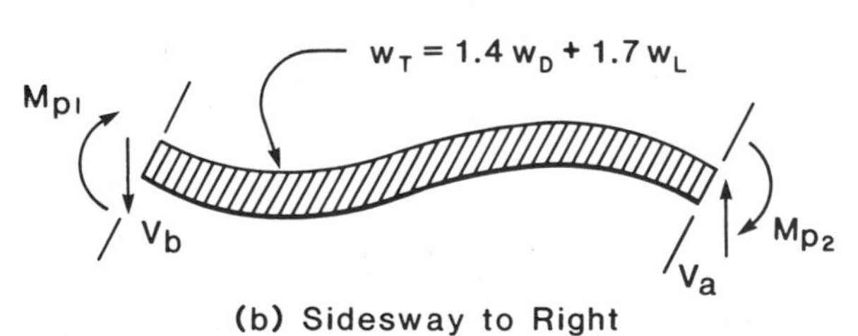

Figure 9-15 Loading cases for shear design of beams—uniformly distributed gravity loads.

Columns The current approach to the design of earthquake-resistant reinforced concrete rigid (i.e., moment-resisting) frames is to have most of the significant inelastic action or plastic hinging occur in the beams rather than in the columns. This is referred to as the "strong-column, weak-beam" concept and is intended to help insure the stability of the frame while undergoing large lateral displacements under earthquake excitation. Plastic hinging at both ends of most of the columns in a story can precipitate a story-sidesway mechanism leading to collapse of the structure at and above the story.

ACI Appendix A requires that the sum of the flexural strengths of the columns meeting at a joint, under the most unfavorable axial load, be at least equal to 1.2 times the sum of the design flexural strengths of the girders in the same plane framing into the joint. Where this requirement is satisfied, closely spaced transverse reinforcement need be provided only over a short distance near the ends of the columns where potential hinging can occur. Otherwise, closely spaced transverse reinforcement is required over the full height of the columns.

The requirements associated with the strong-column, weak-beam concept, however, do not insure that plastic hinging will not occur in the columns. As pointed out in Reference 9-5, a bending-moment distribution among frame members such as is shown in Figure 9-16, characterized by points of inflection being located away from the mid-height of columns, is not uncommon. This condition, which has been observed even under static lateral loading, occurs when the flexural mode of deformation (as contrasted with the shear–beam component of deformation) in tall frame structures becomes significant and may also arise as a result of higher-mode response under dynamic loading. As Figure 9-16 shows, a major portion of the girder moments at a joint is resisted (assuming the columns remain elastic) by one column segment, rather than being shared about equally (as when the points of inflection are located at mid-height of the columns) by the column sections above and below a joint. In extreme cases, such as might result

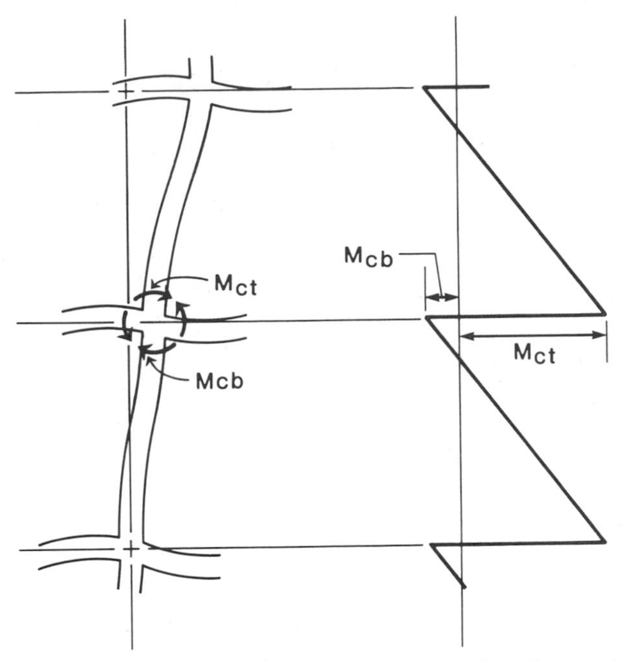

Figure 9-16 Distribution of bending moments in columns at a joint when point of inflection is located away from midheight.

from substantial differences in the stiffnesses of adjoining column segments in a column stack, the point of contraflexure can be outside the column height. In such cases, the moment resisted by a column segment may exceed the sum of the girder moments. In recognition of this, and the likelihood of the hinging region spreading over a longer length than would normally occur, the recent revision of the *Tentative Recommended Lateral Force Requirements*[9-25] of the Structural Engineers Association of California (SEAOC Recommendations adapted by UBC-88) require that where the calculated point of contraflexure in a column lies outside the middle half of the clear height of the member, confinement reinforcement is to be provided over the full height of the column.

Tests on beam–column specimens incorporating slabs,[9-26, 9-27] as in normal monolithic construction, have shown that slabs significantly increase the effective flexural strength of the beams and hence reduce the column-to-beam flexural strength ratio, if the beam strength is based on the bare beam section. Reference 9-26 recommends consideration of the slab reinforcement over a width equal to at least the width of the beam on each side of the member when calculating the flexural strength of the beam.

Another phenomenon that may lead to plastic hinging in the columns occurs in two-way (three-dimensional rigid) frames subjected to ground motions along a direction inclined with respect to the principal axes of the structure. In such cases, the resultant moment from girders lying in perpendicular planes framing into a column will generally be greater than that corresponding to either girder considered separately.[9-5] (Current practice allows consideration of design earthquake loads along each principal axes of a structure separately, as nonconcurrent loadings.) Furthermore, the biaxial moment capacity of a reinforced-concrete column under skew bending will generally be less than the larger uniaxial moment capacity. Tests reported in Reference 9-27 indicate that where bidirectional loading occurs in rectangular columns, the decrease in strength of the column due to spalling of the corner cover concrete, and bond deterioration along the column longitudinal bars at and near the corner can be large enough to shift the hinging from the beams to the columns. Thus, under concurrent bidirectional loading, columns in two-way frames designed according to the strong-column, weak-beam concept mentioned above can either yield before the framing girders or start yielding immediately following yielding of the girders.

It is worth noting that the 1985 report of ACI–ASCE Committee 352 on beam–column joints in monolithic reinforced concrete structures[9-28] recommends a minimum overstrength factor of 1.4, instead of the 1.2 given in ACI 318–83, for the flexural strength of columns relative to that of beams meeting at a joint when the beam strength is based only on the bare beam section (excluding slab). A design procedure (*capacity design*), based on the work of Paulay,[9-29, 9-30] that attempts to minimize the possibility of yielding in the columns of a frame due to the above factors has been adopted in New Zealand.[9-24] The procedure prescribes comparable overstrength factors of 1.5 or greater[9-31] for determining the flexural strength of columns relative to beams. In capacity design, the flexural strength of T or inverted-L beams is to be determined by considering the slab reinforcement over a specified width (depending upon column location) beyond the column faces as effective in resisting negative moments. It is clear from the above that the New Zealand capacity design requirements call for greater column strength than is currently used or recommended in U.S. practice.

To safeguard against strength degradation due to hinging in the columns of a frame, codes generally require lateral reinforcement for both confinement and shear in regions of potential plastic hinging. As in potential hinging regions of beams, the closely spaced transverse reinforcement in critically stressed regions of columns is intended to provide confinement for the concrete core, lateral support of the longitudinal column reinforcement against buckling and resistance (in conjunction with the confined core) against transverse shear. The transverse reinforcement can take the form of spirals, circular hoops, or rectangular hoops, the last with crossties as needed.

Early tests[9-32] of reinforced concrete columns subjected to large shear reversals had indicated the need to provide adequate transverse reinforcement not only to confine the concrete but also to carry most, if not all, of the shear in the hinging regions of columns. The beneficial effect of axial load—a maximum axial load of one-half the balance load was used in the tests—in delaying the degradation of shear strength in the hinging region was also noted in these tests. An increase in column strength due to improved confinement by longitudinal reinforcement uniformly distributed along the periphery of the column section was noted in tests reported in Reference 9-33. Tests cited in Reference 9-31 have indicated that under high axial load, the plastic hinging region in columns with confinement reinforcement provided over the usually assumed hinging length (i.e., the longer section dimension in rectangular columns or the diameter in circular columns) tends to spread beyond the confined region. To prevent flexural failure in the less heavily confined regions of columns, the New Zealand Code[9-18] requires that confining steel be extended to 1.5 times the usual assumed plastic-hinge length when the axial load exceeds $0.36\phi f_c' A_g$, where $\phi = 0.75$ and A_g is the gross area of the column section.

The basic intent of the ACI Code provisions relating to confinement reinforcement in potential hinging regions of columns is to preserve the axial-load-carrying capacity of the column after spalling of the cover concrete has occurred. This is similar to the intent underlying the column design provisions for gravity and wind loading. The amount of confinement reinforcement required by these provisions is independent of the level of axial load. Design for shear is to be based on the largest nominal moment strengths at the column ends consistent with the factored design axial compressive load. Some investigators,[9-5] however, have suggested that an approach that recognizes the potential for hinging in critically stressed regions of columns should aim primarily at achieving a minimum ductility in these regions. Studies by Park and associates, based on sectional analyses[9-31] as well as tests,[9-34, 9-35] indicate that although the ACI Code provisions based on maintaining the load-carrying capacity of a column after spalling of the cover concrete has occurred are conservative for low axial loads, they can be unconservative for high axial loads, with particular regard to attaining adequate ductility. Results of these studies indicate the desirability of varying the confinement requirements for the hinging regions in columns according to the magnitude of the axial load, more confinement being called for in the case of high axial loads.

ACI Appendix A limits the spacing of confinement reinforcement to $\frac{1}{4}$ the minimum member dimension or 4 in., with no limitation related to the longitudinal bar diameter. The New Zealand Code requires that the maximum spacing of transverse reinforcement not exceed the least of $\frac{1}{5}$ the minimum column dimension, 7.9 in., or 6 times the diameter of the longitudinal reinforcement. The third limitation is intended to relate the maximum allowable spacing to the need to prevent premature buckling of the longitudinal reinforcement. In terms of shear reinforcement, ACI Appendix A requires that the design shear force be based on the nominal moment strengths (without an increase in f_y) at the ends of the column corresponding to the factored axial force that results in the largest moments. The larger amount of transverse reinforcement required for either confinement or shear is to be used.

One should note the significant economy, particularly with respect to volume of lateral reinforcement, to be derived from the use of spirally reinforced columns.[9-31] The saving in the required amount of lateral reinforcement, relative to a tied column of the same nominal capacity, which has also been observed in designs for gravity and wind loading, acquires greater importance in earthquake-resistant design in view of the superior ductile performance of the spirally reinforced column. Figure 9-17b, from Reference 9-36, shows one of the spirally reinforced columns in the first story of the Olive View Hospital building in California following the February 9, 1971 San Fernando earthquake. A tied corner column in the first story of the same building is shown in Figure 9-17c. The upper floors in the four-story building, which were stiffened by shear walls that were discontinued below the second-floor level, shifted approximately 2 ft horizontally relative to the base of the first-story columns, as indicated in Figure 9-17a.

Beam–Column Joints Except for cases of poor detailing and construction, observations of damage due to earthquakes have indicated few instances of failure of beam–column joints in moment-resisting frames. This contrasts with numerous cases of failure of members, especially columns. However, it is clear that beam–column joints have to be designed so that the connected elements can perform properly. This requires that the joints be proportioned and detailed to allow the columns and beams framing into them to develop and maintain their strength as well as stiffness while undergoing large inelastic deformations. A loss in strength or stiffness in a frame resulting from deterioration in the joints can lead to a substantial increase in lateral displacements of the frame, including possible instability due to P-delta effects.

The design of beam–column joints is primarily aimed at (i) preserving the integrity of the joint so that the strength and deformation capacity of the connected beams and columns can be developed and substantially

(a)

(b)

(c)

Figure 9-17 Damage to columns of the 4-story Olive View Hospital building during the February 9, 1971 San Fernando, California, earthquake. (From Ref. 9-36.) (a) A wing of the building showing approximately 2 ft drift in first story. (b) Spirally reinforced concrete column in first story. (c) Tied rectangular corner column in first story.

maintained, and (ii) preventing significant degradation of the joint stiffness due to cracking of the joint and loss of bond between concrete and the longitudinal column and beam reinforcement or anchorage failure of beam reinforcement. Of major concern here is the disruption of the joint core as a result of high shear reversals. As in the hinging regions of beams and columns, measures aimed at insuring proper performance of beam–column joints have focused on providing adequate confinement and well as shear resistance to the joint.

The forces acting on a typical interior beam–column joint in a frame undergoing lateral displacement are shown in Figure 9-18a. It is worth noting in Figure 9-18a that each of the longitudinal beam and column bars is subjected to a pull on one side and a push on the other side of the joint. This combination of forces tends to push the bars through the joint, a condition that leads to slippage of the bars and even a complete pullthrough in some test specimens. Slippage resulting from bond degradation under repeated yielding of the beam reinforcement is reflected in a reduction in the beam-end fixity and thus increased beam rotations at the column faces. This loss in beam stiffness can lead to increased lateral displacements of the frame and potential instability.

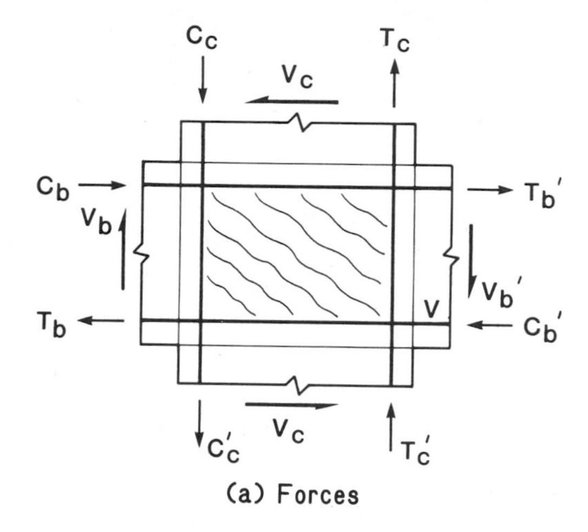

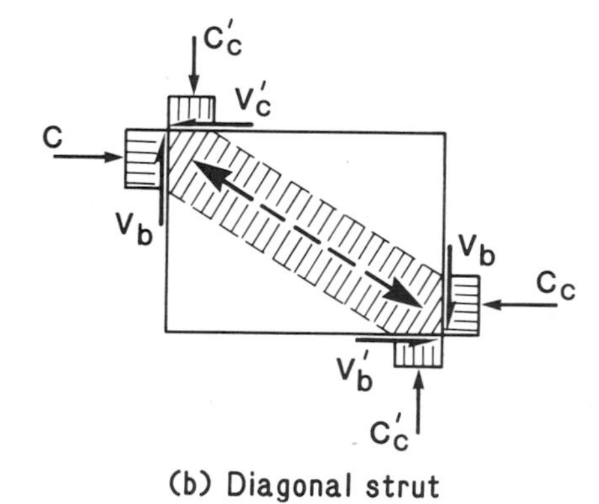

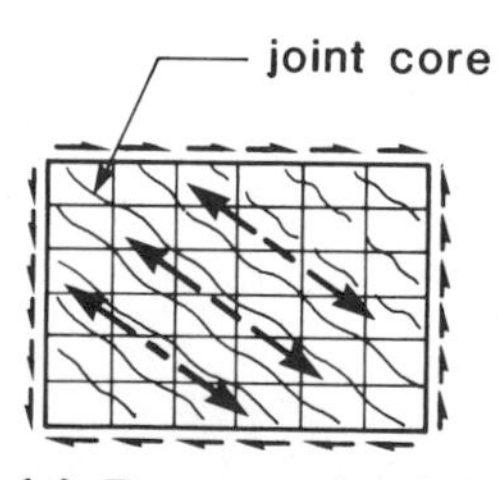

Figure 9-18 Forces and postulated shear-resisting mechanisms in a typical interior beam-column joint. (Adapted from Ref. 9-31.) (a) Forces acting on beam-column joint. (b) Diagonal strut mechanism. (c) Truss mechanism.

Two basic mechanisms have been postulated as contributing to the shear resistance of beam–column joints. These are the diagonal strut and the joint truss mechanisms, shown in Figure 9-18b and c, respectively. After several cycles of inelastic deformation in the beams framing into a joint, the effectiveness of the diagonal strut mechanism tends to diminish as through-depth cracks start to open between the faces of the column and the framing beams and as yielding in the beam bars penetrates into the joint core. The joint truss mechanism develops as a result of the interaction between confining horizontal and vertical reinforcement and a diagonal compression field acting on the elements of the confined concrete core between diagonal cracks. Ideally, truss action to resist horizontal and vertical shears would require both horizontal confining steel and intermediate vertical column bars (between column corner bars). Tests cited in Reference 9-37 indicate that where no intermediate vertical bars are provided, the performance of the joint is worse than where such bars are provided.

Tests of beam–column joints[9-26, 9-38, 9-39] in which the framing beams were subjected to large inelastic displacement cycles have indicated that the presence of transverse beams (perpendicular to the plane of the loaded beams) considerably improves joint behavior. Results reported in Reference 9-26 show that the effect of an increase in joint lateral reinforcement becomes more pronounced in the absence of transverse beams. However, the same tests indicated that slippage of column reinforcement through the joint occurred with or without transverse beams. The use of smaller-diameter longitudinal bars has been suggested[9-37] as a means of minimizing bar slippage. Another suggestion has been to force the plastic hinge in the beam to form away from the column face, thus preventing high longitudinal steel strains from developing in the immediate vicinity of the joint. This can be accomplished by suitably strengthening the segment of beam close to the column (usually a distance equal to the total depth of the beam) using appropriate details. Some of the details proposed include a combination of heavy vertical reinforcement with crossties (see Figure 9-13), intermediate longitudi-

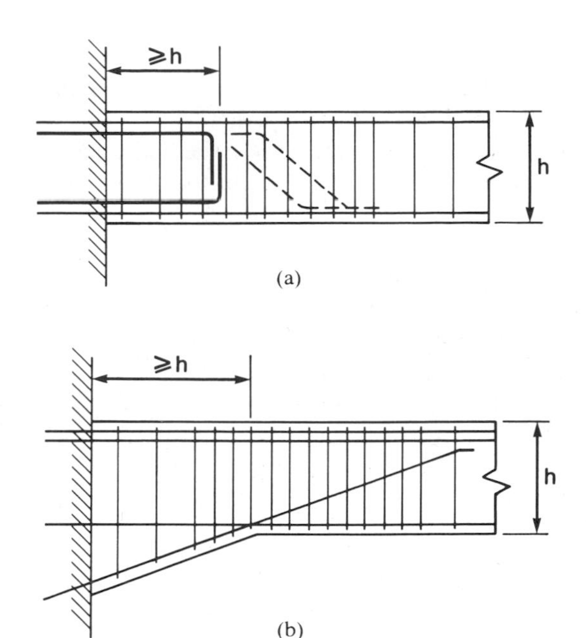

Figure 9-19 Proposed details for forcing beam hinging away from column face. (From Ref. 9-24.) See also Figs. 9-13 and 9-14. (a) Supplementary flexural reinforcement. (b) Haunch.

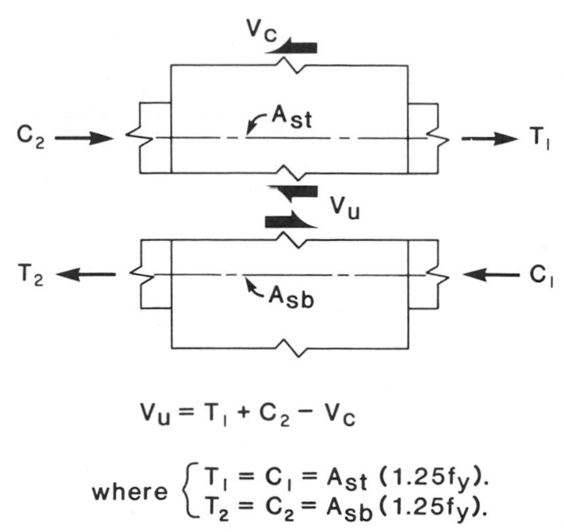

Figure 9-20 Shear force at midheight of beam-column joint—ACI Appendix A design procedure.

nal shear reinforcement (see Figure 9-14),[9-40] and supplementary flexural reinforcement and haunches, as shown in Figure 9-19.[9-31]

The current approach to beam–column joint design in the United States, as contained in ACI Appendix A, is based on providing sufficient horizontal joint cross-sectional area that is adequately confined to resist the shear stresses in the joint. The approach is based mainly on results of a study by Meinheit and Jirsa[9-39] and subsequent studies by Jirsa and associates. The parametric study reported in Reference 9-39 identified the horizontal cross-sectional area of the joint as the most significant variable affecting the shear strength of beam–column connections. Although recognizing the role of the diagonal strut and joint truss mechanisms, the current approach defines the shear strength of a joint simply in terms of its horizontal cross-sectional area. The approach presumes the provision of confinement reinforcement in the joint, unless the joint is confined on all four sides by beams of dimensions satisfying certain requirements. In the ACI Appendix A method, shear resistance calculated as a function of the horizontal cross-sectional area at mid-height of the joint is compared with the total shear across the same mid-height section. Figure 9-20 shows the forces involved in calculating the shear at mid-height of a typical joint. Note that the stress in the yielded longitudinal beam bars is to be taken equal to 1.25 times the specified nominal yield strength f_y of the reinforcement.

The recent ACI–ASCE Committee 352 Recommendations[9-28] have added a requirement relating to the uniform distribution of the longitudinal column reinforcement around the perimeter of the column core, with a maximum spacing between perimeter bars of 8 in. or one-third the column diameter or the cross-section dimension. The lateral confinement, whether from steel hoops or beams, and the distributed vertical column reinforcement, in conjunction with the confinement concrete core, provide the necessary elements for the development of an effective truss mechanism to resist the horizontal and vertical shears acting on a beam–column joint. Results of recent tests on bidirectionally loaded beam–column joint specimens[9-27] confirm the strong correlation between joint shear strength and the horizontal cross-sectional area noted by Meinheit and Jirsa.[9-39]

Some investigators[9-31, 9-37] have suggested that the ACI Appendix A approach does not fully reflect the effect of the different variables influencing the mechanisms of resistance operating in a beam–column joint and have proposed alternative expressions based on idealizations of the strut and joint truss mechanisms.

To limit slippage of beam bars through interior beam–column joints, the ACI–ASCE Committee 352 Recommendations call for a minimum column dimension equal to 20 times the diameter of beam bars passing through the joint. For exterior joints, where beam bars terminate in the joint, the maximum size of beam bar allowed is a No. 11 bar. The more stringent New Zealand Code requires a column size at least 35 times the beam bar diameter.

When the depth of an exterior column is not sufficient to accommodate the required development length for beam bars, a beam stub at the far (exterior) side of the column,[9-31] such as is shown in Figure 9-21, can be used. Embedding the 90° beam bar hooks outside of the heavily stressed joint region reduces the stiffness degra-

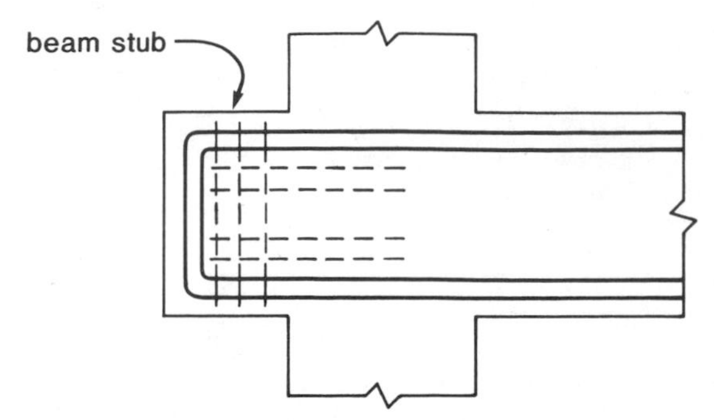

Figure 9-21 Exterior beam stub for anchoring beam bars.

dation due to slippage and improves the overall performance of the connection.

Slab–Column Connections By omitting consideration of the reinforced concrete flat plate in its provisions governing the design of structures in high-seismic-risk areas, ACI Appendix A essentially excludes the use of such a system as part of a ductile frame resisting seismic loads in such areas. Two-way slabs without beams, i.e., flat plates, are, however, allowed in areas of moderate seismic risk.

The flat plate structure is an economical and widely used form of construction in nonseismic areas, especially for multistory residential construction. Its weakest point, as is well known, is its vulnerability to a punching shear failure at the slab-column junctions. The collapse of a number of buildings using such a system during the 1964 Anchorage, Alaska and the 1967 Caracas, Venezuela earthquakes, as well as several buildings using waffle slabs during the September 1985 Mexican earthquake,[9-41, 9-42] clearly dramatized this vulnerability. Although a flat plate may be designed to carry vertical loads only, with structural walls taking the lateral loads, significant shears may still be induced at the slab-column junctions as the structure displaces laterally during earthquake response.

Tests on slab–column connections subjected to reversed cyclic loading[9-43, 9-44] indicate that the ductility of flat-slab–column connections can be significantly increased through the use of stirrups enclosing bands of flexural slab reinforcement passing through the columns. Such shear-reinforced bands essentially function as shallow beams connecting the columns.

Structural Walls Reinforced-concrete structural walls (commonly referred to as shear walls), when properly designed, represent economical and effective lateral stiffening elements that can be used to reduce potentially damaging interstory displacements in multistory structures during strong earthquakes. The structural wall, like the vertical steel truss in steel buildings, has had a long history of use for stiffening buildings laterally against wind forces. The effectiveness of properly designed structural walls in reducing earthquake damage in multistory buildings has been well demonstrated in a number of recent earthquakes.

In earthquake-resistant design, the appreciable lateral stiffness of structural walls can be particularly well utilized in combination with properly proportioned coupling beams in coupled wall systems. Such systems allow considerable inelastic energy dissipation to take place in the coupling beams (which are relatively easy to repair) at critical levels, sometimes even before yielding occurs at the bases of the walls.

Attention in the following discussion will be focused on slender structural walls, i.e., walls with a height-to-width ratio greater than about 2.0, such as are used in multistory buildings. These walls generally behave like vertical cantilever beams. Short or squat walls, on the other hand, resist horizontal forces in their plane by a predominantly truss-type mechanism, with the concrete providing the diagonal compressive strut(s) and the steel reinforcement the equilibrating vertical and horizontal ties. Tests on low-rise walls subjected to slowly reversed horizontal loading[9-45] indicate that for walls with height-to-width ratios of about 1.0, horizontal and vertical reinforcement are equally effective. As the height-to-width ratio of a wall approaches $\frac{1}{2}$ or less, the vertical reinforcement becomes more effective in resisting shear than the horizontal steel.[9-46]

In the following discussion, it will be assumed that the isolated structural wall is loaded by a resultant horizontal force acting at some distance above the base. Under such a loading, flexural hinging will occur at the base of the wall. Where the wall is designed and loaded so that it yields in flexure at the base, as might be expected under strong earthquakes, its behavior becomes a function primarily of the magnitude of the shear force that accompanies such flexural hinging as well as the reinforcement details used in the hinging region near the base. Thus, if the horizontal force acts high above the base (long shear arm), it will take a lesser magnitude of the force to produce flexural hinging at the base than when the point of application of the load is close to the base (short shear arm). For the same value of the base yield moment, the moment-to-shear ratio in the former case is high and the magnitude of the applied force (or shear) is low, while in the latter case the moment-to-shear ratio is low and the applied shear is high. In both cases, the magnitude of the applied shear is limited by the flexural yield strength at the base of the wall.

In this connection, it is of interest to note that dynamic inelastic analyses of isolated walls[9-4] covering a wide range of structural and ground motion parameters have indicated that the maximum calculated shear at the base of walls can be from 1.5 to 3.5 times greater

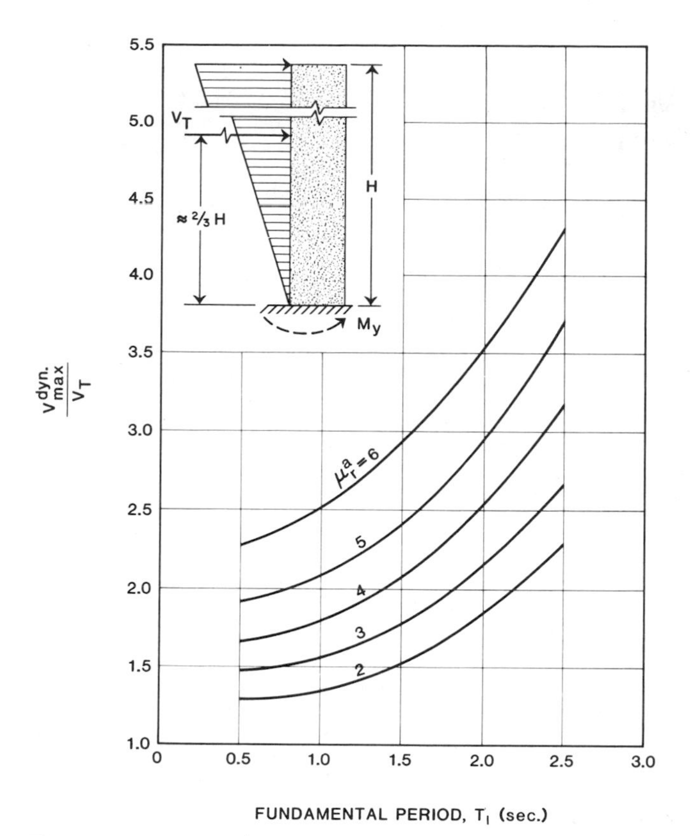

Figure 9-22 Ratio V_{max}^{dyn}/V_T as a function of T_1 and μ_r^a—20-story isolated structural walls. (From Ref. 9-4.)

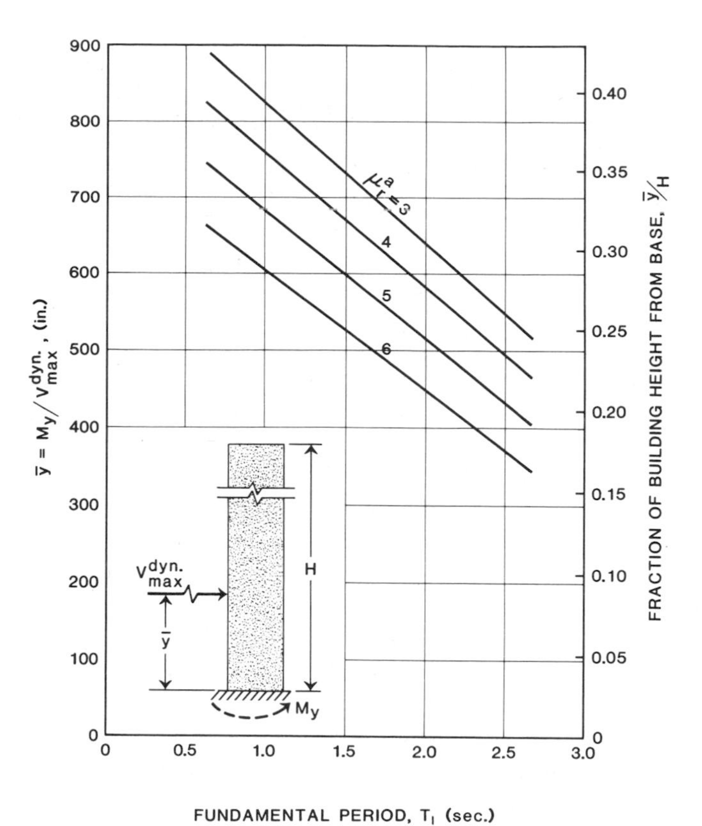

Figure 9-23 Ratio $\bar{y} = M_y/V_{max}^{dyn}$ as a function of T_1 and μ_r^a—20-story isolated structural walls. (From Ref. 9-4.)

than the shear necessary to produce flexural yielding at the base, when such shear is distributed in a triangular manner over the height of the wall, as is prescribed for design in most codes. This is shown in Figure 9-22, which gives the ratio of the calculated maximum dynamic shear, V_{max}^{dyn}, to the resultant of the triangularly distributed shear necessary to produce flexural yielding at the base, V_T, as a function of the fundamental period T_1 and the available rotational ductility μ_r^a. The input accelerograms used in the analyses had different frequency characteristics and were normalized with respect to intensity so that their spectrum intensity (i.e., the area under the corresponding 5%-damped velocity response spectrum, between periods 0.1 and 3.0 sec) was 1.5 times that of the N–S component of the 1940 El Centro record. The results shown in Figure 9-22 indicate that a resultant shear force equal to the calculated maximum dynamic shear need not be applied as high as two-thirds the height of the wall above the base to produce yielding at the base. Figure 9-23, also from Reference 9-4, shows the distance (expressed as the ratio M_y/V_{max}^{dyn}) from the base at which the resultant dynamic force would have to act to produce yielding at the base, as a function of the fundamental period and the available rotational ductility of the wall. The ordinate on the right side of the figure gives the distance above the base as a fraction of the wall height. Note that for all cases, the resultant dynamic force lies below the approximate two-thirds point associated with the triangular loading specified in codes.

These analytical results suggest not only that under strong earthquakes the maximum dynamic shear can be substantially greater than that associated with the lateral loads used to design the flexural strength of the base of the wall, but also, as a corollary, that the moment-to-shear ratio obtaining under dynamic conditions is significantly less than that implied by the code-specified distribution of design lateral loads. These results are important, because unlike beams in frames, where the design shear can be based on the maximum probable flexural strengths at the ends of the member as required by statics (see Figure 9-15), in cantilever walls it is not possible to determine a similar design shear as a function of the flexural strength at the base of the wall using statics alone, unless an assumption is made concerning the height of the applied resultant horizontal force.

Tests on isolated structural walls[9-47, 9-48] have shown that the hinging region, i.e., the region where most of the inelastic deformation occurs, extends a distance above the base roughly equal to the width of the wall. The ductility of the hinging region at the base of a wall,

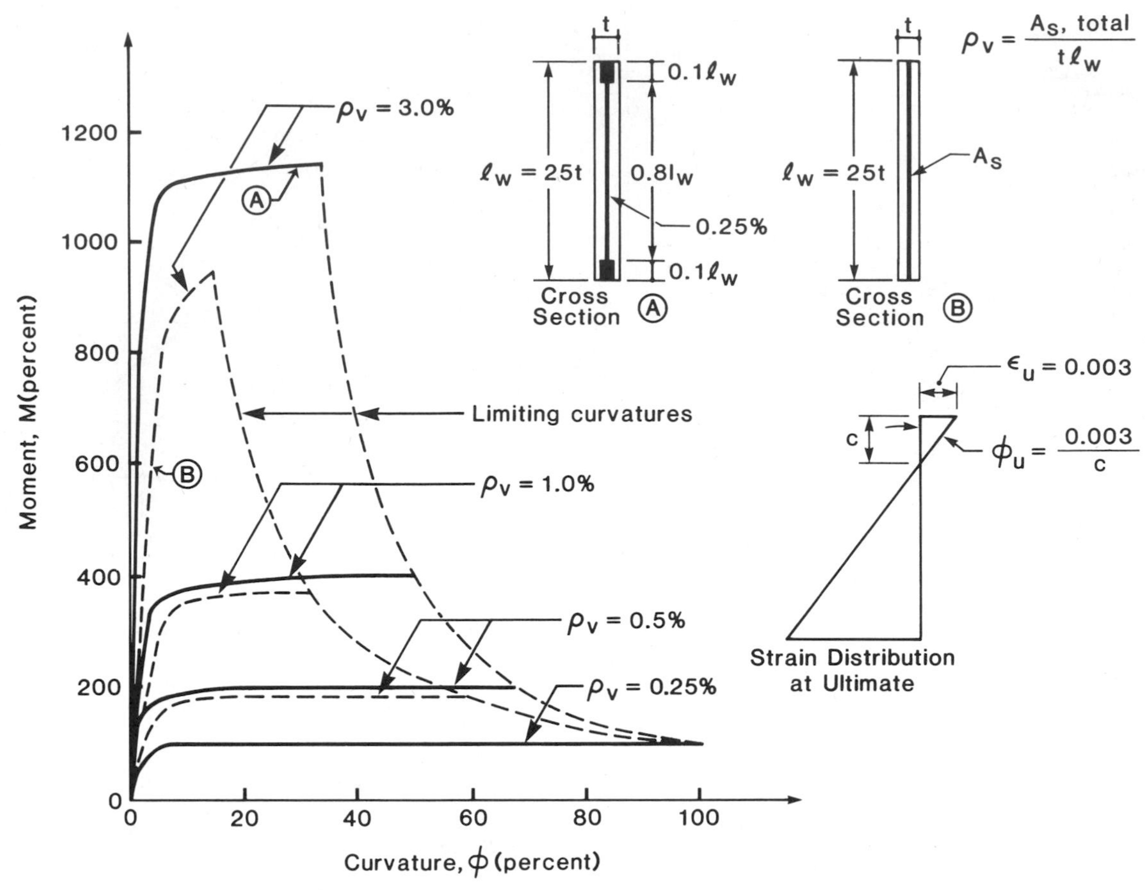

Figure 9-24 Moment-curvature curves for statically loaded rectangular walls as a function of reinforcement distribution. (From Ref. 9-48.)

like the hinging region in beams and columns, is heavily dependent on the reinforcing details used to prevent early disruption of critically stressed areas within the region. As observed in beams and columns, tests of structural walls have confirmed the effectiveness of adequate confinement in maintaining the strength of the hinging region through cycles of reversed inelastic deformation. The adverse effects of high shears, acting simultaneously with the yield moment, on the deformation capacity of the hinging region of walls has also been noted in tests.

Early tests of slender structural walls under static monotonic loading[9-49] have indicated that the concentration of well-confined longitudinal reinforcement at the ends of the wall section can significantly increase the ductility of the wall. This is shown in Figure 9-24 from Reference 9-49. This improvement in behavior resulting from a concentration of well-confined longitudinal reinforcement at the ends of a wall section has also been observed in tests of isolated walls under cyclic reversed loading.[9-47, 9-48] Plain rectangular walls, not having relatively stiff confined boundary elements, are prone to lateral buckling of the compression edge under large horizontal displacements.[9-47, 9-49]

Figure 9-25 shows a sketch of the region at the base of a wall with boundary elements after a few cycles of lateral loading. Several modes of failure have been observed in the laboratory. Failure of the section can occur in flexure by rupture of the longitudinal reinforcement or by a combination of crushing and sliding in a weakened compression flange. Alternatively, failure, i.e., loss of lateral-load-resisting capacity, can occur by sliding along a near-horizontal plane near the base (in rectangular-section walls especially) or by crushing of the web concrete at the junction of the diagonal struts and the compression flange (in walls with thin webs and/or heavy boundary elements).

Since walls are generally designed to be underreinforced, crushing in the usual sense associated with monotonic loading does not occur. However, when the flanges are inadequately confined, i.e., with the longitudinal and lateral reinforcement spaced far apart, concrete fragments within the cores of the flanges that had cracked in flexure under earlier cycles of loading can be lost in subsequent loading cycles. The longitudinal bars can buckle under compression and when subsequently stretched on reversal of the loading can rupture in low-cycle fatigue. It is also worth noting that because of

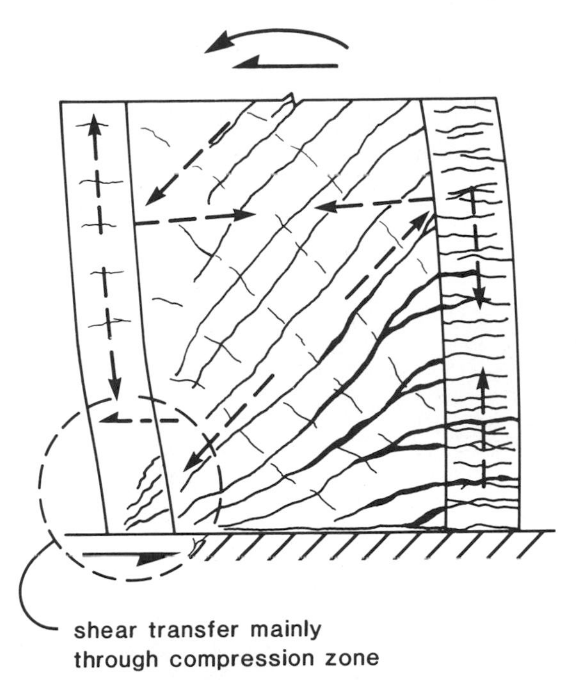

Figure 9-25 Hinging region at base of structural wall with boundary elements, showing typical mechanisms of lateral load resistance. (Adapted from Ref. 9-51.)

the Bauschinger effect (i.e., the early yielding, reflected in the rounding of the stress–strain curve of steel, that occurs during load reversals in the inelastic range and the consequent reduction in the tangent modulus of the steel reinforcement at relatively low compressive stresses), the compression steel in members subjected to reversed cycles of inelastic loading tends to buckle earlier than in comparable monotonically loaded specimens.

As in beams and columns, degradation of strength and ductility of the hinging region of walls is strongly influenced by the magnitude of the shear that accompanies flexural yielding. High shears ($> 6\sqrt{f_c'}$), when acting on a web area traversed by crisscrossing diagonal cracks, can precipitate failure of the wall by crushing of the diagonal web struts or a combined compression–sliding failure of the compression flange near the base. Shear in the hinging region is resisted by several mechanisms, namely, shear-friction along a near-horizontal plane across the width of the wall, dowel action of the tensile reinforcement and to a major extent (as in beams) by shear across the compression flange. After several cycles of load reversals and for moderate moment-to-shear ratios, the flexural cracks become wide enough to reduce the amount of shear carried by shear friction. As suggested by Figure 9-25, the truss action that develops in the hinging region involves a horizontal (shear) component of the diagonal strut that acts on the segment of the compression flange close to the base. If the compression flange is relatively slender and inadequately confined, the loss of core concrete under load reversals results in a loss of stiffness of this segment of the compression flange. The loss of stiffness and strength in the compression flange or its inability to support the combined horizontal (shear) component of the diagonal strut and the flexural compressive force can lead to failure of the wall.

Thus confinement of the flanges of walls, and especially those in the hinging region, is necessary not only to increase the compressive strain capacity of the core concrete but also to delay inelastic bar buckling and, together with the longitudinal reinforcement, prevent loss of the core concrete during load reversals (the so-called "basketing effect"). By maintaining the strength and stiffness of the flanges, confinement reinforcement improves the shear transfer capacity of the hinging region through the so-called "dowel action" of the compression flange, in addition to serving as shear reinforcement. As in beams, the diagonal tension cracking that occurs in walls and the associated truss action that develops induces tensile stresses in the horizontal web reinforcement. This suggests the need for proper anchorage of the horizontal reinforcement in the flanges.

Where high shears are involved, properly anchored crossing diagonal reinforcement in the hinging regions of walls, just as in beams, provides an efficient means of resisting shear and particularly the tendency toward sliding along cracked and weakened planes.

A series of tests of isolated structural wall specimens at the Portland Cement Association[9-47, 9-48] have provided some indication of the effect of several important variables on the behavior of walls subjected to slowly reversed cycles of inelastic deformations. Some results of this investigation have already been mentioned in the preceding. Three different wall cross-sections were considered in the study, namely, plain rectangular sections, barbell sections with heavy flanges (columns) at the ends, and flanged sections with the flanges having about the same thickness as the web. In the following, results for some of the parameters considered will be presented briefly.

1. *Monotonic vs. reversed cyclic loading.* In an initial set of two nominally identical specimens designed to explore the effect of load reversals, a 15% decrease in flexural strength was observed for a specimen loaded by cycles of progressively increasing amplitude of displacement when compared with a specimen the was loaded monotonically. Figures 9-26a and 9-27a show corresponding load–deflection curves for the specimens. A comparison of these figures shows not only a reduction in strength but also that the maximum deflection of the wall subjected to reversed loading was only 8 in., compared to about 12 in. for the monotonically load speci-

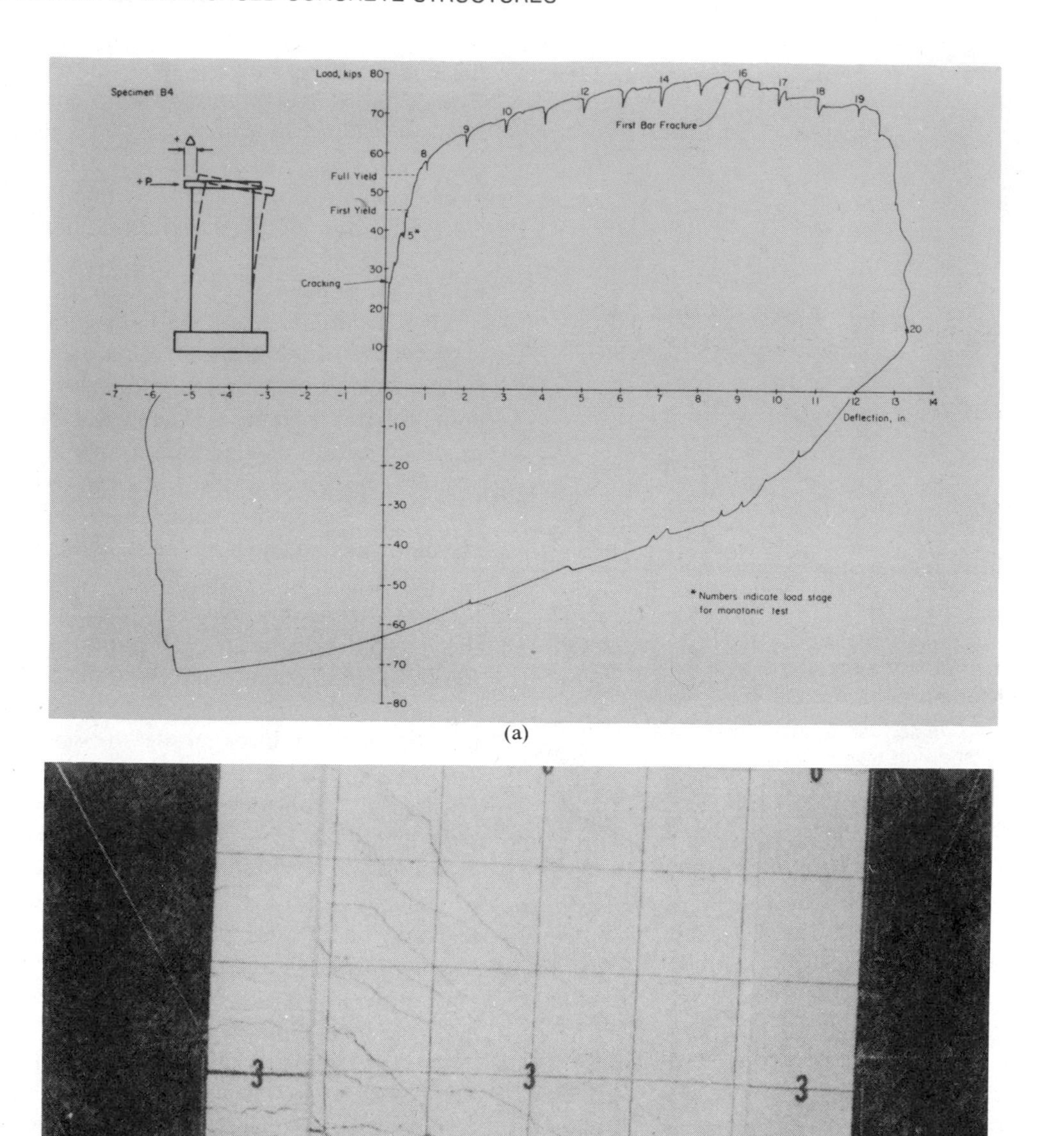

(a)

(b)

Figure 9-26 (a) Load-deflection curve of monotonically loaded specimen. (b) View of specimen at +12 in. top deflection. (From Ref. 9-50.)

men, indicating a reduction in deflection capacity of about 30%. Figure 9-27b, when compared with Figure 9-26a, shows the more severe cracking that results from load reversals.

2. *Level of shear stress.* Figure 9-28 shows a plot of the variation of the maximum rotational ductility with the maximum nominal shear stress in isolated structural wall specimens reported in References 9-47 and 9-48. The decrease in rotational ductility with increasing values of the maximum shear stress will be noted. The maximum rotation used in determining ductility was taken as that for the last cycle in which at least 80% of

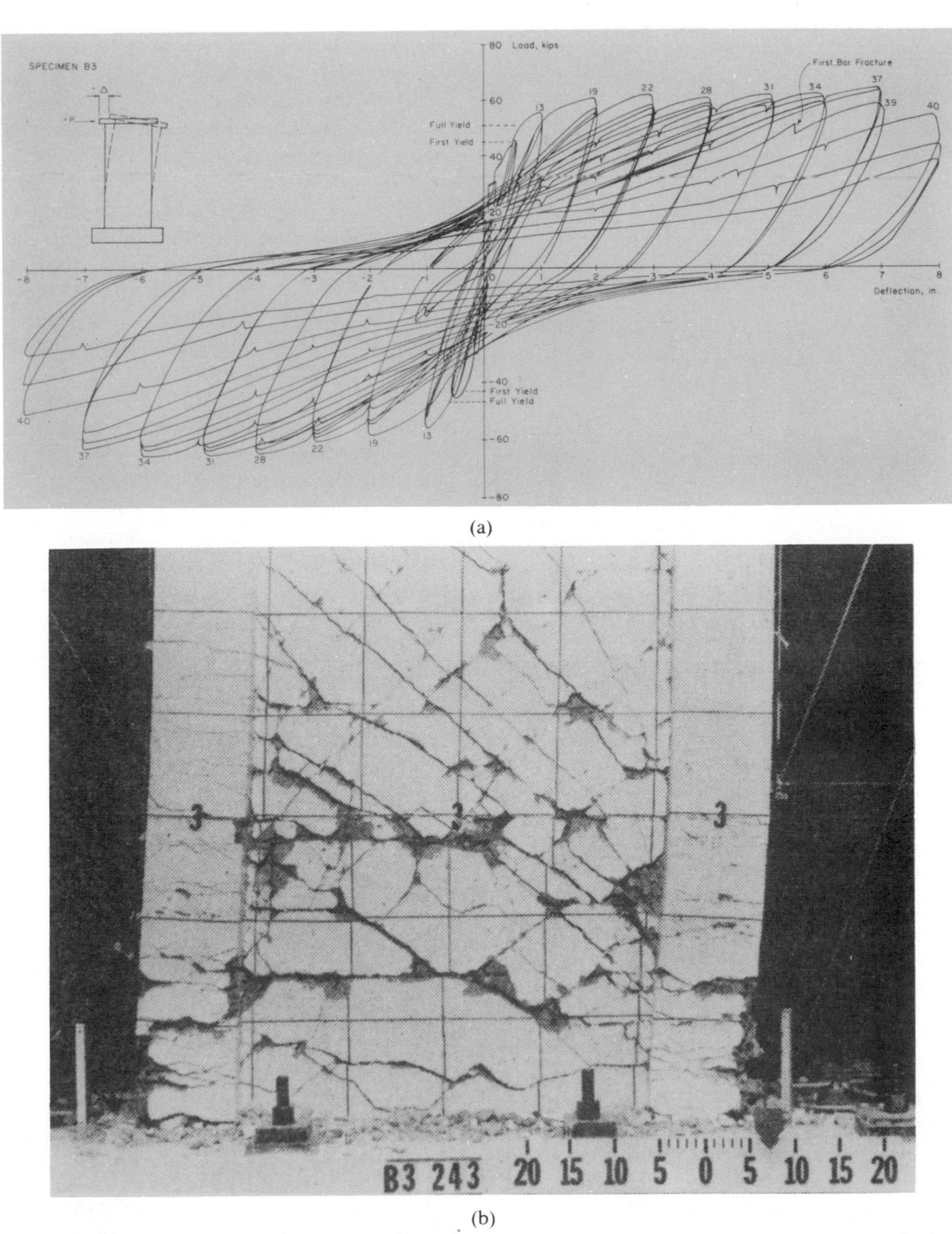

(a)

(b)

Figure 9-27 (a) Load-deflection curve of specimen subjected to load cycles of progressively increasing amplitude. (b) View of specimen at +8 in. top deflection. (From Ref. 9-50.)

the previous maximum observed load was sustained throughout the cycle. The yield rotation was defined as the rotation associated with yielding of all of the tensile reinforcement in one of the boundary elements.

The presencc of axial loads—of the order of 10% of the compressive strength of the walls—increased the ductility of specimens subjected to high shears. In Figure 9-28, the specimens subjected to axial loads are denoted by open symbols. The principal effect of the axial load was to reduce the shear distortions and hence increase the shear stiffness of the hinging region. It may be of interest to note that for walls loaded monotonically,(9-49) axial compressive stress was observed to increase moment capacity and reduce ultimate curvature, results consistent with analytical results from sectional analysis.

3. *Section shape*. As mentioned earlier, the use of wall sections having stiff and well-confined flanges or boundary elements, as against plain rectangular walls, not only allows development of substantial flexural capacity (in addition to being less susceptible to lateral buckling), but also improves the shear resistance and

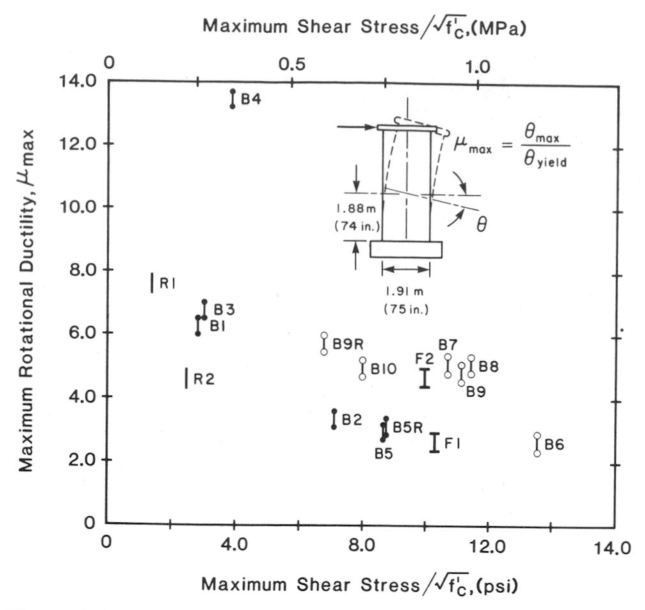

Figure 9-28 Variation of rotational ductility with maximum average shear stress in PCA isolated wall tests. (From Ref. 9-48.)

ductility of the wall. In walls with relatively stiff and well-confined boundary elements, some amount of web crushing can occur without necessarily limiting the flexural capacity of the wall. Corley et al.[9-50] point out that trying to avoid shear failure in walls, particularly walls with stiff and well-confined boundary elements, may be a questionable design objective. Thus, although ACI Appendix A limits the maximum average shear stress in walls to $10\sqrt{f_c'}$ (a value based on monotonic tests) with the intent of preventing web crushing, web crushing occurred in some specimens subjected to shear stresses only slightly greater than $7\sqrt{f_c'}$. However, those specimens where web-crushing failure occurred were able to develop deformations well beyond the yield deformation prior to loss of capacity.

4. *Sequence of large-amplitude load cycles*. Dynamic inelastic analyses of isolated walls[9-7] have indicated that in a majority of cases, the maximum or a near-maximum response to earthquakes occurs early, with perhaps only one elastic response cycle preceding it. This contrasts with the loading program commonly used in quasistatic tests, which consists of load cycles of progressively increasing amplitude. To examine the effect of imposing large-amplitude load cycles early in the test, two nominally identical isolated wall specimens were tested. One specimen was subjected to load cycles of progressively increasing amplitude, as were most of the specimens in this series. Figure 9-29a indicates that specimen B7 was able to sustain a rotational ductility of slightly greater than 5 through three repeated loading cycles. The second specimen (B9) was tested using a modified loading program similar to that shown in Figure 9-6, in which the maximum load amplitude was imposed on the specimen after only one elastic load cycle. The maximum load amplitude corresponded to a rotational ductility of 5. As indicated in Figure 9-29b, the specimen failed before completing the second load cycle. Although results from this pair of specimens cannot be considered conclusive, they suggest that tests using load cycles of progressively increasing amplitude may overestimate the ductility that can be developed under what may be considered more realistic earthquake response conditions. The results do tend to confirm the reasonable expectation that an extensively cracked and "softened" specimen subjected to several previous load cycles of lesser amplitude can better accommodate large reversed lateral deflections than a virtually uncracked specimen that is loaded to near-capacity early in the test. From this standpoint, the severity of the modified loading program, compared to the commonly used progressively-increasing-amplitude loading program, appears obvious.

5. *Reinforcement detailing*. On the basis of the tests on isolated walls reported in References 9-47 and 9-48, Oesterle et al.[9-51] proposed the following detailing requirements for the hinging regions of walls:

- The maximum spacing of transverse reinforcement in boundary elements should be $5d_b$, where d_b is the diameter of the longitudinal reinforcement.
- Transverse reinforcement in the boundary element should be designed for a shear

$$V_{nb} = M_{nb}/1.5l_b,$$

where

M_{nb} = nominal moment strength of boundary element

l_b = width of boundary element (in the plane of the wall)

- No lap splices should be used for crossties in segments of boundary elements within the hinging region.
- A recommendation on anchoring horizontal web reinforcement in the boundary elements, such as is shown in Figure 9-30a, has been adopted by ACI Appendix A. For levels of shear in the range of $5\sqrt{f_c'}$ to $10\sqrt{f_c'}$, the study indicates that alternate 90° and 135° hooks, as shown in Figure 9-30b, can be used.

The specimens tested in this series had special confinement reinforcement only over a length near the base equal to the width of the wall, i.e., the approximate length of the hinging region. Strain readings as well as observations of the general condition of the walls after failure showed that significant inelasticity and damage were generally confined to the hinging region. In view of

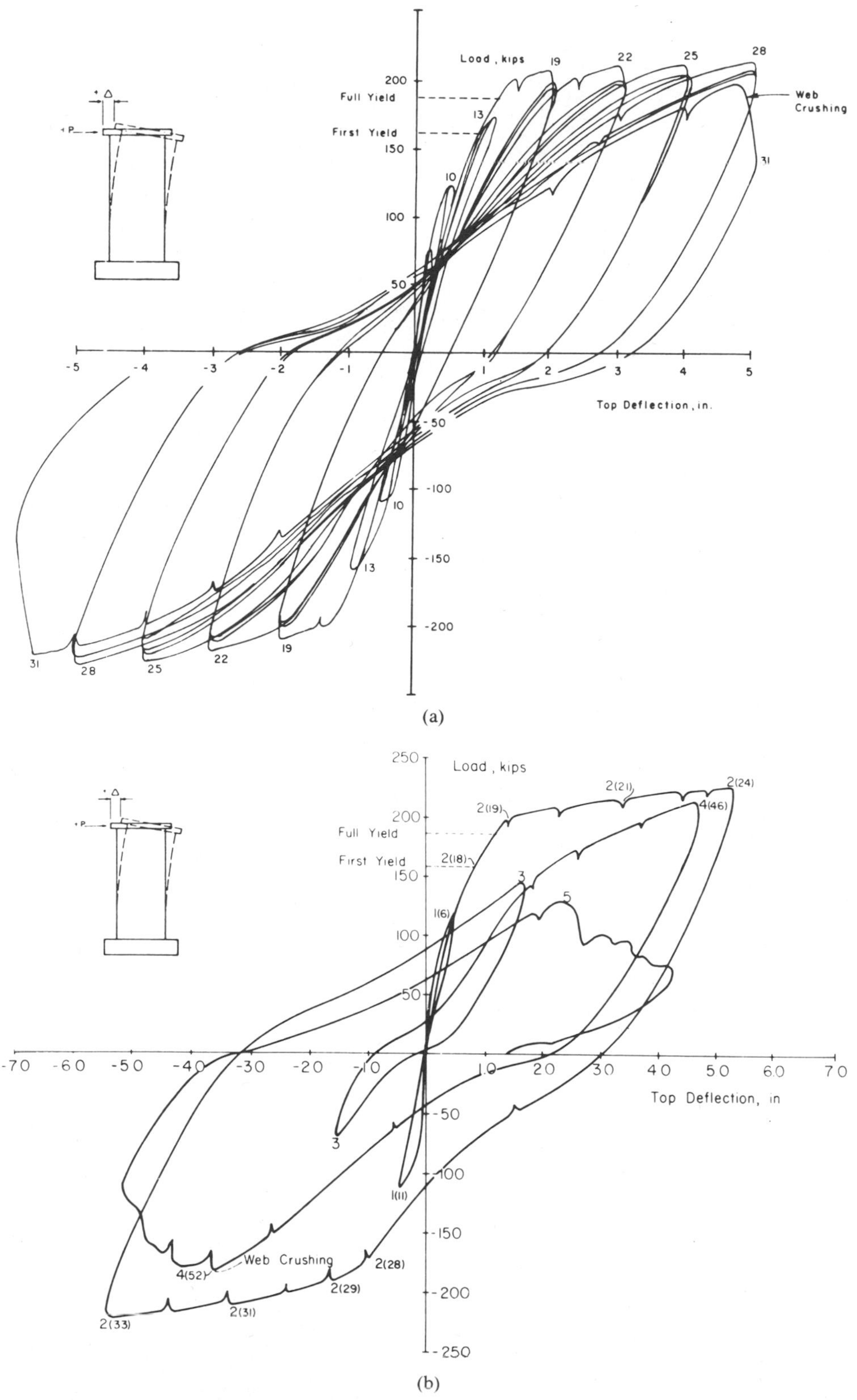

Figure 9-29 Comparison of behavior of isolated walls subjected to different loading histories. (From Ref. 9-50.) (a) Specimen subjected to progressively increasing load amplitudes (see Fig. 9-6a). (b) Specimen subjected to loading history characterized by large-amplitude cycles early in loading (see Fig. 9-6b).

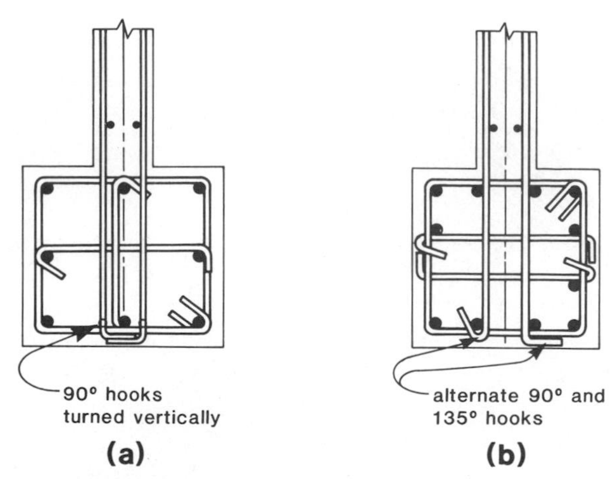

Figure 9-30 Alternative details for anchorage of horizontal web reinforcement in boundary elements. (From Ref. 9-51.) (a) Detail for walls subjected to low-to-moderate stress levels. (b) Detail for walls subjected to high shear stress levels.

this, it has been suggested that special confinement reinforcement for boundary elements need be provided only over the lengths of potential hinging regions. These are most likely to occur at the base and at points along the height of the wall where discontinuities, associated with abrupt and significant changes in geometry, strength, or stiffness, occur.

Coupled Walls As mentioned earlier, a desirable characteristic in an earthquake-resistant structure is the ability to respond to strong ground motion by progressively mobilizing the energy-dissipative capacities of an ascending hierarchy of elements making up the structure.

In terms of their importance to the general stability and safety of a building, the components of a structure may be grouped into primary and secondary elements. *Primary elements* are those upon the integrity of which depend the stability and safety of the entire structure or a major part of it. In this category fall most of the vertical or near-vertical elements supporting gravity loads, such as columns and structural walls, as well as long-span horizontal elements. *Secondary elements* are those components whose failure would affect only limited areas or portions of a structure.

The strong-column, weak-beam design concept discussed earlier in relation to moment-resisting frames is an example of an attempt to control the sequence of yielding in a structure. And the "capacity design" approach adopted in New Zealand—which, by using even greater conservatism in the design of columns relative to beams, seeks to insure that no yielding occurs in the columns (except at their bases)—is yet another effort to achieve a controlled response in relation to inelastic action. By deliberately building in greater flexural strength in the primary elements (the columns), these design approaches force yielding and inelastic energy dissipation to take place in the secondary elements (the beams).

When properly proportioned, the coupled-wall system can be viewed as a further extension of the above design concept. By combining the considerable lateral stiffness of structural walls with properly proportioned coupling beams that can provide most of the energy-dissipative mechanism during response to strong ground motions, a better-performing structural system is obtained. The stiffness of the structural wall makes it a desirable primary element from the standpoint of damage control (by restricting interstory distortions), while the more conveniently repairable coupling beams provide the energy-dissipating secondary elements. Figure 9-31a shows a two-wall coupled-wall system and the forces acting at the base and on a typical coupling beam. A typical distribution of the elastic shear force in the coupling beams along the height of the structure due to a statically applied lateral load is shown in Figure 9-31b. Note that the accumulated shears at each end of the coupling beams, summed over the height of the structure, are each equal to the axial force at the base of the corresponding wall. The height to the most critically stressed coupling beam tends to move downward as the coupling-beam stiffness (i.e., the degree of coupling between the two walls) increases.

In a properly designed earthquake-resistant coupled-wall system, the critically stressed coupling beams should yield first—before the bases of the walls. In addition, they must be capable of dissipating a significant amount of energy through inelastic action. These requirements call for fairly stiff and strong beams. Furthermore, the desire for greater lateral-load-resisting efficiency in the system would favor stiff and strong coupling beams.

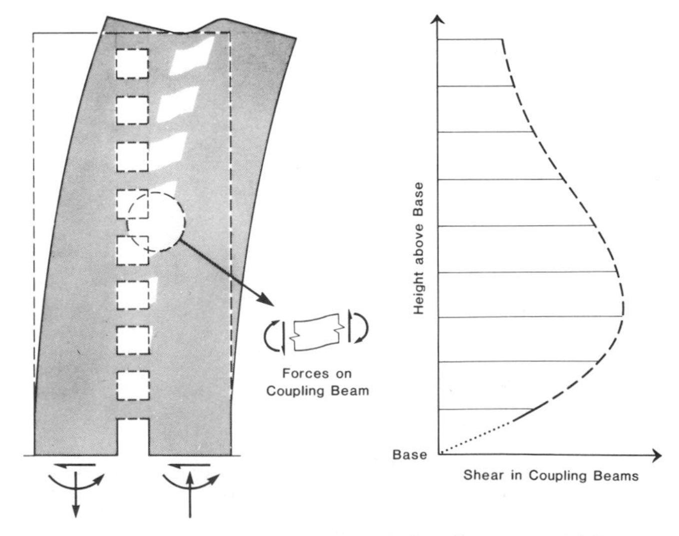

Figure 9-31 Laterally loaded coupled wall system. (a) Forces on walls at base. (b) Typical distribution of shears in coupling beams over height of structure.

However, the beams should not be so stiff or strong flexurally that they induce appreciable tension in the walls, since a net tension would reduce not only the yield moment but also the shear resistance of the wall (recall that a moderate amount of compression improves the shear resistance and ductility of isolated walls). This in turn can lead to early flexural yielding and shear-related inelastic action at the base of the tension wall. Dynamic inelastic analyses of coupled-wall systems[9-53] have shown, and tests on coupled-wall systems under cyclic reversed loading[9-54] have indicated, that when the coupling beams have appreciable stiffness and strength, so that significant net tension is induced in the tension wall, a major part of the total base shear is resisted by the compression wall (i.e., the wall subjected to axial compression for the direction of loading considered), a situation not unlike that which occurs in a beam.

The design of a coupled-wall system would then involve adjusting the wall-to-coupling-beam strength and stiffness ratios so as to strike a balance between these conflicting requirements. A basis for choosing an appropriate beam-to-wall strength ratio, developed from dynamic inelastic-response data on coupled-wall systems, is indicated in Reference 9-55. Once the appropriate relative strengths and stiffnesses have been established, details to insure adequate ductility in potential hinging regions can be addressed.

Because of the relatively large shears that develop in deep coupling beams and the likelihood of sliding shear failures under reversed loading, the use of diagonal reinforcement in such elements has been suggested (see Figure 9-32). Tests by Paulay and Binney[9-56] on diagonally reinforced coupling beams having span-to-depth ratios in the range of 1 to $1\frac{1}{2}$ have shown that this arrangement of reinforcement is very effective in resisting reversed cycles of high shear. The specimens exhibited very stable force–deflection hysteresis loops with

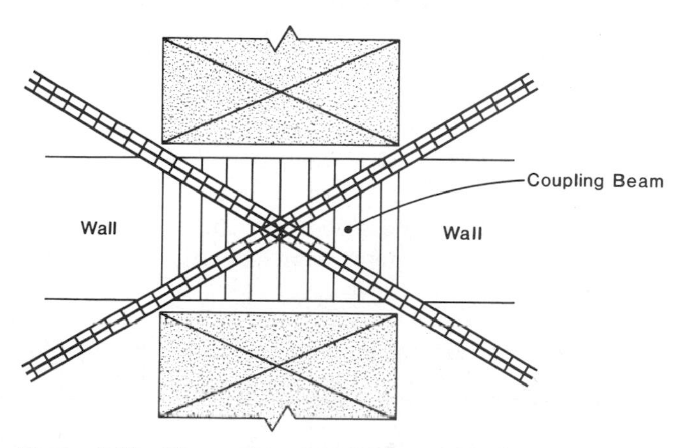

Figure 9-32 Diagonally reinforced coupling beam. (Adapted from Ref. 9-56.)

significantly higher cumulative ductility than comparable conventionally reinforced beams. Tests by Barney et al.[9-57] on diagonally reinforced beams with span-to-depth ratios in the range of 2.5 to 5.0 also indicated that diagonal reinforcement can be effective even for these larger span-to-depth ratios.

In the diagonally reinforced couplings beams reported in Reference 9-57, no significant flexural reinforcement was used. The diagonal bars are designed to resist both shear and bending and assumed to function at their yield stress in both tension and compression. To prevent early buckling of the diagonal bars, Paulay and Binney recommend the use of closely spaced ties or spiral binding to confine the concrete within each bundle of diagonal bars. A minimum amount of "basketing reinforcement," consisting of two layers of small-diameter horizontal and vertical bars, is recommended. The grid should provide a reinforcement ratio of at least 0.0025 in each direction, with a maximum spacing of 12 in. between bars.

9.4 CODE PROVISIONS FOR EARTHQUAKE-RESISTANT DESIGN

9.4.1 Performance Criteria

The performance criteria implicit in most earthquake code provisions require that a structure be able to[9-58]:

1. Resist earthquakes of minor intensity without damage; a structure would be expected to resist such frequent but minor shocks within its elastic range of stresses.
2. Resist moderate earthquakes with minor structural and some nonstructural damage; with proper design and construction, it is expected that structural damage due to the majority of earthquakes will be limited to repairable damage.
3. Resist major catastrophic earthquakes without collapse.

While no clear quantitative definition of the earthquake intensity ranges mentioned above has been given, the criteria provide a general indication of what should be expected in terms of overall performance of structures designed in accordance with the provisions of codes.

It is clear from the above criteria that the principal concern in earthquake-resistant design of buildings is to assure life safety, that is, to prevent collapse under the most intense earthquake that may reasonably be expected at the site during the life of a structure. More recently, however, engineers and public officials have become increasingly aware of the need to insure that buildings housing facilities essential to post-earthquake operations, such as hospitals, power plants, fire stations, and communication centers, not only survive without collapse but remain operational after a severe earthquake. This means that such buildings should suffer a minimum of damage. Thus, for these facilities, damage control has been added to life safety as a major design objective.

Often, damage control becomes desirable from a purely economic standpoint. The extra cost of preventing severe damage to nonstructural components of a building, such as partitions, glazing, ceilings, elevators, and other mechanical and electrical/electronic systems, may be justified by the savings realized in replacement costs and from continued use of a building after a strong earthquake.

9.4.2 Code-Specified Design Lateral Forces

The availability of dynamic analysis programs (see References 9-59 to 9-64) designed for use with large-capacity digital computers has made possible the analytical estimation of earthquake-induced forces and deformations in reasonably realistic models of most structures. However, except perhaps for the relatively simple analysis by modal superposition using response spectra, such dynamic analyses, which can range from a linearly elastic time-history analysis for a single earthquake record to nonlinear analyses using a representative ensemble of accelerograms, are costly and may be economically justifiable as a design tool only for a few large and important structures. At present, when dynamic time-history analyses of a particular building are undertaken for the purpose of design, linear elastic response is generally assumed. Nonlinear (inelastic) time-history analyses are carried out mainly in research work.

For the design of most buildings, reliance will usually have to be placed on the simplified prescriptions found in most codes.[9-1] Although necessarily approximate in character—in view of the need for simplicity and ease of application—the provisions of such codes and the philosophy behind them gain in reliability as design guides with continued application and modification to reflect the latest research findings and lessons derived from observations of structural behavior during earthquakes. Code provisions must, however, be viewed in the proper perspective, that is, as minimum requirements covering a broad class of structures of more or less conventional configuration. Unusual structures must still be designed with special care and may call for procedures beyond those normally required by codes.

The basic form of modern code provisions on earthquake-resistant design has evolved from rather simplified concepts of the dynamic behavior of structures and has been greatly influenced by observations of the performance of structures subjected to actual earthquakes.[9-65] It has been noted, for instance, that many structures built in the 1930s and designed on the basis of more or less arbitrarily chosen lateral forces have successfully withstood severe earthquakes. The satisfactory performance of such structures has been attributed to one or more of the following[9-66, 9-67]: (i) yielding in critical sections of members (yielding not only may have increased the period of vibration of such structures to values beyond the damaging range of the ground motions, but may have allowed them to dissipate a sizable portion of the input energy from an earthquake); (ii) the greater actual strength of such structures resulting from so-called nonstructural elements which are generally ignored in analysis, and the significant energy-dissipation capacity that cracking in such elements represented; and (iii) the reduced response of the structure due to yielding of the foundation.

The distribution of the code-specified design lateral forces along the height of a structure is generally similar to that indicated by the envelope of maximum horizontal forces obtained by elastic dynamic analysis. These forces are considered *service loads*, i.e., to be resisted within a structure's elastic range of stresses. However, the magnitudes of these code forces are substantially smaller than those which would be developed in a structure subjected to an earthquake of moderate-to-strong intensity, such as that recorded at El Centro in 1940, if the structure were to respond elastically to such ground excitation. Thus, buildings designed under the present codes would be expected to undergo fairly large deformations (four to six times the lateral displacements resulting from the code-specified forces) when subjected to an earthquake with the intensity of the 1940 El Centro.[9-2] These large deformations will be accompanied by yielding in many members of the structure, and, in fact, such is the intent of the codes. The acceptance of the fact that it is economically unwarranted to design buildings to resist major earthquakes elastically, and the recognition of the capacity of structures possessing adequate strength and ductility to withstand major earthquakes by responding inelastically to them, lies behind the relatively low forces specified by the codes. These reduced forces are coupled with detailing requirements designed to insure adequate inelastic deformation capacity, i.e., ductility. The capacity of an indeterminate structure to deform in a ductile manner, that is, to deform well beyond the yield limit without significant loss of strength, allows such a structure to dissipate a major portion of the energy from an earthquake without serious damage.

9.4.3. Principal Earthquake-Design Provisions of ANSI A58.1-1982, UBC-85, and Appendix A of ACI 318-83 Relating to Reinforced Concrete

The principal steps involved in the design of earthquake-resistant cast-in-place reinforced concrete buildings, with particular reference to the application of the provisions of nationally accepted model codes or standards, will be discussed below. The minimum design loads specified in ANSI A58.1-1982, "Minimum Design Loads for Buildings and Other Structures" (ANSI-82),[9-68] and the design and detailing provisions contained in Appendix A of ACI 318-83, *Building Code Requirements for Reinforced Concrete*,[9-9] will be used as bases for the discussion. Emphasis will be placed on those provisions relating to the proportioning and detailing of reinforced concrete elements, the subject of the determination of earthquake design forces being treated in Chapter 4. Where appropriate, reference will be made to differences between the provisions of these model codes and those of related codes. Among the more important of these are the *NEHRP Recommended Provisions for the Development of Seismic Regulations for New Buildings* (NEHRP-85),[9-69] which is based on ATC 3-06,[9-70] and the latest edition of the *Recommended Lateral Force Requirements* of the Structural Engineers Association of California (SEAOC-86).[9-25]

The ANSI-82 provisions relating to earthquake design loads are based mainly on those found in the 1979 Edition of the Uniform Building Code (UBC-79; the requirements are essentially the same as those appearing in UBC-85), with a few elements taken from ATC 3-06. The UBC earthquake-design load requirements are generally based on the SEAOC Recommendations. Except for minor modifications, the design and detailing requirements for reinforced concrete members found in UBC-85, NEHRP-85, and SEAOC-86 (UBC-88) are essentially those of ACI Appendix A. The earthquake design provisions of the *National Building Code of Canada—1985* (NBCC-85)[9-71] are, except for minor differences, essentially the same as those found in ANSI-82 and ACI Appendix A.

Although the various code-formulating bodies in the United States tend to differ in what they consider the most appropriate form in which to cast specific provisions and in their judgment of the adequacy of certain design requirements, there has been a tendency for the different codes and model codes to gradually take certain common general features. And while many questions await answers, it can generally be said that the main features of the earthquake-resistant design provisions in most current regional and national codes have good basis in theoretical and experimental studies as well as field observations. As such, they should provide reasonable assurance of attainment of the stated objectives of earthquake-resistant design. The continual refinement and updating of provisions in the major codes to reflect the latest findings of research and field observations[9-72] should inspire increasing confidence in the soundness of their recommendations.

The following discussion will focus on the provisions of ANSI-82 and ACI Appendix A, with occasional references to parallel provisions of NEHRP-85 and SEAOC-86 (UBC-88).

The design earthquake forces specified in ANSI-82 are intended as equivalent static loads. As its title indicates, ANSI-82 is primarily a load standard, defining minimum loads for structures but otherwise leaving out material and member detailing requirements. ACI Appendix A, on the other hand, does not specify the manner in which earthquake loads are to be determined, but sets down the requirements by which to proportion and detail monolithic cast-in-place reinforced concrete members in structures that are expected to undergo inelastic deformations during earthquakes.

Principal Design Steps Design of a reinforced concrete building in accordance with the equivalent static force procedure found in current U.S. seismic codes involves the following principal steps:

1. Determination of design "earthquake" forces:

- Calculation of base shear corresponding to the computed or estimated fundamental period of vibration of the structure. (A preliminary design of the structure is assumed here.)
- Distribution of the base shear over the height of the building.

2. Analysis of the structure under the (static) lateral forces calculated in step (1), as well as under gravity and wind loads, to obtain member design forces. Note that the code-specified forces are to be considered as *service-level* loads (as against factored or ultimate loads). The lateral load analysis can be carried out most conveniently by using a computer program for frame analysis.

Use of dynamic analysis to estimate the seismic design forces in members would take the place of steps (1) and (2) above. ANSI-82, however, requires that the design forces obtained from a dynamic analysis shall not be less than 90% of the corresponding forces obtained using the equivalent static procedure.

3. Designing members and joints for the most unfavorable combination of gravity and lateral loads. The emphasis here is on the design and detailing of members and their connections to insure their ductile behavior.

The above steps are to be carried out in each principal (plan) direction of the building. ANSI-82 allows the design of a structure in each principal direction independently of the other direction on the assumption that the design lateral forces act nonconcurrently in each principal direction. NEHRP-85 and SEAOC-86 (UBC-88) require consideration of orthogonal effects for certain building categories which may be sensitive to torsional oscillations or characterized by significant irregularities. For these cases, the design is to be based on the more severe combination of 100% of the prescribed seismic forces in one direction plus 30% of the forces in the perpendicular direction.

Changes in section dimensions of some members may be indicated in the design phase under step (3) above. However, unless the required changes in dimensions are such as to materially affect the overall distribution of forces in the structure, a reanalysis of the structure using the new member dimensions need not be undertaken. Uncertainties in the actual magnitude and distribution of the seismic forces as well as the effects of yielding in redistributing forces in the structure would make such refinement unwarranted. It is, however, most important to design and detail the reinforcement in members and their connections to insure their ductile behavior and thus allow the structure to sustain without collapse the severe distortions that may occur during a major earthquake. The code provisions intended to insure adequate ductility in structural elements represent the major difference between the design requirements for conventional, non-earthquake-resistant structures and those located in regions of high earthquake risk.

Load Factors, Strength Reduction Factors, and Loading Combinations Used as Bases for Design Codes generally require that the strength or load-resisting capacity of a structure and its component elements be at least equal to or greater than the forces due to any of a number of loading combinations that may reasonably be expected to act on it during its life. In the United States, concrete structures are commonly designed using the ultimate-strength[b] method. In this approach, structures are proportioned so that their (ultimate) capacity is equal to or greater than the required (ultimate) strength. The required strength is based on the most critical combination of factored loads, that is, specified service loads multiplied by appropriate *load factors*. The capacity of an element, on the other hand, is obtained by applying a *strength-reduction factor* ϕ to the nominal resistance of the element as determined by code-prescribed expressions or procedures or from basic mechanics.

Load factors are intended to take account of the different variabilities in the magnitude of the specified loads, lower load factors being used for types of loads that are less likely to vary significantly from the specified values. To allow for the lesser likelihood of certain types of loads occurring simultaneously, reduced load factors are specified for some loads when considered in combination with other loads.

[b]Since ACI 318-71, the term "ultimate" has been dropped, so that what used to be referred to as "ultimate-strength design" is now simply called "strength design."

ANSI-82 requires that structures, their components, and their foundations be designed to have strengths not less than the most severe of the following combinations of loads:

$$U = \begin{cases} 1.4D \\ 1.2D + 1.6L + 0.5(L_r \text{ or } S \text{ or } R) \\ 1.2D + 1.6(L_r \text{ or } S \text{ or } R) + (0.5L \text{ or } 0.8W) \\ 1.2D + 1.3W + 0.5L + 0.5(L_r \text{ or } S \text{ or } R) \\ 1.2D + 1.5E + (0.5L \text{ or } 0.2S) \\ 0.9D - (1.3W \text{ or } 1.5E) \end{cases} \tag{9-1}$$

For garages, places of public assembly, and all areas where the live load is greater than 100 lb/ft^2, the load factor on L in the third, fourth, and fifth combinations in Equation 9-1 is to be taken equal to 1.0. Also, where the effects of F, H, P or T, as defined below, are significant, these are to be considered in design as the following factored loads: $1.3F$, $1.6H$, $1.2P$, and $1.2T$. In Equation 9-1,

U = required strength to resist the factored loads

D = dead load

L = live load

L_r = roof live load

S = snow load

R = rain load

W = wind load

E = earthquake load

F = load due to fluids with well-defined pressures and maximum heights

H = load due to soil pressure

P = load due to ponding

T = load due to effects of temperature, shrinkage, moisture changes, creep, differential settlement, or combinations thereof

ACI 318-83 specifies slightly different load factors for some load combinations, as follows:

$$U = \begin{cases} 1.4D + 1.7L \\ 0.75[1.4D + 1.7L \pm (1.7W \text{ or } 1.87E)] \\ 0.9D + (1.3W \text{ or } 1.43E) \\ 1.4D + 1.7L + (1.7H \text{ or } 1.4F) \\ 0.9D - (1.7H \text{ or } 1.4F) \\ 0.75(1.4D + 1.7L + 1.4T) \geqslant 1.4(D + T) \end{cases} \tag{9-2}$$

For the design of earthquake-resistant structures, UBC-85, as well as SEAOC-86 (UBC-88), modifies the second and third equations in Equation 9-2 to read as follows:

$$U = \begin{cases} 1.4(D + L + E) \\ 0.9D \pm 1.4E \end{cases} \quad (9\text{-}3)$$

NEHRP-85 requires consideration of the following load combinations in addition to those normally required by governing code:

$$U = \begin{cases} (1.1 + 0.5A_v)D + 1.0L + 1.0S \pm 1.0E \\ (0.9 - 0.5A_v)D \pm 1.0E \end{cases} \quad (9\text{-}4)$$

where A_v is the effective peak acceleration for the site as indicated on the corresponding map given in NEHRP-85. Values of A_v corresponding to the seven different map areas range from 0.05 to 0.4, the lowest value being assigned to map area 1 and the highest to map area 7.

As mentioned, the capacity of a structural element is calculated by applying a strength reduction factor ϕ to the nominal strength of the element. The factor ϕ is intended to take account of variations in material strength and uncertainties in the estimation of the nominal member strength, the nature of the expected failure mode, and the importance of a member to the overall safety of the structure. For conventional reinforced concrete structures, ACI 318-83 specifies the following values of the strength reduction factor ϕ:

0.90 for flexure, with or without axial tension

0.90 for axial tension

0.75 for spirally reinforced members subjected to axial compression, with or without flexure

0.70 for other reinforced members (tied columns) subjected to axial compression, with or without flexure (an increase in the ϕ value for members subjected to combined axial load and flexure is allowed as the loading condition approaches the case of pure flexure)

0.85 for shear and torsion

0.70 for bearing on concrete

ACI Appendix A specifies the following exceptions to the above values of the strength-reduction factor as given in the main body of the ACI Code:

1. For structural members other than joints, a value $\phi = 0.60$ is to be used for shear when the nominal shear strength of a member is less than the shear corresponding to the development of the nominal flexural strength of the member. For shear in joints, $\phi = 0.85$.

2. For frame members subjected to flexure and an axial compression exceeding $A_g f_c'/10$—where A_g is the gross cross-sectional area—and the transverse reinforcement provided does not conform to the more stringent requirements for lateral reinforcement for earthquake-resistant structures [see below, "Code Provisions Designed to Insure Ductility in Reinforced Concrete Members, item 3, paragraph (f)], $\phi = 0.50$. Exception 1 above applies mainly to low-rise walls or portions of walls between openings. Exception (2) is intended to discourage the use of ties as lateral reinforcement in columns designed for earthquake resistance.

Code Provisions Designed to Insure Ductility in Reinforced Concrete Members The principal provisions of Appendix A of ACI 318-83[(9-9)] will be discussed below. As indicated earlier, the requirements for proportioning and detailing reinforced concrete members found in UBC-85, NEHRP-85, and SEAOC-86 (UBC-88) are essentially those of ACI Appendix A. Modifications to the ACI Appendix A provisions found in NEHRP-85 and SEOAC-86 (UBC-88) will be referred to where appropriate.

Special provisions governing the design of earthquake-resistant structures first appeared in the 1971 edition of the ACI Code. The provisions of Appendix A supplement or supersede those in the main body of the code and deal with the design of ductile moment-resisting space frames and shear walls of cast-in-place reinforced concrete.

Neither the main body of the code nor Appendix A specifies the magnitude of the earthquake forces to be used in design. The Commentary to Appendix A states that the provisions of the Appendix are intended to enable structures to sustain a series of oscillations in the inelastic range without critical loss in strength. It is generally accepted that the intensity of shaking envisioned by the provisions of Appendix A correspond to those of UBC seismic zones 3 and 4. In the 1983 edition of the ACI Code, a section (Section A.9) was added to cover the design of frames located in areas of moderate seismic risk, roughly corresponding to UBC seismic zone 2. For structures located in areas of low seismic risk (corresponding to UBC seismic zones 0 and 1) and designed for the specified earthquake forces, very little inelastic deformation may be expected. In these cases, the ductility provided by designing to the provisions contained in the main body of the code will generally be sufficient.

A major objective of the design provisions in ACI Appendix A, as well as in the main body of the code, is to have the strength of a structure governed by a ductile type of flexural failure mechanism. Stated another way, the provisions are aimed at preventing the brittle or abrupt types of failure associated with inadequately reinforced and overreinforced members failing in flexure, as well as with shear (i.e., diagonal tension) and anchorage or bond failures. The main difference between Appendix A and the main body of the ACI Code lies in the greater range of deformation, with yielding actually expected at critical locations, and hence the greater ductility required in designs for resistance to major earthquakes. The need for greater ductility follows from the design philosophy that uses reduced forces in proportioning members and provides for the inelastic deformations that are expected under severe earthquakes by special ductility requirements.

A provision unique to earthquake-resistant design of frames is the so-called strong-column, weak-beam requirement. As discussed in Section 9.3.4 under "Beam–Column Joints," this requirement calls for the sum of the flexural strengths of columns meeting at a frame joint to be at least 1.2 times that of the beams framing into the joint. This is intended to force yielding in such frames to occur in the beams rather than in the columns and thus preclude possible instability due to plastic hinges forming in the columns. As pointed out earlier, this requirement may not guarantee nondevelopment of plastic hinges in the columns. The strong-column, weak-beam requirement often results in column sizes that are larger than would otherwise be required, particularly in the upper floors of multistory buildings with appreciable beam spans.

1. *Limitations on material strengths.* ACI Appendix A requires a minimum specified concrete strength f_c' of 3000 lb/in.^{2} and a maximum specified yield strength of reinforcement, f_y, of 60,000 lb/in.^{2}. These limits are imposed with a view to restricting the unfavorable effects that material properties beyond these limits can have on the sectional ductility of members. Appendix A requires that reinforcement for resisting flexure and axial forces in frame members and wall boundary elements be ASTM 706 grade 60 low-alloy steel intended for applications where welding or bending, or both, are important. However, ASTM 615 billet steel bars of grade 40 or 60 may be used provided the following two conditions are satisfied:

$$(\text{actual } f_y) \leqslant (\text{specified } f_y) + 18{,}000 \text{ lb/in.}^2$$

$$\frac{\text{actual ultimate tensile stress}}{\text{actual } f_y} \geqslant 1.25$$

The first requirement helps to limit the increase in magnitude of the actual shears that can develop in a flexural member beyond that computed on the basis of the specified yield stress when plastic hinges form at the ends of a beam. The second requirement is intended to insure reinforcement with a sufficiently long yield plateau.

In the "strong column-weak beam" frame intended by the code, the relationship between the moment capacities of columns and beams may be upset if the beams turn out to have much greater moment capacity than intended by the designer. Thus, the substitution of 60-ksi steel of the same area for specified 40-ksi steel in beams can be detrimental. The shear strength of beams and columns, which is generally based on the condition of plastic hinges forming (i.e., M_y acting) at the member ends, may become inadequate if the actual moment capacity at the member ends is greater than intended as a result of the steel having a substantially greater yield strength than specified.

2. *Flexural members (beams).* These include members having a clear span greater than four times the effective depth that are subject to a factored axial compressive force not exceeding $A_g f_c'/10$, where A_g is the gross cross-sectional area. Significant provisions relating to flexural members of structures in regions of high seismic risk are discussed below.

(a) Limitations on section dimensions:

$$\text{width/depth} \geqslant 0.3$$

$$\text{width} \begin{cases} \geqslant 10 \text{ in.} \\ \leqslant \text{width of supporting column} \\ \quad +1.5 \times (\text{depth of beam}) \end{cases}$$

(b) Limitations on flexural reinforcement ratio (see also Figure 9-33):

$$\rho_{min} = \begin{cases} 200/f_y \\ \text{two continuous bars at both top} \\ \quad \text{and bottom of member} \end{cases}$$

$$\rho_{max} = 0.025$$

(c) Moment capacity requirements:
At beam ends

$$M_y^+ \geqslant 0.50 M_y^-$$

At any section in beam span

$$M_y^+ \text{ or } M_y^- \geqslant 0.25\left(M_y^{max} \text{ at beam ends}\right)$$

(d) Restrictions on lap splices: Lap splices shall not be used

(1) within joints,
(2) within $2h$ from face of support, where h is total depth of beam,
(3) at locations of potential plastic hinging.

Lap splices, where used, are to be confined by hoops or spiral reinforcement with a maximum spacing or pitch of $d/4$ or 4 in.

In addition to the above and the requirements found in Chapter 12 of ACI 318-83, SEAOC-86 specifies a minimum lap length of 24 bar diameters.

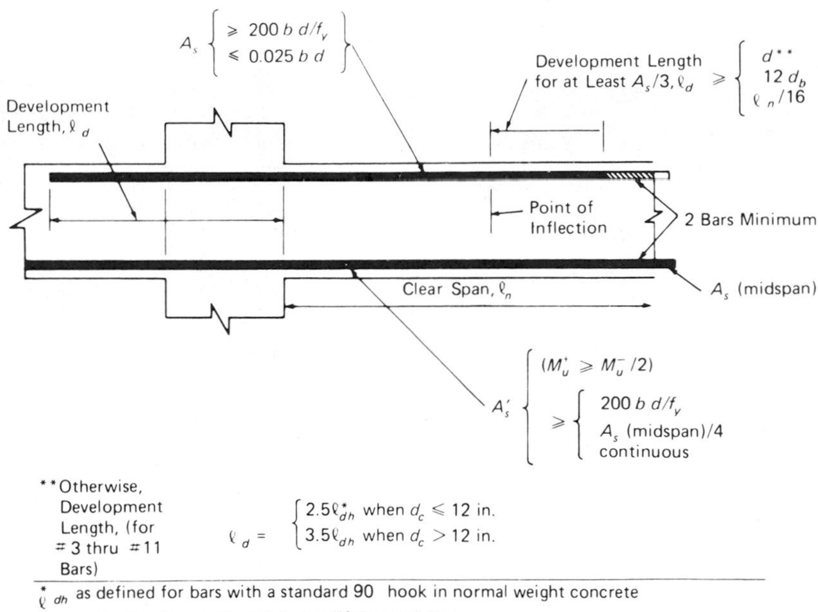

Figure 9-33 Longitudinal reinforcement requirements for flexural members.

(e) Restrictions on welding of longitudinal reinforcement: Welded splices and mechanical connectors may be used provided:

(1) they are used only on alternate bars in each layer at any section;

(2) the distance between splices of adjacent bars is $\geqslant$ 24 in.

It is worth noting that both NEHRP-85 and SEAOC-86 (UBC-88) specifically prohibit the welding of stirrups, ties, inserts, or other similar elements of longitudinal bars.

(f) Development length requirements for longitudinal bars in tension:

(1) For bar sizes 3 through 11 with a standard 90° hook (as shown in Figure 9-34) in normal-weight concrete, the development length

$$l_{dh} \geqslant \begin{cases} \dfrac{f_y d_b}{65\sqrt{f_c'}} \\ 8d_b \\ 6 \text{ in.} \end{cases}$$

(d_b is bar diameter).

(2) When bars are embedded in lightweight-aggregate concrete, the development length is to be at least 1.25 times the values indicated above.

(3) The 90° hook shall be located within the confined core of a column or boundary element.

(4) For straight bars of sizes 3 through 11, the development length

$$l_d \geqslant 2.5 \times (l_{dh} \text{ for bars with } 90° \text{ hooks})$$

when the depth of concrete cast in one lift beneath the bar is $\leqslant$ 12 in., or

$$l_d \geqslant 3.5 \times (l_{dh} \text{ for bars with } 90° \text{ hooks})$$

if the abovementioned depth is > 12 in.

(5) If a bar is not anchored by means of a 90° hook within the confined column core, the portion of the required straight development length not located within the confined core shall be increased by a factor of 1.6.

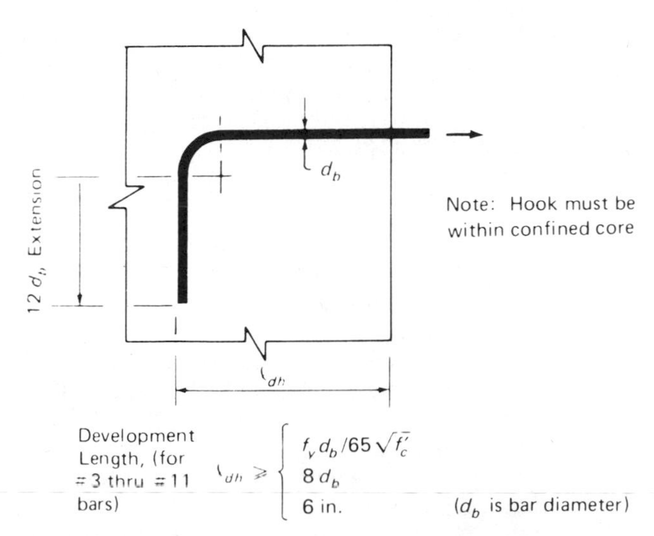

Figure 9-34 Development length for beam bars with 90° hooks.

(g) Transverse reinforcement requirements for confinement and shear: Transverse reinforcement in beams must satisfy requirements associated with

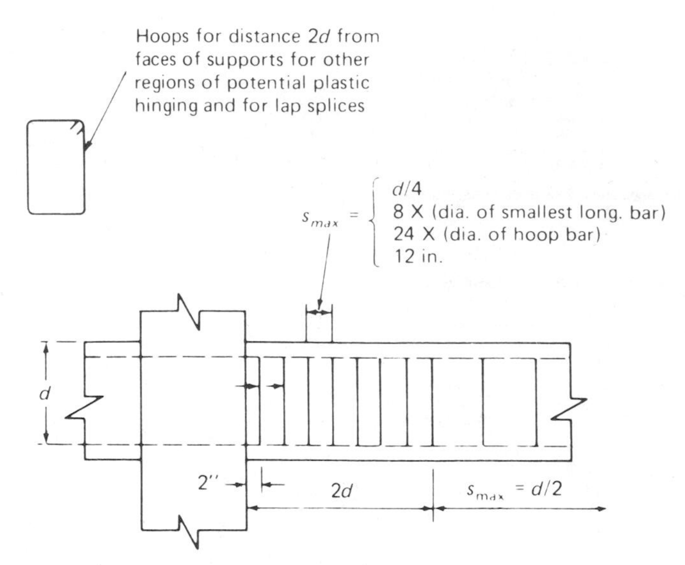

Figure 9-35 Transverse reinforcement limitations for flexural members. Minimum bar size—#3.

their dual function as confinement reinforcement and shear reinforcement (see Figure 9-35).

(1) Confinement reinforcement in the form of hoops is required:

(i) over a distance $2d$ from faces of support (where d is the effective depth of the member);

(ii) over distances $2d$ on both sides of sections within the span where flexural yielding may occur due to earthquake loading.

(2) Hoop spacing:

(i) First hoop at 2 in. from face of support.

(ii) Maximum spacing

$$\leqslant \begin{cases} d/4 \\ 8 \times \text{(diameter of smallest longitudinal bar)} \\ 24 \times \text{(diameter of hoop bars)} \\ 12 \text{ in.} \end{cases}$$

(3) Lateral support for perimeter longitudinal bars where hoops are required: Every corner and alternate longitudinal bar shall be supported by the corner of a hoop with an included angle $\leqslant 135°$, with no longitudinal bar farther than 6 in. along the tie from such a laterally supported bar. Where the longitudinal perimeter bars are arranged in a circle, a circular hoop may be used.

(4) Where hoops are not required, hoop spacing $\leqslant d/2$.

(5) Shear reinforcement—to be provided so as to preclude shear failure prior to development of plastic hinges at beam ends. Design shears for determining shear reinforcement are to be based on a condition where plastic hinges occur at beam ends due to the combined effects of lateral displacement and factored gravity loads (see Figure 9-15). The *probable flexural strength*, M_{pi}, associated with a plastic hinge is to be computed using a strength reduction factor $\phi = 1.0$ and assuming a stress in the tensile reinforcement given by $f_s = 1.25 f_y$.

(6) In determining the required shear reinforcement, the contribution of the concrete, V_c, is to be neglected if the shear associated with the probable flexural strengths at the beam ends is greater than one-half the total design shear and the factored axial compressive force including earthquake effects is less than $A_g f_c'/20$.

NEHRP-85 requires that the concrete contribution to shear resistance, V_c, be neglected if the shear associated with M_{pi} acting at the beam ends exceeds one-fourth (instead of one-half as in ACI Appendix A) the total design shear.

(7) The transverse reinforcement provided must satisfy the requirements for confinement or shear, whichever is more stringent.

Discussion:

(a) Limitations on section dimensions: These limitations have been guided by experience with test specimens subjected to cyclic inelastic loading.

(b) Flexural reinforcement limitations: Because the ductility of a member decreases with increasing tensile reinforcement ratio, ACI Appendix A limits the maximum reinforcement ratio to 0.025. The use of a limiting ratio based on the "balanced condition" as given in the main body of the code, while applicable to members loaded monotonically, fails to describe conditions in flexural members subjected to reversals of inelastic deformation. The limiting ratio of 0.025 is based mainly on considerations of steel congestion and also on limiting shear stresses in beams of typical proportions. From a practical standpoint, low steel ratios should be used whenever possible. The requirement of at least two bars, top and bottom, is intended to insure integrity of the member under reversed loading.

The selection of the size, number, and arrangement of flexural reinforcement should be made with full consideration of construction requirements. This is particularly important in relation to beam–column connections, where construction difficulties can arise as a result of reinforcement congestion. The preparation of large-scale drawings of the connections, showing all beam, column, and joint reinforcements, will help eliminate unanticipated problems in the field. Such large-scale drawings will pay dividends in terms of lower bid prices and a smooth-running construction job. Reference 9-73 provides further recommendations on reinforcement detailing.

(c) Positive moment capacity at beam ends: To allow for the possibility of the positive moment at the end of a beam due to earthquake-induced lateral displacements exceeding the negative moment due to the gravity loads, the code requires a minimum positive moment capacity at beam ends equal to 50% of the corresponding negative moment capacity.

(d) Lap splices: Lap splices of flexural reinforcement are not allowed in regions of potential plastic hinging, since such splices are not considered to be reliable under reversed inelastic cycles of deformation. Hoops are mandatory for confinement of lap splices at any location because of the likelihood of loss of the concrete cover.

(e) Welded splices and mechanical connectors: Welded splices and mechanical connectors are to conform to the requirements given in the main body of the code. A major requirement is that the splice develop at least 125% of the specified yield strength of the bar.

As pointed out, NEHRP-85 and SEAOC-86 (UBC-88) specifically prohibit the welding of stirrups, ties, inserts, or other similar elements to longitudinal bars.

(f) Development length: The expression for l_{dh} given above already includes the coefficients 0.7 (for concrete cover) and 0.80 (for ties) that are normally applied to the basic development length, l_{db}. This is so because ACI Appendix A requires that hooks be embedded in the confined core of a column or boundary element. The expression for l_{dh} also includes a factor of about 1.4, representing an increase over the development length required for conventional structures, to provide for the effect of load reversals.

Except in very large columns, it is usually not possible to develop the yield strength of a reinforcing bar from the framing beam within the width of a column unless a hook is used. Where beam reinforcement can extend through a column, its capacity is developed by embedment in the column and within the compression zone of the beam on the far side of the connection (see Figure 9-33). Where no beam is present on the opposite side of a column, such as in exterior columns, the flexural reinforcement in a framing beam has to be developed within the confined region of the column. This is usually done by means of a standard 90° hook plus whatever extension is necessary to develop the bar, the development length being measured from the near face of the column, as indicated in Figure 9-34. The use of a beam stub at the far (exterior) side of a column may also be considered (see Figure 9-21).

ACI Appendix A makes no provision for the use of size 14 and 18 bars because of lack of sufficient information on the behavior at anchorages of such bars when subjected to load reversals simulating earthquake effects.

(g) Transverse reinforcement: Because the ductile behavior of earthquake-resistant frames designed to current codes is premised on the ability of the beams to develop plastic hinges with adequate rotational capacity, it is essential to insure that shear failure does not occur before the flexural capacity of the beams has been developed. Transverse reinforcement is required for two related functions: (i) to provide sufficient shear strength so that the full flexural capacity of the member can be developed, and (ii) to insure adequate rotation capacity in plastic-hinging regions by confining the concrete in the compression zones and by providing lateral support to the compression steel. To be equally effective with respect to both functions under load reversals, the transverse reinforcement should be placed perpendicular to the longitudinal reinforcement.

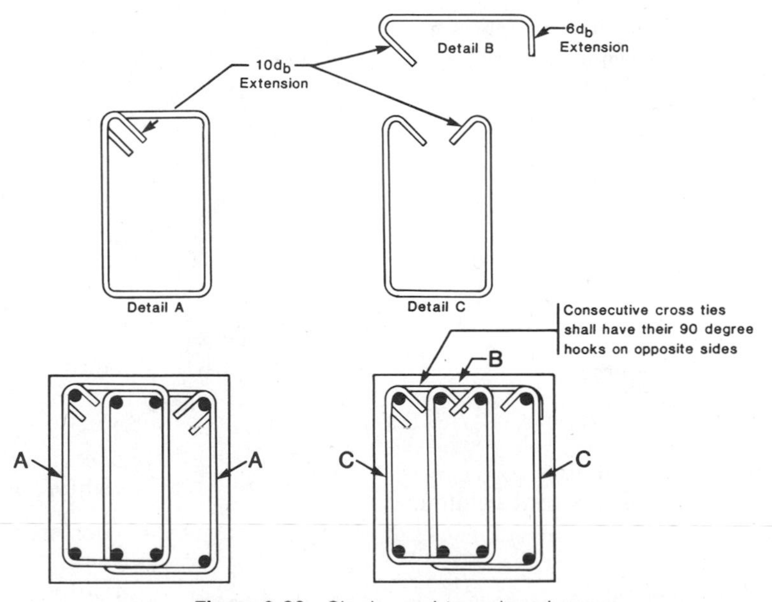

Figure 9-36 Single- and two-piece hoops.

Shear reinforcement in the form of stirrups or stirrup ties is to be designed for the shear due to factored gravity loads and the shear corresponding to plastic hinges forming at both ends of a beam. Plastic end moments associated with lateral displacement in either direction should be considered (Figure 9-15). It is important to note that the required shear strength in beams (as in columns) is determined by the flexural strength of the frame member (as well as the factored loads acting on the member), rather than by the factored shear force calculated from a lateral load analysis. The use of the factor 1.25 on f_y for calculating the probable moment strength is intended to allow for the actual steel strength exceeding the specified minimum and also recognizes that the strain in reinforcement of sections undergoing large rotations can enter the strain-hardening range.

To allow for load combinations not accounted for in design, a minimum amount of web reinforcement is required throughout the length of all flexural members. Within regions of potential hinging, stirrup ties or hoops are required. A hoop may be made of two pieces of reinforcement: a stirrup having 135° hooks with 10-diameter extensions anchored in the confined core and a crosstie to close the hoop (see Figure 9-36). Consecutive ties are to have their 90° hooks on opposite sides of the flexural member.

3. *Frame members subjected to axial load and bending.* ACI Appendix A makes the distinction between columns or beam–columns and flexural members on the basis of the magnitude of the factored axial load acting on the member. Thus, if the factored axial load does not exceed $A_g f_c'/10$, the member falls under the category of flexural members, the principal design requirements for which were discussed in the preceding section. When the factored axial force on a member exceeds $A_g f_c'/10$, the member is considered a beam–column. Major requirements governing the design of such members in structures located in areas of high seismic risk are given below.

(a) Limitations on section dimensions:

$$\text{shortest cross-sectional dimension} \geqslant 12 \text{ in.}$$

(measured on line passing through geometric centroid);

$$\frac{\text{shortest dimension}}{\text{perpendicular dimension}} \geqslant 0.4$$

(b) Limitations on longitudinal reinforcement:

$$\rho_{\min} = 0.01, \qquad \rho_{\max} = 0.06$$

(c) Flexural strength of columns relative to beams framing into a joint (the so-called strong-column, weak-beam provision):

$$\sum M_e \geqslant \tfrac{6}{5} \sum M_g \tag{9-5}$$

where

ΣM_e = sum of the design flexural strengths of the columns framing into joint. Column flexural strength to be calculated for the factored axial force, consistent with the direction of the lateral loading considered, that results in the lowest flexural strength

ΣM_g = sum of design flexural strengths of beams framing into joint

(d) Restriction on use of lap splices: Lap splices are to be used only within the middle half of the column height and are to be designed as tension splices.

In addition to the above, SEAOC-86 (UBC-88) requires that splices in columns be designed as Class A tension splices in accordance with Chapter 12 of ACI 318-83 and that these be enclosed by hoops as for splices in beams.

(e) Welded splices or mechanical connectors for longitudinal reinforcement: Welded splices or mechanical connectors may be used at any section of a column, provided that:

(1) they are used only on alternate longitudinal bars at a section;
(2) the distance between splices along the longitudinal axis of the reinforcement is $\geqslant$ 24 in.

(f) Transverse reinforcement for confinement and shear: As in beams, transverse reinforcement in columns must provide confinement to the concrete core and lateral support for the longitudinal bars as well as shear resistance. In columns, however, the transverse reinforcement must all be in the form of closed hoops or continuous spiral reinforcement. Sufficient reinforcement should be provided to satisfy the requirements for confinement or shear, whichever is larger.

(1) Confinement requirements (see Figure 9-37):

(i) Volumetric ratio of spiral or circular hoop reinforcement:

$$\rho_s \geqslant \begin{cases} 0.12\dfrac{f_c'}{f_{yh}} \\[2ex] 0.45\left(\dfrac{A_g}{A_{ch}} - 1\right)\dfrac{f_c'}{f_{yh}} \end{cases} \tag{9-6}$$

where

f_{yh} = specified yield strength of transverse reinforcement, in lb/in.^{2}

A_{ch} = core area of column section, measured to the outside of transverse reinforcement, in in.^{2}

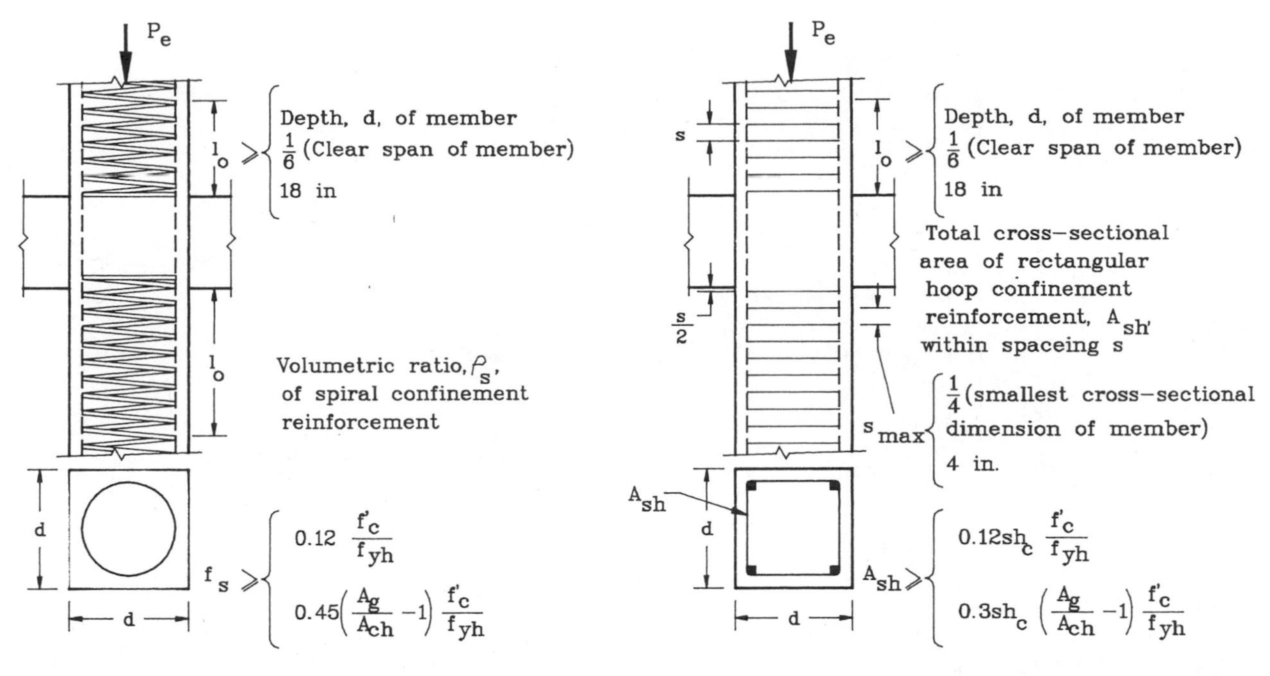

Figure 9-37 Confinement requirements for column ends.

(ii) Rectangular hoop reinforcement, total cross-sectional area, within spacing s:

$$A_{sh} \geqslant \begin{cases} 0.12 s h_c \dfrac{f_c'}{f_{yh}} \\ 0.3 s h_c \left(\dfrac{A_g}{A_{ch}} - 1 \right) \dfrac{f_c'}{f_{yh}} \end{cases} \tag{9-7}$$

where

h_c = cross-sectional dimension of column core, measured center-to-center of confining reinforcement

s = spacing of transverse reinforcement measured along axis of member, in in.

s_{max} = min{$\frac{1}{4}$(smallest cross-sectional dimension of member), 4 in.}

maximum spacing in plane of cross-section between legs of overlapping hoops or cross ties is 14 in.

(2) Confinement reinforcement is to be provided over a length l_0 from each joint face or over distances l_0 on both sides of any section where flexural yielding may occur in connection with lateral displacements of the frame, where

$$l_0 \geqslant \begin{cases} \text{depth } d \text{ of member} \\ \frac{1}{6}(\text{clear span of member}) \\ 18 \text{ in.} \end{cases}$$

NEHRP-85 and SEAOC-86 further require that confinement reinforcement be provided at any section of a column where the nominal axial strength, ϕP_n, is less than the sum of the shears corresponding to the probable flexural strengths of the beams (i.e., based on $f_s = 1.25 f_y$ and $\phi = 1.0$) framing into the columns above the level considered.

(3) Transverse reinforcement for shear in columns is to be based on the shear associated with the largest nominal moment strengths at the column ends (using $f_s = f_y$ and $\phi = 1.0$) corresponding to the factored axial compressive force resulting in the largest moment strengths. SEAOC-86 (UBC-88) requires that the moment strength at the column ends be based on $f_s = 1.25 f_y$ and $\phi = 1.0$. The calculated end moments of columns meeting at a joint, however, need not exceed the sum of the probable moment strengths of the girders framing into the joint.

(g) Column supporting discontinued walls: Columns supporting discontinued shear walls or similar stiff elements are to be provided with transverse reinforcement over their full height below the discontinuity (see Figure 9-38) when the axial compressive force due to earthquake effects exceeds $A_g f_c'/10$.

In addition to the above, NEHRP-85 and SEAOC-86 (UBC-88) require that transverse reinforcement in columns supporting discontinued walls be extended above the discontinuity by at least the development length of the largest vertical bar and below the base by the same amount where the column rests on a wall. Where the column termi-

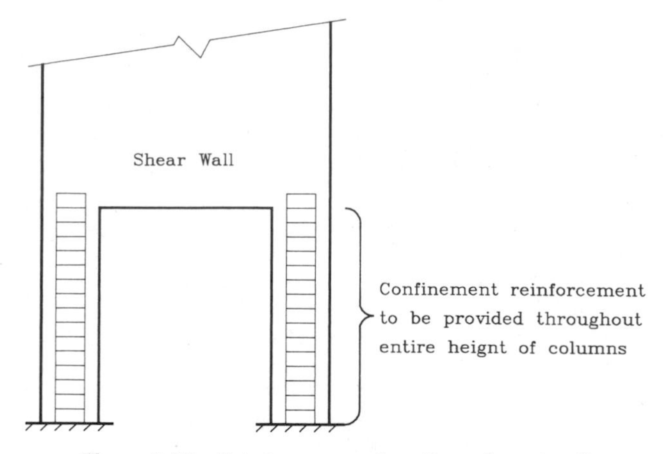

Figure 9-38 Columns supporting discontinued wall.

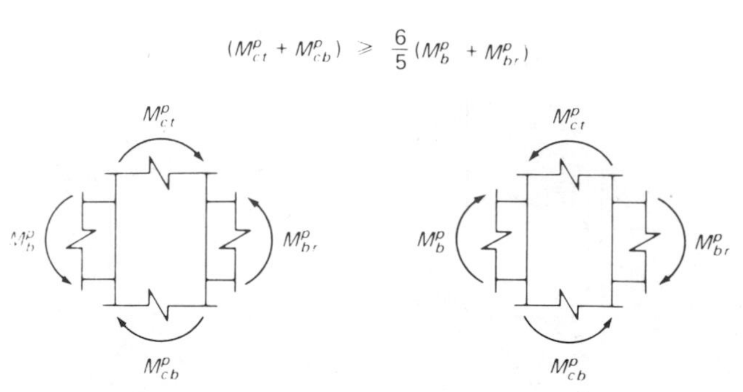

Figure 9-39 Strong column-weak beam frame requirements.

nates in a footing or mat, the transverse reinforcement is to be extended below the top of the footing or mat a distance equal to the compressive development length of the largest vertical bar or the lead length of a standard hook.

SEAOC-86 (UBC-88) also requires the provision of transverse reinforcement over the full height of a column when the calculated point of contraflexure in a column lies outside the middle half of the clear height.

Discussion:

(b) Reinforcement ratio limitation: ACI Appendix A specifies a reduced upper limit for the reinforcement ratio in columns from the 8% of Chapter 10 of the code to 6%. However, construction considerations will in most cases place the practical upper limit on the reinforcement ratio ρ near 4%. Convenience in detailing and placing reinforcement in beam–column connections makes it desirable to keep the column reinforcement low.

The minimum reinforcement ratio is intended to provide for the effects of time-dependent deformations in concrete under axial loads as well as maintain a sizable difference between cracking and yield moments.

(c) Relative column-to-beam flexural strength requirement: To insure the stability of a frame and maintain its vertical-load-carrying capacity while undergoing large lateral displacements, ACI Appendix A requires that inelastic deformations be generally restricted to the beams. This is the intent of Equation 9-5 (see Figure 9-39). As mentioned, formation of plastic hinges at both ends of most columns in a story can precipitate a sidesway mechanism leading to collapse of the story and the structure above it. Also, as pointed out in Section 9.3.4 under "Beam–Column Joints," compliance with this provision does not insure that plastic hinging will not occur in the columns.

If Equation 9-5 is not satisfied at a joint, columns supporting reactions from such a joint are to be provided with transverse reinforcement over their full height. Columns not satisfying Equation 9-5 are to be ignored in calculating the strength and stiffness of the structure. However, since such columns contribute to the stiffness of the structure before they suffer severe loss of strength due to plastic hinging, they should not be ignored if neglecting them results in unconservative estimates of design forces. This may occur in determining the design base shear or in calculating the effects of torsion in a structure. Columns not satisfying Equation 9-5 should satisfy the minimum requirements for members not proportioned to resist earthquake-induced forces, discussed under item 6 below.

(f) Transverse reinforcement for confinement and shear: Sufficient transverse reinforcement in the form of rectangular hoops or spirals should be provided to satisfy the larger requirement for either confinement or shear.

Circular spirals represent the most efficient form of confinement reinforcement. The extension of such spirals into the beam–column joint, however, may cause some construction difficulties.

Rectangular hoops, when used in place of spirals, are less effective with respect to confinement of the concrete core. Their effectiveness may be increased, however, with the use of supplementary crossties. The crossties have to be of the same size and spacing as the hoops and have to engage a peripheral longitudinal bar at each end. Consecutive crossties are to be alternated end for end along the longitudinal reinforcement and are to be spaced no further than 14 in. in the plane of the column cross-section (see Figure 9-40). The requirement of having the crossties engage a longitudinal bar at each end would almost preclude placing them before the longitudinal bars are threaded through.

In addition to confinement requirements, the transverse reinforcement in columns must resist the maximum shear associated with the formation of plastic hinges at the column ends. Although the strong-column, weak-beam provision governing rel-

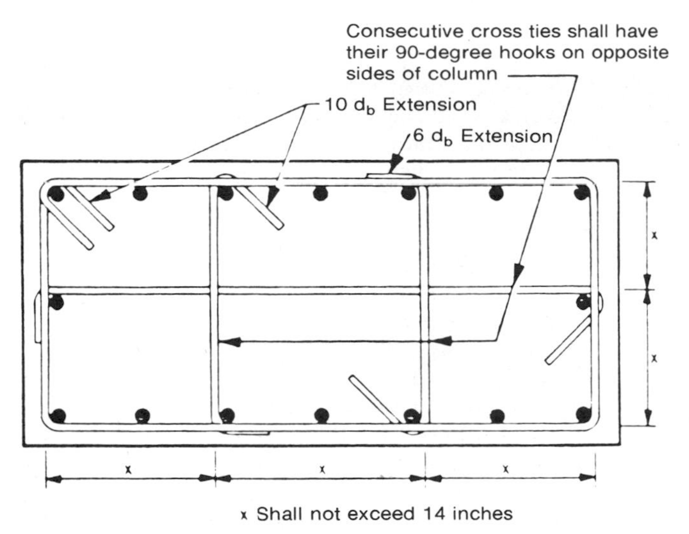

Figure 9-40 Rectangular transverse reinforcement in columns.

ative moment strengths of beams and columns meeting at a joint is intended to have most of the inelastic deformation occur in the beams of a frame, the code recognizes that hinging can occur in the columns. Thus, the shear reinforcement in columns is to be based on the shear corresponding to the development of the nominal moment strengths at the ends of the columns, i.e., the yield moments, with $\phi = 1.0$. The values of these plastic moments —obtained from the P–M interaction diagram for the particular column section considered—are to be the maxima consistent with the possible factored compressive axial forces on the column. Moments associated with lateral displacements of the frame in both directions, as indicated in Figure 9-41, should be considered. The axial load corresponding to the maximum moment capacity should then be used in computing the permissible shear in concrete, V_c.

(g) Columns supporting discontinued walls: Columns supporting discontinued shear walls tend to be subjected to large shears and compressive forces, and can be expected to develop large inelastic deformations during strong earthquakes; hence the requirement for transverse reinforcement throughout the height of such columns. The SEAOC-86 (UBC-88) requirement for transverse reinforcement over the full height of a column when the calculated point of contraflexure lies outside the middle half of the clear height is intended to provide for likely hinging in the column, as discussed in Section 9.3.4 under "Columns."

4. *Beam–column connections.* In conventional reinforced-concrete buildings, the beam–column connections usually are not designed by the structural engineer. Detailing of reinforcement within the joints is normally relegated to a draftsman or detailer. In earthquake-resistant frames, however, the design of beam–column connections requires as much attention as the design of the members themselves, since the integrity of the frame may well depend on the proper performance of such connections.

Because of the congestion that may result from too many bars converging within the limited space of the joint, the requirements for the beam–column connections have to be considered when proportioning the

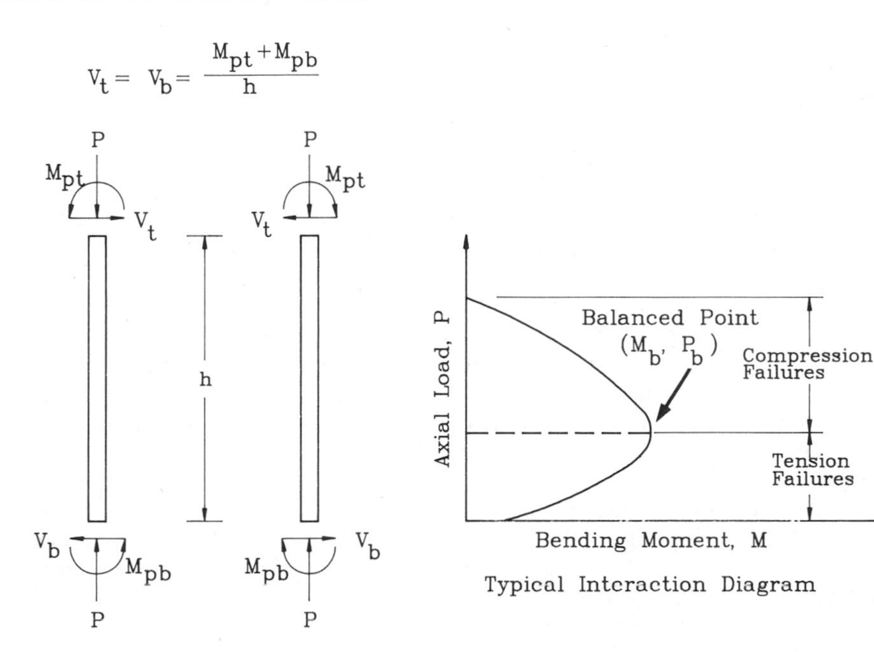

Figure 9-41 Loading cases for design of shear reinforcement for columns.

columns of a frame. To minimize placement difficulties, an effort should be made to keep the amount of longitudinal reinforcement in the frame members on the low side of the permissible range.

The provisions of ACI Appendix A dealing with beam-column joints relate mainly to:

(a) Transverse reinforcement for confinement: Minimum confinement reinforcement, as required for potential hinging regions in columns and defined by Equations 9-6 and 9-7, must be provided in beam–column joints. For "confined" joints, a 50% reduction in the required amount of confinement reinforcement is allowed, the required amount to be placed within the depth of the shallowest framing member. A confined joint is defined as one with beams framing into all four sides and where each beam has a width equal to at least three-fourths of the width of the column face into which it frames.

(b) Transverse reinforcement for shear: The horizontal shear force in a joint is to be calculated by assuming the stress in the tensile reinforcement of framing beams equal to $1.25f_y$ (see Figure 9-20). The shear strength of the connection is to be computed (for normal-weight concrete) as

$$\phi V_c = \begin{cases} \phi 20\sqrt{f_c'}\,A_j & \text{for confined joints} \\ \phi 15\sqrt{f_c'}\,A_j & \text{for unconfined joints} \end{cases}$$

where

$\phi = 0.85$ (for shear)

A_j = minimum cross-sectional area of joint in a plane parallel to the axis of the reinforcement generating the shear force

For the case where the beam width is less than that of the column face into which it frames, SEAOC-86 (UBC-88) requires that the effective width to be used in calculating A_j not exceed the beam width plus twice the least column projection beyond the beam side. NEHRP-85 requires the area A_j to be based on the effective depth to the centroid of the longitudinal column reinforcement, instead of the overall depth used in ACI Appendix A.

For lightweight concrete, V_c is to be taken as three-fourths the value given above for normal-weight concrete.

(c) Anchorage of longitudinal beam reinforcement terminated in a column must be within the confined column core.

Where longitudinal beam bars extend through a joint, SEAOC-86 (UBC-88) requires that the column depth in the direction of loading be not less than 20 times the diameter of the largest longitudinal beam bar.

Discussion:

(a) Transverse reinforcement for confinement: The transverse reinforcement in a beam–column connection helps maintain the vertical-load-carrying capacity of the joint even after spalling of the outer shell. It also helps resist the shear force transmitted by the framing members and improves the bond between steel and concrete within the joint.

The minimum amount of transverse reinforcement, as given by Equations 9-6 and 9-7, must be provided through the joint regardless of the magnitude of the calculated shear force in the joint. The 50% reduction in the amount of confinement reinforcement allowed for joints having beams framing into all four sides recognizes the beneficial confining effect provided by these members.

(b) Results of tests reported in Reference 9-39 indicate that the shear strength of joints is not too sensitive to the amount of transverse (shear) reinforcement. Based on these results, the 1983 edition of ACI Appendix A defines the shear strength of beam–column connections as a function only of the cross-sectional area of the joint, (A_j) and f_c' (see Section 9.3.4 under "Beam–Column Joints").

When the design shear in the joint exceeds the shear strength of the concrete, the designer may either increase the column size or increase the depth of the beams. The former will increase the shear capacity of the joint section, while the latter will tend to reduce the required amount of flexural reinforcement in the beams, with accompanying decrease in the shear transmitted to the joint. Yet another alternative is to keep the longitudinal beam bars from yielding at the faces of the columns by detailing the beams so that plastic hinging occurs away from the column faces.

(c) The anchorage or development-length requirements for longitudinal beam reinforcement in tension have been discussed earlier under flexural members. Splicing of main reinforcement within the joint should be avoided whenever possible.

5. *Shear Walls.* When properly proportioned so that they possess adequate lateral stiffness to reduce interstory distortions due to earthquake-induced motions, shear walls or structural walls reduce the likelihood of damage to the nonstructural elements of a building. When used with rigid frames, walls form a system that combines the gravity-load-carrying efficiency of the rigid frame with the lateral-load-resisting efficiency of the structural wall. In the form of coupled walls linked by appropriately proportioned coupling beams (see Section 9.3.4 under "Coupled Walls"), alone or in combination with rigid frames, structural walls provide a laterally stiff structural system that allows significant energy dissipation to take place in the more easily repairable coupling beams.

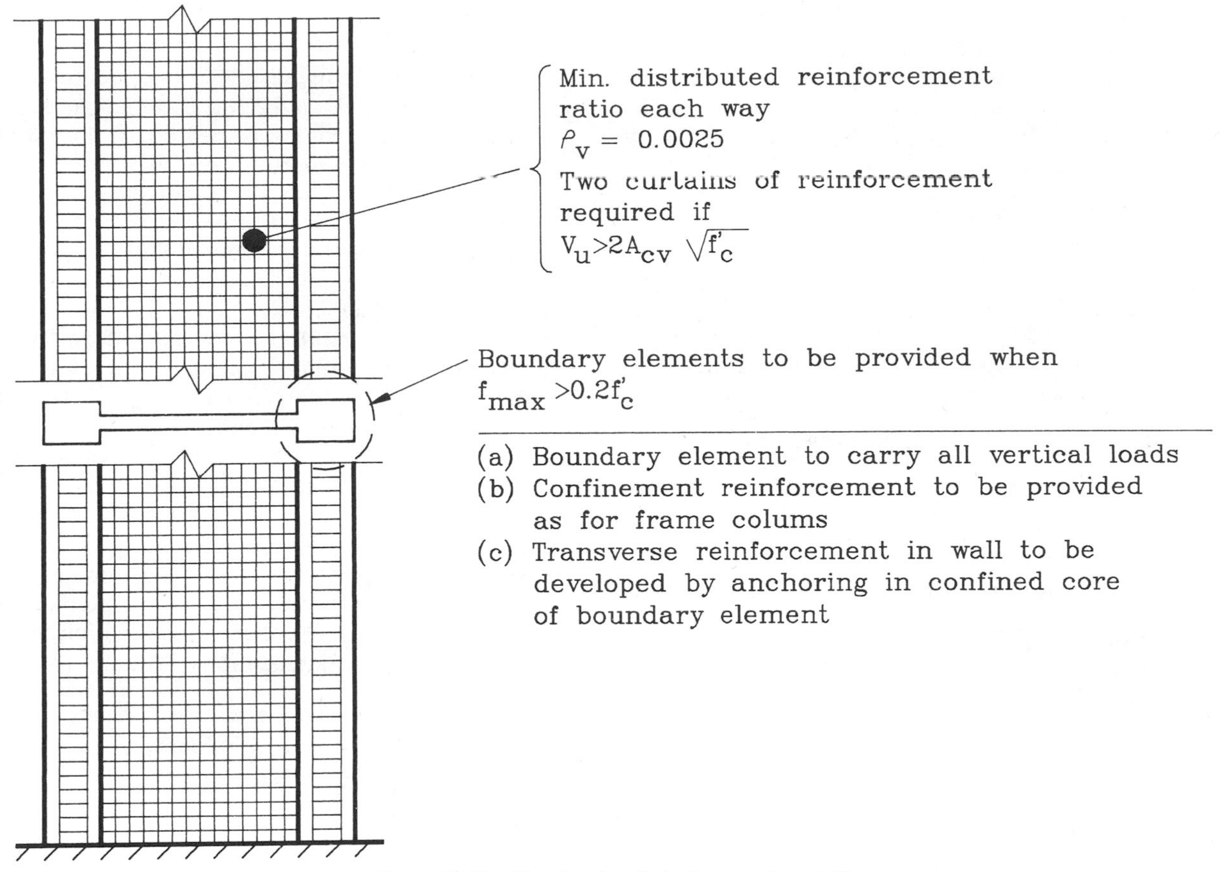

Figure 9-42 Structural wall design requirements.

Observations of the comparative performance of rigid-frame buildings and buildings stiffened by structural walls during recent earthquakes(9-74) have pointed to the consistently better performance of the latter. The performance of buildings stiffened by properly designed structural walls has been better with respect to both life safety and damage control. The need to insure that critical facilities remain operational after a major tremor and the need to reduce economic losses from structural and nonstructural damage, in addition to the primary requirement of life safety (i.e., no collapse), has focused attention on the desirability of introducing greater lateral stiffness in earthquake-resistant multistory buildings. Where acceleration-sensitive equipment is to be housed in a structure, the greater horizontal accelerations that may be expected in laterally stiffer structures should be allowed or provided for.

The principal provisions of ACI Appendix A relating to structural walls (and diaphragms) are as follows (see Figure 9-42):

(a) Walls (and diaphragms) are to be provided with shear reinforcement in two orthogonal directions in the plane of the wall. The minimum reinforcement ratio for both longitudinal and transverse directions is

$$\rho_v = \frac{A_{sv}}{A_{cu}} = \rho_n \geqslant 0.0025$$

where the reinforcement is to be continuous and distributed uniformly across the shear area, and where

A_{cv} = net area of concrete section, i.e., product of thickness and width of wall section

A_{sv} = projection on A_{cv} of area of shear reinforcement crossing the plane of A_{cv}

ρ_n = reinforcement ratio corresponding to plane perpendicular to plane of A_{cv}

The maximum spacing of reinforcement is 18 in. At least two curtains of reinforcement, each having bars running in the longitudinal and transverse directions, are to be provided if the in-plane factored shear force assigned to the wall exceeds $2A_{cv}\sqrt{f_c'}$.

Where boundary elements are not required, SEAOC-86 calls for the provision of a minimum amount of reinforcement parallel to the edges of walls and openings in walls equal to twice the cross-sectional area of the minimum shear reinforcement required per lineal foot of wall.

(b) Shear strength of walls (and diaphragms): For walls with a height-to-width ratio $h_w/l_w \geqslant 2.0$, the shear strength is to be determined using the expression:

$$\phi V_n = \phi A_{cv}\left(2\sqrt{f_c'} + \rho_n f_y\right)$$

where

ϕ = 0.60, unless the nominal shear strength provided exceeds the shear corresponding to development of nominal flexural capacity of the wall

A_{cv} = net area as defined earlier

h_w = height of entire wall or of segment of wall considered

l_w = width of wall (or segment of wall) in direction of shear force

For walls with $h_w/l_w < 2.0$, the shear may be determined from

$$\phi V_n = \phi A_{cv}\left(\alpha_c\sqrt{f_c'} + \rho_n f_y\right)$$

where the coefficient α_c varies linearly from a value of 3.0 for $h_w/l_w = 1.5$ to 2.0 for $h_w/l_w = 2.0$.

Where a wall is divided into several segments by openings, the value of the ratio h_w/l_w to be used in calculating V_n for any segment is not to be less than the corresponding ratio for the entire wall.

The nominal shear strength V_n of all wall segments or piers resisting a common lateral force is not to exceed $8A_{cv}\sqrt{f_c'}$, where A_{cv} is the total cross-sectional area of the walls. The nominal shear strength of any individual segment of wall or pier is not to exceed $10A_{cp}\sqrt{f_c'}$, where A_{cp} is the cross-sectional area of the pier considered.

(c) Development length and splices: All continuous reinforcement is to be anchored or spliced in accordance with provisions governing reinforcement in tension, as discussed for flexural members.

Where boundary elements are present, the transverse reinforcement in walls is to be anchored within the confined core of the boundary element to develop the yield stress in tension of the transverse reinforcement.

SEAOC-86 (UBC-88) further requires that splices in transverse reinforcement be staggered in bars of the same layer as well as with respect to adjacent bars in another layer, where two curtains or reinforcement are used.

For shear walls with minimum edge reinforcement, both NEHRP-85 and SEAOC-86 (UBC-88) require that transverse reinforcement terminating at the edges of the walls be provided with standard hooks enclosing the edge reinforcement. Otherwise, the edge reinforcement is to be enclosed in U-stirrups of the same size and spacing as the transverse reinforcement.

(d) Boundary elements: Boundary elements are to be provided, both along the vertical boundaries of walls and around the edges of openings, if any, when the maximum extreme-fiber stress in the wall due to factored forces including earthquake effects exceeds $0.2f_c'$. The boundary members may be discontinued when the calculated compressive stress becomes less than $0.15f_c'$.

Boundary elements need not be provided if the entire wall is reinforced in accordance with the provisions governing transverse reinforcement for members subjected to axial load and bending, as given by Equations 9-6 and 9-7.

Boundary elements of structural walls are to be designed to carry all the factored vertical loads on the wall, including self-weight and gravity loads tributary to the wall, as well as the vertical forces required to resist the overturning moment due to factored earthquake loads. Such boundary elements are to be provided with confinement reinforcement in accordance with Equations 9-6 and 9-7.

Discussion:

(a) The use of two curtains of reinforcement in walls subjected to significant shears (i.e., $> 2A_{cv}\sqrt{f_c'}$) serves to reduce fragmentation and premature deterioration of the concrete under load reversals into the inelastic range. Distributing the reinforcement uniformly across the height and width of the wall helps control the width of inclined cracks.

(b) ACI Appendix A allows calculation of the shear strength of structural walls using a coefficient $\alpha_c = 2.0$. However, advantage can be taken of the greater observed shear strength of walls with low height-to-width ratios h_w/l_w by using an α_c value of up to 3.0 for walls with $h_w/l_w = 1.5$ or less.

The upper bound on the average nominal shear stress that may be developed in any individual segment of wall ($10\sqrt{f_c'}$) is intended to limit the degree of shear redistribution among several connected wall segments. A wall segment refers to a part of a wall bounded by openings or by an opening and an edge.

It is important to note that Section A.2.3.1 of ACI Appendix A requires the use of a strength-reduction factor ϕ for shear of 0.6 for all members (except joints) where the nominal shear strength is less than the shear corresponding to the development of the nominal flexural strength of the member. In the case of beams, the design shears are obtained by assuming plastic end moments corresponding to a tensile steel stress of $1.25f_y$ (see Figure 9-15). Similarly, for a column the design shears are determined not by applying load factors to shears obtained from a lateral load analysis, but from consideration of the maximum developable moments at the column ends consistent with the axial force on the column. This approach to shear design is intended to insure that even when flexural hinging occurs at member ends due to earthquake-induced deformations, no shear failure would develop. Under the above conditions, ACI Appendix A allows the use of the normal strength-reduction factor for shear of 0.85. When design shears are not based on the condition of flexural strength being developed at member ends, the code requires the use of a lower shear strength-reduction factor to achieve the same result, that is, prevention of premature shear failure.

As pointed out earlier, in the case of multistory structural walls, a condition similar to that used for the shear design of beams and columns is not so readily established. This is so primarily because the magnitude of the shear at the base of a (vertical cantilever) wall, or at any level above, is influenced significantly by the forces and deformations beyond the particular level considered. Unlike the flexural behavior of beams and columns in a frame, which can be considered as close-coupled systems (i.e., with the forces in the members determined by the forces and displacements within and at the ends of the member), the state of flexural deformation at any section of a structural wall (a far-coupled system) is influenced significantly by the displacements of points far removed from the section considered. Results of dynamic inelastic analyses of isolated structural walls under earthquake excitation[9-3] also indicate that the base shear in such walls is strongly influenced by the higher modes of response.

A distribution of static lateral forces along the height of the wall essentially corresponding to the fundamental mode response, such as is assumed by most codes,[9-1] will produce flexural yielding at the base if the section at the base is designed for such a set of forces. Other distributions of lateral forces, with a resultant acting closer to the base of the wall, can produce yielding at the base only if the magnitude of the resultant horizontal force, and hence the base shear, is increased. Results of the study of isolated walls referred to above,[9-3] which would also apply to frame–shear-wall systems in which the frame is flexible relative to the wall, in fact indicate that for a wide range of wall properties and input motion characteristics, the resultant of the dynamic horizontal forces producing yielding at the base of the wall generally occurs well below the two-thirds-of-total-height level associated with the fundamental-mode response (see Figure 9-23). This would imply significantly larger base shears than those due to lateral forces distributed according to the fundamental mode response. The study of isolated walls mentioned above indicates ratios of maximum dynamic shears to "fundamental-mode shears" (i.e., shears associated with horizontal forces distributed according to the fundamental-mode response, as used in codes) ranging from 1.3 to 4.0, the value of the ratio increasing with increasing fundamental period (see Figure 9-22).

(c) Since multistory structural walls behave essentially as vertical cantilever beams, the horizontal transverse reinforcement is called upon to act as web reinforcement. As such, these bars have to be fully anchored in the boundary elements, using standard 90° hooks whenever possible.

(d) ACI Appendix A uses an extreme-fiber compressive stress of $0.2f_c'$, calculated using a linearly elastic model based on gross sections of structural members and factored forces, as indicative of significant compression. Structural walls subjected to compressive stresses exceeding this value are generally required to have boundary elements.

Figure 9-43 illustrates the condition assumed as basis for requiring that boundary elements of walls be designed for all the gravity loads (W) as well as the vertical forces associated with overturning of the wall due to earthquake forces (H). This requirement assumes that the boundary element alone may have to carry all the vertical (compressive) forces at the critical wall section when the maximum horizontal earthquake force acts on the wall. Under load reversals, such a loading condition imposes severe demands on the concrete in the boundary elements. Hence the requirement for confinement reinforcement similar to those for frame members subjected to axial load and bending.

Diaphragms of reinforced concrete, such as floor slabs, that are called upon to transmit horizontal forces through bending and shear in their own plane, are treated in much the same manner as structural walls.

6. *Frame members not forming part of lateral-force-resisting system.* Frame members that are not relied on

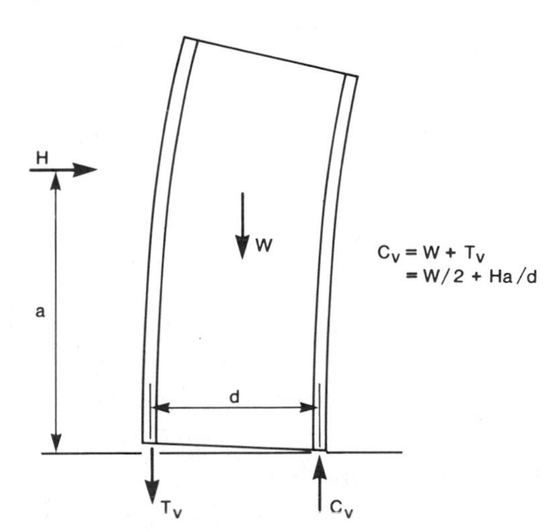

Figure 9-43 Loading condition assumed for design of boundary elements of structural walls.

to resist earthquake-induced forces need not satisfy the stringent requirements governing lateral-load-resisting elements. These relate particularly to the transverse reinforcement requirements for confinement and shear. Non-lateral-load-resisting elements, whose primary function is the transmission of vertical loads to the foundation, need comply only with the reinforcement requirements of ACI Appendix A, in addition to those found in the main body of the code.

A special requirement for non-lateral-load-resisting elements is that they be checked for adequacy with respect to a lateral displacement representing the expected actual displacement of the structure under the design earthquake. For the purpose of this check, ACI Appendix A uses a value of twice the displacement calculated under the factored lateral loads, or $2 \times 1.7 = 3.4$ times the displacement due to the code-specified loads. UBC-85 specifies a lateral displacement equal to $3/K$ times the displacement due to the code forces (K is the horizontal force factor that enters the expression for the design base shear V). With K varying from 0.67 to 2.0, the factor $3/K$ takes on values ranging from 4.5 to 1.5. NEHRP-85 calls for checking non-lateral-load-resisting elements using a lateral displacement equal to C_d (the *deflection amplification factor*) times the displacement calculated under the code-prescribed forces. For concrete structures in regions of high seismicity, C_d varies from 4 to 6.5, depending on the type of structure. The corresponding value in SEAOC-86 (UBC-88) is $3R_w/8$ (R_w is a coefficient analogous to $1/K$ in UBC-85) times the displacements due to the code-specified lateral forces. For regions of high seismicity, R_w varies from 6 to 12, depending on the type of structure. Thus, the factor $3R_w/8$ ranges from 2.25 to 4.5. These factors reflect the maximum displacements that may be expected under a major earthquake which are generally larger than those produced by the code-specified "equivalent lateral forces." Under this requirement, the gravity-load system should be capable of maintaining its vertical-load-carrying capacity, without reduction, under the specified lateral displacement. Where plastic hinging can occur under the specified displacement, confinement reinforcement complying with Equation 9-6 or 9-7 has to be provided. These elements, however, are not required to be designed for the moments associated with the lateral forces.

For gravity-load frame members subjected to factored axial compressive forces exceeding $A_g f_c'/10$, the following requirements relating to transverse reinforcement have to be satisfied: The maximum tie spacing (over the length l_0 from the face of the joint),

$$s_0 \leqslant \begin{cases} 8 \times \text{(diameter of smallest longitudinal bar)} \\ 24 \text{ tie diameters} \\ \frac{1}{2}\text{(least cross-sectional dimension of column)} \end{cases}$$

where

$$l_0 \geqslant \begin{cases} \frac{1}{6}\text{(clear height of column)} \\ \text{maximum cross-sectional dimension of column} \\ 18 \text{ in.} \end{cases}$$

The first tie is to be located within a distance of $s_0/2$ from the face of the joint. The maximum tie spacing in any part of the column is $2s_0$.

7. *Frames in regions of moderate seismic risk.* Although ACI Appendix A does not define "moderate seismic risk" in terms of a commonly accepted quantitative measure, it assumes that the probable ground-motion intensity in such regions would be a fraction of that expected in a high-seismic-risk zone, to which the major part of Appendix A is addressed. By the above description, an area of moderate seismic risk would correspond roughly to zone 2 as defined in ANSI-82.[9-68] For regions of moderate seismic risk, the provisions for the design of structural walls given in the main body of the ACI Code are considered sufficient to provide the necessary ductility. The requirements in ACI Appendix A for structures in moderate-risk areas relate mainly to frames.

The same axial compressive force ($A_g f_c'/10$) used to distinguish flexural members from columns in high-seismic-risk areas also applies in regions of moderate seismicity.

(a) Shear design of beams, columns, or two-way slabs resisting earthquake effects: The magnitude of the design shear is not to be less than either of the following:

   (1) The sum of the shear associated with the development of the nominal moment strength at each restrained end and that due to factored gravity loads, if any, acting on the member. This is similar to the corresponding requirement for high-risk zones and illustrated in Figure 9-15, except that the stress in the flexural tensile reinforcement is taken as f_y rather than $1.25f_y$.

   (2) The maximum factored shear corresponding to the design gravity and earthquake forces, but with the earthquake forces taken as twice the value normally specified by codes. Thus, if the critical load combination consists of dead load (D) + live load (L) + earthquake effects (E), then the design shear is to be computed from

$$U = 0.75[1.4D + 1.7L + 2(1.87E)]$$

(b) Detailing requirements for beams: The positive moment strength at the face of a joint must be at least one-third the negative moment capacity at the same section. (This compares with one-half for high-seismic-risk areas.) The moment strength—positive or negative—at any section is to be no less than one-fifth the maximum moment strength at either end of a member. Stirrup spacing requirements are identical to those for beams in high-seismic-risk areas.

(c) Detailing requirements for columns: The tie spacing requirements for columns are identical to those for gravity-load-carrying members, as given above under item 6, with the additional minimum spacing of 12 in.

(d) Detailing requirements for two-way slabs without beams: As mentioned earlier, requirements for flat plates in ACI Appendix A appear only in the section relating to areas of moderate seismic risk. This suggests that ACI Appendix A considers the use of flat plates as acceptable components of the lateral-load-resisting system only for areas of moderate seismicity.

Specific requirements relating to flat-plate and flat-slab reinforcement for frames in moderate-risk zones are given in ACI Appendix A and illustrated in the Commentary to the Appendix.

9.5 DESIGN EXAMPLES — REPRESENTATIVE ELEMENTS OF A 12-STORY FRAME – SHEAR WALL BUILDING

9.5.1 Preliminaries

A significant part of the damage observed in engineered buildings during earthquakes has resulted from the effects of major structural discontinuities that were inadequately provided for. The message here is clear. Unless proper provision is made for the effects of major discontinuities in geometry, mass, stiffness, or strength, it would be prudent on the part of the engineer to avoid such conditions, which are associated with force concentrations and large ductility demands in localized areas of the structure. Where such discontinuities are unavoidable or desirable from the architectural standpoint, an analysis to obtain estimates of the forces associated with the discontinuity is recommended. SEAOC-86 (UBC-88) provides guidelines for estimating design forces in structures with various types of vertical and plan irregularities.

In addition to discontinuities, major asymmetry, with particular regard to the disposition in plan of the lateral-load-resisting elements, should be avoided whenever possible. Such asymmetry, which can result in a significant eccentricity between the center of stiffness and the center of mass (and hence of the resultant inertial force), can produce appreciable torsional forces in the structure. Torsional effects can be critical for corner columns or end walls, i.e., elements located far from the center of stiffness.

Another important point to consider in the preliminary design of a structure relates to the effectiveness of the various lateral-load-resisting components, particularly where these differ significantly in deformation capacity. Efficient use of structural components would suggest that the useful range of deformation of the principal lateral-load-resisting elements in a structure be of about the same magnitude whenever practicable. This is illustrated in Figure 9-44a, which shows load–deformation curves of representative elements (1) and (2) in a structure. Such a design allows all the resisting elements to participate in carrying the induced forces over the entire range of deformation. In Figure 9-44b, the resisting elements (1) and (2) not only possess different initial stiffnesses but, more importantly, exhibit different ductilities (not ductility *ratios*) or defor-

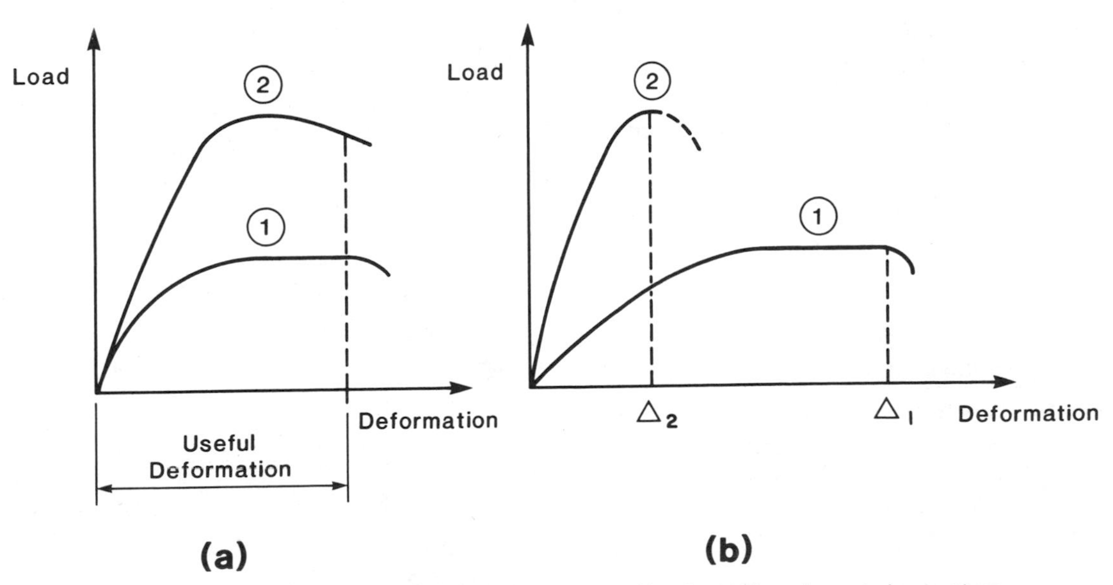

Figure 9-44 Relative deformation capacity in lateral-load-resisting elements in structure.

mation capacities. In such a case, which is typical of a frame–shear-wall structure, the design should be aimed at insuring that the maximum probable deformation or lateral displacement under dynamic conditions does not exceed the deformation capacity Δ_2 of element (2); or, if the maximum expected deformation could exceed Δ_2, then element (1) should be so designed that it can support the additional load that may come upon it when element (2) loses a considerable part of its load-carrying capacity. It is worth noting that, generally, the lateral displacements associated with full mobilization of the ductility of rigid (open) frames are such that significant nonstructural damage can be expected (unless such movement is anticipated and provided for).

The need to tie together all the elements making up a structure or a portion of it that is intended to act as a unit cannot be overemphasized. This applies to the superstructure as well as foundation elements. Where a structure is divided into different parts by expansion joints, as when the various parts differ considerably in height, plan size, shape, or orientation, a sufficient gap should be provided between adjacent parts to prevent their pounding against each other. To avoid pounding between adjacent buildings or parts of the same building when vibrating out of phase with each other, a gap equal to at least six times the sum of the maximum lateral deflections of the two structures under the design (code-specified) lateral forces, or the sum of the maximum deflections of the two structures as indicated by a dynamic analysis, would be desirable.

A good basis for the preliminary design of an earthquake-resistant building is a structure proportioned to satisfy the requirements for gravity and wind loads. The planning and layout of the structure, however, must be undertaken with due consideration of the special requirements for earthquake-resistant design. Thus, modifications in both configuration and proportions to anticipate earthquake-related requirements should be incorporated at the outset into the basic design for gravity and wind. Essential to the finished design is particular attention to details that can often mean the difference between a severely damaged structure and one with only minor, repairable damage.

9.5.2 Example Designs of Elements of a 12-Story Frame – Shear Wall Building[c]

The application of the earthquake-resistant design provisions of ANSI A58.1-1982 (ANSI-82)[(9-68)] with respect to design loads and those of Appendix A of ACI 318-83[(9-9)] relating to proportioning and detailing of members will be illustrated for representative elements of a 12-story frame–shear wall building located in seismic zone 4. Except for minor variations in local areas, the division of the United States into different seismic zones (i.e., 0, 1, 2, 3, and 4), as adopted by ANSI-82 (which is based on ATC 3-06[(9-70)]), is very similar to that used in UBC-85 (or UBC-79, on which most of the other ANSI-82 seismic-design provisions are based). A number of areas classified under zone 1 in UBC-85 are shown as located in zone 0 in ANSI-82.

The typical framing plan and section of the structure considered are shown in Figure 9-45a and b, respectively. The columns and structural walls have constant cross-sections throughout the height of the building. The floor beams and slabs also have the same dimensions at all floor levels. Although the dimensions of the structural elements in this example are within the practical range, the structure itself is hypothetical and has been chosen mainly for illustrative purposes. Other per-

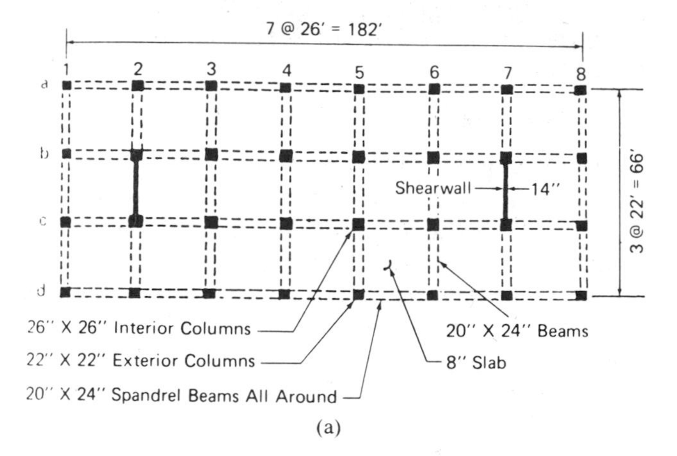

(a)

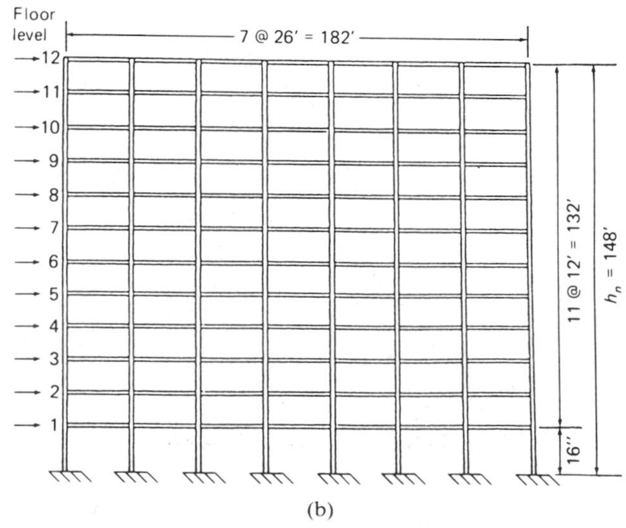

(b)

Figure 9-45 Structure considered in design example. (a) Typical floor framing plan. (b) Longitudinal section.

[c]Reproduced, with minor modifications, from Reference 9-75, with permission from Van Nostrand Reinhold Company.

tinent design data are as follows:

Service loads—vertical:

- Live load:
  Basic, 50 lb/ft^2.
  Additional average uniform load to allow for heavier basic load on corridors, 25 lb/ft^2.
  Total average live load, 75 lb/ft^2.
- Superimposed dead load:
  Average for partitions 20 lb/ft^2.
  Ceiling and mechanical 10 lb/ft^2.
  Total average superimposed dead load, 30 lb/ft^2.

Material properties:

- Concrete:

$$f_c' = 4000 \text{ lb/in.}^2; \qquad w_c = 145 \text{ lb/ft}^3.$$

- Reinforcement:

$$f_y = 60 \text{ ksi}.$$

1. *Determination of design lateral forces.* On the basis of the given data and the dimensions shown in Figure 9-45, the weights that may be considered lumped at a floor level (including that of all elements located between two imaginary parallel planes passing through mid-height of the columns above and below the floor considered) and the roof were estimated and are listed in Tables 9-1 and 9-2. The calculation of the base shear V, as explained in Chapter 4, for the transverse and longitudinal directions is shown at the bottom of Tables 9-1 and 9-2. For this example, the importance factor I and the soil factor S have been assigned values of unity. Note that a value $K = 0.8$ has been used in the transverse direction where the structure consists of a frame–shear wall system, while $K = 0.67$ in the longitudinal direction, where moment-resisting frames make up the structure.

Calculation of the undamped (elastic) natural periods of vibration of the structure in the transverse direction, as shown in Figure 9-46, using the story weights listed in Table 9-1 and member stiffnesses based on gross concrete sections, yielded a value for the fundamental period of 1.34 sec (compared to a T value of 0.91 sec obtained by the approximate formula given in ANSI-82). The mode shapes and corresponding periods of the first five modes of vibration of the structure in the transverse direction are shown in Figure 9-47.

The lateral seismic design forces acting at the floor levels, resulting from the distribution of the base shear in each principal direction, are also listed in Tables 9-1 and 9-2. For example, the seismic lateral force F_x at the 10th floor level in the transverse direction is given by

$$F_{10} = \frac{(V - F_t) w_x h_x}{\sum_{i=1}^{n} w_i h_i} = \frac{(1500 - 96)(124)(2200)}{2{,}144{,}300}$$

$$= 179 \text{ kips}$$

Table 9-1 Design Lateral Forces in Transverse (Short) Direction (Corresponding to Entire Structure)*

			Seismic forces			Wind forces		
Floor level (from base)	Height h_x, ft	Story weight w_x, kips	$w_x h_x$, ft-kips	Lateral force F_x, kips	Story shear ΣF_x, kips	Wind pressure (Average), lb / ft^2	Lateral force H_x, kips	Story shear ΣH_x, kips
12 (Roof)	148	2100	311,000	300*†	300	23.7	25.9	25.9
11	136	2200	299,000	196	496	23.1	50.5	76.4
10	124	2200	273,000	179	675	22.6	49.4	125.8
9	112	2200	246,000	161	836	22.0	48.1	173.9
8	100	2200	220,000	144	980	21.4	46.7	220.6
7	88	2200	193,000	126	1106	20.6	45.0	265.6
6	76	2200	167,000	109	1215	19.8	43.2	308.8
5	64	2200	141,000	92	1307	19.0	41.5	350.3
4	52	2200	114,000	75	1382	18.2	39.8	390.1
3	40	2200	88,000	58	1440	17.1	37.3	427.4
2	28	2200	61,500	40	1480	15.9	34.7	462.1
1	16	2200	30,800	20		14.4	36.7	
—					1500			498.8
		26,300	2,144,300	1500			498.8	

*Base shear, $V = ZIKCSW$, where $C = 1/(15\sqrt{T})$ and $T = 0.05h_n/\sqrt{D}$. In transverse direction, $h_n = 148$ ft and $D = 66$ ft $\Rightarrow$ $T = 0.91$ sec and $C = 0.07$. Thus, $V = (1.0)(1)(0.8)(0.07)(1)W = 0.56W = 0.056(26{,}300) = 1473$, say 1500 kips. $F_t = 0.07TV = (0.07)(0.91)(1500) = 96$ kips.

† Representing the sum $F_t + F_{12}$.

Table 9-2 Design Lateral Forces in Longitudinal Direction (Corresponding to Entire Structure)*

		Seismic forces				Wind forces		
Floor level (from base)	Height h_x, ft	Story weight w_x, kips	$w_x h_x$, ft-kips	Lateral force F_x, kips	Story shear ΣF_x, kips	Wind pressure (Average), lb / ft²	Lateral force H_x, kips	Story shear ΣH_x, kips
12 (Roof)	148	2100	†	239‡	239	19.7	7.8	7.8
11	136	2200	†	149	388	19.3	15.3	23.1
10	124	2200	†	136	524	18.8	14.9	38.0
9	112	2200	†	122	646	18.0	14.3	52.3
8	100	2200	†	109	755	17.2	13.6	65.9
7	88	2200	†	96	851	16.6	13.2	79.1
6	76	2200	†	83	934	16.0	12.7	91.8
5	64	2200	†	70	1104	15.0	11.9	103.7
4	52	2200	†	57	1061	14.0	11.1	114.8
3	40	2200	†	44	1105	13.0	10.3	125.1
2	28	2200	†	30	1135	11.8	9.4	134.5
1	16	2200	†	15	1150	10.1	9.3	143.8
		26,300	2,144,300	1150			143.8	

*In longitudinal direction, $T = C_T h_n^{3/4}$, where C_T(concrete frames) = 0.025: $T = 0.025(148)^{3/4} = 1.06$ sec. $C = 1/(15\sqrt{T}) = 1/(15\sqrt{1.06}) = 0.65$. Base shear, $V = ZIKCSW = (1.0)(1)(0.67)(0.065)(1)W = 0.0436W = 0.0436(26{,}300) = 1146$, say 1150 kips. $F_t = 0.07TV = (0.07)(1.06)(1150) = 85$ kips.

†Same as for transverse direction.

‡$F_t + F_{12}$.

Also listed in the tables are the story shears corresponding to the distributed lateral forces.

For comparison, the wind forces and story shears corresponding to a basic wind speed of 75 mi/h and Exposure B (urban and suburban areas), computed as prescribed in ANSI-82, are shown for each direction in Tables 9-1 and 9-2.

Lateral-load analyses of the structure along each principal direction, under the respective seismic and wind loads, were carried out assuming no torsional effects. The model shown in Figure 9-45 was used for the analysis in the transverse direction. This model consists of three different frames linked by hinged rigid bars at the floor levels. (This device, which imposes

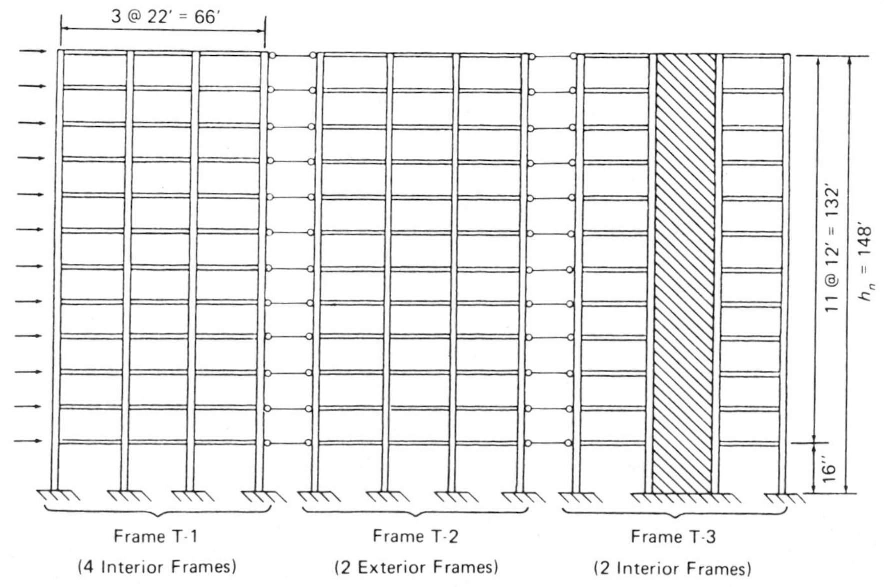

Figure 9-46 Analytical model for lateral load analysis of structure in transverse direction.

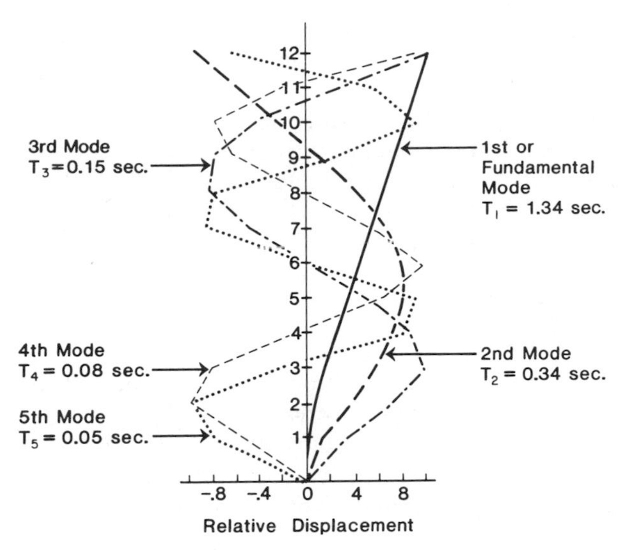

Figure 9-47 Undamped natural modes and periods of vibration of structure in transverse direction.

equal horizontal displacements at each floor level, is used to model the effect of the floor slabs, which generally may be assumed as very stiff in their own planes.) Frame T-1 represents the four identical interior frames along lines 3, 4, 5, and 6, which have been lumped together in this single frame, while Frame T-2 represents the two exterior frames along lines 1 and 8. The third frame, T-3, represents the two identical frame–shear-wall systems along lines 2 and 7. In the longitudinal direction, two linked frames, each similar to the frame shown in Figure 9-45b, were used to represent the two identical exterior frames L-1 along lines *a* and *d* and the two identical frames L-2 along lines *b* and *c* (see Figure 9-46).

The lateral displacements due to both seismic and wind forces listed in Tables 9-1 and 9-2 are shown plotted in Figure 9-48. Although the seismic forces used to obtain the curves of Figure 9-48 are approximate, the results shown still serve to draw the distinction between wind and seismic forces, that is, the fact that the former are external forces the magnitudes of which are proportional to the exposed surface, while the latter represent inertial forces depending primarily on the mass and stiffness properties of the structure. Thus, while the ratio of the total wind force in the transverse direction

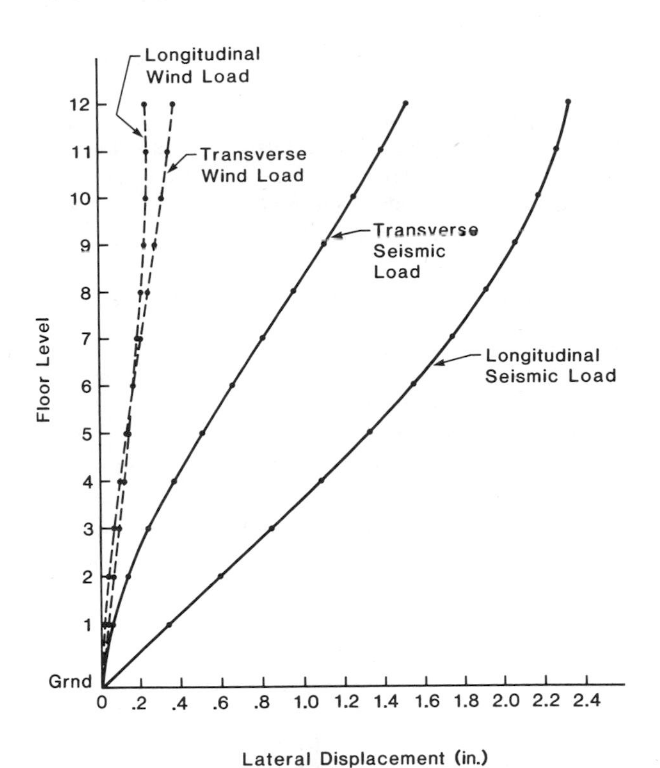

Figure 9-48 Lateral displacements under seismic and wind loads.

Table 9-3 Distribution of Horizontal Seismic Story Shears among the Three Transverse Frames Shown in Figure 9-46

	Frame T-1 (4 interior frames)		Frame T-2 (2 exterior frames)		Frame T-3 (2 interior frames with shear walls)		
Story level	Story shear	% of total	Story shear	% of total	Story shear	% of total	Total story shear, kips
12	364	121	153	51	−217	−72	300
11	255	51	115	23	126	26	496
10	297	44	129	19	249	37	675
9	304	36	133	16	399	48	836
8	316	32	138	14	526	54	980
7	321	29	140	13	645	58	1106
6	319	26	140	12	756	62	1215
5	307	24	134	10	866	66	1307
4	285	21	124	9	973	70	1382
3	242	17	105	7	1093	76	1440
2	205	14	88	6	1187	80	1480
1	48	3	20	1	1432	96	1500

to that in the longitudinal direction (see Tables 9-1 and 9-2) is about 3.5, the corresponding ratio for the seismic forces is only 1.3. As a result of this and the smaller lateral stiffness of the structure in the longitudinal direction, the displacement due to seismic forces in the longitudinal direction is significantly greater than that in the transverse direction. By comparison, the displacements due to wind are about the same for both directions. The typical deflected shapes associated with predominantly cantilever or flexure structures (as in the transverse direction) and shear (open-frame) buildings (as in the longitudinal direction) are evident in Figure 9-48. The average deflection indices, that is, the ratios of the lateral displacement at the top to the total height of the structure, are $\frac{1}{4790}$ for wind and $\frac{1}{1170}$ for seismic loads in the transverse direction. The corresponding values in the longitudinal direction are $\frac{1}{7510}$ for wind and $\frac{1}{760}$ for seismic loads.

An idea of the distribution of lateral loads among the different frames making up the structure in the transverse direction may be obtained from Table 9-3, which lists the portion of the total story shear at each level resisted by each of the three lumped frames. Note that at the top (12th floor level), the lumped frame T-1 takes 121% of the total story shear. This reflects the fact that in frame–shear-wall systems of average proportions, interaction between frame and wall under lateral loads results in the frame "supporting" the wall at the top, while at the base most of the horizontal shear is resisted by the wall. Table 9-3 indicates that for the structure

Table 9-4 Summary of Design Moments for Typical Beams on Sixth Floor of Interior Transverse Frames along Lines 3 through 6 (Figure 9-46)

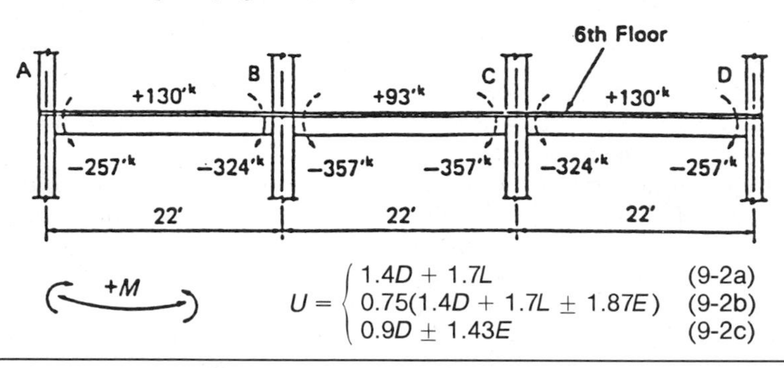

$$U = \begin{cases} 1.4D + 1.7L & \text{(9-2a)} \\ 0.75(1.4D + 1.7L \pm 1.87E) & \text{(9-2b)} \\ 0.9D \pm 1.43E & \text{(9-2c)} \end{cases}$$

Beam *AB*

	Design moment, ft-kips		
	Near *A*	mid-span of *AB*	*B*
9-2a	−86	+130	−179
9-2b:			
Sidesway to right	+127	+126	−324
Sidesway to left	−257	+70	+56
9-2c:			
Sidesway to right	+162	+79	−265
Sidesway to left	−230	+22	+125

Beam *BC*

	Design moment, ft-kips		
	B	Mid-span	*C*
9-2a	−179	+93	−179
9-2b:			
Sidesway to right	+89	+80	−357
Sidesway to left	−357	+60	+89
9-2c:			
Sidesway to right	+157	+47	−297
Sidesway to left	−297	+26	+157

considered, the two frames with walls take 96% of the shear at the base in the transverse direction.

To illustrate the design of two typical beams on the sixth floor of an interior frame, the results of the analysis in the transverse direction under seismic loads have been combined, using Equation 9-2, with results from a gravity-load analysis of a single-story bent including these beams. The results are listed in Table 9-4. Similar values for typical exterior and interior columns on the second floor of the same interior frame are shown in Table 9-5. Corresponding design values for the structural wall section at the first floor of frame T-3 (see Figure 9-46) are listed in Table 9-6. The last column in Table 9-6 lists the axial load on the boundary elements (the 26 × 26-in. columns forming the flanges of the structural walls) calculated according to the ACI requirement that these be designed to carry all factored loads on the walls, including self-weight, gravity loads, and vertical forces due to earthquake-induced overturning moments. The loading condition associated with this requirement is illustrated in Figure 9-43. In both Tables 9-5 and 9-6, the additional forces due to the effects of horizontal torsional moments corresponding to the minimum ANSI-prescribed eccentricity of 5% of the building dimension perpendicular to the direction of the applied forces have been included.

It is pointed out that for buildings located in seismic zones 3 and 4 (i.e., high-seismic-risk areas), the detailing requirements for ductility prescribed in ACI Appendix A have to be met even when the design of a member is governed by wind loading rather than seismic loads.

2. *Design of flexural member AB.* The aim is to determine the flexural and shear reinforcement for the beam AB on the sixth floor of a typical interior transverse frame. The beam carries a dead load of 3.7 kips/ft of span and a live load of 1.95 kips/ft. The critical design (factored) moments are shown circled in Table 9-4. The beam has dimensions $b = 20$ in. and $d = 21.5$ in. The slab is 8 in. thick. $f_c' = 4000$ lb/in.^{2} and $f_y = 60{,}000$ lb/in.^{2}

In the following solution, the boxed-in section numbers at the right-hand margin correspond to those in ACI 318-83.

(a) Check satisfaction of limitations on section dimensions:

$$\frac{\text{width}}{\text{depth}} = \frac{20}{21.5} = 0.93 > 0.3 \quad \text{O.K.}$$

A.3.1.3
A.3.1.4

$$\text{width} = 20 \text{ in.} \begin{cases} \geqslant 10 \text{ in.} \quad \text{O.K.} \\ \leqslant (\text{width of supporting column} \\ \quad +1.5 \times \text{depth of beam}) \\ = 26 + 1.5(21.5) = 58.25 \text{ in.} \quad \text{O.K.} \end{cases}$$

Table 9-5 Summary of Design Moments and Axial Loads for Typical Columns on Second Floor of Interior Transverse Frames along Lines 3 through 6 (Figure 9-45a)

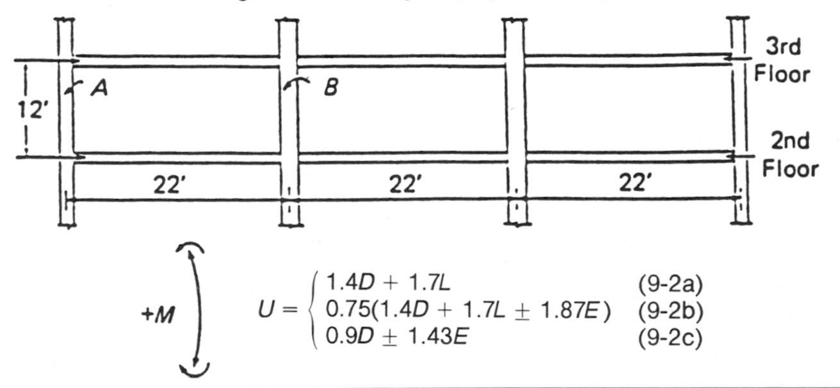

+M

$$U = \begin{cases} 1.4D + 1.7L & \text{(9-2a)} \\ 0.75(1.4D + 1.7L \pm 1.87E) & \text{(9-2b)} \\ 0.9D \pm 1.43E & \text{(9-2c)} \end{cases}$$

	Exterior column *A*			Interior column *B*		
	Axial load, kips	Moment, ft-kips*		Axial load, kips	Moment, ft-kips*	
		Top	Bottom		Top	Bottom
9-2a	(−930)	−95	+95	(−1678)	+53	−53
9-2b:						
Sidesway to right	−526	−15	−9	−1225	(+173	−207)
Sidesway to left	870	(−133	+151)	−1293	−93	+127
9-2c:						
Sidesway to right	−328	+25	−44	−857	+146	−181
Sidesway to left	−680	−101	+120	−925	−126	+161

*Including moments corresponding to horizontal torsional shears due to minimum story-shear eccentricity.

Table 9-6 Summary of Design Loads on Structural Wall Section at First Floor Level of Transverse Frame along Line 2 (or 7) (Figure 9-46)

$$U = \begin{cases} 1.4D + 1.7L & \text{(9-2a)} \\ 0.75(1.4D + 1.7L \pm 1.87E) & \text{(9-2b)} \\ 0.9D \pm 1.43E & \text{(9-2c)} \end{cases}$$

	Design forces acting on entire structural wall			
	Axial load, kips	Bending* (overturning) moment, ft-kips	Horizontal* shear, kips	Axial† on boundary element, kips
10-2a	(4598)	Nominal	Nominal	2326
10-2b	3449	53,280	1129	(4166)*
10-2c	2506	(54,464)	(1153)	3744*

*Includes effect of horizontal torsion due to accidental eccentricity of story shear (ecc. = 0.05 × 182 = 9.1 ft).
† Based on loading condition illustrated in Figure 9-42.

(b) Determine required flexural reinforcement:

(1) Negative moment reinforcement at support B: Since the negative flexural reinforcement for both beams AB and BC at joint B will be provided by the same continuous bars, the larger negative moment at joint B will be used. In the following calculations, the effect of any compressive reinforcement will be neglected. From $C = 0.85f_c'ba = T = A_s f_y$,

$$a = \frac{A_s f_y}{0.85 f_c' b} = \frac{60A_s}{(0.85)(4)(20)} = 0.882A_s$$

$$M_u \leqslant \phi M_n = \phi A_s f_y\left(d - \frac{a}{2}\right)$$

$$(357)(12) = (0.90)(60)A_s \times [21.5 - (0.5)(0.882A_s)]$$

$$A_s^2 - 48.75A_s + 179.9 = 0$$

or

$$A_s = 4.02 \text{ in.}^2$$

Alternatively, convenient use may be made of design charts for singly reinforced flexural members with rectangular cross-sections, given in standard references.(9-76) *Use four No. 9 bars, $A_s = 4.0$ in.²* This gives a negative moment capacity at support B of $\phi M_n = 355$ ft-kips.

Check satisfaction of limitations on reinforcement ratio:

$$\rho = \frac{A_s}{bd} = \frac{4.0}{(20)(21.5)} \qquad \boxed{\text{A.3.2.1}}$$

$$= 0.0093 > \rho_{min} = \frac{200}{f_y} = 0.0033$$

$$\text{and} < \rho_{max} = 0.025 \qquad \text{O.K.}$$

(2) Negative moment reinforcement at support A:

$$M_u = 257 \text{ ft-kips}$$

As at support B, $a = 0.882A_s$. Substitution into

$$M_u = \phi A_s f_y\left(d - \frac{a}{2}\right)$$

yields $A_s = 2.82$ in.². *Use three No. 9 bars, $A_s = 3.0$ in.²*. This gives a negative moment capacity at support A of $\phi M_n = 272$ ft-kips.

(3) Positive moment reinforcement at supports: A positive moment capacity at the supports equal to at least 50% of the corresponding negative moment capacity is required, i.e., $\boxed{\text{A.3.2.2}}$

$$\min M_u^+(\text{at support } A) = \frac{272}{2} = 136 \text{ ft-kips}$$

which is less than $M^+_{max} = 162$ ft-kips at A (see Table 9-4);

min M^+_u (at support B for both spans AB and BC)

$$= \frac{355}{2} = 178 \text{ ft-kips}$$

Note that the above required capacities are greater than the design positive moments near the mid-spans of both beams AB and BC.

(4) Positive moment reinforcement at mid-span—to be made continuous to supports:

$$a = \frac{A_s f_y}{0.85 f_c' b} = \frac{60A_s}{(0.80)(4)(60)} = 0.294A_s$$

Substituting into

$$M_u = (162)(12) = \phi A_s f_y \left(d - \frac{a}{2}\right)$$

yields A_s required at $A = 1.7$ in.^{2}. Similarly, corresponding to the required capacity at support B, $M^+_u = 178$ ft-kips, we have A_s(required) $= 1.85$ in.^{2}. *Use three No. 7 bars continuous through both spans.* $A_s = 1.80$ *in.*^{2} This provides a positive moment capacity of 172 ft-kips—about 3% less than that required at support B. Check:

$$\rho = \frac{1.8}{(20)(21.5)} = 0.0042$$

$$> \rho_{min} = \frac{200}{f_y} = 0.0033 \quad \text{O.K.} \quad \boxed{10.5.1}$$

(c) Calculate required length of anchorage of flexural reinforcement in exterior column:

$$\text{Development length } l_{dh} \geqslant \begin{cases} f_y d_b / 65\sqrt{f_c'} \\ 8d_b \\ 6 \text{ in.} \end{cases} \quad \boxed{\text{A.6.4.1}}$$

(plus standard 90° hook located in confined region of column). For the No. 9 (top) bars (bend radius $\geqslant 5d_b$),

$$l_{dh} \geqslant \begin{cases} \dfrac{(60{,}000)(1.128)}{65\sqrt{4000}} = \underline{17 \text{ in.}} \\ (8)(1.128) = 9.0 \text{ in.} \\ 6 \text{ in.} \end{cases}$$

For the No. 7 bottom bars (bend radius $\geqslant 4d_b$),

$$l_{dh} \geqslant \begin{cases} \dfrac{(60{,}000)(0.875)}{65\sqrt{4000}} = \underline{13 \text{ in.}} \\ (8)(0.875) = 7 \text{ in.} \\ 6 \text{ in.} \end{cases}$$

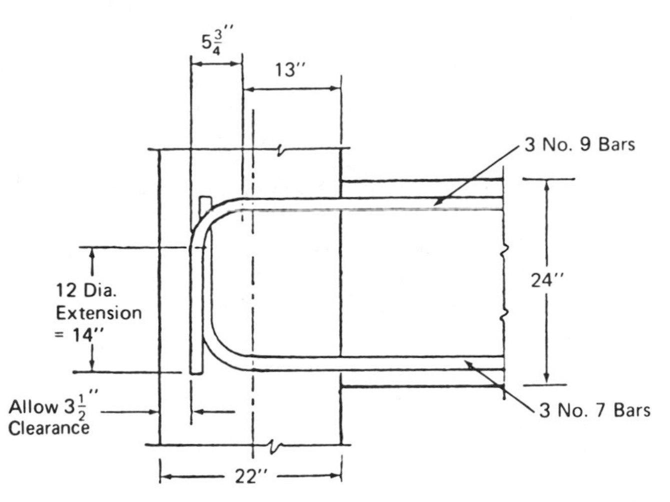

Figure 9-49 Detail of anchorage of flexural reinforcement in exterior column.

Figure 9-49 shows the detail of flexural reinforcement anchorage in the exterior column. Note that the development length l_{dh} is measured from the near face of the column to the far edge of the vertical 12-bar-diameter extension (see Figure 9-34).

(d) Determine shear-reinforcement requirements: Design for shears corresponding to end moments obtained by assuming the stress in the tensile flexural reinforcement equal to $1.25f_y$ and a strength reduction factor $\phi = 1.0$, plus factored gravity loads (see Figure 9-15). Table 9-7 shows values of design end shears corresponding to the two loading cases to be considered. In the table,

$$w_u = 0.75(w_D + w_L) = 0.75[1.4(3.7) + 1.7(1.95)]$$
$$= 6.37 \text{ kips/ft}$$

ACI Appendix A requires that the contribution of concrete to shear resistance, V_c, be neglected if the earthquake-induced shear force (corresponding to the probable flexural strengths at beam ends calculated using $1.25f_y$ instead of f_y and $\phi = 1.0$) is greater than one-half the total design shear and the axial compressive force including earthquake effects is less than $A_g f_c'/20$. $\boxed{\text{A.7.2.1}}$

For sidesway to the right, the shear at end B due to the plastic end moments in the beam (see Table 9-7) is

$$V_B = \frac{238 + 482}{20} = 36 \text{ kips}$$

which is less than 50% of the total design shear, $V_u = 99.7$ kips. Therefore, the contribution of concrete to shear resistance can be considered in determining shear reinforcement requirements.

At right end B, $V_u = 99.7$ kips. Using

$$V_c = 2\sqrt{f_c'}\, b_w d = \frac{2\sqrt{4000}\,(20)(21.5)}{1000} = 54.4 \text{ kips}$$

Table 9-7 Determination of Design Shear Forces for Beam Spans

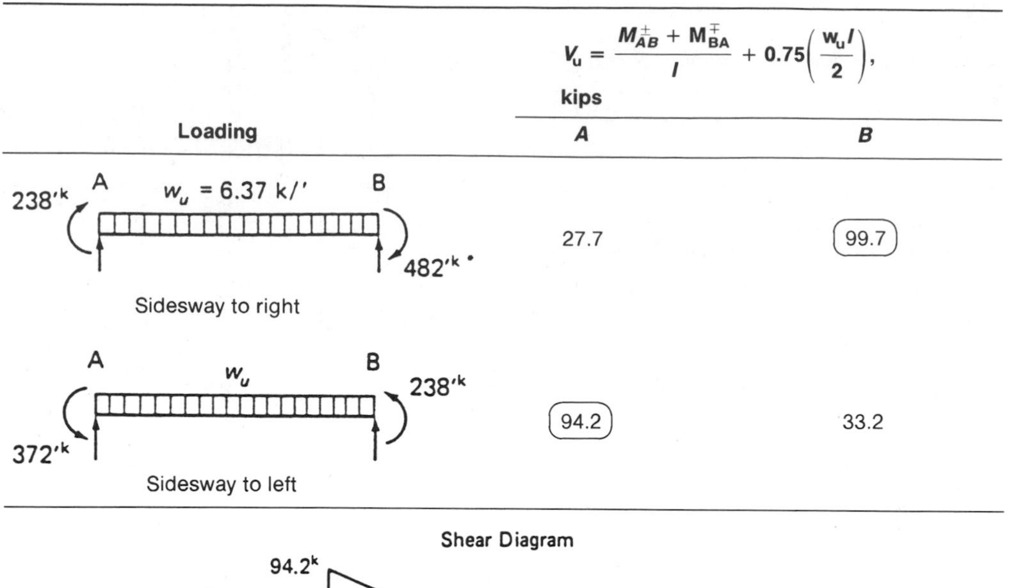

Loading	$V_u = \frac{M_{AB}^{\pm} + M_{BA}^{\mp}}{l} + 0.75\left(\frac{w_u l}{2}\right)$, kips A	B
Sidesway to right	27.7	(99.7)
Sidesway to left	(94.2)	33.2

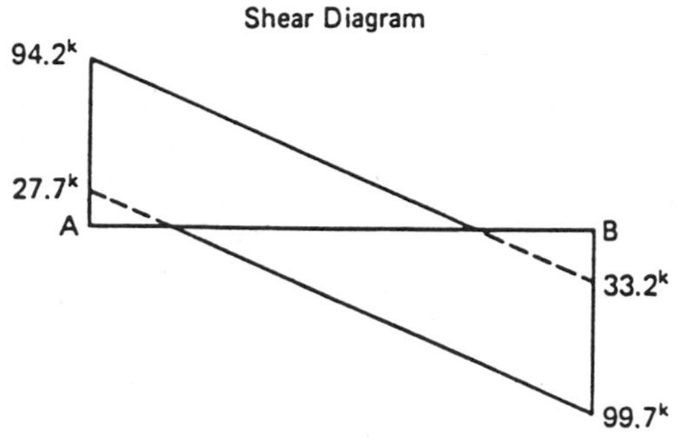

[a]When compression reinforcement is considered, a value about 0.5% greater than the above value is obtained.

we have

$$\phi V_s = V_u - \phi V_c = 99.7 - 0.85 \times 54.4 \quad \boxed{11.3.1.1}$$
$$= 53.5 \text{ kips}$$
$$V_s = 62.9 \text{ kips}$$

Required spacing of No. 3 closed stirrups (hoops), since A_v(2 legs) = 0.22 in.^{2}:

$$s = \frac{A_v f_y d}{V_s} = \frac{(0.22)(60)(21.5)}{62.9} \quad \boxed{11.5.6.2}$$
$$= 4.5 \text{ in.}$$

Maximum allowable hoop spacing within distance $2d = 2(21.5) = 43$ in. from faces of supports:

$$s_{max} = \begin{cases} d/4 = 21.5/4 = 5.4 \text{ in.} \\ 8 \times (\text{dia. of smallest long bar}) \\ \quad = 8(0.875) = 7 \text{ in.} \\ 24 \times (\text{dia. of hoop bars}) = 24(0.375) = 9 \text{ in.} \\ 12 \text{ in.} \end{cases}$$

$\boxed{A.3.3.2}$

Beyond distance $2d$ from the supports, maximum spacing of stirrups:

$$s_{max} = d/2 = 10.5 \text{ in.} \quad \boxed{A.3.3.4}$$

Use No. 3 stirrups spaced as shown in Figure 9-50.

Where the loading is such that inelastic deformation may occur at intermediate points within the span (e.g., due to concentrated loads near mid-span), the spacing of hoops will have to be determined in a manner similar to that used above for regions near supports. In the present example, the maximum positive moment near mid-span (see Table 9-4) is much less than the positive moment capacity provided by the three No. 7 continuous bars. $\boxed{A.3.3.1}$

(e) Negative-reinforcement cutoff points: For the purpose of determining cutoff points for the negative reinforcement, a moment diagram corresponding to plastic end moments and 0.9 times the dead load will be used. The cutoff point for two of the four No. 9 bars at the top, near support B of beam AB, will be determined.

With the negative moment capacity of a section with two No. 9 top bars equal to 186 ft-kips (calculated using $f_s = 60$ ksi and $\phi = 0.9$), the distance

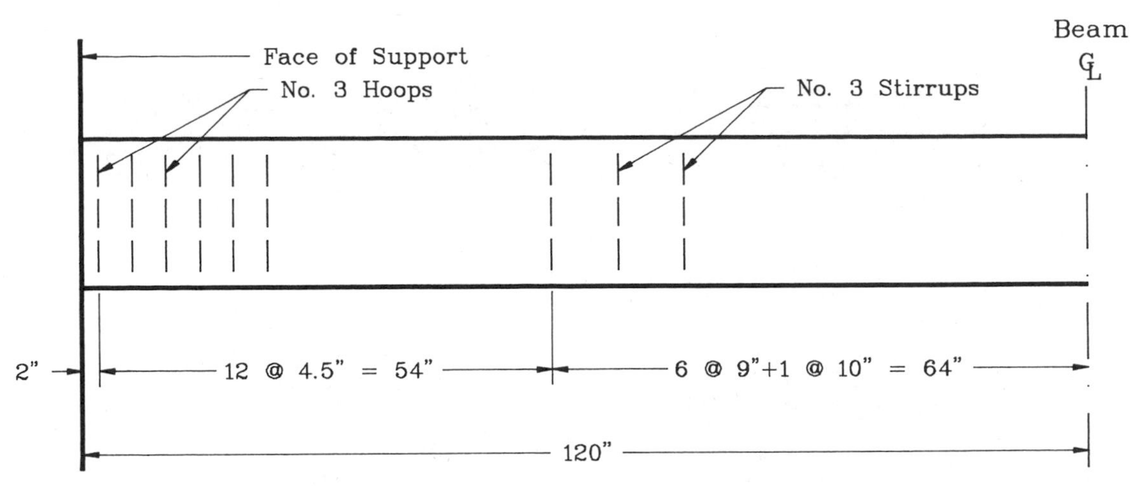

Figure 9-50 Spacing of hoops and stirrups in beam *AB*.

from the face of the right support B to where the moment under the loading considered equals 186 ft-kips is readily obtained by summing moments about section a–a in Figure 9-51 and equating these to -186 ft-kips. Thus,

$$69.4x - 482 - 3.34\frac{x^2}{2} = -186$$

Solution of the above equation gives $x = 4.82$ ft. Hence, two of the four No. 9 bars near support B may be cut off (noting that $d = 21.5$ in. $> 12d_b = 12 \times 1.128 = 13.5$ in.) at [12.10.3]

$$x + d = 4.82 + \frac{21.5}{12} = 6.6 \text{ ft}$$

from the face of the right support B. With l_{dh} for a No. 9 top bar equal to 16.5 in., the required development length for such a bar with respect to the tensile force associated with the negative moment at support B is $l_d = 3.5l_{dh} = 3.5 \times 16.5/12 = 4.8$ ft

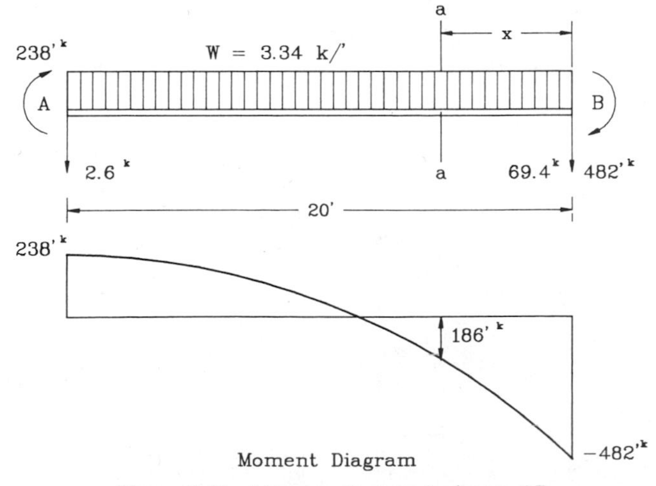

Figure 9-51 Moment diagram for beam *AB*.

< 6.6 ft. Thus, *the two No. 9 bars can be cut off 6.6 ft from the face of the interior support B.* [A.6.4.2] At end A, one of the three No. 9 bars can also be cut off at a similarly computed distance from the (inner) face of the exterior support A.

(f) Flexural reinforcement splices: Lap splices of flexural reinforcement should not be placed within a joint, within a distance $2d$ from faces of supports, or at locations of potential plastic hinging. Note that all lap splices have to be confined by hoops or spirals with a maximum spacing or pitch of $d/4$, or 4 in., over the length of the lap. [A.3.2.3]

(1) Bottom bars, No. 7: The bottom bars along most of the length of the beam may be subjected to maximum stress. Use class C splice. Required length of splice $= 1.7l_d \geqslant 12$ in., where [12.16.1]

$$l_d = 0.04A_b f_y/\sqrt{f_c'}$$

$$\geqslant 0.0004d_b f_y = \frac{(0.04)(0.60)(60{,}000)}{\sqrt{4000}}$$

$$\geqslant (0.0004)(0.875)(60{,}000)$$

$$= 22.8 \text{ in. (governs)} > 21 \text{ in.}$$ [12.2.2]

Class C splice length $= (1.7)(22.8) = 39$ in.

(2) Top bars, No. 9: Since the mid-span portion of the span is always subject to a positive bending moment (see Table 9-4), splices in the top bars should be located at or near midspan. Required length of class A splice $= 1.4l_d \geqslant 12$ in., where, by using the same expression given above, one obtains $l_d = 38$ in. Required spice length $= (1.4)(38) = 53.$ in.

(g) Detail of beam. See Figure 9-52.

3. *Design of frame column A.* The aim here is to design the transverse reinforcement for the exterior tied

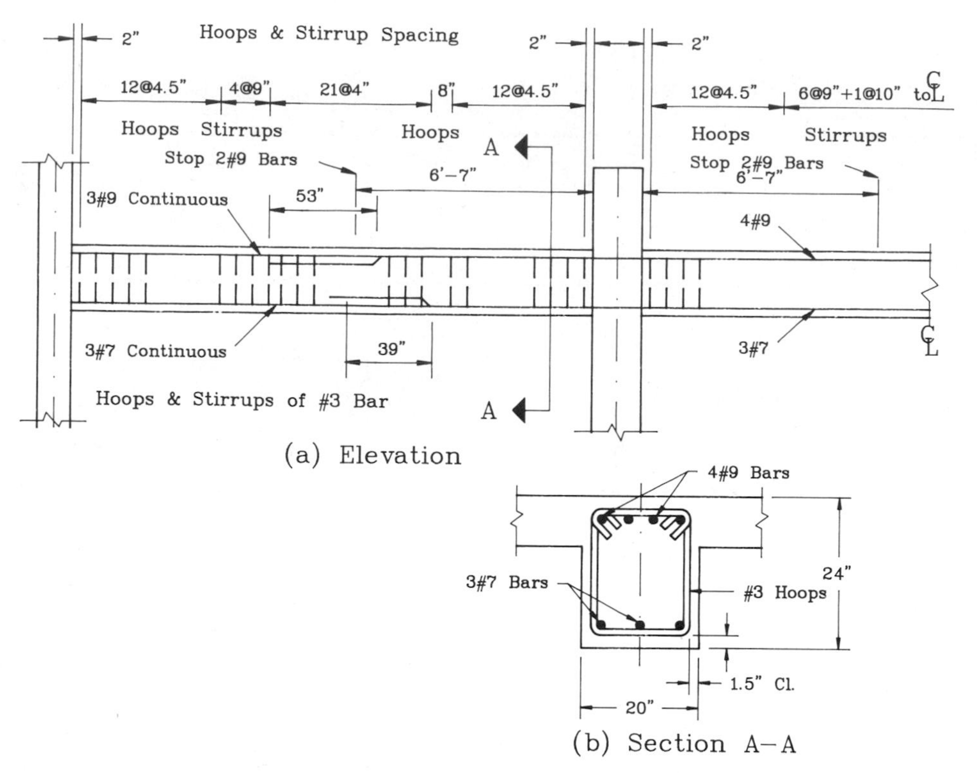

Figure 9-52 Detail of reinforcement for beam *AB*.

column on the second floor of a typical transverse interior frame, that is, one of the frames in frame T-1 of Figure 9-46. The column dimension has been established as 22 in. square and, on the basis of the different combinations of axial load and bending moment corresponding to the three loading conditions listed in Table 9-5, *eight No. 7 bars arranged in a symmetrical pattern* have been found adequate.[9-77, 9-78] Assume the same beam section framing into the column as considered in the preceding section. $f_c' = 4000$ lb/in.^{2} and $f_y = 60{,}000$ lb/in.^{2}.

From Table 9-5, $P_u(\max) = 930$ kips:

$$P_u(\max) = 930 \text{ kips} > \frac{A_g f_c'}{10} = \frac{(22)^2(4)}{10} = 194 \text{ kips} \qquad \boxed{\text{A.4.1}}$$

Thus, ACI Appendix A provisions governing members subjected to bending and axial load apply.

(a) Check satisfaction of vertical reinforcement limitations and moment capacity requirements:

(1) Reinforcement ratio:

$$0.01 \leqslant \rho \leqslant 0.06$$

$$\rho = \frac{A_{st}}{A_g} = \frac{8(0.60)}{(22)(22)} = 0.01 \qquad \text{O.K.} \qquad \boxed{\text{A.4.3.1}}$$

(2) Moment strength of columns relative to that of framing beam in transverse direction (see Figure 9-53):

$$\sum M_e(\text{columns}) \geqslant \tfrac{6}{5}\sum M_g(\text{beams}) \qquad \boxed{\text{A.4.2.2}}$$

From Section 9.5.2, item 2, ϕM_n^- of the beam at A is 272 ft-kips, corresponding to sidesway to the left. From Table 9-5, the maximum axial load on column A at the second floor level for sidesway to the left is $P_u = 870$ kips. Using the P–M interaction charts given in ACI SP-17A,[9-78] the moment capacity of the column section corresponding to $P_u = \phi P_n = 870$ kips, $f_c' = 4$ ksi, $f_y = 60$ ksi, $\gamma = 0.75$ (γ = ratio of distance between centroids of outer rows of bars to dimension of cross-section in the direction of bending, and $\rho = 0.01$ is obtained as $\phi M_n = M_e = 284$ ft-kips. With the same size

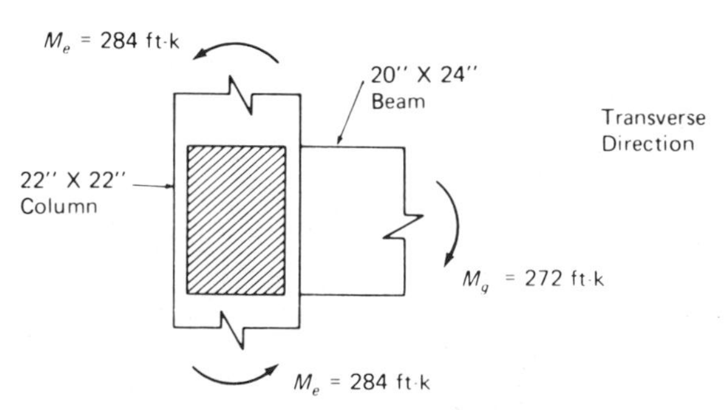

Figure 9-53 Relative flexural strengths of beam and columns at exterior joint — transverse direction.

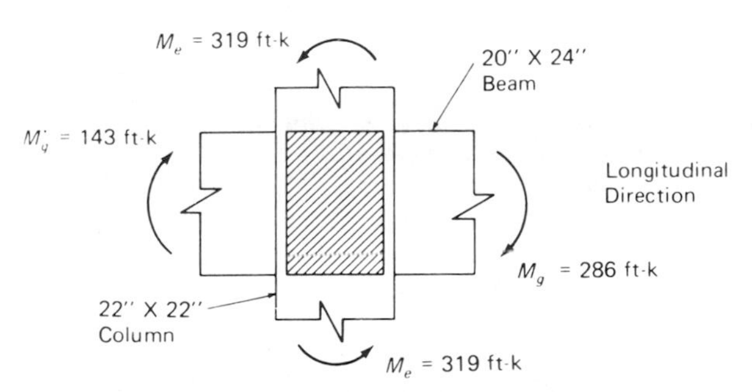

Figure 9-54 Relative flexural strengths of beams and columns at exterior joint — longitudinal direction.

column above and below the beam, total moment capacity of columns = 2(284) = 568 ft-kips. Thus,

$$\sum M_e = 568 > \tfrac{6}{5} M_g = \frac{(6)(272)}{5}$$
$$= 326 \text{ ft-kips} \qquad \text{O.K.}$$

(3) Moment strength of columns relative to that of framing beams in longitudinal direction (see Figure 9-54): Since the columns considered here are located in the center portion of the exterior longitudinal frames, the axial forces due to seismic loads in the longitudinal direction are negligible. (Analysis of the longitudinal frames under seismic loads indicated practically zero axial forces in the exterior columns of the four transverse frames represented by frame T-1 in Figure 9-46.) Under an axial load of (see Table 9-5)

$$0.75[1.4D + 1.7L + 1.87] \qquad (E = 0)$$
$$= (0.75)(930) = 698 \text{ kips},$$

the moment capacity of the column section with eight No. 7 bars is obtained as $\phi M_n = M_e = 319$ ft-kips. If we assume a ratio for the negative moment reinforcement of about 0.0075 in the beams of the exterior longitudinal frames (b_w = 20 in., d = 21.5 in.), then

$$A_s = \rho b_w d \approx (0.0075)(20)(21.5)$$
$$= 3.23 \text{ in.}^2$$

Assume four No. 8 bars, A_s = 3.16 in. Negative moment capacity of beam:

$$a = \frac{A_s f_y}{0.85 f_c' b_w} = \frac{(3.16)(60)}{(0.85)(4)(20)} = 2.79 \text{ in.}$$

$$\phi M_n^- = M_g^- = \phi A_s f_y \left(d - \frac{a}{2} \right)$$
$$= (0.90)(3.16)(60) \frac{21.5 - 1.39}{12}$$
$$= 286 \text{ ft-kips}$$

Assume a positive moment capacity of the beam on the opposite side of the column equal to one-half the negative moment capacity calculated above, or 143 ft-kips. Total moment capacity of beams framing into joint in longitudinal direction, for sidesway in either direction:

$$\sum M_g = 284 + 143 = 429 \text{ ft-kips}$$
$$\sum M_e = 2(319) = 638 \text{ ft-kips}$$
$$> \tfrac{6}{5} \sum M_g = \tfrac{6}{5}(429) = 515 \text{ ft-kips} \qquad \text{O.K.}$$

A.4.2.2

(b) Determine transverse reinforcement requirements:

(1) Confinement reinforcement (see Figure 9-37). Transverse reinforcement for confinement is required over a distance l_0 from column ends, where

$$l_0 \geqslant \begin{cases} \text{depth of member} = \underline{22 \text{ in.}} \text{ (governs)} \\ \tfrac{1}{6}(\text{clear height}) = \dfrac{10 \times 12}{6} = 20 \text{ in.} \\ 18 \text{ in.} \end{cases}$$

A.4.4.4

Maximum allowable spacing of rectangular hoops:

$$s_{max} = \begin{cases} \tfrac{1}{4}(\text{smallest dimension of column}) \\ \quad = \tfrac{22}{4} = 5.5 \text{ in.} \\ \underline{4 \text{ in.}} \text{ (governs)} \end{cases}$$

A.4.4.2

Required cross-sectional area of confinement reinforcement in the form of hoops:

$$A_{sh} \geqslant \begin{cases} 0.12 s h_c \dfrac{f_c'}{f_{yh}} \\ 0.3 s h_c \left(\dfrac{A_g}{A_{ch}} - 1 \right) \dfrac{f_c'}{f_{yh}} \end{cases}$$

A.4.4.1

where the terms are as defined for Equation 9-6. For a hoop spacing of 4 in., f_{yh} = 60,000 lb/in.^{2}, and tentatively assuming No. 4 bar hoops (for the purpose of estimating h_c and A_{ch}), the required cross-sectional area is

$$A_{sh} \geqslant \begin{cases} \dfrac{(0.12)(4)(18.4)(4000)}{60{,}000} \\ \quad = \underline{0.59 \text{ in.}^2} \text{ (governs)} \\ (0.3)(4)(18.4)\left(\dfrac{484}{357} - 1 \right) \dfrac{4000}{60{,}000} \\ \quad = 0.52 \text{ in.}^2 \end{cases}$$

A.4.4.3

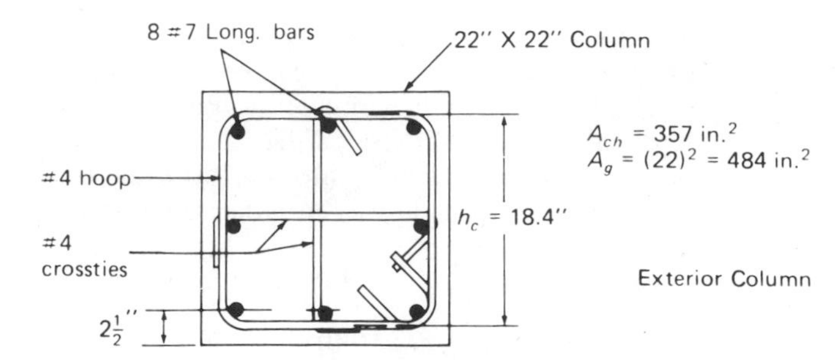

Figure 9-55 Detail of column transverse reinforcement.

No. 4 hoops with one crosstie, as shown in Figure 9-55, provide $A_{sh} = 3(0.20) = 0.60$ in.^{2}.

(2) Transverse reinforcement for shear: As in the design of shear reinforcement for beams, the design shear in columns is based not on the factored shear forces obtained from a lateral-load analysis, but rather on the nominal flexural strength provided in the columns. ACI Appendix A requires that the shear be determined from the largest nominal moment strength consistent with the estimated axial forces on the column. [A.7.1.2]

Assume that an axial force close to $\phi P_b = 484$ kips (corresponding to the "balanced point" on the interaction diagram for the column section considered—which would yield close to if not the largest moment strength) can occur (see Table 9-5). On this basis,

$$M_u(\text{at column ends}) = \phi M_b = 355 \text{ ft-kips}$$

from which (See Figure 9-41)

$$V_u = \frac{2M_u}{l} = \frac{2(355)}{10} = 71 \text{ kips}$$

Assume, for convenience,

$$V_c = 2\sqrt{f_c'}\,bd = \frac{2\sqrt{4000}\,(22)(19.5)}{1000} = 54 \text{ kips}$$

Required spacing of No. 4 hoops with $A_v = 2(0.20) = 0.40$ in.^{2} (neglecting crossties) and $V_s = (V_u - \phi V_c)/\phi = 29.5$ kips:

$$s = \frac{A_v f_y d}{V_s} = \frac{(2)(0.20)(60)(19.5)}{29.5} = 15.9 \text{ in.}$$

[11.5.6.2]

Thus, the transverse reinforcement spacing over the distance $l_0 = 22$ in. near the column ends is governed by the requirement for confinement rather than shear.

Maximum allowable spacing of shear reinforcement: $d/2 = 9.7$ in. [11.5.4.1]

Use No. 4 hoops and crossties spaced at 4 in. within a distance of 24 in. from the columns ends and No. 4 hoops spaced at 9 in. or less over the remainder of the column.

(c) Minimum length of lap splices for column vertical bars: [A.4.3.2]

ACI Appendix A limits the location of lap splices in column bars within the middle half of the member length, the splices to be designed as tension splices. [12.16.2]

Since generally all of the column bars will be spliced at the same location, a Class C splice will be required. [12.16.1]

The required length of splice is $1.7l_d$, where

$$l_d = \frac{0.04A_b f_y}{\sqrt{f_c'}}$$

$$\geqslant 0.0004 d_b f_y = \frac{(0.04)(0.60)(60{,}000)}{\sqrt{4000}}$$

$$\geqslant (0.0004)(0.875)(60{,}000)$$

$$= 23 \text{ in. (governs)} > 21 \text{ in.}$$

[12.2.2]

Thus, required splice length $= 1.7(23) = 39$ in.

Use 40-in. lap splices.

(d) Detail of column. See Figure 9-56.

4. *Design of exterior beam–column connection.* The aim is to determine the coefficient and shear-reinforcement requirements for the exterior beam–column connection between the beam considered in item 2 above and the column in item 3. Assume the joint to be located at the sixth floor level.

(a) Transverse reinforcement for confinement: ACI Appendix A requires the same amount of confinement reinforcement within the joint as for the length l_0 at column ends, unless the joint is confined by beams framing into all vertical faces of the column. In the latter case, only one-half the transverse reinforcement required for unconfined joints need be provided. [A.6.2.1] [A.6.2.2]

In the case of the beam–column joint considered here, beams frame into only three sides of the column, so that the joint is considered unconfined.

In item 4 above, confinement requirements at column ends were satisfied by No. 4 hoops with crossties, spaced at 4 in.

(b) Check shear strength of joint: The shear across section x–x (see Figure 9-57) of the joint is obtained as the difference between the tensile force from the top flexural reinforcement of the framing beam (stressed to $1.25f_y$) and the horizontal shear

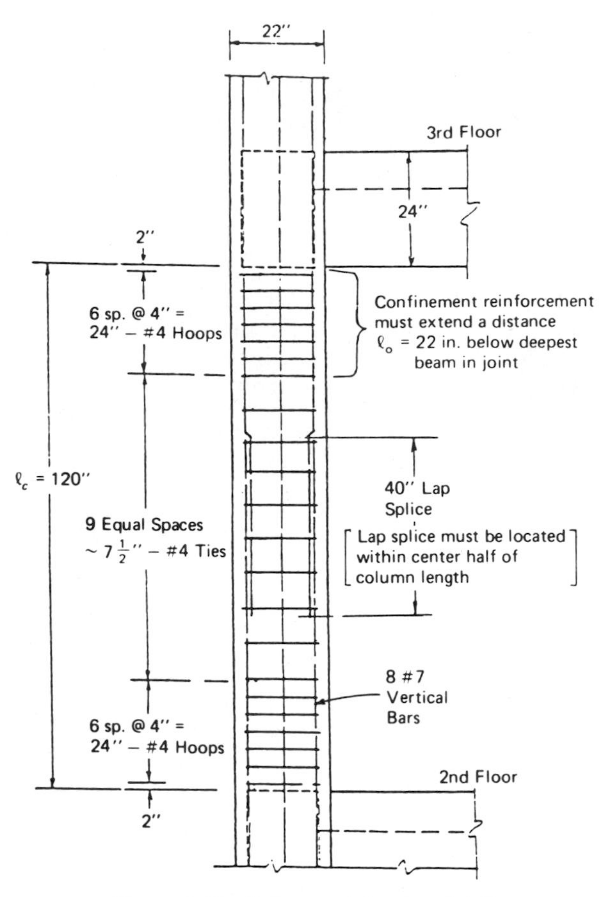

Figure 9-56 Column reinforcing details.

from the column above. The tensile force from the beam (three No. 9 bars), 3.0 in.^{2}) is

$$(3)(1.25)(60) = 225 \text{ kips}$$

An estimate of the horizontal shear from the column, V_h can be obtained by assuming that the beams in the adjoining floors are also deformed so that plastic hinges form at their junctions with the column, with M_p(beam) = 372 ft-kips (see Table

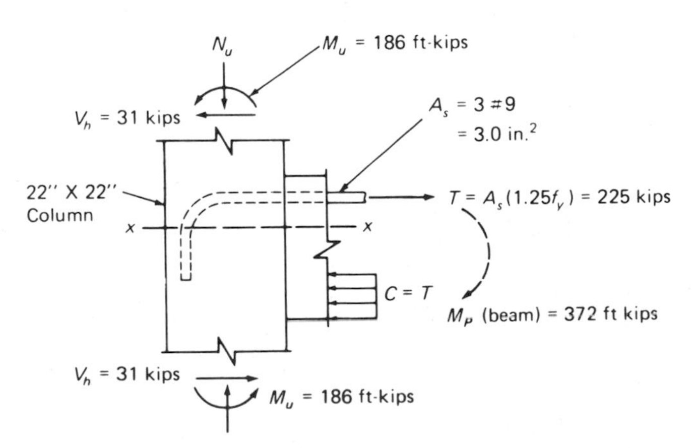

Figure 9-57 Horizontal shear in exterior beam-column joint.

9-7, for sidesway to left). By further assuming that the plastic moments in the beams are resisted equally by the columns above and below the joint, one obtains for the horizontal shear at the column ends

$$V_h = \frac{M_p(\text{beam})}{\text{story height}} = \frac{372}{12} = 31 \text{ kips}$$

Thus, the net shear at section x–x of joint is 225 − 31 = 194 kips. ACI Appendix A gives the nominal shear strength of a joint as a function only of the gross area of the joint cross-section A_j and the degree of confinement by framing beams. For the unconfined joint considered here,

$$\phi V_c = \phi 15\sqrt{f_c'}\,A_j$$

$$= \frac{(0.85)(15)(\sqrt{4000})(22)^2}{1000}$$

$$= 390 \text{ kips} > V_u = 194 \text{ kips} \qquad \text{O.K.}$$

A.2.3.1
A.6.3.1

Note that if the shear strength of the concrete in the joint as calculated above were inadequate, any adjustment would have to take the form (since transverse reinforcement above the minimum required for confinement is considered not to have a significant effect on shear strength) of either an increase in the column cross-section (and hence A_j) or an increase in the beam depth (to reduce the amount of flexural reinforcement required and hence the tensile force T).

(c) Detail of joint. See Figure 9-58. (The design should be checked for adequacy in the longitudinal direction.)

Note: The use of crossties within the joint may cause some placement difficulties. To relieve the congestion, No. 6 hoops spaced at 4 in. but without crossties may be considered as an alternative. Although the cross-sectional area of confinement reinforcement provided by No. 6 hoops at 4 in. (A_{sh} = 0.88 in.^{2}) exceeds the required amount (0.59 in.^{2}), the requirement of A.4.4.3 of ACI Appendix A relating to a maximum spacing of 14 in. between crossties or legs of overlapping hoops (see Figure 9-40) will not be satisfied. However, it is believed that this will not be a serious shortcoming in this case, since the joint is restrained by beams on three sides.

5. *Design of interior beam–column connection.* The objective is to determine the transverse reinforcement and shear requirements for the interior beam–column connection at the sixth floor of the interior transverse frame considered in previous examples. The column is 26 in. square and is reinforced with eight No. 11 bars.

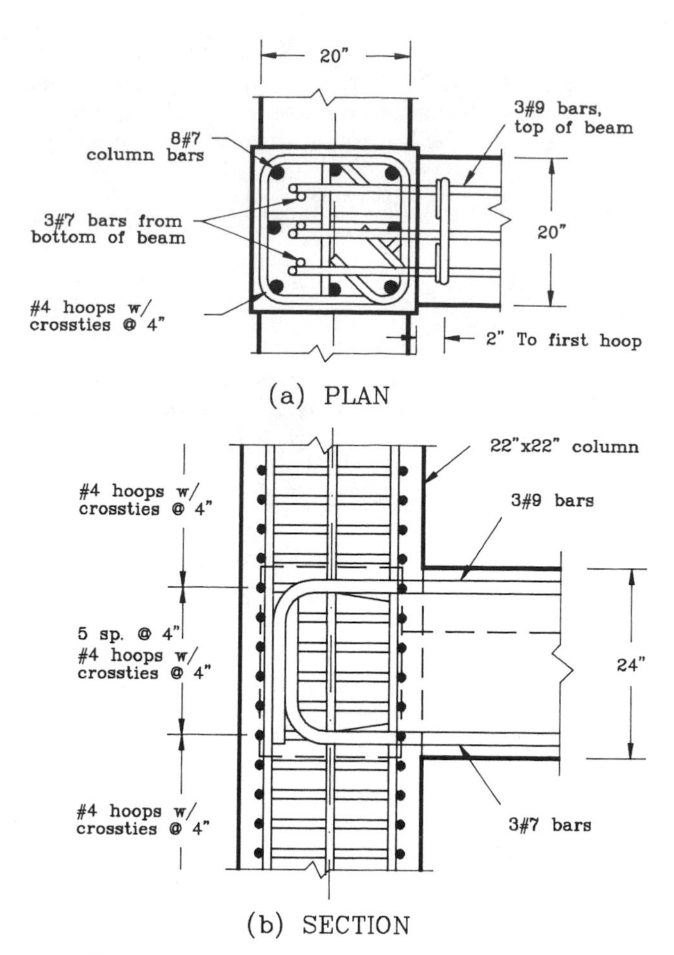

Figure 9-58 Detail of exterior beam-column joint.

The beams have dimensions $b = 20$ in. and $d = 21.5$ in. and are reinforced as noted in Section item 2 above (see Figure 9-52).

(a) Transverse reinforcement requirements (for confinement): Maximum allowable spacing of rectangular hoops,

$$s_{max} = \begin{cases} \frac{1}{4}(\text{smallest dimension of column}) \\ \quad = 26/4 = 6.5 \text{ in.} \\ \underline{4 \text{ in.}} \text{ (governs)} \end{cases}$$

A.4.4.2

For the column cross-section considered and assuming No. 4 hoops, $h_c = 21.9$ in., $A_{ch} = (22.4)^2 = 502$ in.^{2}, and $A_g = (26)^2 = 676$ in.^{2}. With a hoop spacing of 4 in., the required cross-sectional area of confinement reinforcement in the form of hoops is

$$A_{sh} \geqslant \begin{cases} 0.12sh_c\dfrac{f_c'}{f_{yh}} = \dfrac{(0.12)(4)(21.9)(4000)}{60{,}000} \\ \qquad = \underline{0.70 \text{ in.}^2} \quad \text{(governs)} \\ 0.3sh_c\left(\dfrac{A_g}{A_{ch}} - 1\right)\dfrac{f_c'}{f_{yh}} \\ \quad = (0.3)(4)(21.9)\left(\dfrac{676}{502} - 1\right)\dfrac{4000}{60{,}000} \\ \quad = 0.61 \text{ in.}^2 \end{cases}$$

A.4.4.1

Since the joint is framed by beams (having widths of 20 in., which is greater than $\frac{3}{4}$ of the width of the column) on all four sides, it is considered confined, and a 50% reduction in the amount of confinement reinforcement indicated above is allowed. Thus, A_{sh}(required) $\geqslant 0.35$ in.^{2}.

No. 4 hoops with crossties spaced at 4 in. o.c. provide $A_{sh} = 0.60$ in.^{2}. (See Note at end of item 4.)

(b) Check shear strength of joint: Following the same procedure used in item 4, the forces affecting the horizontal shear across a section near mid-depth of the joint shown in Figure 9-59 are obtained:

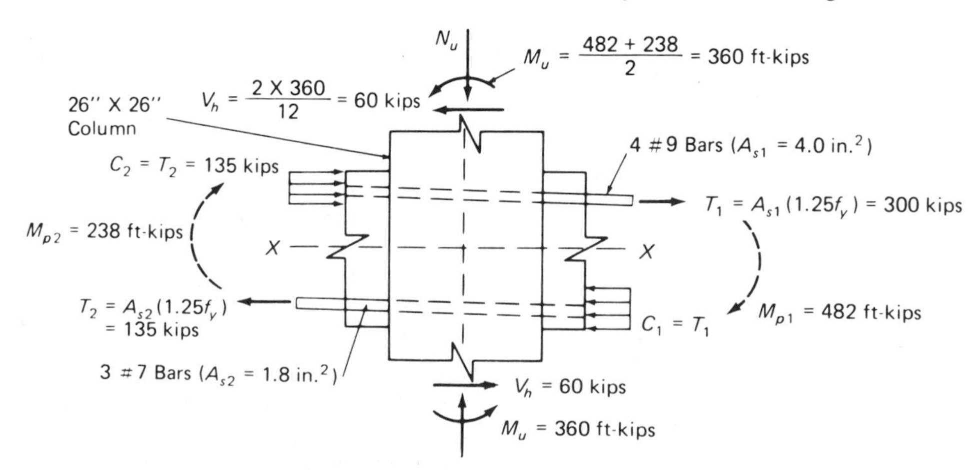

Figure 9-59 Forces acting on interior beam-column joint.

the joint shown in Figure 9-59 are obtained:

$$(\text{Net shear across section } x\text{–}x) = T_1 + C_2 - V_h = 300 + 135 - 60 = 375 \text{ kips} = V_u$$

Shear strength of joint, noting that joint is confined:

$$\phi V_c = \phi 20\sqrt{f_c'}\,A_j = \frac{(0.85)(20)\sqrt{4000}\,(26)^2}{1000} \quad \boxed{\text{A.6.3.1}}$$

$$= 726 \text{ kips} > V_u = 375 \text{ kips} \qquad \text{O.K.}$$

6. *Design of structural wall (shearwall).* The aim is to design the structural wall section at the first floor of one of the identical frame–shear wall systems shown in frame T-3 (see Figure 9-46). The preliminary design, as shown in Figure 9-45, is based on a 14-in.-thick wall with 26-in.-square vertical boundary elements, each of the latter being reinforced with eight No. 11 bars.

Preliminary calculations indicated that the cross-section of the structural wall at the lower floor levels needed to be increased. In the following, a 20-in.-thick wall section with 32 × 50-in. boundary elements reinforced with 24 No. 11 bars is investigated, and other reinforcement requirements determined.

The design forces on the structural wall at the first floor level are listed in Table 9-6. Note that because the axis of the shear wall coincides with the center line of the transverse frame of which it is a part, lateral loads do not induce any vertical (axial) force on the wall.

The calculation of the maximum axial force on the boundary element corresponding to Equation 9-2b in Table 9.6, $P_u = 4166$ kips, shown in Table 9-6, involved the following steps: At base of wall:

- Dead load, $D = 2823$ kips.
- Live load, $L = 412$ kips.
- Moment at base of wall due to seismic load (from lateral load analysis of transverse frames, including a moment of 1806 ft-kips due to accidental eccentricity), $M_b = 38{,}056$ ft-kips.

Referring to Figure 9-43, and noting the load factors used in Equation 9-2b of Table 9.6,

$$W = 0.75(1.4D + 1.7L) = (0.75)[(1.4)(2823) + (1.7)(412)] = 3489 \text{ kips}$$

$$Ha = 1.4M_b = (1.4)(38{,}056) = 53{,}278 \text{ ft-kips}$$

$$C_v = \frac{W}{2} + \frac{Ha}{d} = \frac{3489}{2} + \frac{53{,}278}{22} = 4166 \text{ kips}$$

(a) Check whether boundary elements are required: ACI Appendix A requires boundary elements to be provided if the maximum compressive extreme-fiber stress under factored forces exceeds $0.2f_c'$, unless the entire wall is reinforced to satisfy Sections A.4.4.1 through A.4.4.3 of the Appendix (relating to confinement reinforcement).

It will be assumed that the wall will not be provided with confinement reinforcement over its entire height. For a homogeneous rectangular wall 26.17 ft long (horizontally) and 20 in. (1.67 ft) thick,

$$I_{n.a.} = \frac{(1.67)(26.17)^3}{12} = 2494 \text{ ft}^4$$

$$A_g = (1.67)(26.17) = 43.7 \text{ ft}^2$$

Extreme-fiber compressive stress under $M_u = 53{,}280$ ft-kips and $P_u = 3449$ kips (see Table 9-6):

$$f_c = \frac{P_u}{A_g} + \frac{M_u h_w/2}{I_{n.a.}} = \frac{3449}{43.7}\,\frac{(53{,}280)(26.17)/2}{2494}$$

$$= 358.4 \text{ ksf} = 2.49 \text{ ksi} > 0.2f_c' = (0.2)(4) = 0.8 \text{ ksi}.$$

Therefore, *boundary elements are required*, subject to the confinement and special loading requirements specified in ACI Appendix A.

(b) Determine minimum longitudinal and transverse reinforcement requirements for wall:

(1) Check whether two curtains of reinforcement are required: ACI Appendix A requires that two curtains of reinforcement be provided in a wall if the in-plane factored shear force assigned to the wall exceeds $2A_{cv}\sqrt{f_c'}$, where A_{cv} is the cross-sectional area bounded by the web thickness and the length of section in the direction of the shear force considered. From Table 9-6, the maximum factored shear force on the wall at the first floor level is $V_u = 1153$ kips: $\boxed{\text{A.5.2.2}}$

$$2A_{cv}\sqrt{f_c'} = \frac{(2)(20)(26.17 \times 12)\sqrt{4000}}{1000} = 839 \text{ kips} < V_u = 1153 \text{ kips}$$

Therefore, *two curtains of reinforcement are required.*

(2) Required longitudinal and transverse reinforcement in wall: Minimum required reinforcement ratio,

$$\rho_v = \frac{A_{sv}}{A_{cv}} = \rho_n \geqslant 0.0025$$

(max. spacing = 18 in.) $\boxed{\text{A.5.2.1}}$

With $A_{cv} = (20)(12) = 240$ in.^{2}, (per foot of wall) the required area of reinforcement in each direction per foot of wall is $(0.0025)(240) = 0.60$ in.^{2}/ft. Required spacing of No. 5 bars [in two curtains, $A_s = 2(0.31) = 0.62$ in.^{2}]:

$$s(\text{required}) = \frac{2(0.31)}{0.60}(12) = 12.4 \text{ in.} < 18 \text{ in.}$$

(c) Determine reinforcement requirements for shear. [Refer to discussion of shear strength design for structural walls in Section 9.4.3, under "Code Provisions to Insure Ductility in Reinforced Concrete Members," item 5, paragraph (b).] To allow for increased shear reinforcement requirements above the minimum indicated above, assume two curtains of No. 6 bars spaced at 14 in. o.c. both ways. Shear strength of wall ($h_w/l_w = 148/26.17 = 5.66 > 2$):

$$\phi V_n = \phi A_{cv}\left(2\sqrt{f_c'} + \rho_n f_y\right) \quad \boxed{\text{A.7.3.2}}$$

where

$$\phi = 0.60$$

$$A_{cv} = (20)(26.17 \times 12) = 6281 \text{ in.}^2 \quad \boxed{\text{A.2.3.1}}$$

$$\rho_n = \frac{2(0.44)}{(20)(12)}\left(\frac{12}{14}\right) = 0.00314$$

Thus,

$$\phi V_n = \frac{(0.60)(6281)\left[2\sqrt{4000} + (0.00314)(60{,}000)\right]}{1000}$$

$$= \frac{3768.6[126.4 + 188.4]}{1000} = 1186 \text{ kips}$$

$$> V_u = 1153 \text{ kips} \quad \text{O.K.}$$

Therefore, *use two curtains of No. 6 bars spaced at 14 in o.c. in both horizontal and vertical directions.* $\boxed{\text{A.7.3.5}}$

(d) Check adequacy of boundary element acting as a short column under factored vertical forces due to gravity and lateral loads (see Figure 9-43): From Table 9-6, the maximum compressive axial load on boundary element is $P_u = 4166$ kips. $\boxed{\text{A.5.3.3}}$

With boundary elements having dimensions 32 in. × 50 in. and reinforced with 24 No. 11 bars,

$$A_g = (32)(50) = 1600 \text{ in.}^2$$

$$A_{st} = (24)(1.56) = 37.4 \text{ in.}^2$$

$$\rho_{st} = 37.4/1600 = 0.0234$$

Axial load capacity of boundary element acting as a short column:

$$\phi P_n(\max) = 0.80\phi\left[0.85 f_c'(A_g - A_{st}) + f_y A_{st}\right]$$

$$= (0.80)(0.70)[(0.85)(4)(1600 - 37.4) + (60)(37.4)]$$

$$= (0.56)[5313 + 2246] = 4233 \text{ kips}$$

$$> P_u = 4166 \text{ kips} \quad \text{O.K.} \quad \boxed{\text{9.3.5.2}}$$

(e) Check adequacy of structural wall section at base under combined axial load and bending in the plane of the wall: From Table 9-6, the following combinations of factored axial load and bending moment at the base of the wall are listed, corresponding to Eqs. 9-2a, b and c:

9-2a: $P_u = 4598$ kips, M_u small

9-2b: $P_u = 3449$ kips, $M_u = 53{,}280$ ft-kips

9-2c: $P_u = 2506$ kips, $M_u = 54{,}464$ ft-kips

Figure 9-60 shows the ϕP_n–ϕM_n interaction diagram (obtained using a computer program for generating P–M diagrams) for a structural wall section having a 20-in.-thick web reinforced with two curtains of reinforcement each having No. 6 horizontal and vertical bars spaced at 14 in o.c. and 32 × 50-in. boundary elements reinforced with 24 No. 11 vertical bars, with $f_c' = 4000$ lb/in.^{2}, $f_y = 60{,}000$ lb/in.^{2} (see Figure 9-61). The design load combinations listed above are shown plotted in Figure 9-60. The point marked a represents the P–M combination corresponding to Equation 9-2a, with similar notation used for the other two load combinations.

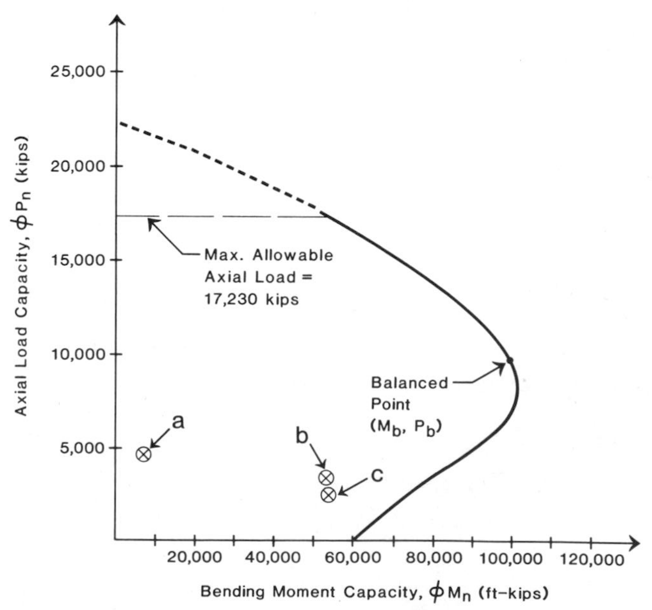

Figure 9-60 Axial load-moment interaction diagram for structural wall section.

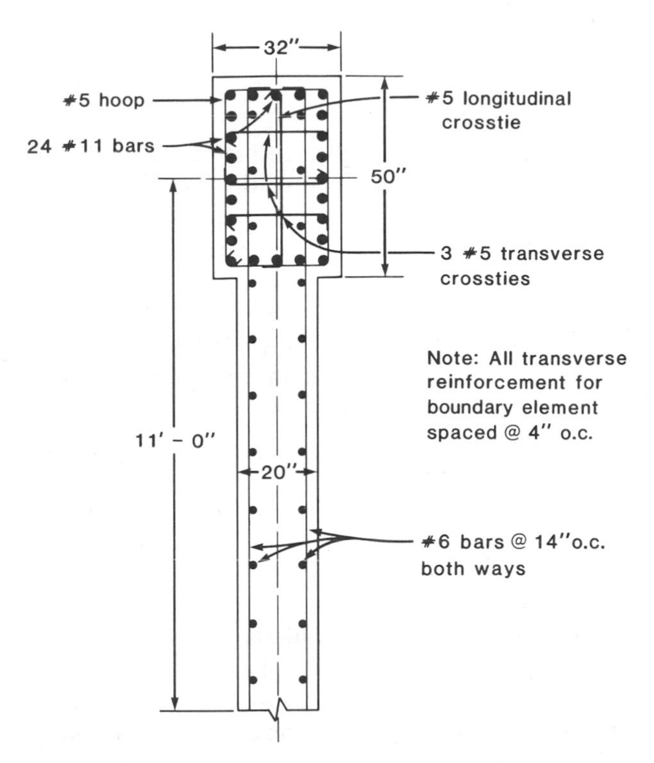

Figure 9-61 Half section of structural wall at base.

It is seen in Figure 9-60 that the three design loadings represent points inside the interaction diagram for the structural wall section considered. Therefore, *the section is adequate with respect to combined bending and axial load.*

Incidentally, the "balanced point" in Figure 9-60 corresponds to a condition where the compressive strain in the extreme concrete fiber is equal to $\epsilon_{cu} = 0.003$ and the tensile strain in the row of vertical bars in the boundary element farthest from the neutral axis (see Figure 9-61) is equal to the initial yield strain, $\epsilon_y = 0.00207$.

(f) Determine lateral (confinement) reinforcement required for boundary elements (see Figure 9-61): The maximum spacing is

$$s_{max} = \begin{cases} \frac{1}{4}(\text{smallest dimension of boundry element}) = \frac{32}{4} = 8 \text{ in.} \\ \underline{4 \text{ in.}} \text{ (governs)} \end{cases}$$ A.5.3.2, A.4.4.2

(1) Required cross-sectional area of confinement reinforcement in short direction:

$$A_{sh} \geqslant \begin{cases} 0.12 s h_c \dfrac{f'_c}{f_{yh}} \\ 0.3 s h_c \left(\dfrac{A_g}{A_{ch}} - 1\right) \dfrac{f'_c}{f_{yh}} \end{cases}$$ A.4.4.1

Assuming No. 5 hoops and crossties spaced at 4 in. o.c. and a distance of 3 in. from the center line of the No. 11 vertical bars to the face of the column, we have

$$h_c = 44 + 1.41 + 0.625 = 46.04 \text{ in.}$$

(for short direction),

$$A_{ch} = (46.04 + 0.625)(26 + 1.41 + 1.25) = 1337 \text{ in.}^2$$

$$A_{sh} \geqslant \begin{cases} (0.12)(4)(46.04)(4/60) = \underline{1.47 \text{ in.}^2} \text{ (governs)} \\ (0.3)(4)(46.04)\left(\dfrac{(32)(50)}{1337} - 1\right)\left(\dfrac{4}{60}\right) = 0.72 \text{ in.}^2 \end{cases}$$

(required in short direction).
With three crossties (five legs, including outside hoops),

$$A_{sh}(\text{provided}) = 5(0.31) = 1.55 \text{ in.}^2 \qquad \text{O.K.}$$

(2) Required cross-sectional area of confinement reinforcement in long direction:

$$h_c = 26 + 1.41 + 0.625 = 28.04 \text{ in.}$$

(for long direction),

$$A_{ch} = 1337 \text{ in.}^2$$

$$A_{sh} \geqslant \begin{cases} (0.12)(4)(28.04)(4/60) = \underline{0.90 \text{ in.}^2} \text{ (governs)} \\ (0.3)(4)(28.04)(1.196 - 1)(4/60) = 0.44 \text{ in.}^2 \end{cases}$$

(required in long direction).
With one crosstie (i.e., three legs, including outside hoop,

$$A_{sh}(\text{provided}) = 0.93 \text{ in.}^2 \qquad \text{O.K.}$$

(g) Determine required development and splice lengths: ACI Appendix A requires that all continuous reinforcement in structural walls be anchored or spliced in accordance with the provisions for reinforcement in tension as given in the Appendix. A.5.2.4

(1) Lap splice for No. 11 vertical bars in boundary elements (the use of mechanical connectors may be considered as an alternative to lap splices for these large bars): Assuming that 50% or less of

the vertical bars are spliced at any one location, a class B splice may be used. 12.16.2
Required length of splice = $1.3l_d$, where 12.16.1

$$l_d \geqslant 2.5 \times \begin{cases} f_y d_b / 65\sqrt{f_c'} = \dfrac{(60{,}000)(1.41)}{65\sqrt{4000}} \\ \qquad = \underline{21 \text{ in.}} \text{ (governs)} \\ 8d_b = (8)(1.41) = 12 \text{ in.} \\ 6 \text{ in.} \end{cases}$$

A.6.4.2

Thus the required splice length is (1.3)(2.5)(21) = 68 in.

(2) Lap splice for No. 6 vertical bars in wall "web": Again assuming no more than 50% of bars spliced at any one level, so that a class B splice may be used, and using the same expression for l_d as above, $l_d = 11$ in. Hence, the required length of splice is (1.3)(2.5)(11) = 36 in.

(3) Development length for No. 6 horizontal bars in wall, assuming no hooks are used within the boundary element: Since it is reasonable to assume that the depth of concrete cast in one lift beneath a horizontal bar will be greater than 12 in., the required factor of 3.5 to be applied to the development length, l_{dh}, required for a 90°-hooked bar will be used [Section 9.4.3, under "Code Provisions Designed to Insure Ductility in Reinforced-Concrete Members," item 2, paragraph (f)]: A.6.4.2

$$l_d = 3.5 l_{dh}$$

$$\geqslant 3.5 \times \begin{cases} \dfrac{f_y d_b}{65\sqrt{f_c'}} = \dfrac{(60{,}000)(7.5)}{65\sqrt{4000}} \\ \qquad = \underline{11 \text{ in.}} \text{ (governs)} \\ 8d_b = (8)\ (0.75) = 6 \text{ in.} \\ 6 \text{ in.} \end{cases}$$

Thus, the required development length is l_d = 3.5(11) = 39 in. This length can be accommodated within the confined core of the boundary element, so that no hooks are needed, as assumed. However, because of the likelihood of large horizontal cracks developing in the boundary elements, particularly in the potential hinging region near the base of the wall, the horizontal bars will be provided with 90° hooks engaging a vertical bar, as recommended in the Commentary to ACI Appendix A and as shown in Figure 9-61.

Required lap splice length for No. 6 horizontal bars, assuming 50% or less of bars are spliced at any one location: $1.3l_d$ = (1.3)(39) = 51 in.

(h) Detail of structural wall: See Figure 9-61. It will be noted there that the No. 6 vertical-wall "web" reinforcement, required for shear resistance, has been carried into the boundary element. The Commentary to ACI Appendix A specifically states that the concentrated reinforcement provided at wall edges for bending shall not be included in determining shear-reinforcement requirements. The area of vertical shear reinforcement located within the boundary element could, if desired, be considered as contributing to the axial load and bending capacity.

EPILOGUE TO CHAPTER 9

Since submission of the manuscript for this chapter, material representing deliberations of ACI Committee 318—Standard Building Code, relating to proposed revisions to the current ACI 318-83 Appendix A, has become available. These revisions, if finalized and approved, will form part of the next edition of the ACI Building Code Requirements for Reinforced Concrete, which is scheduled for publication in 1989. Incidentally, the Special Provisions for Seismic Design, which form Appendix A in ACI 318-83, will become Chapter 21 of ACI 318-89.

The more significant changes to ACI 318-83 Appendix A that are being considered relate to the following:

1. The 10-diameter extension for a 135° hook in crossties (Section A.1) has been reduced to 6 diameters, with a minimum of 3 in. An explicit requirement to alternate the 90° hooks end for end in successive crossties engaging the same longitudinal bars has also been added.

2. In the design of shear reinforcement for members subjected to combined bending and axial load (columns—Section 7.1.2), the design shear force is to be based on the probable moment strength M_{pr} (using $f_s = 1.25f_y$ and $\phi = 1.0$) of the column end sections corresponding to the factored axial compressive force (within the probable range of such force) which will result in the largest moment strength. However, the design shear need not exceed that determined on the basis of the probable moment strengths of beams framing into the joints at the ends of the column, nor should it be less than the factored shear determined from analysis of the structure.

At a typical column-end joint with a beam framing on each side, the upper limit based on the probable beam moment strength is obtained by summing the positive moment strength of the beam on one side and the negative moment strength of the beam on the other

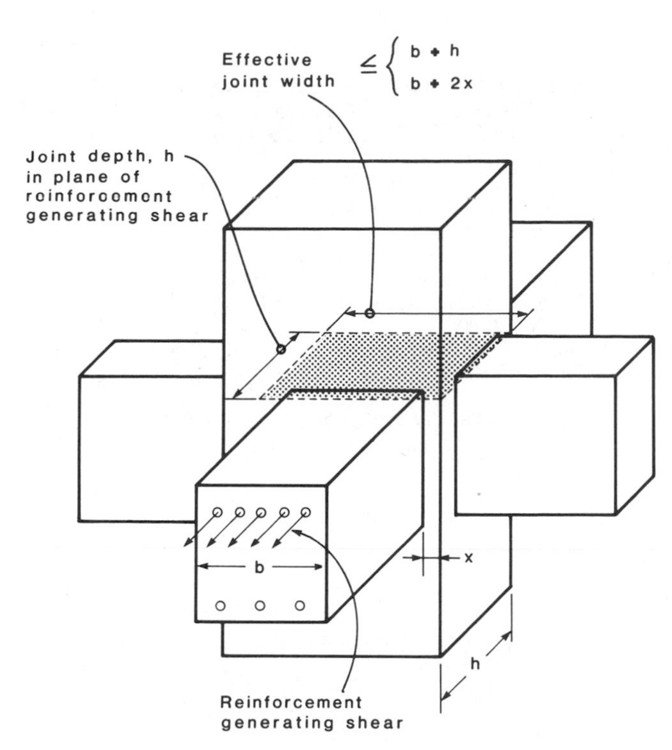

Figure 9-62 Beam-column panel zone.

side and distributing the sum to the column segments above and below the joint in accordance with linear-frame-analysis principles.

The current ACI Appendix A provision relating to columns requires the design of shear reinforcement to be based only on the nominal ($f_s = 1.0 f_y$ and $\phi = 1.0$) moment strength of a column. The proposed revisions are more in accord with SEAOC-86 (and UBC-88).

3. A clarification of the definition of the effective shear area with a joint, A_j, used in calculating the shear strength of beam–column joints, is provided in the proposed revision. As illustrated in Figure 9-62, the effective area is the product of the joint depth and the effective width of the joint. The joint depth is taken as the overall depth of the column (in the direction of the shear considered), while the effective width of the joint is to be taken equal to the width of the column if the beam and column are of the same width, or, where the column is wider than the framing beam, shall not exceed the smaller of:

- beam width plus joint depth, and
- beam width plus twice the least column projection beyond the beam side (the distance x in Figure 9-62)

Here again, these changes are more in conformance with SEAOC-86 (UBC-88).

4. For beam–column joints that may be considered confined by virtue of beams framing on all four sides (Section A.6.2.2), the spacing of transverse reinforcement within the joint is not to exceed either (i) one-quarter of the minimum member dimension, or (ii) 6 in. (instead of 4 in. in the current code—Section 4.4.2).

5. The factor 0.12 appearing in the alternate expression (Equation A-4) for determining the required cross-sectional area of rectangular-hoop (confinement) reinforcement in columns has been reduced to 0.09 on the basis of the satisfactory performance of columns reinforced with the lesser amount of properly detailed hoops and crossties. The revised expression now reads

$$A_{sh} = 0.09 s h_c f_c' / f_{yh}.$$

6. A statement specifically prohibiting the welding of stirrups, ties, inserts, or other similar elements to longitudinal reinforcement required by design load combinations including earthquake effects has been added to Section A.2.5. This change also conforms to requirements found in SEAOC-86 (UBC-88).

REFERENCES

The following abbreviations will be used to denote commonly occurring reference sources:

- Organizations and conferences:

EERI	Earthquake Engineering Research Institute
WCEE	World Conference on Earthquake Engineering
ASCE	American Society of Civil Engineers
ACI	American Concrete Institute
PCA	Portland Cement Association
PCI	Prestressed Concrete Institute

- Publications:

JEMD	*Journal of Engineering Mechanics Division, ASCE*
JSTR	*Journal of the Structural Division, ASCE*
JACI	*Journal of the American Concrete Institute*

9-1 International Conference of Building Officials, 5360 South Workman Mill Road, Whittier, CA 90601, *Uniform Building Code*. The latest edition of the Code is the 1985 Edition.

9-2 Clough, R. W. and Benuska, K. L., "FHA Study of Seismic Design Criteria for High-Rise Buildings," Report HUD TS-3, Federal Housing Administration, Washington, Aug. 1966.

9-3 Derecho, A. T., Ghosh, S. K., Iqbal, M., Freskakis, G. N., and Fintel, M., "Structural Walls in Earthquake-Resistant Buildings, Dynamic Analysis of Isolated Structural Walls—Parametric Studies," Report to the National Science Foundation, RANN, Construction Technology Laboratories, PCA, Skokie, IL, Mar. 1978.

9-4 Derecho, A. T., Iqbal, M., Ghosh, S. K., Fintel, M., Corley, W. G., and Scanlon, A., "Structural Walls in Earthquake-Resistant Buildings, Dynamic Analysis of Isolated Structural Walls—Development of Design Procedure, Design Force Levels," Final Report to the National Science Foundation, ASRA, Construction Technology Laboratories, PCA, Skokie, IL, July 1981.

9-5 Park, R. and Paulay, T., *Reinforced Concrete Structures*, John Wiley & Sons, New York, 1975.

9-6 Paulay, T., "Earthquake-Resisting Walls—New Zealand Design Trends," *JACI*, 144–152, May–June 1980.

9-7 Derecho, A. T., Iqbal, M., Fintel, M., and Corley, W. G., "Loading History for Use in Quasi-static Simulated Loading Test," *Reinforced Concrete Structures Subjected to Wind and Earthquake Forces*, ACI Special Publication SP-63, 329–344 1980.

9-8 Oesterle, R. G., Aristizabal-Ochoa, J. D., Fiorato, A. E., Russell, H. G., and Corley, W. G., "Earthquake-Resistant Structural Walls—Tests of Isolated Walls—Phase II," Report to the National Science Foundation, ASRA, Construction Technology Laboratories, PCA, Skokie, IL, Oct. 1979.

9-9 American Concrete Institute, Detroit, Michigan, "Building Code Requirements for Reinforced Concrete—ACI 318-83." The latest edition of the code is the 1983 Edition.

9-10 Iyengar, K. T. S. R., Desayi, P., and Reddy, K. N., "Stress–Strain Characteristics of Concrete Confined in Steel Binders," *Mag. Concrete Res.* 22, No. 72, Sept. 1970.

9-11 Sargin, M., Ghosh, S. K., and Handa, V. K., "Effects of Lateral Reinforcement upon the Strength and Deformation Properties of Concrete," *Mag. Concrete Res.* 75–76, June–Sept. 1971.

9-12 Sturman, G. M., Shah, S. P., and Winter, G., "Effects of Flexural Strain Gradients on Microcracking and Stress–Strain Behavior of Concrete," Title No. 62-50, JACI, July 1965.

9-13 Clark, L. E., Gerstle, K. H., and Tulin, L. G., "Effect of Strain Gradient on the Stress–Strain Curve of Mortar and Concrete," Title No. 64-50, JACI, Sept. 1967.

9-14 Mattock, A. H., "Rotational Capacity of Hinging Regions in Reinforced Concrete Beams," *Proc. Intl. Symposium on Flexural Mechanics of Reinforced Concrete*, ASCE, 1965, 143–181, 1965. Also PCA Development Dept. Bulletin 101.

9-15 Corley, W. G., "Rotational Capacity of Reinforced Concrete Beams," *JSTR Proc.* 92 (ST5), 121–146, Oct. 1966. Also PCA Development Dept. Bulletin 108.

9-16 Naaman, A. E., Harajli, M. H., and Wight, J. K., "Analysis of Ductility in Partially Prestressed Concrete Flexural Members," *PCI J.*, 64–87, May–June 1986.

9-17 Bertero, V. V. and Fellippa, C., "Discussion of 'Ductility of Concrete,' by Roy, H. E. H. and Sozen, M. A.," *Proc. Intl. Symp. on Flexural Mechanics of Reinforced Concrete*, ASCE, 227–234, 1965.

9-18 Standard Association of New Zealand, *Code of Practice for General Structural Design and Design Loadings for Buildings—NZS 4203:1984*, Wellington, 1984.

9-19 Bertero, V. V., "Seismic Behavior of Structural Concrete Linear Elements (Beams and Columns) and Their Connections," *Proc. of the A.I.C.A.P.-C.E.B. Symposium on Structural Concrete under Seismic Actions*, Rome, I, 123–212, 1979.

9-20 Popov, E. P., Bertero, V. V., and Krawinkler, H., "Cyclic Behavior of Three Reinforced Concrete Flexural Members with High Shear," Report No. EERI 72-5, Univ. of California, Berkeley, Oct. 1972.

9-21 Brown, R. H. and Jirsa, J. O., "Shear Transfer of Reinforced Concrete Beams Under Reversed Loading," Paper No. 16, *Shear in Reinforced Concrete*, Vol. 1, ACI Publication SP-42, 347–357, 1974.

9-22 Bertero, V. V. and Popov, E. P., "Hysteretic Behavior of R. C. Flexural Members with Special Web Reinforcement," *Proc. U.S. National Conference on Earthquake Engineering—1975*, Ann Arbor, MI, 316–326, 1975.

9-23 Scribner, C. F. and Wight, J. K., "Delaying Shear Strength Decay in Reinforced Concrete Members under Large Load Reversals," Report UMEE 78R2, Dept. of Civil Engineering, Univ. of Michigan, Ann Arbor, 1978.

9-24 Standards Association of New Zealand, *Code of Practice for the Design of Concrete Structures, NZS 3101, Part 1:1982,* Wellington, 1982.

9-25 Seismology Committee, Structural Engineers Association of California (SEAOC), *Tentative Recommended Lateral Force Requirements*, Dec. 15, 1986.

9-26 Ehsani, M. R. and Wight, J. K., "Effect of Transverse Beams and Slab on Behavior of Reinforced Beam-to-Column Connections," *JACI* 82, No. 2, 188–195, Mar.-Apr. 1985.

9-27 Leon, R. and Jirsa, J. O., "Bidirectional Loading of R. C. Beam–Column Joints," *EERI Earthquake Spectra* 2, No. 3, 537–564, May 1986.

9-28 ACI–ASCE Committee 352, "Recommendations for Design of Beam–Column Joints in Monolithic Reinforced Concrete Structures," *ACI J. Proc.* 82, No. 3, 266–283, May–June 1985.

9-29 Paulay, T., "Deterministic Design Procedure for Ductile Frames in Seismic Areas," Paper No. 15, *Reinforced Concrete Structures Subjected to Wind and Earthquake Forces*, ACI Publication SP-63, 357–381, 1980.

9-30 Paulay, T., "Developments in Seismic Design of Reinforced Concrete Frames in New Zealand," *Can. J. Civil Eng.* 8, No. 2, 91–113, June 1981.

9-31 Park, R., "Ductile Design Approach for Reinforced Concrete Frames," *EERI Earthquake Spectra* 2, No. 3, 565–619, May 1986.

9-32 Wight, J. K. and Sozen, M. A., "Strength Decay of RC Columns under Shear Reversals," *JSTR* 101, No. ST5, 1053–1065, May 1975.

9-33 Sheikh, S. and Uzumeri, S. M., "Strength and Ductility of Tied Concrete Columns," *JSTR* 106, No. ST5, 1079–1102, May 1980.

9-34 Park, R., Priestley, M. J. N., and Gill, W. D., "Ductility of Square-Confined Concrete Columns," *JSTR* 108, No. ST4, 929–950, Apr. 1982.

9-35 Priestly, M. J. N. and Park, R., "Strength and Ductility of Concrete Bridge Columns under Seismic Loading," *ACI Structural J.*, 61–76, Jan.–Feb. 1987.

9-36 Jennings, P. C. (ed.), "Engineering Features of the San Fernando Earthquake, February 9, 1971," Earthquake Engineering Research Laboratory, California Institute of Technology, Pasadena, June 1971.

9-37 Paulay, T., Park, R., and Priestley, M. J. N., "Reinforced Concrete Beam–Column Joints under Seismic Actions," *JACI Proc.* 75, No. 11, 585–593, Nov. 1978.

9-38 Hanson, N. W. and Conner, H. W., "Seismic Resistance of Reinforced Concrete Beam–Column Joints," *JSTR* 93, ST5, 533–560, Oct. 1967.

9-39 Meinheit, D. F. and Jirsa, J. O., "Shear Strength of R/C Beam–Column Connections," *JSTR* 107, ST11, 2227–2244, Nov. 1982.

9-40 Abdel-Fattah, B. and Wight, J. K., "Study of Moving Beam Plastic Hinging Zones for Earthquake-Resistant Design of R/C Buildings," *ACI Structural J.*, 31–39, Jan.–Feb. 1987.

9-41 Rosenblueth, E. and Meli, R., "The 1985 Earthquake: Causes and Effects in Mexico City," *ACI Concrete Int.* 8, No. 5, 23–34, May 1986.

9-42 Mitchell, D., Adams, J., DaVall, R. H., Lo, R. C., and Weichert, D., "Lessons from the 1985 Mexican Earthquake," *Can. J. Civil Eng.* 13, No. 5, 535–557, 1986.

9-43 Carpenter, J. E., Kaar, P. H., and Corley W. G., "Design of Ductile Flat Plate Structures to Resist Earthquakes," *Proc. 5th WCEE*, Rome, 1973.

9-44 Symonds, D. W., Mitchell, D., and Hawkins, N. M., "Slab–Column Connections Subjected to High Intensity Shears and Transferring Reversed Moments" SM 76-2, Division of Structures and Mechanics, Univ. of Washington, Oct. 1976.

9-45 Cardenas, A. E., Russell, H. G., and Corley, W. G., "Strength of Low-Rise Structural Walls," Paper No. 10, *Reinforced Concrete Structures Subjected to Wind and Earthquake Forces*, ACI Publication SP-63, 221–241, 1980.

9-46 Barda, F., Hanson, J. M., and Corley, W. G., "Shear Strength of Low-Rise Walls with Boundary Elements," *Reinforced Concrete Structures in Seismic Zones*, ACI Publication SP-53, 149–202, 1977.

9-47 Oesterle, R. G., Fiorato, A. E., Johal, L. S., Carpenter, J. E., Russell, H. G., and Corley, W. G., "Earthquake-Resistant Structural Walls—Tests of Isolated Walls," Report to the National Science Foundation, Portland Cement Association, Nov. 1976.

9-48 Oesterle, R. G., Aristizabal-Ochoa, J. D., Fiorato, A. E., Russell, H. G., and Corley, W. G., "Earthquake Resistant Structural Walls—Tests of Isolated Walls—Phase II," Report to the National Science Foundation, Portland Cement Association, Oct. 1979.

9-49 Cardenas, A. and Magura, D. D., "Strength of High-Rise Shear Walls—Rectangular Cross Section," *Response of Multistory Concrete Structures to Lateral Forces*, ACI Publication SP-36, American Concrete Institute, 1973.

9-50 Corley, W. G., Fiorato, A. E., and Oesterle, R. G., "Structural Walls," Paper No. 4, *Significant Developments in Engineering Practice and Research*, Sozen, M. A. (ed.), ACI Publication SP-72, 77–130, 1981.

9-51 Oesterle, R. R., Fiorato, A. E., and Corley, W. G., "Reinforcement Details for Earthquake-Resistant Structural Walls," *ACI Concrete Int.*, 55–66, Dec. 1980.

9-52 Paulay, T., "The Design of Ductile Reinforced Concrete Structural Walls for Earthquake Resistance," *EERI Earthquake Spectra* 2, No. 4, 783–823, Oct. 1986.

9-53 Saatcioglu, M., Derecho, A. T., and Corley, W. G., "Dynamic Inelastic Response of Coupled Walls as Affected by Axial Forces," *Nonlinear Design of Concrete Structures*, Proc. of CSCE–ASCE–ACI–CEB International Symposium, Univ. of Waterloo, Ontario, 639–670, Aug. 1979.

9-54 Shiu, K. N., Takayanagi, T., and Corley, W. G., "Seismic Behavior of Coupled Wall Systems," *JSTR* 110, No. 5, May 1051–1066, 1984.

9-55 Saatcioglu, M., Derecho, A. T., and Corley, W. G., "Parametric Study of Earthquake-Resistant Couple Walls," *JSTR* 113, No. 1, 141–157, Jan. 1987.

9-56 Paulay, T. and Binney, J. R., "Diagonally Reinforced Coupling Beams of Shear Walls," Paper No. 26, *Shear in Reinforced Concrete*, ACI Publication SP-42, Vol. 2, 579–598, 1974.

9-57 Barney, G. B., Shiu, K. N., Rabbat, B., Fiorato, A. E., Russell, H. G., and Corley, W. G., "Behavior of Coupling Beams under Load Reversal," PCA Res. & Dev. Bulletin No. 68, 1980.

9-58 Seismology Committee, Structural Engineers Association of California, *Recommended Lateral Force Requirements and Commentary*, San Francisco, 1980.

9-59 Bathe, K-J., Wilson, E. L., and Peterson, F. E., "SAP IV—A Structural Analysis Program for Static and Dynamic Response of Linear Systems," Report No. EERC 73-11, June 1973 (revised Apr. 1974), Univ. of California, Berkeley.

9-60 "NASTRAN (NASA Structural Analysis Computer System)", Computer Software Management and Information System (COSMIC), Univ. of Georgia, Athens, Georgia.

9-61 "ANSYS—Engineering Analysis Systems," Swanson Analysis Systems, Inc., Houston, PA.

9-62 Wilson, E. L., Hollings, J. P., and Dovey, H. H., "ETABS—Extended Three-Dimensional Analysis of Building Systems," Univ. of California, Berkeley, Apr. 1975.

9-63 Guendelman, R. and Powell, G. H., "DRTABS (DRAIN-TABS)—Inelastic Earthquake Response of Three-Dimensional Buildings," Univ. of California, Berkeley, Oct. 1976.

9-64 *Computer Software for Earthquake Engineering*, National Information Service for Earthquake Engineering (NISEE), Earthquake Engineering Research Center, Univ. of California, Berkeley, July 1984.

9-65 Clough, R. W., "Dynamic Effects of Earthquakes," *Trans. ASCE* 126, Part II, Paper No. 3252, 1961.

9-66 Blume, J. A., "Structural Dynamics in Earthquake-Resistant Design," *Trans. ASCE* 125, Part I, Paper No. 3054, 1960.

9-67 Berg, G. V., "Response of Multistory Structures to Earthquakes," Paper No. 2790, JEMD, Apr. 1961.

9-68 American National Standard Institute, Inc., *Minimum Design Loads for Buildings and Other Structures* (ANSI A58.1-1982), New York. The latest edition is the 1982 Edition.

9-69 Building Seismic Safety Council, "NEHRP Recommended Provisions for the Development of Seismic Regulations for New Buildings" (NEHRP—National Earthquake Hazards Reduction Program), 1985 Edition, Part 1: Provisions, Part 2: Commentary, Washington, DC 20005, Feb. 1986.

9-70 Applied Technology Council, "Tentative Provisions for the Development of Seismic Regulations for Buildings," ATC Publication 3-06, U.S. Government Printing Office, Washington, 505 pp., 1978.

9-71 Associate Committee on the National Building Code, *National Building Code of Canada—1985*, National Research Council of Canada, Ottawa. The latest edition of the Code is the 1985 Edition.

9-72 Earthquake Engineering Research Institute, "Reducing Earthquake Hazards: Lessons Learned from Earthquakes," EERI Publication No. 86-02, Nov. 1986.

9-73 ACI Committee 315, "Details and Detailing of Concrete Reinforcement (ACI 315-80)," *JACI* 83, No. 3, 485–512, May–June 1986.

9-74 Fintel, M., "Ductile Shear Walls in Earthquake-Resistant Multistory Buildings," *JACI* 71, No. 6, 296–305, June 1974.

9-75 Derecho, A. T., Fintel, M., and Ghosh, S. K., "Earthquake-Resistant Structures," Chapter 12, *Handbook of Concrete Engineering*, 2nd Edition, M. Fintel (ed.), Van Nostrand Reinhold, 411–513, 1985.

9-76 ACI Committee 340, "Design Handbook in Accordance with the Strength Design Method of ACI 318-83: Vol. 1—Beams, One-Way Slabs, Brackets, Footings, and Pile Caps" (ACI 340.1R-84), Publication SP-17(84), American Concrete Institute, 1984.

9-77 Concrete Reinforcing Steel Institute, *CRSI* Handbook, Schaumburg, IL. The latest edition of the handbook is the 1984 Edition.

9-78 ACI Committee 340, *Design Handbook in Accordance with the Strength Design Method of ACI 318-77: Vol. 2—Columns,* Publication SP-17A (78), American Concrete Institute, Detroit, 1978.

Chapter 10

Seismic Design of Masonry Structures

*Gary C. Hart, Ph.D.**

10.1 INTRODUCTION

The seismic design of masonry structures has made significant advances in the last decade. Initially the lead was provided by New Zealand and Canadian structural engineers, and their contributions can be noted in the proceedings of the first three North American Masonry Conferences[10-1, 10-2, 10-3] plus the sixth and seventh Canadian Masonry Symposia.[10-4, 10-5] In the United States the work of the Masonry Society in the development of the 1985 Uniform Building Code[10-6] provided a point at which we can mark the change in attitude and direction of seismic masonry design. While notable earlier masonry research efforts by Hegemier[10-7] and Mayes[10-8] were directed at seismic design considerations, it was the development of the 1985 UBC code, the Structural Engineers Association of California (SEAOC) review of the proposed code, and finally the adaptation in the 1985 by International Conference of Building Officials that started the new direction for seismic design of masonry structures.

This new direction is most often called "strength design," after the same rename from concrete design. However, as the next section will indicate, we are really advancing towards "Limit-State Design." The development of this new seismic design approach from the design implementation perspective is documented by approval by the International Conference of Building Officials (ICBO) of three design standards. They are:

1. The strength-design criteria for slender walls in Section 2411 of the 1985 UBC.
2. The strength-design criteria for one- to four-story buildings in ICBO Evaluation Services, Inc., Evaluation Report Number 4115, first published in 1983.[10-9]
3. The strength-design criteria for shear walls in Section 2412 of the 1988 UBC.[10-10]

It is the purpose of this chapter to indicate how these new seismic criteria can be used by structural engineers.

10.2 BEHAVIOR AND LIMIT STATES

The behavior of a masonry component or system when subjected to loads can be described in terms of behavior and limit states. For illustrative purposes, we will use the slender wall shown in Figure 10-1.

As indicated in this figure, the slender wall can be idealized for structural design as evolving through several identifiable states of behavior prior to reaching its final deformed position. We can define this evolution in terms of *behavior states*. Table 10-1 defines the behavior states for the slender wall. For example, the first behavior state corresponds to the stress condition where the load-induced tensile stress is less than the modulus of

*Professor of Engineering, University of California at Los Angeles, and President, Englekirk and Hart Consulting Engineers, Los Angeles, California.

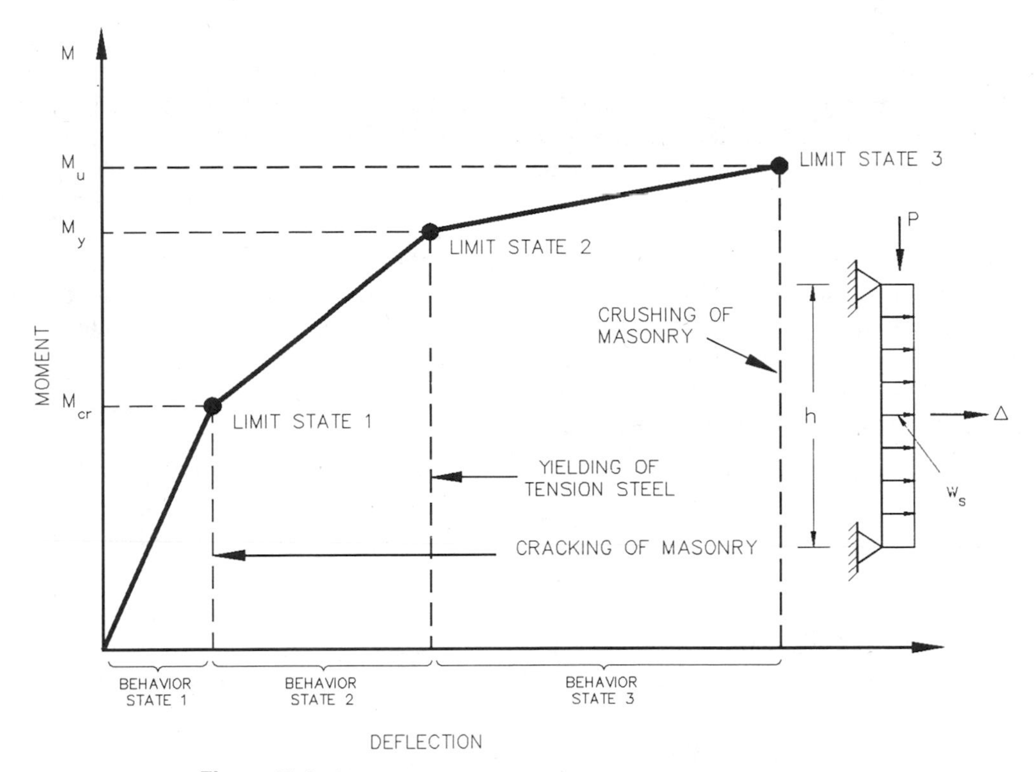

Figure 10-1 Moment-deflection curve for a typical slender wall.

rupture. In this state, the wall cross-section is uncracked and the load-induced moment is less than the cracking moment capacity of the wall cross-section.

A *limit state* exists at the end of each behavior state (see Table 10-1). For example, at the end of the first behavior state, we have the first limit state, and it exists when the lateral load on the wall produces a tensile stress equal to the modulus of rupture.

Table 10-1 Behavior and Limit States for a Ductile Slender Wall

State	Description
Behavior state 1	Uncracked cross-section and $M < M_{cr}$
Limit state 1	$M = M_{cr}$ and stress in the masonry equal to the modulus of rupture.
Behavior state 2	Cracked cross-section with strain in the steel less than its yield strain and $M_{cr} < M < M_y$.
Limit state 2	$M = M_y$ and strain in the steel equal to its yield strain.
Behavior state 3	Cracked cross-section with strain in the steel greater than its yield strain but the maximum strain in the masonry less than its maximum usable strain and $M_y < M < M_u$.
Limit state 3	$M = M_u$ and strain in masonry equal to maximum usable strain.

The slender wall goes through several behavior states prior to reaching its final or *ultimate limit state*. For example, if we consider the load-induced moment as a measurable variable, it can be used to define the existence of the first limit state. In this case, the load-induced moment M will be equal to the cracking moment of the cross-section (M_{cr}). The second limit state exists when the moment M is equal to the yield moment (M_y), and the third limit state exists when M is equal to the moment capacity of the wall (M_u). Therefore, we have identified three limit states whose existence can be numerically quantified as follows:

Limit State	$M =$
1	M_{cr}
2	M_y
3	M_u

Each of these limit states can be the focus of concern for the structural engineer, depending on the client or the design requirements. For example, the first limit state relates to the cracking of the cross-section, and thus to possible water penetration. It can be viewed as a *serviceability limit state*. The second limit state defines the start of permanent steel deformation or significant structural damage. It can be viewed as either a *serviceability* or a *structural-damage limit state*. Finally, the third limit state defines the limit of our acceptable wall performance from a life-safety perspective. Therefore, it is an *ultimate* or *strength limit state*. Typically, it is this limit state that we are concerned with when we use the design approach called strength design. Limit state design can be thought of as a generalization of strength design where we leave open the possibility of addressing limit states other than the strength limit state.

The structural engineer must review the limit states that can exist for the structure he or she is designing. Then, a design criterion must be established that insures, with an acceptable level of safety, that the limit states that structural engineer has identified as undesirable do not exist. For example, current slender-wall design criteria adopted by the International Congress of Building Officials (ICBO) in the 1985 and 1988 Uniform Building Codes (UBC) identify an ultimate or strength design limit state that corresponds to limit state 3 in Table 10-1.[10-6, 10-10] For this example, the *limit-state equation* is

$$M_L \leqslant \phi M_n \tag{10-1}$$

where

M_L = load-induced moment obtained from factored design loads

M_n = nominal (or design) moment capacity of the wall

ϕ = capacity reduction factor that is intended to insure that an acceptable level of safety exists in the final design

The design criteria must address both sides of Equation 10-1. The load-induced moment is obtained from a structural analysis using factored deterministic design loads. We calculate the nominal moment capacity of the wall using the nominal design values of the structural parameters, e.g., specified compressive strength, modulus of elasticity, etc., and the equations of structural engineering.

10.3 LIMIT STATES AND STRUCTURAL RELIABILITY

One task in the United States–Japan coordinated research program under the direction of the Technical Coordinating Council for Masonry Research (TCCMAR) was the evaluation of available approaches whereby masonry design could incorporate the analytical method of structural reliability into *limit-state design*.[10-11] These reliability methods ranged from the very direct to the extremely sophisticated. It is the conclusion of the TCCMAR Category 8, Task 8.1 research that it is possible to significantly extend the rigor of today's masonry code to incorporate structural reliability. The new steel-design criterion accepted for the 1988 Uniform Building Code is *load and resistance factor design* (LRFD) and is based on structural reliability (see References 10-12 to 10-15). LRFD will, in all probability, be the basis of modern reinforced-masonry design. The remainder of this section presents the basics of the LRFD approach and indicates why the identification and quantification of behavior and limit states is so important.

A limit state occurs when a load Q on a structural component equals the resistance R of the component. The occurrence of the limit state exists when $F = 0$, where

$$F = R - Q \tag{10-2}$$

Consider our slender-wall example and the third (or strength) limit state. We can consider R to be the moment capacity of the wall and Q to be the dead plus live plus seismic moment demand. If we denote the moment demand as M, and the moment capacity as M_u, then Equation 10-2 can be written as

$$F = M_u - M \tag{10-3}$$

This equation is called the limit-state design equation. The strength limit state exists when $M = M_u$ or, alternatively, $F = 0$. Stated differently, if F is greater than zero, we know that one of the first three behavior states exists and that the third limit state does not exist.

The economics of building design and construction requires us to have a balance between safety (assurance that a limit state will not exist or be violated) and construction costs. This, historically, has been attained by using a *factor of safety*. In structural reliability, the parallel quantity is referred to as the *reliability index* associated with the limit state under consideration.

Because M_u and M are not known with certainty, they are called random variables. F is a function of M_u and M. Hence, it is also a random variable, with a mean $\mathbf{F}$ and standard deviation σ_F. The reliability index is defined in terms of the statistical moments of F. The reliability index can be defined as

$$\beta = \mathbf{F}/\sigma_F \tag{10-4}$$

Structural reliability theory and the associated mathematics are typically too complex for most design applications. Therefore, for design purposes, we must develop more direct design criteria. Ideally, it is based on structural reliability concepts. This can be accomplished using the *first-order second-moment* structural reliability theory. This theory first performs a Taylor series expansion of F in terms of the random variables, for example R and Q. This expansion is done about the mean value of the random variables, and only the first-order partial derivatives are retained—whence the name first-order.

Next, the mean and standard deviation of F in its Taylor-series expanded form are calculated in terms of the mean and standard deviation (or, alternatively, coefficient of variation) of R and Q. Thus, the "second-order" in the name "first-order second-moment" refers to second-order statistical moments. With the mean and standard deviation of F so calculated, the relability index can be expressed in terms of a constant α and the means (**R** and **Q**) and coefficient of variations (V_R and V_Q) of the random variables. So doing, we can write

$$\mathbf{Q}\, e^{\mu\beta V_Q} = \mathbf{R}\, e^{-\mu\beta V_R} \qquad (10\text{-}5)$$

Note that the right side of the equation relates to the resistance, and the left side to the load effect. If we again consider the slender wall example, we can express this equation as

$$\mathbf{M}\, e^{\mu\beta V_M} = \mathbf{M}_u\, e^{-\mu\beta V_{M_u}} \qquad (10\text{-}6)$$

where $\mathbf{M}_u$ and $\mathbf{M}$ are the means of M_u and M, and V_{M_u} and V_M are their coefficients of variation.

The left-hand side of Equation 10-6 is the *design moment demand* and ideally is equal to the left-hand side of Equation 10-1. The ANSI A58.1-1982 load factors or similar reliability-based factored loads define this design moment demand.

The right-hand side of Equation 10-6 is the *design moment capacity* and contains the structural reliability or safety, β. This can be written as

$$M_D = \mathbf{M}_u\, e^{-\mu\beta V_{M_u}} \qquad (10\text{-}7)$$

If we recall the right-hand side of the limit-state design equation for moment capacity given in Equation 10-1, it follows that

$$M_D = \phi M_n = M_u\, e^{-\mu\beta V_{M_u}} \qquad (10\text{-}8)$$

Therefore, the capacity reduction factor ϕ for this limit state is

$$\phi = \frac{M_u}{M_n}\, e^{-\mu\beta V_{M_u}} \qquad (10\text{-}9)$$

Equation 10-9 shows the dependence of the capacity reduction factor ϕ on (i) the ratio of the mean to nominal design moment, (ii) the uncertainty or quality of construction and analytical modeling as manifested in the value of V_{M_u}, and (iii) the level of safety (β value) that the design criterion seeks to attain. These three items must be the focus of discussion among those involved in the development of future masonry-design criteria.

10.4 SLENDER-WALL DESIGN

10.4.1 General

The appropriate method to model the capacity of a member subjected to both bending and axial loads is an interaction approach which accounts for the relationship between the stresses caused by bending and axial loads. An *interaction diagram*, such as that shown in Figure 10-2, may be constructed by establishing the capacity of the member under various combinations of axial and flexural loads. Although any number of points may be calculated, the critical points identified by numbers 1 through 6 in Figure 10-2 should be more than sufficient to construct an accurate interaction diagram. Each point is described by the axial capacity P_n and moment capacity M_n. Thus, M_n can be computed for a given P_n, or vice versa.

For example, at one extreme, point 1, where no externally applied moment is imposed on the wall, the nominal axial capacity of the wall is

$$P_n = 0.85 f'_m (A_n - A_s) + A_s F_y \qquad (10\text{-}10)$$

The other extreme, point 6, is where the capacity of the member is the pure bending nominal flexural capacity of the wall, or

$$M_n = 0.85 f'_m ab\left(d - \frac{a}{2}\right) \qquad (10\text{-}11)$$

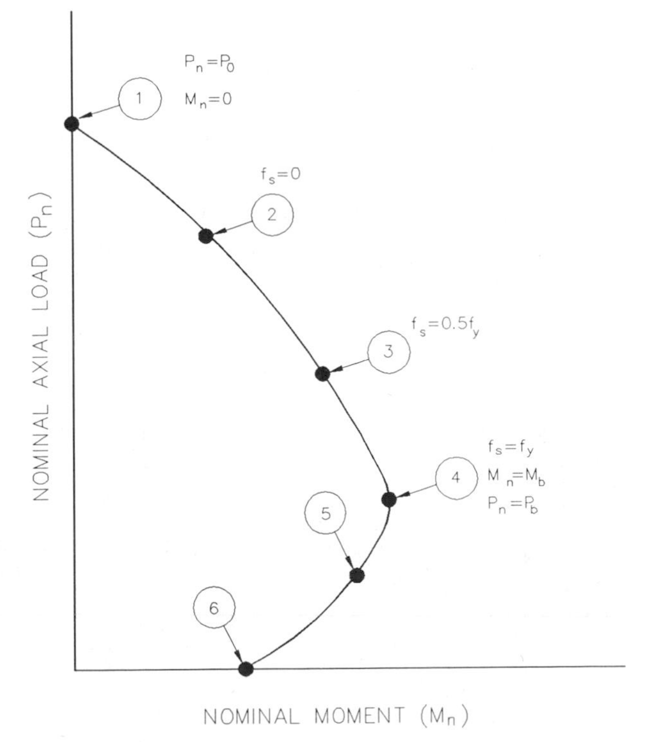

Figure 10-2 Interaction diagram for an eccentrically loaded member.

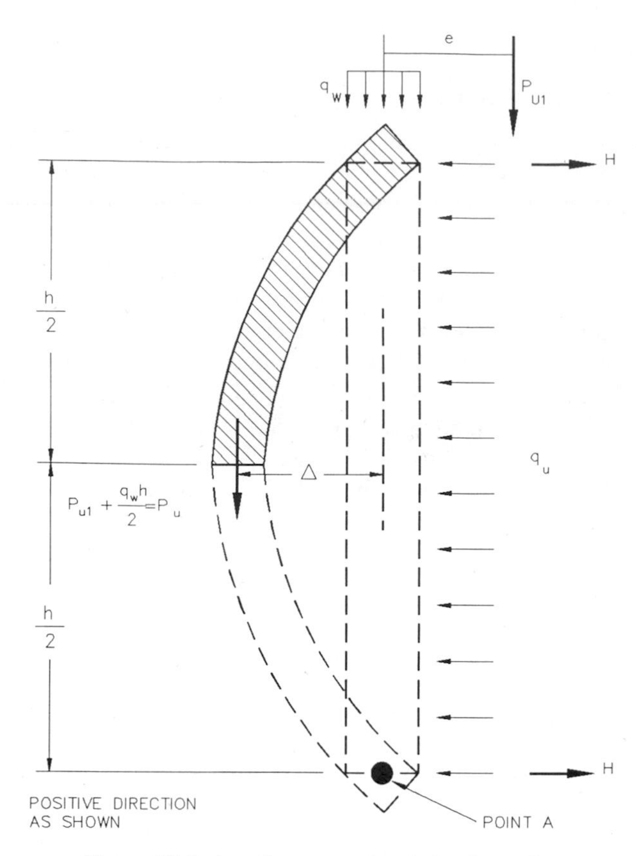

Figure 10-3 Loading geometry of slender wall.

The intermediate points may be established by choosing several condition of strain and using the force–equilibrium and stress–strain relationships developed in Reference 10-16 for calculating P_n and M_n.

10.4.2 Structural Mechanics

The load-induced moment on a wall is a function of lateral wall deflection. If the wall has a height-to-thickness ratio of 25 or more (herein referred to as a *slender wall*), the lateral deflection can produce moments that are significant relative to the moment obtained using small deflection theory.

Figure 10-3 shows the forces acting on a slender wall with a pin connection at each end. The summation of moments about the bottom of the wall, point A, gives the equation for the horizontal force at the upper wall support. That is,

$$P_{ul}e + Hh - q_u\frac{h^2}{2} - q_w h(\Delta_a) = 0 \quad (10\text{-}12)$$

where

P_{ul} = vertical load on wall

e = eccentricity of vertical load

q_u = uniform lateral load on wall per linear foot

q_w = weight of wall per linear foot

Δ_a = "effective" lateral deflection used to estimate dead-load moment

If we assume that

$$\Delta_a = 2\Delta/3 \quad (10\text{-}13)$$

where Δ is the wall's mid-height lateral deflection, then

$$H = \frac{q_u h}{2} - \frac{P_{ul}e}{h} + \frac{2q_w\Delta}{3} \quad (10\text{-}14)$$

The first term corresponds to the classical small-deflection reaction, the second to the change in the magnitude of the force due to an eccentric wall loading, and the third to the lateral wall deflection.

If we take the moment about the mid-height of the wall, the moment induced on the cross-section from the external loads is

$$M = H\frac{h}{2} + P_{ul}(\Delta + e) + (q_w h)\Delta_b - \frac{q_u h}{2}\frac{h}{4} \quad (10\text{-}15)$$

where Δ_b is the "effective" lateral deflection used to estimate the dead-load moment. If we assume that

$$\Delta_b = \Delta/3 \quad (10\text{-}16)$$

which is consistent with Δ_a above, and substitute H into the moment equation, it follows that

$$M = \frac{q_u h^2}{8} + \frac{P_{ul}e}{2} + \left(P_{ul} + \frac{q_w h}{2}\right)\Delta \quad (10\text{-}17)$$

The first term corresponds to the moment due to the classical small-deflection moment from the uniform lateral load, the second term to the moment due to the eccentric vertical load on the wall, and the third term to the moment due to large lateral deflections. This last term can be referred to as the P-delta load.

The moment M and lateral force H are a function of Δ, which in turn is a function of the wall's cross-sectional properties and steel reinforcement as well as the moment M and the lateral load H. Therefore, the calculation of the moment M is iterative.

The ultimate axial load computed using the factored axial forces must be less than the evaluated nominal capacity:

$$\phi P_n \geqslant P_u \quad (10\text{-}18)$$

The slender wall must have a capacity equal to the sum of the superimposed factored axial dead and live loads,

P_{u1}, and the factored wall dead load for the upper half, P_{u2}, along with the factored lateral load from the wall and/or loading above (see Figure 10-3). The moment capacity of a wall section is calculated, assuming that axial strength does not govern the design, and it is checked against the moment generated under the applied lateral load and by the P-delta effect.

Although most walls are loaded at a level which is considerably less than their axial load strength, a check can be made to determine if flexure controls the design, that is,

$$\phi P_b \geqslant P_u \tag{10-19}$$

in which

$$P_b = 0.85 f'_m b a_b - A_s f_y$$

where

$$a_b = \left(\frac{58{,}000}{58{,}000 + f_y} \right) \beta d$$

The nominal moment capacity of the wall section loaded with a concentrically applied load may be determined from force and moment equilibrium (see Figures 10-4 and 10-5). The axial load is

$$P_u = C - T$$

or

$$0.85 f'_m b a = P_u + A_s f_y \tag{10-20}$$

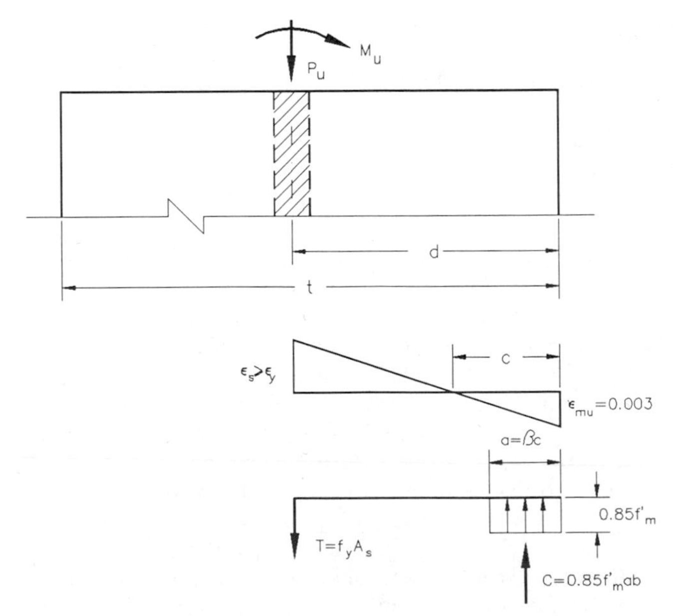

Figure 10-4 Stress and strain diagrams for steel at center of wall.

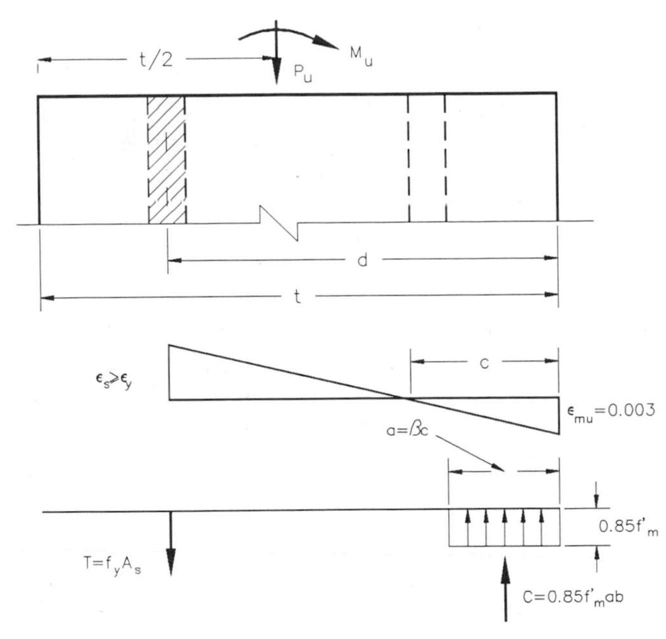

Figure 10-5 Stress and strain diagrams for steel at two faces (ignoring compression steel).

and solving for a yields

$$a = \frac{P_u + A_s f_y}{0.85 f'_m b} \tag{10-21}$$

Summing the internal and external moments about the tension steel yields

$$M_u + P_u\left(d - \frac{t}{2}\right) - C\left(d - \frac{a}{2}\right) = 0$$

Substituting Equation 10-20 for C, and assuming $M_n = M_u$, the nominal moment capacity of a member with steel at two faces (Figure 10-5) is

$$M_n = \left(P_n + A_s f_y\right)\left(d - \frac{a}{2}\right) - P_n\left(d - \frac{t}{2}\right) \tag{10-22}$$

In the more typical case with steel in one layer of reinforcement at the center line of the wall (Figure 10-4), the nominal moment capacity is

$$M_n = \left(P_n + A_s f_y\right)\left(d - \frac{a}{2}\right) \tag{10-23}$$

If the imposed moment, M_u, is less than the reduced moment capacity, ϕM_n, the wall section is acceptable:

$$\phi M_n \geqslant M_u \tag{10-24}$$

This may be determined by comparing Equation 10-17 with Equation 10-22 or 10-23, multiplied by the appropriate ϕ factor.

In 1981, the Structural Engineers Association of Southern California (SEAOSC) tested 32 slender concrete, brick, and concrete masonry panels subjected to a constant axial and increasing lateral load.[10-17] Panel

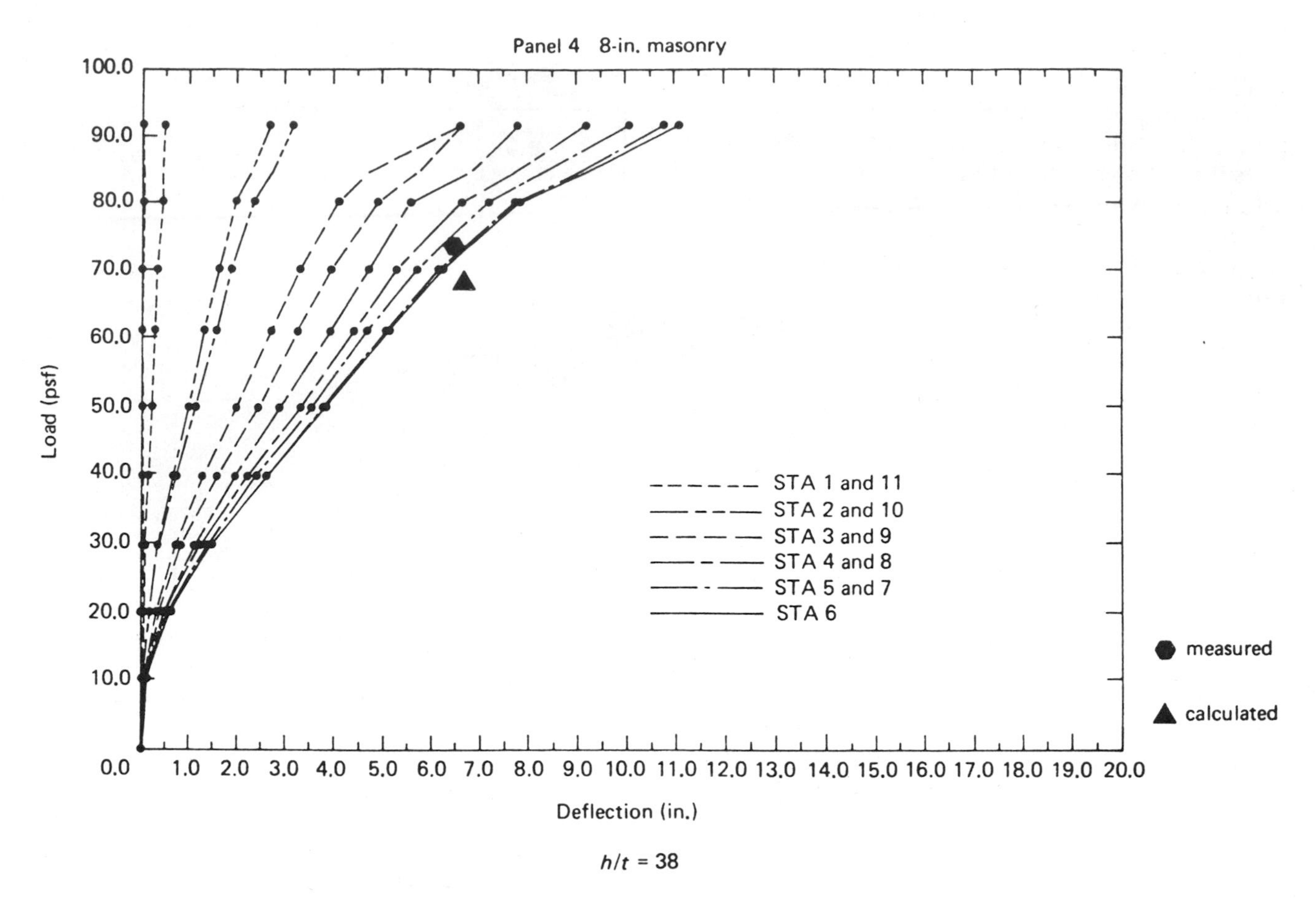

Figure 10-6 Load deflection curves (slender walls).

capacities were predicted using the strength method developed by SEAOSC. The procedure for calculating ultimate moments and deflections is presented in Equations 10-23 and 10-24. Load deflection results of these tests for 8-in.-thick concrete masonry walls are presented in Figure 10-6. A close correlation was obtained between calculations and test data.

10.4.3 State-of-the-Art Limit-State Design Criteria

The limit-state design procedure concerns reinforced hollow-unit concrete masonry slender walls subjected to vertical and horizontal forces causing out-of-plane flexure. The design procedure and criteria described herein differ slightly from the SEAOC criteria in Section 2411 of the 1988 UBC, but the author believes the presented performance criteria better represents the current state of the art.

Conditions for the Design Procedure

1. The minimum nominal thickness of the masonry wall shall be 6 in.

2. The ratio of unsupported height to nominal wall thickness may not exceed 36 unless the axial stress at the location of maximum moment is equal to or less than $0.04 f'_m$.

3. The minimum reinforcement ratio shall be 0.0007 in either direction and 0.002 total.

4. The maximum reinforcement shall not exceed 50% of the balanced steel ratio, ρ_b. The maximum steel in each cell shall not exceed 0.03 times the cell area unless the reinforcing steel is lap spliced, and then it is to be 0.06 times the cell area.

5. The principal wall reinforcement in the direction of span shall not be spliced within the middle third of the span.

6. All units shall be laid in running bond unless the wall is grouted solid.

7. Masonry walls at corners and intersecting cross walls shall be effectively anchored to each other or separated to prevent seismic batter.

8. All grouts shall have a minimum compressive strength f'_c not less than 2000 lb/in.^{2} nor greater than 4000 lb/in.^{2} f'_c shall be determined by prism tests.

9. All grouts shall be consolidated by mechanically vibrating over the height of pour (vibration shall be performed after the initial loss of water and before initial set). Grout space shall be not less than the minimum necessary for mechanical vibration.

10. The specified compressive strength f'_m shall not be less than 1500 lb/in.^{2} nor greater than 3000 lb/in.^{2}. f'_m shall be determined by prism tests.

11. An inspector shall provide continuous inspection during all key phases of wall construction as identified on the structural plans.

Design Procedures The design of hollow unit reinforced concrete masonry shall be based on forces and moments determined from analysis. An analysis which considers slenderness of walls by representing effects of axial load and deflection in the calculation of required moments must be used. This design procedure must satisfy both strength and deflection limit states. The slender-wall design procedures given herein may be used when the ratio of unsupported height to nominal wall thickness is equal to or less than 36 or when the vertical load stress at the location of the maximum moment does not exceed $0.04f'_m$:

$$\frac{P_w + P_0}{A_g} \leqslant 0.04f'_m \qquad (10\text{-}25)$$

where

P_0 = unfactored axial load from tributary floor and/or roof area, pounds

P_w = unfactored weight of the wall tributary to section under consideration, pounds

f'_m = specified compressive strength, lb/in.^{2}

A_g = gross area of wall, square inches

Design Load Factors

1. *General.* Strength required by a masonry wall shall be based on factored loads.

2. *Basic load combinations.* Loading combinations shall be based on the selected loading criteria shown below. The required strength U to resist factored loads and forces shall be as follows:

$$U = 0.75(1.4D + 1.7L + 1.87E) \qquad (10\text{-}26a)$$

$$U = 0.75(1.4D + 1.7L + 1.7W) \qquad (10\text{-}26b)$$

$$U = 0.9D + 1.43E \qquad (10\text{-}26c)$$

$$U = 0.9D + 1.3W \qquad (10\text{-}26d)$$

where

D = dead loads or related internal moments and forces

L = live loads or related internal moments and forces

E = load effects of earthquake or related internal moments and forces

W = wind loads or related internal moments and forces

U = required strength to resist factored loads or related internal moments and forces

Design Assumptions for Nominal Strength

1. The nominal strength of singly reinforced concrete masonry wall cross-sections subject to combined flexural and axial loads shall be based on applicable conditions of equilibrium and compatibility of strains. Strain in reinforcement and masonry shall be assumed directly proportional to the distance from the neutral axis.

2. The maximum usable strain at extreme masonry compression fiber shall be assumed equal to 0.003.

3. For steel strains less than the steel yield strain, the stress in reinforcement shall be taken as E_s times the steel strain. For steel strains greater than the steel yield strain, the stress in the reinforcement shall be considered independent of strains and equal to f_y. Here

f_y = Specified yield strength of the reinforcement, lb/in.^{2}

E_s = Modulus of elasticity of reinforcement = 29,000,000 lb/in.^{2}

4. The tensile strength of masonry shall be neglected in flexural calculations of strength, except when computing the nominal cracking moment strength.

5. In the calculation of nominal moment strength the relationship between masonry compressive stress and masonry strain may be assumed to be rectangular. Masonry stress of $0.85f'_m$ shall be assumed uniformly distributed over an equivalent compression zone bounded by the edges of the cross-section and a straight line located parallel to the neutral axis at a distance a from the fiber of maximum compressive strain.

Design Strength The required moment strength, M_u, shall be equal to or less than the nominal moment strength multiplied by a strength reduction factor:

$$M_u \leqslant \phi M_n \qquad (10\text{-}27)$$

where

M_n = nominal moment strength

ϕ = strength-reduction factor for nominal strength as given in Table 10-2

Modulus of Elasticity The nominal value of the modulus of elasticity of the masonry, E_m, shall be

Table 10-2 Strength-Reduction Factor for Nominal Moment Strength

Nominal Thickness of Wall	ϕ
$\geqslant$ 8 in.	0.80
< 8 in.	0.65

Table 10-3 Nominal Value of Modulus of Rupture for Wall System

Type of Wall	f_r, lb / in.^{2}
Solid grouted	$4.5\sqrt{f'_m}$
Partial grouted	$3.0\sqrt{f'_m}$

Table 10-4 Deflection Limits for Unfactored Loads

Nominal Thickness of Wall	Maximum Wall Deflection, in.
⩾ 8 in.	$0.007h$*
< 8 in.	$0.005h$*

*h is the height of wall between supports, in inches.

assumed as follows:

$$E_m = 750 f'_m \tag{10-28}$$

Modulus of Rupture The nominal value of the modulus of rupture of the partially grouted or solid grouted hollow-unit masonry wall system, f_r, shall be assumed as in Table 10-3.

Deflection Limitations The maximum wall deflection relative to the support, Δ_s, under unfactored lateral and vertical loads shall be limited as noted in Table 10-4.

Design Equations

1. *Deflections*. The mid-height deflection for simple wall support conditions top and bottom due to the unfactored loads, Δ_s, shall be computed using either of the following equations:

$$M_s \leqslant M_{cr} \quad \Rightarrow \quad \Delta_s = \frac{5M_s h^2}{48E_m I_g} \tag{10-29}$$

$$M_{cr} < M_s < M_n \quad \Rightarrow \quad \Delta_s = \frac{5M_{cr} h^2}{48E_m I_g} + \frac{5(M_s - M_{cr})h^2}{48E_m I_{cr}} \tag{10-30}$$

where

I_g = moment of inertia of the uncracked-wall cross-section, in.^{4}

I_{cr} = moment of inertia of the cracked-wall cross-section, in.^{4}

M_{cr} = cracking moment strength

$$= \phi\left(f_r + \frac{P}{A_n}\right)S_m \tag{10-31}$$

Table 10-5 Strength-Reduction Factor for Cracking Moment Strength

Nominal Thickness of Wall	ϕ
⩾ 8 in.	0.90
< 8 in.	0.75

where:

S_m = section modulus of the uncracked wall cross-section, in.^{3}

P = unfactored axial load, pounds

$$= P_0 + P_w \tag{10-32}$$

A_n = net area of cross-section, in.^{2}

ϕ = strength-reduction factor applied to cracking moment strength as given in Table 10-5

M_s = moment due to unfactored loads for a simple wall support condition top and bottom

$$= \frac{wh^2}{8} + P_0\frac{e}{2} + (P_w + P_0)\Delta_s \tag{10-33}$$

where

w = distributed lateral load

e = eccentricity of the vertical load P_0

For other wall support conditions the maximum wall deflection shall be calculated using the equations of structural mechanics.

2. *Required moment strength*. The required moment strength M_u for a simple wall support conditions top and bottom is the moment given by

$$M_u = \frac{w_u h^2}{8} + P_{0u}\frac{e}{2} + (P_{wu} + P_{0u})\Delta_u \tag{10-34}$$

where

w_u = factored distributed lateral load

Δ_u = horizontal deflection at mid-height of wall calculated using Equation 10-30 for factored loads and $M_s = M_u$

P_{wu} = factored weight of the wall tributary to the section under consideration

P_{0u} = factored axial load on the wall from tributary floor and/or roof loads

e = eccentricity of the factored axial load P_{0u}

3. *Nominal moment strength*. The nominal moment strength M_n of the wall is as follows:

$$M_n = A_{se} f_y\left(d - \frac{a}{2}\right) \tag{10-35}$$

where:

$$a = \frac{P + A_s f_y}{0.85 f'_m b} \tag{10-36}$$

$$A_{se} = \frac{P + A_s f_y}{f_y} \tag{10-37}$$

b = tributary width

d = distance from extreme compression fiber to centroid of tension reinforcement

10.4.4 Comments on the State-of-the-Art Limit-State Design Criteria

Reinforced hollow-unit masonry that is constructed with good quality control and has its grout vibrated has been shown through experimental measurements to perform in flexure in a very similar fashion to reinforced concrete. The slender-wall test conducted by the Structural Engineers Association of California and presented in Section 2411 of the 1985 UBC is developed with recognition of this similarity of basic engineering-mechanics performance.

One basic assumption of the existing working-stress design approach for axial load and flexure is that plane cross-sections remain plane during axial load and bending-moment deformations. Alternatively stated, this means that the variation of strain is a linear function of the distance from the neutral axis. The proposed strength-design approach for masonry shear walls makes the same assumption. This assumption is consistent with the assumption used in the strength design of reinforced concrete and is supported by experiments on masonry shear walls such as those presented for a 6-m-tall wall in Figure 10-7.[10-18]

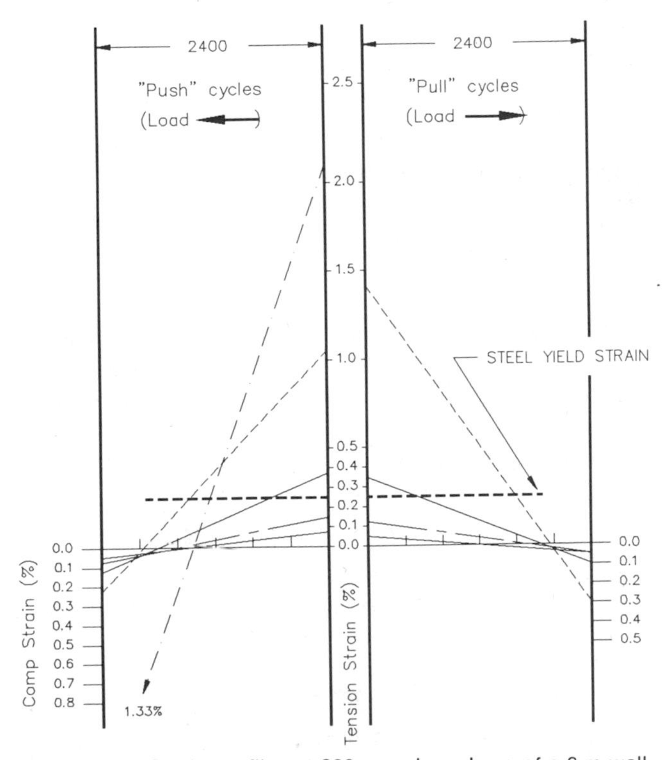

Figure 10-7 Strain profiles at 200 mm above base of a 6 m wall for different deformations.

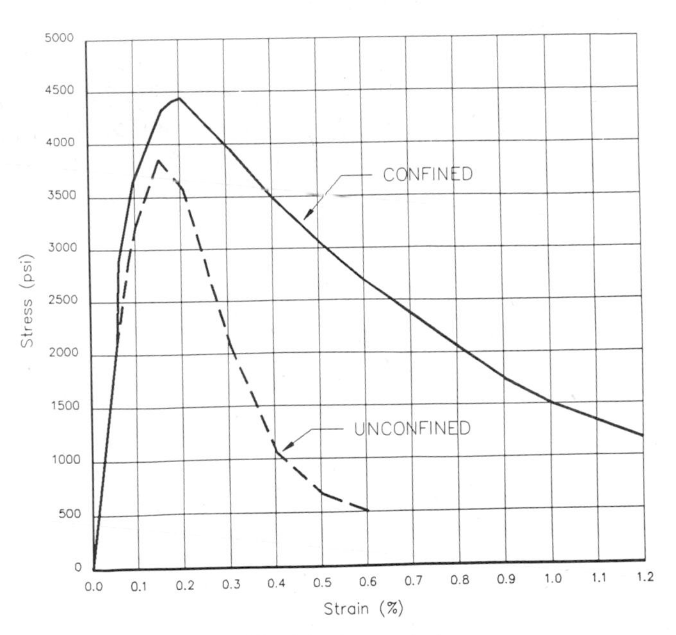

Figure 10-8 Priestly's stress–strain curves.

The assumption is made in the proposed design criteria that a rectangular stress block can be used to calculate the flexural capacity of shear walls. Stress–strain curves such as those presented in Figure 10-8 indicate that the stress–strain curve for masonry is not rectangular in shape but follows more closely a parabolic form. The reason for the selection of the rectangular stress block is one of convenience, and also, the recognition that the moment capacity of a section with a rectangular stress block closely approximates the moment capacity obtained using the more accurate representation of the stress–strain curve.

Figure 10-9 shows the results of tests conducted in Canada for beams in flexure.[10-21] The test results are compared with the estimated nominal moment capacity using a rectangular stress block and with the design value using a strength-reduction factor of 0.86.

Figure 10-10 shows an idealized stress–strain curve with the parameters defined in Table 10-6 identified on the curve. Based on the TCCMAR data, the value of 0.003 for the maximum usable strain is slightly less the average value or nominal value obtained from the test results.

The value of the maximum usable strain selected as part of this criterion is equal to the value most often cited for the design of reinforced-concrete members. One might be inclined to be concerned with our selection of 0.003 because it is the same value as used for

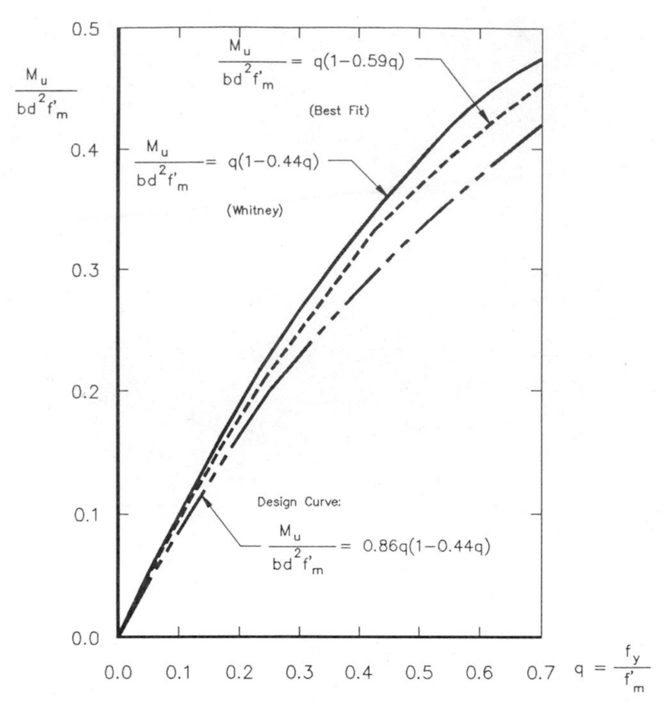

Figure 10-9 Tension controlled flexural test results.

reinforced concrete. However, as indicated in Figure 10-11, the maximum usable strain value for concrete with maximum compressive value comparable to those values specified in the criteria for masonry far exceeds the 0.003 value. In particular, as reinforced concrete can attain significantly higher maximum compressive values, it is only at these maximum compressive values where the 0.003 limitation is reasonable.

The maximum strain can be increased where confinement is provided (see Figure 10-8). Experimental evidence indicate that confinement increases the maximum usable strain, and therefore the component curvature ductility.

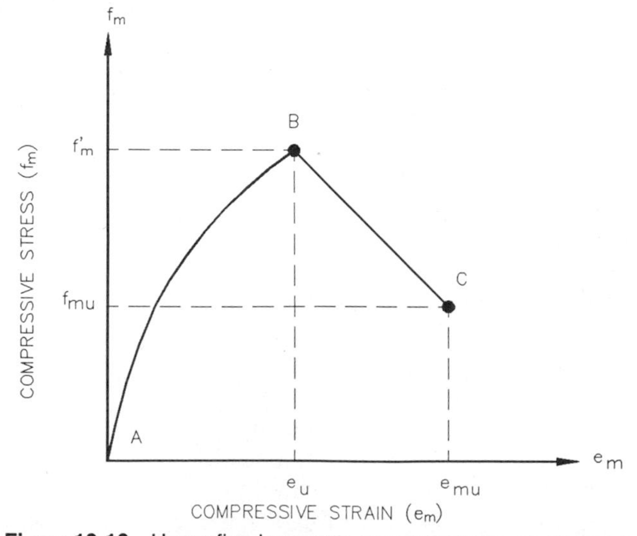

Figure 10-10 Unconfined concrete masonry stress–strain curve.

Table 10-6 Design Parameters for the Unconfined Concrete Masonry Stress–Strain Curve

Parameter	Comment
f'_m	Ultimate compressive stress. Nominal design value is specified by design engineer.
ϵ_u	Strain corresponding to f'_m. We recommend a nominal design value of 0.0020 to 0.0025.
f_{mu}	The minimum usable compressive stress in the strain region defined by strain values greater than the strain at ultimate compressive stress, i.e., ϵ_{mu}. We recommend a nominal design value of $0.5f'_m$.
ϵ_{mu}	Maximum usable unconfined strain. Alternately stated, it is the strain corresponding to the minimum usable compressive stress. We recommend a nominal design value of 0.0030.

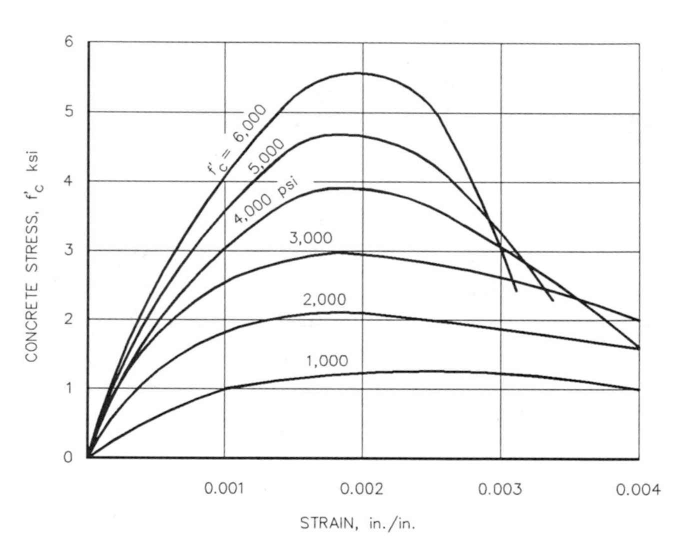

Figure 10-11 Typical stress–strain curves for concrete under short-time loading.

10.4.5 Example

The following example is worked using the state-of-the-art limit-state design criteria described in Section 10.4.4. The wall geometry and loading are shown in Figure 10-12.

Material Properties

$$f'_m = 3.0 \text{ ksi}$$
$$f_r = 4.5\sqrt{f'_m} \text{ lb/in.}^2$$
$$E_m = 750 f'_m \text{ ksi}$$
$$f_y = 60 \text{ ksi}$$
$$A_s = 0.22 \text{ in.}^2/\text{ft}$$
$$e = 5.5 \text{ in.}$$
$$d = 3.81 \text{ in.}$$

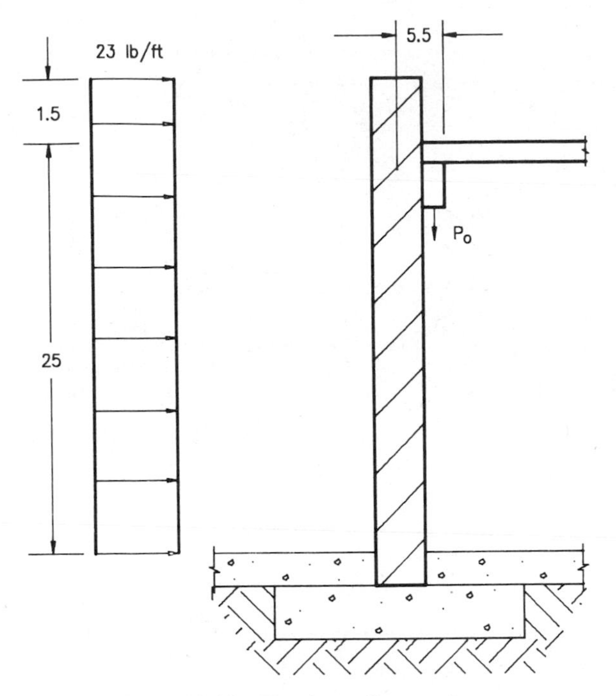

Figure 10-12 Slender wall example.

Self-Weight of the Wall

$$P_w = (25/2 + 1.5)80$$
$$= 1120 \text{ lb/ft}$$
$$= 1.12 \text{ kips/ft}$$

Maximum Axial Load Check

$$0.04f'_m A_g = 0.04 \times 3 \times 7.62 \times 12$$
$$= 10.47 \text{ kips/ft}$$
$$> P_0 + P_w = 1.66 \text{ kips/ft O.K.}$$

$$f_r = 4.5\sqrt{3000} = 246 \text{ lb/in.}^2$$

$$I_g = 12 \times 7.63^3/12 = 443 \text{ in.}^4$$

$$S = 12 \times 7.63^2/6 = 116 \text{ in.}^3$$

$$M_{cr} = \phi\left(\frac{P}{A} + f_r\right)S$$

$$= 0.9\left(\frac{(1.12 + 0.3) + (0.7 \times 0.24)}{7.625 \times 12} + 0.246\right)$$

$$\times \left(\frac{116}{12}\right)$$

$$= 2.29 \text{ kip-ft} = 27.5 \text{ kip-in.}$$

Transformed Moment of Inertia

$$k = -n\rho + \left[(n\rho)^2 + 2n\rho\right]^{0.5}$$

$$n = \frac{E_s}{E_m} = \frac{29 \times 10^3}{750 \times 3} = 12.88$$

$$A_{se} = \frac{A_s f_y + P}{f_y} = \frac{0.22 \times 60 + 1.66}{60} = 0.247 \text{ in.}^2\text{/ft}$$

$$\rho = \frac{A_{se}}{bd} = \frac{0.247}{12 \times 3.81} = 0.0054$$

$$k = 0.3099$$

$$I_{cr} = d^3\left[4k^3 + 12n\rho(1-k)^2\right] = 28.56 \text{ in.}^4\text{/ft}$$

$$M = \frac{wh^2}{8} + P_0\frac{e}{2} + (P_w + P_0)\,\Delta_s$$

and

$$\Delta = \begin{cases} \dfrac{5M_{cr}h^2}{48E_mI_g} + \dfrac{5(M - M_{cr})h^2}{48E_mI_{cr}} & \text{for } M_{cr} < M < M_n \\[2ex] \dfrac{5Mh^2}{48E_mI_g} & \text{for } M < M_{cr} \end{cases}$$

Unfactored (Service) Moments and Displacements

$$\Delta_1 = 0$$

$$M_1 = \frac{23 \times 25^2}{8} + \frac{540 \times 5.5}{2 \times 12} + 1660 \times 0$$

$$= 1.92 \text{ kip-ft} = 23.0 \text{ kip-in.} < M_{cr} = 27.5 \text{ kip-in.}$$

$$\Delta_2 = \frac{5 \times 1.92 \times 25^2}{48 \times 750 \times 3 \times 144 \times 443} \times 12^5$$

$$= 0.216 \text{ in.}$$

Table 10-7 provides the remainder of the solution.

Table 10-7 Convergence of Deflections and Moments

Iteration	Δ_s, in.	M_s, kip-in.	Δ_s, in.	M_u, kip-in.
1	0.00	23.04	0.00	29.17
2	0.26	23.40	0.50	30.05
3	0.22	23.41	0.59	30.27
4	0.22	23.41	0.62	30.32
5	0.22	23.41	0.64	30.34
6	0.22	23.41	0.64	30.35

Factored (Ultimate) Moments and Displacements Using UBC load factors,

$$U = 0.75(1.4D + 1.7L + 1.7W)$$

$$\Delta_1 = 0$$

$$M_1 = \frac{29.3 \times 25^2}{8} + \frac{621 \times 55}{2 \times 12} + 1747 \times 0$$

$$= 2.43 \text{ kip-ft} = 29.2 \text{ kip-in.}$$

$$> M_{cr} = 27.5 \text{ kip-in.}$$

$$\Delta_2 = \left[\frac{5 \times 2.29 \times 25^2}{48 \times 750 \times 3 \times 144 \times 433} + \frac{5(2.43 - 2.29) \times 25^2}{48 \times 750 \times 3 \times 144 \times 26.13} \right] \times 12^5$$

$$= 0.50 \text{ in.}$$

Table 10-7 provides the remainder of the solution.

Nominal Moment Capacity

$$a = \frac{0.247 \times 60}{0.85 \times 3 \times 12} = 0.48 \text{ in.}$$

$$M_n = 0.8 \times 60 \times 0.247(3.81 - 0.48/2)$$

$$M_n = 42.3 \text{ kip-in.}$$

$$= 3.50 \text{ kip-ft}$$

10.4.6 SLWALL Computer Program

A computer program called SLWALL has been developed with funding from the National Science Foundation and the Concrete Masonry Association of California and Nevada as part of the U.S.–Japan masonry research program called TCCMAR. SLWALL requires the user to have LOTUS 1-2-3 (Version 2) and is available for a nominal charge from the Concrete Masonry Association of California and Nevada.

Figure 10-13 shows the spreadsheet and the output of the SLWALL computer program for the example wall of Section 10.4.5. With this very flexible format the user can benefit from being able to perform spreadsheet "what if" studies.

10.5 SHEAR-WALL DESIGN

10.5.1 General

Over 100 masonry shear walls with different steel ratios, axial load levels, and sizes have been tested in the last decade. Therefore, it is possible to develop a design criteria which is based on good-quality (typically, cyclic load reversal) test data. The design criteria for reinforced hollow-unit concrete masonry shear walls in many respects follow the design criteria for reinforced-con-

```
               *************************************
               * TALL SLENDER CONCRETE MASONRY WALL *
               *************************************

                      SLWALL VERSION (1.01)
                        SEPTEMBER, 1987
                   ---------------------------

*****************************************************************
*                                                               *
*  THIS CONCRETE MASONRY SLENDER WALL PROGRAM (SLWALL) WAS WRITTEN *
*  AS PART OF THE TCCMAR MASONRY RESEARCH PROGRAM AND WAS FUNDED  *
*  BY THE NATIONAL SCIENCE FOUNDATION AND THE CONCRETE MASONRY    *
*  ASSOCIATION OF CALIFORNIA AND NEVEDA.  THE WORK WAS CONDUCTED  *
*  AS PART OF THE CATEGORY 2 RESEARCH WORK UNDER THE JOINT        *
*  RESEARCH EFFORT OF EKEH (EWING AND ASSOCIATES, KARIOTIS AND    *
*  ASSOCIATES, AND ENGLEKIRK AND HART INC.) AND WAS WRITTEN BY THE *
*  STAFF OF ENGLEKIRK AND HART, INC.                              *
*                                                               *
*  SLWALL IS WRITTEN AS A PART OF THE NATIONAL EARTHQUAKE HAZARD  *
*  REDUCTION PROGRAM (NEHRP) AND IS INTENDED TO BE BOTH A RESEARCH *
*  AID AND DESIGN AID FOR CONCRETE MASONRY SLENDER WALLS.         *
*                                                               *
*  NEITHER THE NATIONAL SCIENCE FOUNDATION, THE CONCRETE MASONRY  *
*  ASSOCIATION OF CALIFORNIA AND NEVEDA NOR ENGLEKIRK AND HART,   *
*  INC. ASSUMES ANY RESPONSIBILITY FOR ANY ERRORS, MISTAKES, OR   *
*  MISREPRESENTATIONS THAT MAY RESULT FROM THE USE OF SLWALL.     *
*  SLWALL IS PROVIDED IN AN "AS IS" CONDITION.  NO WARRANTIES OF  *
*  ANY KIND, WHETHER STATUTORY, WRITTEN, ORAL, EXPRESSED, OR      *
*  IMPLIED (INCLUDING WARRANTIES OF ANY FITNESS AND               *
*  MERCHANTABILITY) SHALL APPLY.                                  *
*                                                               *
*  THE COPY OF THIS PROGRAM MAY BE OBTAINED FROM :                *
*  CONCRETE MASONRY ASSOCIATION                                   *
*  OF CALIFORNIA AND NEVEDA                                       *
*  6060 SUNRISE VISTA DR. SUITE 1875                              *
*  CITRUS HEIGHTS, CA 95610                                       *
*  TEL: (916) 722-1700                                            *
*                                                               *
*****************************************************************

                                        JOB #   :    85-G142
                                        DATE    :    9-87
                                        SHEET   :      1
                                        TITLE 1: EXAMPLE # 1
                                        TITLE 2: SOLID GROUTED WALL


 1.  NOMINAL DESIGN LOADS

        DEAD ......................................    300.00 PLF
        LIVE.......................................    240.00 PLF
        WIND.......................................     23.00 PSF
        SEISMIC....................................     10.00 PSF

 2.  WALL PROPERTIES

        NOMINAL WALL THICKNESS.(t).................      8.00 IN
        WALL HEIGHT.(hw)...........................     25.00 FT
        HEIGHT/THICKNESS RATIO.(hw/teff)...........     39.34
        PARAPET HEIGHT.(hp)........................      1.50 FT
        THICKNESS OF FACE SHELL....................      1.25 IN

 3.  MATERIAL PROPERTIES

  3.1: CONCRETE MASONRY
        SPECIFIED COMPRESSIVE STRENGTH.............      3.00 KSI
        MODULUS OF ELASTICITY......................       750 f'm
        MODULUS OF RUPTURE.........................      4.50 f'm^1/2

  3.2: REINFORCING STEEL
        SPECIFIED YIELD STRENGTH...................        60 KSI
        MODULUS OF ELASTICITY......................     29000 KSI

 4.  VERTICAL REINFORCING STEEL

        AREA OF STEEL..............................      0.22 IN^2/FT
        CODE MINIMUM AREA OF STEEL (0.0007Ag)......      0.06 IN^2/FT OK
        CODE MAXIMUM AREA OF STEEL (0.0032Ag)......      0.29 IN^2/FT OK
        DISTANCE FROM CENTROID OF STEEL TO
        EXTREME COMPRESSION FIBER..................      3.81 IN

 5.  DISTANCE FROM VERTICAL LOAD POINT TO CENTER OF      5.50 IN
     WALL

 6.  MAXIMUM ALLOWABLE MID-HEIGHT DEFLECTION.......     0.007 h

 7.  PARTIALLY GROUTED WALLS

        7.1: IS WALL PARTIALLY GROUTED?
                               (YES=1, NO=0).....          0
        7.2: IF PARTAILLY GROUTED, THEN C/C DISTANCE
                OF GROUTED CELLS?..................        0 IN

 8.  DO YOU WANT TO INCLUDE AXIAL LOAD IN CRACKING MOMENT?
                               (YES=1, NO=0).....          1

 9.  DO YOU WANT TO USE EFFECTIVE AREA OF STEEL IN
     CALCULATING DESIGN MOMENT CAPACITY?
                               (YES=1, NO=0).....          1

10.  PERCENTAGE OF LIVE LOAD TO BE USED TO CALCULATE
     CRACKING MOMENT?..............................       70 %
```

Figure 10-13 SLWALL worksheets and results for the slender wall example.

```
11. CAPACITY REDUCTION FACTORS (PHI FACTORS)

      DESIGN CRACKING MOMENT CAPACITY............      0.90
      DESIGN STRENGTH MOMENT CAPACITY............      0.80

12. LOAD FACTORS
                          DEAD     LIVE     WIND       QUAKE     RED.FAC
      CASE 1              1.00     1.00     1.00       1.00         1.00
      CASE 2              1.40     1.70     1.70       1.87         0.75
      CASE 3              0.90     0.00     1.30       1.40         1.00

13. CALCULATED RESULTS / SECTION PROPERTIES

      GROSS CROSS-SECTIONAL AREA OF THE WALL.....        92 IN^2/FT
      SELF WEIGHT OF THE WALL....................        80 PSF
      GROSS MOMENT OF INERTIA ...................       443 IN^4/FT
      GROSS SECTION MODULUS......................       116 IN^3/FT
      DISTANCE OF NEUTRAL AXIS FROM EXTREME
      COMPRESSION FIBER..........................      1.18 IN
      CRACKED MOMENT OF INERTIA .................        29 IN^4/FT
      MODULUS OF ELASTICITY OF MASONRY...........      2250 KSI
      MODULUS OF RUPTURE OF MASONRY..............       246 PSI

14. MAXIMUM ALLOWABLE AXIAL LOAD CHECK (LOAD CASE 1 ONLY)

      DESIGN AXIAL LOAD                                1660 PLF
      MAXIMUM ALLOWABLE AXIAL LOAD (0.04Agf'm)        10980 PLF
      CHECK                                            OK

15. SERVICIBILITY LIMIT STATE DEFLECTION CHECK (LOAD CASE 1 ONLY)

      DESIGN CRACKING MOMENT CAPACITY                  2.30 FT-KIP/FT
      DEFLECTION CORRESPONDING TO DESIGN
      CRACKING MOMENT                                  0.26 IN
      SERVICIBILITY DEMAND MOMENT                      1.95 FT-KIP/FT
      DEFLECTION                                       0.22 IN
      MAXIMUM ALLOWABLE DEFLECTION                     2.10 IN
      CHECK                                             OK

16. STRENGTH LIMIT STATE DESIGN MOMENT CAPACITY CHECK (LOAD CASES 2 & 3)

                                         CASE 2     CASE 3
      STRENGTH MOMENT DEMAND              2.53       2.46 FT-KIP/FT
      DESIGN STRENGTH MOMENT CAPACITY     3.56       3.45 FT-KIP/FT
      DEFLECTION                          0.64       0.59 IN
      CHECK                                OK         OK
```

Figure 10.13 *Continued*

crete shear walls. However, as we shall later discuss, a major area of disagreement exists between many engineers who design concrete shear walls and many masonry designers over the use of highly reinforced boundary members. With that issue put aside it is possible, as this section will illustrate, to design ductile masonry shear walls that will perform well during seismic loading.

10.5.2 Structural Mechanics

The reader is referred to Volume II of the book entitled *Earthquake Design of Concrete Masonry Buildings* by Englekirk and Hart,[10-16] and to *Design of Reinforced Masonry* by Schneider and Dickey,[10-23] for excellent discussions of the structural mechanics of reinforced masonry design. In most respects it parallels the standard development of structural engineering design we are familiar with. For example, plane cross sections are assumed to remain plane and a rectangular (Whitney) stress block replaces a more complex stress strain curve. The reader may wish to refer to these two references prior to reading the next subsection.

10.5.3 State-of-the-Art Limit-State Design Criterion

The following design criterion is very similar to the 1988 UBC design criterion. The reader is referred to Reference 10-24 for a history of that development.

Notation

A_n = net cross-sectional area perpendicular to axial load, in.^{2}

Table 10-8 Nominal Shear-Strength Coefficient

M/Vd*	C_d
$\leqslant 0.25$	2.4
$\geqslant 1.00$	1.2

* M is the maximum bending moment that occurs simultaneously with the shear load V at the section under consideration. Interpolation by straight line for M/Vd values between 0.25 and 1.00 is permitted.

A_{mv} = net area of masonry section bounded by wall thickness and length of section in the direction of shear force considered, in.^{2}

A_s = area of tension reinforcement, in.^{2}

a_b = length of compressive stress block, inches

b = effective width of wall, inches

C_d = masonry shear-strength coefficient as obtained from Table 10-8

d = distance from extreme compression fiber to centroid of tension reinforcement, inches

D = dead loads, or related internal moments and forces

E = load effects of earthquake, or related internal moments and forces

E_s = modulus of elasticity of steel, 29,000,000 lb/in.^{2}

ϵ_{mu} = maximum usable compressive strain of masonry

F_s = allowable stress in reinforcement, lb/in.^{2}

f_s = computed stress in reinforcement, lb/in.^{2}

f'_m = specified compressive strength of masonry at the age of 28 days, lb/in.^{2}

f_y = specified yield strength of reinforcement, lb/in.^{2}

L = live loads, or related internal moments and forces

L_w = length of wall

P_b = nominal balanced-design axial strength

P_0 = nominal axial strength without bending loads

P_u = required axial strength

U = required strength to resist factored loads, or related internal moments and forces

V_n = nominal shear strength

V_m = nominal shear strength provided by masonry

V_s = nominal shear strength provided by shear reinforcement

ρ_n = ratio of distributed shear reinforcement on a plane perpendicular to plane of A_{mv}

ϕ = strength-reduction factor

Quality-Control Provision

1. Special inspection during construction of the shear wall is required, especially after placement of the steel and prior to the pouring of the grout.
2. f'_m shall not be less than 1500 lb/in.^{2} nor greater than 4000 lb/in.^{2}. However, in concrete masonry a limit of 3000 lb/in.^{2} is recommended unless special quality-control measures are taken or recommended by the engineer.
3. f'_m shall be verified with prism testing.

Design Procedure

1. *Required strength.*

- For earthquake loading, the load factors shall be

$$U = 1.4(D + L + E) \qquad (10\text{-}38)$$

$$U = 0.90 \pm 1.4E \qquad (10\text{-}39)$$

- The required strength U to resist dead load D and live load L shall be at least equal to

$$U = 1.4D + 1.7L \qquad (10\text{-}40)$$

2. *Design strength.* The design strength provided by the shear-wall cross-section in terms of axial force, shear, and moment shall be computed as the nominal strength multiplied by the strength reduction factor ϕ. Shear walls shall be proportioned so that the design strength exceeds the required strength. The strength-reduction factor ϕ shall be as follows:

- Axial load and axial load with flexure: 0.65.
For members in which f_y does not exceed 60,000 lb/in.^{2}, with symmetrical reinforcement, ϕ may be increased linearly to 0.85 as ϕP_n decreases from $0.65P_b$ to zero. (Note that the 1988 UBC has 0.25 instead of 0.65.) For solid grouted wall P_b may be calculated by

$$P_b = 0.85 f'_m b a_b \qquad (10\text{-}41)$$

where

$$a_b = 0.85 \frac{\epsilon_{mu}}{\epsilon_{mu} + f_y/E_s} d \qquad (10\text{-}42)$$

- Shear: 0.60.
The shear-strength reduction factory may be 0.80 for any shear wall when its nominal shear strength exceeds the shear corresponding to development of its nominal flexural strength for the factored-load combination.

3. *Design assumptions for nominal strength.* The nominal strength of shear-wall cross-sections shall be based on the assumptions prescribed in Section 10-4. The maximum usable strain, ϵ_{mu}, at the extreme masonry compression fiber shall not exceed 0.003 unless compression tests on prisms indicate higher values are justified.
4. *Reinforcement.*

- Minimum reinforcement shall be 0.0007 in either direction and 0.002 total.
- When the shear-wall failure mode is in flexure, the nominal flexural strength of the shear wall shall be at least three times the cracking moment strength of the wall from Equation 10-31.
- All continuous reinforcement shall be anchored or spliced in accordance with 1985 UBC Section 2409(e)1, 2, 3A (with $F_s = 0.5f_y$), 3B, 3D, 3F, and 3G (with $F_s = f_y$).
- The minimum amount of vertical reinforcement shall not be less than one-half the horizontal reinforcement.
- Maximum spacing of horizontal reinforcement within the region defined in item 6, paragraph (c)(1) below shall not exceed three times nominal wall thickness or 24 in., whichever is less.

5. *Axial strength.* The nominal axial strength of the shear wall supporting axial loads only shall be calculated by

$$P_0 = 0.85 f'_m (A_n - A_s) + f_y A_s \qquad (10\text{-}43)$$

The axial design strength provided by the shear-wall cross-section shall satisfy the equation

$$P_u \leq \phi(0.80) P_0 \qquad (10\text{-}44)$$

6. *Shear strength.*

(a) The nominal shear strength shall be determined using either paragraph (b) or (c) below. Table 10-9 gives the maximum nominal shear strength values.

Table 10-9 Maximum Nominal Shear-Strength Values

M/Vd*	$V_n/(A_{mv}\sqrt{f'_m})$
≤ 0.25	6.0
≥ 1.00	4.0

* M is the maximum bending moment that occurs simultaneously with the shear load V at the section under consideration. Interpolation by straight line for M/Vd values between 0.25 and 1.00 is permitted.

(b) The nominal shear strength of the shear wall shall be determined, except as provided in paragraph (c) below, from

$$V_n = V_m + V_s \tag{10-45}$$

where

$$V_m = C_d A_{mv} \sqrt{f'_m} \tag{10-46}$$

and

$$V_s = A_{mv} \rho_n f_y \tag{10-47}$$

(c) For a shear wall whose nominal shear strength exceeds the shear corresponding to development of its nominal flexural strength, two shear regions exist:

(1) For all cross-sections within the region defined by the base of the shear wall and a plane at a distance L_w above the base of the shear wall, the nominal shear strength shall be determined from

$$V_n = A_{mv} \rho_n f_y \tag{10-48}$$

The required shear strength for this region shall be calculated at a distance $L_w/2$ above the base of the shear wall but not to exceed one-half story height.

(2) For the other region the nominal shear strength of the shear wall shall be determined from Equation 10-45.

7. *Confinement of vertical steel.* All vertical reinforcement whose corresponding masonry compressive stress, corresponding to factored forces, exceeds $0.4f'_m$ shall be confined when the failure mode is flexure. When vertical steel needs to be confined, that shall be done with a minimum of No. 3 bars at a maximum of 8-in. spacing or equivalent within the grouted core and within the region defined as the base of the shear wall. When confinement is needed, the vertical steel confined shall be at least from the end of the wall to a lateral distance three times the thickness of the wall.

10.5.4 Comments on State-of-the-Art Design Criteria for Shear Walls

The design strength is obtained by multiplying the nominal strength by a strength-reduction factor. The nominal strength is ideally the best professional estimate of the true strength of the member. The strength-reduction factor is selected to take account of the uncertainty of the value of the parameters in the nominal strength equation, the workmanship in the field, and the general confidence in the equation's ability to predict the actual performance of the member.

For walls subjected to flexure and axial load the variation in the numerical value of the strength reduction factor is a function of the axial load on the shear wall. The primary reason for this is to insure that the wall's performance is that of an underreinforced flexural member. Therefore, we have divided the interaction diagram for the shear wall into two zones for the purpose of setting a value of the strength reduction factor. Zone 1 corresponds to sufficiently low axial loads to insure a very ductile shear-wall performance. We have provided an axial-load limit of less than 65% of an approximate calculation of the balanced-design axial load P_b. This alternative approach, by including the balance design axial load, places a stronger emphasis on the importance of quantifying the intensity of the axial load as a function of the balanced-design axial load in order to promote ductility. The value of 65% P_b is reasonable based on a reliability analysis which incorporated uncertainty in material properties and the design equation.[10-25] To provide a straightforward calculation of the balanced-design axial load, we have provided an equation which is a good approximation for the use here (i.e., typically less than 10% error). This approximation assumes that the positive and negative forces from the steel in tension and compression balance each other in the equilibrium equation.

Zone 2 is for values of axial load greater than 65% of the balanced-design axial load. The numerical value of the strength-reduction factor in zone 2 is equal to 0.65. To insure that the quality of the masonry is consistent with the engineering design assumptions, the minimum value of f'_m is set at 1500 lb/in.^{2}. The maximum recommended value for f'_m is 3000 lb/in.^{2} unless a special level of quality control is used for the concrete masonry. Unless the engineer has checked with his local block supplier, it is reasonable to assume that 3000 lb/in.^{2} is a practical limit.

The strength-reduction factor for shear walls where the mode of failure is shear is equal to 0.60. This typically represents shear walls that are long compared to their height.

For walls where flexure is a possible failure mode, the shear resistance that is provided is checked to insure that the shear corresponding to the development of the full nominal flexural strength of the wall is provided. This approach is consistent with the approach taken for reinforced concrete in the 1985 UBC. In this situation, the strength-reduction factor for shear is equal to 0.80.

The equation used to calculate the axial strength of the wall is equal to the specified compressive strength times the net area of the wall times an effective-stress parameter value of 85% plus the yield stress of the steel times the area of the steel. This equation is directly consistent with the equation used in reinforced-concrete design.

For pure axial-load design, the strength-reduction factor is equal to 0.65 and was discussed in Section 10-4. A further reduction is made to reduce the axial load by multiplying the nominal strength by 0.8 in order to allow for accidental eccentricities.

The shear strength of shear walls can be determined using either of two alternative approaches. The first approach is used for shear walls where the failure mode is shear. In this situation, the strength reduction factor is equal to 0.60 and the nominal shear strength is obtained by adding two terms. The first term is the shear strength assumed to be provided by the masonry in a reinforced masonry wall. The second term is the shear strength provided by the shear reinforcement.

The second approach used to calculate the nominal shear strength of a wall is appropriate for shear walls where a flexural mode of failure is possible. The intent of this approach is to require that sufficient shear reinforcement be placed in the wall to insure a ductile flexural failure. In this situation, the strength-reduction factor for shear is equal to 0.80. The flexural failure mode will result in a shear wall where the region near the base will be called upon to undergo an inelastic moment-curvature response. Therefore, we have identified two shear regions for such a shear wall. Shear region 1 extends up from the base of the wall a distance equal to the length of the wall and is a plastic hinge region. In this region because of the inelastic cyclic response, only the shear resistance provided by the steel is considered in the design. In region 2, above the plastic hinge, the masonry and the steel are both used to calculate the shear strength of the wall.

Let us now return to the maximum nominal shear strength and nominal shear-strength coefficients in Tables 10-8 and 10-9. The design values presented in these tables are reduced values obtained in part from published research.[10-8, 10-24]

The use of boundary members in shear walls is a highly controversial topic in masonry design. The New Zealand building code does not allow boundary members to be used in masonry shear walls.[10-26] The New Zealand approach is to encourage the designer to distribute the vertical steel uniformly along the length of the wall. This, it is argued, provides a more consistent distribution of shear stress between the wall and the foundation. The counter to this argument is the current approach taken by reinforced-concrete design criteria. In essence, the current approach for reinforced-concrete walls is to design the shear wall as if it were essentially a second-class ductile frame and discount the concrete between the boundary members. The net result of this design is high axial loads at the ends of the wall. The author's position is that the New Zealand approach results in a much better shear-wall design.

The approach we recommend for steel-confinement determination specifies that the factored loads are applied to the shear wall; then, using the principles of mechanics, the compressive stress in the masonry immediately adjacent to the vertical reinforcing bars is calculated. If this stress exceeds 40% of the maximum specified compressive stress, the vertical reinforcement must be confined. The figure 40% is based on an approximate unconfined-masonry-prism strain value of 0.001 for a stress–strain curve that is parabolic between zero stress and maximum compressive stress (see Figure 10-12). If the strains are below 0.001 then based on observations of prism tests we can expect no significant loss of strength or stiffness due to cyclic loading and small internal masonry cracking.

10.5.5 Shear-Wall Design Examples

This example problem is for an 8-in. block wall that is 24 ft long and 57 ft 6 in. tall. The wall lateral loading corresponds to UBC seismic zone 4, which produces a design base shear and overturning moment of 120 kips and 4608 kip-ft, respectively. The unfactored dead and live loads at the base of the wall are 486 and 84 kips, respectively. This example was suggested by Mr. James Lai and was worked and reviewed as part of the Structural Engineers Association of California Seismology Committee review of the 1988 UBC Code, Section 2412, Shear Wall Design Criteria.

Nominal Design Loads

Dead load = 486 kips

Live load = 84 kips

Lateral unfactored Chapter 23

seismic shear force = 120 kips

Seismic moment M_s = 4608 kip-ft

Wall Properties

Nominal block thickness = 8 in.

Length of wall = 24 ft

Specified compressive strength = 3.0 ksi

Modulus of rupture = $4.5\,(f'_m)^{0.5}$

Maximum usable masonry strain = 0.003

Load Factors (1988 UBC)

$$U = 1.4D + 1.7L$$
$$U = 1.4D + 1.4L + 1.4E$$
$$U = 0.9D - 1.4E$$

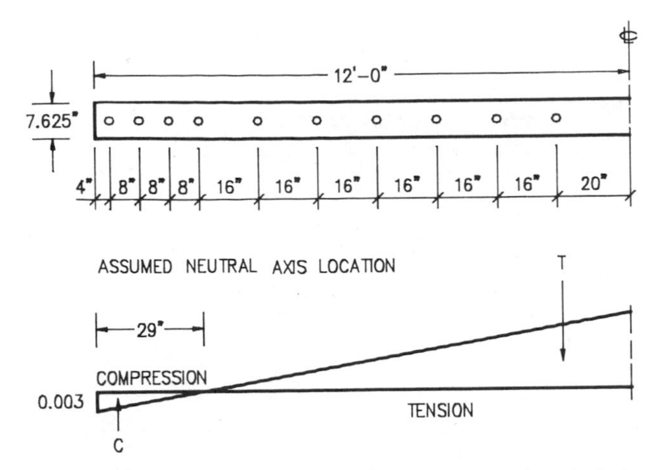

Figure 10-14 Strain profile with maximum compression strain in masonry equal to 0.003.

Factored Axial Dead and Live Load

$$P_u = 1.4 \times 486 + 1.7 \times 84 = 823 \text{ kips}$$

Nominal Pure Axial Load Capacity

$$P_{0n} = 0.85 f'_m l_w t$$
$$= 0.85 \times 3.0 \times 24 \times 12 \times 7.625 = 5600 \text{ kips}$$

Design Pure Axial Load Capacity

$$P_{0u} = 0.65 P_{0n} = 3640 \text{ kips} > P_u \qquad \text{O.K.}$$

Nominal Pure Bending-Moment Capacity

Specified yield strength = 60 ksi

Vertical steel: see Figure 10-14

$$A_s = 12 \text{ in.}^2$$

Assume that the neutral is located at a distance of 29 in. from the compression face (see Figure 10-15). Then it follows that

$$T = A_s f_s = 0.6 \times 45 + 0.6 \times 15 \times 60$$
$$C = A_s f_s + 0.85 \times 0.85 \times f'_m \times b \times c$$
$$= (0.6 \times 60 + 0.6 \times 51 + 0.6 \times 27 + 0.6 \times 3) + (0.72 \times 3 \times 7.625 \times 29)$$

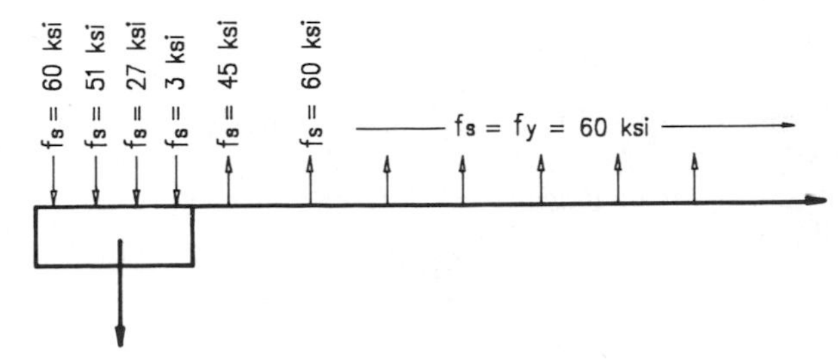

Figure 10-15 Force equilibrium.

So

$$T - C = 3 \approx 0 \qquad \text{O.K.}$$

Taking moment about extreme compression fiber,

$$M_{0n} = -(0.6 \times 60 \times 4 + 0.6 \times 51 \times 12 + 0.6 \times 27 \times 20 + 0.6 \times 3 \times 28) - \frac{0.722 \times 3 \times 7.625 \times 29 \times 29}{2 \times 0.85} + (0.6 \times 45 \times 44 + 0.6 \times 60 \times 60 + \cdots + 0.6 \times 60 \times 284)$$

$$M_{0n} = -\frac{885 - 5903 + 100{,}980}{12} = 7849 \text{ kip-ft}$$

Factored Required Moment Demand

$$M_u = 1.4 M_s = 1.4 \times 4608 = 6451 \text{ kip-ft}$$

Nominal Cracking Moment Capacity

$$\text{Gross section modulus} = \frac{7.625 \times 288^2}{6} = 105{,}408 \text{ in.}^3$$

$$\text{Modulus of rupture} = 4.5 \times (3000)^{0.5} = 246 \text{ lb/in.}^2$$

$$M_{cr} = \left(\frac{P}{A} + f_r\right) S$$
$$= \left(\frac{570 \times 10^3}{2196} + 246.4\right) \frac{105{,}408}{12{,}000}$$
$$= 4445 \text{ kip-ft}$$

Is Confinement of Vertical Steel Needed? Compressive stress of masonry,

$$0.4 f'_m = 120 \text{ ksi}$$

Corresponding strain in masonry when we assume that the maximum stress f'_m corresponds to a strain of 0.0025:

$$0.001$$

- Load case 1: Assume that the neutral axis is located at a distance of 160 in. from the extreme compression face (see Figure 10-16)

$$T = 0.6(0.7 + 3.6 + \cdots + 22.5) = 77 \text{ kips}$$
$$C = 0.6(28.3 + 26.8 + \cdots + 6.5) + \tfrac{1}{2} \times 160 \times 7.625 \times 1.2 = 844 \text{ kips}$$
$$C - T = 767 \approx P_u = 1.40 + 1.4L = 798 \qquad \text{O.K.}$$

Taking moments about plastic centroid, we ob-

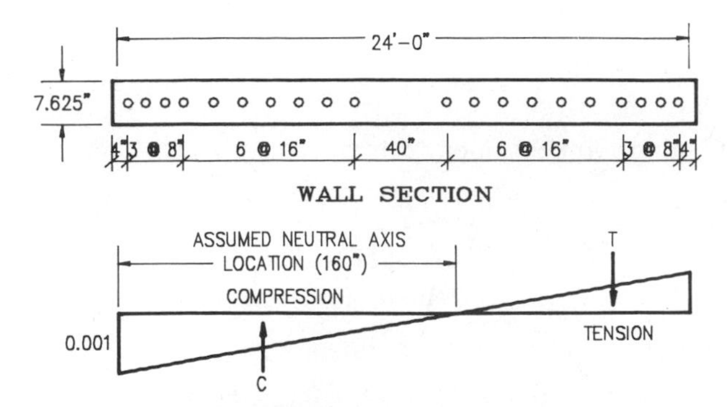

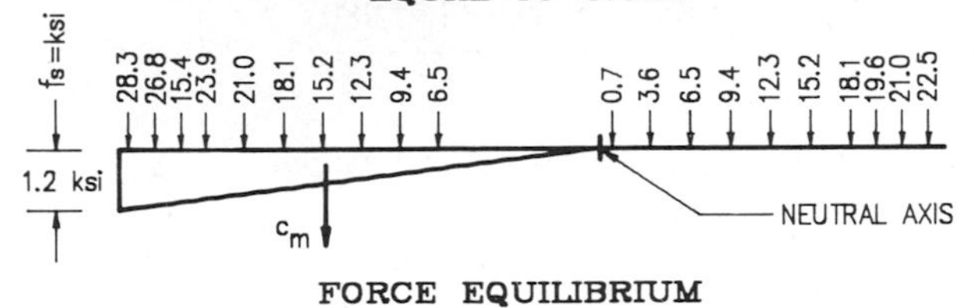

Figure 10-16 Strain profile and force equilibrium.

tain the moment corresponding to $0.4f_m'$,

$$
\begin{aligned}
&0.6(0.7 \times 10 + 3.6 \times 36 + \cdots + 22.5 \times 140) \\
&\quad +0.6(28.3 \times 140 + 26.8 \times 132 + \cdots \\
&\qquad +6.5 \times 20) \\
&\quad +\tfrac{1}{2} \times 160 \times 7.625 \times 1.2 \times \left(\tfrac{2}{3} \times 160 - 16\right) \\
&\qquad = +\frac{8475.6 + 11{,}510.0 + 66{,}392.4}{12} \\
&\qquad = 7198 \text{ kip-ft}
\end{aligned}
$$

Thus confinement of vertical compression steel is not needed, because this moment (7198 kip-ft) is greater than the factored required moment demand (6451 kip-ft).

- Load case 2:

$$P_u = 0.9D = 437 \text{ kips}$$

Assume that the neutral axis is located at a distance of 115 in. from the extreme compression face:

$$C - T = 438 \text{ kips} \approx P_u = 437 \text{ kips} \qquad \text{O.K.}$$

Therefore, moment corresponding to $0.4f_m'$ is

$$6951 \text{ kip-ft}$$

Thus confinement of vertical compression steel is not needed, because 6951 kip-fit is greater than 6451 kip-ft.

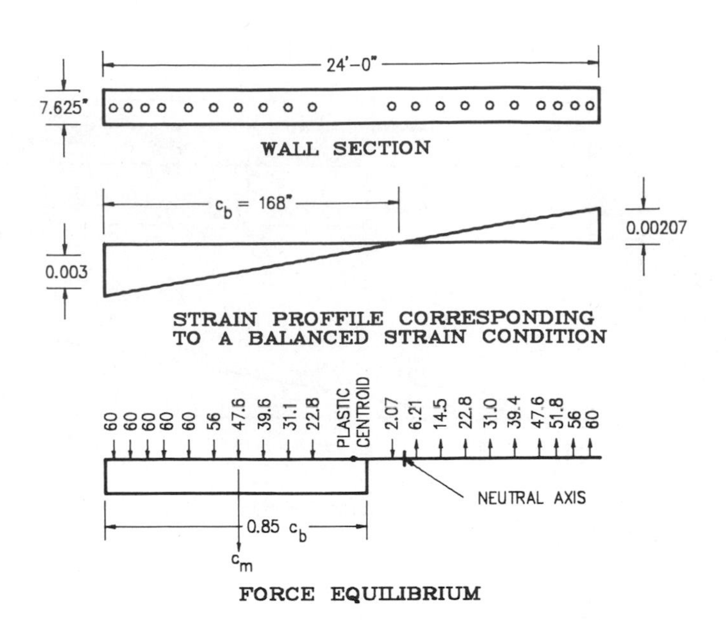

Figure 10-17 Strain profile and force equilibrium.

Approximate Design Balanced Axial Load P_{bu} See Figure 10-17. We have

$$C_b = \frac{\epsilon_{mu}}{\epsilon_s + \epsilon_{mu}} d = \frac{0.003}{0.003 + 0.00207} \times 284 = 168.01$$

$$
\begin{aligned}
T &= 0.6[6.21 + 14.5 + \cdots + 60] = 197.5 \text{ kips} \\
C &= 0.6[60 \times 60 + \cdots + 2.07] \\
&\quad + 0.85 \times 0.85 \times 168 \times 7.625 \times 3.0 = 30{,}755 \text{ kips} \\
P_{bn} &= C - T = 2878 \text{ kips} \approx 2880 \text{ kips}
\end{aligned}
$$

Design balanced axial load:

$$P_{bu} = 0.65P_{bn} = 1872 \text{ kips}$$

Note: This approximation to calculate P_{bn} assumes that the force from the compression steel cancels that from the tension steel.

Compare Ultimate and Balanced Axial Loads

1. Ultimate load:

$$
\begin{aligned}
P_u &= 1.4D + 1.4L = 798 \text{ kips} \\
&= 28\% \text{ of } P_{bn} \\
&= 43\% \text{ of } P_{bu}
\end{aligned}
$$

$$0.1f_m'A_n/P_{bn} = 0.23 < 0.25P_{bn}$$

Capacity reduction factor 0.65

So use 0.23 for interpolation of strength-reduction factor.

2. Balanced load:

$$
\begin{aligned}
P_u &= 0.9D = 437 \text{ kips} \\
&= 15\% \text{ of } P_{bn} \\
&= 23\% \text{ of } P_{bu}
\end{aligned}
$$

```
                 ********************************
                 * CONCRETE MASONRY SHEAR-WALL *
                 ********************************

                     SHWALL VERSION (1.01)
                       SEPTEMBER, 1987
                 ------------------------

*****************************************************************
*                                                               *
*  THIS CONCRETE MASONRY SHEAR WALL PROGRAM (SHWALL) WAS WRITTEN *
*  AS PART OF THE TCCMAR MASONRY RESEARCH PROGRAM AND WAS FUNDED *
*  BY THE NATIONAL SCIENCE FOUNDATION AND THE CONCRETE MASONRY   *
*  ASSOCIATION OF CALIFORNIA AND NEVADA.  THE WORK WAS CONDUCTED *
*  AS PART OF THE CATEGORY 2 RESEARCH WORK UNDER THE JOINT       *
*  RESEARCH EFFORTS OF EKEH (EWING AND ASSOCIATES, KARIOTIS AND  *
*  ASSOCIATES, AND ENGLKERIK AND HART INC.) AND WAS WRITTEN BY   *
*  THE STAFF OF ENGLEKIRK AND HART, INC.                         *
*                                                               *
*  SHWALL IS WRITTEN AS A PART OF THE NATIONAL EARTHQUAKE        *
*  HAZARD REDUCTION PROGRAM (NEHRP) AND IS INTENDED TO BE BOTH A *
*  RESEARCH AID AND DESIGN AID FOR CONCRETE MASONRY SHEAR WALLS. *
*                                                               *
*  NEITHER THE NATIONAL SCIENCE FOUNDATION, THE CONCRETE MASONRY *
*  ASSOCIATION OF CALIFORNIA AND NEVEDA NOR ENGLEKIRK AND HART,  *
*  INC. ASSUMES ANY RESPONSIBILITY FOR ANY ERRORS, MISTAKES, OR  *
*  MISREPRESENTATIONS THAT MAY RESULT FROM THE USE OF SHWALL.    *
*  SHWALL IS PROVIDED IN AN "AS IS" CONDITION.  NO WARRANTIES OF *
*  ANY KIND, WHETHER STATUTORY, WRITTEN, ORAL, EXPRESSED, OR     *
*  IMPLIED (INCLUDING WARRANTIES OF ANY FITNESS AND              *
*  MERCHANTABILITY) SHALL APPLY.                                 *
*                                                               *
*  THE COPY OF THIS PROGRAM MAY BE OBTAINED FROM :               *
*  CONCRETE MASONRY ASSOCIATION                                  *
*  OF CALIFORNIA AND NEVADA                                      *
*  6060 SUNRISE VISTA DR. SUITE 1875                             *
*  CITRUS HEIGHTS, CA 95610                                      *
*  TEL: (916) 722-1700                                           *
*                                                               *
*****************************************************************

                                   JOB #   :    85-G142
                                   DATE    :    11-87
                                   SHEET   :     1
                                   TITLE 1:    SHEAR WALL
                                   TITLE 2:    EXAMPLE 1

1.  NOMINAL DESIGN LOADS

       DEAD LOAD.................................     486.0  KIPS
       LIVE LOAD.................................      84.0  KIPS
       LATERAL SHEAR FORCE (Vs)..................     120.0  KIPS
       MOMENT(Ms)................................    4608.0  K-FT

2.  WALL PROPERTIES

       NOMINAL BLOCK THICKNESS.(t)...............       8.0  IN.
       LENGTH OF WALL.(Lw).......................      24.0  FT
       SPECIFIED COMPRESSIVE STRENGTH............       3.0  KSI
       MODULUS OF RUPTURE........................       4.5  f'm^0.5
       MAXIMUM USABLE MASONRY COMPRESSIVE STRAIN..    0.003  -/-

3.  LOAD FACTORS

       ENTER 1 FOR ANSI OR 2 FOR U.B.C LOAD FACTOR       2
       U =  1.4 D +  1.7 L
       U =  1.4 D +  1.4 L +  1.4 E
       U =  0.9 D -  1.4 E

4.  FACTORED AXIAL DEAD AND LIVE LOAD

       Pu=  1.4 D +  1.7 L.......................       823  KIPS

5.  NOMINAL AXIAL STRENGTH (Pon = 0.85 f'mLwt)....     5600  KIPS

6.  DESIGN AXIAL STRENGTH   (Pou = 0.65 Pon).......     3640  KIPS

       Pu =  15  % OF NOMINAL AXIAL STRENGTH (Pon)           O.K.
       Pu =  23  % OF DESIGN AXIAL STRENGTH  (Pou)           O.K.

7.  NOMINAL PURE BENDING MOMENT CALCULATION

       SPECIFIED YIELD STRENGTH..................      60.0  KSI
       ESTIMATED NEUTRAL AXIS LOCATION ..........      29.0  IN.

       VERTICAL STEEL LOCATION

                    AS1  =   0.60  IN^2        D1 =     4.00  IN.
                    AS2  =   0.60  IN^2        D2 =    12.00  IN.
                    AS3  =   0.60  IN^2        D3 =    20.00  IN.
                    AS4  =   0.60  IN^2        D4 =    28.00  IN.
                    AS5  =   0.60  IN^2        D5 =    44.00  IN.
                    AS6  =   0.60  IN^2        D6 =    60.00  IN.
                    AS7  =   0.60  IN^2        D7 =    76.00  IN.
                    AS8  =   0.60  IN^2        D8 =    92.00  IN.
                    AS9  =   0.60  IN^2        D9 =   108.00  IN.
                    AS10 =   0.60  IN^2        D10=   124.00  IN.
                    AS11 =   0.60  IN^2        D11=   164.00  IN.
                    AS12 =   0.60  IN^2        D12=   180.00  IN.
                    AS13 =   0.60  IN^2        D13=   196.00  IN.
                    AS14 =   0.60  IN^2        D14=   212.00  IN.
                    AS15 =   0.60  IN^2        D15=   228.00  IN.
                    AS16 =   0.60  IN^2        D16=   244.00  IN.
```

```
                    AS17 =   0.60  IN^2        D17=   260.00  IN.
                    AS18 =   0.60  IN^2        D18=   268.00  IN.
                    AS19 =   0.60  IN^2        D19=   276.00  IN.
                    AS20 =   0.60  IN^2        D20=   284.00  IN.
       AREA OF STEEL      =    12.0  IN^2
       SUM OF TENSION AND COMPRESSION FORCES......         3  KIPS O.K.
       NEUTRAL AXIS MUST BE  O.K.      ...........
       NEW NEUTRAL AXIS LOCATION..................     29.00  IN.
       NOMINAL PURE BENDING MOMENT (Mon)..........      7849  K-FT

8.  FACTORED REQUIRED MOMENT

       Mu = 1.4 Ms................................      6451  K-FT

9.  NOMINAL CRACKING MOMENT CAPACITY OF SECTION (Mcr)

       DO YOU WANT TO INCLUDE AXIAL LOAD IN
       CRACKING MOMENT CALCULATION?  (YES=1,NO=0).         1
       PERCENTAGE OF LIVE LOAD TO BE USED TO
       CALCUTLATE NOMINAL CRACKING MOMENT.........       100  %
       GROSS SECTION MODULUS......................    105408  IN^3
       MODULUS OF RUPTURE.........................       246  PSI
       NOMINAL CRACKING MOMENT CAPACITY...........      4445  K-FT

10. EVALUATE NECESSITY OF VERTICAL STEEL CONFINEMENT (BOUNDARY MEMBER)

       COMPRESSIVE STRESS AT 0.4 f'm   ...........      1.20  KSI

       ESTIMATED NEUTRAL AXIS LOCATION............    160.00  IN.
       Pu = 1.4 D +  1.4 L........................       798  KIPS
       SUMMATION OF TENSION AND COMPRESSION FORCES       767  KIPS O.K.
       MOMENT CORRESPONDING TO STRESS   0.4 f'm...      7197  K-FT
       THEREFORE :          BOUNDARY ELEMENT IS   NOT NEEDED
       THE N.A. MUST BE  O.K......................

       ESTIMATED NEUTRAL AXIS LOCATION............    115.00  IN.
       Pu = 0.9 D.................................       437  KIPS
       SUMMATION OF TENSION AND COMPRESSION FORCES       438  KIPS O.K.
       MOMENT CORRESPONDING TO STRESS   0.4 f'm...      6951  K-FT
       THEREFORE :          BOUNDARY ELEMENT IS   NOT NEEDED
       THE N.A. MUST BE  O.K......................

11. CALCULATE DESIGN BALANCED AXIAL LOAD (Pbu)

       BALANCED NEUTRAL AXIS (Cb).................    168.08  IN.
       NOMINAL BALANCED AXIAL LOAD (Pbn)..........      2880  KIPS
       DESIGN BALANCED AXIAL LOAD (Pbu=0.65 Pbn)..      1872  KIPS

12. COMPARE ULTIMATE AXIAL LOAD AND BALANCED AXIAL LOAD

       Pu = 1.4 D +  1.4 L........................       798  KIPS
       Pu =  28  % OF NOMINAL BALANCED LOAD (Pbn).
       Pu =  43  % OF DESIGN BALANCED LOAD (Pbu)..
       FLEXURAL PHI FACTOR........................      0.65

       Pu = 0.9 D.................................       437  KIPS
       Pu =  15  % OF NOMINAL BALANCED LOAD (Pbn).
       Pu =  23  % OF DESIGN BALANCED LOAD (Pbu)..
       FLEXURAL PHI FACTOR........................      0.72

13. CALCULATE DESIGN BALANCED MOMENT CAPACITY (Mbu)

       NOMINAL BALANCED MOMENT (Mbn)..............     21048  K-FT
       DESIGN BALANCED MOMENT (Mbu =    0.65 Mbn)..     13681  K-FT

14. CALCULATE DESIGN PURE BENDING MOMENT CAPACITY (Mou)

       NOMINAL PURE BENDING MOMENT(Mon)...........      7849  K-FT
       DESIGN PURE BENDING MOMENT (Mou=0.85 Mon)        6672  K-FT

15. CALCULATE DESIGN PURE AXIAL LOAD CAPACITY (Pou)

       NOMINAL PURE AXIAL LOAD (Pon)..............      6289  KIPS
       DESIGN PURE AXIAL LOAD (Pou =    0.65 Pon)..      4088  KIPS
       MAXIMUM USABLE AXIAL LOAD (Pmax=0.80 Pou)..      3270— KIPS

16. DESIGN MOMENT CAPACITY (M)

       Mu = 1.4 Ms................................      6451  K-FT

       DESIGN MOMENT CAPACITY CORRESPONDING TO

            1.4 D +  1.4 L........................      8759  K-FT O.K.
            M/Mcr.................................       2.0

            0.9 D.................................      7634  K-FT O.K.
            M/Mcr.................................       1.7

17 SHEAR DEMAND

       Vu = 1.4 Vs  ..............................       168  KIPS
       REQUIRED FLEXURAL DUCTILE SHEAR (Vdu)......       351  KIPS

18 SHEAR CAPACITY

       MAXIMUM NOMINAL SHEAR STRENGTH TABLE 24-K..       481  KIPS

       PLASTIC HINGE REGION ENTER 1 OTHERWISE 2...         1
       Vm = Cd Ag SQRT(f'm) TABLE 24-L............         0  KIPS
       Av = 0.6 IN^2      S = 24.0 IN.^2...........
       Vs =.......................................       432  KIPS
       Vn =Vm + Vs................................       432  KIPS
       DUCTILE SHEAR RATIO (Vn/Vdu)...............      1.23
       SHEAR PHI FACTOR...........................      0.80
       Vu    <<<     0.80 Vn = 346 KIPS O.K. ......
```

Figure 10-19 SHWALL worksheets and results for the shear wall example.

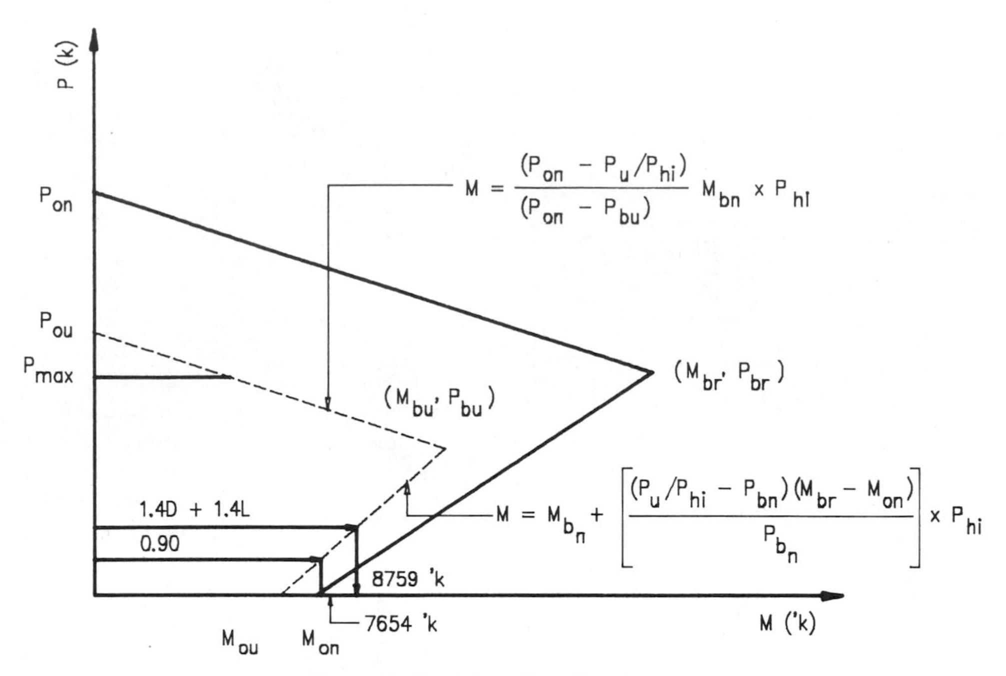

Figure 10-18 Interaction diagram.

Capacity reduction factor

$$= 0.85 - \left(\frac{0.85 - 0.65}{0.23} \times 0.15\right) = 0.72$$

Design Balanced Moment Capacity Taking moment about plastic centroid,

$$M_{\mathrm{bn}} = [0.6 \times 60 \times 140 + 0.6 \times 132 + \cdots + 22.8 \times 0.6 \times 10] - 2.07 \times 0.6 \times 20 + 0.852 \times 168 \times 7.625 \times 3.0 \times 72.6 - [0.6 \times 60 \times 140 + \cdots + 0.6 \times 6.21 \times 36]$$

$$= \frac{28{,}973.6 - 24.84 + 201{,}581 + 22{,}044.8}{12}$$

$$M_{\mathrm{bn}} = 21{,}047.8 \text{ kip-ft}$$

$$M_{\mathrm{bu}} = 0.65 M_{\mathrm{bn}} = 13{,}681 \text{ kip-ft}$$

Design Pure Bending-Moment Capacity M_{0u} Nominal pure bending moment:

$$M_{0n} = 7849 \text{ kip-ft}$$

Design pure bending moment:

$$M_{0u} = (0.85 M_{0n}) = 6672 \text{ kip-ft}$$

Design Pure Axial-Load Capacity Nominal pure axial load:

$$P_{0n} = 0.85 f'_m A_c + f_y A_s$$

$$= 0.85 \times 3.0(7.625 \times 288 - 12.0) + 60 \times 12.0$$

$$= 6289 \text{ kips}$$

Design pure axial load:

$$P_{0u} = 0.65 P_{0n} = 4088 \text{ kips}$$

Maximum usable axial load:

$$P_{\max} = 0.8 P_{0u} = 3270 \text{ kips}$$

Design Moment Capacity M See Figure 10-18. We have

$$M_u = 1.4 M_s$$

- Case 1:

$$P_u = 1.4D + 1.4L = 798 \text{ kip} < P_{\mathrm{bu}} = 2880$$

$$M = \left(\frac{(P_u/\phi - P_{\mathrm{bn}})(M_{\mathrm{bn}} - M_{0n})}{P_{\mathrm{bn}}} + M_{\mathrm{bn}}\right)\phi$$

$$= \left(\frac{(798/0.65 - 2880)(21048 - 7849)}{2880} + 21048\right)0.65 = 8759 \text{ kip-ft}$$

$$\frac{M}{M_{cr}} = \frac{8759}{4445} = 1.97$$

- Case 2:

$$P_u = 0.9D = 437 < P_{\mathrm{bu}} = 2880$$

$$M = \left(\frac{(437/0.72 - 2880)(21{,}048 - 7849)}{2880} + 21{,}048\right)0.72 = 7654 \text{ kip-ft}$$

$$\frac{M}{M_{cr}} = \frac{7654}{4445} = 1.7$$

Shear Demand

$$V_u = 1.4V_s = 1.4 \times 120 = 168 \text{ kips}$$

$$V_{du} = \frac{(V_s/M_s)M_{max}}{\phi}$$

$$= \frac{(120/4608) \times 8759}{0.65} = 351 \text{ kips}$$

Shear Capacity Maximum nominal shear (1988 UBC Table 24-K):

$$\frac{M}{V_d} = \frac{4608}{120 \times 24} = 1.6 > 0.25$$

$$V_{n\max} = 4.0 \times A_{mv}(f_m')^{0.5}$$

$$= \frac{4.0 \times 24 \times 12 \times 7.625(3000)^{0.5}}{1000} = 481 \text{ kips}$$

Assuming the system is in the plastic-hinge region, then

$$V_m = 0$$

$$V_s = A_v f_y L_w / s$$

$$= 0.6 \times 60 \times 288/24 = 432 \text{ kips}$$

$$V_n = V_m + V_s = 432 \text{ kips}$$

Ductile shear ratio:

$$V_n/V_{du} = 1.23$$

$$\text{Shear strength-reduction factor} = 0.80$$

$$\phi = \begin{cases} 0.8 & \text{if} \quad V_n \geqslant V_{du} \leqslant V_{n\max} \\ 0.6 & \text{if} \quad V_n < V_{du} > V_{n\max} \end{cases}$$

$$V_u = \phi V_n = 0.8 \times 432 = 345.6 > V_u = 168 \text{ kips} \quad \text{O.K.}$$

10.5.6 SHWALL Computer Program

A computer program called SHWALL was developed as part of the TCCMAR program and is available from the Concrete Masonry Association of California and Nevada. Figure 10-19 provides a solution to the example worked in the previous section using SHWALL.

REFERENCES

10-1 First North American Masonry Conference, Univ. of Colorado, Boulder, CO, Aug. 1978.

10-2 Second North American Masonry Conference, Univ. of Maryland, College Park, MD, Aug. 1982.

10-3 Third North American Masonry Conference, Univ. of Texas, Arlington, TX, June 1985.

10-4 Third Canadian Masonry Symposium, Edmonton, Alberta, Canada, June 1982.

10-5 Fourth Canadian Masonry Symposium, Univ. of New Brunswick, New Brunswick, Canada, June 1986.

10-6 International Conference of Building Officials, *Uniform Building Code, 1985 Edition*, Whittier, CA, 1985.

10-7 Hegemier, G., *Various Contributions to Second North American Masonry Conference*, Univ. of Maryland, College Park, MD, Aug. 1982.

10-8 Mayes, R., *Various Contributions to Second North American Masonry Conference*, Univ. of Maryland, College Park, MD, Aug. 1982.

10-9 International Conference of Building Officials, "Strength Design of One to Four Story Concrete Masonry Building," Report No. 4115, ICBO, Whittier, CA, Feb. 1985.

10-10 International Conference of Building Officials, *Uniform Building Code, 1988 Edition*, Whittier, CA, 1988.

10-11 Noland, J., *Various Contributions to Fourth North American Masonry Conference*, Univ. of California at Los Angeles, Los Angeles, CA, Aug. 1987.

10-12 Hart, G. C., *Uncertainty Analysis, Loads and Safety in Structural Engineering*, Prentice-Hall, Englewood Cliffs, NJ, 1982.

10-13 Galambos, T. V., "Probability Based Load Criteria, Assessment of Current Design," *Proc. J. Structural Eng. ASCE* 108, No. ST5, May 1982.

10-14 Cornell, C. A., "Probability Based Load Criteria, Load Factors, and Load Combinations," *Proc. J. Structural Eng., ASCE* 108, No. ST5, May 1982.

10-15 Englekirk, R. E. and Hart, G. C., *Earthquake Design of Concrete Masonry Buildings, Volume I*, Prentice-Hall, Englewood Cliffs, NJ, 1982.

10-16 Englekirk, R. E. and Hart, G. C., *Earthquake Design of Concrete Masonry Buildings, Volume II*, Prentice-Hall, Englewood Cliffs, NJ, 1984.

10-17 ACI–SEAOSC Task Committee on Slender Walls, *Test Report on Slender Walls*," Los Angeles, CA, Feb. 1980–Sept. 1982.

10-18 Priestly, M. and Elder, D., "Seismic Behavior of Slender Concrete Masonry Shear Walls," *Bull. New Zealand Nat. Soc. Earthquake Eng.*, 15, No. 1, Mar. 1982.

10-19 Priestly, M. and Elder, D., "Stress-Strain Curves for Unconfined and Confined Concrete Masonry," *ACI J.* 80, No. 7, May/June 1983.

10-20 Wang, C. K. and Salmon, C. G., *Reinforced Concrete Design*, 3rd Edition, Harper and Row Publishers, New York, 1979.

10-21 Keller, G. R. and Sutler, G. T., "Variability of Reinforced Concrete Masonry Beam Strength in Flexure and Shear," Second North American Masonry Conference, College Park, MD, Aug. 1982.

10-22 Kingsley, G. R. and Atkinson, R. H., "Stress–Strain Behavior of Grouted Hollow Unit Masonry," Fourth Canadian Masonry Symposium, Univ. of New Brunswick, New Brunswick, Canada, June 1986.

10-23 Schneider, R. R. and Dickey, W. L., *Reinforced Masonry Design*, 2nd Edition, Prenctice-Hall, Englewood Cliffs, NJ, 1987.

10-24 Hart, G. C., *Various Contributions to Fourth North American Masonry Conference*, Univ. of California at Los Angeles, Los Angeles, CA, Aug. 1987.

10-25 Sajjad, N. A., "Reliability Analysis of Selected Concrete Masonry Design Equation," Masters Project, Univ. of California at Los Angeles, Los Angeles, CA 1987.

10-26 Priestly, M., "New Zealand Seismic Design Philosophy for Masonry Structures," Fourth North American Masonry Conference, Los Angeles, CA, Aug. 1987.

Chapter 11

Geotechnical and Foundation Design Considerations

Marshall Lew Ph.D., P.E., * *and John Nissen, S.E.*†

11.1 INTRODUCTION

Structures come in different shapes, forms, and sizes. However, all structures have at least one feature in common: they all have a foundation. A foundation is the means by which the superstructure interfaces with the underlying soil or rock. Under static conditions, generally only the vertical loads of a structure need be transferred to the supporting soil or rock. In a seismic environment, the loads imposed on a foundation from a structure under seismic excitation may greatly exceed the static vertical loads; in addition, there will be horizontal forces and possibly some moments at the foundation level.

Consideration must also be given to what could happen to the supporting soil or rock under seismic excitation. For example, an earthquake might cause liquefaction to occur in loose sandy soils, which would cause a virtually complete loss of all bearing capacity of the soil; needless to say, a structure founded on such soils would suffer great distress and upset.

This chapter will attempt to identify those phenomena which would affect the design of foundations in a seismic environment. Some of these phenomena can be effectively designed for by structural detailing, but some of these phenomena are beyond the wizardry of the structural engineering profession, and geotechnical wizardry may also be needed. In some instances, there may not be an economical engineering solution for the problem.

This chapter will be different from other chapters in this handbook in the respect that not all of the solutions to the seismic problems will be engineering solutions. This just points out the limitations of the science and art we know as engineering. We as engineers must be able to recognize our limitations and realize that we cannot always be the knight that saves the damsel in distress. If we can attain at least this little enlightenment, we will all be better engineers.

In a seismic environment, there may exist a potential for ground failures. It is obvious that if the ground should fail beneath a structure, the structure could be severely damaged or even destroyed. Such an event would threaten property and life. Several different ground failure mechanisms will be discussed in this chapter.

11.2 SITE AND SOIL CONDITIONS

Because a foundation must be capable of adequately supporting a structure in an economical manner, it is imperative that there be a proper geotechnical investigation. This geotechnical investigation should provide information about the soil types beneath the site and their

*Director of Earthquake Engineering, LeRoy Crandall and Associates, Glendale, California.
† Project Manager, John A. Martin and Associates, Los Angeles, California.

physical characteristics (strength, compressibility, permeability, etc.). The investigation should also provide economical and feasible alternatives for the support of the structure. These recommendations should take into account the functionality and purpose of the structure. In a seismic environment, the geotechnical investigation would also need to evaluate the behavior of the supporting soils under earthquake excitation, predict the consequences for the structure, and recommend foundation types.

Not only is it important to investigate the soil conditions, the general site conditions also merit deep scrutiny. This investigation should include features near the building area and also distant features. Important nearby site features include water levels, topographic features, and the presence of other structures both above and below ground. Offsite and even distant features could have some influence upon the proposed structure, especially in a seismically active area. For example, there could be large bodies of water retained by earth dams that could fail in an earthquake; if the structure is in the path of this potential inundation, the consequences could be very grave indeed.

11.3 SOIL LIQUEFACTION

11.3.1. Causes of Liquification

Soil liquefaction during an earthquake leads to loss of strength or stiffness of the soil. This could result in the settlement of structures, cause landslides, precipitate failures of earth dams, or cause other types of hazards. Soil liquefaction has been observed to occur most often in loose saturated sand deposits.

During strong earthquake shaking, a loose saturated sand deposit will have a tendency to compact and thus decrease in volume. If this deposit cannot drain rapidly, there will be an increase in the pore water pressure. The effective stress in the sand deposit is equal to the difference between the overburden pressure and the pore water pressure. With increasing oscillation, the pore water pressure can increase to the value of the overburden pressure. Since the shear strength of a cohesionless soil is directly proportional to the effective stress, the sand will not have any shear strength: it is now in a liquefied state. *Sand boils* appearing at the ground surface during an earthquake are evidence that liquefaction has occurred. Greater detail describing the liquefaction phenomenon and methods of analysis to determine the potential for liquefaction may be found in several references.[11-1, 11-2, 11-3]

11.3.2 Liquefaction-Induced Ground Failures and Effects on Structures

If a soil becomes liquefied and loses its shear strength, ground failures may result. If there are structures founded over or near these soil deposits, they may be damaged. Youd[11-4] has classified ground failures caused by liquefaction into three categories:

1. lateral spreading,
2. flow failures, and
3. loss of bearing capacity.

Lateral spreading is the movement of surficial soil layers in a direction parallel to the ground surface which occurs when there is a loss of shear strength in a subsurface layer due to liquefaction (see Figure 11-1). Lateral spreading usually occurs on very gentle slopes with a slope of less than 5%. If there is differential lateral spreading under a structure, there can be sufficient tensile stresses developed in the structure to tear it apart. Flexible buildings have been observed to withstand extensional displacement better than stiff or brittle buildings.[11-5]

Lateral spreading can have a catastrophic impact upon long, linear buried utilities ("lifelines"). During the great 1906 San Francisco earthquake, it is believed that every break in the water supply pipeline was caused by lateral spreading. This, of course, severely hampered efforts against the fires that were triggered by the earthquake, which eventually destroyed much of San Francisco.

Flow failures occur when large zones of soil become liquefied or blocks of unliquefied soils flow over a layer of liquefied soils. Flow slides can develop where the slopes are generally greater than 5%. This phenomenon had tragic effects during the 1964 Alaska earthquake.

Liquefaction can also result in the loss of bearing capacity usually accompanied by large soil deformations. Structures supported on these soils may settle, tilt, or even overturn. Buried structures have even been floated out of the ground. In extreme cases, where the thickness of the liquefied soils is large, tilting or overturning failures can occur, such as those observed in Niigata, Japan during the 1964 earthquake (Figure 11-2). Where the liquefied soil is thin, or where there is relatively thick nonliquefied soil overlying a liquefied soil deposit, severe tilting or overturning of structures will not occur, but differential vertical settlements can.

Buried structures, such as underground tanks, may be subject to excess buoyancy because of the increase in the pore water pressure associated with liquefaction. Retaining structures, such as retaining walls or port structures, may also be subjected to an increase in the lateral pressures should liquefaction occur in the adjacent soils. The formation of sink holes (when sand boils occur) may cause differential settlement or tilting of structures established on shallow foundations.

Of course, the degree to which structures are affected directly or indirectly by liquefaction-caused failures will depend upon how extensive the liquefaction is. If the

Figure 11-1 Liquefaction-caused lateral spreading at the Civil Court Building in Purnea, India from the 1934 Bihar-Nepal earthquake. (Photo courtesy of Professor Les Youd, Brigham Young University.)

Figure 11-2 Liquefaction-induced loss of bearing capacity of apartment buildings during the 1964 Niigata, Japan, earthquake. (Photo courtesy of United States Geological Survey.)

liquefaction occurs in a thick and horizontally extensive layer of sand, the effects on structures may be expected to be very great. If, in contrast, the liquefaction is isolated to very thin and discontinuous layers (lenses) of soils, structures may suffer minimal or even no noticeable damage.

11.3.3 Facing the Liquefaction Problem

If liquefaction is identified as a hazard, there are choices that must be made. For new construction, the available choices are:

1. design for liquefaction by modifying the site soil conditions or strengthening the structure,
2. abandon or move the project, or
3. accept the risks by proceeding without designing for liquefaction.

Obviously, economics will influence the choice in a major way. The second choice would be dependent on whether there was an alternate site without the liquefaction problem. The third choice could invite unwanted liability exposure and problems of uninsurability, or even jeopardize future property values and the viability of the project. The second and third choices could be subject of much discourse but are outside the intention and scope of this work, which will be concerned with designing for liquefaction.

11.3.4. Mitigation of Liquefaction Hazard by Site Modification

There are site-modification methods which can reduce the susceptibility of the soils beneath a site to liquefaction.[11-2] These methods are summarized in Table 11-1.

The first general category of site-modification methods involves the excavation of the potentially liquefiable soils. This soil may then be recompacted as an engineered fill to a higher density, so that it will have less potential to liquefy. Alternatively, the native soils may be improved with additives and then properly compacted as an engineered fill. Another solution would be to waste the excavated material and replace it completely with properly compacted imported material that would be nonliquefiable.

The second general category of site improvement methods is in-situ densification of the liquefiable soils. By densification, the soils will have less potential to liquefy because a dense soil does not tend to decrease in volume when subjected to earthquake shaking; indeed, it may even tend to become less dense, thus reducing the possibility of excess pore pressures developing. Vibroflotation is one process by which a machine is lowered into the ground and compacts loose soils by simultaneous vibration and saturation.[11-6] As the machine vibrates, water is pumped in faster than it can be absorbed by the soil. The more granular particles are vibrated into a more dense state while the excess water carries off the finer particles to the ground surface (see Figures 11-3 and 11-4). Granular soils are added from the ground surface to compensate for the loss of the finer particles and the increased density. It has been reported by Ishihara et al.[11-7] that oil tanks supported on sand soils compacted by the vibroflotation technique suffered little damage and settlement in the 1978 Miyagiken-Oki earthquake, while nearby similar facilities supported on loose sand deposits that were not densified suffered considerable damage and significant settlement.

In-situ densification may also be accomplished by dynamic compaction (which is also referred to as impact densification or heavy tamping). Dynamic compaction utilizes a heavy falling weight to produce a

Table 11-1 Methods to Mitigate Soil Liquefaction by Site Modification[11-2]

I. Excavation and replacement of liquefiable soils
   A. Excavation and engineered compaction of existing soil
   B. Excavation and engineered compaction of soils improved with additives
   C. Excavation of existing soils and replacement with properly compacted nonliquefiable soils

II. Densification of in-situ soils
   A. Compaction piles
   B. Vibratory probes
   C. Vibroflotation
   D. Compaction grouting
   E. Dynamic compaction or impact densification

III. In-situ improvement of soils by alteration
   A. Mixing in-situ soils with additives
   B. Removing in-situ soils by jetting and replacement with nonliquefiable soils

IV. Grouting or chemical stabilization

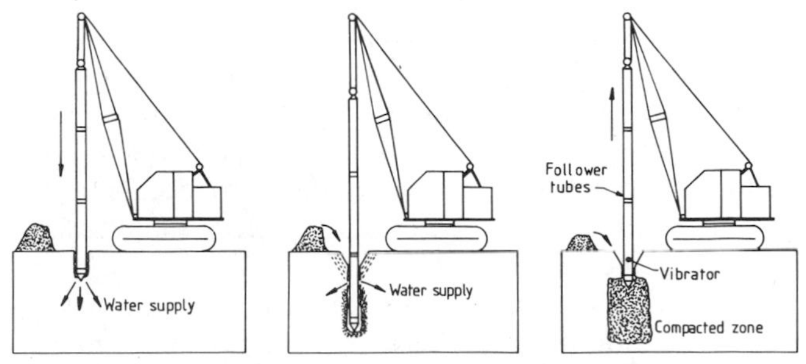

Figure 11-3 Vibroflotation technique. (Illustration courtesy of GKN Hayward Baker, Inc.)

Figure 11-4 Water being pumped during vibroflotation. (Illustration courtesy of GKN Hayward Baker, Inc.)

Figure 11-5 Dynamic Compaction technique. (Illustration courtesy of GKN Hayward Baker, Inc.)

shock wave which is propagated to some depth in the ground (Figure 11-5). The effect of this compaction in granular soils is to generate high pore water pressures. As these pressures are dissipated by drainage, compaction (or more correctly, consolidation) occurs, and the soils become more dense and therefore more resistant to liquefaction.

The third major category of soil improvement methods is alteration of the soil. The soil may be made more resistant by the construction of mixed-in-place solidified piles or walls. Lime, cement, or asphalt may be mixed in place to create piles or walls to provide shear resistance which will confine an area of liquefiable soils to prevent flow. Vibroreplacement is one such process by which soils can be improved; it is especially suitable when there are significant amounts of fine soils which do not readily respond to vibratory compaction. In vibroreplacement, a vibrator is used to penetrate the soil to a desired depth, and the resulting cavity is filled with coarse-grained material such as gravel or crushed rock. This material is then compacted and forms a *stone column* (see Figure 11-6). Stone columns are installed

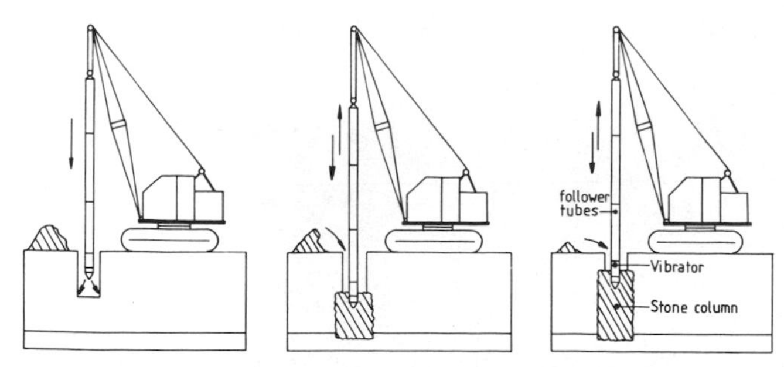

Figure 11-6 Vibro-Replacement. (Illustration courtesy of GKN Hayward Baker, Inc.)

on a predetermined grid pattern. They have low compressibility and high shear strength. Because they are coarse-grained, excess pore pressures in the surrounding soils can be quickly dissipated. Another alteration process is soil freezing.

The fourth category of soil improvement methods is soil grouting or chemical stabilization. These methods improve the shear resistance of the soils by the injection of particulate matter, resins, or chemicals into the voids.

11.3.5 Mitigation of Liquefaction Hazard by Structural Design

In designing a structure to resist liquefaction one must take into account the deformations of the soil that could occur during liquefaction. This will greatly affect the foundation design of the building.

Designing for liquefaction may be accomplished by the use of piles or caissons which rely upon the soil or rock beneath the potentially liquefiable soil layers for support. Such designs need to take account of possible downdrag forces on the piles or caissons because of the settlement of the upper soils. Also, special design for the lateral forces or base shear may be needed because there could be a significant loss of the ability to transfer horizontal forces to the liquefied soils; this may require the use of battered piles or the design of caissons as unsupported columns through the liquefied zones. The foundation system will need to be tied together quite well to resist possible lateral spreading of the soils. Floor slabs on grade could be subject to settlement or differential movement and may have to be structurally supported to minimize the effects of liquefaction.

For structures of low profile and uniform mass distribution, a mat foundation may be feasible. The mat should be able to bridge the local areas of settlement so that the structure can act more or less as a rigid body. Any permanent deformations of the structure could then be corrected by grouting or mud-jacking the structure to its proper level.

Wall structures retaining potentially liquefiable soils, such as those at port and harbor facilities may be subjected to greater than normal lateral earth pressures should liquefaction occur. Earth pressures may increase from an at-rest or active condition to a condition where the pressure distribution is equivalent to that imposed by a fluid having a density equal to the total density of the soil.

Even with a structural solution to mitigate against liquefaction, there will remain some risk that the structure will be damaged and that some remedial work will be needed after the liquefaction event.

11.3.6 Mitigation of Liquefaction Hazard by Drainage

Dewatering systems may reduce the potential for liquefaction by removing the water from those layers which could liquefy. Also, the resulting increase in effective overburden pressure will add to the resistance of the soils against liquefaction. If total dewatering of a site is not practical, providing some means of drainage may mitigate the problem. Drainage allows for the rapid dissipation of excess pore pressures in the potentially liquefiable soil layers. If the pore pressures can be relieved quickly, the effective stresses will not decrease significantly and the soil will retain most of its shear strength, not allowing liquefaction to occur. Vertical gravel drains placed in a grid pattern may be able to accomplish this. Vibroreplacement also utilizes this principle, as the coarse-grained stone columns will be very permeable in comparison with the surrounding soils.

There are methods under development to utilize prefabricated drainage material similar to conventional vertical wick drains to control the effects of liquefaction. These drains would be of sufficient size to accommodate the large volumes of water generated during a liquefaction event without undue head loss. An integral water reservoir would allow water to be stored during an earthquake; the water would be gradually drained back into the surrounding soils. A water outlet is not required for this system.

11.4 SEISMIC SETTLEMENT, SUBSIDENCE, AND DIFFERENTIAL COMPACTION

Seismic settlement and *subsidence* are two terms used to describe surface subsidence resulting from compaction or densification of granular soils by earthquake-induced vibrations, which may occur over large areas. Although this phenomenon produces a result which is similar to what occurs from liquefaction, it occurs in dry or partially saturated soils or in saturated soils which have good drainage, that is, soils that do not liquefy.[11-8]

During an earthquake, a granular soil is subjected to cyclical shear from horizontal and vertical accelerations. In a strong earthquake, the horizontal motions can cause densification because of the numerous shear cycles that occur. Whitman and DePablo[11-9] have

Figure 11-7 Differential compaction between an area with older natural soils and an area with loose fill soils from the 1986 San Salvador earthquake. (Photo courtesy of Robert Chieruzzi, LeRoy Crandall and Associates.)

demonstrated that vertical accelerations greater than the acceleration of gravity (g) are required to cause significant densification of granular soils. However, it has been reported that over 1 m (about 3 ft) of ground subsidence due to densification was experienced in Valdivia, Chile during the 1960 earthquake.[11-10] It has also been reported that there was ground subsidence on the order of 5 to 7 m over a very large area in the Mississippi Valley as a result of the New Madrid earthquakes of 1811 and 1812.[11-8] It is difficult to determine whether some of these reported instances of subsidence were at least partially due to liquefaction or some tectonic movement, or if they were totally a result of seismic settlement.

Differential compaction occurs when there is marked difference in the density of the soils in a horizontal sense. Such a phenomena was observed during the San Salvador earthquake of 1986 (Figure 11-7).

11.5 LANDSLIDING AND LURCHING

11.5.1 Landsliding

Earthquakes may trigger landsides or other forms of slope instability. Slope failures may occur as a result of the development of excess pore pressures which will reduce the shear strength of the soils or cause loss of strength along bedding or joints in rock materials. The Turnagain Heights landslide occurred as a result of the 1964 Alaska earthquake (Figure 11-8). The epicenter of the earthquake (Richter magnitude 8.5) was about 130 km from Anchorage, but the strong ground motion lasted more than 3 minutes. Seed and Wilson[11-11] believed that the long duration of the ground motion caused the pore water pressures to continually increase, causing liquefaction of silt and fine-sand lenses which led to the landslide. Earthquake-caused liquefaction within the Lower Van Norman Dam during the February 9, 1971 San Fernando earthquake nearly resulted in the overtopping of the dam (see Figure 11-9), which would have threatened thousands of people who lived below the dam in the populous San Fernando Valley.

Earthquakes may also cause shallow debris slides in areas with high, steep slopes. These slides can be minor or quite major. The 1970 debris avalanche triggered by the Peruvian earthquake of May 31, 1970 buried the towns of Yungay and Ranrahirca, in which 18,000 lives were lost (see Figure 1-19).

Careful consideration should be given to structures which are sited in a location that could directly or indirectly be affected by some form of slope instability. A very careful geotechnical and geologic investigation will be needed to determine if such hazards exist and if there are any practical means of mitigation.

11.5.2 Lurching

Lurching means movement of soil or rock masses at right angles to a cliff or steep slope. Structures founded either in part or whole on such masses may experience significant lateral and vertical deformations.

Figure 11-8 The Turnagain Heights landslide occurred as a result of the 1964 Great Alaska earthquake and had a length of about 1.5 mi and width from $\frac{1}{4}$ to $\frac{1}{2}$ mi. (Photograph by United States Geological Survey.)

Figure 11-9 Van Norman Dam after the 1971 San Fernando earthquake. (Photo by the United States Geological Survey.)

11.6 FLOODING, TSUNAMIS, AND SEICHES

11.6.1 Flooding

Seismic activity may cause some calamity elsewhere which results in flooding at the site under consideration. An important part of the site investigation is the identification of any bodies of water or structures that contain water and that are located above or upstream of the site. The consequences of failure of these bodies or structures should be evaluated to determine the probable flood limits and depths of inundation. The effect of this potential flooding on structure and function could affect the siting of a building. It may be practical to raise the finished floor elevation to be reasonably above the maximum expected flood elevation; if this is not possible, re-siting of the proposed building may be necessary. Otherwise, emergency procedures will need to be established in the event that the flood hazard becomes a reality.

In some regions of the United States, studies of flood hazard have been performed and flood maps are available. Some of these studies have been performed by the United States Army Corps of Engineers; others are available from the Federal Emergency Management Agency (FEMA).

11.6.2 Tsunamis

A tsunami is a long sea wave that can be generated by a rapidly occurring change in seafloor topography caused by tectonic displacement. Such tectonic displacement may be caused by earthquakes, undersea landslides, and volcanic eruptions. Among earthquakes, it seems that strike-slip earthquakes are less likely to cause tsunamis and that a substantial vertical offset caused by a dip-slip mechanism is necessary to generate large tsunamis. A tsunami may be caused by a nearby fault rupture, or by distant earthquakes which may be thousands of miles away. In the open sea, these waves travel at great velocities; however, their amplitudes are quite small, and their wavelengths very long. The velocity of a tsunami water wave is approximately given by the relationship

$$v = \sqrt{gD}$$

where g is the acceleration of gravity and D is the water depth.

As the wave approaches a coastline, the shallower depth of water will cause the amplitude of the water wave to become greater. The wave may become even more accentuated where there are topographic features such as narrow bays and very shallow waters. In fact, the meaning of the word tsunami is literally "harbor wave" in the Japanese language. The wavefront may crash on shore and move inland.

Tsunamis do not result from every earthquake whose source is beneath the seafloor. They do not occur very often, but they can cause significant damage and loss of life. Tsunamis have occurred most frequently in the Pacific Ocean. Japan has been the victim of numerous tsunamis throughout recorded history. The city of Hilo on the big island of Hawaii has been devastated several times by tsunamis; the offshore topography channels the destructive energy into waves that were estimated to be from 21 to 26 feet high in the 1946 tsunami. Extensive damage from tsunamis occurred during the so-called Great Alaska Earthquake of March 27, 1964 (see Figures 11-10 and 11-11). Tsunamis have also occurred in the Atlantic and Indian Oceans, although not as frequently. They have even been reported in the Mediterranean Sea.[11-8, 11-12]

Structural damage from tsunamis is caused by the force of the water and the impact of boats and other objects that may be carried by the water. Structures with open fronts or large areas of glass with continuous rear walls have been found to be more likely to be damaged. Observations made after several tsunamis suggest that light frame buildings are subject to very severe damage or total destruction because of their flexible construction. Heavy timber construction is also found to be very susceptible to damage from tsunamis. If not firmly anchored to the foundations, such structures have a tendency to float if the water level is high enough. Heavier buildings constructed of structural steel or reinforced concrete tend to be less damaged. In some structures which seem to have withstood tsunami forces, the structural elements at the lower levels turn out to have sustained significant damage from the passage of the water.

It has been suggested that a structure could be designed to resist tsunami waves.[11-13] Special consideration would be needed to minimize the effects of tsunami. First, the major axis (long dimension) of the building should be oriented parallel to the expected direction of the wave. A building with this orientation would have a minimal surface area that could be attacked by the oncoming waves. In addition, such a building will have greater strength to resist the wave forces because of the greater amount of structural elements providing resistance. Consideration should also be given to leaving the lower portion of the building completely open, which would greatly reduce the total load applied to the structure from the tsunami wave.

The forces exerted on a structure by a tsunami are not easy to predict. The horizontal fluid pressure exerted by flowing waters, p, can be estimated by the equation

$$p = 0.5 C_D \rho V_s^2$$

where C_D is a coefficient of drag for submerged objects, which is a function of the shape of the object (which may be a wall or a column), ρ is the mass density of water, and V_s is the speed of the water surge, which is approximated by

$$V_s = 2\sqrt{g d_s}$$

where g is the acceleration of gravity and d_s is the height of the water surge.[11-13] In addition to the dy-

Figure 11-10 Waterfront at Seward, Alaska, looking south, before the 1964 Great Alaska earthquake-generated underwater landslides, surge-waves, and tsunami waves devastated the waterfront. (Photo by the United States Geological Survey.)

Figure 11-11 Waterfront at Seward a few months after the earthquake, looking north. (Photo by the United States Geological Survey.)

namic fluid pressures, there could be impact loading from objects carried by the tsunamis.

11.6.3 Seiches

A seiche may occur when earthquake ground motion causes water in closed or almost closed body (such as a bay, a lake, a reservoir, or even a swimming pool) to oscillate from one side to the other.(11-8) Large seiches may occur when the frequency of the incoming earthquake waves is the same as the natural frequency of the water body and causes resonant oscillation. This oscillation can cause overtopping of dams and damage to structures located near the water, and can continue for hours.

11.7 SOIL-STRUCTURE INTERACTION

11.7.1 Conventional Structural-Dynamic Analysis and Soil-Structure Interaction

In the normal dynamic analysis of a building, the usual method is to determine the free-field ground motion at the site of the building, and to apply that motion at the base of the building, assuming that the base is fixed. This may be valid for the case where the building is founded on rock. However, if the building is founded on soft soils, the earthquake motion at the base of the building is not likely to be identical to the free-field ground motion. The presence of the structure will modify the free-field motions, because the soil and structure will interact to create a new dynamical system. This *soil–structure interaction* will result in a structural response that may be quite different from the response computed from a fixed-base building subjected to a free-field ground motion.

Certainly it is a simpler problem when one can separate the determination of the design ground motion from the dynamic analysis of the building, as in a conventional dynamic analysis. This uncoupling of the soil system from the building system will usually give a conservative predicted response. This may be a rationale for using a fixed-base model, for convenience sake, rather than a soil–structure interaction model. Another reason is that soil–structure interaction involves two distinct disciplines (as practiced in the United States), namely geotechnical and structural engineering. However, a fixed-base model may not be able to take into account all of the possible modes of response such as rocking of the structure or deformation of its base. Additionally, the periods of vibration of the structure may be longer because of the interaction. In critical structures, such as nuclear reactors, some of these other modes of response may be just as important as the primary translation modes of vibration. The change in period may also affect the response of the overall structure and its substructures or components.

It turns out that in soil–structure interaction analysis, the whole is greater than the sum of the two parts. There needs to be an understanding of both soil dynamics and structural analysis and an ability to combine these two. Because of the interaction of the soil and the structure, both need to be modeled. However, it should be recognized that, in comparison to the structure, the soil is essentially a semi-infinite medium. Thus the soil–structure interaction model subjected to dynamic loading cannot be treated in the same way one would consider static loading. When analyzing a soil–structure system under static loading, it is sufficient to model the structure on the soil system within fixed or semifixed boundaries at a sufficient distance from the structure so that boundary conditions do not affect the static response of the structure. Under dynamic loading, the fictitious boundaries could not be far enough away from the structure to avoid affecting the structural response; i.e., the boundaries would reflect the traveling waves within the soil mass and not allow the energy to pass through to infinity. In an attempt to model boundaries properly, special techniques such as the boundary-element method have been developed.

11.7.2 Elements of Soil-Structure Interaction Analysis

Consider two identical structures with rigid foundations, one founded on stiff rock and the other founded on soft soil, as shown in Figure 11-12. The soil layer overlies the rock, and the distance between the two structures is small, so that it may be reasonably assumed that the incident earthquake waves arriving from the earthquake source are identical for the two structures. For illustration purposes, we will consider only a vertically propagating shear wave which produces only

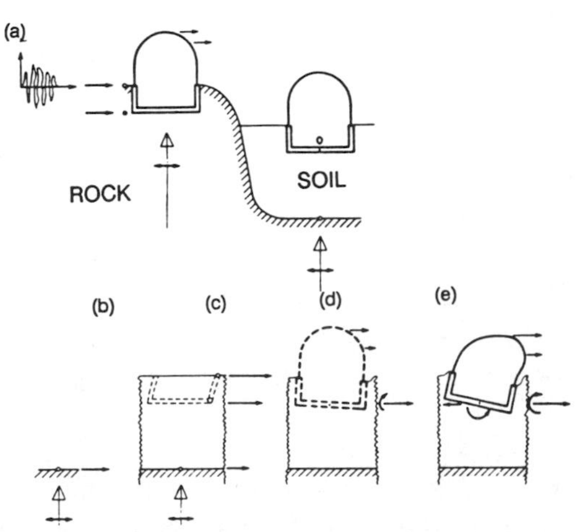

Figure 11-12 Seismic response of structure founded on rock and on soil. (a) Sites; (b) outcropping rock; (c) free field; (d) kinematic interaction; (e) inertial interaction.(11-15)

horizontal motions. A control motion may be defined on the free ground surface of the rock, say at point A. As the rock is stiff, it may be reasonably assumed that the motion at any point in the rock, say at point B, is the same as the control motion at point A.

For the structure founded on the rock, a fixed-base condition will exist and the horizontal ground motion applied to the base of the structure by the earthquake will be equal to the control motion. Rocking of the structure will not develop in a fixed-base condition.

For the structure founded on soft soil, the earthquake motion at the base of the structure will not be the same as for the structure founded on rock, and neither will the base be fixed. The motion at point C at the top of the rock will not be as great as the motion at point A, because of the presence of the overlying soil layer. As the wave propagates upward in the soil layer, the motion may be amplified or deamplified—usually the former. The frequency content of the motion will also change. The rigid base of the structure will also modify the motion. The motions will undergo a kinematic interaction which results in the base being subject to some average horizontal displacement and also some rocking. These rigid-body motions in the base will apply inertial loading on the superstructure, which will excite the structure. The excited structure will then cause a demand on the supporting soils to resist transverse shear and overturning moment. These demands on the soils cause *inertial interaction*, which results in deformations of the soils, which in turn ultimately cause further modification of the motion at the base of the structure.

11.7.3 Limitations of Soil–Structure Interaction Analyses

Implicit in the formulation of present soil–structure interaction analyses is the assumption that the principle of superposition is valid. A result of this assumption is that the response that is computed is for a linear system. However, soils are notoriously nonlinear when subjected to strong ground motions at the levels of engineering interest.(11-14) Although it may be possible to use material properties that are compatible with the strain levels produced during an earthquake, this is still far from being a true nonlinear analysis.

Because a significant mass of soil must be modeled around the structure, there will be a large number of degrees of freedom, which usually make large demands on computational storage and run time. This may be alleviated by substructuring the problem into two parts. The first part of the analysis is to compute the free-field response of the site (without the structure present). The motions are determined at the nodes where the structure is attached. The force–displacement relationships of these nodes are also determined. The second part of the

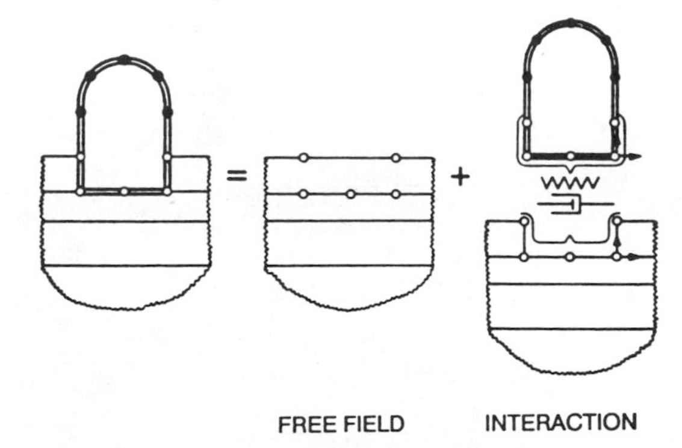

Figure 11-13 Seismic soil-structure interaction with substructure method.(11-15)

analysis is the study of the superstructure mounted on spring–dashpot systems subjected to the free-field motions determined from the first part of the analysis (Figure 11-13).

Great care must be exercised in soil–structure interaction analyses. The basic assumptions show that, although this type of analysis is more sophisticated than a conventional rigid-base analysis, the current state of the art still falls far short of modeling reality. Such analyses should be tempered with much engineering judgment. For more detailed information on the theory of soil–structure interaction, the reader is referred to the work by Wolf.(11-15)

11.8 FAULT RUPTURE

If ground-surface rupture due to earthquake faulting were to occur beneath a structure, there would be a significant threat to the integrity of the structure and to its contents and inhabitants (see Figure 11-14). The structure will in all probability be a substantial, if not a total, loss due to the differential ground displacement between the two sides of the fault. This displacement will be mostly lateral if the fault is a strike-slip fault, and mostly vertical if it is a thrust-type fault; some faults will exhibit a combination of these two types of movement.

The state of California has enacted legislation, known as the "Alquist–Priolo Special Studies Zones Act," which specifies the process to define fault-rupture hazard zones in California. One of the specific criteria given in this act provides that "No structure for human occupancy shall be permitted to be placed across the trace of an active fault." A structure for human occupancy is any structure that has an occupancy rate of 2000 person-hours per year. In addition, the act states that "... the area within fifty feet of such active faults shall be presumed to be underlain by active branches of that fault unless proven otherwise." This last provision essentially defines a setback distance from an active

Figure 11-14 Home damaged by faulting during the 1971 San Fernando earthquake. (Photo by the United States Geological Survey.)

fault within which structures with significant human occupancy should not be built.

11.9 SELECTION OF A FOUNDATION SYSTEM

11.9.1 General

The selection of a foundation system for a structure subject to seismic loading will not, in most cases, be different from that for static loading only. As seismic forces will ultimately translate into vertical and horizontal forces and perhaps a moment at the foundation level, no special procedures are needed to select a foundation system.

There may be some instances where the seismic forces in some foundation elements will be so great that a special design may be required. In buildings with shear walls to resist lateral loads, there could be great overturning loads, especially if the shear walls are short in comparison with the lateral plan dimension of the building. Extensive grade beam systems may be utilized to distribute the loads to other foundations. Deep piling or caissons may be needed to resist uplift loads. If such deep foundations are used in combination with shallow spread-footing foundations, attention will be needed to accommodate possible large differential settlements between adjacent foundations under static loading conditions as well as seismic conditions.

There are two major foundation systems utilized for the majority of projects being designed and built today: spread footings and pile foundations.

The spread or pad footing is the most widely used foundation system, due primarily to its ease of construction and cost effectiveness. It distributes the superimposed vertical loads more or less uniformly to the soils below in direct bearing. The footing is subject to flexural bending and beam (or one-way) shear resulting from this soil reaction, and also to punching shear, as the column above will want to shear through the footing, in a localized manner, at the structure–footing interface (Figure 11-15).

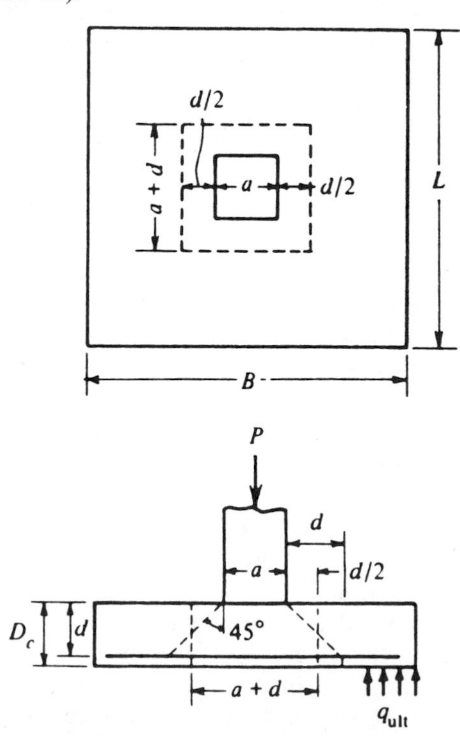

Figure 11-15 Assumptions used for finding depth of footing to satisfy punching shear requirements. (From *Foundation Engineering Handbook,* Winterkom and Fang, eds., Van Nostrand Reinhold, 1975.)

One of the major drawbacks to a spread-footing foundation is its inability to handle heavy loading, such as those encountered in high-rise construction or industrial applications. The size of the footing is in direct proportion to the maximum allowable soil bearing pressure. There is a point of diminishing return that must be evaluated for each specific project. This is a result of the fact that as the superimposed load increases, the pad size will increase, thereby becoming less economical and negating one of its major advantages.

The other important disadvantage of the pad footing system is the fact that the competent bearing material must be located close to the soil–structure interface. This inhibits its use on sites with substantial amounts of fill material or sites where the bearing material is not readily accessible or at a constant elevation, as with hillside sites.

Lateral loads are resisted by a combination of passive soil resistance and frictional resistance developed at the soil–structure interface. It is important to note that the footing is subject to forces in addition to those generated by the superimposed vertical loads.

The pile or caisson technique is used primarily on sites with an overlying layer of fill material, where the competent soil is at a variable depth below the structure–foundation interface, or when there are especially heavy loads to be resisted. It distributes loads to the soil by way of frictional resistance developed along the shaft of the pile or by direct end bearing. Vertical loads are distributed to individual piles by means of a pile cap, which functions similarly to a spread footing and is subject to essentially the same shears and bending moments. Figure 11-16 illustrates a major difference between a spread footing and pile foundation system. As one can see from this example, the same load-carrying capacity is achieved by the pile foundation in approximately 25% of the area required by the spread footings. The main drawback to a pile foundation is its relatively high cost of construction. This is due to the amount of time required to drill and pour or to drive the piles.

In pile systems the lateral loads are resisted by a combination of passive soil resistance against the pile cap, flexural resistance of the pile or caisson, and frictional resistance developed at the structure–soil interface. The piles or caissons must be designed to resist the shear and moment induced by a seismic event. Piles may also be battered, that is, drilled or driven at an angle to the vertical axis of the pile cap so as to introduce a horizontal component for lateral load resistance, although its vertical load carrying capacity is then reduced as the angle of the batter is increased (see Figure 11-17).

There are other foundation systems in use which may be considered when special design circumstances occur. These systems, which will not be fully explored in this chapter, include strap footings, combined footings, and

AXIAL FORCE = 1700 kips

ALLOWABLE SOIL PRESSURE = 5 kips/ft^2

SPREAD FOOTING

Required Area = 1700/5 = 340 ft^2

Use 18'-6" X 18'-6" Pad

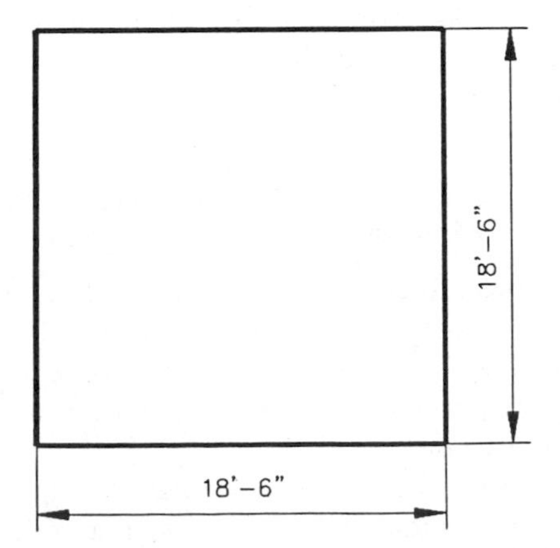

PILES

Using 100 Ton Capacity Piles and 40 ft Embed.

Minimum 3 ft Spacing

Number of Piles = 1700/200 = 8.5

Use 9 Piles, Pile Cap Size = 9' X 9'

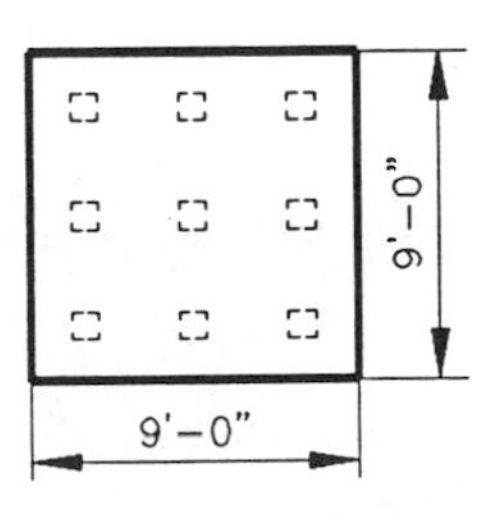

Figure 11-16 Comparison of surface area required by a spread footing and a pile foundation.

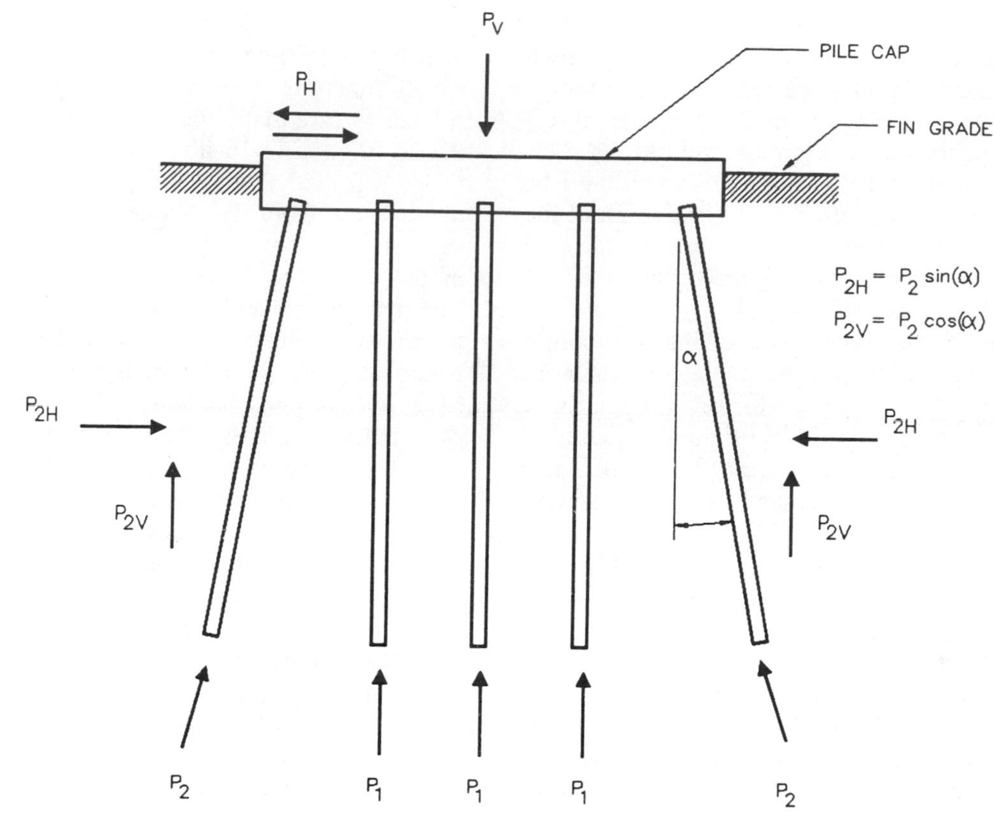

Figure 11-17 Pile foundation forces.

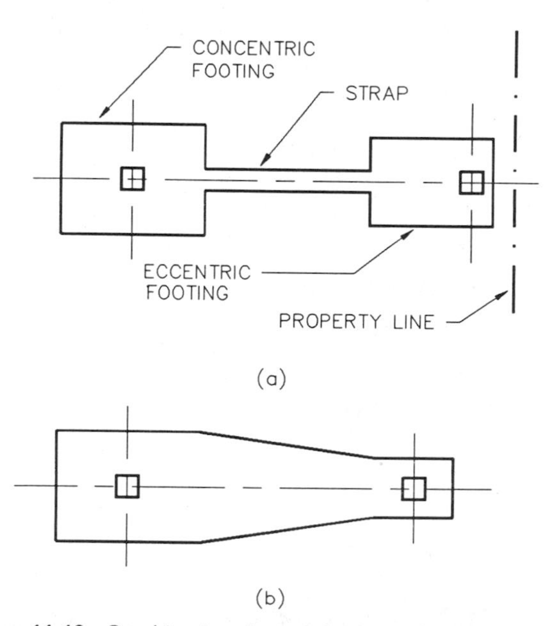

Figure 11-18 Combined and strap footings. (a) Strap footing. (b) Combined footing.

mat (or raft) foundations (see Figure 11-18). In strap footings, a reinforced-concrete grade beam, or *strap*, is used to resist a moment induced by a spread footing being eccentrically loaded. This situation arises at footings located near a property line or existing construction. Another, similar system is that of the *combined footing*, which occurs when two vertical loads are located so that separate spread footings under each would overlap each other. The design is based on the soil reaction resultant being concentric to the resultant of the vertical loads.

The last main-foundation system to be discussed is the mat, or raft, foundation. The idea is to adequately distribute all superimposed loads, generally from a series of closely space column loads above, by means of a thick reinforced-concrete mat (Figure 11-19). This system has been used successfully on high-rise structures where the majority of the vertical, lateral, and overturning forces are confined to a small building footprint, rendering other foundations systems unbuildable or uneconomic.

11.9.2 Comparison of Code Requirements for Foundation Design

The codes which are commonly used as minimum design standards are UBC-85,[11-16] NEHRP-85,[11-17] and SEAOC-86,[11-18] which has been adopted as the seismic design requirement for the 1988 edition of the Uniform Building Code (UBC-88).

All three codes noted above attempt to establish minimum standards for both design and detailing of the two most common foundations systems, the spread-footing and pile foundations. There are considerable

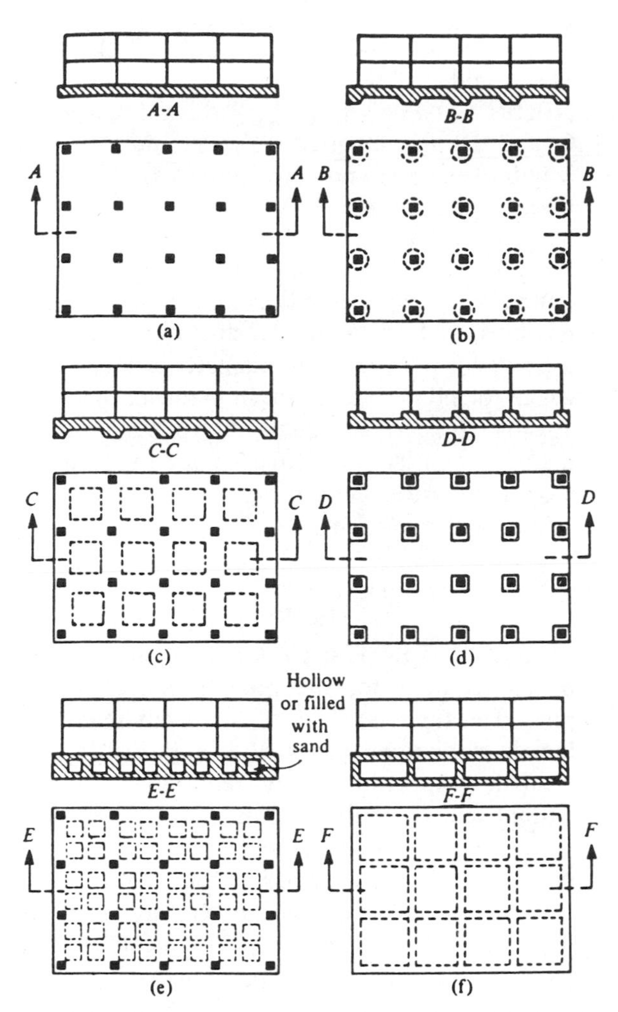

Figure 11-19 Common types of mat foundations. (a) Flat plate. (b) Flat plate thickened under columns. (c) Two-way beam and slab. (d) Flat plate with pedestals. (e) Cellular construction. (f) Basement walls as rigid frame. (From *Foundation Engineering Handbook*, Winterkom and Fang, eds., Van Nostrand, Reinhold, 1975.)

differences between the three, which will be highlighted both in this section and in the design examples which follow. It is important to note that while all codes provide guidelines for design and detailing, they are not intended to replace engineering judgment and should never be the sole basis for design decision making.

The SEAOC-86 (UBC-88), the UBC-85, and the NEHRP-85 each determine the lateral forces by considering the relative structural merits of a specific lateral-load-resisting system, the geographic location of the site with respect to known seismically active areas, the relative importance of each structure, and the anticipated response of the structure based on its geometry and the classification of the soil. The SEAOC-86 (UBC-88) and NEHRP-85 also provide minimum recommendations for response-spectrum and time-history analyses.

Within the context of the foundation design in a seismic environment, the following comparison will be amplified in the design examples:

Spread Footings All three codes allow the soil bearing pressure to be increased 33% when dead plus live plus seismic loads are considered.

- SEAOC-86 (UBC-88): Strut ties or the equivalent are required to interconnect the spread footings, in tension or compression, for a force equal to 10% of the larger of the footing and the column load. The strut ties are not required if the pad footings are founded on and embedded into approved competent soil or rock.
- NEHRP-85: Strut ties or the equivalent are required to interconnect the spread footings, in tension or compression, for a force equal to 25% of the larger of the footing and the column load multiplied by the effective peak velocity-related acceleration (A_v), which is dependent on the specific site's geographic location (see Chapter 4). The strut ties are not required in Seismic Performance Categories *A* or *B* or if the pad footings are founded directly on rock.
- UBC-85: No strut-tie requirements.

Piles

1. Strut-Tie Requirements.

- SEAOC-86 (UBC-88): Pile caps, tops of foundation piers, and caissons shall be interconnected by strut ties or the equivalent capable of resisting, in tension or compression, 10% of the larger of the footing and the column load.
- NEHRP-85: Ties are required as noted for spread footings except in seismic performance category A.
- UBC-85: Pile caps and individual caissons shall be interconnected by ties capable of resisting in tension and compression a minimum horizontal force equal to 10% of the larger of the pile-cap and the caisson loading.

2. Pile-to-Pile-Cap Connection.

- SEAOC-86 (UBC-88): The pile reinforcement must be embedded for the tensile development length without reduction for the excess area of steel. This requirement may be modified to the compression development length when 1.4 times the earthquake force is less than or equal to 0.9 times the dead load. The pile must extend 4 in. minimum into the pile cap. Special reinforcing is required and is noted in the comparison of precast prestressed piles of Example 11.2.
- NEHRP-85: Requirements are the same as the SEAOC-86 (UBC-88) except that the develop-

ment length is that required for compression unless there is a net tensile load that must be resisted by the piles.

- UBC-85: No special requirements.

3. Reinforcing-steel requirements.

(A) Uncased, poured-in-place piles:

- SEAOC-86 (UBC-88): Longitudinal reinforcing is required for 15 ft minimum, or the full length if that is less than 15 feet:

$$A_s(\min) = 0.005A_g$$

or

$$A_s(\min) = 0.01A_{g1}$$

or

$$A_s(\min) = 4 \text{ No. 6 bars}$$

where A_{g1} is the minimum shaft diameter for structural consideration. Ties shall be spaced no further apart than 8 bar diameters and spaced at 3 in. on center in the top 4 ft. All ties shall be No. 4 bars unless the pile diameter is less than 20 in., in which case No. 3 ties may be used.

- NEHRP-85: Four seismic performance categories C and D, the longitudinal reinforcing is required for a distance of ten pile diameters in the top of the pile:

$$A_s(\min) = 0.005A_g$$

or

$$A_s(\min) = 4 \text{ No. 6 bars}$$

Tie size and spacing are the same as required for the SEAOC-86.

- UBC-85: Same requirements as for column reinforcing.

(B) Precast, nonprestressed:

- SEAOC-86 (UBC-88): The longitudinal and tie reinforcing are as required for the uncased, poured-in-place piles for the top half of the pile, and

$$A_s(\min) = 0.01(A_g)$$

Nonprestressed piles shall not resist flexure induced by a seismic event unless the upper two-thirds of the pile contains special transverse reinforcing as defined in Appendix A of the ACI 318-83 code.[11-19]

- NEHRP-85: For seismic performance categories C and D, the longitudinal steel must be full length of the pile and equal to $0.01A_g$. Ties shall be as noted for the SEAOC-86 over the upper half of the pile.
- UBC-85: The longitudinal steel shall be as required by the applied forces and resultant internal stresses. Ties (or spirals) shall be at least No. 5 gauge for piles 16 in. in diameter or less, No. 4 gauge for piles 17 to 19 in. in diameter, and No. 3 gauge (or $\frac{1}{4}$-inch round) for piles 20 in. in diameter and larger. All ties shall be spaced not more than 3 in. on center for the top and bottom 2 ft, and 8 in. elsewhere.

(C) Precast, prestressed:

- SEAOC-86 (UBC-88): The longitudinal steel shall be as required for the precast nonprestressed piles. Ties (spirals) shall be No. 3 minimum and have a maximum spacing of 4 in. in the upper 2 ft of the pile. Additionally, reinforcing equal to $0.01A_g$ and extending a minimum of 8 in. into the pile cap is required.
- NEHRP-85: This code has various requirements for the longitudinal and tie reinforcing which depend on the magnitude and type of service loading present, the anticipated use of the pile, the specific type of soil, and whether the pile is hollow core or marine piling. The code itself should be consulted for specific design requirements.
- UBC-85: All longitudinal reinforcement shall be high-tensile seven-wire strands. Ties (spirals) shall not be less than No. 5 gauge for piles 24 in. to 35 in. in diameter. All other piles shall have No. 3 gauge or $\frac{1}{4}$-inch round ties. Ties shall be spaced as noted for precast nonprestressed piles, except that the first five ties at each end of the pile shall be spaced 1 in. on center.

It should be noted that SEAOC-85 (UBC-88) and NEHRP-85 both recommend that certain types of piles not resist earthquake-induced flexural loads unless deformation of the soil profile is taken into account, although they are not in agreement on the pile types.

There are two additional types of piles, not discussed here, which have limited use in seismic design. They are metal-cased concrete piles and steel or concrete-filled steel piles. These are generally special systems whose use is dictated, among other things, by unusual soil conditions or weather exposures.

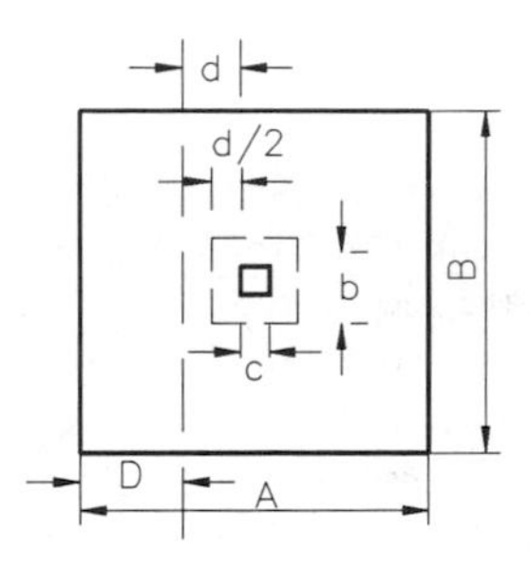

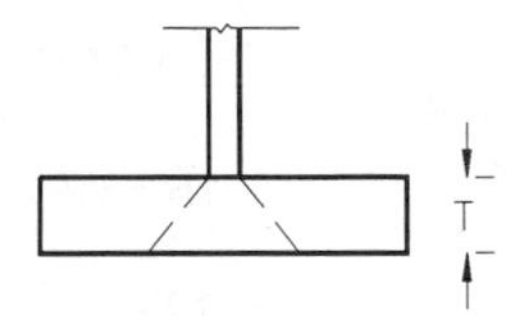

Figure 11-20 Spread footing design (Example 11-1).

11.9.3 Design Examples

Example 11-1

Design a spread footing to support a 14-in. square column (see Figure 11-20) under the following working stress loads:

Dead	$P_D = 405$ kips
Live	$P_L = 130$ kips
Seismic	$P_S = 235$ kips

The allowable soil pressure is established by a geotechnical engineer to be 3000 lb/ft², and we have $f_c' = 3000$ lb/in.², $f_y = 60$ ksi. For strut tie design according to the NEHRP-85 requirements assume $A_v = 0.53g$ and seismic performance category C.

Determine governing load. Working load:

$$P_W = P_D + P_L = 405 + 130 = 535 \text{ kips}$$

Ultimate load:

$$P_U = 1.4P_D + 1.7P_L = 567 + 221 = 788 \text{ kips}$$

Thus

$$\text{ultimate load factor} = 788/535 = 1.47$$

Ultimate soil pressure:

$$q_u = (3000)(1.47) = 4410 \text{ lb/ft}^2$$

Ultimate load including seismic (ACI 318-83,[11-19] Equation 9-2):

$$P_{us} = 0.75[1.4P_D + 1.7P_L + (1.7)(1.1)P_s]$$
$$= 0.75(567 + 221 + 439.5) = 920.6$$
$$> 788 \text{ kips} \qquad \text{(governs)}$$

Note: The ACI formula reduces the load by $\frac{1}{3}$ rather than increasing the soil pressure by $\frac{1}{3}$.

Determine required footing size:

$$A_F = \frac{P_u}{q_u} = \frac{920.6}{4.41} = 208.75 \text{ ft}^2$$
$$A_F = AB, \qquad \therefore \quad B = A$$
$$\therefore \quad A = \sqrt{208.75} = 14.45 \text{ ft}$$

where A_F, A, and B are the required area and dimensions of the footing, respectively. Try 15-ft 0-in. × 15-ft 0-in. square pad footing:

$$\text{trial thickness} = 2 \text{ ft } 9 \text{ in.}$$

Check soil pressure, assuming 50-lb/ft³ difference in the weight between concrete and soil:

$$\frac{920.6}{15^2} + (2.75)(0.05)(1.4) = 4.28 \text{ ksf}$$
$$< 4.41 \text{ ksf} \qquad \text{O.K.}$$

Check punching shear:

$$v_c = 4\sqrt{f_c'} = 4\sqrt{3000} = 219 \text{ lb/in.}^2$$
$$d = (2.75)(12) - 3.1 = 29 \text{ in.}$$
$$b = C + 2(d/2) = 14 + 29 = 43 \text{ in.}$$
$$V_u = q_u\left[AB - (d/12)^2\right]$$

where d is the distance from the extreme compression fiber to the centroid of the reinforcing steel, b is the dimension of one side of the critical-section perimeter, and C is the plan dimension of the column. q_u is the ultimate soil pressure less the footing weight, the term $(d/12)^2$ represents the soil pressure that does not contribute to the punching shear, and V_u is the design ultimate shear at the critical section,

$$V_u = 4.09\left[(15 \times 15) - (29/12)^2\right]$$
$$= 4.09[225 - 5.84] = 896.4 \text{ kips}$$
$$v_u = V_u \times 1000/\phi b_0 d$$

where v_u is the factored ultimate shear stress in psi and b_0 is the perimeter of critical section.

$$\phi = 0.85 \qquad \text{for shear}$$
$$b_0 = 4b = 4 \times 43 = 172 \text{ in.}$$
$$d = 29 \text{ in.}$$
$$v_u = \frac{896.4 \times 1000}{0.85 \times 172 \times 29} = 211.42$$
$$< 219 \qquad \text{O.K.}$$

Therefore no shear reinforcing is required.

Check beam shear:

$$v_c = 2\sqrt{f_c'} = 2\sqrt{3000} = 109 \text{ lb/in.}^2$$

$$V_u = (D - d)(A \text{ or } B)\frac{q_u}{12}$$

$$D = \left(\tfrac{15}{2} - \tfrac{7}{12}\right)12 = 83 \text{ in.}$$

(D is the shear span, which is the distance from the critical section to the edge of the footing.)

$$d = 29 \text{ in.}$$

$$V_u = \frac{(83 - 29)(15)(4.09)}{12} = 276.08 \text{ kips}$$

$$v_u = \frac{V_u \times 1000}{\phi(A \text{ or } B)d} = \frac{276.08 \times 1000}{0.85 \times 15 \times 12 \times 29} = 62.22 \text{ lb/in.}^2 < 109 \text{ lb/in.}^2 \qquad \text{O.K.}$$

Determine flexural reinforcing: at face of column,

$$M_u = q_u(A \text{ or } B)L^2/2$$

where $L = D = 83$ in. $= 6.92$ ft. Hence

$$M_u = \frac{4.09(15)(6.92)^2}{2} = 1468.9 \text{ ft-kip}$$

Assume No. 8 bars: $d = 33.3 - 1.0 - 0.5 = 28.5$ in. Then

$$M_u = \phi A_s f_y\left(d - \frac{a}{2}\right)$$

Try 15 No. 8 bars: $A_s = 11.85$ in.^{2}. Then

$$a = \frac{A_s f_y}{0.85 f_c'(A \text{ or } B)} = \frac{11.85 \times 60}{0.85 \times 3 \times 15 \times 12} = 1.55 \text{ in.}$$

$$\therefore \qquad A_s = \frac{M_u \times 12}{\phi f_y(d - a/2)}, \qquad \phi = 0.90$$

$$A_s(\text{required}) = \frac{1468.9 \times 12}{0.9 \times 60(28.5 - 1.55/2)} = 11.7 \text{ in.}^2 < A_s(\text{provided}) \qquad \text{O.K.}$$

Therefore use 15-ft 0-in × 15-ft 0-in. × 2-ft 9-in.-thick pad footing with 15 No. 8 bars each way at the bottom.

Strut-tie check:

- SEAOC-86: Since the footing is founded on competent soil, approved by the building official, no strut ties are required.
- NEHRP-85: Strut ties are required, since the footing is not founded on rock and the structure is located in seismic performance category C. We have

$$\frac{\text{tension}}{\text{compression}} = 0.25(A_v)(P_u) = 0.25(0.53)(920.6) = 122 \text{ kips}$$

  Tension bars: $A_s = 122/\phi f_y = 122/(0.9 \times 60) = 2.26$ in^2, minimum of four bars: $2.26/4 = 0.56$ in.^{2}; therefore try four No. 7 bars ($A_s = 2.40$ in.^{2}) in a 12 × 12-in. beam. Compression bars:

$$P_A(\text{allowable}) = A_N(\phi)f_c'$$

  where the net compression area is

$$A_N = (12 \times 12) - 2.4 = 141.6 \text{ in.}^2$$

  We have $\phi = 0.70$ for compression. Thus

$$P_A = 141.6 \times 0.70 \times 3 = 297.3 \text{ kips} < 122 \text{ kips} \qquad \text{O.K.}$$

  The strut beam shall be reinforced as required by the ACI 318-83 code for compression and tension members.
- UBC-85: Strut ties are not needed unless specifically required by the soils engineer or the building official.

Example 11-2

Design a pile foundation for the problem of Example 11-1. Assume a 100-kip pile capacity and 14-in. square precast prestressed piles (Figure 11-21). The imposed ultimate moment is 80 ft-kips per pile. $f_c' = 6000$ lb/in.^{2}, $f_{ci}' = 4000$ lb/in.^{2} (at transfer), and $\frac{1}{2}$-in.-diameter 270-kip seven-wire strands are used as prestressing tendons.

Pile design:

$$f_c' = 6000 \text{ lb/in.}^2$$

$$f_{ci} = 4000 \text{ lb/in.}^2 \qquad \text{(at transfer)}$$

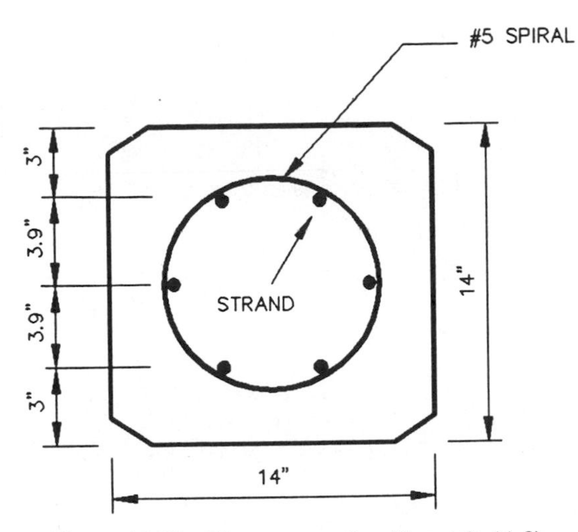

Figure 11-21 Pile cross section (Example 11-2).

Prestressing steel: $\frac{1}{2}$-in. diameter, 270 kips, seven-wire strand.

$$\text{Area of each strand} = 0.153 \text{ in.}^2$$
$$\text{Ultimate tensile strength per strand} = 41{,}310 \text{ lb}$$
$$\text{Initial force per strand} = 28{,}915 \text{ lb}$$
$$\text{Losses per strand } (35{,}000 \text{ lb/in.}^2) = 5{,}355 \text{ lb}$$
$$\text{Working force per strand} = 23{,}560 \text{ lb}$$
$$\text{Effective prestress} = 721 \text{ lb/in.}^2$$

$$f_{se} = 0.8 \times 0.7 \times f_{pu} = 0.8 \times 0.7 \times 270$$
$$= 151.2 \text{ ksi}$$

where f_{se} is the effective stress in the prestressed reinforcement after allowance for all losses, and f_{pu} is the specified tensile strength of the prestressing tendons.

$$\text{Effective strain} = 151.2 \times 0.153 = 22.13 \text{ kips/strand}$$

Compression block (see Figure 11-22):

$$C = (0.8)(14)(0.85 f_c')(c)$$
$$= 0.8 \times 14 \times 0.85 \times 6c = 57.12 \text{ kips}$$

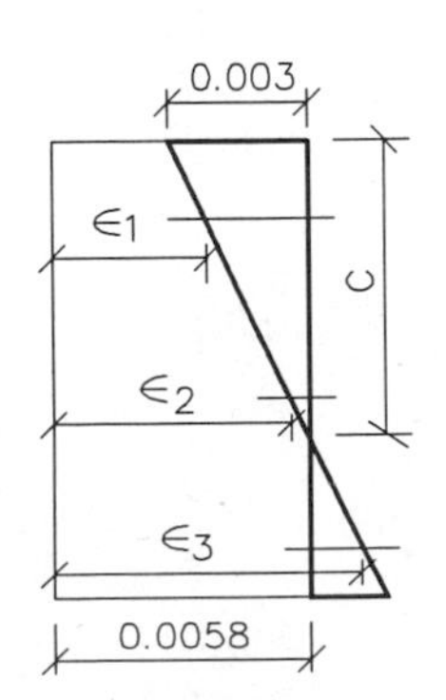

Figure 11-22 Strain diagram.

Try $c = 9.4$ in.; then $C = 536.9$ kips:

$$\epsilon_1 = 0.0058 - \frac{6.3}{9.4} \times 0.003 = 0.0038$$
$$f_1 = 106.4 \text{ ksi}$$
$$T_1 = 106.4 \times 2 \times 0.153 = 32.56 \text{ kips}$$
$$\epsilon_2 = 0.0058 - \frac{3.8}{9.4} \times 0.003 = 0.0046$$
$$f_2 = 128.8 \text{ ksi}$$
$$T_2 = 39.42 \text{ kips}$$
$$\epsilon_3 = 0.0058 - \frac{1.5}{9.4} \times 0.003 = 0.0063$$
$$f_3 = 176.4 \text{ ksi}$$
$$T_3 = 53.9 \text{ kips}$$
$$\sum T = 32.56 + 39.42 + 53.9 = 125.88 \text{ kips}$$
$$M_u = \frac{0.70}{12}[(536.9 \times 3.11) + 3.9(53.9 - 32.56)]$$
$$= \frac{0.70}{12}[1669.76 + 83.2]$$
$$= 102.3 \text{ ft-kips}$$
$$> 80 \text{ ft-kips} \quad \text{O.K.}$$

Sketches of an individual pile detailed per SEAOC-86 (UBC-88) and UBC-85 codes are shown in Figure 11-23.

Pile-cap design (see Figure 11-24): Determine number of piles (N):

$$N_1 = \frac{P_D + P_L}{100} = \frac{405 + 130}{100} = 5.35$$
$$N_2 = \frac{P_D + P_L + P_S}{100 \times \frac{4}{3}} = 5.78$$

Use six 14-in. square by 40-ft-long piles.

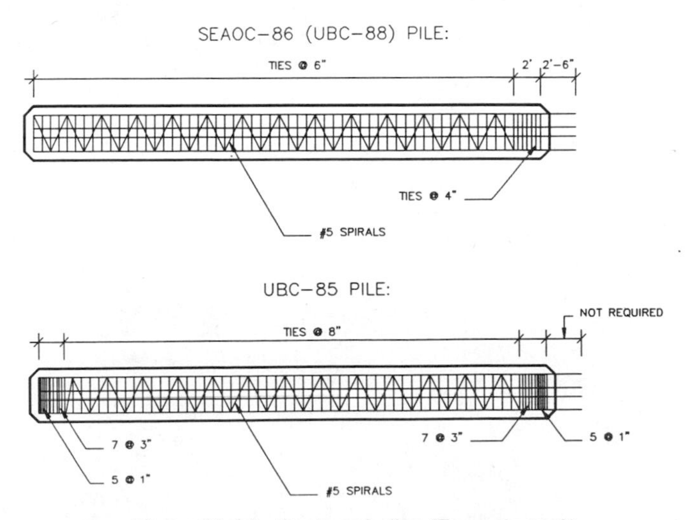

Figure 11-23 Designed piles (Example 11-2).

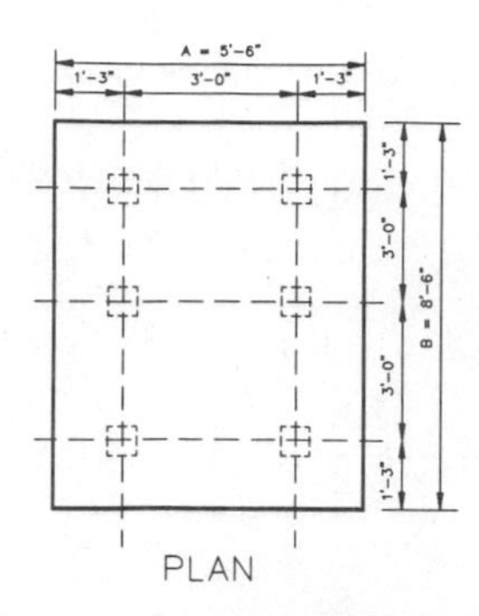

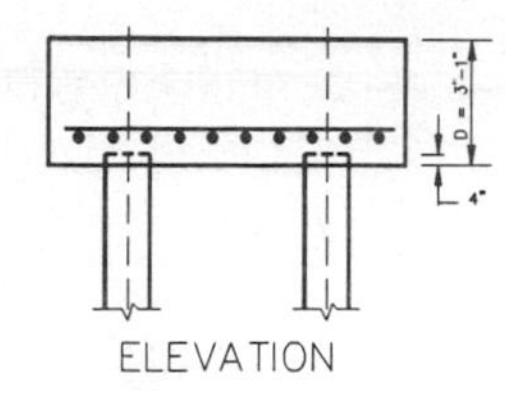

Figure 11-24 Pile cap design (Example 11-2).

Based on a 3-ft center-to-center pile spacing and 1 ft 3 in. of edge distance,

$$A = 3 + (2 \times 1.25) = 5.5 \text{ ft}$$

$$B = (2 \times 3) + (2 \times 1.25) = 8.5 \text{ ft}$$

(Minimum and maximum pile spacings are established by the soil engineer to insure that individual friction capacities are maintained.)

Establish minimum depth of the pile cap: Ultimate load per pile (P_u) is 920.6 kips from Example 11-1. Hence

$$\frac{920.6}{N} = \frac{930.6}{6} = 153.43 \text{ kips/pile}$$

Check punching shear at column:

$$v_c = 4\sqrt{f_c'} = 219 \text{ lb/in.}^2$$

$$v_u = V_u \times 1000/\phi b_0 d$$

$$\phi = 0.85 \quad \text{and} \quad b_0 = 4(14 + d)$$

$$v_u = \frac{v_u \times 1000}{0.85(56 + 4d)}$$

Set $v_u = v_c$. Then

$$219 = \frac{920.6 \times 1000}{0.85(56d + 4d^2)}$$

$$\therefore \quad d_{\min 1} = 29 \text{ in.}$$

$$\therefore \quad D = 29 + 1 + 3 + 4 = \underline{37 \text{ in.}}$$

Check punching shear at pile:

$$v_u = \frac{153.43 \times 1000}{0.85 b_0 d}$$

$$b_0 = 4(14 + 29) = 172 \text{ in.}$$

$$d = 29 \text{ in.}$$

$$v_u = 36.19 \text{ lb/in.}^2 < 219 \text{ lb/in.}^2 \qquad \text{O.K.}$$

Check beam shear (short direction):

$$v_c = 2\sqrt{f_c'} = 2\sqrt{3000} = 109 \text{ lb/in.}^2$$

$$V_u = 2 \times 153.43 = 306.86 \text{ kips}$$

$$v_u = \frac{V_u \times 1000}{0.85Bd}$$

$$B = 5.5 \times 12 = 66 \text{ in}$$

Set $v_u = v_c$. Then

$$v_c = \frac{306.86 \times 1000}{0.85 \times 66 \times d}$$

$$d_{\min} = \frac{306.86 \times 1000}{0.85 \times 66 \times 109} = 50.18 \text{ in.}$$

By inspection, this cone lies outside of the pile cap and is therefore is not applicable.

Establish depth of pile cap by deep beam shear analysis as specified in the ACI 318-83 Code:

$$v_c = \left(3.5 - \frac{2.5M_u}{V_u d}\right)\left(1.9\sqrt{f_c'} + \frac{2500\rho_w V_u d}{M_u}\right)$$

where

$$M_u = (2 \times 153.43)[(36 - 7) \times 0.5] = 4450 \text{ in.-kips}$$

$$V_u = 306.86 \text{ kips}$$

the minimum steel ratio per ACI Code is

$$\rho_w = 0.0018$$

and

$$d = 29 \text{ in.}$$

Thus

$$v_c = \left(3.5 - 2.5\frac{4450}{8899}\right) \times \left(1.9\sqrt{f_c'} + \frac{2500 \times 0.0018 \times 8899}{4450}\right)$$

$$= (2.25)(104 + 9.0)$$

$$= 254.25 \text{ lb/in.}^2 < 6\sqrt{f_c'} \qquad \text{O.K.}$$

Recheck v_u:

$$v_u = \frac{306.86 \times 1000}{0.85 \times 66 \times 29} = 188.62 \text{ lb/in.}^2 < 254 \text{ lb/in.}^2$$

Flexural reinforcing, longitudinal direction:

$$M_u = (2 \times 153.43)(3.0 - 0.58) = 742.6 \text{ ft-kips}$$

$$\text{M.S.} = \frac{742.6 \times 12}{\phi f_c' b d^2}$$

where

$$\phi = 0.90$$
$$b = B = 5.5 \times 12 = 66 \text{ in.}$$
$$d = 29 \text{ in.}$$

Thus

$$\text{M.S.} = 0.0595 \quad \therefore \quad \omega = 0.062$$

$$\rho = \frac{\omega f_c'}{f_y} = \frac{0.062}{30} = 0.0030$$

$$A_{Sf} = 0.003 \times 66 \times 29 = 5.69 \text{ in.}^2$$

Use eight equally spaced No. 8 bars.

Transverse direction:

$$M_u = (3 \times 153.43)(1.5 - 0.58) = 423.5 \text{ ft-kips}$$

$$\text{M.S.} = \frac{M_u \times 12}{0.9 \times 3 \times 8.5 \times 12 \times 29^2} = 0.0219$$

$$\omega = 0.022, \quad \therefore \quad \rho = 0.0011 < 0.0018$$

$$A_S = 0.0018 \times 8.5 \times 29 = 5.33 \text{ in.}^2$$

Use seven equally spaced No. 8 bars.

Strut ties:

- SEAOC-86 (UBC-88): Strut ties are required:

  $$P = (0.10)(920.6) = 92.06 \text{ kips}$$

  Note: This force may be resisted by strut ties, slab on grade, or any other equivalent method.
- NEHRP-85: Same as required in Example 11-1.
- UBC-85: Same as SEAOC-86.

11.10 LATERAL SEISMIC EARTH PRESSURES

Lateral earth pressures are imposed on retaining structures. Under static conditions, retaining structures would be subjected to active lateral earth pressures. These pressures are normally computed utilizing the classical theories developed by Coulomb and Rankine. The methodologies to determine the active lateral earth pressures on retaining walls for static conditions may be found in most geotechnical references, such as the United States Navy *Design Manual DM-7.2*.

When there is an earthquake, one can see that inertial forces from the ground shaking will impose additional load on a retaining wall. The most commonly used formulation to calculate the seismic lateral earth pressure on a retaining-wall structure is the Mononobe–Okabe formulation, which has been described in detail by Seed and Whitman.[11-20] This method is an extension of Columb earth-pressure theory with the addition of horizontal and vertical forces to account for the earthquake loads. This method assumes that there is sufficient wall movement to produce the minimum wall pressures and that the backfill material consists of dry cohesionless materials.

For cantilevered retaining structures, most investigators have agreed that the point of application of the resultant of the dynamic earth pressure should be at a height of $0.5H$ to $0.67H$ above the base of the wall.[11-20] Prakash[11-3] has recommended that the point of application of the resultant be taken at $0.55H$ above the base of a flexible wall end at $0.45H$ above the base of a rigid wall.

The Hospital Code of the State of California[11-21] contains a section which provides for a seismic increment of earth pressure. This provision requires that where buildings provide lateral support for walls that retain soil, and where the exterior grades on the opposite sides of the building differ by more than 6 ft, the earth pressure on the higher side should also include a seismic increment of earth pressure. In lieu of a determination by a qualified geotechnical engineer, this code provision requires that the seismic increment of earth pressure should be considered as an uniformly distributed pressure P_d determined as

$$P_d = 0.03 Z \Delta h$$

where

P_d = seismic increment of active earth pressure due to earthquake, lb/ft^2

Δ = soil density, lb/ft^3

h = height of retained earth above the bottom lateral support for the wall, ft

Z = seismic-hazard-zone coefficient

The seismic-hazard-zone coefficient Z is the same as that found in the UBC-85 Code. The value of Z for zone 4 is 1.0.

REFERENCES

11-1 Seed, H. B. and Idriss, I. M., *Ground Motions and Soil Liquefaction during Earthquakes*, Earthquake Engineering Research Institute, Berkeley, CA, 1982.

11-2 National Research Council, *Liquefaction of Soils during Earthquakes*, National Academy Press, Washington, 1985.

11-3 Prakash, S., *Soil Dynamics*, McGraw-Hill Book Company, New York, 1981.

11-4 Youd, L. T., "Major Cause of Earthquake Damage is Ground Failure," *Civil Eng.* 48, No. 4, 47–51, Apr. 1978.

11-5 Youd, L. T., "Ground Failure Displacement and Earthquake Damage to Buildings," *Proceedings, 2nd ASCE Conference on Civil Engineering and Nuclear Power*, Knoxville, TN, Sept. 1980.

11-6 Brown, R. E., "Vibroflotation Compaction of Cohesionless Soils," *J. Geotech. Div. ASCE* 103, No. GT12, 1437–1451, Dec. 1977.

11-7 Ishihara, K., Iwasake, Y., and Nakajima, M., "Liquefaction Characteristics of Sand Deposits at an Oil Tank Site During the 1978 Miyagiken-Oki Earthquake," *Soil and Foundations* 20, No. 2, 97–111, June 1980.

11-8 Hunt, R. E., *Geotechnical Engineering Investigation Manual*, McGraw-Hill Book Company, New York, 1984.

11-9 Whitman, R. V. and DePablo, P. O., "Densification of Sand by Vertical Vibrations," *Proceedings, 4th World Conference on Earthquake Engineering*, Santiago, Chile, 1980.

11-10 Lambe, W. T. and Whitman, R. V., *Soil Mechanics*, John Wiley & Sons, New York, 1969.

11-11 Seed, H. B. and Wilson, S. D., "The Turnagain Heights Landslide, Anchorage, Alaska," *J. Soil Mech. and Foundation Eng. Div. ASCE* 93, No. SM4, 325–353, July 1967.

11-12 Legget, R. F. and Karrow, P. F., *Handbook of Geology in Civil Engineering*, McGraw-Hill Book Company, New York, 1983.

11-13 Wiegel, R. E. (ed.), *Earthquake Engineering*, Prentice-Hall, Englewood Cliffs, NJ, 1970.

11-14 Seed, H. B., "Earthquake Effects on Soil-Foundation Systems," *Foundation Engineering Handbook*, H. F. Winterkorn and H. Y. Fang (eds.), Van Nostrand Rienhold Company, New York, 1975.

11-15 Wolf, J. P., *Dynamic Soil-Structure Interaction*, Prentice-Hall, Englewood Cliffs, NJ, 1985.

11-16 International Conference of Building Officials, *Uniform Building Code—1985*, Whittier, CA, 1985.

11-17 Federal Emergency Management Agency, *1985 Edition of NEHRP Recommended Provisions for the Development of Seismic Regulations for New Buildings*, FEMA-96, Feb. 1986.

11-18 Seismology Committee of Structural Engineers Association Of California, "Tentative Lateral Force Requirements," Nov.–Dec. 1986.

11-19 American Concrete Institute, *Building Code Requirements for Reinforced Concrete (ACI 318-83)*, ACI, Detroit, Michigan, 1983.

11-20 Seed, H. B., and Whitman, R. V., "Design of Earth Retaining Structures for Dynamic Loads," Proc. ASCE Spec. Conf. on Lateral Stresses in the Ground and Design of Earth Retaining Structures, Ithaca, New York, pp 103–147, 1970.

11-21 State of California, *California Administrative Code*, *Title 24*, Part II, Sacramento, California, 1985.

Chapter 12

Design of Nonstructural Systems and Components

*Thomas A. Sabol, Ph.D., S.E.**

12.1 INTRODUCTION

Prior to the 1964 Alaska earthquake, little attention had been paid to the problem of damage to nonstructural components and the contents of structures. Nearly all of the designer's effort was expended on mitigating damage to the lateral-force-resisting system, while seismic restraint of mechanical, electrical, plumbing, and architectural systems was left to the individual manufacturers or the installation contractors. However, the 1964 Alaska earthquake, and subsequent earthquakes, focused attention on the need to include explicit consideration of the seismic design for nonstructural elements to mitigate property losses and life–safety hazards resulting from damage to these systems.

Nonstructural systems consist of architectural, mechanical, and electrical elements. These elements include such components as nonbearing walls, exterior wall panels, suspended ceilings, air-conditioning equipment, boilers, elevators, electrical switch gear, transformers, and lighting fixtures.

In many cases, the monetary value of the damage associated with the nonstructural elements exceeds the value of the damage to the structural components.

Past earthquakes have provided ample evidence of damage to nonstructural systems. For example, library shelves are capable of sustaining the downward load of several thousand pounds of books, yet are easily toppled when subject to even moderate earthquake, as shown in Figure 12-1. Mechanical equipment is often mounted on vibration isolators to prevent the transmission of annoying or damaging vibrations caused by the normal operation of the equipment. However, the potential for shifting or toppling of heavy equipment during an earthquake is seldom considered. Figure 12-2 shows a heavy chiller shifted off its vibration isolators during an earthquake in southern California. Everyday objects such as lighting fixtures can be excited by earthquake motion and fall from the suspended ceiling or hung by their electrical connections as shown in Figure 12-3.

The intent of this chapter is to provide an understanding of the necessity for seismic design of nonstructural systems. We will also examine the available procedures and guidelines for seismic design of such elements. This will be accomplished by presenting a discussion of current design standards in Section 12.2.

Section 12.3 discusses design considerations for architectural components by presenting examples of well-performing systems and details to prevent earthquake-related damage. Similar information is presented in Section 12.4 for mechanical and electrical systems. Examples to illustrate the above concepts are presented throughout the chapter.

*Project Director, Englekirk and Hart Consulting Engineers, Inc., Los Angeles, California.

Figure 12-1 Overturned library shelves, Lawerence Livermore Laboratory.

Figure 12-2 Chiller shifted from vibration isolation mounts, Olive View Hospital, San Fernando Valley earthquake of 1971.

Figure 12-3 Damaged suspended ceiling and light fixtures, Olive View Hospital, San Fernando Valley earthquake of 1971.

12.2 DESIGN STANDARDS

12.2.1 General

The design standards for nonstructural systems cannot be expected to address every situation. However, if this limitation is recognized, these standards do provide insight and needed guidance to help mitigate the damage to nonstructural systems. Included in this section are discussions of the design standards sponsored by the Uniform Building Code, the United States military, and

Table 12-1 Horizontal Factor C_p According to UBC-85(12-1)

Part of building	Direction of horizontal force	C_p*
1. Exterior bearing and nonbearing walls, interior bearing walls and partitions, interior nonbearing walls and partitions — see also Section 2312 (j) 3 C. Masonry or concrete fences over 6 ft high	Normal to flat surface	0.3†
2. Cantilever elements:		
a. Parapets	Normal to flat surfaces	0.8
b. Chimneys or stacks	Any direction	
3. Exterior and interior ornamentations and appendages	Any direction	0.8
4. When connected to, part of, or housed within a building:		
a. Penthouses, anchorage and supports for chimneys, stacks and tanks, including contents‡	Any direction	0.3
b. Storage racks with upper storage level at more than 8 ft in height, plus contents§		
c. All equipment or machinery‡		
d. Fire sprinkler system		
e. Supports and bracing, equipment racks and piping for hazardous production material		0.45
5. Suspended ceiling framing systems (applies to seismic zones 2, 3, and 4 only) — see also Section 4701 (e)	Any direction	0.3**,††
6. Connections for prefabricated structural elements other than walls, with force applied at center of gravity of assembly	Any direction	0.3‡‡
7. Access floor systems	Any direction	0.3§§

* C_p for elements laterally self-supported only at the ground level may be two-thirds of value shown.

† See Also Section 2309 (b) for minimum load and deflection criteria for interior partitions.

‡ For flexible and flexibly mounted equipment and machinery, the appropriate values of C_p shall be determined with consideration given both to the dynamic properties of the equipment and machinery and to the building or structure in which it is placed but shall be not less than the listed values. The design of the equipment and machinery and their anchorage is an integral part of the design and specification of such equipment and machinery.

For essential facilities and life-safety systems, the design and detailing of equipment which must remain in place and be functional following a major earthquake shall consider drifts in accordance with Section 2312 (k).

§ W_p for storage racks shall be the weight of the racks plus contents. The value of C_p for racks over two storage support levels in height shall be 0.24 for the levels below the top two levels. In lieu of the tabulated values, steel storage racks may be designed in accordance with U.B.C. Standard No. 27-11.

Where a number of storage rack units are interconnected so there are a minimum of four vertical elements in each direction and each column line designed to resist horizontal forces, the design coefficients may be as for a building with K values from Table No. 23-I, $CS = 0.2$ for use in the formula $V = ZIKCSW$ and W equal to the total dead load plus 50% of the rack-rated capacity. Where the design and rack configurations are in accordance with this paragraph, the design provisions in U.B.C. Standard No. 27-11 do not apply.

** Ceiling weight shall include all light fixtures and other equipment which is laterally supported by the ceiling. For purposes of determining the lateral force, a ceiling weight of not less than 4 lb/ft^2 shall be used.

†† Does not apply to ceilings constructed of lath and plaster or gypsum board screw or nail attached to suspended members that support a ceiling at one level extending from wall to wall.

‡‡ The force shall be resisted by positive anchorage and not by friction.

§§ W_p for access floor systems shall be the dead load of the access floor system plus 25% of the floor live load and a 10-lb/ft^2 partition load.

the National Earthquake Hazards Reduction Program. Applications of floor response spectra in evaluation of forces acting on nonstructural systems and components are also discussed.

Most existing design codes assume that the behavior of the nonstructural components in a building can be uncoupled from the response of the primary structural components. The nonstructural systems are then designed for earthquake resistance using equivalent static lateral forces. The accuracy of such an approach is limited. Certain codes have attempted to address the problem by explicitly considering the dynamic response of coupled systems. Furthermore, there are several specialized design procedures for various types of electrical, mechanical, and plumbing systems.

Nearly all of the contemporary design standards use the same basic formula to establish the design lateral force:

$$F_p = ZC_pW_c \tag{12-1}$$

where

F_p = lateral force applied to the nonstructural component

Z = numerical coefficient that depends on the seismic zone in which the structure is located

C_p = horizontal force factor that varies with the type of nonstructural component

W_c = weight of the whole, or a part, of the nonstructural component

The essential difference between the model standards is in the value of the coefficient C_p.

The purpose of the analysis of nonstructural components is to verify that they would withstand, without collapse or excessive deflection, the forces and deformations caused by a major seismic event. The determination of a realistic loading level represents a difficult problem. Various techniques that utilize equivalent static lateral forces, acceleration time histories, or floor response spectra are available to estimate these forces. However, in all but the most comprehensive analyses, neither the time nor the money is available to use the latter two methods. This leaves the equivalent-static-lateral-force method as the most practical alternative for a wide variety of situations. If a floor response spectrum is available, it can be used instead of the forces described in this section.

Nonstructural components whose weights are large in comparison with the structural weight tend to have significant effects upon the overall response of the structure. Furthermore, the response of these components is highly dependent upon the response of the building. As a consequence, the components that are not covered by our discussion are those whose weights:

1. exceed 20% of the total dead weight of the floor where the component is located or
2. exceed 10% of the total dead weight of the structure.

Such components should be analyzed together with the structure and are not suitable for the methods presented here.

Table 12-2 Horizontal Factor C_p According to UBC-88[(12-2)]

Elements of structures and nonstructural components	C_p*
I. Part of Structure	
1. Walls, including the following:	
a. Unbraced (cantilevered) parapets.	2.0
b. Other exterior walls above the ground floor.	0.75
c. All interior bearing and nonbearing walls and partitions.[†]	0.75
d. Masonry or concrete fences over 6 ft high.	0.75
2. Penthouse (except where framed by an extension of the building frame).	0.75
3. Connections for prefabricated structural elements other than walls, with force applied at center of gravity.[‡]	0.75
4. Diaphragms.[§]	—
II. Nonstructural Components	
1. Exterior and interior ornamentations and appendages.	2.0
2. Chimneys, stacks, trussed towers, and tanks on legs.	
a. Supported on or projecting as an unbraced cantilever above the roof more than one-half its total height.	2.0
b. All others, including those supported below the roof with unbraced projection above the roof less than one-half its height, or braced or guyed to the structural frame at or above its center of mass.	0.75
3. Signs and billboards.	2.0
4. Mechanical, plumbing, and electrical equipment and machinery and associated piping.[**]	0.75
5. Tanks and vessels (plus contents), including support systems and anchorage.	0.75
6. Storage racks (include contents).	0.75
7. Anchorage for permanent floor-supported cabinets and bookstacks more than 5 ft in height (includes contents).	0.75
8. Anchorage for suspended ceilings and light fixtures — see also Section 4701 (e) and Footnote 3 after the item.[††,‡‡]	0.75
9. Access floor systems.[‡,§§]	0.75

* See Section 2312(g)2 for items supported at or below grade.
[†] See Section 2312(h)2D.
[‡] Applies to seismic zones 2, 3, and 4 only.
[§] See Section 2312(h)2I.
[**] Equipment and machinery shall include but not be limited to such items as boilers, heat exchangers, chillers, pumps, motors, air-handling units, cooling towers, transformers, switch gear, control panels, and life-safety equipment. It shall include sprinkler systems, other major piping, and the ducting serving such equipment and machinery.
[††] Ceiling weight shall include all light fixtures and other equipment or partitions which are laterally supported by the ceiling. For purposes of determining the lateral seismic force, a ceiling weight of not less than 4 lb/ft^2 shall be used.
[‡‡] Ceilings constructed of lath and plaster or gypsum board screw- or nail-attached to suspended members that support a ceiling at one level extending from wall to wall need not be analyzed, provided the walls are not over 50 ft apart.
[§§] W_p for access floor systems shall be the dead load of the access floor systems plus 25% of the floor live load plus a 10-lb/ft^2 partition load allowance.

12.2.2 The Uniform Building Code Provisions

According to both 1985 and 1988 editions of the Uniform Building Code,[12-1, 12-2] the design lateral force on a nonstructural component is computed as

$$F_p = ZIC_pW_p \tag{12-2}$$

where Z and I are the seismic-zone factor and the importance factor, respectively (see Section 4.3). W_p is the weight of the nonstructural component. The UBC-85 values for the coefficient C_p are given in Table 12-1.

UBC-88 differentiates between the rigid and nonrigid items. The rigid components are defined as those having a fixed base period less than or equal to 0.06 sec. The UBC-88 values of C_p for rigid items are given in Table 12-2. For nonrigid items or flexibly supported items, the 1988 edition of the code requires that the values of C_p be determined considering dynamic properties of both the equipment and the structure which supports it. In the absence of a detailed analysis, the value of C_p for a nonrigid component may be taken as twice the value listed in Table 12-2. However, it need not exceed 2.0. The value of C_p for elements or components supported at or below the ground level may be two-thirds of the value shown in Table 12-2. However, the design force should not be less than that obtained by treating the item as an independent structure.

12.2.3 The *Tri-Services Manual* Provisions

The *Tri-Services Manual*[12-3] contains the seismic-design requirements for United States military installations. As in the UBC formulation, an equivalent-static-force approach is used for most components. The recommended design forces for architectural components are identical to those in the UBC. However, the design forces for mechanical and electrical equipment are obtained using a response spectrum.

Two general classifications of equipment are defined:

1. Equipment at the ground or on the seismic base of the structure.
2. Equipment above the seismic base.

Equipment Installed at the Ground For equipment installed at the ground the seismic force F_p, is given by

$$F_p = ZCW_p \tag{12-3}$$

where Z is the seismic-zone coefficient as described by the UBC code, C is the seismic-force coefficient, and W_p is the weight of equipment.

The value of C is based on a site-independent response spectrum shown in Figure 12-4 as well as a soil constant C_s such that

$$C = S_aC_s \tag{12-4}$$

The soil constant C_s is dependent on the allowable

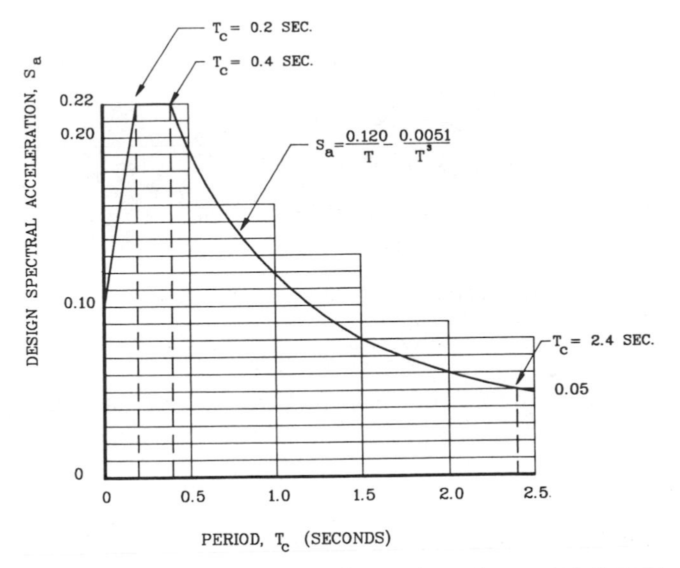

Figure 12-4 Design spectrum for equipment mounted directly on the ground.

bearing value f_p (ksf) of the soil at the site:

$$C_s = 2.25 - 0.125f_p \geqslant 1.0 \tag{12-5}$$

A graphic representation of Equation 12-5 is shown in Figure 12-5. For sites with engineered fill of only moderate depth, the value of f_p should be based on the native soil, not the engineered fill. For values of f_p greater than or equal to 10.0 ksf the response is assumed to be equal to that of the underlying bedrock ($C_s = 1.0$).

If the equipment is *rigidly* mounted on the ground, then

$$C = 0.10C_s \tag{12-6}$$

Equipment Installed in the Building The design lateral force for equipment installed in the building is

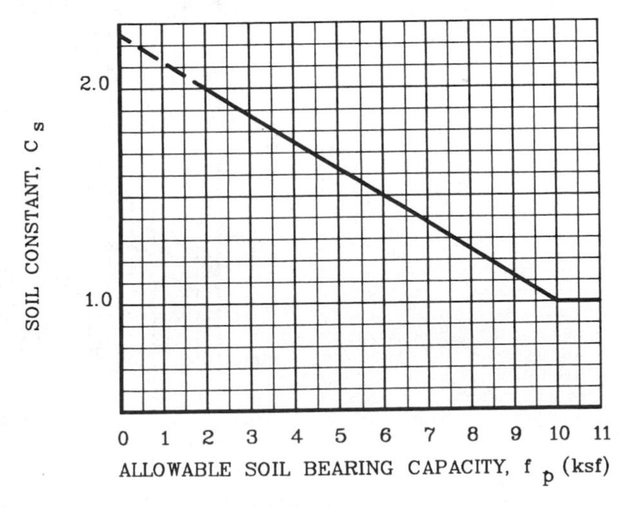

Figure 12-5 Soil constant (C_s) versus allowable soil bearing capacity (f_p).

given by

$$F_p = ZKCSW_p \qquad (12\text{-}7)$$

where

Z = seismic-zone coefficient

K = numerical coefficient for the selected structural system (see Table 4-1)

C = seismic-force coefficient

W_p = weight of equipment

The seismic-force coefficient C, in this case, is defined as

$$C = C_s A_h \text{M.F.} \qquad (12\text{-}8)$$

where C_s is as described previously, and

A_h = design acceleration for the equipment

M.F. = dynamic magnification factor

A_h is intended to describe an envelope of maximum accelerations based on building height and location of the equipment within the building. As the equipment is installed at higher locations within the structure, the maximum accelerations will increase due to dynamic amplification caused by the seismic response of the building. For one- to three-story buildings this amplification is usually less than for buildings of greater height. Consequently, two equations are used to define A_h (see Figure 12-6):

$$A_h = 0.10 + 0.05\frac{h_c}{h_n} \qquad (12\text{-}9)$$

for buildings of one to three stories, and

$$A_h = 0.10 + 0.15\frac{h_c}{h_n} \qquad (12\text{-}10)$$

for multistory buildings, where

h_c = height at which the equipment is installed

h_n = total height of the building above the seismic base

If the input motion is approximated by a harmonic function, the magnification factor M.F. can be obtained from

$$\text{M.F.} = \frac{1}{\sqrt{\left[1 - (T_c/T)^2\right]^2 + (2\beta T_c/T)^2}} \qquad (12\text{-}11)$$

where β is the ratio of critical damping. Equation 12-11 is presented in graphic form in Figure 12-7.

The period of the building, T, can be estimated from information presented in Chapters 3 and 4. The fundamental period of the equipment, T_c, is to be obtained from:

$$T_c = 0.32(W_p/K_c)^{0.5} \qquad (12\text{-}12)$$

where K_c is the stiffness of the resilient equipment mounts. The stiffness of the equipment support attachment is determined in terms of the load per unit deflection of the center of gravity (lb/in.). For stable resilient attachments, K_c is taken as the spring constant. For other resilient attachments, K_c is taken as the slope of the load–deflection curve at the point of interest.

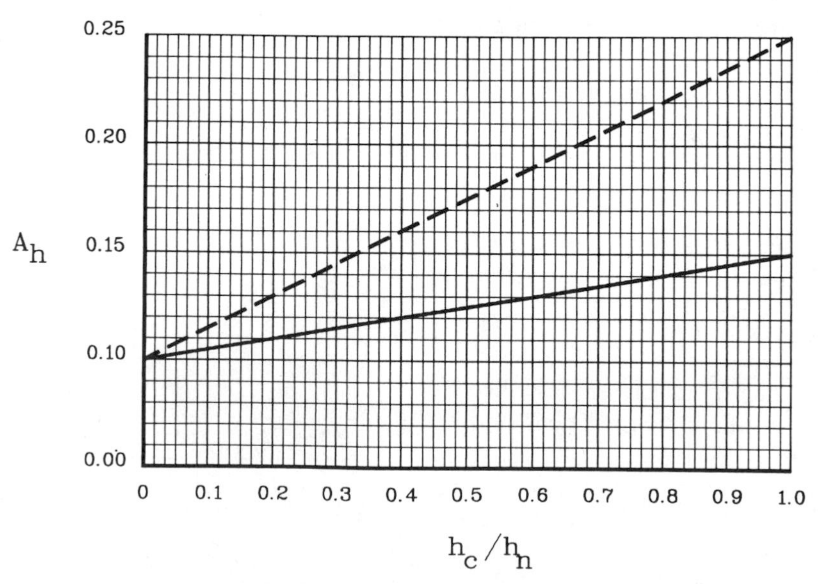

Figure 12-6 Design accelerations for mechanical and electrical equipment.

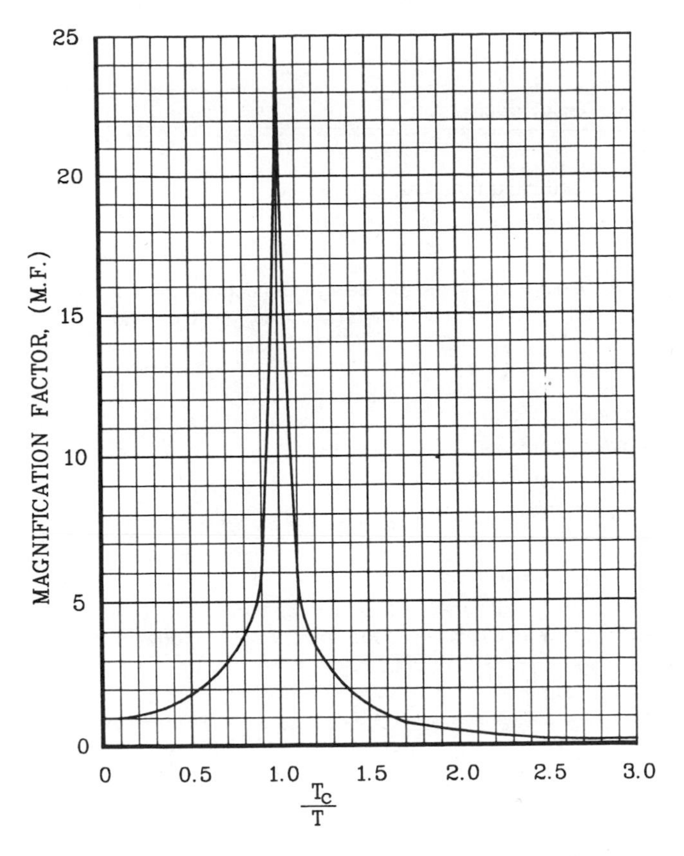

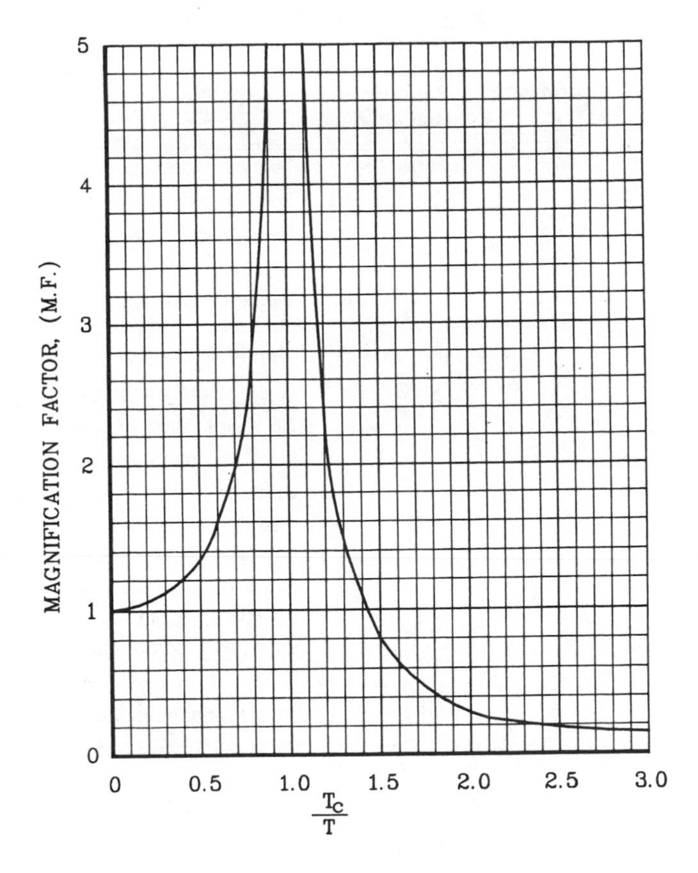

Figure 12-7 Dynamic magnification factor.

12.2.4 The NEHRP-85 Provisions

Design provisions for nonstructural damage mitigation with a level of sophistication similar to that in the *Tri-Services Manual* are contained in the document developed by the Federal Emergency Management Agency as part of the National Earthquake Hazards Reduction Program (NEHRP-85).[12-4] These design criteria are based directly upon a study sponsored by the Applied Technology Council and embodied in the document known as ATC 3-06.[12-5] The NEHRP-85 employs a static type of analysis similar to that in the UBC. However, for evaluation of design lateral forces, the nonstructural components are divided into two categories: (i) architectural components, and (ii) mechanical/electrical components.

A factor P is defined to distinguish three levels of performance characteristics: *superior* ($P = 1.5$), *good* ($P = 1.0$), and *low* ($P = 0.5$). The performance-criterion factor P and the corresponding design goals for various nonstructural components are summarized in Tables 12-3 and 12-4. Based on the seismic-hazard exposure group assigned to the building (see Section 4.6.10), individual nonstructural elements are required to satisfy a designated minimum performance criterion. The performance-characteristic levels required for various nonstructural components are given in Tables 12-5 and 12-6.

The NEHRP-85 design forces correspond to the ultimate limit state of the system (factored loads). Should design comparisons involve working loads, the forces calculated in this subsection should be reduced by a factor of 1.2.[12-4]

Seismic Forces on Architectural Components Seismic forces acting on architectural systems, their components, and their attachments are estimated as follows:

$$F_p = A_v C_c P W_c \tag{12-13}$$

where

F_p = seismic force applied to a component of a building or equipment at its center of gravity

A_v = seismic coefficient determined from Figure 4-7

C_c = seismic coefficient for components of architectural systems as given in Table 12-5

P = performance criteria factor

W_c = weight of component

Table 12-3 Definition of Performance Criteria for Architectural Components[12-4]

Performance P	Performance characteristic	Design goal
1.5	Superior (S)	Maximum resistance to lateral forces. Damage limited to cosmetic damages. All operating functions unimpaired. Minimal glass breakage (safety glass may crack). No loss of any fire rating or protection. System or component shall be able to handle 1.5 times the design deflections of any structural member to which it is attached or could have loads imposed on it due to structural member design movement.
1.0	Good (G)	Average resistance to lateral forces. No major falloff of wall or ceiling components allowed. No glass fallout except for tempered-glass fragments. All operating functions normally operable or readily repaired on site. Fire rating 75% intact. This does not mean 75% of unit is intact; it means that a 4-h wall shall have a 3-h, etc. Minor damage to system or component structure allowed. System or component shall be able to handle 1.0 times the design deflections of any structural member to which it is attached or could have loads imposed on it due to structural member design movement.
0.5	Low (L)	Low resistance to lateral forces. Glass fallout permitted. Ceilings and lighting fixtures fall down. Major components must substantially stay in place but are not operable and must be repaired. System or component structural damage may occur. Fire ratings impaired. System or component shall be able to handle 0.5 times the design deflections of any structural member to which it is attached or could have loads imposed on it due to structural member design movement.

Table 12-4 Definition of Performance Criteria for Mechanical and Electrical Equipment[12-4]

Performance P	Performance characteristic	Design goal
1.5	Superior (S)	High resistance to static and dynamic seismic forces. All operating functions unimpaired. No broken piping regardless of size. No interruptions of utility services other than normal transfer of functions to alternative sources.
1.0	Good (G)	Moderate resistance to static and dynamic forces. All major equipment operates normally or is easily repaired at site. No broken main distributing piping or vessel. No short circuits or broken electrical circuits.
0.5	Low (L)	Low resistance to static and dynamic seismic forces. Major equipment must substantially stay in place. Broken main distribution piping and vessels tolerated. Fallout of lighting fixtures tolerated. Short circuits or broken electrical circuits tolerated.

Table 12-5 Seismic Coefficient C_c and Performance-Characteristic Levels Required for Architectural Systems and Components[(12-4)]

Architectural Components	Factor C_c	Required Performance-characteristic levels by seismic-hazard exposure group* III	II	I
Appendages				
Exterior nonbearing walls	0.9	S	G†	L‡
Wall attachments	3.0	S	G†	L‡
Connector fasteners	6.0			
Veneer attachments	3.0	G	G§	L
Roofing units	0.6	G	G†	NR
Containers and miscellaneous components (free-standing)	1.5	G	G	NR
Partitions				
Stairs and shafts	1.5	S	G**	G
Elevator shafts	1.5	S	L**	L††
Vertical shafts	0.9	S	L**	L‡‡
Horizontal exists including ceilings	0.9	S	S	G
Public corridors	0.9	S	G	L
Private corridors	0.6	S	L	NR
Full-height area separation partitions	0.9	S	G	G
Full-height other partitions	0.6	S	L	L
Partial-height partitions	0.6	G	L	NR
Structural fireproofing	0.9	S	G**	L‡‡
Ceilings				
Fire-rated membrane	0.9	S	G**	G
Non-fire-rated membrane	0.6	G	G	L
Architectural equipment — ceiling, wall, or floor mounted	0.9	S	G	L
Architectural components — elevator and hoistway structural systems				
Structural frame providing supports for guiderail brackets	1.25	S	G	G
Guiderails and brackets	1.25	S	G	G
Car and counterweight guiding members	1.25	S	G	G

* NR = not required.
† May be reduced one performance level if the area facing the exterior wall is normally inaccessible for a distance of 10 ft and the building is only one story.
‡ Shall be raised one performance level if the area facing the exterior wall is normally accessible within a distance of 10 ft plus 1 ft for each floor height.
§ May be reduced one performance level if the area facing the exterior wall is normally inaccessible for a distance of 10 ft plus 1 ft for each floor of height.
** Shall be raised one performance level if the building is more than four stories or 40 ft in height.
†† May be reduced to NR if the building is less than 40 ft in height.
‡‡ Shall be raised one performance level for an occupancy containing flammable gases, liquids, or dust.

Seismic Forces on Mechanical / Electrical Equipment Seismic forces acting on mechanical/electrical systems, their components, and their attachments can be estimated as follows:

$$F_p = A_v C_c P a_c a_x W_c \qquad (12\text{-}14)$$

where

F_p = seismic force applied to a component of a building or equipment at its center of gravity

A_v = seismic coefficient determined from Figure 4-7

C_c = seismic coefficient for components of mechanical or electrical systems as given in Table 12-6

P = performance-criterion factor

a_c = amplification factor related to the response of a system or component as affected by the type of attachment as determined from Table 12-7

a_x = amplification factor at level x related to the variation of response in height of the building

$$= 1.0 + \frac{h_x}{h_n}$$

Table 12-6 Seismic Coefficient C_c and Performance-Characteristic Levels Required for Mechanical / electrical components(12-4)

Mechanical / Electrical Components†	Factor‡ C_c	Required performance-characteristic levels by seismic-hazard exposure group*		
		III	II	I
Emergency electrical systems (code-required) Fire and smoke detection system (code-required) Fire suppression systems (code-required) Life-safety systems components	2.00	S	S	S
Elevator machinery and controller anchorage	1.25	S	G	G
Boilers, furnaces, incinerators, water heaters, and other equipment using combustible energy sources or high-temperature energy sources, chimneys, flues, smokestacks, and vents Communication systems Electrical bus ducts and primary cable systems Electrical motor control centers, motor control devices, switchgear, transformers, and unit substations Reciprocating or rotating equipment Tanks, heat exchangers, and pressure vessels Utility and service interfaces	2.00	S	G	L
Machinery (manufacturing and process)	0.67	S	G	L
Lighting fixtures	0.67§	S	G	L
Ducts and piping distribution systems:				
Resiliently supported	2.00	S	G	NR
Rigidly supported	0.67**	S	G	NR
Electrical panelboards and dimmers	0.67	S	G	NR
Conveyor systems (nonpersonnel)	0.67	S	NR	NR

*NR = not required.

†Where mechanical or electrical components are not specifically listed in this table, the designer shall select a similarly listed component, subject to the approval of the authority having jurisdiction, and shall base the design on the performance and C_c values for the similarl component.

‡C_c values listed are for horizontal forces. C_c values for vertical forces shall be taken as one-third of the horizontal values.

§Hanging- or swinging-type fixtures shall use a C_c value of 1.5 and shall have a safety cable attached to the structure and the fixture at each support point, capable of supporting four times the vertical load.

**Seismic restraints may be omitted from the following installations: (1) gas piping of less than 1-in. inside diameter, (2) piping in boiler and mechanical rooms of less than $1\frac{1}{4}$ in. inside diameter, (3) all other piping of less than $2\frac{1}{2}$-in. inside diameter, (4) all electrical conduit of less than $2\frac{1}{2}$-in. inside diameter, (5) all rectangular air-handling ducts less than 6 ft^2 in cross-sectional area, (6) all round air-handling ducts less than 28 in. in diameter, (7) all piping suspended by individual hangers 12 in. or less in length from the top of the pipe to the bottom of the support for the hanger, and (8) all ducts suspended by hangers 12 in. or less in length from the top of the duct to the bottom of the support for the hanger.

where

h_x = height above the seismic base at level x

h_n = total height of the building above the shear base

W_c = weight of component or equipment

In Table 12-7, T_c is the fundamental period of the equipment and T is the fundamental period of the building. T_c is determined using Equation 12-12.

Example 12-1

A reciprocating chiller is mounted on a concrete housekeeping pad in the mechanical area of a ductile frame building. The chiller weights 15.0 kips. The building is classified in seismic-hazard exposure group II and is located in Los Angeles, California. The vibration engineer has recommended that resilient isolators be used to damp the vibration created during the operation of the unit. These isolators have a total stiffness K_c

Table 12-7 Amplification Factor a_c for Mechanical / Electrical Equipment[12-4]

Equipment attachment	a_c
For fixed or direct attachment to buildings	1.0
For resilient mounting systems:	
with seismic activated restraining devices	1.0
with elastic restraining device:	
if $T_c/T < 0.6$ or $T_c/T > 1.4$	1.0
if $T_c/T \geqslant 0.6$ or $T_c/T \leqslant 1.4$	$\geqslant 2.0$
if mounted on the ground or on a slab in direct contact with the ground	2.0

equal to 6.6 kips/in. The dimensions of the chiller are shown in Figure 12-8. The building's height is 180 ft, and its estimated period is 1.2 sec.

1. Using the NEHRP design criteria, establish the design lateral force on the anchors if the mechanical area is located 170 ft above the shear base.
2. Using the NEHRP design criteria, establish the design lateral force on the chiller if the mechanical area is located at the ground floor (0-ft level).
3. Using the provisions of the UBC-85 code, establish the design lateral force on the chiller based on the parameters given in part 1.
4. Using the *Tri-Services Manual* criteria, establish the design lateral force on the chiller based on the parameters given in part 1. Assume that the allowable soil bearing value f_p is 5.0 ksf.

- Part 1:

$$A_v = 0.4 \quad \text{(Figure 4-7)}$$

$$C_c = 2.0 \quad \text{(Table 12-6)}$$

$$P = 1.0 \quad \text{(Table 12-4)}$$

$$T_c = 0.32\left(\frac{W_c}{K_c}\right)^{0.5} = 0.32\left(\frac{15 \text{ kips}}{6.6 \text{ kips/in.}}\right)^{0.5} = 0.48 \text{ sec}$$

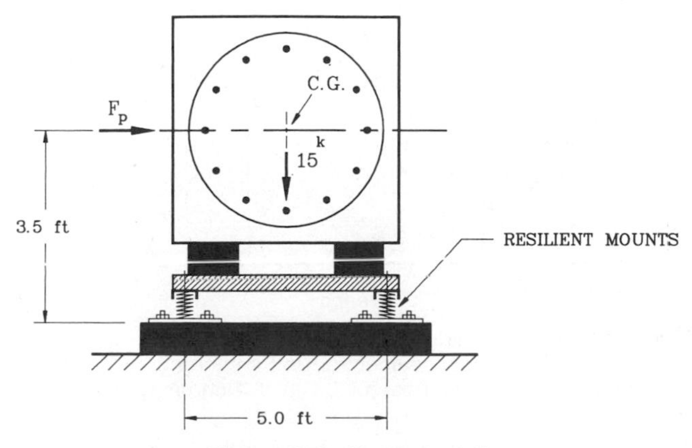

Figure 12-8 Example chiller.

$$\frac{T_c}{T} = \frac{0.48 \text{ sec}}{1.2 \text{ sec}} = 0.4 \quad \Rightarrow \quad a_c = 1.0$$

$$a_x = 1.0 + \frac{h_x}{h_n} = 1.0 + \frac{170}{180} = 1.94$$

Substitute the information gathered above into Equation 12-14 to obtain the value of the seismic force on the chiller:

$$F_p = A_v C_c P a_c a_x W_c = 0.4(2.0)(1.0)(1.0)(1.94)(15.0 \text{ kips}) = 23.3 \text{ kips}$$

Assuming four anchor points, the shear force applied to the anchor points is

$$F_v = \frac{23.3 \text{ kips}}{4} = 5.8 \text{ kips/anchor}$$

Thus the overturning moment is

$$M_{ot} = (23.3 \text{ kips})(3.5 \text{ ft}) = 81.5 \text{ kip-ft}$$

and the resisting moment is

$$M_r = (15.0 \text{ kips})(2.5 \text{ ft}) = 37.5 \text{ kip-ft}$$

Summing moments about a corner of the base, we obtain the uplift force F_t in the anchors

$$F_t = \frac{81.5 - 37.5}{5.0} = 8.8 \text{ kips}$$

Since there are two anchor groups per side, the uplift force at the anchor point is 4.4 kips/group. Some designers would recommend that a reduced value of the resisting moment be used to take account of vertical acceleration that would reduce the effectiveness of the equipment weight. A value of 90% of the calculated resisting moment seems reasonable.

- Part 2:

$$a_x = 1.0 + \frac{0 \text{ ft}}{180 \text{ ft}} = 1.0$$

The revised value of F_p is

$$F_p = 0.4(2.0)(1.0)(1.0)(1.0)(15.0 \text{ kips}) = 12.0 \text{ kips}$$

The advantage of locating heavy equipment such as this chiller on the lower floors of the building is evident when one observes the reduction in lateral force from 23.3 to 12.0 kips.

- Part 3: Using the criteria contained in the UBC-85 code, the force applied on the chiller is given by Equation 12-2:

$$F_p = ZC_pW_p$$

The value of Z for zone 4 is 1.0, and the value $C_p = 0.3$ is obtained from Table 12-1. Thus

$$F_p = 1.0(0.3)(15.0 \text{ kips}) = 4.5 \text{ kips}$$

- Part 4: Using Equation 12-7,

$$F_p = ZKCW_p$$
$$Z = 1.0 \qquad \text{(from part 3)}$$
$$K = 0.67$$

for a ductile frame (see Section 4.3), and

$$C = C_sA_h\text{M.F.}$$

From Equation 12-5,

$$C_s = 2.25 - 0.125f_p = 2.25 - 0.125(5.0 \text{ ksf})$$
$$= 1.63$$

For multistory buildings such as this one, the envelope of maximum accelerations, A_h, is given by Equation 12-10:

$$A_h = 0.10 + 0.15\frac{h_x}{h_n} = 0.10 + 0.15\frac{170}{180} = 0.24$$

The magnification factor M.F. can be obtained from Equation 12-11 or, more quickly, from Figure 12-7. In part 1 we established the value of T_c/T to be equal to 0.4, which provides us with M.F. = 1.2. Thus,

$$C = 1.63(0.24)(1.2) = 0.47$$

Substituting the required values, we obtain the lateral force on the chiller:

$$F_p = 1.0(0.67)(0.47)15.0 \text{ kips} = 4.7 \text{ kips}$$

If we compare the values of the design lateral forces obtained in parts 1, 2, and 4, we notice that the NEHRP provisions result in the highest design force level. Note that the NEHRP force from part 1 has been converted into a working stress load using the factor 1.2 discussed previously:

- Part 1: NEHRP criteria, $F_p = 19.3$ kips.
- Part 3: UBC criteria, $F_p = 4.5$ kips.
- Part 4: Tri-Services criteria, $F_p = 4.7$ kips.

12.2.5 Approach Using Floor Response Spectrum

The ground-motion response spectrum is commonly used to estimate the seismic response of a building. Similarly, the floor response spectrum can be used to evaluate the seismic forces acting on sensitive equipment installed within a building. It is evident that for a given ground motion, the motion experienced by various floors in the building, and as a result their response spectra, are not the same. Figure 12-9 shows typical floor response spectrum curves for the first level and the roof of a ten-story steel moment frame.

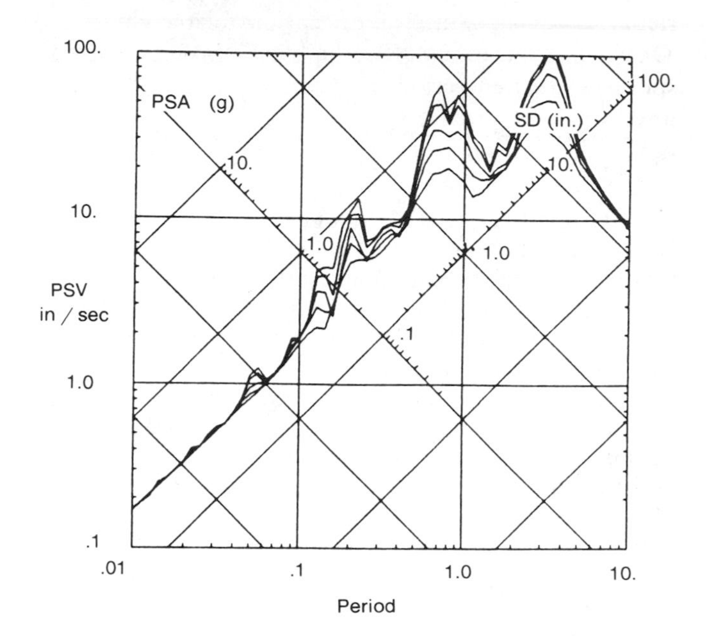

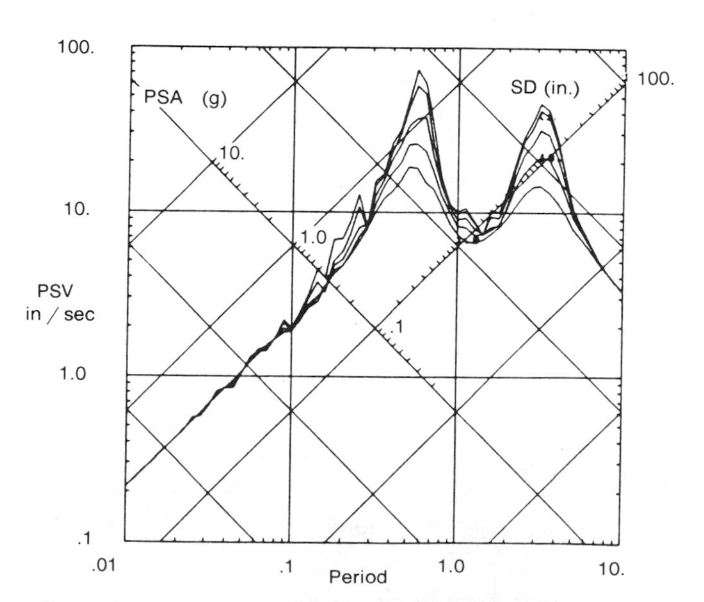

Figure 12-9 Typical floor response spectra for a 10 story steel frame. Top: Roof response spectra. Bottom: First floor response spectra. (Illustration Courtesy of Dr. T. J. Dehghahnyar, John A. Martin & Associates.)

The traditional methods of generating floor response spectra rely upon developing a time history of the response at a given level in a structure. This requires that a time-history analysis based on an actual earthquake record (or an artificial record) be performed to obtain the time history of the response of the floor level. This time history is then used to generate the floor response spectrum. Using multiple time histories and applying statistical techniques to smooth out the results is both time-consuming and expensive. Because of this drawback, techniques have been developed to generate floor response spectra without having to perform a series of time-history analyses.[12-6]

One of these methods, which uses the ground motion response spectrum to estimate accelerations at different points in the structure, was suggested by M. P. Singh.[12-6] This method, however, is an approximate one and lacks a complete mathematical justification. Other methods, discussed by Biggs and Roesset,[12-7] attempt to overcome some of the deficiencies of this approximate approach without resorting to the laborious procedures required by the time-history analysis method.

In spite of the deficiencies associated with the approximate method outlined in Reference 12-6, it does allow the designer to consider the response of the equipment and the structure using more than one period of vibration. It also permits the designer to explore the changes in response caused by changes in the location of the equipment in the building.

In the following discussion, the subscripts "s" and "e" represent parameters pertaining to the structure and the equipment, respectively. Modes corresponding to the structure and to the equipment are represent using the subscripts "n" and "m", respectively.

The maximum acceleration, A_{sne}, is given by

$$A_{sne} = A_{0n}\Gamma_{sn}\phi_{sne} \tag{12-15}$$

where

A_{0n} = acceleration ordinate obtained from ground-motion response spectrum for period T_n and the corresponding damping ratio

Γ_{sn} = modal participation factor for mode n

ϕ_{sne} = mode-shape ordinate from the eigenvector for mode n for the point under consideration

From the manufacturer of the equipment, one can obtain the natural periods of the equipment, T_{em}, its mode shapes ϕ_{emr}, and its modal participation factors Γ_{em}. If this information is not available, one can perform a dynamic analysis of a rough model of the equipment to estimate the required information.

The modal accelerations at the location of the equipment, A_{emr}, are obtained from:

$$A_{emr} = A_{em}\phi_{emr}\Gamma_{em}$$

where

A_{em} = total modal acceleration

$$= \left[\sum_{n'}(A'_{emn})^2 + \frac{\Sigma_{n''}(\Gamma_{sn}\phi_{sne}A''_{emn})^2}{\Sigma_{\text{all } n}(\Gamma_{sn}\phi_{sne})^2}\right]^{1/2} \tag{12-16}$$

The first summation includes those modes for which $T_{em}/T_{sn} \leqslant 1.25$. The numerator of the second term includes those modes for which $T_{em}/T_{sn} > 1.25$ (i.e. the remaining modes). The denominator includes all of the modes.

The values of A'_{emn} and A''_{emn}, the maximum accelerations of the equipment in mode m which is associated with mode n of the structure, depend on the ratio of T_{em}/T_{sn}. For values of $T_{em}/T_{sn} \leqslant 1.25$,

$$A_{emn} = \frac{A_{emn}}{A_{sne}}A_{sne} \tag{12-17}$$

and for $T_{em}/T_{sn} > 1.25$,

$$A''_{emn} = \frac{A_{emn}}{A_{emg}}A_{emg} \tag{12-18}$$

Figures 12-10 and 12-11 provide the value of A_{emn}/A_{sne} if $T_{em}/T_{sn} < 1.25$ for damping values of 4% and 7% for the building. A_{emg} is the spectral acceleration ordinate from the ground response spectrum for the given mode of the equipment. Figure 12-12 supplies the values of A_{emn}/A_{emg} if $T_{em}/T_{sn} > 1.25$.

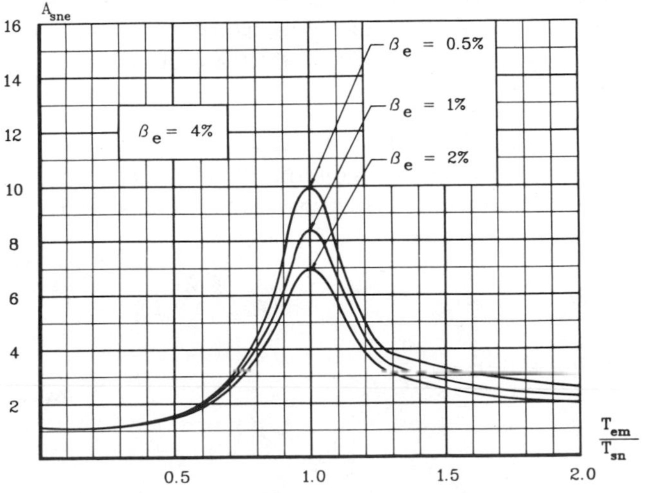

Figure 12-10 Magnification factor based on response of structure, β_s = 4%.

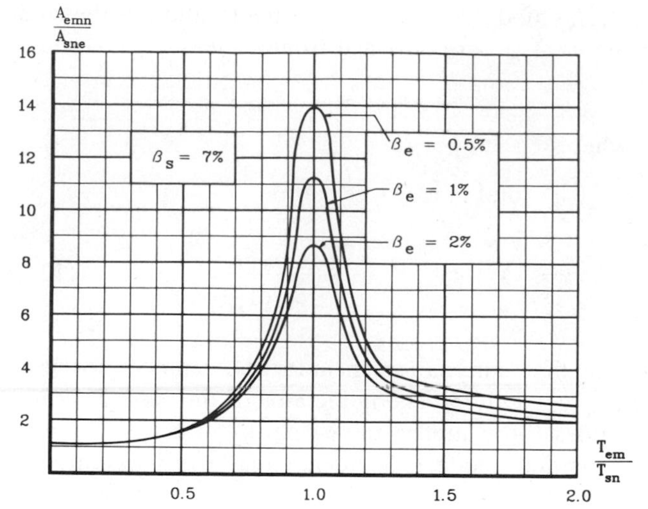

Figure 12-11 Magnification factor based on response of structure, $\beta_s = 7\%$.

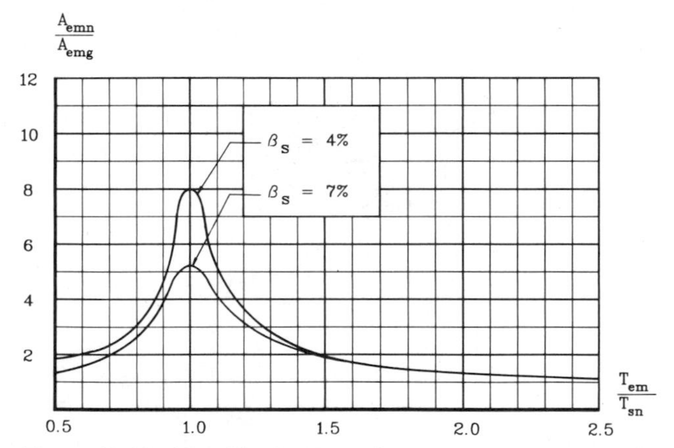

Figure 12-12 Magnification factor based on ground motion.

The above calculations are performed for each mode of interest for both the equipment and the structure. The total acceleration for all of the modes considered can be obtained using the root-mean-square procedure. The following example illustrates the procedure.

Example 12-2

Determine the maximum acceleration of a piece of mechanical equipment mounted on the floor slab of a building. Pertinent physical parameters of the structure for the first three modes are as follows:

$$T_{s1} = 0.25 \text{ sec}, \quad \phi_{1e} = 0.4, \quad \Gamma_{s1} = 1.5$$

$$T_{s2} = 0.05 \text{ sec}, \quad \phi_{2e} = 0.5, \quad \Gamma_{s2} = 1.0$$

$$T_{s3} = 0.03 \text{ sec}, \quad \phi_{3e} = -0.6, \quad \Gamma_{s3} = 0.4$$

The damping β_s is 4%. Based on information from the manufacturer, pertinent physical parameters for the two modes of response of the equipment are as follows:

$$T_{e1} = 0.08 \text{ sec} \quad \phi_{e11} = 1.0, \quad \phi_{e12} = 0.5 \quad \Gamma_{e1} = 1.5$$

$$T_{e2} = 0.025 \text{ sec} \quad \phi_{e21} = 1.0, \quad \phi_{e22} = -0.5 \quad \Gamma_{e2} = -0.5$$

The damping β_e is 1%. Review of the ground response spectrum for the first three natural periods of the building has yielded the following spectral ordinates:

$$A_{01} = 0.15g, \qquad A_{02} = 0.05g, \qquad A_{03} = 0.05g$$

The spectral ordinates associated with the equipment and the ground motion response spectrum are $A_{e1g} = 0.13g$ and $A_{e2g} = 0.05g$.

Based on the information presented above, the maximum accelerations at the equipment for the first three modes are

$$A_{s1e} = A_{01}\Gamma_{s1}\phi_{1e} = (0.15g)(1.5)(0.4) = 0.090g$$

$$A_{s2e} = A_{02}\Gamma_{s2}\phi_{2e} = (0.05g)(1.0)(0.5) = 0.025g$$

$$A_{s3e} = A_{03}\Gamma_{s3}\phi_{3e} = (0.05g)(0.4)(-0.6) = -0.012g$$

- Equipment mode 1: The ratio $T_{e1}/T_{s1} = 0.08/0.20 = 0.40$. Thus, from Figure 12-10 we get $A_{emn}/A_{sne} = 1.2$. Hence

$$A'_{emn} = \frac{A_{emn}}{A_{sne}} A_{sne} = 1.2(0.090) = 0.11g$$

Similarly, for the other two modes of the structure,

$$\frac{T_{e1}}{T_{s2}} = \frac{0.08}{0.5} = 1.6$$

$$\frac{A_{e12}}{A_{s2e}} = 1.7 \qquad \text{(from Figure 12-12)}$$

$$A''_{e12} = \frac{A_{e12}}{A_{e1g}} A_{e1g}$$

$$= 1.7(0.13g)$$

$$= 0.22g$$

and

$$\frac{T_{e1}}{T_{s3}} = \frac{0.08}{0.03} = 2.7$$

$$\frac{A_{e13}}{A_{s3e}} = 1.1 \qquad \text{(from Figure 12-12)}$$

$$A''_{e13} = \frac{A_{e13}}{A_{e1g}} A_{e1g}$$

$$= 1.1(0.13g)$$

$$= 0.14g$$

Solving for A_{e1}, we obtain

$$A_{e1} = \left((0.11g)^2 + \frac{[1.0(0.5)(0.22g)]^2 + [0.4(-0.6)(0.14g)]^2}{[1.5(0.4)]^2 + [1.0(0.5)]^2 + [0.4(-0.6)]^2} \right)^{1/2} = 0.18g$$

Thus, the maximum modal acceleration at the equipment is

$$A_{e11} = A_{e1}\phi_{e11}\Gamma_{e1} = (0.18g)(1.0)(1.5) = 0.27g$$

- Equipment Mode 2: The same procedure is used. We obtain the value $A_{e2} = 0.112g$, and

$$A_{e21} = A_{e2}\phi_{e21}\Gamma_{e2} = 0.112g(1.0)(-0.5) = 0.056g$$

Using the root-mean-square technique to combine the first and second modes of equipment response, we obtain the maximum acceleration of the equipment, A_e:

$$A_e = \left[(0.27g)^2 + (0.056g)^2\right]^{0.5} = 0.28g.$$

Note that, at least in this case, the final answer ($A_e = 0.28g$) is nearly equal to that obtained by considering only the first mode of response of the equipment ($A_{e11} = 0.27g$).

12.3 DESIGN CONSIDERATIONS FOR ARCHITECTURAL COMPONENTS

12.3.1 General

Architectural components and systems include nonbearing interior partitions, exterior wall panels, suspended ceilings, shelves, and the like. Past earthquakes have provided numerous examples of damage to these systems. In many cases the pre-earthquake remedy for this damage would have consisted of a more thorough lateral bracing, often at little additional cost. However, in some other cases significant costs would be incurred in limiting nonstructural damage to similarly low levels.

There are two general strategies that one might follow to respond to the seismic forces:

1. An isolation approach, in which the elements are provided with sufficient separation from the structure so that the deformation of the structure will not produce appreciable stress on the elements.
2. A deformation approach, in which the elements are designed with the intention that they will be able to undergo the required deformation.

The seismic force used in the design is dependent upon the weight of the system or component, the seismic coefficient for the site, the seismic coefficient for the component, and the required performance characteristics. The amplification effects, due to the location of a component with respect to the building height and type of structural system, are not considered explicitly in many of the nonstructural design standards. The designer must recognize where these factors might be relevant and consider their influence accordingly.

Seismic excitation produces different motions in various structural members. If nonstructural components are connected to these members, they will be forced to undergo the deformation resulting from the net deflection of these structural members. For example, all of the components in a structure that connect the floor slabs together must be designed to accommodate the level of drift anticipated for the structure.

Using the anticipated story drifts, one can establish the degree of separation required to prevent or limit the damage sustained by some architectural elements. In providing these separations, however, one must not forget that the nonstructural components must be restrained against the forces generated by themselves. These forces arise, for example, from the dynamic response of a masonry wall or a suspended ceiling.

A second method is to provide the architectural element with sufficient strength to withstand the loads generated by the structural deformation. In most cases this approach will not yield acceptable results. If one could provide the nonstructural elements with the strength required to resist the forces and deformations produced by the structural response, then why were not these components included in the seismic-force-resisting system?

It therefore seems difficult to justify this strength oriented approach from an economic standpoint. In addition, the designer must recognize that nonstructural elements possessing significant strength and stiffness can have an unintended effect on the structural response of the building itself. Nonbearing walls with insufficient separations can cause the accidental failure of primary structural elements. Figure 12-13 illustrates what can happen if a partial-height concrete-block partition is connected directly to adjacent structural columns. The strength and stiffness of the wall produces an unintended shortening of the clear height of the column and permits the column to fail in shear rather than in the anticipated flexural mode.

Another consideration that might guide one's approach to nonstructural seismic design is a recognition of the consequences of failure of the nonstructural system or component. There are some parts of a system whose failure during an earthquake causes no more

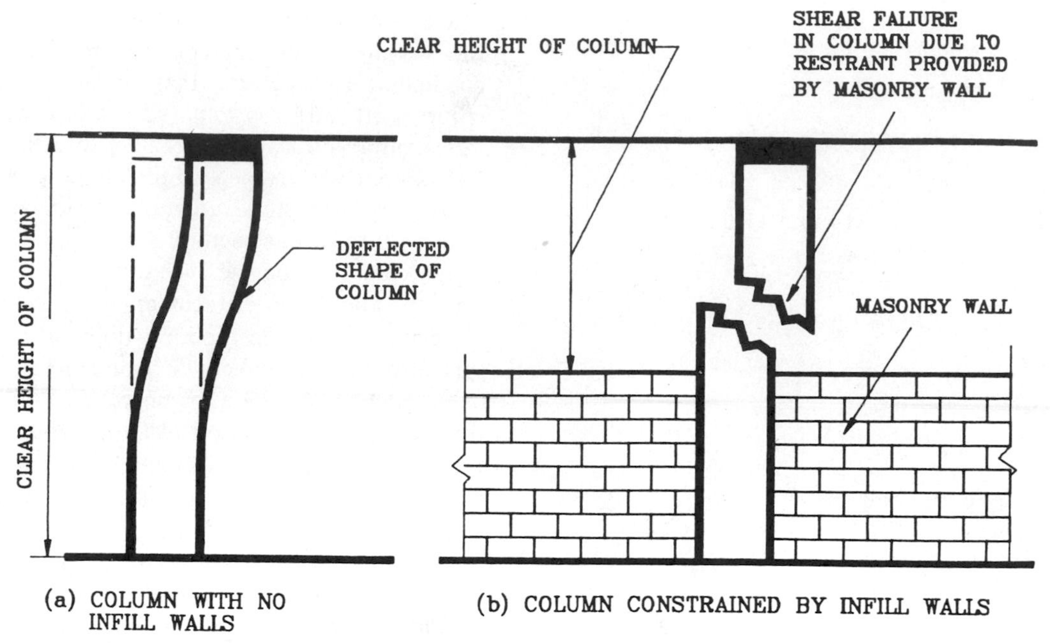

Figure 12-13 Effect of infill walls on column behavior.

than a temporary inconvenience. However, the failure of other nonstructural components can result in injury or death. Clearly, the failure of these components must be prevented.

The building location is one criterion that might be used to evaluate the consequences of nonstructural damage. Buildings located along continuously occupied sidewalks or streets would present a significantly greater threat to life safety due to architectural facade failures than would buildings with ample, landscaped separations from pedestrian and vehicular flow. What might be an acceptable level of nonstructural damage in the isolated structure becomes an element of sizable concern for the building surrounded by unprotected people. The adjacency of nonstructural elements with high damage potentials to areas of high occupancy within the building might also provide guidance as to the level of structural integrity required. The performance criteria contained in the NEHRP provisions and, to a lesser extent, in the UBC code quantify the relevance of occupancy to the design force levels.

12.3.2 Examples of Damage to Architectural Systems

Many components are inherently subject to damage that is uncontrollable or can be controlled only at great cost. Experience from past earthquakes provides a range of examples of this type of nonstructural damage. While it would be desirable to eliminate this damage completely, frequently it is expensive to do so. This uncontrollable damage is generally minimal in effect: plaster cracking, and minor damage to such items as laboratory glassware or material, file drawers, office equipment, and the like. Further, the failure to identify certain items as hazards will result in some damage. The specific types of damage that appear to represent the accepted threshold of nonstructural damage together with indications as to the alternatives in improvement options are presented in this section to help the designer in evaluating the benefits gained in preventing various kinds of nonstructural damage. Many of the examples of damage thresholds for nonstructural damage in this section are based on the work performed by the Veterans Administration.[12-8]

- *Plaster* is a very brittle finish material and fails when subjected to even minor bending or shearing forces. It will crack or spall out of the supporting lath. This is particularly true for plaster surfacing on structural elements such as shear walls, columns, and beams. The structural elements, twisting or bending through designed limits, will cause cracking in the surface plaster. Such cracking is not necessarily expensive to repair, and further, is not a major hazard to life safety except on ceilings, where spalls can cause injury. Isolation of the furring from structural elements may be helpful. However, since the repair costs are minimal, a decision to have existing plaster work remain unchanged until after an earthquake may be a more practical approach. On ceilings, where the hazard is greater, replacement may be warranted.
- *Tile and masonry joints:* Damage to these elements, in most cases, is due to unreinforced

partitions. If other seismic resistance characteristics are accommodated (including adequate compression control and expansion joints), the cracking will be more of a nuisance than a hazard. Repair of mortar or grout joints may be necessary following an earthquake.

- *Partitions:* Some modern interior-planning approaches utilize nonanchored partition systems that rely upon the weight, corners, or spread bases to supply stability. These partitions are more susceptible to overturning than anchored systems when subjected to seismic forces. Decisions on their use should take into account the flexibility of such systems and their ease of installation, as opposed to the possible danger of overturning during an earthquake. Of particular importance in this regard are those systems that utilize hanging furniture or storage systems as part of the partition system. Some partitions are lightweight screens and would not cause injury or significant damage if overturned. However, other systems are more dense—heavier even than full-height stud and gypsum board walls. The weight and stability of these systems must be given careful consideration in high-seismic-exposure areas.
- *Ceiling tiles:* Many types of ceiling tile are very brittle and will crack if stressed by deformation of the ceiling suspension system during an earthquake. Further, deformation can also cause separations in the suspension system, and this, in turn, may cause tiles to fall. Bracing of the ceiling system can reduce these effects, but may not eliminate them altogether (particularly in the case of exposed spline systems).
- *Shear cracking of surface finishes near doors and windows:* Although this type of damage is most common in plaster surfaces, other materials such as gypsum board and ceramic tile are also subject to cracking from movement of walls in relation to openings. Post-earthquake repairs are inexpensive and usually unavoidable.

Experience obtained from past earthquakes provides valuable information. Table 12-8 lists the damage to architectural components and systems that has been identified after past earthquakes. As can be seen, the most frequently damaged components included ceiling systems, light fixtures, and partition systems. Of all ceiling systems, suspended plaster systems performed the worst. Better performance was obtained with concealed spline systems and surface-mounted ceilings. For all suspended systems, damage was most noticeable at room perimeters, where the enclosing partitions acted against the ceiling system. Perimeter damage included buckling of the suspension and falling ceiling tiles, plaster, or gypsum boards.

All partition systems sustained some damage. Block-type partitions, including clay tile, concrete masonry, and gypsum block, were damaged by joint failures in the partition or by compression failures at the top due to action of the structure on the partition. Other damage included horizontal movement and corner failures. Metal-stud partitions, when carried full height between structural floors, were damaged similarly in that structural vibration caused bending in the studs. Studs, carried to or just above the ceiling system, were overturned by lateral action of the ceiling system on the partitions. Some separation was evident between the studs and carrying channels at both the top and bottom. The partition systems that performed well included metal demountable partition systems and drywall partitions. Partitions that crossed building seismic joints were particularly susceptible to damage due to differential structural movement across the joints. Partitions that contained buried piping or rigid conduit also sustained some damage due to the interaction of the piping and the partition body.

12.3.3 Characteristics of Well-Performing Architectural Systems

A variety of measures to prevent nonstructural damage to architectural systems have been successful. This section discusses some of these measures and illustrates certain details. Much of the information in this section is adapted from the work performed by the Veterans Administration.[12-8]

Facades and Glazing Good seismic performance has been observed in architectural facades and glazing systems that exhibited the following characteristics:

Heavy rigid facades are used only on rigid structural systems and are not attached to relatively flexible building frames. Concrete masonry unit filler walls are not installed in a manner that would restrain the lateral deflection of the building frame. A gap, with an adequately sized resilient filler (if necessary for sound or fire control), separates the structural frame from the nonstructural filler walls. Figure 12-14 illustrates one method of providing this separation while at the same time bracing the wall against out-of-plane motion.

Curtain walls are securely attached to the building frame. Flexible gaskets in curtain walls are designed and installed so that they do not come loose when the wall is subjected to repeated racking. All glass panels are set in resilient mounts with sufficient space for in-plane motion and are supported by mullions designed to withstand earthquake forces. Tempered glass is used in exits or where large glazed areas front public walks.

Table 12-8 Damage to Architectural Systems and Components During Past Earthquakes(12-8)

Systems	Components	Recorded Damage
Partitions	Permanent — masonry and tile	Cracking of units; horizontal drift; unit losses or compression failures at top of partitions; joint failures; overturning.
	Permanent — stud and gypsum board or plaster	Overturning associated with ceiling failures adjacent to partitions; finish cracking; horizontal drift; delamination of finish from studs.
	Demountable — metal, wood, metal and glass	Separations at top and bottom channel; compression breaks; overturning; cracking or separation of fixed glass from partition body.
Furring	Plaster or gypsum board	Cracks in finish; separation failures from furred structural element due to movement of structure.
Ceilings	Suspended lay-in tile system — exposed splines	Unwinding or breakage of hangers; separation of tiles from suspension system; compression bending of system at room perimeters; breakage at building seismic joints; shear breakage in suspension interconnections.
	Suspended concealed spline systems	Failures similar to exposed splines, except less tiles separate from suspension system.
	Suspended plaster or gypsum board	Plaster spalls from lath; shear cracks in finish; suspension system sustains similar damage to other suspended systems; gypsum board separates from supports.
	Surface-applied tile, plaster, or gypsum board	Generally better performance than suspended systems; plaster cracks and spalls due to structural movement; adhesive failures in ceiling tile.
Light fixtures	Lay-in fluorescent (recessed and semirecessed)	Racking of ceiling suspension causes fixtures to separate from suspension system and fall. Where fixtures are supported separately from ceiling system, performance is better. Failures within fixtures included separation of diffusers, lenses, and lamps from housings.
	Stem-hung and chain-hung fluorescent	Separation of stem at structural connection point; twisting of fixture causes breakage in stems and chain breakage. Multiple fixture installations, end to end, experience most common damage due to interaction of fixtures with each other. Long-stem fixtures sustain more damage than short-stem. Internal damage similar to lay-in fixtures.
	Surface-mounted fluorescent	Ceiling fixtures perform similarly to lay-in fixtures. Wall fixtures perform better than ceiling fixtures except in instances of wall failure. Internal damage similar to others.
	Stem-hung incandescent	Performance similar to stem-hung fluorescent fixtures except that incandescent fixtures are usually hung with a single flexible stem. Damage due primarily to fixture swaying and encountering other structural and nonstructural components. Internal damage consists of lens and globe separation as well as lamp breakage.
	Surface-mounted incandescent	Ceiling fixtures perform similarly to surface-mounted fluorescent. Wall fixtures perform well.
	Ornamental fixtures	Chandeliers and other fixtures of a similar nature fall similarly to stem-hung fixtures. Internal damage due to multiple moving elements interacting with each other.
Doors and frames	Wood, hollow metal, metal, and glass	Frames warp from enclosing wall movement; doors occasionally deform hinges.

Brick veneer facades on steel-frame buildings are avoided unless the brick veneer is securely tied to a separate wall that is independent of the steel frame. Otherwise, adequate provisions for the resulting deflections must be made. Wire or straight-rod ties are not used to anchor face brick to a wall, especially when a layer of insulation or an air gap separates the two elements. Large masonry facades are designed as part of the structural system and are not considered as nonstructural ornamentations unless they are properly attached to a structural wall.

Partitions Partitions in buildings with flexible structural frames are anchored to only one structural element, such as a floor slab, and separated by a physical gap from all other structural elements. Reinforced-concrete masonry unit partitions tied to more than one structural element are considered as part of the structural system. Unreinforced concrete masonry units are not used for partitions or filler walls.

Conduits and piping in partitions are tied to the structural element to which the partition is anchored. Openings in partitions for pipes, conduits, and ducts are

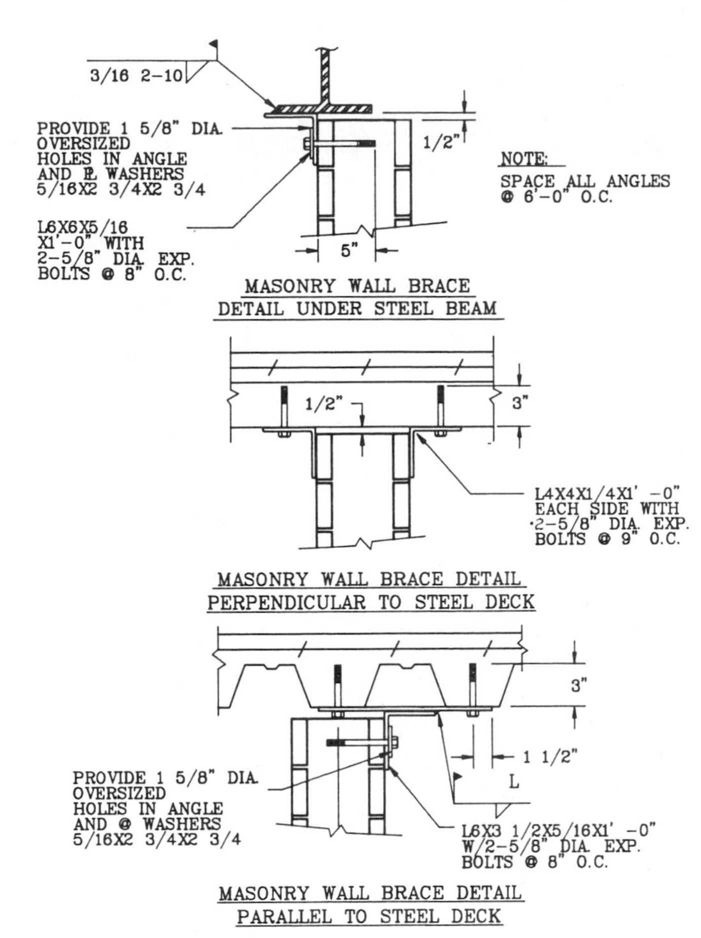

Figure 12-14 Nonbearing masonry wall details.

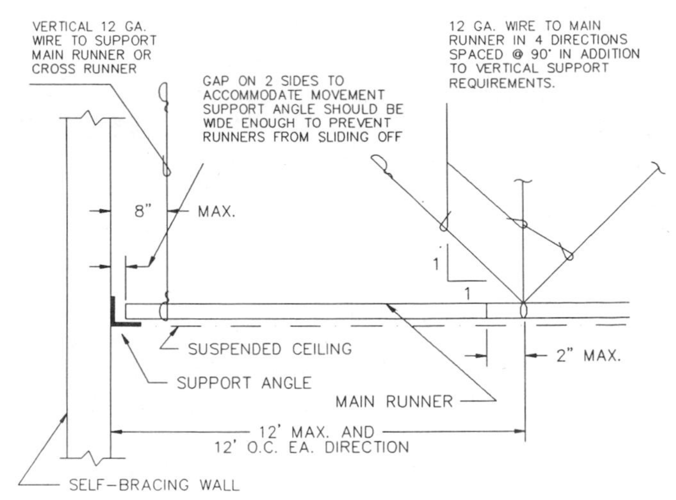

Figure 12-15 Suspended ceiling bracing detail.

properly reinforced and made large enough to preclude direct contact with fixtures.

Ceilings Fluorescent lighting fixtures are not installed in or on exposed tee-grid or concealed-spline suspended ceilings unless the ceiling suspension system is designed to carry the added weight of the fixtures during an earthquake or the fixtures are independently supported and laterally braced. Exposed tee-grid or concealed-spline suspended ceilings are laterally braced and provided with a physical separation at the walls, particularly in large rooms with high ceilings and deep attic spaces. Figure 12-15 illustrates general provisions to brace ceilings and prevent light fixtures from falling.

Exposed tee-bar, luminous, and concealed-spline ceiling systems must satisfy all of the requirements of Reference 12-9, and in particular:

1. Ceiling grids are braced at regular intervals against lateral and vertical movements.
2. The ceiling system is not fastened to the surrounding walls or partitions. Use soffits to return the ceiling system to the supporting slab. Where the ceiling must join a wall or partition, an angle wall trim, wide enough to allow for differential movements, is provided. Main and cross runners have hangers at the perimeter so that wall trims do not support the ceiling.
3. Cross runners are fastened to the main runners using locking clips or similar devices to prevent cross tees pulling or twisting out of the main runners.

Gypsum board with glue-on tiles and lath and plaster ceiling systems must satisfy all of the requirements of Reference 12-9, and in particular:

1. Ceiling systems are braced at regular intervals against vertical movement and are provided with lateral braces at the perimeter.
2. Gypsum-board ceilings are reinforced at nail points by the use of steel nailing strips. Nails with large heads are used to install gypsum-board ceilings.
3. Furring channel joints in irregular-shaped ceilings are made up using rivets, bolts, and welds. Corners are braced so that they do not pivot.
4. Large ceiling areas separated by rows of linear diffusers or light fixtures are held together with rigid ties, and the diffusers and light fixtures are secured to the ceiling system.

Exits Unreinforced-concrete wall units are not used in any exit walls. Brittle veneers, such as tiles, are not applied directly to the inside of concrete stairways. If they are used, they should be mounted on a separate stud wall or furring strip.

Floor covers for seismic joints in corridors are designed to take three-dimensional movements. Special attention should be given to movement parallel to the joint.

All exit door openings are structurally sound, to prevent jumping of the doors in their frames.

Free-standing or glass lay-in shelves are not placed in public areas, particularly near exit doors. Displays in wall-mounted or recessed showcases are tied down so that they cannot come loose and break the glass front during an earthquake. Where this is impractical, tempered glass is used for greater reliability.

Pendant-mounted light fixtures are not placed in exits. Recessed or surface-mounted lights are preferable.

Heavy ornamentation, such as marble veneer, is avoided in exit lobbies. If a marble veneer is used, it should be securely fastened to structural elements with dovetails and clip angles, or it should be mounted on furring walls.

Storage Racks and Cabinets Storage racks are designed to withstand earthquake forces and are anchored to the floor or laterally braced from the top to the structural elements. Racks are designed with lateral bracing and anchor bolts and can withstand anticipated lateral and uplift loads. Rows of racks have rigid ties installed at the top of the racks to brace and stabilize the entire installation. Racks placed along walls are anchored to the wall to avoid battering between the wall and the rack.

Filing cabinets and map or plan drawers are anchored to the floor or walls, and all drawers are fitted with positive-locking safety latches.

Vital furniture and equipment are anchored to the floor or wall. Loose materials stored on high shelves are held in place by face bars. Medical supplies are kept in cabinets that are anchored to the floor or walls and fitted with latched doors.

Roof Tiles Heavy roof tiles usually will perform satisfactorily if they are anchored to the roof diaphragm. Figure 12-16 illustrates one method of securing mission tiles with corrosion-resistant wire ties. Consideration of corrosion resistance is important because there are examples of tile ties having rusted from exposure and failed when quickly loaded during an earthquake.

12.4 DESIGN CONSIDERATIONS FOR MECHANICAL / ELECTRICAL EQUIPMENT

12.4.1 General

Mechanical and electrical equipment is subject to earthquake damage, as are the architectural components discussed in Section 12.3. Often, since the structural engineer has little influence over the methods used to install these systems, the potential for significant nonstructural damage is quite large. Most electrical and mechanical equipment is purchased as premanufactured items rather than being manufactured specifically for a project. The particular characteristics of the equipment

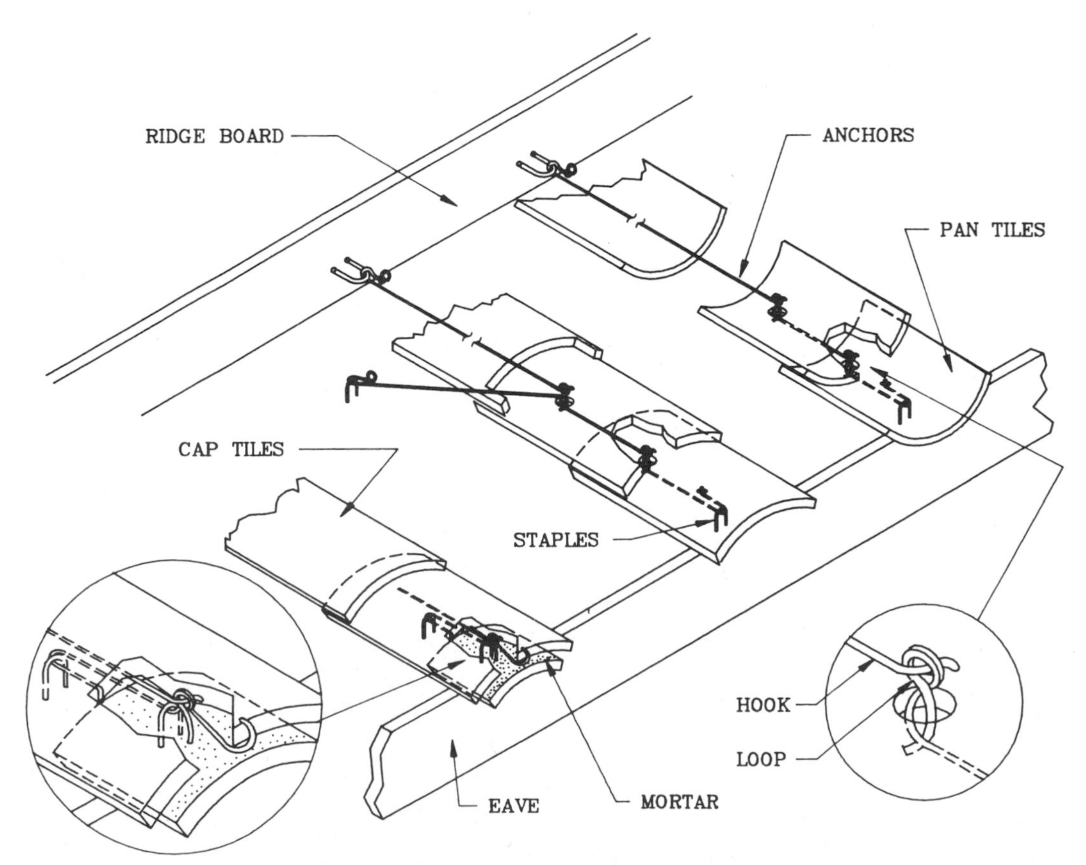

Figure 12-16 Roof tile anchors detail.

Table 12-9 Damage to Mechanical Components during Past Earthquakes[12-8]

Systems	Components	Recorded damage
Mechanical	Rigidly mounted equipment such as boilers, chillers, generators, tanks	Generally perform well where there is no damage to structural base. Some shearing of attachment devices and corresponding horizontal displacement; tall tanks overturn; supports fail. Greatest damage is to equipment that rested on structural base without positive anchorage; overturning and horizontal movement sever connected lines and pipes.
	Vibration-isolation-mounted fans, pumps, air handlers, etc.	Devices fail and cause equipment to fall. Some damage due to unrestrained shaking on vibration-isolation device. Suspended equipment fails more often than floor-mounted equipment.
Piping	Water, stream, sprinkler, gas, waste, etc.	Large-diameter rigid piping fails at elbows and bends. Joint separations; hanger failures. Small-diameter piping performs better than larger piping due to bending without breaking. Single failures of hanger assemblies frequently causes progressive overloading and failures at other hangers and piping supports. Piping performs better in vertical runs where there are lateral restraints than in horizontal runs where there is no lateral bracing. Failures at building seismic joints due to differential movements.
Ducts	Rectangular, square, and round ducts	Breakage most common at bends. Supporting yokes fail; long runs fail as a result of large-amplitude swaying.
Elevators	Counterweights, guiderails	Separation of counterweights from rails. Damage from counterweights includes structural beams, cables, and cabs.
	Motor–generator sets	Sheared-off vibration-isolation devices.
	Controller panels	Overturning when unanchored at bases. Hinged panels thrown open.
	Cars — guiding systems	Generally perform well.
	Hoistway doors	Some doors jam or fall outwards.
	Hydraulic elevator systems	Generally perform well. Some cylinders shift out of plumb.
Escalators		Generally perform well. Some treads are damaged by falling debris.

are established by the manufacturers, and these, in turn, determine damage potential. Earthquakes have caused a considerable amount of damage to such items as motor rotors, fan blades, switch assemblies, housings, and internal wires or pipes. For computer equipment, the possibilities of expensive damage are very great, although many internal items are mounted on rocker-arm assemblies that allow for some vibration damping.

There are generally two classes of equipment installations of interest with respect to seismic design:

1. equipment rigidly anchored to the building structure or the ground and
2. equipment mounted on vibration isolators.

Table 12-9 illustrates several examples of damage to various mechanical systems.

Failures of equipment rigidly mounted to the ground or building typically reflect a lack of strength in the anchors or an inability of connecting service piping to reconcile the differential displacements.

Vibrating mechanical equipment, such as chillers or emergency generators, has traditionally been mounted on resilient mounting systems, particularly when the equipment is on the upper floors of a structure. These resilient mounting devices include stable springs, pneumatic restraining devices, and elastic restraining devices. Any such device should be capable of withstanding the calculated forces.

Service conditions require this type of mounting to limit the transmission of vibration and sound to the rest of the structure. Past earthquakes have provided several examples of failures of these vibration-type mounts because of their inability to undergo the large lateral displacements common in such long period systems. These large displacements also cause failures in the connecting service piping because of the differential movement at the points of connection. The resilient mounting devices act as filters and damp out or decelerate the high-frequency vibration characteristic of the equipment during normal operation. Their long periods, often in the range of the building period or that of the input ground motion, often results in forces and displacements significantly greater than those contemplated by the manufacturer.

In addition to the heavy types of equipment common to mechanical and electrical installations, light fixtures have sustained considerable damage during past earthquakes. Table 12-10 illustrates several example of nonstructural damage to various types of electrical systems. Light-fixture types that sustained the most frequent damage included pendant- or chain-hung fluorescent fixtures that were not directly supported from the struc-

Table 12-10 Damage to Architectural Systems and Components During Past Earthquakes(12-8)

Systems	Components	Recorded damage
Emergency equipment	Generators	Generally perform well when bolted securely to structural bases.
	Communications and lighting equipment	Perform similarly to other electrical equipment. Some battery racks collapse. Unsecured battery-powered emergency lighting falls.
	Exit corridors, doors, lighting	Many exit doors become deformed and jam. Exit corridors are blocked with debris. Exit lights perform well.
	Battery packs	Most remain in place where strapped to walls.
Electrical equipment	Panels, transformers, ducts, switchboards, distribution systems	Tall equipment overturns where not bolted at base or braced at top. Many instances of panels performing better than enclosing partitions. Horizontal movement of large equipment. Rigid conduits with structure support fail.

ture. Fixtures that were installed in long rows end to end were damaged more frequently than those that were installed separately, as damage was caused by the swaying of the fixtures against each other during an earthquake. The light fixtures that performed well included surface-mounted fixtures and recessed fixtures that were directly supported from the structure or from rigid, permanent partitions.

12.4.2 Characteristics of Well-Performing Mechanical Systems

The following paragraphs illustrate characteristics of installation that have exhibited good performance during past earthquakes.

Equipment with Vibration Isolation Heavy mechanical equipment is not mounted on the upper floors of tall buildings unless all vibration-isolation mounts are carefully analyzed for earthquake resistance. This is of particular importance in view of the amplification of acceleration at the upper stories of tall buildings. Floor-mounted vibration-isolation devices are bolted to the equipment base and to the structural slab. The use of heavy bases under equipment mounted on vibration isolators is avoided, where possible, to reduce inertial forces.

Lateral and vertical restraining devices are provided around the base of vibration-isolated, floor-mounted equipment to restrict the displacements. Resilient material is provided on the contact surface of the restraining devices to retard impact loads. Figures 12-17 illustrates one such type of device.

Vibration-isolation hangers for suspended equipment are tightly installed against the supporting structural member. A structural restraining frame is provided around suspended heavy equipment. Cross bracing is provided between hanger rods on all four sides of suspended lightweight equipment.

Equipment without Vibration Isolators Supports for tanks and heavy equipment are designed to withstand earthquake forces and are anchored to the floor or otherwise secured. Suspended tanks and heavy equipment are strapped to their hanger systems and provided with lateral bracing. Domestic water heaters are provided with legs that can withstand earthquake forces and are anchored to the floor and/or strapped to a structurally sound wall.

Threaded pipe is not used for tank or equipment legs. The weakened plane created by the pipe threads has lead to several failures of these supporting pipes.

All horizontal tanks are strapped to their saddles, lugs are welded to the tanks at support points to prevent horizontal movement, and saddles are bolted to the structural slab.

Ceiling-mounted expansion tanks are tightly installed against the supporting structural member. All suspended equipment is braced on all four sides.

Any frame supporting elevated tanks or equipment is provided with adequate bracing and is anchored to the structural slab and walls.

All floor-mounted equipment and tanks are bolted to the structural slab.

Piping Systems Pipelines are tied to only one structural system. Where structural systems change and relative deflections are anticipated, movable joints are installed in the piping to allow for such movement. Figure 12-18 illustrates such an arrangement.

Suspended piping systems should have consistent freedom throughout. For example, branch lines are not

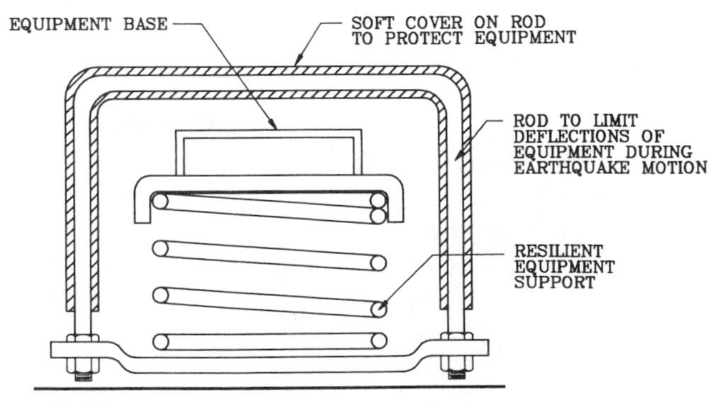

Figure 12-17 Resilient mounting device with restraint against excessive lateral deflection.

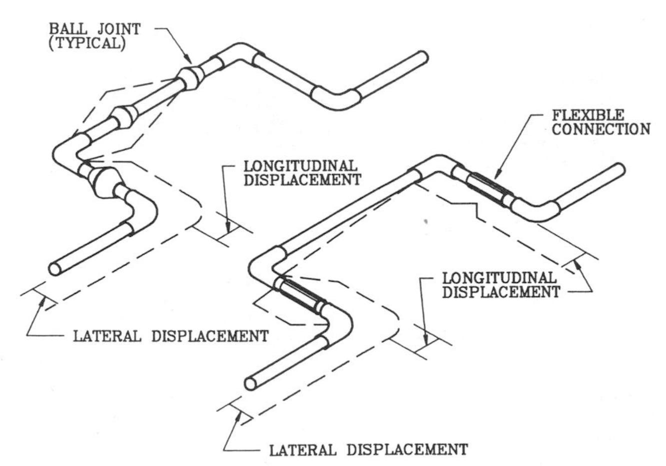

Figure 12-18 Pipe-work details at a seismic gap.

anchored to structural elements if the main line is allowed to sway. If the piping system is allowed to sway, movable joints are installed at equipment connections.

Pipelines leading to thermal-expansion loops or flexible pipe connections are guided to confine the degree of pipe movement.

Pipes should not cross seismic joints. If they must do so, the crossing is made at the lowest floor possible, and all pipe deflections and stresses induced by the deflections are carefully evaluated. Flexible joints are provided where pipes pass through seismic or expansion joints, or where rigidly supported pipes connect to equipment with vibration isolation. Figures 12-19 and 12-20 provide some suggested guidelines for piping that crosses seismic joints.

Sway bracing is provided in both longitudinal and transverse directions on all pipes $2\frac{1}{2}$ in. and larger to limit the stress on pipes. Figure 12-21 illustrates one method of providing this sway bracing. Spans for various pipes under different support conditions are given in Tables 12-11, 12-12, and 12-13.

Branch lines do not support or brace larger piping. No rigid piping system is fastened to two dissimilar parts of the building that may move differently during an earthquake, such as a wall and a roof.

The weight of the vertical pipe risers is supported at a point or points above the center of gravity of the riser, and lateral guides are provided at regular intervals. Large enough pipe sleeves are provided through walls or floors to allow for anticipated differential movements.

Air-Distribution Systems Long hangers and supports for ductwork are provided with lateral bracing. Flexible duct connections are installed in a semifolded condition with enough material to allow for the expected differential movement between fans and the ductwork. Pipe sleeves or duct openings through walls or floors are large enough to allow for the anticipated movement of

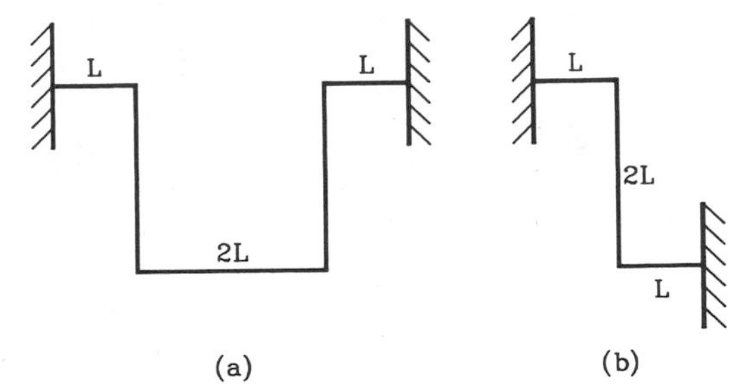

Figure 12-19 Schematic piping layout at seismic gaps.

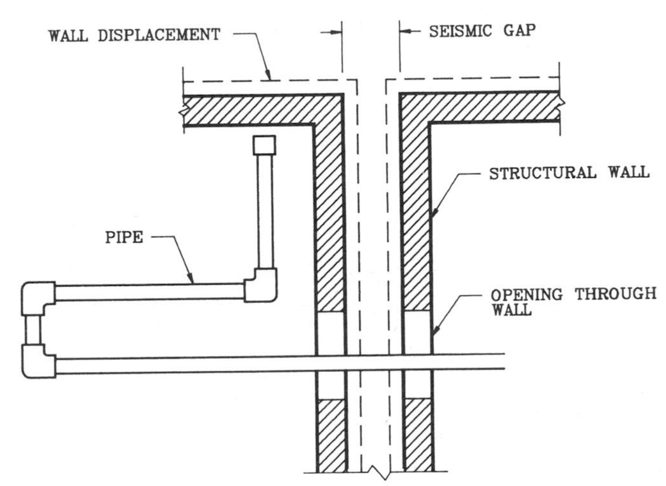

Figure 12-20 Plan of piping crossing a seismic gap.

the pipes and ducts. Horizontal ducts are supported as close as possible to the supporting structural member. Ceiling diffusers and registers are secured to ductwork with sheet-metal screws. Diffusers connected to the flexible ducts have positive ties to the ductwork and/or the wall opening.

Elevators Vibration isolators under the motor–generators are bolted to the floor and to the legs of the motor–generators. The isolators should have sufficient strength to withstand the earthquake forces. Selector and controller panels are bolted to the floor and, if

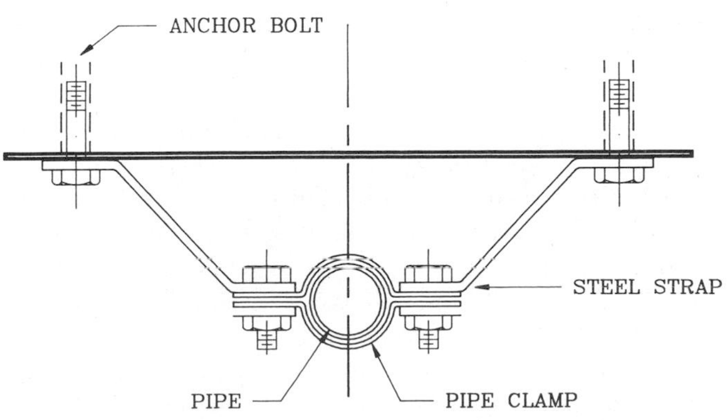

Figure 12-21 Pipe clamp and sway bracing.

Table 12-11 Maximum Allowable Spans for Pipes with Clamped End Conditions(12-3)

Diameter, in.	Std. wt. steel pipe 40 s	Ex. strong steel pipe 80 s	Copper tube Type K	Copper tube Type L	Copper tube Type M	85 Red brass & SPS copper pipe
$\frac{1}{2}$	7-9	—	5-6	5-6	5-3	6-6
$\frac{3}{4}$	8-6	8-6	6-6	6-6	6-3	7-3
1	9-6	9-6	7-3	7-3	7-0	8-0
$1\frac{1}{2}$	11-6	11-6	8-6	8-6	8-3	9-9
2	12-9	13-0	9-9	9-6	9-6	10-9
$2\frac{1}{2}$	14-0	14-3	10-9	10-6	10-6	11-9
3	15-6	15-9	11-9	11-6	11-3	13-0
$3\frac{1}{2}$	16-6	16-9	12-6	12-3	12-0	14-0
4	17-3	17-9	13-6	13-0	13-0	14-9
5	19-0	19-6	15-0	14-6	14-3	16-0
6	20-9	21-3	16-3	15-9	15-6	17-3
8	23-3	24-3				
10	25-9	26-6				
12	27-6	28-6				

Maximum span, ft-in.

Table 12-12 Maximum Allowable Spans for Pipes with Simple End Conditions(12-3)

Diameter, in.	Std. wt. steel pipe 40 s	Ex. strong steel pipe 80 s	Copper tube Type K	Copper tube Type L	Copper tube Type M	85 Red brass & SPS copper pipe
$\frac{1}{2}$	5-0	—	3-9	3-6	3-6	4-3
$\frac{3}{4}$	5-9	5-9	4-3	4-3	4-0	5-0
1	6-6	6-6	5-0	4-9	4-6	5-6
$1\frac{1}{2}$	7-6	7-9	5-9	5-6	5-6	6-6
2	8-6	8-6	6-6	6-6	6-3	7-0
$2\frac{1}{2}$	9-3	9-6	7-3	7-0	7-0	8-0
3	10-3	10-6	7-9	7-6	7-6	8-9
$3\frac{1}{2}$	11-0	11-0	8-3	8-3	8-0	9-3
4	11-6	11-9	9-0	8-9	8-6	9-9
5	12-9	13-0	10-0	9-6	9-6	10-9
6	13-9	14-0	10-9	10-6	10-3	11-6
8	15-6	16-0				
10	17-0	17-6				
12	18-3	19-0				

Maximum span, ft-in.

possible, provided with sway braces at the top. All electrical components within the panels are secured to the panel fame, and all doors and hinged panels are fitted with positive-locking latches.

Counterweight guide rails serving buildings of five or more stories are 15 lb/ft or heavier, and their supports should be designed to withstand earthquake forces. A safety shoe is used for counterweight guide-rail bracket design. The type of bracket used depends on the building height and location in the hoistway. Properly designed safety shoes for roller guides are provided to protect the roller assemblies from being damaged by the counterweights. Counterweight guide rails are strengthened by using a section heavier than the typical 8-lb/ft rolled section. Brackets are strengthened using gusset plates or ties placed at frequent intervals. The car guide rails on long spans are strengthened by installing spacers between the back-to-back rails at midpoints between the separator beams. This increases the rigidity of both rails.

Ventilation, communications, and lighting systems are connected to emergency power systems and are designed to operate when the normal power fails. Seismic switches to shut down the elevator during an earthquake and then lower the cars to the nearest floors are provided. Manual override switches are needed for low-speed operation by rescue and maintenance personnel after the earthquake.

Table 12-13 Maximum Allowable Spans for Pipes with Fixed-Pin End Conditions[12-3]

	Maximum span, ft-in.					
Diameter, in.	Std. wt. steel pipe 40 s	Ex. strong steel pipe 80 s	Copper tube Type K	Copper tube Type L	Copper tube Type M	85 Red brass & SPS copper pipe
$\frac{1}{2}$	6-6	—	4-6	4-6	4-6	5-3
$\frac{3}{4}$	7-0	7-0	5-6	5-3	5-3	6-0
1	8-0	8-0	6-0	6-0	5-9	6-9
$1\frac{1}{2}$	9-6	9-6	7-3	7-0	7-0	8-0
2	10-6	10-9	8-0	8-0	7-9	9-0
$2\frac{1}{2}$	11-9	11-9	9-0	8-9	8-6	9-9
3	12-9	13-0	9-9	9-6	9-3	10-9
$3\frac{1}{2}$	13-6	14-0	10-6	10-3	10-0	11-6
4	14-6	14-9	11-0	11-0	10-9	12-3
5	16-0	16-3	12-3	12-0	11-9	13-3
6	17-0	17-9	13-6	13-0	12-9	14-3
8	19-3	20-0				
10	21-3	22-0				
12	23-0	23-6				

The elevator hoistway and the surrounding structural system are adequately reinforced and braced to prevent distortion at the doors and to prevent debris from falling into the shaft.

Hydraulic elevator equipment is secured to floors and/or walls, and splashproof oil tanks are used.

12.4.3 Characteristics of Well-Performing Electrical Systems

Damage to electrical equipment is generally confined to light fixtures, building electrical distribution systems, and emergency power and lighting systems. This section discusses the characteristics of electrical systems that have performed well during past earthquake.

Light Fixtures Pendant-hung fluorescent fixtures, especially when mounted end to end in long rows, are provided with flexible lateral bracing at both the ceiling supports and the bottom connections to the fixtures. Lighting fixtures supported by flexible hangers are located so that they will not collide with other building elements. Extreme flexibility in hangers is less desirable than damped flexibility. Support systems designed for pendant mounting from a horizontal surface are not used on a sloping surface, because some of the freedom of movement is used up in vertical alignment. No pendant light fixtures are located below high ceilings. They should be surface-mounted and secured to a supporting grid system that meets the supporting and bracing requirements for suspended ceilings.

Recessed lighting fixtures are secured to and supported by a ceiling suspension system designed to carry the weight of the ceiling fixtures. Alternatively, the fixtures can be provided with independent safety supports. All hooks used to hang lighting fixtures should have safety latches. Recessed lighting fixtures are supported by and secured directly to the main runners of the ceiling support system. They are not supported by furring or cross runners or nailing bars in the case of gypsum-board ceilings. When the locations of the runners are not compatible with the lighting patterns, auxiliary support members of equal strength are provided. All recessed lighting fixtures are provided with independent secondary supports attached to the fixture housing and the building structure. These supports should be a minimum of two 12-gauge wires placed on diagonal corners of each fluorescent fixture, and each wire should be capable of supporting four times the weight of the fixture. Although this requirement may appear redundant after the fixture is secured to a ceiling support system, it is desirable in the event of ceiling failure of improperly installed fixture-to-ceiling attachments.

Grilles, diffusers, and lenses are not made of glass or metal and are permanently fastened to the fixture or provided with safety chains or adequate hinges.

Preferred surface-mounted fixture installations are direct attachments to the building structure. However, suspended installations that use positive locking devices are an acceptable alternative.

Building Electrical Systems All electrical equipment such as transformers, switch gear, and control panels must be anchored to the building. Flexible braided connections are used in place of rigid copper bus whenever relative movement may occur between switchboard components. Additional pull boxes with slack conductors are provided in long conduit runs to avoid tensioning of conductors.

Crossing seismic joints with conduits and bus ducts is avoided where possible. When seismic joints are crossed, arrangements that permit the required deflections are used. The crossing should be made at the lowest possible floor. Separate ground conductors are provided in all conduit runs that cross seismic joints and elsewhere in the electrical system where grounding systems could be broken.

Emergency Power and Lighting Systems Emergency power generators installed in building are mounted on adequately designed vibration isolators. The vibration isolators and the connecting service piping are provided with horizontal restraints.

Starter battery racks are adequately secured to the structure. The batteries are each attached to the rack with a positive mounting to restrain movement.

Battery-powered emergency lighting units are securely tied to the building.

REFERENCES

12-1 International Conference of Building Officials, *Uniform Building Code*, 1985 Edition, Whittier, CA, 1985.

12-2 International Conference of Building Officials, *Uniform Building Code*, 1988 Edition, Whittier, CA, 1988.

12-3 Departments of Navy, Army, and Air Force, *Tri-Services Manual: Seismic Design of Buildings*, Navy NAV FAC P-355, Apr. 1973.

12-4 Federal Emergency Management Agency, *NEHRP Recommended Provisions for the Development of Seismic Regulations for New Buildings, Part 1, Provisions*, Building Seismic Safety Council, 1985.

12-5 Applied Technology Council, "Tentative Provisions for the Development of Seismic Regulations for Buildings," ATC 3-06, National Bureau of Standards, Washington, 1978.

12-6 Singh, M. P., "Generation of Seismic Floor Spectra," *J. Eng. Mech. Div. ASCE* 101, No. EM5, Oct. 1975.

12-7 Biggs, J. M. and Roesset, J. M., "Seismic Analysis of Equipment Mounted on a Massive Structure," *Seismic Design of Nuclear Power Plants*, R. J. Hanson (ed.), M.I.T. Press, Cambridge, MA, 1970.

12-8 Veterans Administration, *Earthquake Resistive Design of Nonstructural Elements of Buildings*, VA Construction Standard CD-55, Veterans Administration, Washington, Nov. 1973.

12-9 City of Los Angeles, "Rule of General Application RGA 4-74, Minimum Requirements for Ceiling Suspension Systems Supporting Luminous Ceilings and Acoustic Tile Ceilings with and without Light Fixtures," City of Los Angeles, Department of Building and Safety, 1974.

Chapter 13

Design of Structures with Seismic Isolation

*Ronald L. Mayes, Ph.D.**

13.1 INTRODUCTION

Because of today's concern for liability, engineering innovations must be exhaustively tested and analytically proven to a degree unknown in the past. Early engineers were respected for their ability to design from first principles and produce designs that were conceptually right even though analytical or laboratory methods did not exist that would remove all doubt. For the most part, the great early engineers removed doubt by force of their personality and confidence. They took risks that would be unthinkable today.

The field of seismic design is, as perhaps befits a subject directly concerned with both life safety and uncertainty, cautious and slow to innovate. In practice, improved seismic design does not represent a market opportunity because seismic safety is generally taken for granted. Like other code-dominated issues, and like airplane safety, seismic safety has never been much of a selling point. Money diverted to improve seismic resistance is often seen as a detraction from more visible and enjoyable attributes.

Improvement in seismic safety, since about the time of the San Francisco earthquake of 1906, has been due primarily to acceptance of ever-increasing force levels to which buildings must be designed. Innovation has been confined to the development and acceptance of economical structural systems that perform reasonably well, accommodate architectural demands such as open exteriors and the absence of interior walls, and enable materials such as steel and reinforced concrete to compete in the marketplace on near-equal terms.

The vocabulary of seismic design is limited. The choices for lateral resistance lie among shear walls, braced frames, and moment-resistant frames. Over the years, these have been refined and their details developed, and methods of analysis and modeling have improved and reduced uncertainty. But the basic approach has not changed: construct a very strong building and attach it securely to the ground. This approach of arm wrestling with nature is neither clever nor subtle, and it involves considerable compromise.

Although codes have mandated steadily increasing force levels, in a severe earthquake a building, if it were to remain elastic, would still encounter forces several times above its designed capacity. This situation is quite different from that for vertical forces, in which safety factors insure that actual forces will not exceed 50% of designed capacity unless a serious mistake has been made. For vertical forces, this is easy to do. But to achieve similar performance for seismic forces, the structure would be unacceptably expensive and its architectural impact would be extreme. This discrepancy between seismic demand and capacity is traditionally

*President, Dynamic Isolation Systems, Berkeley, California.

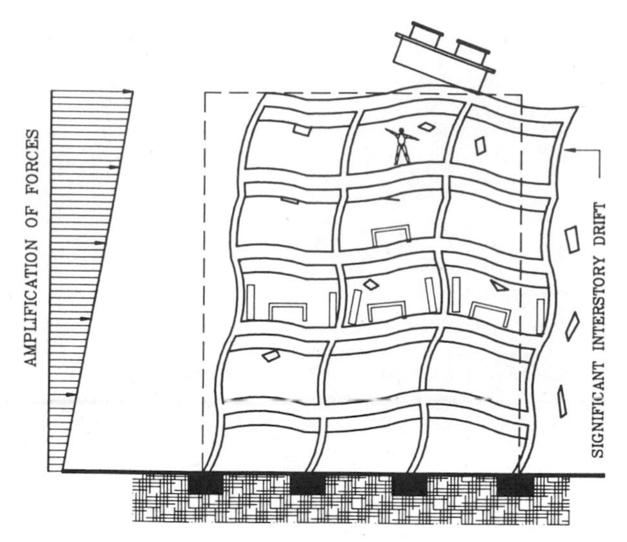

Figure 13-1 Conventional structure.

accommodated by reserve capacity, which includes uncalculated additional strength in the structure and often the contribution of partitions and exterior cladding to the strength and stiffness of the building. In addition, the ability of materials such as steel to dissipate energy by permanent deformation—which is called ductility—greatly reduces the likelihood of total collapse.

Modern buildings contain extremely sensitive and costly equipment that has become vital in business, commerce, education and health care. Electronically kept records are essential to the proper functioning of our society. These building contents frequently are more costly and valuable than the buildings themselves. Furthermore, hospitals, communication and emergency centers, and police and fire stations must be operational when needed most: immediately after an earthquake.

Conventional construction can cause very high floor accelerations in stiff buildings and large interstory drifts in flexible structures. These two factors cause difficulties in insuring the safety of the building components and contents (Figure 13-1).

In the last few years, an alternative to the brute-force response to nature has finally reached a stage, if not of fruition, at least of application. This approach is obvious and easily explainable at the cocktail-party level: Why not detach the building from the ground in such a way that the earthquake motions are not transmitted up through the building, or are at least greatly reduced? This conceptually simple idea has required much research to make it feasible, and only with modern computerized analysis has it become possible. Application has depended on very sophisticated materials research into both natural and composite materials in order to provide the necessary performance.

This new concept, now generally termed *seismic isolation*, meets all the criteria for a classic modern technological innovation. Imaginative advances in conceptual thinking were necessary, as were materials new to the industry, and ideas have developed simultaneously on a worldwide basis. But the method threatens conventional and established design procedures, so the road to seismic-isolation innovation is paved with argument, head shaking, and bureaucratic caution—all, to some extent, well-intentioned and necessary, given our litigious society.

Mounting buildings on an isolation system will prevent most of the horizontal movement of the ground from being transmitted to the buildings. This results in a significant reduction in floor accelerations and interstory drifts, thereby providing protection to the building contents and components (Figure 13-2).

The principle of seismic isolation is to introduce flexibility at the base of a structure in the horizontal plane, while at the same time introducing damping elements to restrict the amplitude of the motion caused by the earthquake. The concept of isolating structures from the damaging effects of earthquakes is not new. The first patent for a seismic isolation scheme was taken out in 1909,[13-1] and since that time several proposals with similar objectives have been made (see References 13-2 to 13-7). Nevertheless, until the last decade, few structures have been designed and built using these principles.

However, new impetus was given to the concept of seismic isolation by the successful development of mechanical-energy dissipators and elastomers with high damping properties (see References 13-8 to 13-14). Mechanical-energy dissipators, when used in combination with a flexible isolation device, can control the response of the structure by limiting displacements and forces, thereby significantly improving seismic performance. The seismic energy is dissipated in components specifically designed for that purpose, relieving structural elements such as beams and columns from energy-dissipation roles (and thus damage). There are over 100 civil-engineering structures that have now been constructed using the principles of seismic isolation, and Kelly[13-6] and Buckle[13-7] provide an excellent history and world overview.

The advantages of seismic isolation include the ability to eliminate or very significantly reduce structural and nonstructural damage, to enhance the safety of the building contents and architectural facades, and to reduce seismic design forces. The factor-of-5-to-10 reduction in elastic-force reductions achieved with seismic isolation can be expressed in simple terms with regard to building performance as a reduction of a Richter-magnitude-8 event to an event in the 5-to-6 magnitude range. Clearly, this is a very significant reduction. These potential benefits are greatest for stiff structures fixed rigidly to the ground, such as low- and medium-rise buildings, nuclear power plants, bridges, and many types of equipment. Some tectonic and soil-foundation conditions may, however, preclude the use of seismic isolation.

13.1.1 An Idea Whose Time Has Come

The elastomeric bearing and the mechanical damper are fundamental components in any seismic-isolation scheme. But it is not just the invention of the elastomeric bearing and the energy dissipator which has made seismic isolation a practical reality. Three other parallel, but independent, developments have also contributed to its recent success.

The first of these was the development of reliable software for the computer analysis of structures so as to predict their performance and determine design parameters. Work has been in progress for more than 15 years on the software for inelastic analysis of structural systems, and there are many available programs. Application to seismically isolated structures is straightforward, and correlation studies with model tests show the software to be soundly based.

The second development was the use of shaking tables which are able to simulate the effects of real recorded earthquake ground motions on different types of structures. The results of shaking-table tests over the last 12 years (see References 13-15 to 13-21) have provided another mechanism to enhance confidence in the way buildings respond during real earthquakes. In addition, the results provide an opportunity to validate computer modeling techniques which are then used on full-size structures.

A third important development is in the skill of the engineering seismologist in estimating ground motions at a particular site. Recent advances in seismology have given confidence in site-specific ground motions which take into account fault distances, local and global geology, and return period. These design motions are basic input to the computer modeling of seismically isolated systems and are a vital step in the estimation of system performance.

In summary then, five recent developments are together responsible for elevating seismic isolation from fantasy to practical reality:

1. The design and manufacture of high-quality elastomeric (rubber) pads, frequently called bearings, that are used to support the weight of the structure but at the same time protect it from earthquake-induced forces.
2. The design and manufacture of mechanical-energy dissipators (absorbers) and high-damping elastomers that are used to reduce the movement across the bearings to practical and acceptable levels and to resist wind loads.
3. The development and acceptance of computer software for the analysis of seismically isolated structures which includes nonlinear material properties and the time-varying nature of the earthquake loads.
4. The ability to perform shaking-table tests using real recorded earthquake ground motions to evaluate the performance of structures and provide results to validate computer modeling techniques.
5. The development and acceptance of procedures for estimating site-specific earthquake ground motions for different return periods.

13.2 CONSIDERATIONS FOR SEISMIC ISOLATION

The need for seismic isolation of a structure may arise if any of the following situations apply:

- Increased building safety and postearthquake operability are desired.
- Reduced lateral design forces are desired.
- Alternate forms of construction with limited ductility capacity (such as precast concrete) are desired in an earthquake region.
- An existing structure is not currently safe for earthquake loads.

For new structures current building codes apply in all seismic zones, and therefore many designers may feel that the need for seismic isolation does not exist because the code requirements can be satisfied by current designs. However, the commentary to the Structural Engineers Association of California (SEAOC) *Recommended Lateral Force Requirements*[13-22] states that buildings designed in accordance with its provisions will

- resist minor earthquakes without damage,
- resist moderate earthquakes without structural damage but with some nonstructural damage,
- resist major earthquakes without collapse but with structural and nonstructural damage.

These principles of performance also apply to buildings that are rehabilitated to code-level design forces.

Seismic isolation promises the capability of providing a building with better performance characteristics than our current code approach and thus represents a major step forward in the seismic design of civil engineering structures. In the case of a building retrofit, the need for isolation may be obvious: the structure may simply not be safe in its present condition should an earthquake occur. In such cases, if seismic isolation is suitable, its effectiveness compared with alternative solutions such as strengthening should be examined.

13.2.1 Solutions for Nonstructural Damage

One of the more difficult issues to address from a conventional design viewpoint is that of reducing nonstructural and building-content damage. This is very often ignored, and when addressed, can be very expensive to incorporate in conventional design. In fact, the cost of satisfying the more stringent bracing requirements of nonstructural elements in a California hospital is on the order of $2 to $4 per square foot more than for the conventional building requirements.

There are two primary mechanisms that cause nonstructural damage. The first is related to interstory drift between floors, and the second to floor accelerations. Interstory drift is defined as the relative displacement that occurs between two floors divided by the story height. Floor accelerations are the absolute accelerations that occur as a result of the earthquake, and in conventional construction they generally increase up the height of the building. Together, these two components cause damage to the building contents, architectural facades, partitions, piping and ductwork, ceilings, building equipment, and elevators (Figure 13-1).

There are two different design philosophies that are debated within the structural-engineering profession with regard to minimizing nonstructural damage. One argues that stiff buildings are the best solution. Stiff buildings reduce interstory drift, but they produce high floor accelerations. The other school of thought argues that flexible buildings are the solution, since they attract less force and tend to reduce floor accelerations. Although this is true, flexible buildings have much higher interstory drifts, and this accentuates damage to components that are sensitive to drift.

Clearly, a design concept that reduces both interstory drift and floor accelerations combines the best aspects of these two current design philosophies. Seismic isolation is such a concept (Figure 13-2), since it significantly reduces both floor accelerations and interstory drift and thus provides a viable economic solution to the difficult problem of reducing nonstructural earthquake damage.

13.3 BASIC ELEMENTS OF SEISMIC ISOLATION SYSTEMS

There are three basic elements in any practical seismic isolation system. These are:

1. a flexible mounting so that the period of vibration of the total system is lengthened sufficiently to reduce the force response;
2. a damper or energy dissipator so that the relative deflections between building and ground can be controlled to a practical design level; and
3. a means of providing rigidity under low (service) load levels such as wind and minor earthquakes.

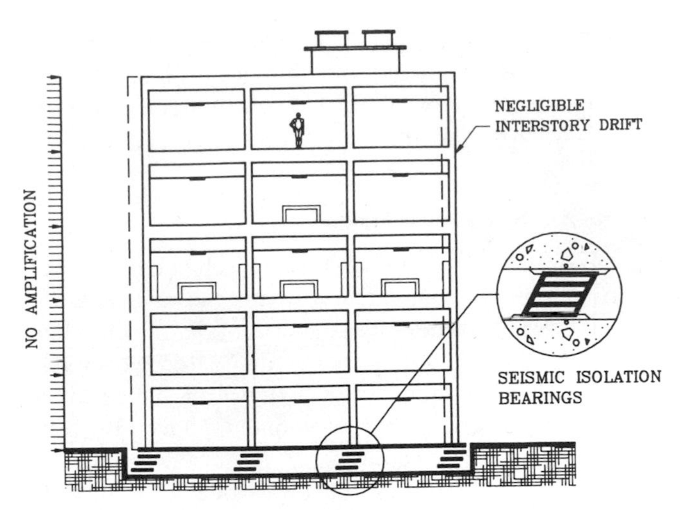

Figure 13-2 Base isolated structure.

Flexibility Bridge structures have for a number of years been supported on elastomeric bearings,[3-23] and as a consequence have already been designed with a flexible mount. It is equally possible to support buildings on elastomeric bearings, and numerous examples exist where buildings have been successfully mounted on pads. To date this has been done primarily for vertical-vibration isolation rather than seismic protection. Over 100 buildings in Europe and Australia have been built on rubber bearings to isolate them from vertical vibrations from subway systems below, and are performing well more than 40 years after construction. By increasing the thickness of the bearing, additional flexibility and period shift can be attained.

While the introduction of lateral flexibility may be highly desirable, additional vertical flexibility is not. Vertical rigidity is maintained by constructing the rubber bearing in layers and sandwiching steel shims between layers. The steel shims, which are bonded to each layer of rubber, constrain lateral deformation of the rubber under vertical load. This results in vertical stiffnesses several hundred times the lateral stiffness and of a similar order of magnitude to conventional building columns.

An elastomeric bearing is not the only means of introducing flexibility into a structure, but it appears to be the most practical and the one with the widest range of application. Other possible devices include rollers, friction slip plates, capable suspensions, sleeved piles, and rocking (stepping) foundations (Figures 13-3 to

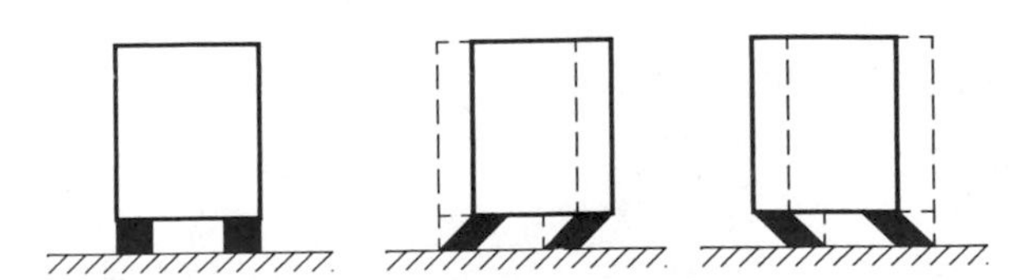

Figure 13-3 Elastomeric bearings.

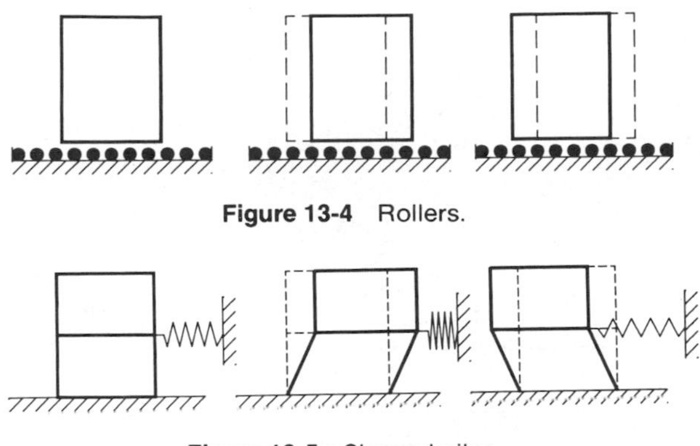

Figure 13-4 Rollers.

Figure 13-5 Sleeved piles.

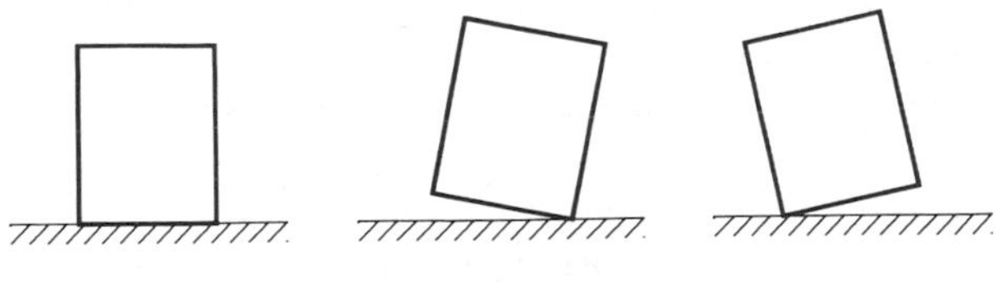

Figure 13-6 Rocking.

13-7). The reduction in force with increasing period (flexibility) is shown schematically in the force-response curve of Figure 13-8. Substantial reductions in base shear are possible if the period of vibration of the structure is significantly lengthened.

The reduction in force response illustrated in Figure 13-8 is primarily dependent on the nature of the earthquake ground motion and the period of the fixed-base structure. Further, the additional flexibility needed to lengthen the period of the structure will give rise to large relative displacements across the flexible mount. Figure 13-9 shows an idealized displacement response curve from which displacements are seen to increase with increasing period (flexibility). However, as shown in Figure 13-10, if substantial additional damping can be introduced into the structure, the displacement problem can be overcome. It is also seen that increasing the damping reduces the forces at a given period and removes much of the sensitivity to variations in ground-motion characteristics, as indicated by the smoother force response curves at higher damping levels.

Energy Dissipation One of the most effective means of providing a substantial level of damping is through hysteretic energy dissipation. The term "hysteretic" refers to the offset in the loading and unloading curves under cyclic loading. Work done during loading is not

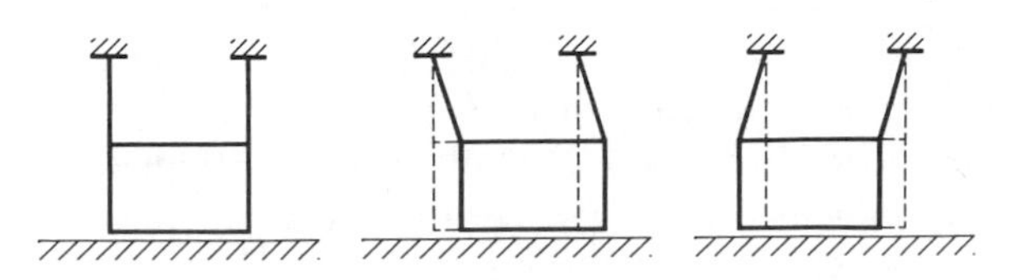

Figure 13-7 Cable suspension.

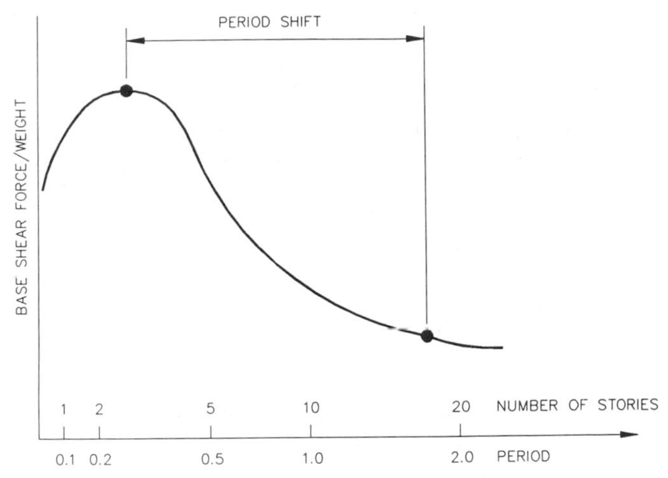

Figure 13-8 Idealized force response spectrum.

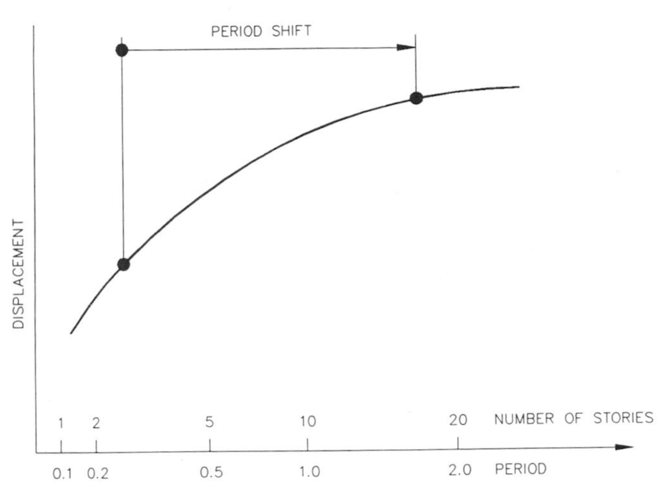

Figure 13-9 Idealized displacement response spectrum.

completely recovered during unloading, and the difference is lost (dissipated) as heat. Figure 13-11 shows an idealized force–displacement loop, where the enclosed area is a measure of the energy dissipated during one cycle of motion. Mechanical devices which use the plastic deformation of either mild steel or lead to achieve this behavior have been developed (References 13-8 to 13-13), and several mechanical-energy dissipation devices developed in New Zealand are shown in Figure 13-12.

Many engineering materials are hysteretic by nature, and all elastomers exhibit this property to some extent. By the addition of special-purpose fillers to elastomers, it is possible to increase their natural hysteresis without unduly affecting their mechanical properties.[13-9] Such a technique gives a useful source of damping, but so far it has not been possible to achieve the same level of energy dissipation as is possible with, say, a lead–rubber elastomeric bearing.

Friction is another source of energy dissipation which is used to limit deflections. However, it can be a difficult

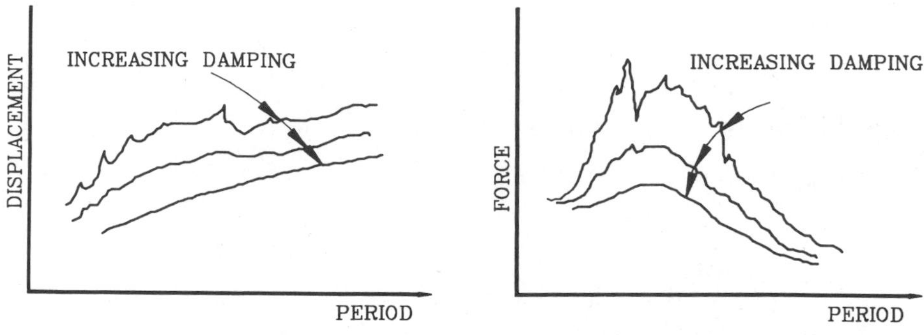

Figure 13-10 Response spectra for increasing damping.

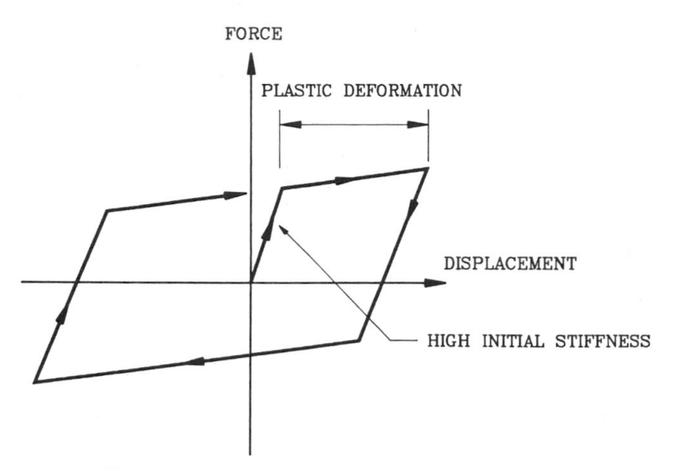

Figure 13-11 Hysteretic force-deflection curve.

source to quantify, and reliable systems tend to be an order of magnitude more expensive than with either of the above mechanisms. A further disadvantage is that most frictional devices are not self-centering, and a permanent offset between the sliding parts may result after an earthquake.

Hydraulic damping has been used successfully in some bridges and a few special-purpose structures.[13-7] Potentially high damping forces are possible from viscous fluid flow, but maintenance requirements and high initial cost have restricted the use of such devices.

Rigidity for Low Lateral Loads While lateral flexibility is highly desirable for high seismic loads, it is clearly undesirable to have a structural system which will vibrate perceptibly under frequently occurring loads such as minor earthquakes or wind loads. Lead–rubber bearings (and other mechanical-energy dissipators) provide the desired low load rigidity by virtue of their high elastic stiffness (Figure 13-11). Some other seismic isolation systems require a wind-restraint device for this purpose—typically a rigid component designed to fail under a given level of lateral load. This can result in a shock loading being transferred to the structure due to the sudden loss of load in the restraint. Nonsymmetrical failure of such devices can also introduce undesirable torsional effects in a building. Further, such devices will need to be replaced after each failure.

Table 13-1 summarizes the sources of flexibility and energy dissipation that have been discussed above. A more detailed explanation of these concepts can be found in the proceedings of a Workshop on Base Isolation and Passive Energy Dissipation that was conducted by Applied Technology Council.[13-24]

13.4 FORCE–DEFLECTION CHARACTERISTICS

Conceptually, there are four basic types of force–deflection relationships for isolation systems. These idealized relationships are shown in Figure 13-13, with each idealized curve having the same design displacement D for the design-level earthquake.

A *linear* isolation system is represented by curve A and has the same isolated period for all earthquake load levels. In addition, the force generated in the superstructure is directly proportional to the displacement across the isolation system. A linear isolation system will require some form of wind-restraining mechanism to be added to the system.

A *hardening* isolation system is represented by curve B. This system is soft initially (long effective period) and then stiffens (effective period shortens) as the earthquake load level increases. When the earthquake load level induces displacements in excess of the design displacement in a hardening system, the superstructure is subjected to higher forces and the isolation system to lower displacements than in a comparable linear system. Like a linear system, a hardening system will also require some form of additional wind-restraining mechanism.

A softening isolation system is represented by curve C. This system is stiff initially (short effective period) and softens (effective period lengthens) as the earthquake load level increases. When the earthquake load level induces displacements in excess of the design displacement in a softening system, the superstructure is subjected to lower forces and the isolation system to

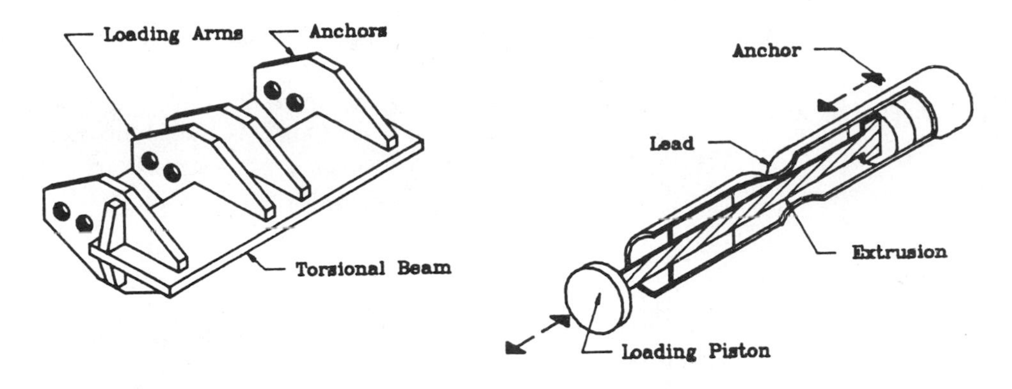

TORSIONAL BEAM DEVICE LEAD EXTRUSION DEVICE

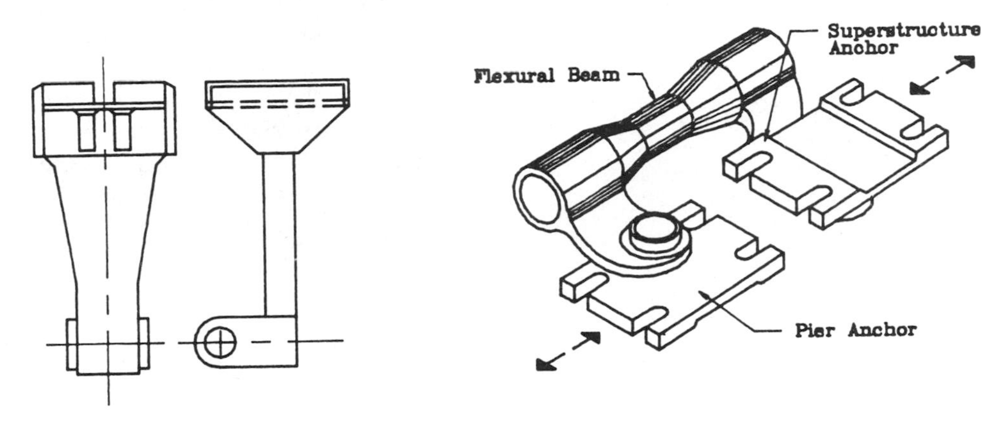

FLEXURAL PLATE DEVICE FLEXURAL BEAM DEVICE

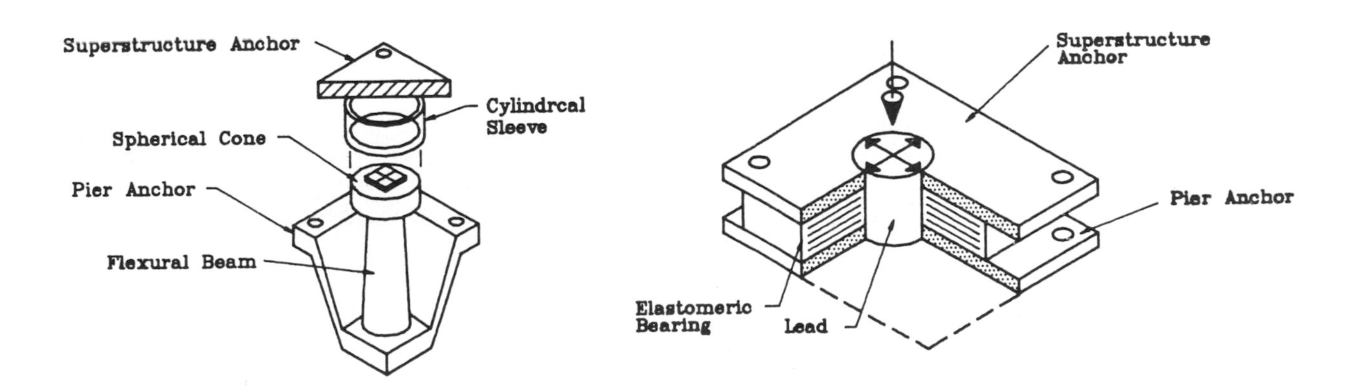

FLEXURAL BEAM DEVICE LEAD – RUBBER DEVICE

Figure 13-12 Various mechanical energy dissipators.

Table 13-1 Alternative Sources of Flexibility and Energy Dissipation

Flexible Mounting Systems
Unreinforced rubber blocks
Elastomeric bearings (reinforced rubber blocks)
Sliding plates
Roller and / or ball bearings
Sleeved piles
Rocking systems
Suspended floors
Air cushions
Slinky springs
Damping Devices / Mechanisms
Plastic deformation of a metal
Friction
High-damping elastomers
Viscous fluid damping
Tuned mass damping

higher displacements than in a comparable linear system. The high initial stiffness of a softening system is the wind-restraining mechanism.

A *sliding* isolation system is represented by curve *D*. This system is governed by the friction force of the isolation system. As in the softening system, the effective period lengthens as the earthquake load level increases, and the loads on the superstructure remain constant. The displacement of the sliding isolation system after repeated earthquake cycles is highly dependent on the vibratory characteristics of the ground motion and may exceed the design displacement. Consequently, minimum design requirements do not adequately define the peak seismic displacement for seismic isolation systems governed solely by friction forces. The value of the coefficient must be high enough to resist the wind forces.

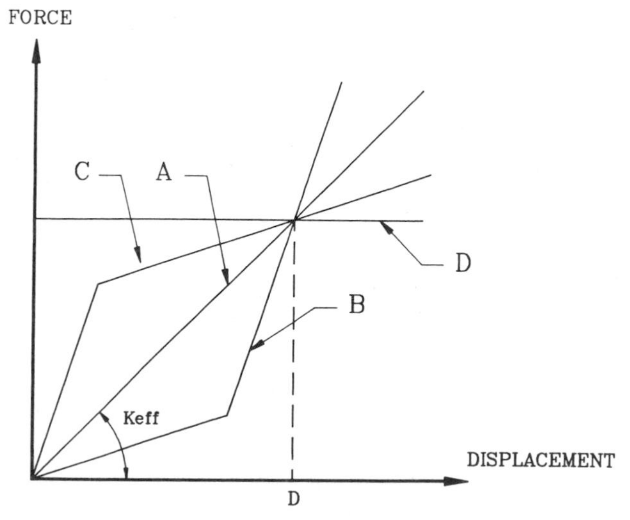

Figure 13-13 Idealized force-displacement relationships for isolation systems.

13.5 SEISMIC-ISOLATION DESIGN PRINCIPLES

The design principles for seismic isolation are illustrated in Figure 13-14. The top curve on this figure shows the realistic elastic forces based on a 5% ground response spectrum which will be imposed on a nonisolated structure from the new SEAOC "blue book."[13-25] The spectrum shown is for a rock site *if* the structure has sufficient elastic strength to resist this level of load. The lowest curve shows the forces which the Uniform Building Code[13-27] requires a structure to be designed for, and the second-lowest curve shows the probable strength assuming the structure is designed for the UBC forces. The probable strength is 1.5 to 2.0 times higher than the design strength because of the design load factors, actual material strengths which are greater in practice than those assumed for design, conservatism in structural design, and other factors. The difference between the maximum elastic force and the probable yield strength is an approximate indication of the energy which must be absorbed by ductility in the structural elements.

When a building is isolated, the maximum elastic forces are reduced considerably due to period shift and energy dissipation, as shown in Figures 13-8 and 13-10. The elastic forces on a seismically isolated structure are shown by the dashed curve in Figure 13-14. This curve corresponds to a system with approximately 30% equivalent viscous damping.[13-28]

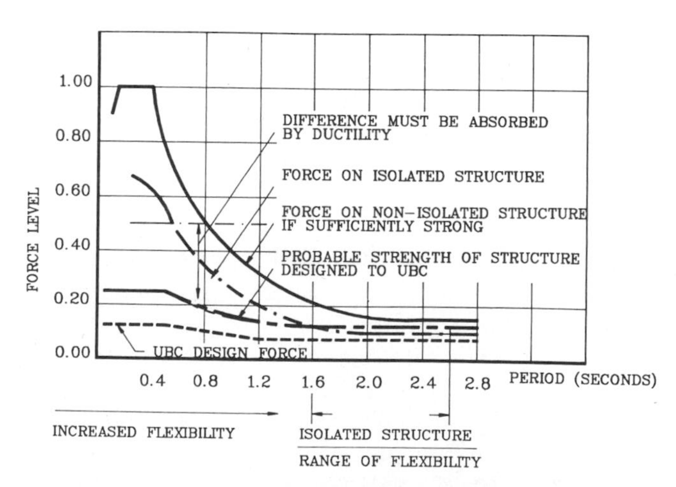

Figure 13-14 Design principles of seismic isolation.

If a stiff building, with a fixed-base fundamental period of 1.0 sec or less, is isolated, then its fundamental period will be increased into the 1.5- to 2.5-sec range (Figure 13-8). This results in a reduced UBC design force (Figure 13-14), but more importantly in the 1.5- to 2.5-sec range the probable yield strength of the isolated building is approximately the same as the maximum forces to which it will be subjected. Therefore, there will be little or no ductility demand on the structural system, and the lateral design forces are reduced by approximately 50%.

13.6 FEASIBILITY OF SEISMIC ISOLATION

Structures are generally suitable for seismic isolation if the following conditions exist:

- The subsoil does not produce a predominance of long period ground motion such as that obtained in Mexico City.
- The structure has two stories or more (or is unusually heavy).
- The site permits horizontal displacements at the base of the order of 6 in.
- The structure is fairly squat.
- Wind lateral loads and other nonearthquake loads are less than approximately 10% of the weight of the structure.

Each project must be assessed individually and early in the design phase to determine its suitability for seismic isolation. For this assessment, there are differences between new construction and the retrofit of existing structures. The following sections provide some guidelines for each of the situations.

13.6.1 New Construction

Structure The first consideration in assessing the suitability of a new project is the structure itself. Seismic isolation achieves a reduction in earthquake forces by lengthening the period of vibration at which the structure responds to the earthquake motions. The most significant benefits obtained from isolation are in structures for which the fundamental period of vibration without base isolation is short—less than 1 sec. The natural period of a building generally increases with increasing height. Taller buildings reach a limit at which the natural period is long enough to attract low earthquake forces without isolation.

Therefore, seismic isolation is most applicable to low-rise and medium-rise buildings and becomes less effective for high-rise ones. The cutoff depends mainly on the type of framing system. Shear-wall structures and braced-frame structures are generally stiffer than moment frames of equivalent height, and so for shear walls and braced frames isolation may be effective up to 12 to 15 stories, whereas with moment frames the cutoff is generally about 8 to 10 stories. These numbers are only generalizations, and there are of course exceptions. Most isolation systems do not perform well under tension, and the installation procedures are such that tension is prevented from being transferred into the bearing should it occur. A small amount of uplift is acceptable, and connection details accommodate this. The isolation system must also resist maximum lateral loads from other sources without yielding in order to avoid unacceptable displacements and vibrations under service loads, such as wind. Therefore, if these service lateral loads exceed about 10% of the structure's weight, the building should not be isolated.

Soil Conditions The second consideration when assessing the suitability of a structure for seismic isolation is the soil condition and the geology of the site. Generally, the stiffer the soil, the more effective the isolation.

The flexibility of the structure determines how it will respond to a given earthquake motion. However, the form of the earthquake motion as it arrives at the base of a structure may be modified by the properties of the soil through which the earthquake waves travel. If the soil underlying the structure is very soft, the high-frequency content of the motion may be filtered out, and the soil may produce long-period motions. An extreme example of this was seen in the 1985 Mexico City earthquake. Lengthening the period of a stiff structure in these lake-bed soil conditions will amplify rather than reduce the ground motions, and hence for Mexico City seismic isolation should not be considered.

Another geologic consideration is the distance from a major fault. If a structure is located within 10 km of a major fault, then it is recommended that the design displacements be multiplied by a factor of 1.5 to allow for the recently observed fling or pulse effect of near-fault ground motions.

Adjacent Structures A third consideration in assessing suitability is any constraints imposed by adjacent structures at the proposed site. As discussed earlier, the basic concept of seismic isolation requires a tradeoff between reduced forces and increased displacements. Practical isolation systems minimize these displacements, but nevertheless base displacements of the order of 3 to 12 in. generally occur. If the site is very confined due to neighboring buildings built on the boundary, it may not be possible to accommodate these displacements.

13.6.2 Retrofit of Existing Structures

Retrofit of existing structures to improve their earthquake safety involves additional considerations, compared with new construction, because of the constraints already present. Some structures are inherently more suitable for retrofit of seismic isolation than others. For example, bridge superstructures are generally supported on steel bearings. Replacement of these bearings with elastomeric ones is a fairly simple, low-cost operation that will lead to a reduction in earthquake forces and allow the option of redistributing forces away from the weak substructures into abutments more capable of sustaining them.

Buildings are often more difficult to retrofit than bridges. However, seismic isolation may often be an effective solution for increasing the earthquake safety of existing buildings without the addition of new structural elements which detract from the features which originally make the building worth preserving. Although seismic isolation reduces earthquake forces, it does not eliminate them. Consequently, the strength and ductility of an existing structure must at least be sufficient to resist the reduced forces that result from isolation. If the strength of the existing structure is extremely low (less than 0.05 of the weight of the building), then additional strengthening of the structure will be required. In such cases, the economics of large additional strengthening vs. some strengthening and the provision of isolation will need to be studied.

In addition to the conditions discussed above for new buildings, the issues to be addressed in the seismic isolation retrofit of an existing structure are:

- Is there sufficient clearance with adjacent buildings to permit a movement of 3 to 12 in.?
- Do the building and its existing foundations have sufficient strength and ductility to resist the reduced seismic forces?
- What is the appropriate level for the plane of isolation—foundation level, basement level, ground level, or the top, bottom, or mid-height of the columns?

The pros and cons with regard to the plane of isolation are:

- Any structure with a full subbasement or basement that can be temporarily disrupted is a good isolation candidate, since the work can be confined to that area.
- A structure with piled foundations can be more easily retrofitted at the foundation level than one with spread footings.
- Provision for the zone of isolation at the top, bottom, or mid-height of the basement-, first-, or second-level columns requires a detailed evaluation of the column capacities. If the strength of the column is not sufficient to resist the reduced isolation forces, three potential options exist. First, the column may be strengthened and act as a cantilever. Second, a new framing system with stiff beams may be developed at the plane of isolation to reduce the column forces. Third, the mid-height column solution may be considered, since it reduces the column moments by a factor of 2.

In summary, seismic rehabilitation of an existing structure provides the ability to confine most of the construction work to the level where the plane of isolation is to be provided, whereas conventional methods generally require the addition of structural elements to all levels of the building. This tradeoff can be very important if continued use of the facility is desired, as in hospitals.

13.6.3 Uplift and Overturning

In many types of structural systems increasing lateral forces will induce net tensions in elements once the axial loads caused by the overturning moment exceeds the gravity loads. This may occur for example at the edges of shear walls or the columns in braced or moment-resisting frames.

In conventional design this tension is resisted in the base connections and foundations, although only if it occurs under the code levels of earthquake lateral loads. The more severe loading occurring under actual maximum earthquakes will produce overturning moments much greater than the design value, and therefore tension forces will be induced even where none are indicated under code loading. In this case, it is assumed that the structural detailing and redundancies are sufficient to prevent failure due to the uplift.

More recent studies[13-15] have indicated that uplift may in fact be beneficial in reducing earthquake forces in conventional structures. In fact, at least two actual structures in New Zealand have been explicitly designed for uplift as a form of seismic isolation: a stepping bridge and a chimney stack.

For a structure isolated on elastomeric bearings the effects of uplift must be examined more carefully, since the elastomeric bearing is not suitable for resisting large tensile loads. The connection detail at the bearing-to-structure interface is designed so as not to transmit tension, using either dowels or keeper plates. These connections typically allow up to 1 in. of uplift before disengaging. As for a conventional frame, uplift in itself is not necessarily detrimental to the building response. However, excessive amounts of uplift could cause the bearing to disengage from its connection with a loss of performance.

Therefore, if uplift is indicated in an isolated structure, detailed analysis must be performed to quantify the vertical displacements for connection design. This involves a nonlinear analysis with realistic maximum credible earthquake records and requires significant analytical effort.

To avoid this, the optimum strategy is to avoid or minimize uplift. This is done by careful configuration of the lateral load-resisting elements. The important parameters are the height-to-width ratio of the lateral load-resisting system and the amount of gravity load carried by these elements.

13.7 DESIGN GUIDELINES

Recognizing the growing importance of seismic isolation as an approach to reducing structural response due to earthquakes, the Structural Engineers Associations of Northern and Southern California created committees to develop necessary provisions for guidance in the design of seismically isolated buildings. The northern group (SEAONC) completed tentative design guidelines[13-26] in September, 1986 for conventional buildings and these are now being reviewed by the State Seismology Committee for possible adoption in the Uniform Building Code. In January, 1987, the Building Safety Board of California Office of Statewide Health Planning and Development adopted "An Acceptable Method for Design and Review of Hospital Buildings Utilizing Base Isolation."

The background of the approach and objectives of the SEAONC document for conventional buildings is provided in the document's introduction, an extract of which is given below:

> The advantages of seismic isolation and the recent advancements in isolation system products have already led to the design and construction of a number of seismically-isolated buildings and bridges in California and Salt Lake City. This activity has, in turn, identified a need to supplement existing codes with design requirements developed specifically for seismically-isolated buildings. This need is shared by the public which requires assurance that seismically-isolated buildings are "safe" and by the engineering profession which requires a minimum standard upon which design and construction can be based. Accordingly, the Base-Isolation Subcommittee of the Seismology Committee of the Structural Engineers Association of Northern California has developed the following seismic isolation design requirements to supplement the "Recommended Lateral Force Requirements and Commentary" (Blue Book) document.[13-26]
>
> Rather than addressing a specific method of seismic isolation this document provides general design requirements applicable to a wide range of possible seismic isolation schemes. In remaining general, the design requirements rely on mandatory testing of isolation system hardware to confirm the engineering parameters used in the design and to verify the overall adequacy of the isolation system. Some systems may not be capable of demonstrating acceptability by test and, consequently, would not be permitted. In general, acceptable systems will:
>
> - remain stable for required design displacements,
> - provide increasing resistance with increasing displacement,
> - not degrade under repeated cyclic load, and
> - have quantifiable engineering parameters (e.g., force–deflection characteristics and damping).
>
> The design requirements are based on a severe level of earthquake ground motion which corresponds, approximately, to a 500-year return period event as described by the recommended ground motion spectra of the Blue Book.[13-26] The isolation system, including all connections and supporting structural elements, is required to be designed (and tested) for the full response effects of this level of ground motion. Structural elements above the isolation system, however, are required to be designed for slightly reduced loads (i.e., loads reduced by a factor of up to 2.7) if the structural system has sufficient ductility, etc., to respond inelastically. The above limits on isolation system and superstructure response are intended to ensure that seismically isolated buildings will be at least as safe as conventional buildings during extreme events considering the uncertainties in the new and developing technology of seismic isolation.

The text of the document is divided into the following five sections:

A. Lateral-Force Design Requirements
B. Minimum Lateral Earthquake Displacements and Forces
C. Alternate Determination of Seismic Displacement and Force
D. Required Tests of Isolation System
E. Additional Requirements

Each of these sections specifies design requirements for seismically isolated structures in a manner consistent with new "blue book" requirements[13-26] for conventional fixed-base buildings. In addition, a full running commentary is included with each section to describe the background and explain the basis for each design requirement. The significant aspects of each of the five sections are discussed in the following subsections.

13.7.1 Lateral-Force Design Requirements

Section A provides the link between the document's design requirements for seismically isolated buildings and the general design requirements specified by the new "blue book" for conventional structures. Section A also provides all necessary defintions, symbols, and notation specific to seismic-isolation design requirements.

13.7.2 Minimum Lateral Earthquake Displacements and Forces

Section B of the document specifies minimum earthquake displacements and forces on the basis of the properties of the isolation system and the vibratory characteristics of the ground. Design parameters are defined by simple prescriptive formulas which are similar to those used by the new "blue book" to prescribe forces for design of conventional buildings. In this manner, design engineers and building officials alike have a clearly defined basis to evaluate the design adequacy of isolated structures. Although the more complex methods of Section C are recommended, and in some cases required, the use of simple prescriptive formulas is considered essential for the unambiguous description of minimum design requirements.

Inherent uncertainty in design parameters is significantly reduced by the document's required use of isolation-system properties which must be based on test results, as described below:

> Minimum earthquake displacements and forces on seismic-isolated structures shall be based on the true deformational characteristics of the isolation system. . . .
>
> The deformational characteristics to be used in the determination of the displacements and forces shall be based on existing test data of the isolation system so that a reasonable level of confidence is obtained in the analysis results. . . .

Design-Displacement Formula Section B of the document prescribes minimum design displacements for the isolation system by the formula

$$D = \frac{10ZNST}{B}$$

where

Z = seismic-zone coefficient (e.g., 0.3 for zone 3, and 0.4 for zone 4)

S = soil-type coefficient (i.e., 1.0 for type S1, 1.5 for type S2, and 2.0 for type S3)

T = isolated-building period

N = near-field coefficient (i.e., 1.0 for sites more than 10 km from an active fault, 1.25 for sites within 10 km of an active fault, and 1.5 for sites within 5 km of an active fault)

B = damping coefficient (e.g., 1.0 for 5%-damped systems, 1.2 for 10%-damped systems, 1.5 for 20%-damped systems, and 1.7 for 30%-damped systems)

Close examination of this formula reveals that displacements required for design of 5%-damped isolated buildings located at least 10 km from an active fault are consistent with the shape of the spectra recommended by the new "blue book"[13-26] for design of conventional buildings with periods greater than 1 sec. Two additional terms (i.e., the near-field coefficient N and the damping coefficient B) have been added to the displacement formula to allow for increased ground-motion uncertainty of near fault sources and for the energy-dissipation characteristics of systems which have effective damping other than 5% of critical.

Building-Period Formula Section B of the document states that the period T of the isolated building shall be calculated from the minimum effective stiffness of the isolation system as follows:

$$T = 2\Delta\sqrt{\frac{W}{k_{\min} g}}$$

where

W = building weight

g = gravity constant

$k_{\min}$ = minimum effective stiffness of the isolated system, determined by test

By basing the period T on the minimum effective stiffness of the isolation system, this formula describes the longest period of the building, considering stiffness variability. In this manner, the design displacement D represents the maximum excursion of the isolated structure for the design-basis event.

Minimum Lateral Forces Section B of the document specifies minimum lateral forces in terms of the design displacement D, and prescribes a different level of design for elements at or below the isolation interface and for elements above.

1. *On structural elements at or below the isolation interface.* For working-stress design of elements at or

below the isolation interface, the minimum lateral force or shear, V_b, is prescribed by

$$V_b = \frac{k_{max} D}{1.5}$$

where

k_{max} = maximum effective stiffness of the isolation system, determined by test

D = design displacement.

This formula is intended to keep foundation elements, isolation-system components, and connections to structural elements fully elastic for the design-basis event by basing the peak shear force on the maximum effective stiffness of the isolation system multiplied by worst-case displacements. A factor of 1.5 is used to conservatively approximate the relationship between nominal yield of the material and working-stress design allowables.

2. *On structural elements above the isolation interface.* For working-stress design of the structure above the isolation interface, the minimum lateral force or shear, V_s, is prescribed by

$$V_s = \frac{2k_{max} D}{R_w}$$

where

k_{max} = maximum effective stiffness of the isolation system, determined by test

D = design displacement

R_w = numerical coefficient specified in Section 1-G of the new "blue book" for various structural systems, but not greater than 8

In all cases, the value of V_s shall not be less than any one of the following:

- the lateral force required by governing building codes or applicable design documents for a structure with an empirical period equal to the isolated period T,
- the ultimate capacity of a wind-restraint system, if one is used, and
- the yield level of a softening system or the static-friction level of a sliding system.

The formula, and limitations, for the force on structural elements above the isolation interface is intended to permit a limited amount of inelastic response for ductile structures. For most structural systems, inelastic response of the seismically isolated superstructure is limited to one-half that allowed for a conventional building. For very ductile structures (i.e., $R_w \geqslant 8$) inelastic response is not permitted to exceed a maximum of 2.7 (i.e., 4 divided by the working-stress-design reduction factor 1.5) times the nominal yield of the system. This will, in effect, limit actual inelastic response to a factor of about 2, or less, for the design basis event.

Building-Period Limits and Vertical Distribution of Force For use of the Section B requirements of the document, the isolated period T must be at least 5 times greater than the period of the building on a fixed base. This requirement insures that lateral displacement will occur primarily in the isolation system, rather than in the superstructure of the building. Provided the period limits are met, then Section B prescribes the vertical distribution of seismic acceleration over the height of the structure above the isolation interface. If the period limits of Section B are not met, a dynamic analysis must be performed in accordance with Section C provisions to allow for structure–isolation-system interaction.

13.7.3 Alternate Determination of Seismic Displacement and Force

Section C of the document provides for a more rigorous analysis method than the simple-formula approach of Section B. Section C provisions can be used for the design of isolated buildings subject to the following document limitations:

> Section B defines the minimum seismic isolation base shear forces and design displacements. As an alternate the following procedures may be utilized, but the resulting base shear forces and design displacements shall not be less than 90% of those calculated in Section B. If a dynamic analysis is required as per Section B.4, then the resulting base shear forces and design displacements shall not be less than 100% of those required by Section B.

A brief listing of Section C provisions is given below:

Analysis Procedures

- Dynamic analysis shall be based on isolation-system properties determined by test (i.e., Section D),
- either response-spectrum or time-history analysis methods shall be used, and
- the seismic input shall be based on 500-year return-period spectra (i.e., either the design spectra corresponding to Section B requirements or those determined by a site-specific hazard analysis).

Design Parameters

- Structural elements at or below the isolation interface shall be designed for forces from the dynamic analysis divided by 1.5, and
- structural elements above the isolation interface shall be designed for forces obtained from the dynamic analysis divided by $R_w/2$, subject to the limitations of Section B.

13.7.4 Required Tests of Isolation Systems

Because accurate representation of the isolation system's properties is essential for reliable design, detailed testing requirements are specified in Section D. General test criteria are described in the document by the following paragraph:

> The deformational characteristics and damping values used in the design and analysis shall be based on existing test data of the system and confirmed by the following tests on a selected sample of the components prior to construction. The isolation system to be tested shall include the ultimate restraint system, the wind restraint system, as well as individual isolators if such systems are used in the design. The tests specified are for validating the properties of the seismic isolation system. They should not be considered as manufacturing quality control requirements.

A brief list of Section D test requirements is given below:

Sequence of Tests

- 20 fully reversed cycles shall be used at the design wind level,
- 3 fully reversed cycles of loading shall be used at each of 5 increments of the design displacement, and
- 10 fully reversed cycles of loading shall be used at the full design displacement.

Determination of Design Properties

- The maximum and minimum effective stiffness shall be determined at the design displacement, considering both positive and negative loadings, and
- the equivalent viscous damping shall be calculated from the areas of test hysteresis loops.

System Adequacy

- The isolation system shall have positive incremental stiffness,
- The isolation system shall have less than 25% deterioration of stiffness and damping, and
- The isolation system shall have residual displacement less than 25% of the design displacement.

13.7.5 Additional Requirements

Additional performance, detail, and quality-control requirements are covered in the last section of the draft requirements. Because of their importance, the document's text is given here in full.

Environmental Conditions In addition to the requirements for vertical and lateral loads induced by wind and earthquake, the isolation system shall be designed with consideration given to other environmental conditions including aging effects, creep, fatigue, operating temperature, and exposure to moisture or damaging substances.

Wind Loads Isolated structures shall resist design wind loads at all levels above the isolator level in accordance with the general wind-design provisions. At the isolator level, a wind-restraint system shall be provided as necessary to avoid human discomfort within the building and as necessary to limit lateral displacement in the isolation system to a value equal to that required between floors.

Fire Resistance Fire resistance for the isolation system shall meet that required for the building columns, walls, or other structural elements.

Lateral Restoring Force The isolation system shall be configured to produce a lateral restoring force as necessary to meet the requirements of Section D.5(3).

Vertical-Load Stability The isolation system shall provide a factor of safety of 3 for vertical loads (dead load plus live load) in its laterally undeformed state. It shall also be designed to be stable under the full design vertical loads at a horizontal displacement of 1.5 times the design displacement for softening systems, or 1.5 times the design force for hardening systems.

Overturning The factor of safety against global structural overturning at the isolation level shall be not less than 1.0 for the load combinations defined by the recommended lateral-force requirements. All gravity and seismic loading conditions shall be investigated, except that seismic forces for overturning calculations shall be based on $k_{max}D$, the maximum base shear force, not V_s, the design base shear force for the superstructure; and W shall be used for the vertical restoring force.

Local uplift of individual elements is permitted provided the resulting deflections do not cause overstress or instability of building elements.

Design Review Engineering review of the isolation-system concept and design is required.

Lateral Drift The structure above the isolation system shall conform to conventional drift criteria.

Horizontal Distribution of Force A horizontal diaphragm or other structural elements shall provide continuity immediately above or below the individual isolators at the base of a building and shall have adequate rigidity to insure that the building structure moves as a rigid body on top of the isolators, and shall be strong enough to transmit forces (due to nonuniform ground motion) from one part of the building to another.

Separations Minimum separations between the isolated building and surrounding retaining walls or other fixed obstructions shall be not less than either 1.5 times the design displacement or the minimum distance required for conventional structures.

Components above the Isolation Interface Design of nonstructural component anchorage or bracing shall be consistent with the response of the particular structure under consideration, as substantiated by analysis and/or tests. Alternatively, anchorage or bracing may be designed using the force requirements for non-base-isolated structures.

Components Which Cross the Isolation Interface All architectural, equipment, and utility components which cross the seismic interface shall be designed to accommodate displacement of the isolation system as defined by Section B.3.3.

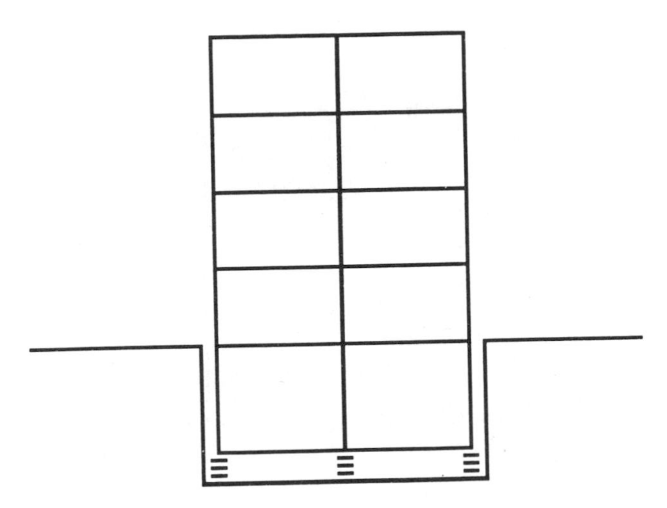

ADVANTAGES

- No special detailing required for separation of internal services such as elevator and stairways.
- No special cladding separation details.
- Base of columns connected by diaphragm at isolation level.
- Simple to incorporate back-up system for vertical loads.

DISADVANTAGES

- Added structural costs unless sub-basement required for other purposes.
- Requires a separate (independent) retaining wall.

Figure 13-15 Bearings located in subbasement.

13.8 SEISMIC-ISOLATION CONFIGURATIONS

The seismic-isolation configuration, including the layout and the installation details for the isolation system, depends on the site constraints, type of structure, construction, and other related factors. The following details are provided as an aid in determining appropriate layouts for particular projects and are not intended to restrict the designer in individual cases.

13.8.1 Bearing Location

Figures 13-15 to 13-18 provide typical planes of isolation for elastomeric bearings in buildings both with and without separate basement levels. Some of the advantages and disadvantages associated with each layout are listed in the figures. The following general guidelines are considerations for determining a suitable layout:

- The bearing location should permit access for inspection and replacement, should this become necessary.
- A full diaphragm above or below the isolators to distribute lateral loads uniformly to each bearing is preferable. If distribution is by tie beams only, the bearings should be arranged in proportion to the lateral load taken by each element, i.e., larger bearings under stiffer elements.
- Free movement for the maximum predicted horizontal displacement must be available.
- A layout which allows stub walls or columns as a backup system for vertical loads should be used wherever possible.
- Consideration must be given to the continuity of services, stairways, and elevators at the plane of isolation.
- Consideration must be given to details for cladding if it will extend below the plane of isolation.

13.8.2 Connection Details

Although connection details vary for each project, the design principles remain the same:

1. The bearing must be free to deform in shear between the outer shims; i.e., the upper surface of the bearing must be able to move freely horizontally.
2. The connections must have the capacity for transferring maximum seismic forces between the substructure and the superstructure.
3. Ease of construction must be kept in mind to insure access for installation and, in the case of a retrofit, temporary support for the superstructure.

The most common bearing construction has outer load

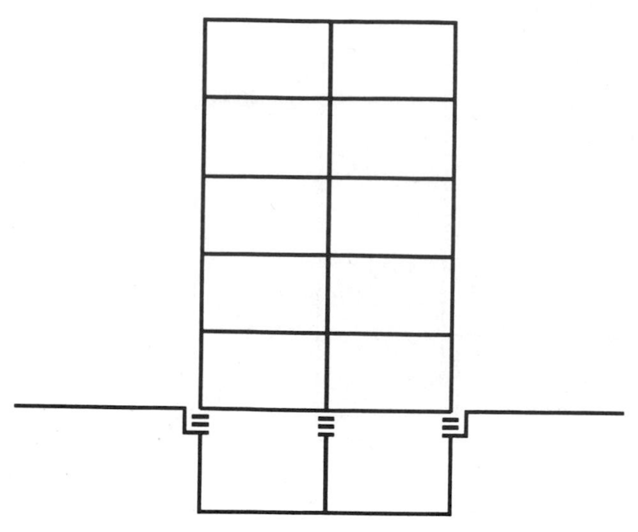

ADVANTAGES

- No Sub–basement Requirement.
- Minimal added structural costs.
- Base of columns connected by diaphragm at isolation level.
- Backup system for vertical loads provided by columns.

DISADVANTAGES

- May require cantilevered elevator shaft below first floor level.
- Special treatment required for internal stairways below first floor level.

Figure 13-16 Bearings located at top of basement columns.

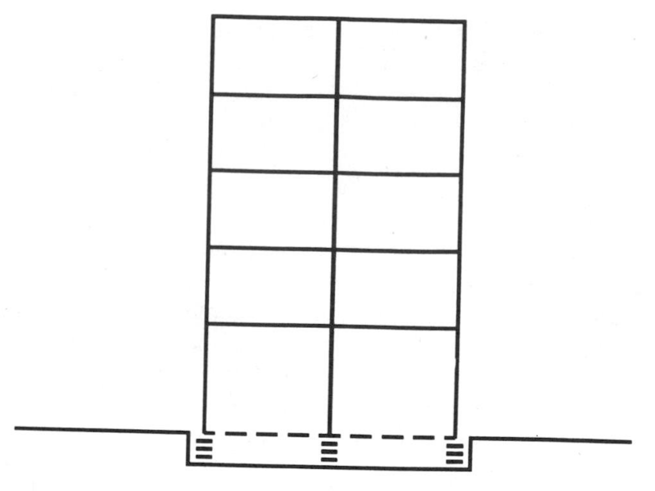

ADVANTAGES

- Minimal added structural costs.
- Separation at level of base isolation is simple to incorporate.
- Base of columns may be connected by diaphragm.
- Easy to incorporate back–up system for vertical loads.

DISADVANTAGES

- May require cantilever pit.

Figure 13-17 Bearings located at bottom of first story columns.

plates of $\frac{3}{4}$- to $1\frac{1}{2}$-in. steel covered by $\frac{1}{8}$-in. rubber layers. During the manufacture, holes for dowels are formed through the outer rubber layers and load plates. Exterior cover plates with dowels are then added to the bearing prior to installation. These exterior plates may be either welded or bolted to the structure. It is important to insure that the dowels do not intrude into the internal rubber layers. Figure 13-19 is an example connection detail using dowels.

13.8.3 Provision for Bearing Removal

Where practical, provision should be made to ease removal and replacement of the bearings should this ever become necessary. This requires two things: (i) a means of supporting the building weight while the bearing is removed, and (ii) a means of removing the bearing without undue damage to the connections.

The ease of meeting this first requirement will depend on the location of the bearings and type of backup safety system used. In a subbasement jacks can generally be used between the foundation and basement floor to support the bearing load. If a backup safety system is used (as described in the following section), provision for jacking may be incorporated into the design. Bear-

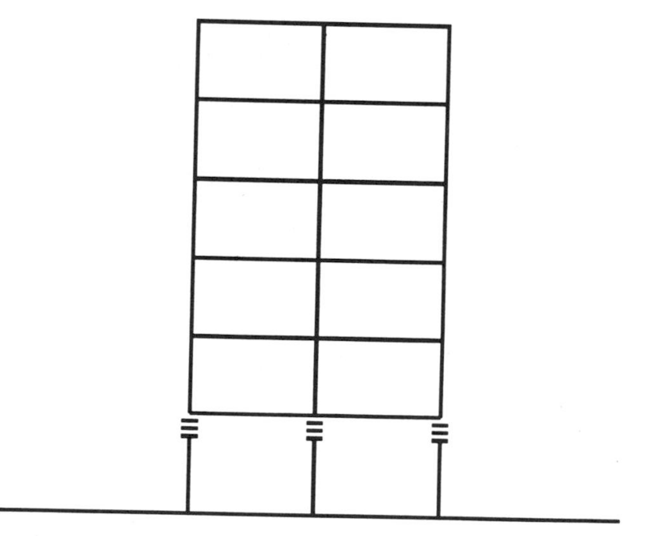

ADVANTAGES

- Minimal added structural costs.
- Economic if first level is for parking.
- Backup system for vertical loads provided by columns.

DISADVANTAGES

- Special detail required for elevators and stairs.
- Special cladding details required if first level is not open.
- Special details required for vertical services.

Figure 13-18 Bearings located at top of first story columns.

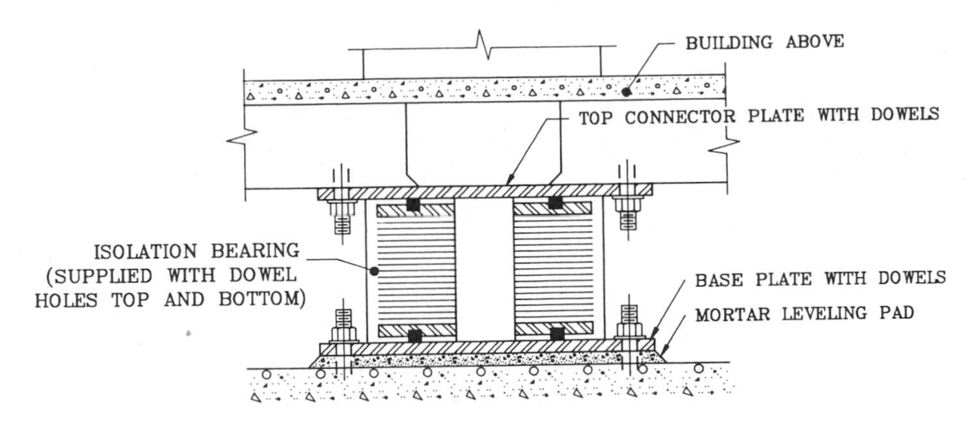

Figure 13-19 Installation using dowels.

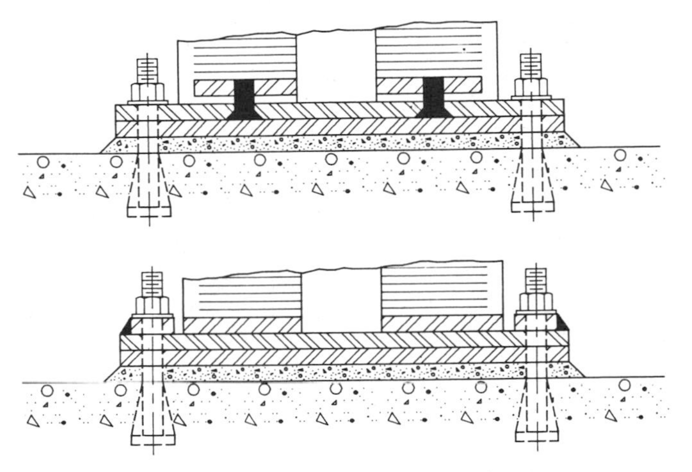

Figure 13-20 Details for replacement bearings.

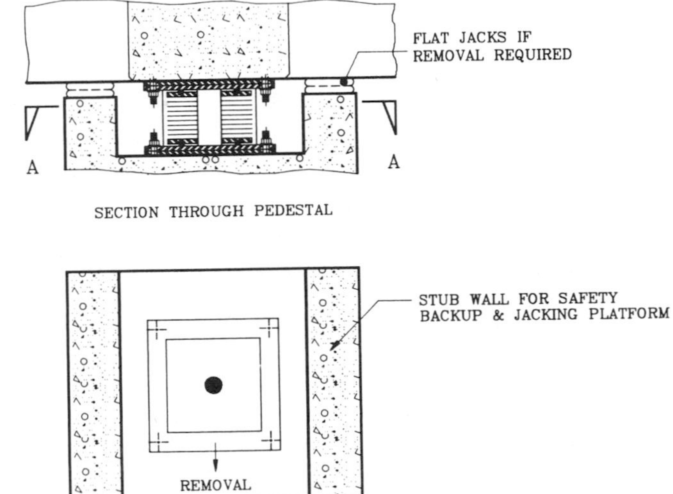

Figure 13-21 Backup and removal detail.

ing locations at the top of columns will require shoring to be erected around columns to provide a jacking platform if a backup system has not been provided.

The removal of the bearing once the load is removed will be simplified if bolted connections are used to connect to the structure. For example, the connection detail shown in Figure 13-19 could be modified to simplify bearing removal. In this modification, double plates would be added at the bottom of the bearing as shown in Figure 13-20. The bearing complete with dowel plates could then be removed. For a welded connection, removal would entail cutting the welds.

A combination of a removal and backup safety-system detail is shown in Figure 13-21.

13.8.4 Backup Safety System

Depending on the importance of the building, it may be considered desirable to incorporate an alternate means of supporting the vertical loads should bearing failure occur.

As for removal, the ease in incorporating such a system depends on the bearing location and configuration. For bearing locations at the top of columns a layout is shown schematically in Figure 13-22. This

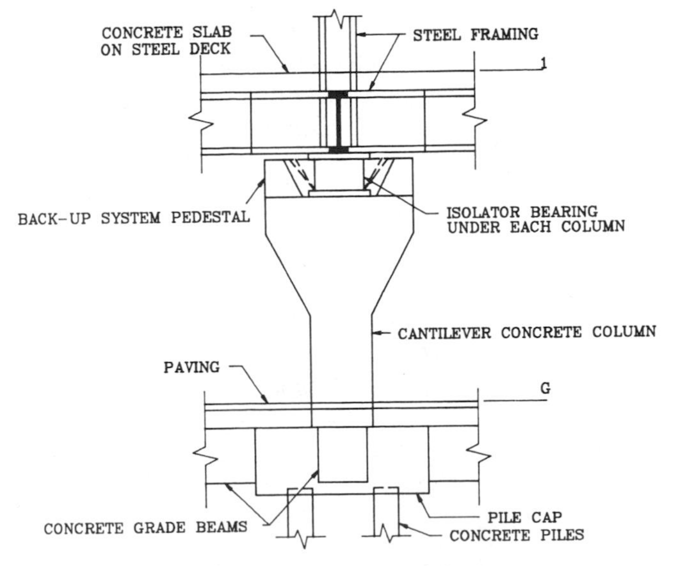

Figure 13-22 Bearings at top of columns.

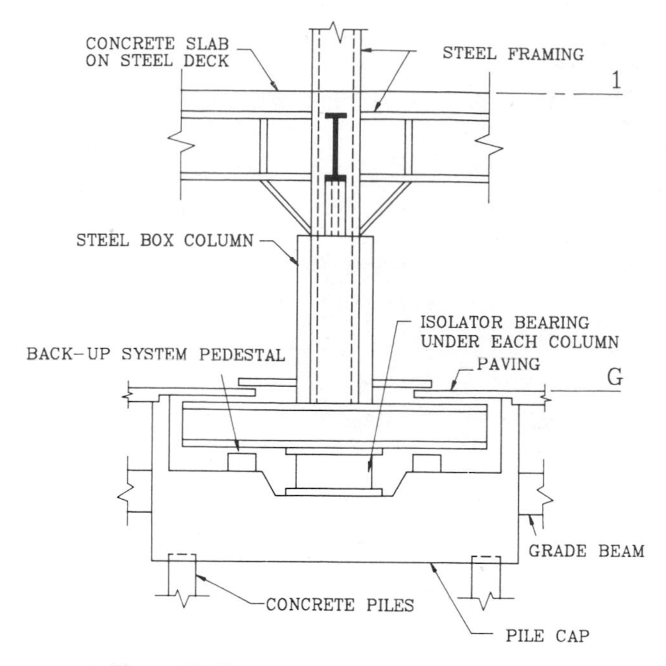

Figure 13-23 Bearings at base of columns.

provides a means of supporting the vertical load, and a lateral displacement limiter. An alternate to the scheme of locating bearings at the top of columns is to locate them at the base of the columns as shown in Figure 13-23.

13.9 DESIGN PROCEDURES

Design procedures have been developed and published for the lead–rubber isolation system.(13-27) These procedures are applicable to any system with similar force–deflection characteristics. The step-by-step procedures that follow are those developed for the lead–rubber system and are similar to those required for any isolation system. The procedures will insure that (i) the bearings will safely support the maximum gravity service loads throughout the life of the structure and (ii) the bearings will provide a period shift and hysteretic damping during one or more design earthquakes. The steps to achieve these aims are:

1. The minimum required plan size is determined for the maximum gravity loads at each bearing location.
2. The total rubber thickness is computed to give the period shift during earthquake loadings.
3. The diameter of the lead plug is calculated to provide the hysteretic damping and wind resistance required.
4. The performance of the bearing as designed is checked under gravity, wind, thermal, earthquake, and any other load conditions.

The need for an isolation system which is stiff under low levels of lateral load (e.g. wind) but flexible under higher levels (i.e. earthquakes) necessarily leads to a nonlinear system. The lead–rubber bearing is a nonlinear system which may be idealized in terms of a bilinear force–deflection curve with constant values throughout many cycles of loading (Figure 13-24). The design procedure takes advantage of two aspects of the nonlinear isolated response:

1. The period-shift effect of seismic isolation provides for a fundamental mode of response which is defined by the effective stiffness of the bearings at the design displacement while the supported structure remains essentially rigid.
2. The force–deflection curve of the lead–rubber bearing may be accurately defined by a stable bilinear hysteresis with predictable stiffness and yield levels.

These two aspects lead to the development of nonlinear response spectra based on the single-degree-of-freedom response developed using a bilinear force–deflection pattern rather than the linearly elastic system used for conventional response spectra. To develop such nonlinear spectra, the concept of an *effective period* is used to take account of the variation in the stiffness and hence period of response of the nonlinear system. The effective period is computed from an effective stiffness, which is the maximum force attained by a system divided by the displacement at the time of occurrence of this force. This effective-period concept is also used in the SEAONC Design Guidelines (Section 13.7).

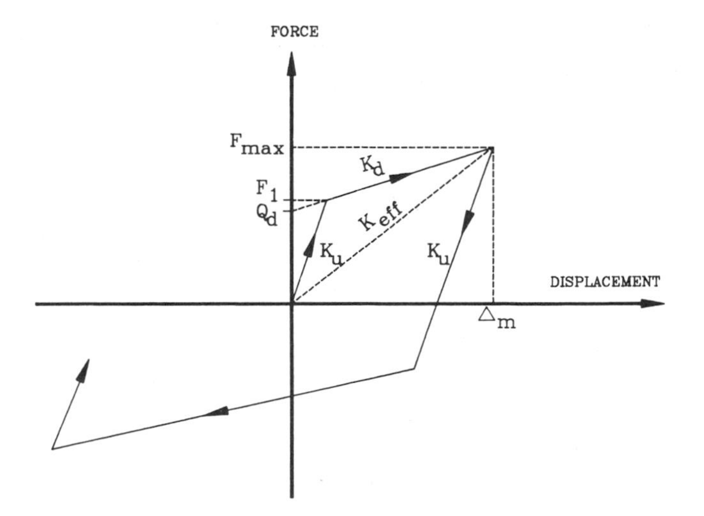

Q_d = Characteristic strength (kips)
F_1 = Yield force (kips)
F_{max} = Maximum force (kips)
K_d = Post–elastic stiffness (kip/inch)
K_u = Elastic (unloading) stiffness (kip/inch)
K_{eff} = Effective stiffness
Δ_m = Maximum bearing displacement

Figure 13-24 Typical hysteresis loop.

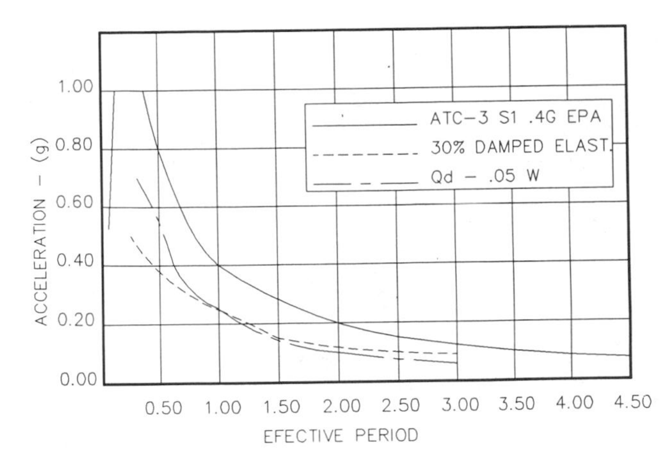

Figure 13-25 Nonlinear response spectrum.

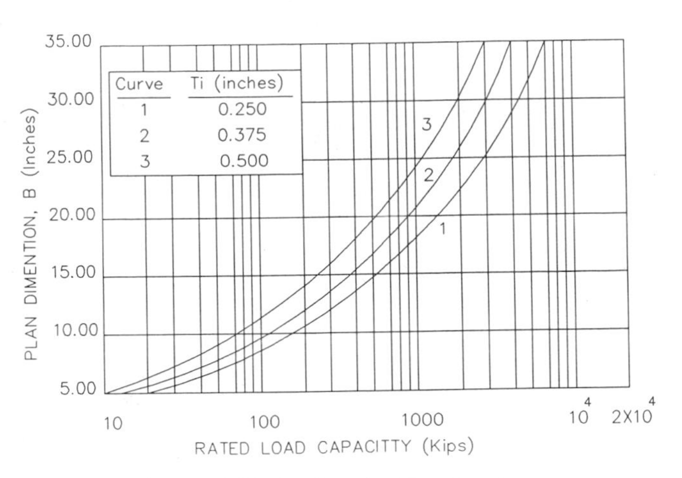

Figure 13-26 Design chart for plan size.

The nonlinear spectra are obtained by subjecting a single-degree-of-freedom system to seven time histories that envelop the various ground-response spectra. The maximum forces and isolation displacements obtained from the seven time histories are averaged and used as one point on the nonlinear spectrum.

The spectra obtained are relatively smooth and similar to the elastic curves for high viscous damping. Figure 13-25 compares the ATC 0.4g spectrum for soil type 1 (5%-damped elastic spectrum) with the nonlinear curve produced using the above procedure and the 30%-damped elastic spectrum. The 30%-damped curve was obtained from the seven time histories used for the nonlinear spectrum and is the mean of the spectra obtained from these motions.

Figure 13-25 shows that for periods less than 1 sec, the hysteretic damping of the lead–rubber bearing is equivalent to less than 30% viscous damping; for 1.0 to 1.5 sec it is approximately constant, and for periods beyond 1.5 sec, isolation not only provides the greatest benefits from the period-shift effect but also provides the most effective levels of hysteretic damping in the bearings.

13.9.1 Plan Size of Bearing

Ultimate failure of an elastomeric bearing under vertical loads is by tensile failures in the layers. The rated-load method[13-23] of computing vertical load capacity applies a factor of safety to the maximum tensile strains. For gravity-load effects, a factor of safety of 3 is used for the specified minimum tensile strength. For a given rubber formulation the rated load is a function of the plan dimension B and the rubber-layer thickness t_i. A design chart may be produced as shown in Figure 13.26 where a family of curves represents various layer thicknesses. This chart enables the required plan dimension to be determined for a given gravity load.

13.9.2 Rubber Thickness

As the rated load is dependent on the individual layer thickness and not the number of layers, the plan dimension may be determined independently of the total rubber thickness (i.e. number of layers), which determines the extent of isolation. The computation of the total rubber thickness is based on the effective-stiffness concept and the use of the nonlinear response spectrum.

The *effective stiffness* K_{eff} is defined as

$$K_{eff} = F_{max}/\Delta_{max} \tag{1}$$

where F_{max} is the maximum force in the bearing and Δ_{max} is the maximum displacement. The maximum force may be expressed as

$$F_{max} = Q_d + K_d \, \Delta_{max} \tag{2}$$

where Q_d is the characteristic strength of the bearing and K_d is the postelastic stiffness. The effective period as defined above is

$$T_{eff} = 2\pi \sqrt{\frac{M}{K_{eff}}} \tag{3}$$

By substituting for F_{max} in (1) and the resulting expression for K_{eff} in (3), the effective period may be expressed as a function of Q_d, K_d, the weight of the structure, and the maximum displacement. By expressing Q_d as a fraction of the weight and normalizing the weight by using the average compressive stress on the bearings, design charts may be produced which relate the rubber thickness T_r to the average compressive stress f_b, as shown in Figure 13-27. These charts are earthquake-specific, and Figure 13-27 is for the ATC 0.4g spectrum, soil type S1.

Figure 13-27 allows for a quick assessment of the range of rubber thicknesses appropriate for a given earthquake and illustrates the seismic-isolation design

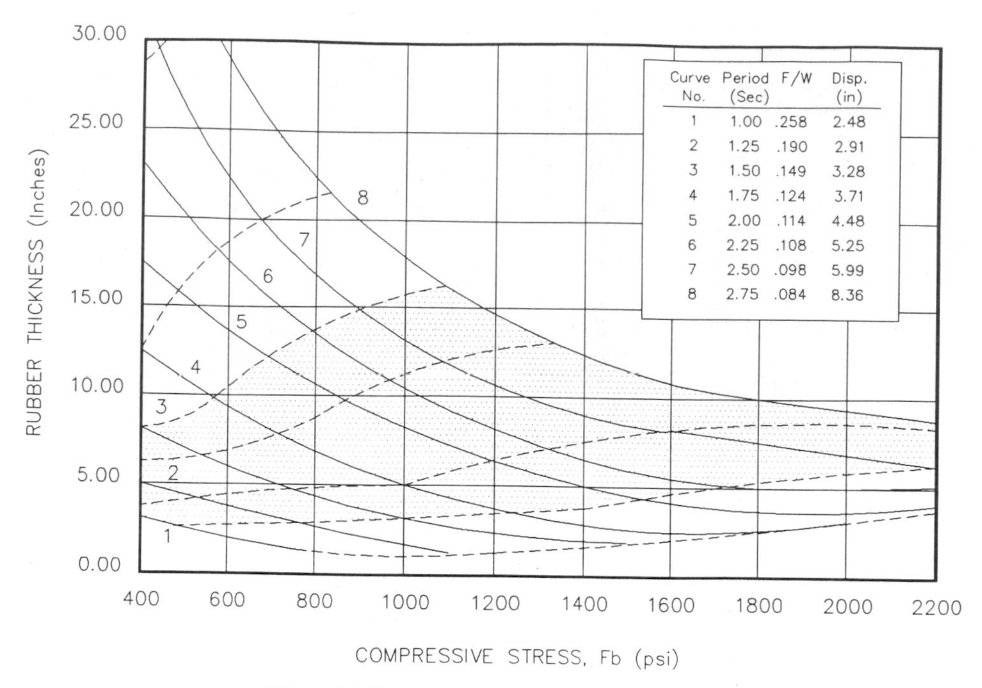

Figure 13-27 Design chart for rubber thickness.

tradeoff between decreased forces and increased displacements as the rubber thickness, and thus the effective period, increases.

13.9.3 Lead-Plug Size

Charts such as the one in Figure 13-27 are for a characteristic strength Q_d that is a given fraction of the structure weight, in this case $0.05W$.

13.9.4 Performance Check

The bearing is checked to insure that the vertical load capacity exceeds the applied loads under the maximum displacement. This check is performed using charts similar to those in Figure 13-25, but each chart is for a specific thickness and the family of curves represent various maximum displacements.

Although the procedure above is used for preliminary designs, the same principles used in the chart development have been used to develop a computer program which allows flexibility in selecting variables (e.g. range of Q_d, rubber formulations, and dimensional restrictions) which cannot be accommodated in a limited series of charts. This program is used for the detailed bearing-design phase but is also suitable for preliminary design.

13.9.5 Application of Design Procedure

The use of the preliminary design procedures is illustrated by design of seismic isolation for two structures:

1. a two-story, ductile, moment-resisting frame,
2. a twelve-story, eccentrically braced frame.

For each structure the charts are used to design seismic isolation for ATC $0.4g$ spectra, using soil type S1. The maximum force levels and displacements are compared with the mean results from three time-history analyses performed using time histories which were frequency scaled to match the design response spectra. As a check, the maximum displacements computed using the proposed SEAONC guidelines for isolated structures are also computed.

SEAONC Guidelines The SEAONC seismic-isolation guidelines (Section 13.7) require that the bearings be designed for a displacement D computed as

$$D = 10ZNST/B$$

where Z is the seismic-zone coefficient (corresponding to A_a in ATC-3), S is the soil-type factor (1.0, 1.5, and 2.0 for ATC soil types S1, S2, and S3 respectively), T is the effective period, and N is a near-field pulse factor. N has a value of 1.0 to 1.5 depending on the distance from an active fault. B is a factor reflecting the damping level in the particular isolation system, ranging from 0.8 for less than 2% damping to 2.0 for 50% or more damping.

For the lead–rubber bearings, Figure 13-25 shows that the equivalent damping for periods of 1.5 sec or longer is approximately 30%. Therefore, a B factor of 1.7 is used.

Two-Story Ductile Frame The two-story ductile frame was designed with 12-in. square bearings under each column to carry gravity loads. This corresponds to an average bearing stress of 540 lb/in.^{2}. A rubber thickness of 2.5 to 9 in. could be used, from Figure 13-27, and 7.5 in. was selected from the design charts to produce an effective period of 1.6 sec, corresponding to a force coefficient of 0.14 at 3.5 in. maximum displacement. The three time-history analyses produced displacements of 3.81, 4.67, and 3.92 in. and base shear coefficients of 0.14, 0.16, and 0.14 respectively, for mean values of 4.13 in. and 0.14. For the nonisolated structure the corresponding base shear coefficient was 0.86. For this structure, the SEAONC displacement based on $Z = 0.4$, $S = 1.0$, $N = 1.0$, $B = 1.7$, and an effective period of 1.6 sec is 3.76 in.

Twelve-Story Braced Frame The twelve-story braced frame was designed with 23-in. square bearings under each interior column and 18-in. square bearings under the exterior columns. This corresponds to an average bearing stress of 1157 lb/in.^{2}. A rubber thickness 3 to 15 in. could be used from Figure 13-27, and 7.0 in. was selected from the design charts to produce an effective period of 2.15 sec corresponding to a force coefficient of 0.11 at 4.95 in. maximum displacement. The three time-history analyses produced displacements of 4.87, 4.85, and 5.00 in. and base shear coefficients of 0.11, 0.11, and 0.11 respectively, for mean values of 4.91 in. and 0.11. For the fixed-base structure the corresponding base shear coefficient was 0.34. For this structure, the SEAONC displacement based on $Z = 0.4$, $S = 1.0$, $N = 1.0$, $B = 1.7$, and an effective period of 2.15 sec is 5.06 in.

Summary of Results The displacements and forces in the isolators for the two structures may be summarized as follows:

Quantity	Value 2-Story	Value 12-Story
Displacement, in.:		
Design procedure	3.5	4.95
Nonlinear analysis	4.1	4.9
SEAONC guidelines	3.8	5.1
Base Shear Coefficient:		
Design procedure	0.14	0.11
Nonlinear analysis	0.15	0.11
Nonisolated	0.86	0.34

For the two-story building, the design procedure gives a good estimate of the base shear coefficient but underestimates the displacements by 15% compared to the nonlinear analyses and 8% compared to the SEAONC guidelines. This building is rather light and at the borderline of suitability for seismic isolation—as seen from Figure 13-27, a stress of 540 lb/in.^{2} is at the lower end of the chart, where the results are very sensitive to the stress level.

The twelve-story structure is more suited for seismic isolation and less sensitive to the stress on the bearings. Therefore, the preliminary design procedure gives results in excellent agreement with the nonlinear analyses and the proposed SEAONC guidelines.

13.10 DESIGN EXAMPLE

A proposed four-story structure is V-shaped in plan with approximate dimensions of 300 by 300 ft (Figure 13-29). The building has a parking level at grade with three floors above. There is a partial penthouse on each building, and a portion of each building is open, forming an atrium.

The structural system consists of a steel frame for the upper three levels with concentric bracing to provide lateral stiffness. At the lowest level, concrete cantilever columns, interconnected by grade beams, support a system of lead–rubber bearings, which serve to isolate the structure above from the ground-level garage.

Typical elevations are shown in Figure 13-28, and the first-floor plan is shown in Figure 13-29.

The site is located 11.5 mi (18.5 km) from the San Andreas Fault, 6.6 mi (11 km) from the Calaveras Fault, and 6 mi (10 km) from the Hayward Fault, exposing the facility to ground motion from several sources.

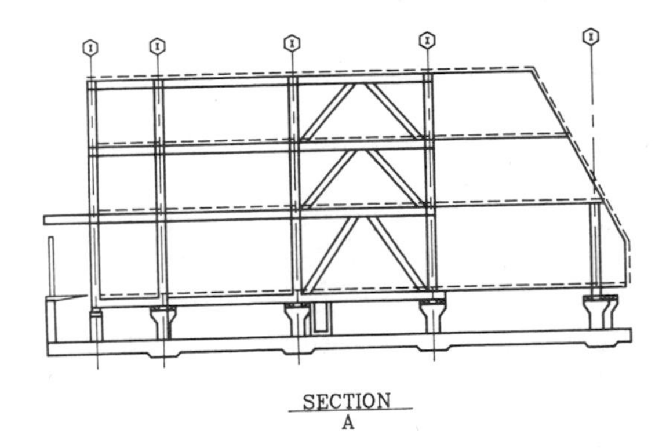

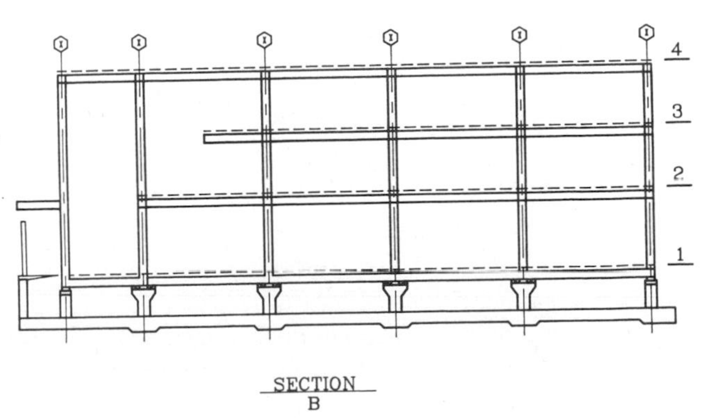

Figure 13-28 Typical structural section.

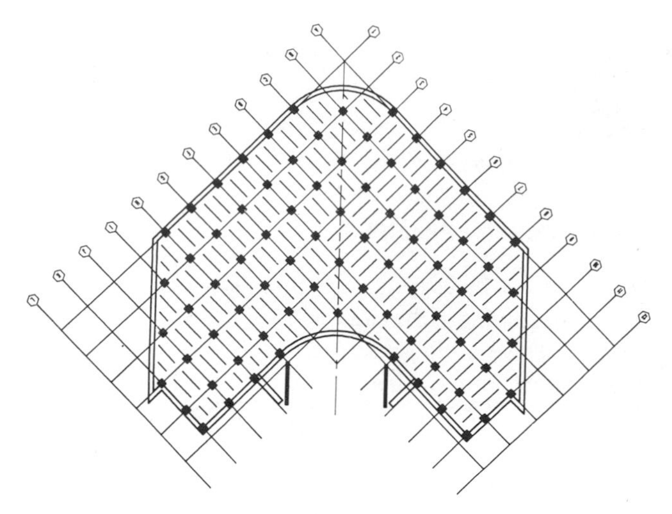

Figure 13-29 First floor plan.

For the purpose of this example, the analyses that follow are based on requirements of zone 4, soil type S2, with a distance greater than 10 km from the fault.

13.10.1 SEAONC Design Requirements

The objective of this example is to compare the results obtained from two methods of analysis with the SEAONC Tentative Seismic Isolation Design Requirements. These are summarized as follows:

SEAONC Section B—Minimum Displacements and Forces The requirements of this section are proposed for use for regular buildings provided the fixed base period of the building does not exceed 20% of the period T of the isolation.

Using Equation 1-3 of the new "blue book," T of the fixed-base building is

$$T = \frac{0.05h_n}{D^{0.5}}$$

where

$$h_n = 52 \text{ ft}$$

$$D = 300 \text{ ft}$$

Thus

$$T = \frac{0.5 \times 52}{300^{0.5}} = 0.15 \text{ sec}$$

For an isolated period of 2 sec,

$$T_{\text{fixed base}} < 20\% \, T_{\text{isolated}}$$

and therefore Section B requirements are applicable for the building, since it is also regular in configuration.

SEAONC Section C—Alternate Determination of Displacements and Forces The specifications of this section either require or permit a more detailed or refined study. There are two methods of analysis permitted: a response spectrum analysis and a time-history analysis. For this project, the time-history method of analysis was used, incorporating the appropriate seismic input. The details of these methods are given in the following subsections.

13.10.2 SEAONC Section B Calculations

Based on the method specified in Section B, the following design parameters are applicable for zone 4 and the site soil conditions:

- Site coefficient,

$$S = 1.5 \qquad \text{(Table 1-BB)}$$

- Near-field response,

$$N = 1.0 \qquad \text{(Table 1-J,} \quad d > 1)$$

- Seismic-zone factor,

$$Z = 0.4 \qquad \text{(Table 1-A)}$$

- Weight (above the isolators),

$$W = 18{,}000 \text{ kips}$$

Equations 1 and 2 of Section B are

$$D = \frac{10ZNST}{B} \tag{1}$$

and

$$T = 2\pi\sqrt{\frac{W}{K_{\min}g}} \tag{2}$$

There are two ways of using these two equations for preliminary design:

- Method I: Perform a preliminary design of the isolation system to obtain a practical isolation system, and obtain a preliminary value of $K_{\min}$, the effective stiffness, and B, the equivalent viscous damping, for the isolation system. With these values, T, and then D, can be determined.
- Method II: Assume an initial isolation period, T. With this value of T, a value for $K_{\min}$ can be determined. This will then set the sizes of the isolation system, and a value of B for the system can then be obtained. D can then be calculated.

In this example, Method I was used. The preliminary design of the isolation system was performed using the design procedures described in Section 13.9. As a consequence of this preliminary design, there were 21 14-in.-diameter bearings, 37 16-in.-diameter bearings, and 26 17-in.-diameter bearings (all 13 in. high). The K_{min} for these bearings was 460 kips/in. Then

$$T = 2\pi\sqrt{\frac{18{,}000}{460 \times 386.4}} = 2.00 \text{ sec}$$

For the lead–rubber bearings used on this project, an equivalent viscous-damping value of 30% was determined. Thus

$$B = 1.7 \qquad \text{(Table 1-K)}$$

and

$$D = \frac{10 \times 0.4 \times 1.0 \times 1.5 \times 2}{1.7}$$
$$= 7.05 \text{ in.}$$

The maximum shear force that the structure will be subjected to is given by the use of Equation 4 (Section B.3.2) without the reduction factor of $R_w/2$:

$$V_{max} = K_{max} D$$

Therefore

$$V_{max} = 460 \times 7.05 = 3243$$
$$= \frac{3243}{18{,}000} \times W = 0.18W.$$

The design base shear force V_s is

$$V_s = \frac{K_{max} D}{R_w/2} \qquad \text{with} \quad R_w = 8$$
$$= \frac{460 \times 7.05}{4} \text{ kips}$$
$$= 811 \text{ kips}$$
$$= \frac{811}{18{,}000} W = 0.045W.$$

However, V_s shall not be less than

1. the lateral seismic force required by governing building codes or applicable design documents for a fixed-base structure with an empirical period equal to the isolated period T,
2. the base shear corresponding to the design wind load, and
3. the yield level of a softening system, the ultimate capacity of a sacrificial wind-restraint system, and the static friction level of a sliding system.

The numerical values of these limitations are as follows:

1. SEAOC "Blue Book":

$$\frac{ZIC}{R_w} W$$

where

$$Z = 0.4$$
$$I = 1$$
$$C = \frac{1.25S}{T^{2/3}}$$
$$R_w = 8$$
$$S = 1.2$$
$$T = 2.0 \text{ sec} \qquad \text{(isolated period)}$$

Therefore

$$C = \frac{1.25 \times 1.2}{2^{2/3}} = 0.94$$

and

$$V = \frac{0.4 \times 1.0 \times 0.94}{8} W$$
$$= 0.047W$$

2. Wind design force:

$$550 \text{ kips} = \frac{550}{18{,}000} W$$
$$= 0.030W$$

3. Yield level of isolators: $0.05W$.

Therefore, the minimum design force of the structure above the isolators is

$$V_s = 0.05W$$

The design shear force for the cantilever columns below the isolation level is given by Equation 3 (Section B.3.1) of SEAONC:

$$V_b = \frac{K_{max} D}{1.5}$$
$$= \frac{460 \times 7.05}{1.5} Z = 2162 \text{ kips}$$
$$= \frac{2162}{18{,}000} W$$
$$= 0.12W$$

13.10.3 SEAONC Section C Calculations

Based on the period-separation requirement of Section B.5, a time-history analysis would not be explicitly required by SEAONC. However, an extensive series of nonlinear time histories was performed, and the modeling and seismic input were compatible with Section C. Thus, the results are of interest for a comparison with Section B requirements.

The results are summarized first, followed by a description of the analytical model and the seismic input and then the detailed results.

Nonlinear Time-History Results

Zone-4 S_2 Spectrum-compatible time histories (section C.3)	Displacement in.	Maximum base shear V_{max}, kips	V_{max}/W
El Centro	6.2	2580	0.14
Olympia	6.1	2650	0.15
Taft	5.6	2450	0.14

Model for Analysis (Section C.2) The building was modeled with the computer program ANSR-II, a general-purpose three-dimensional nonlinear-analysis program. This program has a library of finite elements allowing specification of various types of material nonlinearity and incorporates solution schemes for static and dynamic analyses including geometric nonlinearities as required.

The building was modeled as an assemblage of three-dimensional beam–column members with lumped masses at the nodal points. A series of time-history analyses were performed using ground-acceleration time histories as input.

The cantilever columns in the garage level had not yet been finally sized, and for the purpose of this analysis, it was assumed that they were 30-in. square reinforced-concrete members. The total height was taken to be 7 ft 9 in. with a rigid end block 1 ft 9 in. long at the top of the column, giving an effective length of 6 ft 0 in. The stiffness was based on 4000-lb/in.^{2} concrete, and it was assumed that yielding would not occur in these columns.

The lead–rubber bearings were modeled explicitly as nonlinear elements with their bilinear hysteresis loop.

The equivalent-stick models of the concentric braced frames were developed from a linear elastic model of a single frame. The computer program SAP was used to obtain deflections under a series of unit loads applied at each floor in turn. This effectively provided the stiffness matrix of the individual stories, and this matrix was then converted into the properties of three beam–column elements, one representing each story of the braced frame.

For this study it was assumed that the diaphragms were rigid. This is a reasonable assumption given the width-to-length ratios of the floors. The diaphragms were modeled as beam elements with very high flexural properties, effectively preventing any distortions in the plane of the floors.

Masses were lumped at the five nodes lying along the center line of the first floor and at the three center-line nodes at the upper floor levels. It was assumed that the penthouse loads were at the third floor level, and no attempt was made to model the penthouse itself. This is a reasonable approximation, as the penthouse weight is only about 6% of the third-floor weight and so its effect on response would be minimal.

To allow for any accidental eccentricity in the floor weights, the masses at each node were adjusted to provide an eccentricity, normal to the direction of major earthquake loading, of +5% and −5% of the building length, i.e., 15 ft 0 in.

The hysteretic damping in the lead–rubber bearing is modeled explicitly with a bilinear force–deflection curve in the yielding elements. Viscous damping of 5% was used for the structure. This was achieved by specifying mass- and stiffness-proportional factors so as to achieve 5% damping at periods of 0.10 and 2.0 sec. For periods between these values, the damping will be less than 5%. At the likely natural period of the superstructure (0.4 sec) the damping will be approximately 2% of critical, a conservative value for steel at high stress levels.

Seismic Input (Section C.3) The analyses were performed using the ATC-3-06 0.4g response spectrum for soil type S2. A suite of time histories to match each of the spectral curves meeting the Section C.3 requirements was developed.

The time histories used for the analyses were all frequency-scaled to match the target spectrum and were based on the recorded motions for El Centro 1940 (N–S and E–W components), Olympia 1949 (N04W component), and Taft 1952 (S69E component).

Analyses Performed A total of three different configurations were considered, each differing only in the position of the center of mass:

- Configuration 1: The center of mass lay on the axis of symmetry, i.e., no eccentricity.
- Configuration 2: The center of mass was offset +5% normal to the X axis.
- Configuration 3: The center of mass was offset −5% normal to the X axis.

Configuration 1 was analyzed for a single time history, the El Centro 1940 N–S frequency-scaled record, applied along the axis of symmetry, i.e., at 45° to the X and Z axes of the model. This analysis was mainly to verify the model by examining the response and insuring symmetry.

Table 13-2

Spectrum-compatible earthquake*	Base shear, kips	Base-shear coefficient	Bearing displacement, in.	Frame displacement, in.
El Centro	2580	0.143	6.2	0.6
Olympia	2650	0.147	6.1	0.7
Taft	2450	0.136	5.6	0.6

*ATC 0.4g S2.

Configuration 2 was analyzed three times with different spectrum-compatible time histories. In each case the major loading, 100% of one component, was applied parallel to the X axis, and a load of 40% of the other component was applied concurrently along the Z axis. The X-axis loadings were, respectively, the El Centro 1940 N–S, the Olympia 1949, and the Taft 1952 frequency-scaled records. For all three analyses the concurrent Z or orthogonal loading was 40% of the El Centro 1940 E–W record, frequency-scaled to match the ATC-3 spectrum.

Configuration 3 was analyzed once using the same loading case as the first analysis of configuration 2, i.e., 100% of El Centro 1940 N–S along X, and 40% of El Centro 1940 E–W along Z. The purpose of this analysis was to determine whether the response was sensitive to the direction in which the center of mass was shifted.

Results of Analyses The main items of interest from the analysis are the base shear forces above the isolators, the bearing displacements, and the frame displacements.

The maximum displacements from the spectrum-compatible time histories were 0.25 in. in the garage columns, 6.0 in. in the bearings, and 0.7 in. total in the frames.

The maximum bearing displacement from the analysis is in excellent agreement with the results of the design procedures of Section 13.9, which also predicted 6 in.

The maximum base shear values, both in kips and as a fraction of the weight of the building, are also listed. Maximum base shears for the spectrum-compatible time histories reached 0.15g in the direction of primary loading. The values in the direction of the 40% earthquake loading are generally about one-half those in the direction of 100% loading.

The final results are shown in Table 13-2.

13.10.4 Summary

This example summarizes the results of the analyses that were performed on a proposed four-story structure. Results are obtained using Section B and Section C of the SEAONC Tentative Seismic Isolation Design Requirements, and these are compared with the preliminary design procedures of Section 13.9. The proposed site falls within zone 4 and S2 soil conditions of the code and is just outside of the 10-km near fault zone.

For this structure, the Section B requirements are applicable, and it can be seen from the summary of results that the method produces conservative results with respect to both the isolation displacement and the maximum base shear.

13.11 CONCLUSIONS

Several practical systems of seismic isolation have been developed and implemented in recent years, and interest in the application of this technique continues to grow. Although seismic isolation offers significant benefits, it is by no means a panacea. Feasibility studies are required early in the design phase of a project to evaluate both the technical and the economic issues. If its inclusion is appropriate from a technical and first-cost perspective, then significant life-cycle cost advantages can be achieved. Thus, seismic isolation represents an important step forward in the continuing search for improved seismic safety.

REFERENCES

13-1 Calantarients, J. A., "Improvements in and Connected with Building and Other Works and Appurtenances to Resist the Action of Earthquakes and the Like," Paper No. 325371, Engineering Library, Stanford Univ., Stanford, CA, 1909.

13-2 deMontalk, Robert Wladislas, "Shock Absorbing or Minimizing Means for Buildings," U.S. Patent No. 1,847,820, 1932.

13-3 Bechtold, Jacob, "Earthquake-Proof Building," U.S. Patent No. 845,046, 1907.

13-4 Wright, F. L., *An Autobiography: Frank Lloyd Wright*, Horizon Press, New York, 1977.

13-5 Green, N. B., "Flexible First Story Construction for Earthquake Resistance," *Trans. Amer. Soc. Civil Eng.* 100, 645, 1935.

13-6 Kelly, J. M. "Aseismic Base Isolation: Its History and Prospects," Joint Sealing and Bearing Systems for Concrete Structures, Publication SP-70, American Concrete Institute, 1982.

13-7 Buckle, I. G., "Development and Application of Base Isolation and Passive Energy Dissipation: A World Overview," Applied Technology Council Report 17, Palo Alto, CA, Mar. 1986.

13-8 Skinner, R. E., Tyler, R. G., Heine, A. J., and Robinson, W. H., "Hysteretic Dampers for the Protection of Structures from Earthquakes," *Bull. New Zealand Nat. Soc. Earthquake Eng.* 13, No. 1, Mar. 1980.

13-9 Way, D. and Lew, M., "Design and Analysis of a High Damping Rubber Isolation System," Applied Technology Council Report No. 17, Palo Alto, CA, 1986.

13-10 Jolivet, F. and Richli, M., "Aseismic Foundation System for Nuclear Power Stations," *Transactions of the Fourth Conference on Structural Mechanics in Reactor Technology*, San Francisco, Vol. K, No. 9/2, 1977.

13-11 Castiglioni, A., Urbano, C., and Stupazzini, B., "Seismic Design of Bridges in High Activity Region," *Proceedings of the Seventh European Conference on Earthquake Engineering*, Athens, Vol. 6, 187–203, 1982.

13-12 Ikonomou, A. S., "Seismic Isolation of Bridges with the Alexisismon," *Proceedings of the Conference on Short and Medium Span Bridges*, Toronto, 141–153, 1982.

13-13 Robinson, W. H., "Lead–Rubber Hysteretic Bearings Suitable for Protecting Structures During Earthquakes," *J. Earthquake Eng. and Structural Dynamics* 10, 593–604, 1982.

13-14 Blakeley, R. W. G., et al., "Recommendations for the Design and Construction of Base Isolated Structures," *Bull. New Zealand Nat. Soc. Earthquake Eng.* 12, No. 2, 1979.

13-15 Kelly, J. M. and Tsztoo, D., "Earthquake Simulation Testing of a Stepping Frame with Energy-Absorbing Devices," Report No. UCB/EERC-77/17, Earthquake Engineering Research Center, Univ. of California, Berkeley, 1977.

13-16 "Earthquake Simulator Tests of a Nine-Story Steel Frame with Columns Allowed to Uplift," Report No. UCB/EERC-77/23, Earthquake Engineering Research Center, Univ. of California, Berkeley, 1977.

13-17 Kelly, J. M., Eidinger, J. M., and Derham, C. J., "A Practical Soft Story System," Report No. UCB/EERC-77/27, Earthquake Engineering Research Center, Univ. of California, Berkeley, 1977.

13-18 Kelly, J. M., Beucke, K. E., and Skinner, M. S., "Experimental Testing of a Friction Damped Aseismic Base Isolation System with Fail-Safe Characteristics," Report No. UCB/EERC-80/18, Earthquake Engineering Research Center, Univ. of California, Berkeley, 1980.

13-19 Kelly, J. M., Beucke, K. E., and Skinner, M. S., "Experimental Testing of an Energy-Absorbing Base Isolation System," Report No. UCB/EERC-80/35, Earthquake Engineering Research Center, University of California, Berkeley, 1980

13-20 Kelly, J. M. and Hodder, S. B., "Experimental Study of Lead and Elastomeric Dampers for Base Isolation Systems," Report No. UCB/EERC-81/16, Earthquake Engineering Research Center, Univ. of California, Berkeley, 1981.

13-21 Kelly, J. M., Buckle, I. G., and Tsai, H. C., "Earthquake Simulator Testing of a Base Isolated Bridge Deck," Report No. UCB/EERC-85/09, Earthquake Engineering Research Center, Univ. of California, Berkeley, 1985.

13-22 Structural Engineers Association of California, *Recommended Lateral Force Requirements and Commentary*, San Francisco, 1983.

13-23 Stanton, J. F. and Roeder, C. W., "Elastomeric Bearings: Design, Construction and Materials," NCHRP Report 248, Transportation Research Board, Washington, 1982.

13-24 Applied Technology Council, "Proceedings of a Seminar and Workshop on Base Isolation and Passive Energy Dissipation," ATC Report No. 17, Palo Alto, CA, 1986.

13-25 Structural Engineers Association of California, *Tentative Lateral Force Requirements*, San Francisco, 1985.

13-26 Structural Engineers Association of Northern California, *Tentative Seismic Isolation Design Requirements*, San Francisco, CA, Sept. 1986.

13-27 International Conference of Building Officials, *Uniform Building Code*, Whittier, CA, 1985.

13-28 Kelly, T. E., Mayes, R. L., and Jones, L. R., "Preliminary Design Procedures for Seismically Isolated Structures," Proceedings of a Seminar on Base Isolation and Passive Energy Dissipation, Report No. 17, Applied Technology Council, Palo Alto, CA, 1986.

Conversion Factors*

LENGTH

To Convert from	To	Multiply by
Inches	feet	0.083333
	angstrom units	2.54×10^8
	micrometers	25400
	millimeters	25.4
	centimeters	2.54
	meters	0.0254
Feet	inches	12.0
	angstrom units	3.048×10^9
	micrometers	304800
	millimeters	304.80
	centimeters	30.48
	meters	0.3048
Millimeters	inches	3.9370079×10^{-2}
	feet	3.2808399×10^{-3}
	angstrom units	1×10^7
	micrometers	1×10^3
	centimeters	1×10^{-1}
	meters	1×10^{-3}
Centimeters	inches	0.39370079
	feet	0.032808399
	angstrom units	1×10^8
	micrometers	1×10^4
	millimeters	10
	meters	1×10^{-2}
Meters	inches	39.370079
	feet	3.2808399
	angstrom units	1×10^{10}
	micrometers	1×10^6
	millimeters	1×10^3
	centimeters	1×10^2
Yard	feet	3
	centimeter	91.44
	meter	0.9144
Angstrom units	inches	3.9370079×10^{-9}
	feet	3.28084×10^{-10}
	micrometers	0.0001
	millimeters	1×10^{-7}
	centimeters	1×10^{-8}
	meters	1×10^{-10}
Micrometers	inches	3.9370079×10^{-5}
	feet	3.2808399×10^{-6}
	angstrom units	1×10^4
	millimeters	1×10^{-3}
	centimeters	1×10^{-4}
	meters	1×10^{-6}

*For additional information, see American Society for Testing and Materials, *Metric Practice Guide*, ASTM E 380-72, 34 pp., 1973; International Organization for Standardization "International Standard, SI Units, and Recommendations for the Use of their Multiples and Certain Other Units," Ref. No. ISO 1000-1973(E), 21 pp., 1973; National Bureau of Standards, *Engineering Standards, U.S. Metric Study*, NBS SP 345-11, 250 pp., 1971.

AREA

To Convert from	To	Multiply by
Square meters	square feet	10.76387
	square centimeters	1×10^4
	square inches	1550.0031

To Convert from	To	Multiply by
Square feet	square meeters	9.290304×10^{-2}
	square centimeters	929.0304
	square inches	144
Square centimeters	square meters	1×10^{-4}
	square feet	1.076387×10^{-3}
	square inches	0.155
Square inches	square meters	6.4516×10^{-4}
	square feet	6.9444×10^{-3}
	square centimeters	6.4516
Square yard	acres	2.066×10^{-4}
	square feet	9
	square centimeters	8361.273
	square miles	3.228×10^{-7}
Acres	square meters	4046.849
	square feet	43560
	yards	4840

VOLUME

To Convert from	To	Multiply by
Cubic centimeters	cubic meters	1×10^{-6}
	cubic feet	3.5314667×10^{-5}
	cubic inches	0.061023744
Cubic meters	cubic feet	35.314667
	cubic centimeters	1×10^{6}
	cubic inches	61023.74
Cubic inches	cubic meters	1.6387064×10^{-5}
	cubic feet	5.7870370×10^{-4}
	cubic centimeters	16.387064
Cubic feet	cubic meters	0.028316847
	cubic centimeters	28316.847
	cubic inches	1728
U.S. gallons (gal)	cubic centimeters	3785
	cubic meters	3.785×10^{-3}
	cubic feet	0.133680
	cubic inches	231
	cubic yards	4.951×10^{-3}
	British Imperial gallons	0.833
	liters	3.785

FORCE

To Convert from	To	Multiply by
Pounds	dynes	4.44822×10^{5}
	grams	453.59243
	kilograms	0.45359243
	tons (long)	4.464286×10^{-4}
	tons (short)	5×10^{-4}
	kips	1×10^{-3}
	tons (metric)	4.5359243×10^{-4}
Kips	pounds	1000
	tons (short)	0.500
	kilograms	453.59243
	tons (metric)	0.45359243
Kilograms	dynes	980665
	grams	1000

To Convert from	To	Multiply by
	pounds	2.2046223
	tons (long)	9.8420653×10^{-4}
	tons (short)	11.023113×10^{-4}
	kips	2.2046223×10^{-3}
	tons (metric)	0.001
Tons (short)	kilograms	907.18474
	pounds	2000
	kips	2
	tons (metric)	0.907185
Tons (metric)	grams	1×10^{6}
	kilograms	1000
	pounds	2204.6223
	kips	2.2046223
	tons (short)	1.1023112
Pound force (lbf)	newtons	4.45
Kilogram force (kgf)	newtons	9.81

VELOCITY

To Convert From	To	Multiply by
Centimeters/second	microns/second	10,000
	meters/minute	0.600
	feet/minute	1.9685
	miles/hour	0.022369
	feet/year	1034643.6
Microns/second	centimeters/second	0.0001
	meters/minute	0.000060
	feet/minute	0.00019685
	miles/hour	0.0000022369
	feet/year	103.46436
Feet/minute	centimeters/second	0.508001
	microns/second	5080.01
	meters/minute	0.3048
	miles/hour	0.01136363
	feet/year	525600
Feet/year	microns/second	0.009665164
	centimeters/second	0.0000009665164
	meters/minute	5.79882×10^{-7}
	feet/minute	1.9025×10^{-6}
	miles/hour	2.16203×10^{-8}
Miles/hour	centimeters/second	44.7041
	meters/minute	26.82
	feet/hour	5280
	feet/minute	88
	feet/second	1.467
	miles/second	2.778×10^{-4}

TIME

To Convert from	To	Multiply by
Milliseconds	seconds	10^{-3}
	minutes	1.66666×10^{-5}
	hours	2.777777×10^{-7}
	days	1.1574074×10^{-8}
	months	3.8057×10^{-10}
	years	3.171416×10^{-11}

To Convert from	To	Multiply by
Seconds	milliseconds	1000
	minutes	1.66666×10^{-2}
	hours	2.777777×10^{-4}
	days	1.1574074×10^{-5}
	months	3.8057×10^{-7}
	years	3.171416×10^{-8}
Minutes	milliseconds	60000
	seconds	60
	hours	0.0166666
	days	6.944444×10^{-4}
	months	2.283104×10^{-5}
	years	1.902586×10^{-6}
Hours	milliseconds	3600000
	seconds	3600
	minutes	60
	days	0.0416666
	months	1.369860×10^{-3}
	years	1.14155×10^{-4}
Days	milliseconds	86400000
	seconds	86400
	minutes	1440
	hours	24
	months	3.28767×10^{-2}
	years	0.0027397260
Months	milliseconds	2.6283×10^{9}
	seconds	2.6283×10^{6}
	minutes	43800
	hours	730
	days	30.416666
	years	0.08333333
Years	milliseconds	3.1536×10^{10}
	seconds	3.1536×10^{7}
	minutes	525600
	hours (mean solar)	8760
	days (mean solar)	365
	months	12

STRESS

To convert from	To	Multiply by
Pounds/square inch	pound/square foot	144
	feet of water	2.3066
	kips/square foot	0.144
	kilograms/square centimeter	0.070307
	tons/square meter	0.70307
	atmospheres	0.068046
	kilopascals	6.9
Pounds/square foot	pounds/square inch	0.0069445
	feet of water	0.016018
	kips/square foot	1×10^{-3}
	kilograms/square centimeter	0.000488243
	tons/square meter	0.004882
	atmospheres	4.72541×10^{-4}
	pascals	47.9
Feet of water (at 39.2°F)	pounds/square inch	0.43352
	pounds/square foot	62.427
	kilograms/square centimeter	0.0304791
	tons/square meter	0.304791
	atmospheres	0.029499
	inches of Hg	0.88265
Kips/square foot	pounds/square inch	6.94445
	pounds/square foot	1000
	tons (short)/ square foot	0.5000
	kilograms/square centimeter	0.488244
	tons (metric)/square meter	4.88244
Kilograms/ square centimeter	pounds/square inch	14.223
	pounds/square foot	2048.1614
	feet of water (39.2°F)	32.8093
	kips/square foot	2.0481614
	tons/square meter	10
	atmospheres	0.96784
Tons (short)/ square foot	atmospheres	0.945082
	kilograms/square meter	9764.86
	tons (metric)/ square meter	9.76487
	pounds/square inch	13.8888
	pounds/square foot	2000
	kips/square foot	2.0
Tons (metric)/ square meter	kilograms/square centimeter	0.10
	pounds/square foot	204.81614
	kips/square foot	0.20481614
	tons (short)/ square foot	0.102408
Atmospheres	bars	1.0133
	centimeters of mercury at 0°C	76
	millimeters of mercury at 0°C	760
	feet of water at 39.2°F	33.899
	kilograms/square centimeter	1.03323
	grams/square centimeter	1033.23
	kilograms/square meter	10332.3
	tons (metric)/ square meter	10.3323
	pounds/square foot	2116.22
	pounds/square inch	14.696
	tons (short)/square foot	1.0581
Newtons/square meter	pascals	1.00

UNIT WEIGHT

To Convert from	To	Multiply by
Grams/cubic centimeter	tons (metric)/cubic meter	1.00
	kilograms/cubic meter	1000.00
	pounds/cubic inch	0.036127292
	pounds/cubic foot	62.427961
kilograms/cubic meter	grams/cubic centimeter	0.001
	tons (metric)/cubic meter	0.001
	pounds/cubic inch	3.6127292×10^{-5}
	pounds/cubic foot	0.062427961
Pounds/cubic inch	grams/cubic centimeter	27.679905
	tons (metric)/cubic meter	27.679905
	kilograms/cubic meter	27679.905
	pounds/cubic foot	1728
Pounds/cubic foot	grams/cubic centimeter	0.016018463
	tons (metric)/cubic meter	0.016018463
	kilograms/cubic meter	16.018463
	pounds/cubic inch	$5.78703704 \times 10^{-4}$
Tons (metric)/cubic meter	grams/cubic centimeter	1.00
	kilograms/cubic meter	1000.00
	pounds/cubic inch	0.03612722
	pounds/cubic foot	62.427961

COEFFICIENT OF CONSOLIDATION, C_v

To Convert from	To	Multiply by
Square centimeters/second	square centimeters/year	3.1536×10^{7}
	square meters/year	3.1536×10^{3}
	square inches/second	0.155
	square inches/year	4.8881×10^{6}
	square feet/year	3.39447×10^{4}
Square inches/second	square inches/year	3.15368×10^{7}
	square feet/year	2.1900×10^{5}
	square centimeters/second	6.4516
	square centimeters/year	2.0346×10^{8}
	square meters/year	2.0346×10^{4}

COEFFICIENT OF THERMAL CONDUCTIVITY, *k*

To Convert from	To	Multiply by
Btu/hr-ft^2-(deg F/ft)	cal/sec-cm^2-(deg C/cm)	4.134×10^{-3}
W/cm^2-(deg C/cm)	Btu/hr-ft^2-(deg F/ft)	57.780
W/cm^2-(deg C/cm)	cal/sec-cm^2-(deg C/cm)	0.239

TEMPERATURE

$T(°C) = \frac{5}{9}[T(°F) - 32°]$

$T(°F) = \frac{9}{5}T(°C) + 32°$

$T(K) = T(°C) + 273.18°$

Index